THEORETICAL AND APPLIED RHEOLOGY

Proceedings of the
XIth INTERNATIONAL CONGRESS ON RHEOLOGY
Brussels, Belgium, August 17-21, 1992

VOLUME 1

Edited by

P. Moldenaers
Department of Chemical Engineering,
Katholieke Universiteit Leuven, Leuven, Belgium

and

R. Keunings
Division of Applied Mechanics,
Université Catholique de Louvain, Louvain-la-Neuve, Belgium

1992

ELSEVIER

AMSTERDAM • LONDON • NEW YORK • TOKYO

ELSEVIER SCIENCE PUBLISHERS
Sara Burgerhartstraat 25
P.O. Box 211, 1000 AE Amsterdam
The Netherlands

ISBN: 0-444-89007-6

This book is printed on acid-free paper.

Printed in The Netherlands.

XIth INTERNATIONAL CONGRESS ON RHEOLOGY

Brussels, Belgium, August 17-21, 1992

Organized on behalf of the *INTERNATIONAL COMMITTEE ON RHEOLOGY*

CONGRESS SPONSORS

Commission of the European Communities
FNRS-NFWO, Belgian National Science Foundation

Agfa-Gevaert
BASF
Carri-Med
Dow Benelux
DSM
Elsevier Science Publishers
Polyflow
Rheometrics
Solvay
Unilever U.K.

ORGANIZING COMMITTEE

M.J. Crochet (co-Chairman)	Université Catholique de Louvain
J. Mewis (co-Chairman)	Katholieke Universiteit Leuven
P. Moldenaers (Secretary)	Katholieke Universiteit Leuven
A. Cardon	Vrije Universiteit Brussel
B. Crochet	
B. Goetmaeckers	Agfa-Gevaert
R. Keunings	Université Catholique de Louvain
J. Leblanc	EniChem
C. Marco	Université de Mons
R. Mewis	
G. Vancoppenolle	Agfa-Gevaert

PREFACE

The International Congress on Rheology is held every four years under the auspices of the International Committee on Rheology. Brussels is the venue for the 1992 Congress, organized by the Belgian Group of Rheology. The year of 1992 is of special importance for the European Community as it signals the creation of a single market of 320 million people, with Brussels as its administrative centre.

The programme of the First International Congress on Rheology, held in Scheveningen, the Netherlands (1948), consisted of 9 general lectures and 38 contributed papers. Rheology has since developed into a broad, multidisciplinary field of recognized academic and industrial relevance. More than 900 authors from over 35 countries, including a large representation from Eastern Europe, have contributed to the 1992 Congress in Brussels. The programme of the 1992 Congress is organized in 17 plenary and keynote papers, an industrial panel session, 250 oral contributions, and about 200 poster presentations. This is supplemented by an exhibition of rheological equipment, software, and literature.

On behalf of the Organizing Committee, we wish to express our gratitude to the numerous individuals and organizations that have contributed to the success of the Congress. All credit for the intellectual content of these Proceedings should of course be given to the authors themselves, whether they be eminent personalities, established specialists, or promising Ph.D. students. We have appreciated the care they have exercised in the preparation of the camera-ready manuscripts. The selection of contributed papers has benefited greatly from the expertise of numerous colleagues worldwide. These dedicated individuals have spent a considerable amount of time and effort on such a delicate task. We also wish to thank our sponsors for their financial and organizational support. Finally, it is our pleasure to acknowledge the contribution of the administrative, scientific, and technical staff of the Université Catholique de Louvain and of the Katholieke Universiteit Leuven. In particular, their help in editing these Proceedings is greatly appreciated.

Paula Moldenaers
Roland Keunings

INTERNATIONAL COMMITTEE ON RHEOLOGY

Secretary's Report - August 1988

Since the IXth International Congress on Rheology took place in Acapulco, Mexico, in October 1984, two new rheological societies asked for, and were admitted to membership in ICR. These are the Chinese and the Slovene Societies of Rheology. The first is active in what is without doubt the most populous country on Earth. Slovenia is one of the constituent republics of the Federation of Yugoslavia, and currently is surely the smallest political unit boasting a member society of ICR. The Slovene society has 50 members compared with the Chinese society's 500.

The total number of member societies thus stands at 20. The 20 members represent close to 4740 rheologists. At the time of the last Congress the number of rheologists represented by the then 18 member societies was 3960. The same societies now report 4190 members. Most of the member societies have slightly increased their membership during the past four years. Significant drops were reported only by the Belgian and Dutch societies.

The membership, listed in Table 1 below, extends over four of the six land areas recognized as separate continents : North America, Europe, Asia, and South America. There is no society as yet in Africa nor is there one in Antartica. The chances of holding an International Congress on Rheology (of Ice ?) in Antartica in the near future appear rather slim.

Table 1

Member Societies of the International Committee on Rheology (with foundation date and approximate membership in July, 1988)

1.	The Society of Rheology (USA) (1929)	1150
2.	The British Society of Rheology (1940)	650
3.	Nederlandse Reologische Vereniging (1950)	80
4.	Deutsche Rheologische Gesellschaft (DRG) (1951)	280
5.	Swedish Society of Rheology (1956)	150
6.	Australian Society of Rheology (1959)	100
7.	Arbeitsgruppe Rheologie Gesellschaft Österreichischer Chemiker (1959)	80
8.	Groupe Français de Rhéologie (1964)	200
9.	Rheology Group of the Czechoslovak Chemical Society (1968)	110
10.	Societa Italiana de Reologia (1971)	100
11.	The Society of Rheology, Japan (1973)	720
12.	Belgian Group of Rheology (1974)	50
13.	Sociedad Mexicana de Rheologia (1975)	100
14.	Israël Society of Rheology (1981)	60
15.	Spanish Group of Rheology (1981)	80
16.	Canadian Rheology Group/Groupe Canadien de Rhéologie (1982)	115
17.	Association Argentina de Reologia (1983)	120
18.	Indian Society of Rheology (1984)	40
19.	Chinese Society of Rheology (1985)	500
20.	Slovene Society of Rheology (1987)	50

There has not been much change in the number and type of publications sponsored by member societies. The only major loss is Documentation Rheology which the Federal Institute for Materials Research and Testing (BAM) of the Federal Republic of Germany had produced in collaboration with the German Society of Rheology. The publication has proved too costly under the current economic climate. Some back issues (volume 1959 to 1096/87) are still available (see Newsletter of April 15, 1988).

Table 2 lists the current publications sponsored by member societies. Almost all produce at least occasional circulars or newsletters to keep in touch with their members. These may or may not be mentioned explicitly in the Table. The British and USA societies sponsor major publications which are of interest to members of other societies. These are marked with an asterisk below.

Table 2

Publications by Member Societies of ICR (listed by country)

Argentina	Boletin de la AAR (quarterly)
Australia	Newsletter (quarterly); Proc. Biennial Conference; (every second year)
Belgium	Bulletin (3 times/year)
Canada	Newsletter (2 times/year)
P.R. of China	Proceedings/Abstracts of yearly national congresses
Czechoslovakia	Abstracts of the Triennial National Industrial Rheology Conferences; Circulars (3-4 times/year)
France	Proceedings of the Annual Meeting
Italy	Bolletino (3 times/year)
Japan	Nihon Reoroji Gakkaishi (J. Soc. Rheol. Japan) (quarterly)
Mexico	Bulletin (yearly)
Netherlands	Publications in "Materialen", a monthly journal of the (Dutch) Materials Science Association
United Kingdom	*Rheology Abstracts (quarterly); Bulletin (quarterly)
USA	*Journal of Rheology (8 times/year); Rheology Bulletin (2 times/year)

*Publication of interest to all Member Societies

Virtually all member societies hold Annual Meetings. Only one society (Israël) reports that it does not have regular annual meetings. The larger societies hold meetings which feature the presentations of major scientific programs. These often attract visitors from other countries. Scientific programs are often arranged as joint programs with other (not necessarily rheological) societies in the same country or with participation of others.

Table 3 reports these activities of the member societies.

Table 3

Meetings of the Member Societies of ICR

Argentina	Annual Meeting (December) (ca. 100); Argentine Congress of Rheology (every second year) (ca.50)
Australia	Annual Meeting (February) (ca. 30); Biennial Conference (ca. 70)
Austria	Annual Meeting (Autumn) (ca. 50)
Belgium	Annual Meeting (Fall) (ca.25); 2 or 3 Meetings/year
Canada	Annual Meeting (October); Local meetings in Montreal and Toronto (ca. 20-30); Participation in Meetings of The Society of Rheology in USA
P.R. of China	Annual Meeting (ca. 300); Joint meetings with other Chinese societies
Czechoslovakia	Annual Meeting (Autumn) (ca.50); National Industrial Rheology Conference (triennial) (ca.60)
France	Annual Meeting (Autumn) (50-100); Half-day Seminars (15-40)
German Fed. Rep.	Annual Meeting (Spring) (ca. 150); Occasional joint meetings with other societies
India	Annual Meeting (Winter) (ca. 20-25)
Israël	Occasional meetings (20-30)
Italy	Annual Meeting (Spring of alternate years) (ca.100)
Japan	Annual Meeting (May) (ca.100); Rheology Symposium (Autumn) (ca.150)
Mexico	Annual Meeting (Winter) (ca.60)
Netherlands	Annual Meetings (each Spring and each Fall) (ca.50)
Slovenia	Annual Meeting (September) (40-50)
Spain	Annual Meeting (October) (ca. 30)
Sweden	Annual Meeting (Autumn) (ca.120); Other occasional meetings
United Kingdom	Annual Meeting (September) (ca.150): Spring Meeting (March-April) (ca.50)
USA	Annual Meeting (October) (ca.200)

Officers of the member societies are generally elected at the annual meetings, but not necessarily for one year. Tenures range from one to four years.

Particular mention must be made of the Rheological Conferences which are arranged by the European member societies (including Israël). These take place every four years spaced so that they fall between two International Congresses. They are sponsored by a committee consisting of the European (and the Israeli) delegates to ICR. The activities are ably coordinated by the European Secretary, Dr. J.C.S. Petrie, who also takes care of the timely dissemination of news about scientific meetings of interest to the European and Israeli membership.

Various honors and awards currently bestowed by member societies are listed in Table 4 below. Eight member societies confer honorary memberships. The British, Japanese, and USA societies award major awards, generally to their members only.

Table 4

Honors and Awards Bestowed by Member Societies of ICR (listed by country)

Country	Honors and Awards
Argentina	Honorary Membership
Australia	Honorary Membership
France	Honorary Membership
Germany	Honorary Membership
Japan	Honorary Membership; Society Award; Society Awards for Younger Rheologists
Mexico	Honorary Membership
Slovenia	Honorary Membership
U.K.	Honorary Membership; Gold Medal; Annual Award; Scott Blair Scholarship
U.S.A.	Bingham Medal

The last item I wish to report on are certain ongoing efforts to develop a consistent nomenclature and use of symbols, and the construction of rheological glossaries. These efforts have been slow and sporadic. Nevertheless, some progress has been made . Several glossaries are now available. The next Newsletter will contain a list of these glossaries and of published recommendations relating to nomenclature and symbols, as well as a detailed report on current moves to coordinate the efforts undertaken by member societies of ICR in these areas.

It seems worthwhile to continue these undertakings. A consistent, widely accepted nomenclature and use of symbols is clearly in the interest of all rheologists. Rheological glossaries may, in some sense, be even more important. After all, if we rheologists do not develop them, someone else might, with possibly disastrous results.

Respectfully submitted,

N.W. Tschoegl
Secretary

INTERNATIONAL COMMITTEE ON RHEOLOGY

Secretary's Report - March 1992

Admission of New Member

The first major item to report is the addition of a national society to the International Committee. The new member is the Soviet Society of Rheology, which was admitted in 1991 after receiving unanimous support from ICR Delegates in a mailed ballot. Soviet rheologists have been active and organized for some time : every two years they hold an All-Union Symposium on Rheology, and they organize smaller, more frequent meetings throughout the U.S.S.R. They were formerly part of an umbrella organization, but in 1991 they achieved independence and sought to join the international community. The first President of the Society is Professor Valery G. Kulichikhin and the first Vice-President is Professor Alexander Ya. Malkin, who is also the Delegate to the International Committee. Because the Soviets have long been active in rheology, joining the ICR was long overdue. It might also be noted that, because of the dissolution of the U.S.S.R., the Society will have a new name in the near future.

The Current Roster

At the time of the last Secretary's Report, in 1988, the number of member societies of the ICR was 20. The current roster of 21, including the Soviet Society, is given in Table 1 in order of founding dates. Also listed in Table 1 are memberships, based on a survey completed in January 1992. Summing the figures in the Table, one finds that the number of rheologists in the 21 societies is 4810. Since the comparable number four years ago was 4740, there has been a slight growth in an economically-difficult period. The biggest changes in membership have been the addition of the Soviet Society, with 150 members, and a drop from 500 to 300 in the membership of the Chinese Society.

Table 1
Member Societies of the International Committee on Rheology

In the order founded (year), and approximate membership in 1992.

	Society (year)	Members
1.	The Society of Rheology (USA) (1929)	1100
2.	The British Society of Rheology (1940)	790
3.	Nederlandse Reologische Vereniging (1950)	85
4.	Deutsche Rheologische Gesellschaft (DRG) (1951)	300
5.	Nordic (Swedish) Society of Rheology (1956)	80
6.	Australian Society of Rheology (1959)	60
7.	Arbeitsgruppe Rheologie Gesellschaft Österreichischer Chemiker (1959)	80
8.	Groupe Français de Rhéologie (1964)	260
9.	Rheology Group of the Czechoslovak Chemical Society (1968)	95
10.	Societa Italiana de Reologia (1971)	120
11.	The Society of Rheology, Japan (1973)	750
12.	Belgian Group of Rheology (1974)	120
13.	Sociedad Mexicana de Rheologia (1975)	90
14.	Israël Society of Rheology (1981)	60
15.	Spanish Group of Rheology (1981)	70
16.	Canadian Rheology Group/Groupe Canadien de Rhéologie (1982)	120
17.	Association Argentina de Reologia (1983)	70
18.	Indian Society of Rheology (1984)	60
19.	Chinese Society of Rheology (1985)	300
20.	Slovene Society of Rheology (1987)	50
21.	Soviet Society of Rheology (1991)	150

Table 2

Activities of Member Nations

Country	National Meetings		Local Meetings		Newsletter	Technical Publications
	Frequency	*Attendance*	*No./year*	*Attendance*	*issues/year*	
Argentina	annual	100	-	-	1	-
Australia	biennial	50	6	30	2-3	Abstracts (biennal)
Austria	annual	30	2-3	15	-	-
Belgium	annual	30	-	-	1-2	-
Canada	-	-	2	25	2	-
China	triennial	150	-	-	-	-
Czechoslovakia	triennial	50	3-4	25	-	Abstracts of meetings
France	annual	70	2	20	3	"Les cahiers de Rhéologie" (annual)
Germany	biennial	200	-	-	-	-
India	biannual	30	variable	-	2	-
Israël	biannual	30	occasional	40	-	-
Italy	-	-	2	50	2	-
Japan	biannual	100	12	50	4	"Nihon Reoroji Gakkaishi" 4 issues/year
Mexico	annual	60	-	-	-	-
Netherlands	biannual	30	-	-	-	"Materialen" (monthly)
Slovenia	annual	40	-	-	4	Proc. of national meetings
Spain	annual	20	-	-	1	-
Sweden(Nordic)	annual	50	1	10	4	-
U.K.	biannual	75	4	10	"Bulletin" 4 issues/year	"Rheology Abstracts" 4 issues/year
U.S.A.	annual	300	-	-	2	"J. Rheology" 8 issues/year
U.S.S.R.	biennial	300	1	50	-	-

In listing the national societies in Table 1, the actual name of the society appears wherever possible. For names that cannot be written in Roman script, the English equivalent has been substituted.

It should also be noted that the Swedish Society of Rheology changed its name to the Nordic Society of Rheology.

A New Society in Korea

The Korean Society of Rheology was founded on February 24, 1989, with a membership of about 150 expected from universities, industries and research institutes. The Chairman of the Organising Committee of the new Society was Professor Sang Yong Kim, and the current contact is Prof. Seung Jong Lee, Department of Chemical Engineering, Seoul National University, Seoul 151-742, Korea.

Activities of Member Societies

Some appreciation of rheological activity worldwide may be gained from Table 2. The Table shows the frequency and approximate attendance of national and local meetings, as well as communications from societies to its members. Not included in the Table are many other international meetings, most of which are organized on an ad hoc basis. Some recent examples are : the International Conference on Dynamics of Polymeric Liquids (Capri, 1991), the China-Japan International Conference on Rheology (1991), the International Workshops on Numerical Modelling (the 7th, in Florida, in 1992), the European Conference on Rheology (held every four years between ICR Congresses), the Extensional Flow Workshops (last held at Villard de Lans, 1991). Such meetings are held more and more often, it seems, and the frequency of meetings of all types confirms a diverse and vigorous field. In fact, from the number of meetings, there is on average at least one meeting on rheology every week of the year somewhere in the world.

Honours and Awards

Some societies make awards to their members and it is useful to list these, as is done below in Table 3. The awards which have been established since the last Secretary's Report in 1988 are the A.S.R. Medal (Australia), the Mason Award (Canada) and the Vinogradov Medal (U.S.S.R.).

Table 3
Honours and Awards by Member Societies

Australia :	Honorary Life Membership; A.S.R. Medal
Canada :	CRG/GCR Mason Award
Japan :	SRJ Award, SRJ Young Rheologist Award
Mexico :	Honorary Membership
Slovenia :	Honorary Membership
U.K. :	Annual Award; Gold Medal; Scott Blair Memorial and Biorheology Scholarships
U.S.A. :	Bingham Medal
U.S.S.R. :	Vinogradov Medal

The State of Rheology

The information presented in this Report indicates that the state of rheology worldwide is strong and healthy. Membership is growing, new societies are being formed, meetings and publications are frequent, and more awards for achievement are being established.

David F. James
Secretary

CONTENTS OF VOLUME 1

Molecular Theories : Posters

Constitutive Equations : Oral communications

Constitutive Equations : Posters

Theory : Oral communications

FLUID MECHANICS : Posters

NUMERICAL SIMULATION : Oral communications

NUMERICAL SIMULATION : Posters

MELTS AND POLYMER PROCESSING : **Oral communications**

MELTS AND POLYMER PROCESSING : Posters

POLYMER SOLUTIONS : Oral communications

POLYMER SOLUTIONS : Posters

CONTENTS OF VOLUME 2

CONTRIBUTED PAPERS
(Continued)

LIQUID CRYSTALS : Oral communications

LIQUID CRYSTALS : Posters

SUSPENSIONS : Oral communications

SUSPENSIONS : Posters

FOAMS AND EMULSIONS : Oral communications

FOAMS AND EMULSIONS : Posters

FOOD RHEOLOGY : Oral communications

FOOD RHEOLOGY : Posters

BIORHEOLOGY : Oral communications

BIORHEOLOGY : Posters

ELECTRORHEOLOGY : Oral communications

ELECTRORHEOLOGY : Posters

RHEOLOGY OF SOLIDS : Oral communications

RHEOLOGY OF SOLIDS : Posters

COMPOSITE MATERIALS : Oral communications

COMPOSITE MATERIALS : Posters

RHEOMETRY AND EXPERIMENTAL METHODS : Oral communications

RHEOMETRY AND EXPERIMENTAL METHODS : Posters

INSTRUMENTATION : Oral communications

PLENARY PAPERS

Theoretical and Applied Rheology, edited by P. Moldenaers and R. Keunings
Proc. XIth Int. Congr. on Rheology, Brussels, Belgium, August 17-21, 1992

ROLE OF MOLECULAR MODELING IN POLYMER RHEOLOGY

G. MARRUCCI

Universita' Federico II, Dipartimento di Ingegneria Chimica, Piazzale Tecchio, 80125 Napoli (Italy)

1. INTRODUCTION

Our knowledge of the rheological properties of polymeric liquids has evolved in the course of time by the use of three fundamental tools. The first and most important was (and still is, of course) the experimental investigation. There is no need to elaborate on the central role that "the experiment" plays in rheology, as in all branches of physics for that matter.

The other two tools are conceptual, and have to do with the everlasting attempt at somehow modeling the observed behaviour. When using one of them, the polymeric liquid is looked upon macroscopically, i.e., from the point of view of continuum mechanics. Such an approach is also called "phenomenological". The other tool is usually referred to as "molecular modeling", because the starting point in the analysis is, in some idealized sense at least, the constituent polymeric molecule. The latter approach seems to have recently gained an increasing favour among rheologists, a possible reason being the popularity reached by relatively simple models like the "dumbbell" (ref. 1), or by ingenious ideas like the "reptation" concept (refs. 2-4). (Opposers of this idea also contributed to its popularity.)

In this paper, some general considerations on the usefulness of molecular modeling are presented. First, an analysis of the role played by continuum theories is made, showing which limitations are encountered when applying the classical approach to polymeric liquids. Next, some general concepts of molecular modeling will be briefly presented. It will finally be argued that the information gained by molecular modeling must again merge into continuum mechanics if simulation of complex flows is to be attempted.

2. LINEARITY VS. NONLINEARITY

2.1 Linear constitutive equations

In the history of science, the first investigations of non-equilibrium phenomena led to the discovery of linear "laws" (or constitutive equations). Thus, for example, it was found that the flux of heat was proportional to the gradient of temperature (Fourier law). More importantly for rheology, it was found that, in many viscous fluids at least, the flux of momentum (or, equivalently, the deviatoric part of the stress tensor) was proportional to (the symmetric part of) the velocity gradient.

For the practical use of these linear phenomenological laws, it was only required that the proportionality coefficient (the viscosity in the latter example) be determined, for any given material, as a function of the state variables, prominently temperature. Such a determination was conveniently made experimentally.

It so appears that the possible role of molecular modeling was inevitably meager in this context. At best, it could have been used to somehow "predict" the value of the proportionality coefficient, or its dependence upon the state variables. As already noted, it was simpler, and safer, to obtain this information experimentally.

In this respect, the situation did not substantially change when Boltzmann described viscoelasticity because, again, the corresponding law was formulated in linear terms. Of course, the law of linear viscoelasticity is much more complex than the previous proportionalities. This occurs because the stress σ is now dependent upon the deformation history, i.e., it is obtained through an integral over past time:

$$\sigma(t) = \int_{-\infty}^{t} h(t-t')\,\gamma(t,t')\,dt' \qquad (1)$$

In Eq.(1), $\gamma(t,t')$ is a measure of the deformation occurred between past time t' and present time t, and h(t-t') is a "memory" (or relaxation) function which accounts for the elapsed time t-t'. It is the latter function which characterizes the viscoelastic material at any given temperature.

Now, although finding a molecular model which predicts the function h(t-t') may be attractive, it remains relatively simple (and indeed safer) to obtain h(t-t') experimentally, for example by measuring the frequency response of the material. The temperature dependence of h(t-t') is itself readily obtained experimentally.

2.2 Nonlinearities in classical theories

It will be argued in the following sections that the fact that we are forced to use nonlinear constitutive equations in polymer rheology is the crucial factor that makes molecular modeling extremely useful if not indispensable. This statement might be found confusing at first sight, because nonlinearities are often encountered in classical theories, i.e., in the context of a phenomenological approach. Let us look at this matter of "classical" nonlinearities in some better detail, however, with an emphasis on fluid motion, of course.

A good example of classical nonlinearity is immediately offered by the Navier-Stokes equation. Where is the source of nonlinearity in this equation, however? Certainly not in the constitutive part: indeed, it is assumed that the viscous response of the fluid is linear. The nonlinear terms arise from the fluid inertia; they are straightforwardly linked to the universal Newton's law of motion, i.e., to the momentum conservation principle. In conclusion, no ambiguity can ever be attached to the nonlinearity of the Navier-Stokes equation; nor is molecular modeling needed to predict, or interpret, such nonlinearity.

Another classical example of nonlinearity is provided by non-isothermal flow fields. Assume that a fluid flows under conditions such that inertia effects are negligible, but temperature is not uniform throughout because of a transfer of heat. As is well known, the velocity and temperature fields will obey a set of equations describing conservation of mass, momentum, and energy. This set of equations generates a nonlinear problem even under the most "favourable" assumptions. Indeed, let us assume not only that inertia is negligible but also that heat is transported by conduction only (i.e., coupling terms due to convection are absent), and that heat generated by flow is negligible. Further assume that the constitutive laws relating fluxes to gradients are linear. Yet the problem remains nonlinear because of the temperature dependence of the constitutive parameters.

Many other examples of "classical" nonlinearities could easily be made. In all of them, the constitutive equations *per se* are linear; the source of the nonlinear behaviour is either traced back to "first principles" or else is due to the dependence of constitutive parameters on temperature (or concentration, or, in extreme cases, pressure). In all such examples, the possible role played by molecular modeling is irrelevant or nil; the phenomenological approach is in fact completely satisfactory. We will see that the situation changes drastically when the nonlinearity is constitutive.

2.3 Nonlinear constitutive equations

Soon after polymers were discovered, it was found that they exhibited a viscoelastic response (ref. 5). Unfortunately, it also became apparent that the linear formula of Eq.(1) did not generally apply; rather, linear viscoelasticity could only be recovered asymptotically, i.e., in the limit of small stresses. A prominent example of nonlinearity in polymeric liquids is offered by the so-called non-Newtonian viscosity, i.e., by the fact that, in a steady shear flow, the tangential stress is not proportional to shear rate unless the flow is sufficiently slow.

This intrinsic nonlinearity of polymeric liquids brought up a problem which, even today, is far from being satisfactorily solved. From the phenomenological point of view, there was a growing effort to find a good replacement (or, rather, an extension) of Eq.(1) which could describe the observed nonlinearity. Such a search has generated a multitude of "guesses", i.e., of proposed constitutive equations, which would be impossible to describe here. It can safely be stated that none of these proposals has gained a general consensus, though it cannot be denied that most of them have served (more or less satisfactorily) some specific purpose.

The continuum mechanics of nonlinear viscoelasticity received a formal assessment with the work of Coleman and Noll on so-called "simple" fluids (see ref. 6). These authors generalized the linear functional of the deformation history, as given by Eq.(1), to an arbitrary functional. Thus, the constitutive equation was formally written as:

$$\sigma(t) = \mathop{\mathbf{F}}_{t'=-\infty}^{t'=t} [\gamma(t,t')] \tag{2}$$

The functional **F** was required to obey frame indifference (or material invariance), and to have a topology incorporating the concept of "fading memory". These restrictions, though conceptually important, are too weak to be able to determine the functional "uniquely" (i.e., to within a finite number of material functions, see later). Thus, the only results of this general theory could be obtained from asymptotic expansions. Indeed, it was shown that the functional could be generally expanded as a series of integrals, the first of which is that of Eq.(1), the second is a double integral, and so forth. Significant predictions were obtained from the "slow flow" asymptotic expansion at the second order level, together with the (trivial) result that a simple fluid is indeed Newtonian to within first order in the magnitude of the velocity gradient.

It is time, perhaps, to draw a conclusion on the phenomenological approach to nonlinear viscoelasticity: With the possible exception of asymptotic results, we are actually in the dark as to the "shape" of the nonlinear functional of Eq.(2), i.e., to the form that the constitutive equation of polymeric liquids should take well within the nonlinear range. Nor could the situation be different after all, as discussed in the rest of this section.

Imagine for a moment that someone wants to guess the shape of the nonlinear function y = f(x) (x and y representing physical variables) by only knowing that the curve goes through the origin with some given slope and, maybe, with a given curvature (i.e., by knowing the expansion of f(x) at x=0 up to second order at best). Clearly, any guess of the function far from the origin could not be taken seriously under these conditions. The situation would slightly improve if one or two isolated points far from the origin were known; yet bitter surprises might still arise because of arbitrary interpolations or extrapolations. Of course, this problem is solved phenomenologically if all points of the curve, in the range of interest, are accessible to measurement (even if actual measurements are always limited in number). When dealing with functions, such is ordinarily the case.

Much different is the situation with functionals. A phenomenological determination of the nonlinear functional describing viscoelasticity in polymeric liquids would require measuring stress for arbitrary deformation histories. In spite of the well known ingenuity of experimental rheologists, this is a patent impossibility. What is accessible to measurement is a highly restricted set of possible deformation histories, such as steady-state shear flows, and a few others. Now, the "material functions" which are thus obtained (the non-Newtonian viscosity curve, for example) play the role of isolated points in the functional space. Moreover, these "points" cannot even be ordered: differently from the y = f(x) situation previously considered where it is certainly possible to know how far from the origin is a given x (i.e., how deep it is in the nonlinear range), here we have no easy way of comparing a transient elongational deformation, say, with a steady shear flow, or with a large amplitude oscillation, etc.

In other words, for assigned "values" of the material functions which can be experimentally determined there still exist infinite different functionals (quite apart from the difficulty of inventing some specific mathematical form which agrees with the available data). The conclusion is that there is no real possibility of "determining" the constitutive equation of nonlinear viscoelasticity only from the observed phenomenology (plus the restrictions posed by frame invariance, by the second law of thermodynamics, etc.). The game would always remain a game of guess.

Let us see now how this situation can be improved upon by using molecular modeling.

3. MOLECULAR MODELING

3.1 The basic ingredients

Molecular modeling of polymer properties may mean quite different things depending on the context. For example, when dealing with equilibrium properties, the model of the polymer molecule may even include all of its atoms, and all the interaction potentials. Such a level of detail can be afforded in that context because use can also be made of some general statistical laws which hold at equilibrium. (For example, probabilities are proportional to Boltzmann factors.) Also, very effective numerical algorithms (Monte Carlo simulations) have been developed in the course of many years.

The prediction of dynamic properties is more difficult, especially far from equilibrium. A highly detailed model would require the methods of molecular dynamics which are still essentially inapplicable to polymers; indeed, the size of a representative "cell" of material easily exceeds the capabilities of the largest computers.

Fortunately, for the purpose of rheological predictions the molecular model need not be detailed. Because the time scale of the dynamic response which is of interest in rheology is usually large, the molecular model can be "coarse grained", i.e., there is no need to take as constituent parts of the polymer molecule the monomers in the chemical sense (not to speak of the individual atoms), but rather some larger piece of chain or even, in cruder models, the entire macromolecule.

The basic ingredients of molecular modeling in a rheological context are therefore: i) some appropriate idealization of the polymer chain (the "appropriateness" remains to be specified, of course), and ii) the relevant interactions of a chain with its environment, i.e., with other chains and/or with a solvent if any is present.

The best known example of idealized chain used in rheology is that of Rouse (ref. 7), which models flexible polymeric chains as sequences of beads and springs. The springs are substitutes for chain segments (sub-chains) which are small enough to respond, on the timescale of a rheological experiment, as if they were at equilibrium. Thus the elasticity of the spring is rubberlike. (A deformed rubber is a network of polymer chains in equilibrium under the action of some force.) The beads of the Rouse chain are meant to model the frictional interaction of the chain with its surroundings, the latter being treated as a continuous medium. In other words, the beads are endowed with a friction coefficient (a dynamic parameter of the model).

In fact, in the original Rouse (or Rouse-Zimm) model the friction coefficient completely determines the dynamics, because no other interaction of the chain with its surroundings is assumed to exist. With these simple basic ingredients, the model was found to adequately describe (also accounting for some modern developments on hydrodynamic interactions among the beads, see ref. 8) the behaviour of dilute polymer solutions, where the "surroundings" of a chain is the solvent. The simple Rouse model was also found to describe surprisingly well the dynamics of polymer melts, as long as the molecular weight of the polymer remained below a critical value.

It is worth emphasizing the role of the interactions among the basic ingredients of a molecular model. Indeed, let us now consider a melt of long chains, i.e., polymers with a molecular weight larger than the critical value alluded to in the previous paragraph. In such a melt, the dynamic interactions of a chain with its surroundings are no longer purely frictional; the topological interactions (entanglements) due to uncrossability of the chains are also important. An effective way of accounting for this situation was that of modeling the polymer molecule as a Rouse chain constrained inside a tube (which deforms affinely with the continuum), where the tube constraint takes care of the topological interactions (refs. 4,9). Notice that the idealized chain which mimics the polymer molecule is the same here as before, namely a Rouse chain. The molecular model is quite different, however, as a consequence of the different interactions.

Other popular models idealize the chain in the form of a dumbbell or of a rigid rod. The elastic dumbbell case can be looked upon as a crude Rouse model (only two beads and one spring). By using the dumbbell model, however, it becomes easier to account for deviations from classical rubberlike elasticity due to large chain extensions (non-Gaussian behaviour). In such a case, the molecular model is defined only after the spring elastic law has been specified (ref. 1).

The rigid rod is meant to simulate rigid or semi-rigid polymeric chains. It is well known that these polymers, either in concentrated solutions or in the melt, may form a liquid crystalline phase. The molecular model, therefore, must also account for the thermodynamic interaction which is responsible for the phase change, in particular for the isotropic-nematic transition (ref. 10,11). The interaction is usually taken care of by means of a mean-field potential of the form suggested long ago by Onsager (ref. 12) or by Maier and Saupe (ref. 13).

This rather lengthy exposition of possible "recipes" defining molecular models is justified by the consideration that, in fact, the choice of the basic ingredients represents the crucial part of the game. As discussed in the next sections, the rest of the work, all the way up to the prediction of the stress for a given deformation history, is essentially routine, at least in the sense that no other degrees of freedom are available along the way (except, perhaps, that of adopting some judicious mathematical simplifications).

A conclusion can therefore be reached already. If some of the predictions of a molecular model prove unsatisfactory, i.e., they do not compare favourably with the observations, some change must be made in the basic ingredients, at a level which pertains to the physics of the problem. For example, one could be forced to include in the model some other known physical effect previously left out for simplicity (and in the false hope that it could be neglected).

In comparing molecular modeling to the phenomenological approach, one might say that the guessing game of finding a good constitutive equation of nonlinear viscoelasticity has been moved from the mathematical guess of a suitable functional to the physical guess of acceptable basic ingredients (yet simple ones, see next section). Perhaps, the change is more than merely a matter of taste.

3.2 The "diffusion" equation

Once the idealized chain has been chosen, the coordinates required to specify a chain configuration are known. For example, the orientation in space of a rigid rodlike polymer (which is the only relevant configurational aspect in this case) is described by a unit vector. Similarly, the configuration of an elastic dumbell is specified by the end-to-end vector, whereas a Rouse chain requires a set of vectors.

For reasons which will soon be apparent, it is generally convenient that the number N of these coordinates be small, i.e., that the idealized chain be as simple as possible. (Here, a conflict may arise between simplicity and realism.) Indeed, a function $f(x_1, x_2, .., x_N; t)$ must be introduced which, at any time t, gives the probability density that a molecule be found with assigned values of its coordinates x_i. This function describes how the population of molecules is distributed among the possible configurations available to them. Now, although there are exceptions (the Rouse chain is one of them), the equation which the distribution function obeys (see next paragraphs) is much easier to handle if N is small.

The dynamical equation determining the distribution function is called either "diffusion" or Smoluchowski equation. The structure of this equation is as follows:

$\partial f/\partial t$ = Brownian (or diffusion) term +

contribution of potential +

contribution of velocity gradient (3)

As shown by Eq.(3), the rate of change of the distribution function results from three effects, now briefly described.

The first of them arises from the Brownian thermal motion of the polymer chains (or sub-chains). The random changes in conformation or in orientation of the idealized chains, due to Brownian motion, are described in Eq.(3) by means of a diffusion term which, were f dependent on a single coordinate x, would be of the form $D\,\partial^2 f/\partial x^2$, with D a diffusion coefficient.

The second term describes the effect of the existing potentials, internal and/or external to the chain. For example, for idealized chains containing springs, the spring potential energy (mostly of entropic origin) enters the picture. An external potential arises because of the surrounding chains; it is typically accounted for in the mean-field approximation. Notice that the external potential is related to the second basic ingredient of our molecular model, i.e., to the assumed interactions of the chain with its surroundings. (In some cases, not normally of interest to rheologists, potentials also arise because of fields external to the material like, e.g., a magnetic field.) Be it as it may, if V(x) is the potential (made non-dimensional by means of the Boltzmann factor kT), the second term in Eq.(3) has the structure: $D\,\partial/\partial x\,(\,f\,\partial V/\partial x)$.

Finally, the last term of Eq.(3) is directly linked to the motion. It describes how the chain conformations and/or orientations would be deterministically "dragged along" by virtue of the flow field (were not for the other two effects previously mentioned, of course). If the velocity of this drag motion is symbolically indicated as v (v is linked to the macroscopic velocity gradient through the molecular configuration), the last term has the structure: $-\,\partial/\partial x\,(\,f\,v\,)$.

For any preassigned kinematics of flow, i.e., for any assigned velocity gradient, actual solution of Eq.(3) gives us the most complete "molecular" information. Indeed, it says how the distribution of conformations and orientations of the idealized chains evolves in time and/or achieves a steady state. In the next section, it will be shown how macroscopic "observables" like the stress (or, e.g., the birefringence) are promptly obtained from the detailed information which is contained in the function f. In other words, Eq.(3) implicitly contains all the information required to completely specify the nonlinear functional of Eq.(2), and even more than that.

It is here concluded, however, by noting that Eq.(3) also generally corresponds to a quite complicated differential equation (or even to an integro-differential one, when there is a mean-field potential), which obviously poses mathematical problems. This is in fact the main limitation of molecular modeling: either the physical model is very simple, so that Eq.(3) comes out not too complex, or else solution of Eq.(3) may require drastic (and dangerous) mathematical simplifications. When the solution of Eq.(3) must be obtained numerically, a possible (physically equivalent) alternative is the technique called "Brownian dynamics" (ref. 14), whereby a "sample" of idealized chains is subjected to numerical Brownian motion, as well as to the interaction potentials assumed in the model, and to the assigned flow field.

3.3 The stress and other "observables"

In statistical mechanics, all quantities which can be macroscopically observed are "ensemble averages", i.e., they are obtained by performing an integral of the appropriate molecular quantity over all possible molecular configurations, weighted according to the corresponding probabilities (as described by the distribution function f).

If, by way of example, we assume for simplicity that the molecular configuration is specified by a single coordinate x (as was done in the previous section when showing the structure of the terms in Eq.3), then a macroscopically observable property Q is obtained through the integral

$$Q = \int_a^b f(x)\, q(x)\, dx \tag{4}$$

where q(x) is the value that Q would attain were all molecules in the configuration x, and $a<x<b$ indicates the range of possible configurations. The average in Eq.(4) is usually written $<q>$ for short, i.e., $Q=<q>$ is the "observable".

The stress tensor, and the birefringence, are observables of rheological interest. For any specified molecular model, it is a straigthforward matter to decide which average gives the stress tensor and/or the birefringence. For example, in the dumbell model with a linear spring, the average to which both the stress and the optical tensors are proportional is <**RR**>, where **R** is the end-to-end vector. This explains the so-called stress-optical law, which is in fact based upon the Gaussian behaviour of the polymer chains. If the idealized chain is unextendible, as is true for rigid polymers (or unextended, as is often the case for flexible polymers in concentrated solutions or melts), then only the orientation of the chain is relevant. Thus, if some unit vector **u** measures the chain orientation, the stress can be linked to the average <**uu**>, and sometimes also to the higher rank tensor <**uuuu**>. General methods to obtain the correct expression for the stress are described in (ref. 4).

Once the stress tensor expression which is appropriate to the specific molecular model has been established, the modeling job is formally concluded. Indeed, Eq.(3) provides us with the distribution function f(x) which must be used in an integral such as that in Eq.(4) to obtain the stress. The set of Eqs.(3-4) formally corresponds to the constitutive equation of the polymeric liquid: if the deformation of a material particle as a function of time is known (i.e., the kinematics of motion is specified), then the stress at that material particle is determined (as a function of time) through the intermediation of the distribution function f(x;t) (parametric in time). Notice that f(x;t) represents more information than contained in the final result: the averaging process which is used to calculate the stress also wipes out a lot of "molecular" details.

4. CONCLUDING REMARKS

It has been shown, hopefully in a convincing way, that constitutive nonlinearities cannot be efficiently handled within a phenomenological approach while, at least in principle, molecular modeling is much more suited. The reason is that, once the molecular models have been chosen on physical grounds, the corresponding nonlinear constitutive equations of interest to rheology remain fully determined, without ambiguities.

In actual practice, this seemingly triumphant conclusion must be considerably attenuated. In the first place, the mathematical difficulties encountered in solving Eq.(3) (even for the simplest kinematics) may become overwhelming as soon as the molecular model tries to be somewhat more detailed, i.e., as soon as one tries to account for physical aspects which are a little more than the bare "essential".

Secondly, even if the molecular model is relatively simple, and therefore Eq.(3) can be solved for any preassigned kinematics, in most problems of practical interest the kinematics is in fact unknown beforehand. Applications typically generate boundary-value problems, where velocity and/or stress conditions are assigned at the boundaries, and the flow field must be determined accordingly. The resulting kinematics does not usually correspond to any simple flow field.

In order to solve this kind of problems, one needs an "explicit" constitutive equation, i.e., an equation which links kinematics directly to stress, without the intermediation of a molecular distribution function. In other words, one needs a constitutive equation formulated in the fashion of continuum mechanics.

Such an explicit form of constitutive equation is hardly obtainable, in an exact way, from molecular modeling. Indeed, only few examples are known. More promising appears to be some suitable compromise between exactness (in the sense of exact correspondence to the molecular model adopted) and simplicity, in the sense of somehow eliminating the intermediation of the distribution function.

It so appears that the art of molecular modeling in rheology not only consists in formulating sensible physical models, but also in envisaging judicious mathematical tricks which bring one back, so to speak, from the molecules to the continuum.

REFERENCES

1. R.B. Bird, R.C. Armstrong, O. Hassager and C.F. Curtiss, Dynamics of Polymeric Liquids, Wiley, New York, 1977, Vol. 2.
2. S.F. Edwards, Proc. Phys. Soc., 92 (1967) 9-16.
3. P.G. de Gennes, J. Chem. Phys., 55 (1971) 572-579.
4. M. Doi and S.F. Edwards, The Theory of Polymer Dynamics, Clarendon, Oxford, 1986.
5. J.D. Ferry, Viscoelastic Properties of Polymers, Wiley, New York, 1980.
6. C. Truesdell and W. Noll, The Non-linear Field Theories, Springer, Berlin, 1965.
7. P.E. Rouse, J. Chem. Phys., 21 (1953) 1272-1280.
8. H.C. Öttinger, AIChE J., 35 (1989) 279-286.
9. D. Pearson, E. Herbolzheimer, N. Grizzuti and G. Marrucci, J.Polym.Sci.: Polym.Phys., 29 (1991) 1589-1597.
10. M. Doi, J. Polym. Sci.: Polym. Phys. Ed., 19 (1981) 229-243.
11. G. Marrucci in: A. Ciferri (Ed.), Liquid Crystallinity in Polymers, VCH Publ., New York, 1991, Chapt. 11.
12. L. Onsager, Ann. N. Y. Acad. Sci., 51 (1949) 627-659.
13. W. Maier and A. Saupe, Z.Naturforsch., 15a (1960) 287-291.
14. J. Honerkamp, Stochastische Dynamische Systeme, VCH Verlag, Weinheim, 1990.

Theoretical and Applied Rheology, edited by P. Moldenaers and R. Keunings
Proc. XIth Int. Congr. on Rheology, Brussels, Belgium, August 17-21, 1992

RHEOLOGY, STRUCTURE, BOUNDARY CONDITIONS AND INDUSTRIAL APPLICATIONS

ARTHUR B. METZNER

Department of Chemical Engineering, University of Delaware, Newark, DE 19716 (USA)

Several of the classic studies in rheology which we all appreciate were ones in which visual or optical measurements of structure were coupled with the more traditional measurements of stresses and deformation rates. Speakers at several meetings in the 1970's exhorted rheologists to increase their attention to the evolving material structure as the system was deformed and, very gradually, this has begun to occur. Much remains to be done.

The oral presentation of this material at the Congress will focus upon several areas within the general scope of the title, as follows.

1. Electrorheology is emerging as a very fertile field for research with some possibility for development of new industrial applications (refs. 1-4). Predictions of structure in electrorheological fluid suspensions at rest are impressively able to portray experimental observations; low shear rate rheological responses appear to be in agreement with these predictions. Physically one might expect major changes in fluid behavior with deformation rate and modes (steady or transient, shearing or extensional) and with the scale of the experiment. Additionally, it is not clear that the stress tensor should be symmetric for flows of electrorheological fluids; if it is not then additional effects of the scale of the experiment or industrial flow field are expected to arise (ref. 5). Very few studies to date have addressed these complications and there may be a need for breakthroughs in formulation of new fluids in order to capitalize on the unique fluid properties exhibited by materials in electric fields. There is at least one report of changes in material response with structure of the bounding surfaces (ref. 1).

2. Boundary conditions influence the motion of fluids in manners not appreciated until fairly recently. In the case of polymeric melts, sharkskin and melt fracture phenomena have been demonstrably linked to "slip" or "adherence" conditions at the wall of a duct or die and recent research continues to provide unexpected surprises of great practical significance (refs. 6, 7).

As a second and unrelated example of the importance of boundary conditions we may consider the flow of lubricants. In this case practitioners have long known of apparently-pervasive effects of the surfaces in contact with the lubricant; recent scientific studies (ref. 8) have shown dramatic, order-of-magnitude changes in apparent viscosity of otherwise Newtonian fluids with deformation rate and film thickness when the "fluid" is confined to very thin films.

3. Extraordinarily impressive advances are being made in studying and advancing our knowledge of the structure of concentrated suspensions of particles which are spherical (or nearly so) in shape. Experimental, analytic and numerical tools are all being applied (refs. 9-15). We now know that an apparently random structure of a suspension at rest may be transformed, under shear, into well-defined layers of closely packed particles separated by a suspending liquid which is relatively clear or free of particles. Consequently, these densely packed layers may slide over each other in shearing flows with progressively increasing ease as the duration of an imposed shear field increases - i.e. there is a transition from a solid-like to a fluid-like response, as in thixotropy, even in systems composed of macroscopic particles not normally considered to lead to thixotropy. Disruption of the well-defined layers at high shear rates may lead to the dramatic shear-thickening long known to occur in such systems. Similar disruption of "glide planes" would be expected to occur in all of the more complex flows of industrial significance, with the possibility for both beneficial and adverse changes in suspension structure and rheological response. These latter complexities do not appear to have benefited greatly as yet

from our improved knowledge of suspension structure. A modest beginning has been made in considering how differences in the material structure between dynamic and steady state shearing experiments may be employed to connect results of dynamic measurements to the steady state rheology (refs. 16, 17).

Thus far we have implied that suspensions are spatially homogeneous: they may become layered after large shearing deformations but, within a layer, the suspension everywhere is assumed to possess the same structure. Toy and coworkers (ref. 10) have documented that this may not be so, at least in suspensions of acicular and plate-like particles. This writer is not able to articulate the circumstances under which such inhomogeneities in structure are expected to arise. The publication by Toy et al. does show that suspension inhomogeneities are clearly revealed experimentally by erratic torque signals in rotational viscometers - much as found in viscometric measurements of phase-separated polymer solutions (ref. 18) - and hence a direct and simple experimental test for the presence of inhomogeneities appears to be available.

This author has not cited any of the several available research and review articles, as well as monographs, dealing with suspensions of colloidally-sized particles. Others at this meeting are much more qualified to discuss that area.

4. Flows of polymeric blends and alloys present challenges of very great industrial importance. The rheological properties of the blend and, more importantly, the physical attributes of objects molded from a blend, depend non-linearly upon the phase ratio and the properties of the interfacial regions as well as upon homopolymer properties (refs. 19-23). Again, this author looks for insights from other participants in this Congress(!).

5. Fiber fluids - suspensions of fibrous particles in polymeric melts - provide an opportunity for coupling particle orientation and motion in flows of arbitrary complexity with the rheological stresses. If the suspensions are of moderate concentration (say 30% by volume solids, or less) there may be a fairly dramatic particle drift across streamlines. These motions may result in particle-rich clusters, of varying size, surrounded by regions of low particle concentration (refs. 24, 25). An important observation arising out of these studies is that a perfectly uniform particle distribution throughout a molded object is an unrealizable circumstance in general: the evolution of fiber agglomerates having a distribution of sizes is to be expected even when the original starting material was a suspension having a perfectly uniform distribution of the particle locations within it.

SUMMARY

This very brief manuscript was prepared to draw attention to a number of challenges which, to the writer, appear to be interesting scientifically and significant industrially. Completeness of coverage cannot be claimed: several areas of importance have been omitted to avoid overlap with other contributors and others have undoubtedly been overlooked. The writer apologizes for the latter.

REFERENCES

1. E. A. Collins, C. F. Zukoski and 24 other authors, in an editorial and 10 research papers, J. Rheology, 35 (1991) pp. 1303-1461.
2. Y. Otsubo, M. Sekine and S. Katayama, Electrorheological Properties of Silica Suspensions, J. Rheol., 36 (1992) pp. 000.
3. T. C. Jordan, M. T. Shaw and T. C. B. McLeish, Viscoelastic Response of Electrorheological Fluids. II. Field Strength and Strain Dependence, J. Rheol., 36 (1992) pp. 000.
4. R. T. Bonnecaze and J. F. Brady, Yield Stresses in Electrorheological Fluids, J. Rheol., 36 (1992) pp. 73-115.
5. V. K. Stokes, Theories of Fluids with Microstructure, Springer, Berlin, 1984.
6. S. J. Kurtz and 17 other authors, Workshop on Sharkskin Melt Fracture, Abstracts of 63rd Annual Meeting, Society of Rheology, 1991.
7. S. V. Hatzikiriakos and J. M. Dealy, Wall Slip of Molten High Density Polyethylenes. II. Capillary Rheometry Studies, J. Rheology, 36 (1992) pp. 000.
8. S. Granick, Molecular Tribology, Materials Research Society Bulletin, 16 (1991) pp. 33-35.
9. R. L. Powell, S. Kim and 30 additional authors in an editorial and 9 research papers, J. Rheol., 35 (1991) pp. 719-940.
10. M. L. Toy, L. E. Scriven, C. W. Macosko, N. K. Nelson, Jr. and R. D. Olmsted, Nonhomogeneities in Couette Flow of Ferrite Suspensions, J. Rheol., 35 (1991) pp. 887-899.
11. R. L. Hoffman, Interrelationships of Particle Structure and Flow in Concentrated Suspensions, Materials Research Society Bulletin, 16 (1991) pp. 32-37.

12. B. J. Ackerson, Shear Induced Order and Shear Processing of Model Hard Sphere Suspensions, J. Rheol., 34 (1990) pp. 553-590.
13. J. F. Brady and G. Bossis, Stokesian Dynamics, Ann. Revs. Fluid Mech., 20 (1988) pp. 111-157.
14. H. M. Laun, R. Bung, S. Hess, W. Loose, O. Hess, K. Hahn, E. Hädicke, R. Hingmann, F. Schmidt and P. Lindner, Rheological and Small Angle Neutron Scattering Investigation of Shear-Induced Particle Structures of Concentrated Polymer Dispersions Submitted to Plane Poiseuille and Couette Flows, J. Rheol., 36 (1992) pp. 000.
15. A. J. Poslinski, M. E. Ryan, R. K. Gupta, S. G. Seshadri and F. J. Frechette, Rheological Behavior of Filled Polymeric Systems, Parts I and II, J. Rheol., 32 (1988) pp. 703-735 and 751-771.
16. D. Doraiswamy, A. N. Mujumdar, I. Tsao, A. N. Beris, S. C. Danforth and A. B. Metzner, The Cox-Merz Rule Extended - A Rheological Model for Concentrated Suspensions and Other Materials with a Yield Stress, J. Rheol., 35 (1991) pp. 647-686. See also I. M. Krieger, J. Rheol., 36 (1992) pp. 215-217.
17. A. N. Mujumdar, A. N. Beris and A. B. Metzner, Transient and Steady Behavior of Thixotropic Suspensions With Yield Values, J. Rheol., submitted for publication.
18. C. Rangel-Nafaile, A. B. Metzner and K. F. Wissbrun, Analysis of Stress-Induced Phase Separations in Polymer Solutions, Macromolecules, 17 (1984) pp. 1187-1195.
19. D. Graebling and R. Muller, Rheological Behavior of Polydimethylsiloxane/Polyoxyethylene Blends in the Melt. Emulsion Model of Two Viscoelastic Liquids, J. Rheol., 34 (1990) pp. 193-205 and 1333.
20. J. F. Palierne, Linear Rheology of Viscoelastic Emulsions with Interfacial Tension, Rheol. Acta, 29 (1990) pp. 204-214.
21. L. A. Utracki, On the Viscosity - Concentration Dependence of Immiscible Polymer Blends, J. Rheol., 35 (1991) pp. 1615-1637.
22. E. W. Chan and 35 additional authors, Polymer Processing of Homopolymers and Blends, Abstracts of 63rd Annual Meeting, Society of Rheology, 1991.
23. D. D. Joseph, M. S. Arney, G. Gillberg, H. Hu, D. Hultman, C. Verdiere and T. M. Vinagre, A Spinning Drop Tensioextensometer, J. Rheol., 36 (1992) pp. 000.
24. M. L. Becraft and A. B. Metzner, The Rheology, Fiber Orientation and Processing Behavior of Fiber-Filled Fluids, J. Rheol., 36 (1992) pp. 143-174.
25. T. A. Plumley, Clustering and Migration of Fibers Suspended in Molten Polymers, Ph.D. Thesis, University of Delaware, Newark, DE (1992).

Theoretical and Applied Rheology, edited by P. Moldenaers and R. Keunings
Proc. XIth Int. Congr. on Rheology, Brussels, Belgium, August 17-21, 1992

RHEOLOGY AND COMPUTATION

R. I. TANNER

Department of Mechanical Engineering. The University of Sydney, Sydney, 2006 (Australia).

1. BACKGROUND

A long-term goal of rheology is the reliable prediction, from some prior knowledge of the material's characteristics, of the stresses, temperatures, and flow patterns in new design-driven situations. This program demands, (i) A good mathematical description of the material (rheology), and boundary conditions and (ii) A feasible computational strategy for the system of equations forming the description.

This paper will review some aspects of these twin goals. It is assumed that the use of large or at least medium-scale computation is necessary; experience has shown that even with creeping, Newtonian flows (a linear problem) very few meaningful predictions can be made without computation. Most of the discussion on the rheology will be aimed at polymer melts and solutions. This is done simply to limit the scope of the present paper; the challenges implicit in liquid crystals, electrorheology, concentrated suspensions, and composite materials remain to be discussed elsewhere.

2. RHEOLOGY

Without a reasonable description of the material computational virtuosity is not helpful, so we begin with the question of the constitutive equation. Many materials are nearly incompressible, and this will be assumed here; in problem of die-filling, for example, there are good arguments for relaxing this condition and considering the material as having a density related to pressure and temperature. Errors of order 10% can result if the constant density assumption is made. Examples of computations with compressible media are hard to find; C. R. Beverly (ref. 1) has considered this problem and has computed some results.

Thus we assume

$$\frac{\partial u_i}{\partial x_i} = 0 \qquad (1)$$

where u_i is the component of velocity in the x_i direction..

Similarly, the question of the symmetry of the stress tensor may need careful consideration for some materials. For polymer melts and solutions available evidence from molecular theories (ref. 2) indicates that the assumption of symmetry is satisfactory; it is assumed here; I am not aware of any computations that do not make this assumption.

Given the above constraints, then one uses the decomposition of the stress tensor (σ_{ij}) into

$$\sigma_{ij} = -p\delta_{ij} + \tau_{ij} \qquad (\text{with } \sigma_{ij} = \sigma_{ji}) \qquad (2)$$

where p is the pressure, δ_{ij} is the unit tensor, and τ_{ij} is the extra-stress tensor (ref. 2). The pressure p is not determined by the motion; τ_{ij} is the part of the (symmetric) stress tensor determined by the motion. The balance of linear momentum demands (ref. 2)

$$\frac{\partial \sigma_{ij}}{\partial x_j} + \rho f_i = \rho a_i \qquad (3)$$

where ρ is the density, f_i is the body force component and a_i is the acceleration component. Often we shall ignore ρf_i and ρa_i as they are frequently unimportant in polymeric flows. The effects of temperature on flow are substantial (ref. 2) and generally need to be assessed; however, since it has not proved to be difficult to couple the energy equation with the equations given above, we shall only consider isothermal situations here.

The remaining task in terms of the bulk equations is to connect τ_{ij} to the motion. Sometimes additional structural variables are also involved but ultimately τ_{ij} has to be related to the history of the motion (refs. 2-4).

Constitutive equations that have been used in computational schemes may be classified into various families: Rate-type equations, integral-type expansions and differential models (see ref. 2-4). The scheme is shown below.

2.1 <u>Families of Equations</u>

- Rate-type Equations
  - Generalized Newtonian
  - Viscometric
  - Higher-order equations

- Integral-type Expansions
  - . Green-Rivlin Expansions
  - . KBKZ
  - . Variants of KBKZ
  - . Inelasticity
  - . Larson's Equivalents

- Differential Models
  - . Maxwell, Oldroyd-B
  - . PTT, MPTT
  - . Leonov, modified Leonov
  - . Larson, reversible, irreversible
  - . Giesekus
  - . Others

It has been shown elsewhere (ref. 2) that the Rate-type equations are inadequate for describing general flows; the higher-order models are also difficult to compute with, and they will not be discussed further.

Of the Integral expansions the K-BKZ (ref. 2) is certainly the most realistic, but, as Wagner (ref. 5) has emphasized, it is too "elastic" and an irreversible mechanism has to be introduced into the description in order to describe recoil after release of stress. The theme of irreversibility has also been explored by Larson (ref. 3). Thus, contrary to earlier expectations and beliefs it seems doubtful that the correct limit of polymer behaviour for very fast strains is a purely elastic limit - some loss of elasticity seems to occur, and needs to be accounted for.

For the differential models, a great deal of effort has been expended on single-mode (single relaxation time) models. These models are of the general form

$$\lambda \frac{\Delta \tau_{ij}}{\Delta t} + R_{ij} = 0 \tag{4}$$

where R_{ij} is a function of the velocities, velocity gradients and τ_{ij} itself, λ is a relaxation time and

$$\frac{\Delta \tau_{ij}}{\Delta t} = \frac{\partial \tau_{ij}}{\partial t} + u_k \frac{\partial \tau_{ij}}{\partial x_k} - \tau_{ik} L_{jk} - L_{ik} \tau_{kj} \tag{5}$$

$$\text{where } L_{ij} \equiv \partial u_i / \partial x_j \tag{6}$$

For the PTT model

$$R_{ij} = (1 + \varepsilon \frac{\lambda}{\eta} \tau_{kk}) \tau_{ij} - \eta (L_{ij} + L_{ji}) \tag{7}$$

where η is the viscosity and ε is a dimensionless parameter. When $\varepsilon = 0$ one has the Maxwell model; keeping $\varepsilon = 0$ and adding a solvent viscosity (ηs) gives the Oldroyd-B model. If an appreciable range of stress occurs in the field of flow, then a single mode is unrealistic. Recent trials seem to indicate that the PTT model (refs. 2-3) behaves at least qualitatively correctly in many cases. Mr. C. R. Beverly (University of Sydney) has also shown that an accurate eight-mode description of the IUPAC Low-Density Polyethylene (LDPE) (ref. 6) can be made using this model. However, other models, such as these due to Larson and Giesekus (ref. 3) behave quite similarly.

In summary, we have now got some reasonably good models, but further improvement and a more systematic inclusion of irreversibility would be welcome.

3. BOUNDARY CONDITIONS

The experimental evidence now accumulating suggests that for highly stressed polymers "slip at the wall" often occurs; the usual no-slip boundary condition may not always be adequate. Phan-Thien (ref. 7) has done some exploratory computations with slip included (see also ref. 8).

As an alternative explanation for some of the instabilities seen in flows, it has been postulated that instead of wall slip one is seeing a real effect built into the constitutive model, (as opposed to the wall boundary condition). Decisive evidence to choose between these ideas seems to be lacking now.

Another difficult group of problems arises in connection with the separation of a fluid stream from the lip of a rounded-exit die, or the peeling of a calendared sheet from the rollers. It is not known a priori where the separation paints occur, nor what the criteria for separation are.

4. COMPUTATIONAL SCHEMES

Various reviews of computations have appeared and we refer to Crochet's (ref. 8) as a recent one. Computations at high Weissenberg numbers have proved difficult to achieve in some cases. The Weissenberg number (Wi) is defined for these models in steady flows as

$$Wi = \lambda U/L \tag{8}$$

where λ is the (single) relaxation time, U is a characteristic velocity, and L is a characteristic length. We split the review according to the model type.

4.1 Differential Models

The great majority of computations in this class have been performed with single-mode equations, especially the Maxwell and Oldroyd-B models (ref. 8). Reviewing recent progress with this type of model one can see distinct progress for plane and axisymmetric flows when the geometry is smooth and no free surfaces are present. The flow around a sphere is such a case, and at least two stable, accurate methods have been proposed for it (refs. 9,10) For the flow in a tube with a sinusoidal profile, also a smooth flow, Pilitsis and Beris (ref. 11) have shown that spectral methods are very accurate and stable. For three dimensional cases no feasible methods are yet available for the Maxwell and Oldroyd-B models at high Weissenberg numbers. For problems with free surfaces involving stress singularities, there are many difficulties, especially with the Oldroyd-B and Maxwell models. Rosenberg and Keunings (ref. 12) have discussed this problem carefully. Unsteady flows have also been studied with single-mode models, and recently some work on multimode models has been attempted. However, these computations are quite expensive.

The difficulties with single-mode models of the Maxwell-Oldroyd B type in problems with stress singularities have been mentioned above. One difficulty that has not yet been resolved with these models is the nature of the singularity near a corner or a separation point. With the PTT and similar models, a balance of the first term on the left of eqn. (4) and the term multiplied by ε in the expression (7) seems to occur in regions of very high stress. This indicates that the nature of the singularity is Newtonian, but that there is an increased effective viscosity, of order η/ε. Similar results hold for other differential models. The use of the PTT model, as opposed to the Maxwell, has enabled more stability to be achieved with some computational schemes.

4.2 Integral Expansion Models

Along with increased realism one also has a decrease in computational problems with this type of model. For example in ref. 13, a constitutive equation of the KBKZ type was used, where

$$\tau_{ij} = (1-\theta)^{-1}\int_{-\infty}^{t} M(t-t')\,\phi\,(I_c, I_{c-1}) \times \left\{ C_{ij}^{-1}(t') + \theta Cij(t') \right\} dt' \tag{9}$$

Here

$$M = \sum_{k=1}^{8} \frac{a_k}{\lambda_k} \exp\left(-(t-t')/\lambda_k\right) \tag{10}$$

is a (linear) memory function with 8 modes, ϕ is given by

$$\phi = \frac{\alpha}{(\alpha-3) + \beta_k I_{c-1} + (1-\beta_k) I_c} \tag{11}$$

In (9) C_{ij}^{-1} is the Finger tensor measured relative to the current configuration (ref. 2) and I_{c-1} is the value $(C^{-1})_{ii}$, while $I_c = C_{ii}$.

The α and β_k are parameters chosen to fit the shear and elongational viscosities and θ is related to the ratio of the second normal stress difference to the first normal stress difference; so that

$$\frac{N_2}{N_1} = \frac{\theta}{1-\theta} \tag{12}$$

Luo and Tanner (ref. 13) chose $\theta = -1/9$, as indicated in experiments on solutions (ref. 2).

With this model a reasonably satisfying simulation of IUPAC low-density polyethylene extrusion at 150°C was made (ref. 13). The main problem was that at higher rates of extrusion the computation over-predicted the swelling of the extrudate. Agreement between experiments and computations could be restored by permitting slip at the last two elements on the wall of the die nearest the exit. Attempts to introduce "irreversibility" in the manner of Wagner (ref. 5) did not improve the agreement with experiment, which was surprising. Although the average stress at the die wall was not high (<0.1MPa) it is certain that the lip stresses were considerably higher than the average and it is possible that local, as higher than the average and it is possible that local, as opposed to global, slip occurs near the exit. Clearly, this is a major problem for simulation.

The numerical methods used in ref. 13 involved integration along streamlines and the use of elements whose edges are partly defined by streamlines. While this helps to maintain accuracy in the stress field, it is very inconvenient for problems with recirculating regions. This drawback has now been overcome by using new, false-transient type techniques for integrating the stress equations (9).

These improvements in computational techniques have been made by Luo (refs. 14, 15) and results are able to give reasonable descriptions of some experiments, including 22:1 and 4:1 contraction flows (ref. 15). Although stable, and able to handle free surfaces, these computations are slow and are not able to take advantage of Newton-Raphson schemes. (Boundary element methods also suffer from this problem, see ref. 7). Few unsteady results are available for integral models, and no three-dimensional results are available. Some 3-D boundary element results have been produced (ref. 16) but the computational effort was large.

In fact, the solution of Newtonian and simple variable viscosity (power-law, Carreau model, etc.) problems in three-dimensions (3D) is still excessively time-consuming. For creeping Newtonian flows (ref. 16) the boundary-element method is sometimes convenient, but for non-linear problems it is slow. The new Fortin elements (ref. 8) have speeded up these simpler 3D computations enormously.

5. SUMMARY

There clearly remain a number of unsolved problems in rheology and computation. Some needing urgent solution are

(i) Accurate treatment of singular points on free surfaces
(ii) True 3-D capability
(iii) Special boundary problems, including slip and separation
(iv) Improved constitutive relations
(v) Transient flows

It is hoped that solutions will be forthcoming soon to these major issues.

I thank the University of Delaware (Department of Mechanical Engineering) for hospitality during the preparation of this paper.

REFERENCES

1. C. R. Beverly and R. I. Tanner, in Proceedings 5th National Congress on Rheology, Melbourne, Australia, June 1992, in press.
2. R. I. Tanner, Engineering Rheology, Oxford University Press, Revised Edition, 1988.
3. R. G. Larson, Constitutive Equations for Polymer Melts and Solutions, Butterworths, Boston, 1988.
4. R. B. Bird, R. C. Armstrong and O. Hassager, Dynamics of Polymeric Liquids, Vol. 1, Fluid Mechanics, Wiley, New York, Second Edition, 1987.
5. M. H. Wagner, J. Non-Newtonian Fluid Mechanics, 4 (1978) pp. 39-55.
6. J. Meissner, Pure and Appl. Chem., 42 (1975) pp. 551-612.
7. N. Phan-Thien, J. Non-Newt. Fluid Mech., 26 (1988) pp. 327-340.
8. M. J. Crochet, Rubber Chemistry and Technology, 62 (1989) pp. 426-455.
9. R. C. King, M. N. Apelian, R. C. Armstrong and R. A. Brown J. Non-Newt. Fluid Mech., 29 (1988) pp. 147-216.
10. M. J. Crochet and V. Legat, J. Non-Newt. Fluid Mech., 00 (1992) pp. 000-000.
11. S. Pilitsis and A. N. Beris, J. Non-Newt. Fluid Mech., 31 (1989) pp. 231-287.
12. J. Rosenberg and R. Keunings, J. Non-Newt. Fluid Mech., 40 (1991) pp. 123-151.
13. X-L. Luo and R. I. Tanner, Int. J. for Num. Meth. in Engineering, 25 (1988) pp. 9-22.
14. X-L. Luo and E. Mitsoulis, Int. J. for Num. Meth. Fluids, 11 (1990) pp. 1015-1031.
15. H. J. Park and E. Mitsoulis, J. Non-Newt., Fluid Mech., 00 (1992) pp. 000
16. T. Tran-Cong and N. Phan-Thien, Rheol. Acta, 27 (1988) pp. 639-648.

Theoretical and Applied Rheology, edited by P. Moldenaers and R. Keunings
Proc. XIth Int. Congr. on Rheology, Brussels, Belgium, August 17-21, 1992

RECENT DEVELOPMENTS IN RHEOMETRY

K. WALTERS

Department of Mathematics, University College of Wales, Aberystwyth (U.K.)

SUMMARY

In any detailed study of non-Newtonian Fluid Mechanics, Rheometry must have a role to play. In the present subjective assessment of recent developments in the field, attention is focussed on the following topics:

(i) The antithixotropic shear-thickening instability which appears above a critical value of the Deborah number in rotational shear flows of some elastic liquids.

(ii) The difficulty of measuring the extensional viscosity of mobile elastic liquids.

(iii) The qualitative differences in the rheometrical behaviour of elastic liquids with different microstructures.

(iv) The existence or otherwise of the 'yield stress'.

1. INTRODUCTION

Any comprehensive study of non-Newtonian Fluid Mechanics must involve the following components:

(i) A systematic study of the behaviour of the non-Newtonian liquids involved in simple flow situations, such as those found in conventional rheometers. The discussion in the present paper is biased towards the behaviour of *highly elastic* liquids and the flows of interest include steady simple shear flow, oscillatory shear flow and extensional flow. In the former, the normal stress differences as well as the shear stress are of interest.

(ii) The construction of suitable constitutive equations for the elastic liquids under test. Such equations must necessarily satisfy the well known mathematical constraints arising from a consistent application of the Principles of Continuum Mechanics. Further, their general form can often be deduced from a consideration of the fluid's microstructure. It goes without saying that any proposed equations must be able to simulate, in simple flows, the rheometrical data provided in (i) above.

There is no difficulty in satisfying these conditions *in principle*, but some pragmatism is usually required to meet the dual constraints of tractability and predictive capacity.

(iii) The prediction of the behaviour of the elastic liquids in *complex* flows of practical importance. To all intents and purposes this now involves *Computational* Fluid Dynamics applied within a non-Newtonian framework and the chosen constitutive equations have to be solved in conjunction with the familiar equations of motion and continuity. Nowadays, numerical methods are usually indispensable.

(iv) The final component in the research programme involves an *experimental* study of the behaviour of the elastic liquids in the complex flows already studied theoretically in (iii). A comparison of the predictions of (iii) and the experimental data of (iv) must be seen as an essential part of the Scientific Method.

In general terms, it is true to say that much progress has been made in all four areas, but our major concern in the present communication is with area (i), i.e. Rheometry.

At first sight it may appear that recent developments in Rheometry have been slow and rather low key. Certainly, most of the basic mathematics required for a successful development of a Rheometrical Expert System has been in place for some time and any recent activity has either been of peripheral interest or a matter of 'finite tuning'. There have been one or two notable exceptions. For example, the effect of instrument inertia on the interpretation of data from controlled-stress rheometers formed the basis of a recent Bingham -

award lecture (*1*).

On the instrumentation side, there have been significant developments. Rheometers have become ever more sophisticated and 'user-friendly', reflecting major developments in such areas as electronics and computing. The resulting improvement in instrument performance has made possible the investigation of rheometrical tests which hitherto had been fraught with experimental difficulties. Notable amongst recent advances have been step-change techniques (see, for example, *2*, *3*).

There have also been important advances in our knowledge of the rheometrical behaviour of complicated systems like filled polymer melts, but such developments will not be discussed in the present communication.

It would be wrong to underplay the relevance of all these developments, but here we shall major on fluid-mechanical aspects of Rheometry. The four areas chosen for particular comment reflect some of the author's own interests, but it is to be hoped that the following discussion will not appear too myopic.

2. A SURPRISE - INSTABILITIES IN ROTATIONAL SHEAR FLOWS

Instabilities in shear flows, even for Newtonian liquids, are not uncommon. The appearance of Taylor vortices above a critical set of conditions in some Couette flows is very well documented as is the transition to turbulence in capillary flow. Viscoelasticity invariably introduces extra complications and anyone interested in Birfurcation and Chaos will surely find a fertile field of study in non-Newtonian Fluid Mechanics!

So far as rotational shear flows are concerned, especially those generated in torsional flow between rotating parallel plates or in the corresponding cone-and-plate flow, shear fracture is an important and well documented instability, which severely restricts the attainable shear-rate range for highly elastic liquids (cf *4*, *5*). Here, we report on another instability which apparently was unknown before the mid nineteen eighties. The story is an interesting one, since it displays a productive interaction between Rheometry and Fluid Mechanics.

In 1984, the present author and his colleagues (*6*) observed an unexpected antithixotropic shear thickening in a Boger (*7*) fluid, which was a dilute solution of a high molecular weight polyacrylamide in a maltose syrup/water solvent. Up to that time, the main attraction of Boger fluids had been their (reasonably) constant shear viscosity with shear rate. Jackson et al (*6*) observed that at high shear rates, measurements of the torque and the total normal force in a Weissenberg Rheogoniometer increased steadily *at fixed shear* rate over a considerable period of time. Figure 1 from reference (*6*) illustrates the scale of the effect.

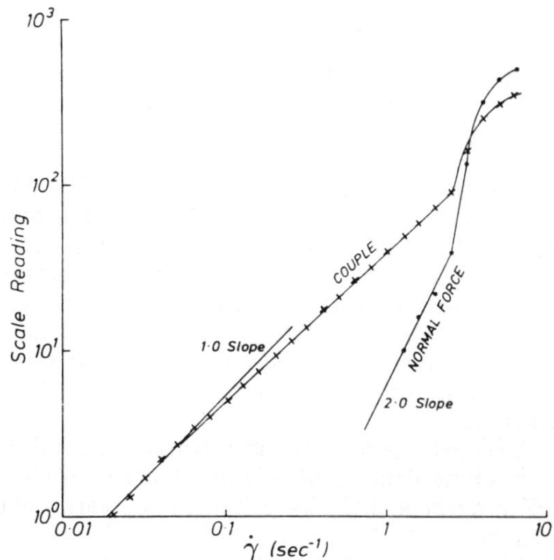

Fig 1. Raw data for a Boger fluid obtained from a torsional flow experiment in a Weissenberg Rheogoniometer. (From Ref *6*.)

The data show the final 'equilibrium' values at each shear-rate for a Boger fluid in a torsional flow. (If the *initial* scale readings had been plotted there would have been no discontinuities in the two curves.) The figure is more than adequate to show the departure from so called 'second-order' behaviour and to illustrate that the *apparent* build up of structure is substantial. Interestingly, Jackson et al noted that halving the rheometer gap at a constant value of shear rate appeared to remove the antithixotropy and they concluded that the rim shear rate in torsional flow could not therefore be the controlling factor of the phenomenon.

At about the same time Phan Thien (*8*, *9*) carried out an instability analysis for the Oldroyd-B constitutive equation in torsional flow and in cone-and-plate flow and predicted critical conditions for the onset of instabilities in both geometries.

If Ω_c is the critical angular velocity at which the instability sets in, it is convenient to define a critical Deborah number D_c by $\lambda\Omega_c$, where λ is the relaxation time. For torsional flow, Phan-Thien (*8*) predicted

$$De_c^{(pp)} = \frac{\pi}{[(1-\beta)(5-2\beta)]^{\frac{1}{2}}}, \tag{1}$$

where β is the ratio of the solvent viscosity to the total viscosity in the Oldroyd-B model. In the corresponding cone-and-plate flow, the Phan Thien (*9*) result is

$$De_c^{(cp)} = \left[\frac{2}{5(1-\beta)}\right]^{\frac{1}{2}}. \tag{2}$$

It is generally agreed that the Oldroyd-B model is a reasonable choice for many Boger fluids and, although it is now conceded that more complicated models are required to capture important aspects of the behaviour of Boger fluids in complex flows (*10*), the Phan-Thien stability analyses must have something important to say to those who study Boger fluids in rotational shear flows, and so it has turned out.

Detailed experiments and a thorough analysis of the resulting data by Magda and Larson (*11*) and McKinley et al (*12*) have helped to clarify the situation.

Magda and Larson (*11*) carried out a series of experiments on Boger fluids and Fig 2 shows the 'long time' shear viscosity of one of the fluids as a function of the 'long-time' value of the shear rate for torsional-flow in a controlled-stress rheometer.

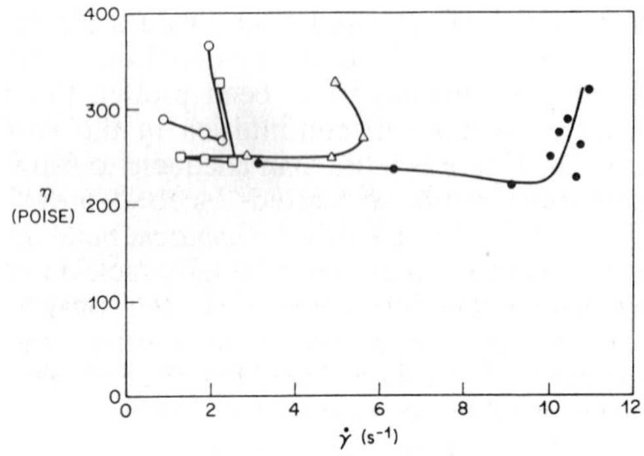

Fig 2. The long time apparent shear viscosity of a Boger fluid as a function of the long time value of the shear rate. Each curve corresponds to a different rheometer gap, and the various points along the curve correspond to different values of the time-dependent stress. Gap: ❍ 2mm, ❑ 1.2mm, Δ 0.6mm and ● 0.3mm. (From Ref *11*.)

Of most importance in the present context is the systematic variation of the critical shear rate with rheometer gap. Indeed, the inverse proportionality between the critical shear rate and the gap led Magda and Larson to view the apparent shear thickening as a manifestation of the Phan-Thien instability, especially since the observed location of the transition was close to that predicted by Phan-Thien.

A later very detailed study of this apparent shear-thickening phenomenon carried out by McKinley et al (*12*) has further elucidated the situation. Their basic findings confirmed the results and conclusions of Magda and Larson that the onset of the instabilities is governed, not by the shear rate or shear stress, but primarily by the rotation rate (i.e. the Deborah number) and that the instability occurs reasonably close to the Phan-Thien prediction. However, McKinley et al (*12*) were able to observe the instability and Fig 3 is a reproduction of their very revealing findings for a glass-based parallel-plate geometry. The figures on the individual pictures relate to the time from start up of the experiment. The rotation rate was rapidly increased at t=1.00 min from a subcritical to a supercritical value of the Deborah number. Fig 3(a) shows the fluid streamlines at 41.9 s before the onset of the instability. In Fig 3(b) a radial banded structure is evident with roll cells that are practically axisymmetric. These streamlines are not steady in time but increase in intensity and propagate both radially outwards from the centre of the disk and radially inwards from the outer rim. When the two sets of rolls meet, the flow undergoes another change of structure (3d).

McKinley et al (*12*) noted that the experimental observations did not match the linear stability requirements of the Phan-Thien theory and were minded to conclude that the good agreement between the experimental observations and the theoretical stability analysis was probably fortuitous. A thorough and very difficult analysis is required to confirm or otherwise this negative conclusion, but time may yet reveal that assigning the Phan-Thien analysis to the graveyard of the 'not relevant' is premature.

One thing is certain. The observed antithixotropic behaviour of Boger fluids is not due to any build up of internal structure in the polymer solutions, but rather the result of an interesting transition from a simple viscometric flow to a time-dependent three-dimensional state in the rheometers. Formal stability analysis (*8*, *9*), routine rheometrical tests (*6*) and detailed study of the flow (*10*, *11*) have all helped to highlight and interpret a provocative finding, which should not be overlooked or ignored by those but simply use Rheometry as a tool (and have no particular interest in Analysis or Fluid Mechanics *per se*.

3. A DISAPPOINTMENT - MEASUREMENT OF EXTENSIONAL VISCOSITY

Figure 4 is an extension (to 1992) of a subjective assessment made by the present author at the 8th International Congress on Rheology in

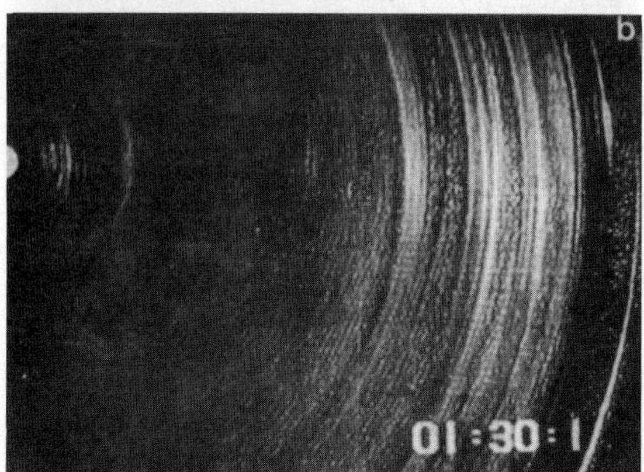

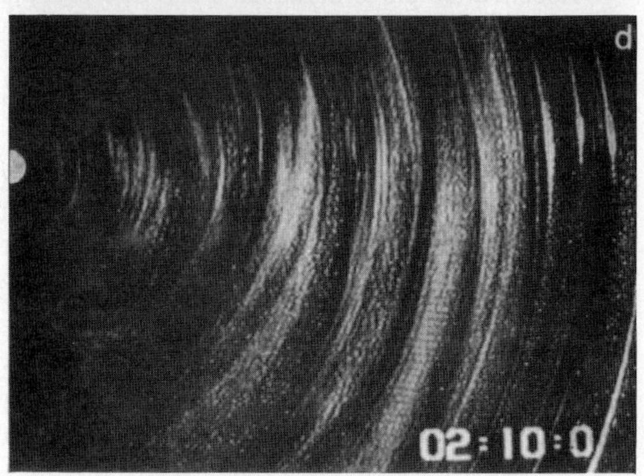

Fig 3. Visualization of the spatial structure of the rotational flow instability in a parallel-plate geometry (a) viscometric flow at a subcritical Deborah number; (b) onset of unstable flow at a supercritical Deborah number showing initial formation of an axisymmetric cellular structure at t = 1:30 min; (c) radial propagation of the secondary flow across the disk at t = 1:40 min; (d) three-dimensional, time-dependent structure ultimately observed in the flow at t = 2:10 min. (From Ref *12*.)

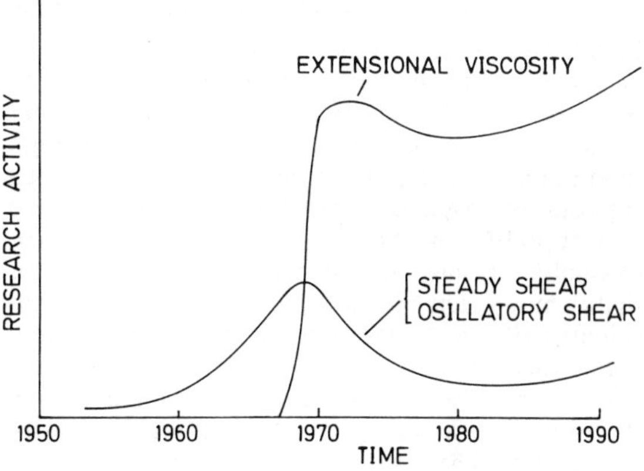

Fig 4. Research activity on an arbitrary scale.

Naples (*13*). It gives an indication, which is necessarily subjective, of the level of research activity, on an arbitrary scale, in various aspects of rheometrical studies. The dominant feature in 1980, and still in 1992, is the significant interest in *extensional* viscosity measurements which shows little sign of abating.

Consider a uniaxial extensional flow field given in Cartesian coordinates by

$$v_x = \dot{\epsilon}x, \quad v_y = -\frac{\dot{\epsilon}}{2}y, \quad v_z = -\frac{\dot{\epsilon}}{2}z, \tag{3}$$

where $\dot{\varepsilon}$ is the (constant) extensional strain rate. The corresponding stress distribution for an elastic liquid can be written in the form (*5*).

$$\sigma_{xx} - \sigma_{yy} = \sigma_{xx} - \sigma_{zz} = \dot{\epsilon}\eta_E(\dot{\epsilon}), \tag{4}$$

where σ_{ik} is the stress tensor and η_E is the extensional viscosity.

It is generally agreed that it is far more difficult to measure η_E than the shear viscosity, especially for mobile liquids. In the case of liquids with relatively high viscosities, like polymer melts, the problem is one of maintaining an extensional flow long enough for the stress (in a controlled strain-rate experiment) or the strain rate (in a controlled stress experiment) to reach a steady state, thus enabling transient effects, not allowed for in (3) and (4), to be ignored.

The problems of determining η_E for mobile liquids like dilute polymer solutions are more acute and are of a different kind to those encountered with polymer melts (cf *5*). However, for many polymer solutions, the extensional viscosities are so high (and potentially important) as to justify a less than rigorous approach, something which would be quite unacceptable in the case of shear flows, for example. It was suggested by the present author that it may often be sufficient to seek simple, but scientifically acceptable, techniques, which yield extensional-viscosity *levels* and are able to indicate whether the liquids are tension thickening or tension thinning, i.e. whether the extensional viscosity increases or decreases with strain rate. The task is therefore to generate a flow which is dominated by extension and then to address the problem of how best to interpret the data in terms of material functions which are rheologically meaningful. Such a philosophy was expanded by the present author at the First European Rheology Congress in Graz in 1986 (cf *14*). To say that it did not meet with unbridled enthusiasm would be an under-statement, but the same basic ideas had already been taken up by other workers in the field and, either knowingly or unknowingly, others were soon to follow. So much so that it took a number of International Workshops on the subject to elucidate the issues (cf *15*, *16*).

The various possibilities have been discussed at length by James and Walters (*17*). Important candidates for study have been

(i) Spinning devices.
(ii) Open-syphon flow.
(iii) Opposing-jet techniques.
(iv) Contraction flows.
(v) Converging channel flows.

When the various devices were first introduced, there were no well characterized test fluids to use in any assessment exercise. This omission was rectified when the test fluids M1 and A1 became available (see, for example, *16*). These fluids have enabled a comparison to be made of the results from each device and Fig 5 is the (disappointing) outcome of one such study. The names in the Figure are associated with the various contributors to the Workshop (*16*).

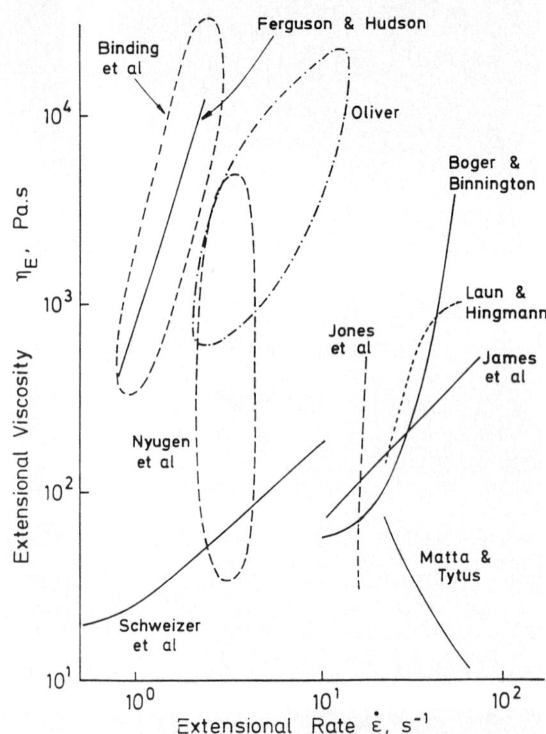

Fig 5. Extensional viscosity measurements on the M1 fluid. The lines and envelopes represent data by the various M1 investigators as published in reference *16*. (Figure taken from Ref *17*.)

James and Walters (*17*) point out that most of the experiments could not be criticised on reproducibility and accuracy and the divergence and spread of the data is due to the interpretation of what are for the most part respectable results.

Recent attempts to understand and interpret the spread of data in Fig 5 using conventional non-Newtonian Fluid Mechanics have been reasonably successful and it is now possible to make some general observations (*17*).

(i) The extensional viscometers should be viewed as providing measurements of *an* extensional viscosity rather than *the* extensional viscosity. In this sense, the various viscometers may have significant utility in Quality Control.

(ii) It is essential to specify accurately the 'prehistory' before any extensional-flow test section and also the duration of the extensional deformation if the resulting interpretation of the data is to have objective relevance.

(iii) Many of the extensional-viscosity techniques may play a positive role in constructing constitutive equations for test fluids.

It may be argued that the above conclusions should have been self evident, given the complicated flow field in the extensional viscometers. However, it would be wrong to deprecate optimism, however misguided, and we should not underestimate the powerful part played by international cooperation in this important area of rheometrical research.

4. A DISCOVERY - THERE IS A QUALITATIVE DIFFERENCE BETWEEN THE BEHAVIOUR OF POLYMERIC AND COLLOIDAL SYSTEMS

Those of us who were nurtured on a diet of Continuum Mechanics might be forgiven the belief that an elastic liquid is an elastic liquid, whatever the detached microstructure. Such a belief might be fuelled by limiting relationships between the various functions like (cf *5*)

$$\eta'(\omega)\big|_{\omega\to 0} = \eta(\dot{\gamma})\big|_{\dot{\gamma}\to 0}, \tag{5}$$

$$\frac{G'(\omega)}{\omega^2}\bigg|_{\omega\to 0} = \frac{N_1}{2\dot{\gamma}^2}\bigg|_{\dot{\gamma}\to 0}, \tag{6}$$

relating the viscometric functions η and N_1 and dynamic functions η' and G'. The preconception is only disturbed when actual rheometrical experiments are carried out on systems with different microstructures. So, for example, a solution of a high-molecular-weight flexible polymer might behave in a qualitatively different way to a colloidal system or even a solution of a semi-rigid polymer. Such a discovery was made by the present author after conducting a series of experiments on aqueous solutions of a flexible polyacrylamide and an aqueous solution of a semi-rigid xanthan gum (cf *18*). The respective concentrations (2% for the polyacrylamide and 3% for the xanthan gum) were chosen such that the shear viscosities were very similar over a reasonable range of shear rates, the idea being that any difference between the two solutions would not be apparent to many with access only to a conventional viscometer. Figure 6 shows the resulting viscometric data for the two solutions. The

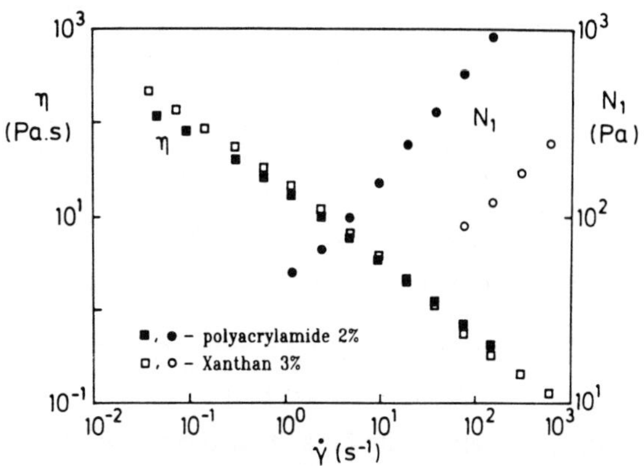

Fig 6. Steady shear data for a 2% polyacrylamide solution and a 3% xanthan gum solution. (From Ref *18*.)

differences in the normal-stress behaviour is clear and is not unexpected. In this mode of deformation, the polyacrylamide solution appears to be more 'elastic'. Such a conclusion is supported and enhanced by use of any of the extensional flow devices discussed in §3 (cf (*5*) p93); the polyacrylamide solution offers far more resistance to an extensional deformation.

The whole question of 'elasticity' in a given liquid is thrown into confusion by the dynamic data for the two solutions shown in Fig 7. Here, the dynamic viscosity η' for the two solutions is very similar, although there is no fundamental reasons why this should have been the case. Of far more interest is the reversal of roles so far as G' is concerned. The gel-like response for the xanthan gum solution yields a higher G' over a substantial frequency range and in this mode of deformation at least, the xanthan gum solution must be seen as being more elastic than its polyacrylamide counterpart. [Of course, the limiting relation (6) would necessitate some reversal of roles in either Fig 6 or Fig 7, but the relevant measurements to elucidate this point are far from easy.]

Many rheologists would find no difficulty accepting the behaviour shown for the xanthan gum solution and would see a correspondence with the behaviour of many colloidal systems. Here, we are looking at a system with a viscoelasticity which is very strong near the rest state, but which is readily broken down by shear and/or extension.

As an aside, we remark that Figs 6 and 7 and the supporting data given in ref (*18*) do illustrate how difficult it is to answer the question " How elastic is a given non-Newtonian liquid?" unless the mode of deformation in mind is made explicit.

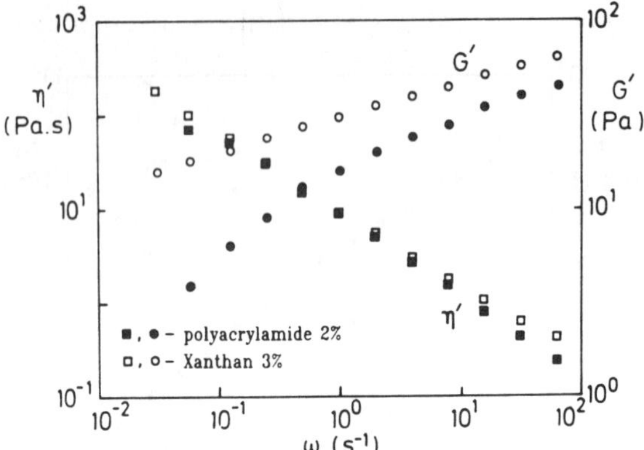

Fig 7. Dynamic data for a 2% acqueous solution of polyacrylamide and a 3% acqueous solution of xanthan gum. (From Ref *18.*)

5. A CONTROVERSY - THE YIELD STRESS MYTH?

At the 9th International Congress on Rheology in Acapulco, Barnes and Walters (*19*) had the effrontery to question the concept of the yield stress, defined in the Glossary of Rheological Terms (*20*) as " that stress below which the substance is an elastic solid and above it a liquid..". Barnes and Walters conjectured that the availability of a new generation of controlled-stress rheometers had made it possible to conclude that the yield stress concept was an idealization and that given accurate measurements, it was possible to show that fluids which flow at high stresses will also flow (perhaps very slowly) at lower stresses, i.e. the viscosity, although (maybe) very high at low shear rates, is nevertheless always finite and there is no yield stress.

To say that the Acapulco lecture and the associated paper (*19*) led to a lively debate would be an understatement and the issue is taken up again at this congress by my co-author H A Barnes (*21*). Much of what has been written on the subject has been less than helpful, but interjections by a number of very respected scientists (see, for example, *22*, *23*) has helped to provide a proper balance. Overstatement is often useful in Science if it helps to elucidate important issues and so it has turned out in the present controversy. As far as the present author is concerned, private discussions with respected colleagues, including one of this Congresses organizers, Professor Mewis, and a study of the relevant literature since 1984 has led to the following conclusions.

(i) It was not inappropriate to raise the issue of the yield stress in 1984 and it is certainly true that, for the majority of cases, the basic conclusion that all materials have a finite low-shear-rate viscosity is correct, although the value of this viscosity may be very high indeed.

(ii) It is conceded that in a limited number of cases, especially in the case of suspensions, sound scientific arguments can be put forward to support the existence of a real yield stress.

(iii) The yield stress is an engineering reality (cf *22*, *23*).

Finally, we remark that those who have argued vociferously and long for the existence of a true yield stress must accept the inevitable corollary - that for such materials, dynamic measurements are of very limited utility. At best, they can only capture and interpret the solid-like response below the yield stress, and the characteristic 'flat topping' of output curves usually encountered in such materials is conclusive evidence of non linearity.

6. CONCLUSIONS

At first sight, Rheometry may not appear to be the most exciting area of non-Newtonian Fluid Mechanics research, but a close study of the relevant literature would suggest that the subject is not without its attractions. The supply of interesting and meaningful problems has not yet been exhausted and significant challenges lie ahead. Most of these are associated with the problem of extending the range of operation of existing rheometers. We close by mentioning one such problem which will require inspired science in its resolution.

Viscous heating in capillary flow will become increasingly important as attempts are made to obtain viscometric data at ever higher shear rates. An important unresolved inverse problem can be simply stated. Given a series of measurements of Pressure drop/Flowrate for different values of the temperature of the capillary walls, is it possible to calculate the viscosity as a unique function of shear rate and temperature?

REFERENCES

1. I.M. Krieger: Bingham Award Lecture 1989, The role of instrument inertia in controlled-stress rheometers, J. of Rheology *34* (1990) 471.
2. W. Gleissle and N. Ohl, On the relaxation of shear and normal stresses of viscoelastic fluids following constant shear rate experiments, Rheological Acta *29* (1990) 261.
3. M.M.K. Khan and R.I. Tanner, Rheology of an LDPE melt in reversing multi step shear and elongational flows, Rheologica Acta *29* (1990) 281.
4. J.F. Hutton, The fracture of liquids in shear: the effect of size and shape, Proc. Roy. Soc. A*287* (1965) 222.

5. H.A. Barnes, J.F. Hutton and K. Walters, An Introduction to Rheology, Elsevier, 1989.
6. K.P. Jackson, K. Walters and R.W. Williams, A rheometrical study of Boger fluids, J. non-Newtonian Fluid Mechanics *14* (1984) 173.
7. D.V. Boger, A highly elastic constant-viscosity fluid, J. non-Newtonian Fluid Mechanics *3* (1977) 87.
8. N.Phan-Thien, Coaxial-disk flow of an Oldroyd-B fluid: Exact solution and stability, J. non-Newtonian Fluid Mechanics *13* (1983) 325.
9. N. Phan-Thien, Cone-and-plate flow of the Oldroyd-B fluid is unstable, J. non-Newtonian Fluid Mechanics *17* (1985) 37.
10. D.V. Boger, M.J. Crochet and R.A. Keiller, On viscoelastic flows through abrupt contractions. To appear in J. non-Newtonian Fluid Mechanics (1992).
11. J.J. Magda and R.G. Larson, A transition occurring in ideal elastic liquids during shear flow, J. non-Newtonian Fluid Mechanics *30* (1988), 1.
12. G.H. McKinley, J.A. Byars, R.A. Brown and R.C. Armstrong, Observations on the elastic instability in cone-and-plate and parallel-plate flows of a polyisobutylene Boger fluid, J. non-Newtonian Fluid Mechanics *40* (1991) 201.
13. K. Walters and H.A. Barnes, Anomolous extensional-flow effects in the use of commercial viscometers, in Rheology Vol 1: Principles: Eds G. Astarita et al. Plenum Press, (1980) 45.
14. D.M. Jones, K. Walters and P.R. Williams, On the extensional viscosity of mobile polymer solutions, Rheological Acta *26* (1987) 20.
15. J. non-Newtonian Fluid Mechanics Vol 30 (1988) 97-368.
16. J. non-Newtonian Fluid Mechanics Vol 35 (1990) 85-470.
17. D.F. James and K. Walters, A critical appraisal of available methods for the measurement of extensional properties of mobile systems. To appear in Techniques in Rheological Measurement, Ed: A.A. Collyer, Eslevier (1992).
18. K. Walters, A.Q. Bhatti and N. Mori, The influence of polymer conformation on the rheological properties of acqueous polymer solutions, in Recent developments in Structured Continua Vol 2, Eds: D.De Kee and P.N. Kaloni, Longman (1990) 182.
19. H.A. Barnes and K. Walters, The yield stress myth? Rheologica Acta *24* (1985) 323.
20. British Standard 5168 (1975) Glossary of Rheological Terms.
21. H.A. Barnes, The yield stress revisited. This Proceedings.
22. G. Astarita, The engineering reality of the yield stress, J. of Rheology *34* (1990) 275.
23. J. Schurz, The yield stress - an empirical reality, Rheological Acta *29* (1990) 170.

KEYNOTE PAPERS

Theoretical and Applied Rheology, edited by P. Moldenaers and R. Keunings
Proc. XIth Int. Congr. on Rheology, Brussels, Belgium, August 17-21, 1992

Measurement of Velocity and Stress Fields in Complex Polymer Flows

R.C. Armstrong[1], R.A. Brown[1], L.M. Quinzani[2], G.H. McKinley[3], and J.A. Byars[1]

[1]Department of Chemical Engineering, Massachusetts Institute of Technology, Cambridge, MA 02139 (USA)
[2]Planta Piloto de Ingenieria Quimica, Universidad Nacional del Sur, 12 de Octubre 1842, Bahía Blanca - 8000 (Argentina)
[3]Division of Applied Sciences, Harvard University, Cambridge MA 02139 (USA)

1. INTRODUCTION

There are a number of interesting challenges that face those working at the interface between rheology and fluid mechanics. Among the most pressing are

i. Determination of the elongational behavior of polymeric liquids in well defined elongational flows.
ii. Development and validation of constitutive equations for nonhomogeneous flows that are neither simple shear nor elongation.
iii. Determination of the rheological behavior of polymeric liquids near solid boundaries and singularities.
iv. Understanding the interrelation between rheological properties of polymers and flow transitions in complex flows.

Much of this agenda has been brought into focus by recent advances in the ability to compute viscoelastic flows numerically (refs. 1-3), though the latter item has been highlighted primarily by the studies of flow through an axisymmetric abrupt contraction by Boger and coworkers (ref. 4) and by our group (refs. 5,6).

In this paper we look at the contributions of velocity and stress measurements in complex flows of viscoelastic liquids toward resolving some of these issues. Because of space limitations, we make no effort to be encyclopedic in our coverage of this problem, but rather we focus on some rather recent measurements in our group. Additional details can be found in ref. 7.

2. EXPERIMENTAL METHODS

2.1 Fluids

The polymer solutions used in this study were made of a high molecular weight polyisobutylene (PIB; Exxon Vistanex L-120, $\bar{M}_w \approx 1.8\times10^6$ g/mol), a low molecular weight polybutene (PB; Amoco L-100, $\bar{M}_w \approx 1000$ g/mol) and a Newtonian solvent, tetradecane (C14). Two types of solutions were prepared: a shear-thinning solution consisting of 5.0 wt% PIB in C14 and a nearly constant viscosity solution or Boger fluid consisting of 0.31 wt% PIB, 94.86 wt% PB, and 4.83 wt% C14. The rheology of these fluids has been characterized extensively in shear flows as described in ref. 8. In this paper we focus on results for the shear thinning fluid. Figures 1 - 3 give the linear viscoelastic and viscometric properties of this fluid. Note that the linear data are well fit by a discrete spectrum with 4 modes. The constants for these modes are shown in Table 1; the approximate value of the solvent viscosity is also shown. The viscometric properties are fit to several nonlinear models (ref. 9) as shown in Figs. 2 and 3. These fits were obtained by using the linear spectrum in Table 1 and the nonlinear parameters in Table 2 (see ref. 8).

Table 1. Linear Viscoelastic Spectrum for the 5.0 wt% PIB/C14 Solution at 25°C.

Mode	λ_k (s)	η_k (Pa·s)
1	0.6855	0.0400
2	0.1396	0.2324
3	0.0389	0.5664
4	0.0059	0.5850
Solvent	---	≈0.002

Table 2. Nonlinear Constitutive Equation Parameters

Model	Parameter	Mode 1	2	3	4
Giesekus	α_k	0.5	0.2	0.3	0.2
Bird-DeAguiar	b_k	10	10	10	20
	σ_k	1.0	0.4	0.5	0.4
Phan-Thien/	ξ	0.13			
Tanner	ε	----			
Acierno et al.	a	----			

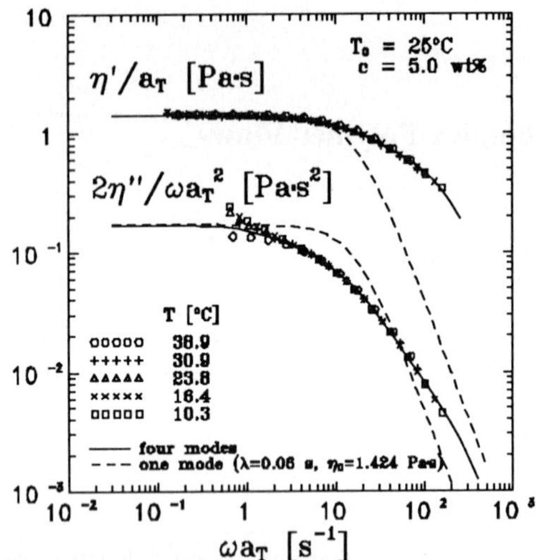

Figure 1. Master curves for the dynamic properties η' and $2\eta''/\omega$ of the 5.0 wt% PIB solution at T_0=25°C. The solid curves are calculated from the spectrum in Table 1.

2.2 Laser Doppler Velocimeter

The LDV apparatus employed in this research is a three-color, six-beam system; it is described in detail in ref. 10. The optical train gives a measuring volume of 38×53×274 μm. By mounting all of the optics on a translating table, this scattering volume can be positioned over a range of 96×48×48 cm with an accuracy of ±4 μm.

2.3 Flow Induced Birefringence (FIB) System

The flow induced birefringence (FIB) system developed in this research is a modification of the two-color flow birefringence system originally developed by Chow and Fuller (refs. 11-12). Our system is shown schematically in Fig. 4, and it is described in detail in ref. 7. In Fig. 4, optical components for the blue and green beams of the Argon laser are denoted by subscripts B and G, respectively. Included in the drawing are the locations of beam splitters (BS), mirrors (M), filters (F), prism polarizers (P), lenses (L), pinholes (Ph), and power meters (D). The flow cell, which sits in the middle of the system is placed on a translating table, so that it can be moved normal to the beams over a 10×20 cm range with a positioning accuracy of ±2 μm.

2.4 Flow Geometry

The flow cell used in obtaining the results reported here is a planar contraction with top and bottom surfaces made of PMMA (for LDV measurements) and side surfaces containing custom made, low birefringence silicate glass windows (for FIB measurements). The width of the system is 25.4 cm, and the internal heights of the up- and down-stream sections are 2.54 and 0.64 cm. Thus the contraction ratio is 3.97:1 for this device. The contraction could be placed in either of two locations within a flow loop (Fig. 5) to allow interrogating the cell with either LDV or FIB.

3. BEHAVIOR NEAR SINGULARITIES

The work of Apelian *et al.* (ref. 13) has shown that the use of a consistent and accurate numerical method is not a sufficient guarantee of convergence for calculations in flows with geometric singularities, e.g. the abrupt contraction flow. Apelian *et al.* found that the convergence of the numerical calculations is only feasible if the stress and velocity fields are mathematically well behaved near the reentrant corner of the contraction. As a consequence, the convergence of the numerical predictions depends on the behavior of the constitutive equation in

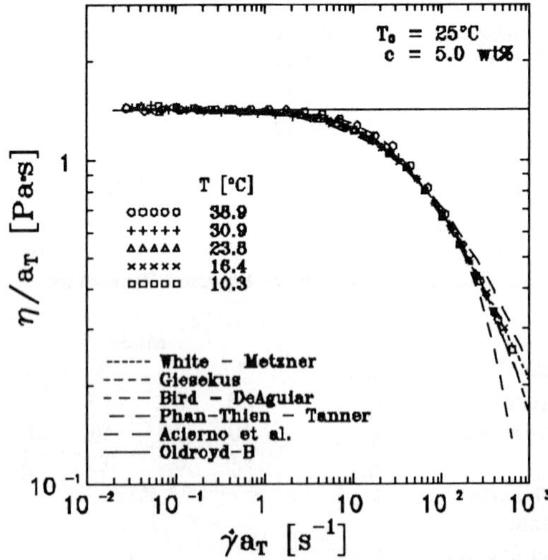

Figure 2. Master curve for the viscosity $\eta(\dot{\gamma})$ and comparison with model predictions.

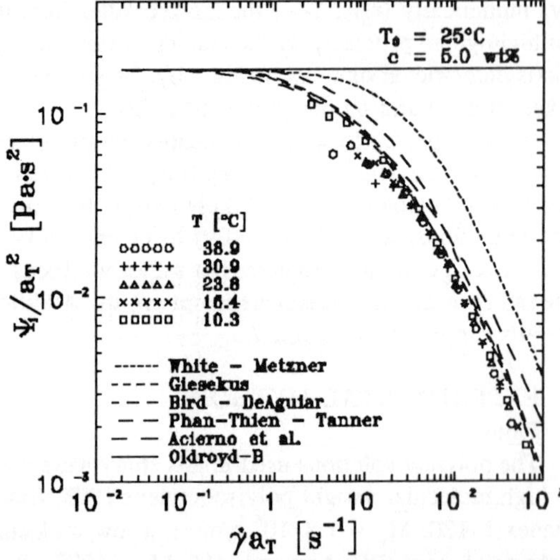

Figure 3. Master curve for the first normal stress coefficient $\Psi_1(\dot{\gamma})$ and comparison with model predictions.

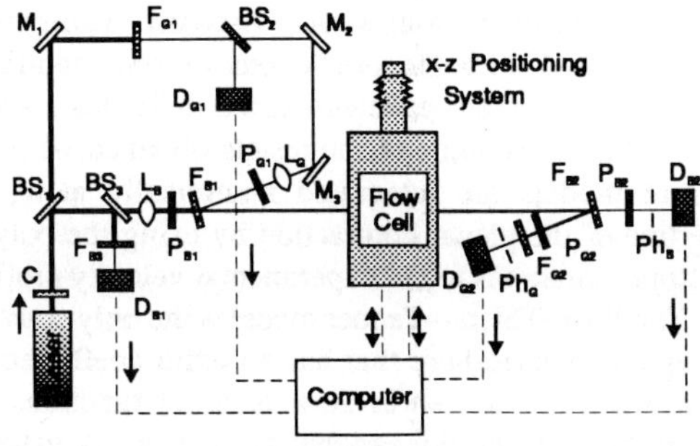

Figure 4. The flow birefringence system.

this region (refs. 13,14).

An analysis of the asymptotic behavior of a Newtonian fluid near a corner singularity patterned after those of Dean and Montagnon (ref. 15) and Moffat (ref. 16) shows that for the Newtonian fluid the stress at the singularity is integrable, i.e. finite forces are predicted about the singularity. Near the corner, the stream function ψ is expressed as an eigenfunction expansion (refs. 15,16)

$$\psi(r,\theta) = \sum r^{\lambda_i} f_i(\theta) \tag{1}$$

where r and θ refer to polar coordinates with origin at the corner, $\theta = 0$ bisecting the fluid region, and θ increasing in the direction of flow. For a Newtonian fluid with no slip boundary conditions, the substitution of Eq. 1 into the equation of motion leads to a series of eigenvalue problems. The velocity gradients and stress scale as r^{λ_i-2} as r $\rightarrow 0$, where $\lambda_1 \cong 1.545$, $\lambda_2 \cong 1.909$, etc. The leading order term ($\tau_{ij} \sim r^{-0.455}$) is dominant at all angles except where $f_1(\theta) = 0$. At those particular angles the leading order term vanishes and the stresses are determined by the lower order contribution ($\tau_{ij} \sim r^{-0.091}$). For the corner flow, the leading order term vanishes at $\theta = 0$ for the normal stresses, and at $\theta = \pm 0.212\,\pi$ for the shear stress.

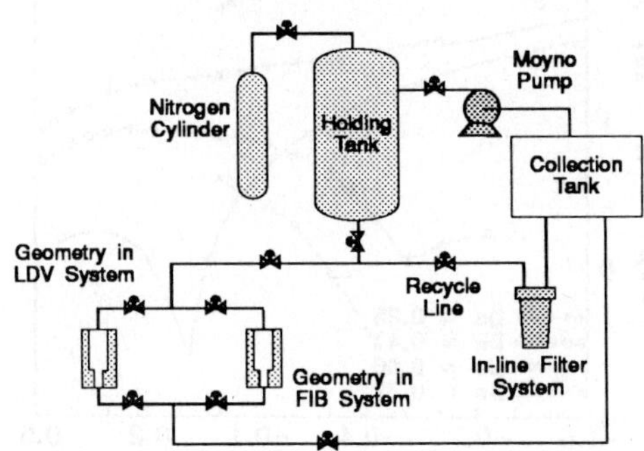

Figure 5. Flow loop used to feed either the LDV or flow birefringence systems.

For a shear thinning generalized Newtonian fluid, Henriksen and Hassager (ref. 17) have shown that the stresses grow less rapidly than the Newtonian rate as r $\rightarrow$ 0.

To allow convergent calculations in flow problems with singularities, Apelian *et al.* proposed the *modified upper-convected Maxwell model* (MUCM) (ref. 13) which predicts constant viscosity and a relaxation time that decreases with increasing magnitude of the trace of the stress tensor. The MUCM model yields stress fields that reduce to the asymptotic expressions for a Newtonian fluid near singularities. The stress has thus the $r^{-0.455}$ Newtonian behavior at leading order with a lower-order $r^{-0.182}$ viscoelastic contribution.

To shed light on the correct behavior of the stress tensor near a singularity, experimental measurements of the shear stress and normal stress difference were made near the reentrant corner of the planar contraction for the angles $\pi/4$, 0, and $-\pi/4$. The size of the beams inside the flow cell provide spatial resolution near the wall of order $r/h \approx 0.2$. This improves somewhat on the previous measurements of Galante and Frattini (ref. 18) which were restricted to $r/h \geq 0.8$. Figures 6 and 7 illustrate the results for the angle $\theta = -\pi/4$. In these and subsequent figures the Deborah number is defined as

$$De = \lambda(\dot{\gamma})\langle v\rangle/h \tag{2}$$

where $\lambda = \Psi_1/2\eta$, $\dot{\gamma} = \langle v\rangle/h$, $\langle v\rangle$ is the average velocity in the downstream slit, and h is the half gap thickness of the downstream slit. It should be noted that the presence of an isoclinic line near the lip, which indicates proximity to a change in sign in the normal stress difference, prevents data from being obtained closer to the singularity than shown.

The full lines in these figures are the predicted stresses for a Newtonian fluid of viscosity $\mu = 1.424$ Pa·s (ref. 14), computed at the smallest flow rate, $Q = 34.8$ cm^3/s. A shear stress growth rate of $r^{-0.23}$ is calculated from the simulations at $\theta = -\pi/4$. Neither the computed slope nor the experimentally measured slope is quite asymptotic. The measured stress growth rate is nearer the asymptotic prediction of the lower-order contribution, $r^{-0.091}$, which dominates at $\theta = \pm 0.212\pi$, than the leading-order rate of $r^{-0.455}$. The locations $\theta = \pm\pi/4$ are very near these angles, and the shear stress shows the effect of that proximity. The agreement between the experimental results and the Newtonian calculations is excellent.

The FIB measurements for the PIB fluid near the reentrant corner of the planar contraction give strong support to the validity of solutions of the form of Eq. 1. Thus there is good reason to require that constitutive equations give velocities and stresses near the singularity of the form $r^p f(\theta)$.

4. ELONGATIONAL PROPERTIES

The capability of measuring both the local velocity and stress profiles in a complex flow such as the abrupt contraction offers the possibility of testing constitutive equations in flows different from the standard shear and shearfree flows used in rheometry. This is done by integrating the constitutive equations along a chosen particle path in conjunction with the measured velocity profile for that path. The resulting stress field is then compared with the experimental results.

The flow along the midplane of the abrupt planar contraction is a transient elongational flow. Measurement of the centerline stress distribution and the velocity profile yields the transient planar elongational viscosity $\bar{\eta}_1^c(z)$ defined by

$$\bar{\eta}_1^{\,c}(z) \equiv -\frac{\tau_{zz} - \tau_{yy}}{\dot{\varepsilon}_{max}} \tag{3}$$

where $\dot{\varepsilon}_{max}$ is the maximum elongation rate that occurs along the centerline. The elongation rate $\dot{\varepsilon}(y = 0,z)$ is calculated from the derivative of the axial velocity along the centerline

$$\dot{\varepsilon}(z) = \frac{\partial v_z(z)}{\partial z} \tag{4}$$

Here we illustrate the use of LDV and FIB to measure the transient planar elongational viscosity $\bar{\eta}_1^c(z)$ of the PIB solution. The evolution of the normal stress difference $-(\tau_{zz} - \tau_{yy})$ and the development of the axial velocity along the centerline of the planar contraction are shown in Figs. 8 and 9. Position is nondimensionalized with the half-width of the downstream channel, h. In order to compute $\bar{\eta}_1^c$, polynomials of degree 10 are fit to the velocity and stress data in Figs. 8 and 9, and these fits are used to calculate $\dot{\varepsilon}$ and $\bar{\eta}_1^c$ (see Figs. 10 and 11).

The measured elongational viscosity distributions were compared with the predictions of five non-linear models by using the coefficients listed in Tables 1 and 2. Figure 12 is an example of the results obtained when the different models are integrated numerically along the centerline of the planar contraction by using the polynomial approximations to the experimental velocity profiles.

The Phan-Thien—Tanner model is the only constitutive equations used here that has an extra coefficient, ε, that was not fit with shear flow material functions and that can be fit from elongational properties. A value of $\varepsilon = 0.25$ was found to give the best fit of the profiles of $\bar{\eta}_1^c(z)$ at large De.

From the above comparison, it is clear that the elongational behavior of the PIB solution is best described by the Phan-Thien—Tanner model. Interestingly, the same qualitative behavior is shown by the apparent uniaxial elongational viscosity measured in the Fluid Analyzer RFX (Rheometrics Inc.) based on opposing jet flow (see Fig. 13). The apparent uniaxial elongational viscosity thins from an overpredicted $\bar{\eta}_{10}$ of approximately 6 Pa·s to approximately 4.5 Pa·s at $\dot{\varepsilon} = 100\,s^{-1}$. The shearfree flow results thus prove to be essential for distinguishing between constitutive equations. Whereas the nonlinear models studied were qualitatively, and sometimes quantitatively, indistinguishable in shear flows, they predict qualitatively different behavior in shearfree flows.

5. CONCLUSIONS

The results presented here suggest the power of combined local velocity and stress measurements in complex viscoelastic flows. In addition to providing

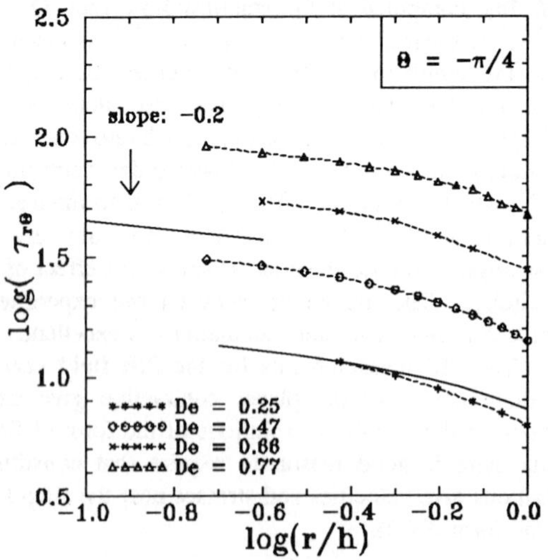

Figure 6. Asymptotic behavior of the shear stress $\tau_{r\theta}$ near the reentrant corner for $\theta = -\pi/4$.

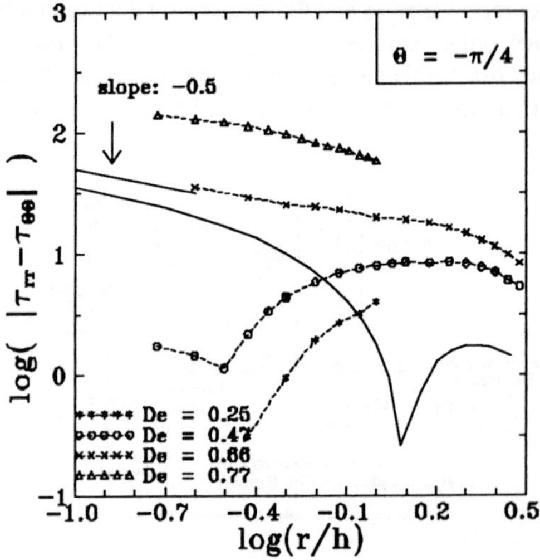

Figure 7. Asymptotic behavior of the normal stress difference $\tau_{rr}-\tau_{\theta\theta}$ near the reentrant corner, $\theta = -\pi/4$.

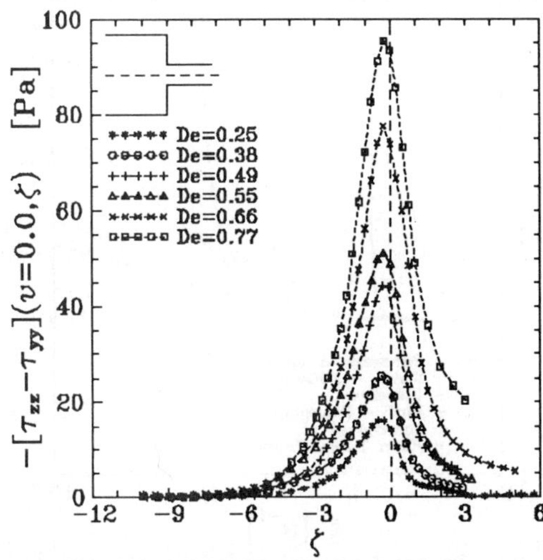

Figure 8. Centerline flow birefringence profiles at six different flow rates.

essential feedback to those performing numerical simulations of viscoelastic flows, they provide otherwise unattainable, fundamental rheological data.

REFERENCES

1. P.J. Coates, R.C. Armstrong, and R.A. Brown, J. Non-Newtonian Fluid Mech., 00 (1992) 000-000.
2. D. Rajagopalan, R.C. Armstrong, and R.A. Brown, J. Non-Newtonian Fluid Mech., 36 (1990) 159-193.
3. R.C. King, M.R. Apelian, R.C. Armstrong, and R.A. Brown, J. Non-Newtonian Fluid Mech., 29 (1988) 147-216.
4. D.V. Boger, Ann. Rev. Fluid Mech., 19 (1987) 157-182.
5. G.H. McKinley, W.P. Raiford, R.A. Brown, and R.C. Armstrong, J. Fluid Mech., 223 (1991) 411-456.

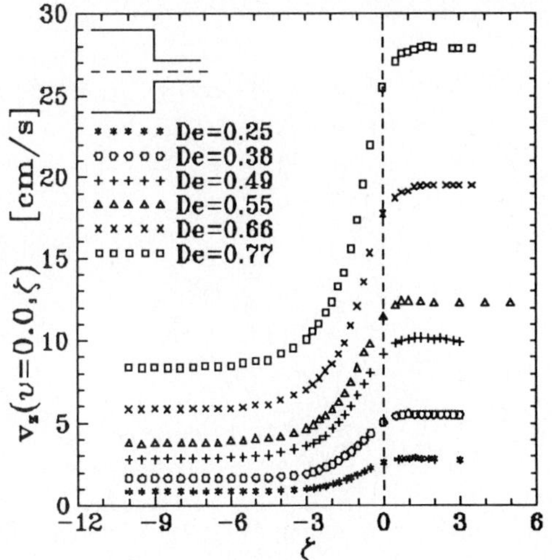

Figure 9. Centerline axial velocity at six different flow rates.

6. J.V. Lawler, S.J. Muller, R.A. Brown, and R.C. Armstrong, J. Non-Newtonian Fluid Mech., 20 (1986) 51-92.
7. L.M. Quinzani, Ph.D. Thesis, Department of Chemical Engineering, MIT, Cambridge, MA (1991).
8. L.M. Quinzani, G.H. McKinley, R.A. Brown, and R.C. Armstrong, J. Rheol., 35 (1990) 705-748.
9. R.B. Bird, R.C. Armstrong, and O. Hassager, Dynamics of Polymeric Liquids. Vol. 1: Fluid Mechanics, Wiley, New York, 1987.
10. G.H. McKinley, Ph.D. Thesis, Department of Chemical Engineering, MIT, Cambridge, MA (1991).
11. A.W. Chow and G.G. Fuller, J. Rheol., 28 (1984) 23-43.
12. A.W. Chow and G.G. Fuller, J. Non-Newtonian Fluid Mech., 17 (1985) 233-243.
13. M.R. Apelian, R.C. Armstrong. and R.A. Brown, J. Non-Newtonian Fluid Mech., 27 (1988) 299-321.
14. P.J. Coates, Ph.D. Thesis, Department of Chemical Engineering, MIT, Cambridge, MA (1992).
15. W.R. Dean and P.E. Montagnon, Proc. Camb. Phil. Soc., 45 (1949) 389-394.
16. H.K. Moffat, J. Fluid Mech., 18, (1964) 1-18.
17. P. Henriksen and O. Hassager, J. Rheol., 33 (1989) 865-879.
18. S.R. Galante and P.L. Frattini, "Spatially Resolved Birefringence Studies of Planar Entry Flow", J. Non-Newtonian Fluid Mech. (1991) submitted.

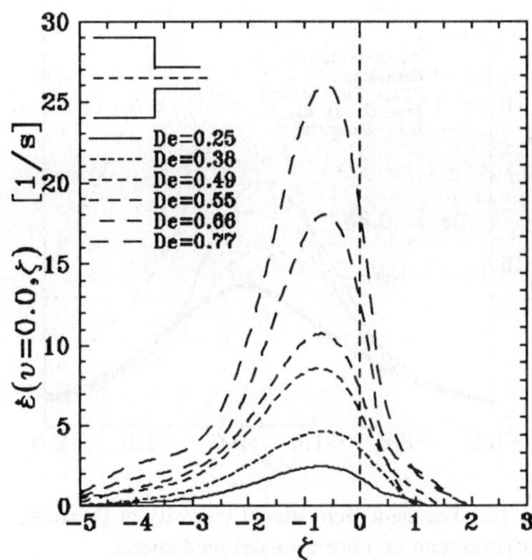

Figure 10. Elongation rate distribution profiles calculated from the velocity data of Fig. 29.

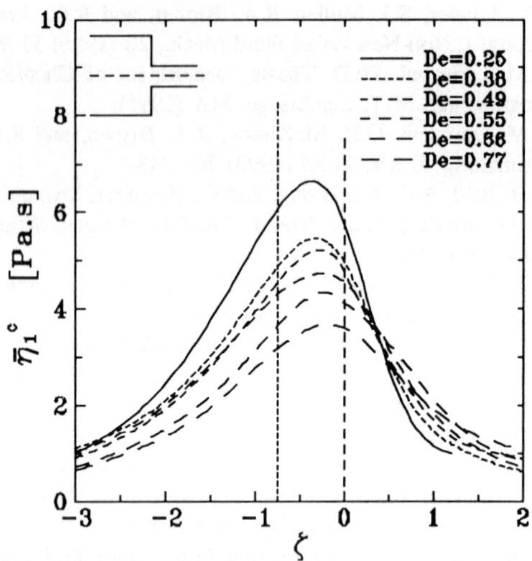

Figure 11. Transient planar extensional viscosity for the 5.0 wt% PIB/C14 at six different De. (---): ζ=-0.75.

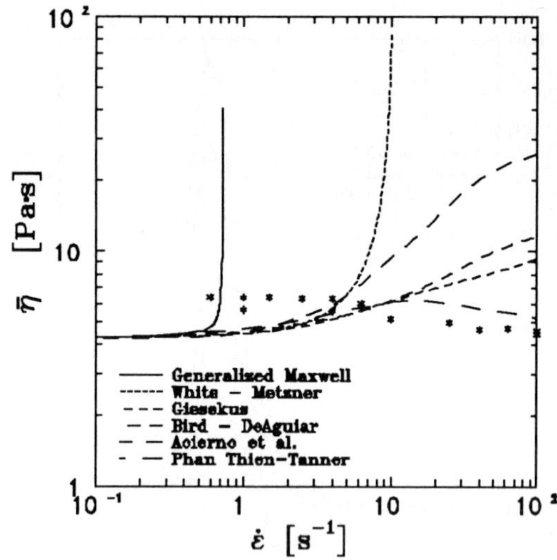

Figure 13. Predicted uniaxial extensional viscosity $\bar{\eta}$. (*): Apparent $\bar{\eta}$ measured in Fluid Analyzer RFX.

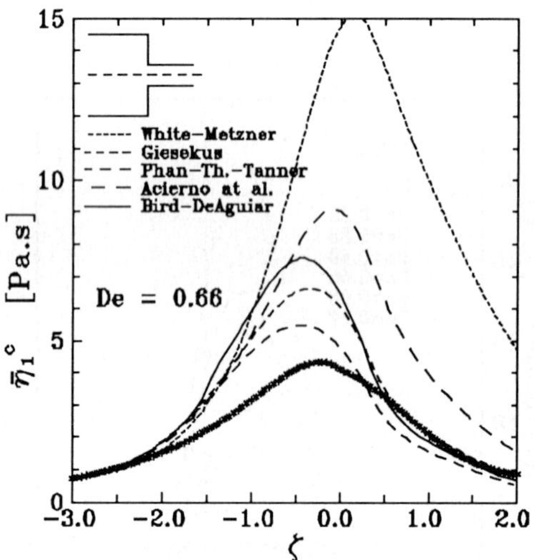

Figure 12. Transient elongational viscosity at De=0.66. Comparison with non-linear model predictions.

Theoretical and Applied Rheology, edited by P. Moldenaers and R. Keunings
Proc. XIth Int. Congr. on Rheology, Brussels, Belgium, August 17-21, 1992

VISCOELASTIC FLOW INSTABILITIES: INCEPTION AND NON-LINEAR EVOLUTION

A.N. BERIS and M. AVGOUSTI

Department of Chemical Engineering, University of Delaware, Newark, DE 19716 (U.S.A.)

1. INTRODUCTION

Viscoelastic flow instabilities represent the limiting factor for enhanced production in many polymer processes, such as extrusion, blow molding etc. Their prevention requires both an adequate understanding of the relevant polymer dynamics and an efficient description of its complex interplay with the corresponding fluid mechanics. Significant advances, in both the above areas, have been recently realized with work which led to the elucidation of the bifurcation diagram of the viscoelastic (for an Upper Convected Maxwell) Taylor-Couette flow problem near the onset of instability for both 2-d (axisymmetric) and 3-d (non-axisymmetric) perturbations. The analysis was made possible by the development of efficient numerical codes, based on spectral collocation methods, which were used in a computer-assisted linear and non-linear analysis as well as in time-dependent viscoelastic flow calculations at both zero and non-zero Reynolds numbers. The present paper overviews these recent developments in a systematic fashion.

The Taylor-Couette flow problem, i.e. the flow between two concentric, independently rotating cylinders, continues to attract considerable attention, even for Newtonian fluids, long after Taylor's first experimental observations (ref. 1). The plethora of pre-chaotic pattern formations observed experimentally prompted the numerous theoretical studies, both analytical and numerical, which attempted to explain most of these phenomena (refs. 2-4). The interest in the Taylor-Couette problem was recently revived by the development of theories concerning bifurcations in the presence of symmetries (ref. 5) which are used to describe the pattern formation observed using various combinations of parameters.

The presence of viscoelasticity in the flow increases the complexity of the mathematical problem. The early theoretical works of Thomas and Walters (refs. 6,7) and Beard et al. (ref. 8) established three observations for the axisymmetric Taylor-Couette flow of an Upper Convected Maxwell fluid. The first is that viscoelasticity acts as a destabilizing agent in the flow. The second is that when the relative rotation between the two cylinders is large enough, overstability (Hopf bifurcation) is found to occur after the onset of instability. The third finding is that the cell size corresponding to the most unstable mode decreases in the presence of elasticity. A very recent reporting by Muller et al. (ref. 9) has shown experimental evidence for the development of purely elastic, inertialess instabilities. Northey et al. (ref. 10) performed creeping flow, time-dependent finite-element simulations with the same conclusions. However, an investigation which could elucidate the transition from a Newtonian-like type of instability, leading to steady-state Taylor vortices, to a purely elastic one, leading to time-periodic secondary vortices, as well as the possible form of the emerging secondary flow patterns, axisymmetric or non-axisymmetric, was lacking.

Thus, the objective of the present work was set to provide a more complete and systematic investigation of the viscoelastic Taylor-Couette flow problem. We achieved that through the coordination of several analytical and computational techniques, involving both standard and more recently developed methods. The problem definition is presented in Section 2. Section 3 presents the results of a computer-assisted linear stability analysis against both 2d and 3d perturbations. The analysis of the bifurcation diagram at the onset of instability (both 2d and 3d) is based on the theory of bifurcation in the presence of symmetries, as explained in Section 4. The results of a non-linear analysis using a multidimensional spectral collocation are presented in Section 5. A sample of the results of a direct time-dependent simulation using the full non-linear equations follows in Section 6. Finally, the conclusions are summarized in Section 7.

2. PROBLEM FORMULATION

Consider two infinitely long, concentric cylinders of radii r_1 and r_2, $r_1<r_2$, independently rotating with angular velocities Ω_1 and Ω_2, respectively, with fluid confined in the annulus between them. A cylindrical coordinate system (r-θ-z) is used with the z-axis the common axis of the two cylinders. The flow problem is characterized by six dimensionless parameters: Three geometric parameters, the ratio of the radii, $\zeta=r_1/r_2$, the dimensionless axial wavelength $L=2\pi/\alpha d$, where d is the gap width, $d=(r_2-r_1)$ and α is the wavenumber corresponding to the assumed periodicity in the axial direction and the azimuthal wavenumber, m, which is an integer number characterizing the assumed periodicity ($\theta\rightarrow\theta+2\pi/m$) in the azimuthal direction (m=0 corresponding to axisymmetric solutions); one kinematic, the ratio of the angular velocities, $\mu=\Omega_2/\Omega_1$; and two flow parameters, the Reynolds number, Re,

$$Re = \frac{d r_1 \Omega_1 \rho}{\eta_p}, \tag{1}$$

where η_p is the polymer viscosity and ρ the fluid density, representing as usual the ratio of inertia to viscous forces, and the Deborah number, De,

$$De = \frac{\lambda_1 r_1 \Omega_1}{d}, \tag{2}$$

indicating the importance of the fluid relaxation time, λ_1, to the flow process characteristic time, $d/(r_1\Omega_1)$.

Alternatively, instead of either the De or Re numbers the elasticity number, ϵ,

$$\epsilon = \frac{De}{Re} = \frac{\eta_p \lambda_1}{\rho d^2} = \frac{\nu_k}{D_\lambda}, \tag{3}$$

can be used, which can be interpreted as a viscoelastic Prandtl number with ν_k and D_λ being an appropriate kinematic viscosity and relaxational diffusivity, respectively. The elasticity number has the advantage to be solely a property of the fluid and not of the flow. As such, it is used extensively for viscoelastic flow calculations involving inertia (refs. 6-8).

The governing equations (i.e. the continuity, momentum and upper convected Maxwell constitutive equation) are expressed in dimensionless form using the gap width d as the scale for length, the linear velocity of the inner cylinder, $r_1\Omega_1$, as the scale for velocity and the mass in unit volume, ρd^3, as the scale for mass. In addition, we assumed non-slip, non-penetration boundary conditions on the cylinder walls, periodic boundary conditions in the axial direction and either axisymmetry or periodic boundary conditions in the azimuthal direction.

3. LINEAR STABILITY ANALYSIS

The classical approach for the study of the stability of the base solution to infinitesimal perturbations involves the evaluation of the eigenvalues associated with the linearized equations of motion around the base solution (linear stability analysis). The eigenvalues of the continuum problem, corresponding to both 2d and 3d perturbations, are estimated in this work from the solution of a generalized algebraic problem obtained by discretizing the continuum one using a spectral collocation method. This approach is very similar to that used by Khorrami et al. (ref. 11) who studied the stability of swirling flows.

The critical Reynolds (or Deborah) number is determined by the minimum in the neutral stability curve representing the Reynolds (Deborah) number as a function of the corresponding wavenumber α, for which the most unstable eigenvalue has a zero real part. The critical wavenumber corresponding to a viscous Newtonian fluid is relatively insensitive to the flow and geometric parameters. For $1\geq\mu\geq0$ the instability corresponds to steady Taylor vortices of nearly square cross section (ref. 12). The opposite is true for viscoelastic flows, for which α_c is fairly sensitive to the flow elasticity, ϵ, and the azimuthal wavenumber, m, as seen in Figure 1. In general, the trend is that the critical wavenumber increases and the critical Reynolds number ($Re_c=De_c/\epsilon$) decreases as the elasticity in the flow increases. This clearly demonstrates that elasticity has a destabilizing effect, in agreement with previous calculations (ref. 6-7) and with experimental data (refs. 9,13).

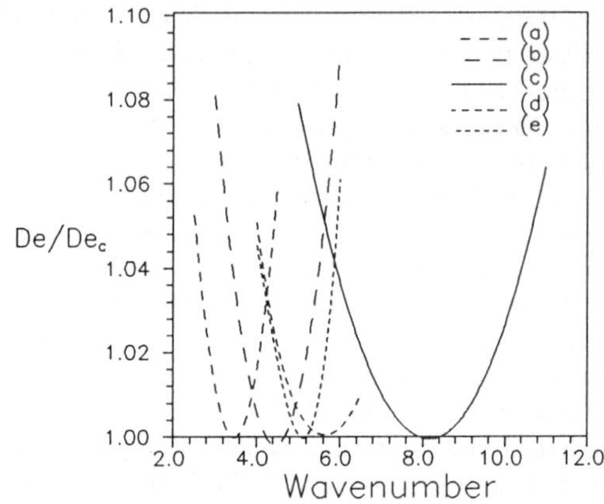

Figure 1. Normalized neutral stability curves for μ=0.5, ζ=0.95 and (a) De_c=2.015; ϵ=0.01, m=0, (b) De_c=18.54; ϵ=0.2424, m=0, (c) De_c=59.3; $\epsilon=5\times10^4$, m=0, (d) De_c=43.0; $\epsilon=5\times10^4$, m=1, (e) De_c=38.4; $\epsilon=5\times10^4$, m=2.

As the elasticity increases, there is a qualitative change in the nature of the instability. Two pairs of complex conjugate eigenfunctions become responsible for the onset of instability, crossing the imaginary axis. In addition, as the elasticity in the flow further increases ($\epsilon \approx 3.3$) there is an exchange between the families corresponding to the most unstable complex eigenvalues . This is represented graphically in Figure 2a depicting the stability limit of the azimuthal Couette flow against 2d (axisymmetric) perturbations. As shown in Figure 1, the stability boundary is not very sensitive to changes in α, which was kept constant within each family in Figure 2, for computational convenience. A similar graph for 3d (non-axisymmetric) perturbations is presented in Figure 2b, corresponding to different critical azimuthal wavenumbers. By direct comparison of Figures 2a and 2b, it is seen that as the elasticity increases, the most unstable perturbation changes from axisymmetric (m=0) to m=3 and finally m=2 non-axisymmetric. The pattern formation corresponding to these instabilities is discussed in the next section.

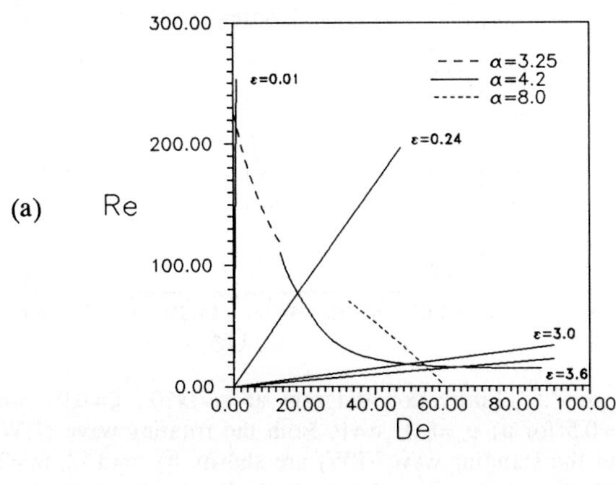

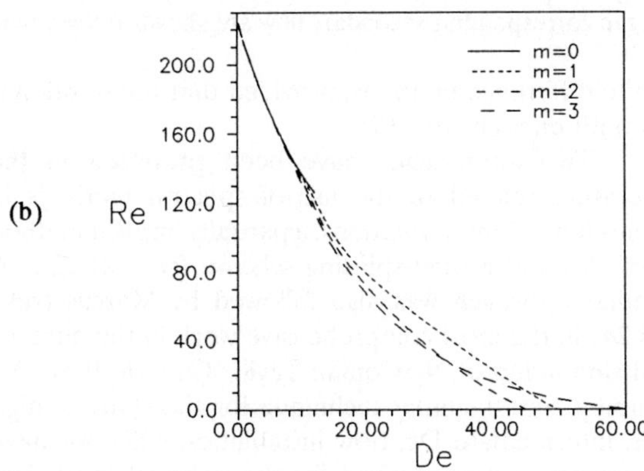

Figure 2. Stability region of an upper convected Maxwell fluid for ζ=0.95 and μ=0.5 for a) axisymmetric (m=0) and b) non-axisymmetric disturbances, α=3.12. For comparison, the stability boundary to steady axisymmetric Taylor vortices (m=0) is repeated in both figures. The straight lines correspond to constant elasticity numbers.

The use of the pseudospectral approach offers an additional advantage besides those of numerical accuracy and efficiency. It allows for the simultaneous calculation of a significant number of the eigenvalues of the continuous system. These eigenvalues correspond to eigenfunctions with the least amount of spatial variations which include, for a fine enough discretization, the most unstable modes of the continuum system. Thereby, one can automatically select the most unstable mode, even when different families switch over as one scans the parameter space. This is a caveat which is often ignored when a parametric study is pursued by following a particular family.

4. BIFURCATION IN THE PRESENCE OF SYMMETRIES

The use of the infinite cylinder approximation, valid for large aspect ratios, is essentially equivalent to the assumption that the solution is spatially periodic in the axial direction. Even so, an axially varying secondary flow solution, corresponds to less symmetry than the original primary (cylindrical Couette) flow which is invariant under axial, azimuthal and temporal translations. Thus at the bifurcation point, a "breaking of symmetries" takes place where the primary family of higher symmetry solutions meets secondary family solutions with lower symmetries. At the bifurcation point, the primary solution ceases to be stable to infinitesimal perturbations. The branches which bifurcate at that point can in principle be stable or unstable. To properly determine the emerging secondary flow patterns and their stability, the theory of bifurcations with symmetry, recently documented by Golubitsky et al. (ref. 5), is used.

Briefly, due to the presence of symmetry, the bifurcation is degenerate, i.e. the eigenvalue of the Jacobian matrix responsible for the onset of instability (which has zero real part) has multiplicity higher than one. Using symmetry arguments (ref. 14), we can group the families of secondary solutions that emerge at the Hopf bifurcation point into two patterns: rotating and standing waves, when axisymmetry is preserved (m=0), or spirals and ribbons when it is not. These generic names reflect the specific spatiotemporal symmetries of the solutions. For example, the rotating wave pattern in 2d and the spirals in 3d remain invariant after an equal phase shift in both time, axial and azimuthal coordinate in space. Whereas, the standing wave in 2d and the ribbon in 3d are invariant under an axial flip. Note, however, that the ribbons still correspond to azimuthal rotating waves.

As nonlinear (Birkhoff normal form) analysis demonstrates (ref. 5), the secondary families of periodic solutions are unstable if at least one of the family branches associated with either one of the two patterns, is subcritical. Note, that in contrast to viscoelasticity, Hopf bifurcations with Newtonian fluids can be observed only when the cylinders are counter-rotating; Chossat et al. (ref. 15) have shown that under certain circumstances ribbons are asymptotically stable. In contrast, with viscoelastic fluids, as seen above, Hopf bifurcations are observed under a much wider range of conditions (at least for the upper convected Maxwell model—see (ref. 16) for the Giesekus model).

5. NONLINEAR ANALYSIS

A nonlinear analysis was performed where the different solution families where traced in the vicinity of the Hopf bifurcation. Again, we used a spectral collocation technique where the solution variables (the velocity, pressure and stress components) were approximated spectrally both in space and time. A truncated Fourier series was used in the axial, (also azimuthal for 3d) and time directions and a truncated Chebyshev series in the radial. A prespecified axial (and azimuthal for 3d) wave number was assumed, whereas the time period was calculated, as well as the Deborah number, using a fixed phase shift and a fixed amplitude of the secondary solution as additional constraints.

To our knowledge, it is the first time that the solution of such high-dimensionality problem is reported, primarily made possible here because of the high accuracy of the spectral approximations and from the fact that next to the bifurcation points only few Fourier modes are necessary. Typical results for both the 2d and 3d problems are shown in Figures 3a and 3b, respectively. It is interesting to note that although both axisymmetric solution families are supercritical (Figure 3a), at least the spiral family bifurcates subcritically (Figure 3b). According to the Golubitski et al. (ref. 5) theory, this implies that both 3d families are unstable and that the emerging solution after the bifurcation might have a considerably different form from the one corresponding to the most unstable eigenfunctions.

6. TIME-DEPENDENT FLOW SIMULATIONS.

Unlike low Reynolds number Newtonian flows, viscoelastic ones can be characterized by extremely steep boundary layers which are developed with increasing elasticity in the flow (ref. 17). The accurate resolution of these boundary layers necessitates the use of powerful numerical methods. Spectral methods have recently been proposed and proven successful in numerical simulations of viscoelastic flows under steady state calculations (ref. 18). The major advantage in adopting spectral and pseudo-spectral (spectral collocation) methods in our numerical discretizations of the partial differential equations is that they enjoy an exponentially fast convergence as compared to low order approximations provided by finite element or finite difference methods, provided that the solution is smooth enough (ref. 19).

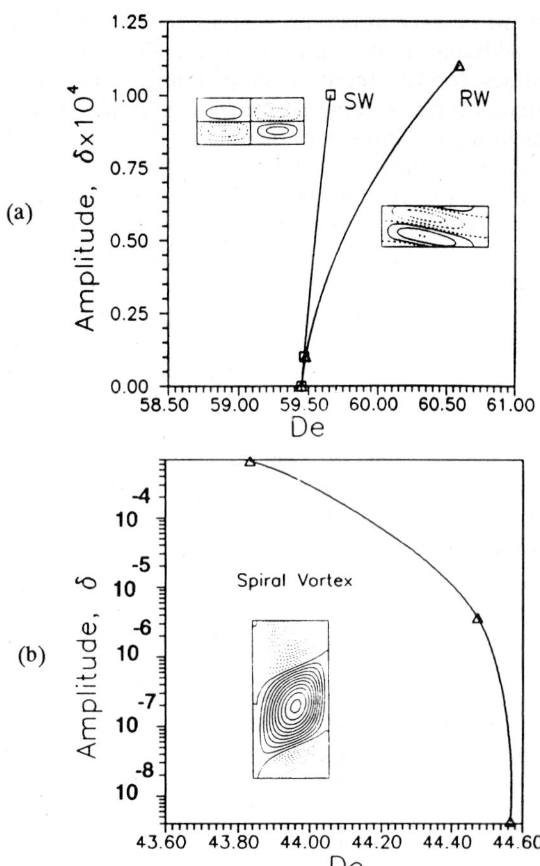

Figure 3 Bifurcation diagram at $\epsilon=1\text{x}10^5$, $\zeta=0.95$ and $\mu=0.5$ for a) $\alpha_c=8.0$, m=0. Both the rotating wave (RW) and the standing wave (SW) are shown. b) $\alpha=3.12$, m=2. Only the spiral wave is shown. In the inserts, the streamlines of the corresponding secondary flow are shown at $\theta=0$, t=0.

Two approaches have been proposed in the literature related to the use of spectral methods in time-dependent simulations: a partially implicit method (ref. 20) and a time-splitting scheme (refs. 21,22). A similar approach was also followed by Marcus (refs. 23,24) in the most comprehensive work in the numerical simulation of Newtonian Taylor-Couette flow. We employ here a similar technique for the study of high Re, intermediate De, flow instabilities, while we have developed a new method for the study of inertialess instabilities. Both methods involve a spectral approximation of the variables, and a hybrid (explicit/implicit) time integration scheme efficiently implemented using fast Poisson solvers and optimal filtering routines (ref. 25). The first method, applicable for finite Re numbers is based on a time-splitting integration, with the divergence-free condition enforced through an influence matrix technique (refs. 26,27). The second one, valid for inertialess creeping flows is based on a semi-implicit time integration of the constitutive equation with both the continuity and the momentum equations enforced as constraints.

6.1 High Reynolds Algorithm

The High Reynolds number algorithm involves a multistage time integration of the governing equations using the velocity, pressure and the elastic stress components as primary variables. In the first stage, the velocity and the stress components are advanced in time, accounting explicitly for the non-linear terms and the elastic component of the stress in the momentum equations and for all the terms in the constitutive equations. In the second stage, a pressure correction to the velocity field is implemented implicitly, where the corresponding pressure is calculated by solving a Poisson equation subject to zero pressure as boundary conditions at the solid wall. In the third stage, the velocity is updated for the viscous terms in the momentum equation implicitly by solving a Helmholtz equation subject to no-slip and non-penetration boundary conditions at the wall.

A final correction is then performed in order to satisfy the incompressibility constraint. This is implemented by adding to the pressure and velocity fields the solution to the homogenous part of the equations solved at stages 2 and 3, with the boundary equations for the pressure so chosen so that the resulting final velocity field is divergence-free everywhere on the solid boundary. Then, it can be proven (ref. 26) that, to within the accuracy of the scheme, the incompressibility constraint is satisfied everywhere within the domain. Note, that in the numerical implementation of the scheme, the final equation for the determination of the appropriate pressure boundary conditions is typically singular due to an incompatibility between the pressure and velocity representations. In our work, this singularity was removed by simply truncating the final system of equations. In consequence, the divergence-free condition was not enforced at a couple of points on the boundary. Although this strategy worked well (the divergence of the velocity was less than 10^{-8}), the more formal projection procedure recently developed by Phillips (ref. 27) is recommended.

6.2 Inertialess Algorithm

The inertialess algorithm, developed for axisymmetric flows, utilizes the streamfunction, azimuthal velocity and extra stress components as primary variables. The stress is updated in time integrating explicitly for the non-linear terms and implicitly for the linear (viscous) ones of the constitutive equation. The new streamfunction is evaluated by solving an inhomogeneous biharmonic equation, obtained by expressing the stresses in terms of the new streamfunction in the equation which results from the elimination of the pressure from the momentum equations. Similarly, the new azimuthal velocity is evaluated from the azimuthal component of the momentum equation. Note, that the resulting equation for the streamfunction in the inertialess algorithm is very similar in spirit to the idea of the reformulation of the momentum equation to an explicitly elliptic form pioneered by King et al. (ref. 18).

The key advantage of both algorithms is a combination of improved numerical stability and computational effectiveness. The numerical stability results from the explicit treatment of the viscous and pressure terms in the equations. This eliminates the severe limitations on the time step size otherwise imposed by a fully explicit algorithm. The computational effectiveness arises from the fact that the only equations required to be solved are (for a cylindrical coordinate system) separable in r and z which allows their efficient spectral solution using direct tensor methods (ref. 28).

6.3 Results

The time-dependent numerical simulation of the Taylor-Couette flow problem was used in order to examine the stability of the emerging secondary flows at the Hopf bifurcation. Indeed, the only other alternative would have been a high order non-linear (Birkhoff normal form) analysis which, however, for the problem at hand, leads to a prohibitively large number of intermediate terms. Two results are reported here in the form of a time-dependent trace of the value of the radial component of the velocity at a given location, $\underline{x}_m$ (close to the center of the Taylor vortices), each one corresponding to a different pattern.

The first graph, shown in Figure 4a, corresponds to a travelling wave (axisymmetric) secondary flow at an intermediate value of ϵ and was obtained with the high Reynolds algorithm using a time step, $\Delta t=0.02$. The second graph, shown in Figure 4b, corresponds to a standing wave (axisymmetric) secondary flow at $\epsilon=\infty$ and was obtained with the creeping flow algorithm using a time step, $\Delta t=1$. In both simulations 16 Fourier and 33 Chebyshev modes were used. The initial perturbations shown in Figure 4 lead eventually to constant amplitude limit cycles. When, for checking purposes, perturbations corresponding to the opposite patterns were used as initial conditions, the amplitude of the solutions increased dramatically till the codes diverged. Thus, the patterns shown in Figure 4 are stable to axisymmetric perturbations at their respective parameter conditions. However, since for Re=0 a 3d Hopf bifurcation occurs earlier, as discussed in the previous sections, the second pattern (at Re=0) must be globally unstable (ref. 29).

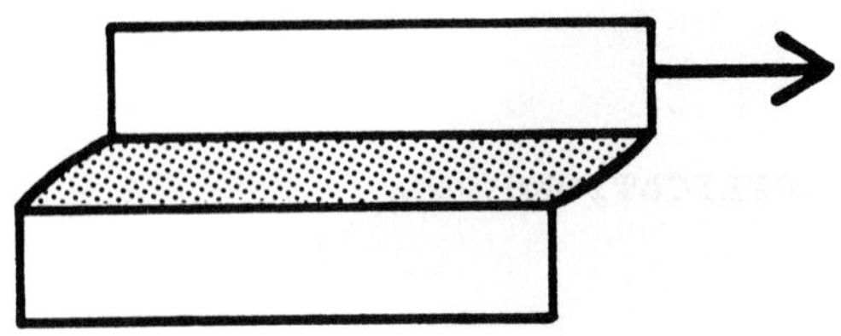

Figure 1 - Sliding Plate Geometry

if the stress could be measured locally rather than being inferred from the total force on one of the plates.

2. THE SHEAR STRESS TRANSDUCER

2.1 The frequency response problem

Devices for detecting tangential forces on flat surfaces have been used in the fields of soil mechanics, aerodynamics, flow of granular materials and rubber elasticity (1). The special challenge that arises in the case of viscous fluids is that the fluid may penetrate the sensor and interfere with its performance. To illustrate this problem it will be convenient to refer to Figure 2, which is a sketch of one type of shear stress transducer.

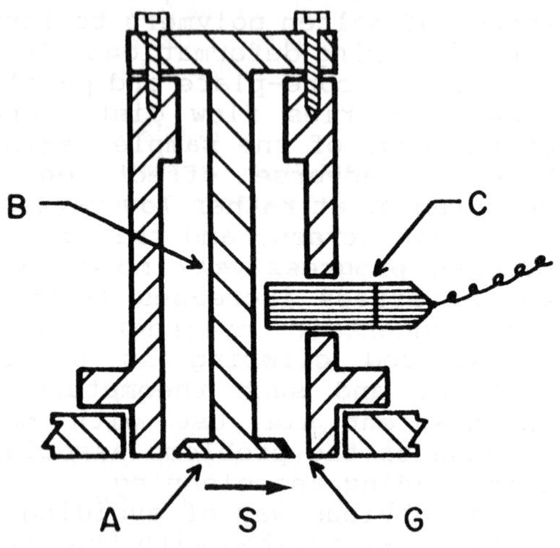

Figure 2 - Shear Stress Transducer with Cantilever Beam

The shear force, **S**, generated by the shearing deformation in the fluid, acts on an active face, **A**, at the end of cantilever beam, **B**, causing the beam to bend. If the deflection of the beam can be monitored by a proximity probe, **C**, then one has a measurement of the shear stress. In order for the beam to respond to the shear force, the active face must be mechanically isolated from the wall in which it is flush mounted, and this implies the existence of a gap, **G**, surrounding the face. Fluid can penetrate into this gap, and this can interfere with the operation of the sensor.

Even if the fluid is not under pressure, there will be some penetration into the transducer due to the compressional stresses caused during the loading of a rheometer or by normal stress differences generated by the shearing of the fluid. However, this penetration will be limited, especially if the gap, **G**, is small. As long as the material inside the transducer does not lose its fluidity, for example by degradation, it will pose no problem for steady state measurements. However, for studies of viscoelasticity, it is essential to measure stress transients, and the damping of the movement of the beam by the fluid in the gap now becomes a serious problem. Several solutions suggest themselves (ref. 4). One can seal the gap by means of a cured elastomer to prevent the ingress of fluid. However, the viscoelastic properties of the elastomer will alter the response of the sensor, and these will change with time at elevated temperature. Another approach is to incorporate a small linear motor into the transducer, activated by a control loop to maintain a constant gap.

It was considered desirable to avoid these complexities by designing the transducer to minimize the effect of fluid on the frequency response. In this regard, it is important to note that it is not the very high frequencies that are of prime interest in nonlinear viscoelastic studies but the plateau and terminal zones.

Our approach to the problem had two objectives:

a) Design the end of the beam to minimize damping due to fluid deformation in the gap.

b) Make the maximum beam deflection as small as possible, while still providing an acceptable signal to noise ratio.

An analysis of the squeezing-shearing flow that occurs in the gap when the beam deflects (ref. 5) revealed that the damping increases markedly as the gap decreases and as the area over which the solid surfaces act on the fluid increases. This suggests a sharp-edged active face, as shown schematically in Figure 2 and in more detail in Figure 3. The question of minimizing the deflection hinges on the sensitivity of the system used to detect the deflection. Various systems were tried (ref. 6), and while strain gauges were found to have insufficient sensitivity and a reflected light method was found to lack stability, the use of a capacitance method was found to yield a satisfactory window of operation (ref. 7). In Meissner's shear stress transducer (ref. 8) LVDTs were used to detect beam deflections in two orthogonal directions.

2.2 <u>The disk-spring transducer</u>

It is desirable to increase the range of the transducer, especially for stress relaxation experiments. In order to achieve this and also to make possible operation at high pressures, the disk-spring transducer was developed (ref. 5). Its principal features are shown in Figure 3. Again, the shear force, **S**, acts on an active face, **A**, but instead of a cantilever beam we have a rigid beam supported by a disk spring, **D**, which deflects in response to the moment on the beam, causing a deflection of the target, **T**, which is grounded and forms one plate of a capacitor. The capacitance probe, **C**, constitutes the other plate. Whereas in the cantilever type there is a loss of sensitivity resulting from the need to position the proximity probe, **C**, some distance from the active face, in the disk-spring type there is an amplification of the deflection of the active face if the upper arm of the beam is larger than the lower arm.

It will be noted that the beam of the disk-spring transducer is subject to axial deflection in response to a normal force acting on the active face. In high pressure applications, it may be desirable to suppress this deflection by incorporating stiffening ribs into the disk spring (refs. 5,9).

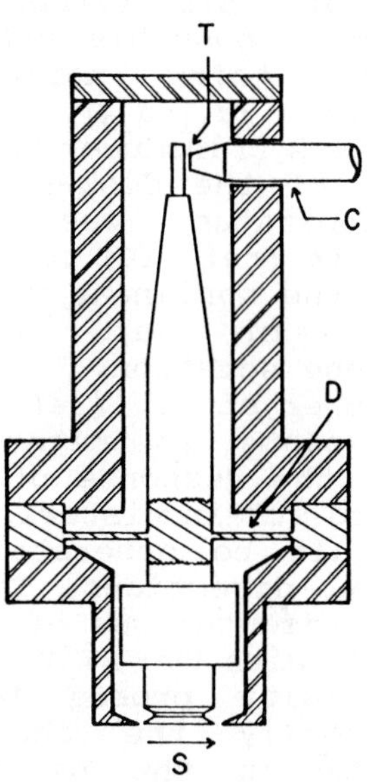

Figure 3 - The Disk-Spring Transducer

3. SLIDING PLATE RHEOMETERS

3.1 <u>Nonlinear viscoelasticity</u>

The use of sliding plate rheometers for the study of a wide range of materials has been reviewed by Dealy and Giacomin (ref. 3), and only instruments incorporating a shear stress transducer will be discussed here. Unless such a transducer is used, the shear stress must be inferred from the total force on one of the plates. This procedure is subject to serious errors, especially for large or rapid deformations, as a result of the following effects:

a) Friction in the mechanisms used to maintain the gap between the plates and keep them parallel,

b) Inhomogeneity of the flow near the edges due to the atypical stress field there,

c) Uncertainties about the exact area of the plates wetted by the sample.

Soong used a shear stress transducer of the cantilever beam type in her sliding plate rheometer, which was designed for use at room temperature (ref. 6). She demonstrated the feasibility of the concept and evaluated several types of proximity probe for detecting the deflection of the beam. This rheometer was later modified by the addition of a birefringence apparatus to make possible the measurement the third normal stress difference, $(\sigma_{11}-\sigma_{33})$. Also, the shear stress transducer was replaced with one of the disk-spring type. This apparatus was used by Demarquette (ref. 10) to carry out a study of the nonlinear viscoelastic behavior of a concentrated polystyrene solution.

Meissner et al. (ref. 8) used a sliding plate rheometer for the study of the response of polymeric liquids to strain histories in which shear could be generated in two, orthogonal directions. Their shear stress transducer was of the cantilever type with two LVDTs serving as the proximity probes to monitor simultaneously the shear stress components in the directions of shear.

Giacomin (refs. 7,11,12) designed a sliding plate rheometer expressly for use with molten plastics. The general features of the rheometer are shown in Figure 4.

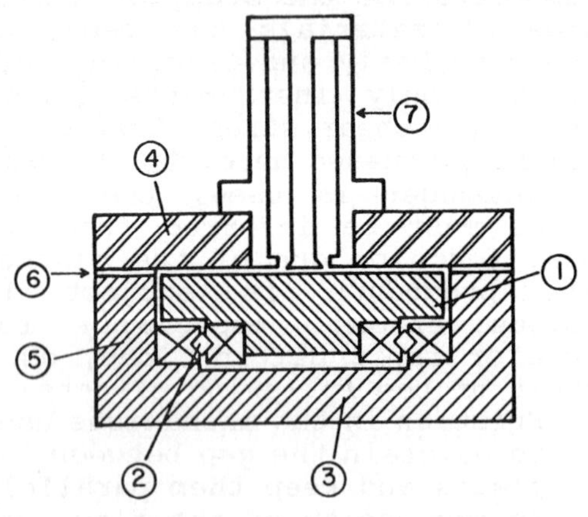

Figure 4 - Cross Section of McGill Sliding Plate Rheometer (Refs. 11,12)

The moving plate, 1, rides on a set of linear bearings, 2, which are mounted on a back plate, 3. The Stationary plate, 4, is fastened by side supports, 6. Shims are inserted at location 6 to fix the gap between the plates. The shear stress in the center of the stationary plate is monitored by a shear stress transducer, 7. The moving plate is driven by a servo-hydraulic linear actuator, which permits large forces to be precisely controlled so that large, rapid, arbitrary strain histories can be generated. A commercial version of this rheometer (ref. 13) is shown in Figure 5.

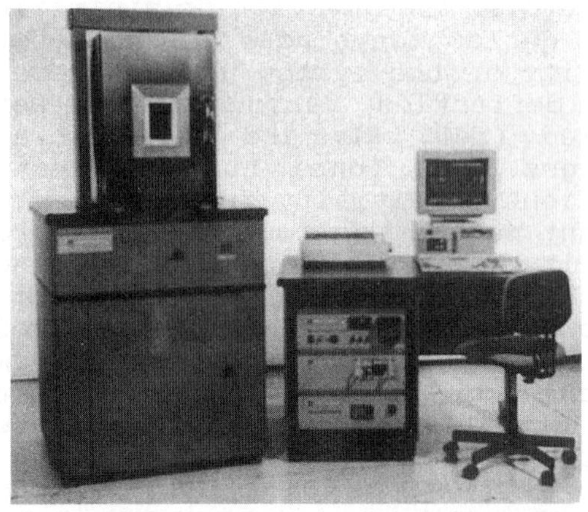

Figure 5 - Commercial Version of the McGill Sliding Plate Rheometer (Ref. 13)

A rheometer of this type has been used in the study of the nonlinear viscoelasticity of several polyethylenes (refs. 11-14).

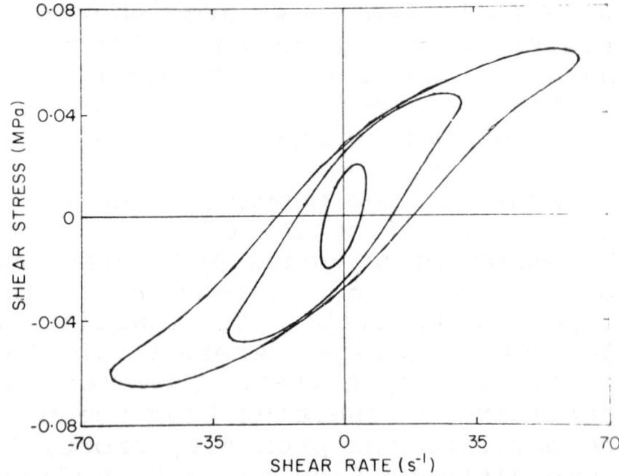

Figure 6 - Shear Stress Versus Shear Rate Loops for Large-Amplitude Oscillatory Shear of a LDPE. Frequency = 1 Hz. Strain amplitudes are 1 (smallest loop), 5 and 10.

Figure 6 shows data from a large amplitude oscillatory shear experiment (ref. 12). This is an interesting strain history in which the Deborah number (**λω**) and the Weissenberg number ($\lambda\dot{\gamma}_o$) can be varied independently. Its use in the study of nonlinear viscoelasticity has been reviewed by Giacomin and Dealy (ref. 15).

3.2 <u>Wall slip studies</u>

Hatzikiriakos used the sliding plate melt rheometer described above to study the slip of molten polymers at solid boundaries (refs. 16,17). Most previous studies of slip in thermoplastics had been carried out using pressure flow devices involving capillaries, slits, or extrusion dies. Because of the complications due to entrance effects, pressure gradients, and nonuniform shear rate, the results of such experiments are subject to certain uncertainties. In the sliding plate rheometer, there are no entrance effects or pressure gradients and the shear rate is constant across the gap. When slip occurs at both plates the true shear rate in the melt is related to the plate velocity, **V**, the gap, **h**, and the slip velocity by the expression:

$$\dot{\gamma}_n \equiv V/h = \dot{\gamma} + 2u_s/h \qquad (1)$$

where $\dot{\gamma}_n$ is the "nominal" shear rate that would be calculated assuming no slip. Assuming that the shear rate and the slip velocity are functions only of the wall shear stress, a plot of $\dot{\gamma}_n$ versus **1/h** at constant shear stress will be a straight line whose slope is $2u_s$. Using this technique Hatzikiriakos found clear evidence of slip in a series of polyethylenes and proposed a model to account for the effects of shear stress and molecular weight on the steady state slip velocity.

He also studied the effect on the slip of materials used as processing aids in the manufacture of blown film (ref. 17). In the sliding plate rheometer, the processing aid can be coated directly onto one or both of the plates.

Hatzikiriakos also studied slip in transient flows including exponential shear and large amplitude oscillatory shear and used the results to evaluate a slip model incorporating a history effect (ref. 16).

4. AN IN-LINE PROCESS RHEOMETER

Most process rheometers for molten plastics are of the "on-line" type, which means that a side stream sample is drawn from the main process flow by means of a gear pump and then passed to the rheometer, which is usually a capillary or a slit. Such a procedure leads to a significant sampling delay, which seriously compromises the ability of the rheometer to be used in a control loop unless the process being control has a very long time constant (ref. 18). To avoid this problem, the rheometer must be of the "in-line" type, that is to say it must be positioned directly in the process flow. It has been proposed (ref. 19) that this be accomplished by use of a shear stress transducer.

An in-line process rheometer that incorporates such a transducer is shown in Figure 7. The arrows show the direction of flow of the polymer. A rotating drum, **A**, drags a small fraction of the polymer into a narrow gap, **B**, between the rotating drum and a slightly larger stationary cylinder. The shear stress exerted by the fluid on the wall of the stationary cylinder is sensed by a shear stress transducer, **C**, which is of the disk-spring type.

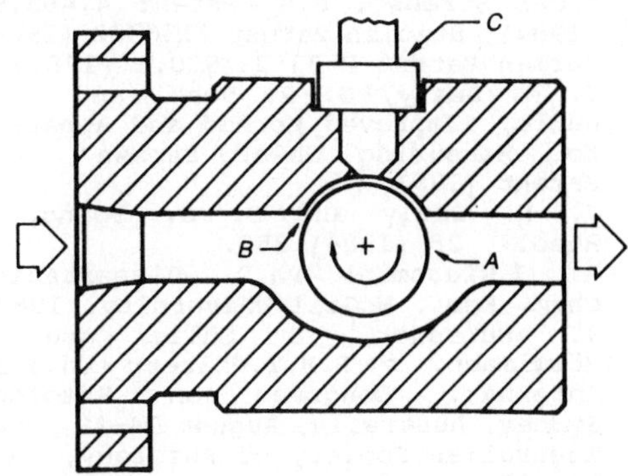

Figure 7 - McGill In-Line Rheometer

The process fluid is allowed to fill the lower portion of the transducer housing (See Figure 3), which is equipped with a vent port so that this fluid can be periodically purged to prevent degradation. Broadhead (refs. 20,21) used a rheometer of this type to control the neutralization of an acid copolymer to produce a range of ionomeric resins. Nelson (ref. 22) used the same rheometer to control the oxidative degradation of polypropylene. He examined the possibility of using the rheometer to determine the linear viscoelastic properties and found that it is impractical to try to generate a small amplitude sinusoidal shear because of the large frictional contribution to the torque on the drive shaft. He examined the possible use of other wave forms (ref. 23) and concluded that by use of a fast Fourier transform one could cope with significant mechanical noise and at the same time obtain very rapidly values of the storage and loss moduli corresponding to a range of frequencies.

REFERENCES

1. J. M. Dealy, Rheometers for Molten Plastics, Van Nostrand Reinhold, New York, 1982.
2. Dealy J.M. and K.F. Wissbrun, Melt Rheology and its Role in Plastics Processing, Van Nostrand Reinhold, NY 1990.
3. J. M. Dealy and A. J. Giacomin, "Sliding plate and sliding plate rheometers", in: Collyer and Clegg (Eds.), Rheological Measurement, Elsevier Applied Science Publishers Ltd., London, 1988, pp. 383-404.
4. J. M. Dealy, "Method of Measuring Shear Stress": U.S. Patent 4,463,928 (1984); British Patent 2130382 (1986); German Patent P 33 33 920.1 (1987)
5. J. M. Dealy, S. R. Doshi and F. R. Bubic, "Improved Method and Apparatus for Measuring Shear Stress", U.S. Patent (1991)
6. J. M. Dealy and S. S. Soong, J. Rheol., 28 (1984) 355.
7. A. J.Giacomin, Ph.D. Dissertation, Chem. Eng., McGill University, 1987.
8. J. Meissner, B Zülle and H. Hürlimann, in: P.H.T.Uhlherr (Ed.),Xth Internat. Congress on Rheology, Sydney, Australia, August 14-19, 1988, Australian Society of Rheology, 1988, Vol. 2, p. 121.
9. M. Sentmanat, M. Eng. Dissertation, Chem. Eng., McGill University, 1992.
10. N. Demarquette and J.M. Dealy, "The use of a new sliding plate rheometer to study the nonlinear viscoelasticity of concentrated polystyrene solutions", J. Rheol., 36 (1992) in press.
11. A.J. Giacomin and J.M. Dealy, "A new rheometer for molten plastics", SPE (ANTEC) Tech. Papers, 32 (1986) 711-714.
12. A.J. Giacomin, T. Samurkas and J.M. Dealy, "A novel sliding plate rheometer for molten polymers", Polym. Eng. Sci., 29 (1989) 499-504.
13. The True Shear Rheometer, Interlaken Technology Corp., Eden Prairie (Minneapolis), Minnesota, U.S.A.
14. T. Samurkas, R.G. Larson and J.M.Dealy, "Strong extensional and shearing flows", J. Rheol., 33 (1089) 559-578.
15. A. J. Giacomin and J. M. Dealy, "Large Amplitude Oscillatory Shear", in: Collyer and Clegg (Eds.), Techniques in Rheological Measurement, Elsevier Applied Science Publishers Ltd., London, 1992.
16. S.G. Hatzikiriakos and J.M. Dealy, J. Rheol., 35 (1991) 497-524.
17. S.G. Hatzikiriakos and J.M. Dealy, "The effect of interface conditions on wall slip and melt fracture of high density polyethylenes", Soc. Plastics Engrs. (ANTEC) Tech. Papers, 37 (1991) 2311-2314.
18. J. M. Dealy and T. O. Broadhead, "Rheometry for Process Control", in: Collyer and Clegg (Eds.), Techniques in Rheological Measurement, Elsevier Applied Science Publishers Ltd., London, 1992.
19. J. M. Dealy, "Method and Apparatus for Measuring Rheological Properties of Fluids": U.S. Patent 4,571,989 (1986); French Patent 85 15252 (1989).
20. T. O. Broadhead, Ph.D. Dissertation, Chem. Eng., McGill University, 1992.
21. T. O. Broadhead, J. M. Dealy and W. I. Patterson, 7th Ann. Mtg, Polym. Proc. Soc., Hamilton, Canada, April 21-24, 1991, pp. 33-34.
22. B. I. Nelson, Ph.D. Dissertation, Chem. Eng., McGill University, 1992.
23. B. I. Nelson and J. M. Dealy,"Dynamic Mechanical Analysis using Complex Waveforms", in: Collyer and Clegg (Eds.), Techniques in Rheological Measurement, Elsevier Applied Science Publishers Ltd., London, 1992.

Theoretical and Applied Rheology, edited by P. Moldenaers and R. Keunings
Proc. XIth Int. Congr. on Rheology, Brussels, Belgium, August 17-21, 1992

SURFACE-INDUCED EFFECTS IN POLYMER MELT FLOW

MORTON M. DENN
Department of Chemical Engineering, University of California at Berkeley, and Center for Advanced Materials, Lawrence Berkeley Laboratory, Berkeley, CA 94720 (USA)

1. INTRODUCTION

It has been a fundamental precept of fluid mechanics for more than a century that the constitution of a solid bounding surface is not relevant to the fluid flow, and that a fluid adopts the velocity of the solid at the surface. This presumption has not always been the case, of course, and Navier himself is responsible for the suggestion that there is a stress-dependent relative velocity between a fluid and the adjacent solid surface. The *no-slip* condition has been reviewed recently by Schowalter (ref. 1). It is pertinent to note in particular that the no-slip condition has been recognized for some time in classical fluid mechanics as a macroscopic approximation which must fail in the neighborhood of the singularity at a contact line when one fluid displaces another at a surface. Rough estimates of the singularities arising in viscoelastic flow near corners and lips predict stresses which exceed the strength of a covalent bond over finite regions (ref. 2), and insistence on the no-slip condition may be one source of the computational difficulties for polymeric liquids described by classical constitutive equations. (The estimates are rough because the form of the singularity near a corner is not known for the viscoelastic liquids of interest.)

The role of the solid surface has come to the fore in recent years in the context of polymer processing because of renewed attention to an extrusion surface defect known as *sharkskin*. Sharkskin manifests itself as a small-amplitude, high-frequency distortion on the surface of extrudates of some polymers, apparently when a critical stress is exceeded, and appears prior to the onset of the grosser instability known as *melt fracture*. Some investigators consider sharkskin to be an early form of melt fracture, while others believe the two phenomena to be distinct. Since the first observations of extrusion instabilities more than thirty years ago there have been suggestions that "wall slip" is a cause of instabilities, or at least an associated phenomenon. The literature through 1976 was reviewed by Petrie and Denn (ref. 3), who concluded at that time that slip was not a cause. Recent experiments require reconsideration of that conclusion. This reconsideration is needed not because of the significance of sharkskin *per se*, since technologies exist for avoiding or delaying the onset, but because a failure of the no-slip condition could have major processing implications in a variety of polymer flows. Slip would introduce the constitution of the shaping surface as a new design variable, hence providing new engineering opportunities for materials fabrication and design.

I have recently reviewed the extant points of view about the cause of extrusion instabilities (ref. 4), which include two types of "constitutive" instabilities: in one the instabilities are a result of multiplicities in the stress constitutive equation akin to ignition-extinction phenomena in combustion, while in the other the mathematical type of the governing equations changes and allows the propagation and growth of high-frequency disturbances. There is some evidence to support the existence of both types of constitutive instabilities, but I believe that the evidence for surface phenomena as the root cause is compelling and has grown since my review was published. Hence this is the topic on which I wish to focus.

Some facts about the onset of sharkskin do

not appear to be controversial, but it should be noted that most data have been obtained with polyethylene. Firstly, there appears to be a critical stress, usually of order 0.2 MPa, below which the surface defect is not observed visually. Secondly, whatever may be happening within the die or at the entrance, the instability occurs at the die *exit*. (There have been many demonstrations of this fact, dating back three decades to work of Howells and Benbow (ref. 5). The experiments I find most convincing are unpublished ones by Nazem, using a computer-controlled extruder in which the screw speed can be increased nearly instantaneously from the stable to sharkskin regime, with an immediate change in the surface character of the extrudate.) Thirdly, the critical conditions for onset are dependent on the die length-to-diameter ratio. Finally, during the transition from sharkskin to full melt fracture there is a regime known as *stick-slip* in which there are alternating regions of smooth and distorted extrudate. There is less agreement about a very important observation first made by Kurtz (ref. 6), who reported a change in the slope of the flow curve which coincides with the visual onset of sharkskin.

2. EXPERIMENTAL BACKGROUND

Adhesive failure between molten polymer and the die wall was proposed in a series of papers more than two decades ago by Vinogradov and his coworkers (c.f. ref 7). (Earlier reports were based on experimental techniques which were flawed.) Vinogradov argued for a transition at high wall stresses to a state in which the polymer melt could be treated as a rubbery solid and adhesive failure could be understood. His data on narrow-distribution polybutadienes and polyisoprenes have been used frequently to test theories of melt fracture because of the range of molecular weights available and the excellent characterization of the polymers.

The experiments which first convinced me of the need for a new look at the possibility of wall slip were reported by Ramamurthy (ref. 8, 9), who demonstrated that the visual onset of sharkskin in linear low-density polyethylene (LLDPE) in a die fabricated of alpha-brass occurred at much higher rates (if at all) than in a chrome-plated die, and that the die surface showed evidence of depletion of zinc. Ramamurthy used the Mooney method to demonstrate the apparent existence of finite slip velocities, and he argued that improved adhesion between the melt and the metal or metal-oxide surface was required to alleviate the distortions. Kalika and Denn (ref. 10) computed slip velocities in the same material in long capillaries by assuming that the the change of slope in the flow curve corresponds to the onset of slip. These experiments are subject to a variety of criticisms because of the use of capillaries, with possible confounding of the data because of end effects and high pressures. (Ramamurthy corrected his short-capillary data for entrance and exit effects, but Kalika and Denn did not.) Recent atmospheric pressure experiments on high-density polyethylene in a parallel-plate Couette rheometer by Hatzikiriakos and Dealy (ref. 11) show convincing evidence of wall slip and seem to validate the conclusions reached in capillaries. Hatzikiriakos and Dealy have since carried out an extensive set of experiments in capillaries (ref. 12) and have correlated their slip-velocity data from the parallel plate and capillary experiments to account for temperature, pressure, and molecular-weight distribution.

Schowalter and coworkers have reported observations of slip in a variety of polymers using a heat transfer technique that records deviations from fully-developed flow. Their data include a polybutadiene (ref. 13) in the molecular-weight range studied by Vinogradov, but they do not observe measurable slip in the sharkskin regime. Piau and coworkers have reported detailed observations of flow transitions, mostly on well-characterized polysiloxanes (ref. 14) but recently on linear polyethylenes as well (ref. 15). They argue that wall slip does not occur with sharkskin, and that the reports of slip are a result of faulty data analysis. It is not obvious that this conclusion can be drawn unambiguously from their experiments, and I find their criticism of other studies to be unconvincing.

3. THEORY

The presumption that flow instabilities are associated with an adhesive failure between the highly-stressed melt and the die wall led Hill, Hasegawa, and Denn (ref. 16) to develop a framework based on the theory of elastomer adhesion. The theory permits a connection between the rate at which new interfacial area is created in a solid peel test, in which the rubbery polymer is stripped from a metal or oxide surface with a constant applied force, and the critical stress and slip velocity in melt flow. The rheology of the bulk fluid enters because of the need to represent stored elastic energy, and pressure and temperature are incorporated through the linear viscoelastic shift factor. The prediction of a critical stress is rather good, as is the dependence of the critical stress on surface energy.

The theory gives an explicit expression for the wall slip velocity, v_s, as follows:

$$v_s = k a_T^{-1} \exp[-ß(4L/D)\tau_w]\tau_w^n \quad (1)$$

τ_w is the wall shear stress. a_T is the temperature-dependent part of the viscoelastic shift factor, which is usually expressed in terms of the W-L-F equation but which is often adequately described by a Van't Hoff exponential. ß is the pressure coefficient of viscosity in the exponential free-volume approximation $\eta \sim \exp(ßp)$, which is also the pressure dependence of the viscoelastic shift factor. n is the power dependence of the peel velocity on the applied force in a solid adhesion experiment. k is proportional to the reciprocal of the surface work of adhesion, which is a thermodynamic quantity, and to a length scale. Hill and coworkers argued that the length scale should be approximately R/4 in a capillary, but a scale characteristic of the surface roughness would seem to be more appropriate and it is not clear how to estimate the length scale in an arbitrary geometry. The analysis takes the *recoverable shear*, the ratio of the wall stress to the shear modulus, to be approximately unity.

Agreement between the theory based on peel experiments and the slip velocities inferred by Kalika and Denn (ref. 10) is remarkably good, giving strong support to the basic approach. The exponential dependence on the pressure (which is represented by the L/D term in Eq. 1) is consistent with the observation that the most critical region is at the die exit, where the pressure is least. The explicit results in the theory assume infinitely-long dies, but Hill in unpublished work has recently shown that accounting for the relaxation of viscoelastic stored energy predicts the L/D dependence of the critical stress for the onset of sharkskin and brings together the results of Ramamurthy and Kalika and Denn. The theory does not contain an obvious mechanism for the observed periodicity of the extrudate distortions.

Hatzikiriakos and Dealy (ref. 12) have formulated a theory of wall slip based on an extension of a kinetic adsorption/desorption analysis first proposed by Lau and Schowalter (ref. 17). The theory fits their data from both capillary and plate rheometers quite well, but it contains a number of *ad hoc* assumptions about functional forms and hence requires experimental data specific to the system being analyzed. The general form for the slip velocity is as follows:

$$v_s = c_1\phi(T)\{1 - c_2 \tanh[(E - c_3\sigma_n/\tau_w)/RT]\}(\tau_w/\tau_{c1}I^{1/4})^n \quad (2)$$

Here, $\phi(T)$ is an arbitrary function of temperature which is assumed to have a W-L-F form, and n is an arbitrary kinetic power-law exponent. E is an activation energy for adsorption/desorption, and the term involving σ_n/τ_w represents a stress-induced correction to the energetics. σ_n is the total normal stress acting on the wall. τ_{c1} is the first critical stress at which surface defects are observed. I is a measure of polydispersity.

Despite a very different physical basis the equation is in fact quite close in form to that developed by Hill and coworkers. The comparison for small L/D is particularly instructive:

$$v_s = ka(T)\exp[-(2c_4L/D)/RT](\tau_w/\tau_{c1}I^{1/4})^n \quad (3)$$

Analysis of Hatzikiriakos and Dealy's reported data shows that the arbitrary temperature term represented here by a(t) is given exactly by their measured viscoelastic shift factor, as predicted by Eq. 1. Hatzikiriakos and Dealy find a stronger pressure dependence at moderate pressures than do Hill and coworkers; the proper comparisons are $4ß\tau_w$ in Eq. 1, which Hill found to be 4×10^{-3}, and $2c_4/RT$ in Eq. 3, for which Hatzikiriakos found 60×10^{-3}. Hatzikiriakos and Dealy found saturation at higher pressures.

Stewart (ref. 18) has recently formulated a theory of slip which combines features of the physical ideas leading up to Eqs. 1 and 2. The major addition is a geometric-mean approximation for the work of adhesion to incorporate the surface tensions of the polymer and presumed coating layers explicitly. Agreement with the data of both Kalika and Denn and Hatzikiriakos and Dealy is quite good, as is the prediction of throughput change in extrusion experiments as a function of the surface tension of fluoropolymer additives.

Two other slip velocity equations which have appeared in the published literature deserve mention. El Kissi and Piau (ref. 19) have reported an empirical equation which does not contain an obvious means of extrapolation to other systems. Leonov (ref. 20) has published an equation based on molecular considerations which requires parameters that do not seem to be readily available.

4. SIMULATION

It has long been the practice in simulations to introduce a small degree of slip at the exit contact line to relieve the singularity, and it is well-known for Newtonian fluids that numerical results are insensitive to this approximation. Limited simulations for viscoelastic liquids in which a simple form of slip velocity is introduced show relative insensitivity (ref. 21). The two theories of slip cited above, although based on very different physical pictures, share common nonlinearities, and it is likely that if slip is a factor in generating surface instabilities and oscillatory flow it is because of nonlinear coupling. The major corrections needed to the adhesion theory of Hill and coworkers are incorporation of a neglected recoverable shear term and replacement of the pressure in the free-volume expression for the shift factor with the average normal stress. The polydispersity factor discovered by Hatzikiriakos and Dealy seems to work well and should be included. We thus obtain the equation

$$v_s = kW_a^{-1}a_T^{-1}\exp[-ß(p-N_1/3)](\tau_w/G)^n\,(\tau_w/I^{1/4})^n \quad (4)$$

where W_a is the work of adhesion, N_1 is the primary normal-stress difference, and G is the shear modulus. Based on Stewart's results the work of adhesion should be taken as proportional to the geometric mean of the surface tensions of polymer and solid surface as long as only dispersive forces control the interaction. n is obtained from adhesion experiments; I from molecular-weight distribution data; and ß, N_1, G, and a_T from rheological data. k is a constant involving a length parameter which is not obvious in an arbitrary flow.

5. SPECTROSCOPY

The notion that physico-chemical interactions between a polymer melt and a surface can influence flow will remain controversial until some direct measurements can be made, and these are likely to be spectroscopic. We have recently completed some attenuated-total-reflectance Fourier-transform infrared spectroscopy measurements in which the rate at which alkanes displace one-another near the surface of a flow cell is monitored (ref. 22). Even with 16-carbon chains we find surprisingly strong surface effects, with stress- and surface-dependent adsorption from the melt operating over time scales of order one minute. These effects are expected to be even more pronounced if the polymers and surfaces experience strong interactions, since in that case there will be surface-induced conformational changes. Such conformational changes have been observed recently in adsorption from solution for poly(methylmethacrylate) on alumina (ref. 23), and we have demonstrated with nuclear magnetic resonance spectroscopy that interactions with a surface can cause the helical conformation of polypeptides to "unwrap" (ref. 24), but comparable measurements from the melt have not been made.

6. ACKNOWLEDGMENT

Preparation of this paper was supported in part by the Director, Office of Energy Research, Office of Basic Energy Sciences, Materials Science Division of the U.S. Department of Energy under Contract No. DE-AC03-76SF00098.

REFERENCES

1. Schowalter, W.R., J. Non-Newtonian Fluid Mech. 29 (1988) 25 - 36.
2. Lipscomb, G.G., R. Keunings, and M.M. Denn, J. Non-Newtonian Fluid Mech. 24 (1987) 85 - 96.
3. Petrie, C.J.S., and M.M. Denn, AIChE J. 22 (1976) 209 - 36.
4. Denn, M.M., Ann. Rev. Fluid Mech. 22 (1990) 13 - 34.

5. Howells, E.R., and J.J. Benbow, Trans. Plastics Inst. 88 (1962) 240 - 53.
6. Kurtz, S.J., in B. Mena et al. (Ed.), Advances in Rheology, vol. 3, Univ. Nat. Auton. Mex., Mexico City, 1984, pp. 399 - 407.
7. Vinogradov, G.V., A.Ya. Malkin, Yu.G. Yanovskii, E.K. Borisenkova, B.V. Yarlykov, G.V. Berezhanya, J. Poly. Sci., Part A-2 10 (1972) 1061 - 84.
8. Ramamurthy, A.V. J. Rheol. 30 (1986) 337 - 57.
9. Ramamurthy, A.V. Adv. Polym. Technol. 6 (1986) 489 - 99.
10. Kalika, D.S., and M.M. Denn, J. Rheol. 31 (1987) 815 - 34.
11. Hatzikiriakos, S.G., and J.M. Dealy, J. Rheol. 35 (1991) 497 - 523.
12. Hatzikiriakos, S.G., and J.M. Dealy, J. Rheol., in press.
13. Lim, F.J., and W.R. Schowalter, J. Rheol. 33 (1989) 1359 - 82.
14. El Kissi, N., and J.M. Piau, J. Non-Newtonian Fluid Mech. 37 (1990) 55 - 94
15. Piau, J.M., Int. Conf. Dynamics Polym. Liqs., Capri, Sept., 1991.
16. Hill, D.A., T. Hasegawa, and M.M. Denn, J. Rheol. 34 (1990) 891 - 918.
17. Lau, H.C., and W.R. Schowalter, J. Rheol. 30 (1986) 193 - 206.
18. Stewart, C.W., SPE ANTEC 92, Detroit, May, 1992.
19. El Kissi, N., and J.M. Piau, C. R. Acad. Sci. Paris, Ser. II 309 (1989) 7 - 9
20. Leonov, A.I., Wear 141 (1990) 137 - 45.
21. Phan-Thien, N., J. Non-Newtonian Fluid Mech. 26 (1988) 327-340.
22. Dietsche, L.R., A.T. Bell, and M.M. Denn, AIChE Annual Meeting, Los Angeles, Nov., 1991.
23. Thakkar, B., F. Konstandinidis, A.K. Chakraborty, R. Tannenbaum, L.W.Potts, J.F. Evans, and M. Tirrell, Langmuir, in press.
24. Fernandez, V.L., J.A. Reimer, and M.M. Denn, submitted for publication.

Theoretical and Applied Rheology, edited by P. Moldenaers and R. Keunings
Proc. XIth Int. Congr. on Rheology, Brussels, Belgium, August 17-21, 1992

RHEOLOGY OF A SYSTEM WITH MESOSCOPIC DOMAIN STRUCTURE

Masao Doi

Department of Applied Physics, Faculty of Engineering Nagoya University, Nagoya 464 (Japan)

1. INTRODUCTION

Polymer blends or liquid crystalline polymers involve structures of large length scale, typically of the order of μm. Thermal effects for such large structures are small. Therefore, these structures are easily changed by flow. Rheological properties of such systems are usually quite complicated. Here I present a theory for such a class of materials. Though the theory does not cover the whole range of materials which involves domains or textures, it indicates an important aspect of such materials, i.e., the lack of a characteristic time and the scaling property. I shall demonstrate this for two systems, the mixture of immiscible fluids (ref.1) and the polydomains in liquid crystalline polymers(ref.2).

2. CONCENTRATED MIXTURE OF IMMISCIBLE FLUIDS

2.1 Interface Pattern and Stress Tensor

When two immiscible fluids are mixed mechanically, droplets of various sizes and shapes are formed. In a flow field, these droplets deform, rupture and aggregate to create rather complex patterns (see Fig.1).

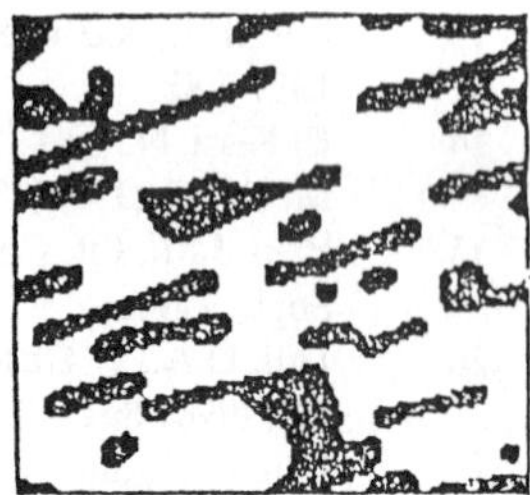

Fig.1 Flow patterns obtained by computer simulation of two immiscible liquids. left:initial state right: after shearing (reproduced from ref.3)

First the dynamics and rheological properties of such systems are discussed. For the sake of simplicity, the two fluids are assumed to have the same viscosity and the same density. A concentrated mixture is considered: the volume ratio is assumed to be about 1:1. The interface of the fluid is regarded as a mathematical surface with no thickness. The Reynolds number is assumed to be small, so that the inertial force is neglected entirely. In such a problem, there are only three parameters characterizing the system, i.e., viscosity η_0,volume fraction ϕ and the interfacial tension Γ.

It is known that the volume average of the stress tensor for such system is given by

$$\sigma_{\alpha\beta} = \eta_0(\kappa_{\alpha\beta}+\kappa_{\beta\alpha})-\frac{\Gamma}{V}\int dS(n_\alpha n_\beta-\frac{1}{3}\delta_{\alpha\beta})-p\delta_{\alpha\beta} \tag{1}$$

where $\kappa_{\alpha\beta} = \partial v_\alpha/\partial r_\beta$ is the macroscopic velocity gradient, and the integral is carried out over all interfaces in the system of volume V. Equation (1) indicates that the stress is determined by the tensor

$$q_{\alpha\beta} = \frac{1}{V} \int dS(n_\alpha n_\beta - \frac{1}{3}\delta_{\alpha\beta}) \qquad (2)$$

which is called the interface tensor.

2.2 Constitutive Equation

Let us consider the time evolution equation for the interfacial tensor under a macroscopic velocity field $\vec{v}(\vec{r},t) = \kappa(t) \cdot \vec{r}$. Crudely speaking, there are two factors influencing the pattern of the interface: one is the flow field which enlarges and orients the interface, and the other is the interfacial tension which opposes these effects. If these effects are accounted for separately, one can write

$$\frac{dq_{\alpha\beta}}{dt} = \left.\frac{dq_{\alpha\beta}}{dt}\right|_{\text{flow}} + \left.\frac{dq_{\alpha\beta}}{dt}\right|_{\text{surface tension}} \qquad (3)$$

The first term can be obtained by a geometrical consideration. In the absence of the interfacial tension, the interface is deformed affinely. Then one can calculate the change of the interfacial tensor(ref.1).

$$\frac{d}{dt}q_{\alpha\beta} = - q_{\alpha\gamma}\kappa_{\gamma\beta} - q_{\beta\gamma}\kappa_{\gamma\alpha} + \frac{2}{3}\delta_{\alpha\beta}\kappa_{\mu\nu}q_{\mu\nu} - \frac{Q}{3}(\kappa_{\alpha\beta} + \kappa_{\beta\alpha}) + \frac{q_{\mu\nu}\kappa_{\mu\nu}}{Q}q_{\alpha\beta} \qquad (4)$$

where Q is the specific interface area $Q = S/V$. The time evolution of Q is given by

$$\frac{d}{dt}Q = -\kappa_{\alpha\beta}q_{\alpha\beta} \qquad (5)$$

On the other hand, the second term in eq.(3) is not easy to assess. Here a simple phenomenological argument is used. Generally speaking, the interfacial tension has two effects:(i) it decreases the area of the interface, and (ii) it makes the system isotropic. The area of the interface is Q, and the degree of the anisotropy can be expressed by $q_{\alpha\beta}/Q$. Thus the simplest relaxation equation is:

$$\left.\frac{d}{dt}Q\right|_{\text{surface tension}} = -r_1 Q \qquad (6)$$

$$\left.\frac{d}{dt}\left(\frac{q_{\alpha\beta}}{Q}\right)\right|_{\text{surface tension}} = -r_2\left(\frac{q_{\alpha\beta}}{Q}\right) \qquad (7)$$

where r_1 and r_2 are the relaxation rates with r_1 representing the decrease of the interfacial area (size relaxation), and r_2 the relaxation of anisotropy (shape relaxation). In general, these relaxation rates are determined by the viscosity η_0 , the interfacial tension Γ and the configuration of the interface. The last can be characterized by Q and $q_{\alpha\beta}$. If we disregard the dependence on $q_{\alpha\beta}$ and assume that r_1 and r_2 are determined by η_0 , Γ and Q, we can show by dimensional analysis,

$$r_1 = c_1\frac{\Gamma Q}{\eta_0} \qquad r_2 = c_2\frac{\Gamma Q}{\eta_0} \qquad (8)$$

where c_1 and c_2 are positive numbers, which may depend on the volume fraction ϕ.

Eq.(8), however, has an unphysical feature. According to Eq.(8), the system approaches the state of $Q = 0$, which is a macroscopically phase separated state. This can happen for the 1:1 mixture which would form a bicontinuous phase. On the other hand, if the volume fraction is not equal to 1:1, the final state may not be the state of $Q = 0$ since once the droplets of the minority phase become spherical, no relaxation would take place. A better model is therefore

$$r_1 = c_1\frac{\Gamma}{\eta_0}(\Sigma q_{\alpha\beta}^2)^{1/2} \qquad (9)$$

which guarantees that the relaxation stops when all the droplets become spherical. Combining these equations we have

$$\frac{d}{dt}q_{\alpha\beta} = - q_{\alpha\gamma}\kappa_{\gamma\beta} - q_{\beta\gamma}\kappa_{\gamma\alpha} + \frac{2}{3}\delta_{\alpha\beta}\kappa_{\mu\nu}q_{\mu\nu}$$

$$- \frac{Q}{3}(\kappa_{\alpha\beta} + \kappa_{\beta\alpha})$$

$$+ \frac{q_{\mu\nu}\kappa_{\mu\nu}}{Q} q_{\alpha\beta} - \lambda(\Sigma q_{\mu\nu}^2)^{1/2} q_{\alpha\beta} \quad (10)$$

$$\frac{dQ}{dt} = -\kappa_{\alpha\beta} q_{\alpha\beta} - \lambda\mu(\Sigma q_{\alpha\beta}^2)^{1/2} Q \quad (11)$$

where $\lambda = c_1 \Gamma/\eta_0$ and μ is a certain numerical constant. Equations (10) and (11) are the time evolution equations for the shape and size of the interface. Given $q_{\alpha\beta}$ and Q, the stress can be calculated by Eq.(1). Thus the set of equations, Eqs.(10) (11) and (1) give the rheological constitutive equation.

2.3 Rheological Properties

The constitutive equation we have derived has properties quite different from those of the usual viscoelastic materials like polymeric liquids. The most important difference is that the present constitutive equation has no intrinsic time constant. This follows from the dimensional analysis: the present system is characterized by Γ, η_0 and ϕ, from which one cannot construct a quantity that has a dimension of time. This leads to various curious rheological properties.

Consider for example, the steady shear flow with shear rate $\dot{\gamma}$. It can be shown from the constitutive equation that the shear stress and the first normal stress differences are proportional to the shear rate:

$$\sigma_{xy} = \eta(\phi)\dot{\gamma} \qquad \sigma_{xx} - \sigma_{yy} = \eta_N(\phi)\dot{\gamma} \quad (12)$$

Thus there is no shear thinning or shear thickening. Although surprising it may seem, this conclusion should hold rigorously since it is a result of dimensional analysis. Notice also that the first normal stress difference is proportional to $\dot{\gamma}$, and is singular at $\dot{\gamma} = 0$. This is like nematic liquid crystals. The singularity comes from the fact that even an infinitesimally small shear rate can change the structure drastically.

2.4 Scaling Property

The lack of the intrinsic time scale leads to an interesting scaling property. Let us write the stress for a given shear rate $\dot{\gamma}(t)$ as

$$\sigma = \sigma(t; [\dot{\gamma}(t)]) \quad (13)$$

then one can easily show

$$\sigma(t; [c\dot{\gamma}(ct)]) = c\sigma(ct; [\dot{\gamma}(t)]) \quad (14)$$

i.e., the stress at time t under the velocity gradient $c\dot{\gamma}(ct)$, is c times larger than the stress at time ct under the velocity gradient $\dot{\gamma}(t)$.

As an example, consider a stepwise change in the shear rate from $\dot{\gamma}_1$ to $\dot{\gamma}_2$ at time $t = 0$ (see Fig.2). Then the scaling relation is written as

$$\sigma(t, c\dot{\gamma}_1, c\dot{\gamma}_2) = c\sigma(ct, \dot{\gamma}_1, \dot{\gamma}_2) \quad (15)$$

Hence, $\sigma(t, \dot{\gamma}_1, \dot{\gamma}_2)$ must be written as

$$\sigma(t, \dot{\gamma}_1, \dot{\gamma}_2) = \dot{\gamma}_2 f\left(\dot{\gamma}_2 t, \frac{\dot{\gamma}_1}{\dot{\gamma}_2}\right) \quad (16)$$

Thus for a fixed ratio of $\dot{\gamma}_1/\dot{\gamma}_2$, the plot of $\sigma(t, \dot{\gamma}_1, \dot{\gamma}_2)/\dot{\gamma}_2$ against the shear strain $\dot{\gamma}_2 t$, becomes independent of the magnitude of $\dot{\gamma}_1$.

Although these predictions are interesting, there has been no detailed report on the rheological properties of the mixture of Newtonian fluids as far as I am aware. However the scaling relation has been found for an entirely different system, i.e., the polydomain liquid crystalline polymers.

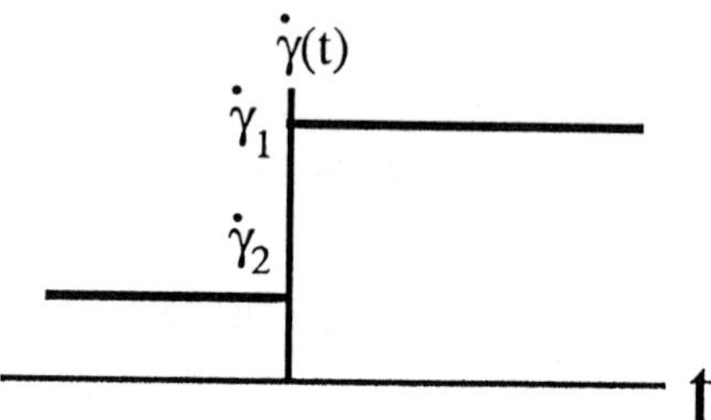

Fig.2 Stepwise change in the shear rate.

3. LIQUID CRYSTALLINE POLYMERS

3.1 Scaling Relation of Liquid Crystalline Polymers

Liquid crystalline polymers have complex domain structures, which causes very complex rheological responses. Several people, however, have found that in the liquid crystalline solution of PBLG, a simple scaling relation exists. The first observation was made by Mead and Larson(ref.4).They measured strain recovery after removal of the shear stress in steady shear flow, and found that the recovery curve becomes independent of the shear rate if the time is scaled by the shear rate in the steady state(Fig.3).

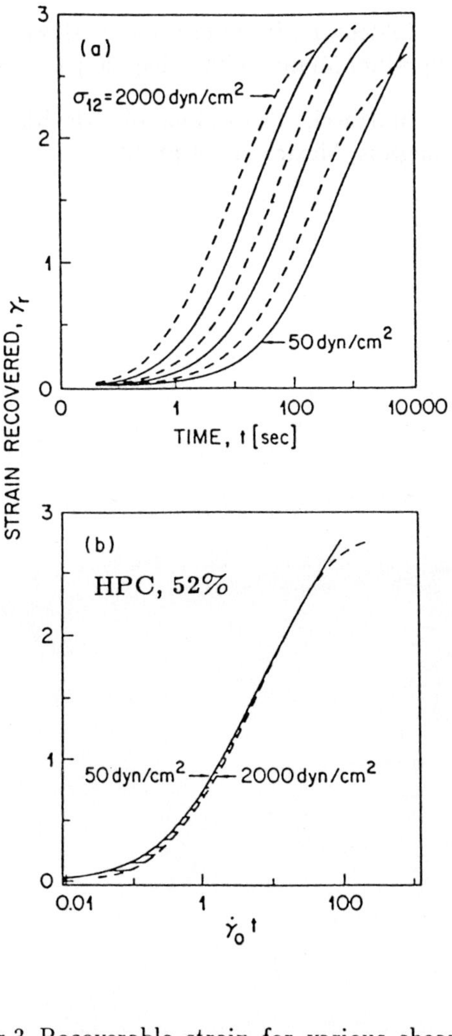

Fig.3 Recoverable strain for various shear stress for a 52:8 % liquid crystalline solution of hydroxypropylcellulose. top: plotted agains time. bottom: plotted against the previous shear rate. Reproduced from ref.4.

A more striking scaling relation was found by Moldenaers, Yanase and Mewis(ref.5). They did experiments with the step change of the shear rate shown in Fig.2 and found that the curves can be superimposed if the shear stress normalized by the steady state value is plotted against the shear strain $\dot{\gamma}_2 t$. Examples of such scaling plots are shown in Fig.4. (Notice that this scaling plot is the same as eq.(17) since the steady shear stress is proportional to the shear rate in the regime they did the experiments.)

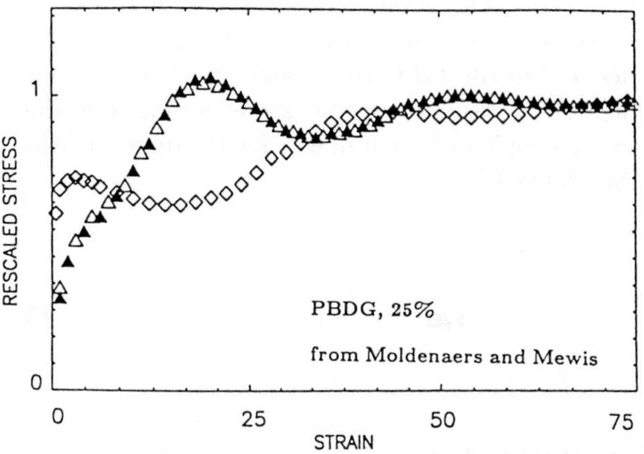

Fig.4 Normalized shear stress after the step chang of the shear rate for a 25% PBDG solution in metacresol. Triangles: flow reversal $-\dot{\gamma}_1 = \dot{\gamma}_2 = 1s^{-1}$ and $-\dot{\gamma}_1 = \dot{\gamma}_2 = 0.4s^{-1}$. Squares: step increase in shear rate $\dot{\gamma}_1 = 0.1s^{-1}, \dot{\gamma}_2 = 1s^{-1}$. Reproduced from ref.5.

3.2 Phenomenological Constitutive Equation

The scaling relation indicates that the polydomain liquid crystalline polymers belong to the same class of materials as the mixture of Newtonian fluids. There is no intrinsic time scale in the system, and the characteristic time of the system is determined by the external conditions such as the pre-shear. Such an idea is not new in liquid crysalline polymers. Marrucci(ref.6) argued that the domain structure of the liquid crystalline polymers is characterized by a single length scale a. For a steady shear flow, a can be estimaed by a dimensional analysis:

$$a \simeq \left(\frac{K}{\eta\dot{\gamma}}\right)^{1/2} \tag{17}$$

where η is the viscosity of the system, and K the Frank elastic constant.

The similarity between the two systems is clear, and one can derive a semi-phenomenological constitutive equation following the same line of argument described in section 2.

The starting point of such a theory is Ericksen-Leslie's theory which describes the motion of the local director field $\vec{n}(\vec{r},t)$ and the stress tensor $\sigma(\vec{r},t)$. The macroscopic stress tensor and the macroscopic order parameter for the director field are defined by

$$\overline{\sigma}(t) = \frac{1}{V}\int d\vec{r}\sigma(\vec{r},t) \tag{18}$$

$$\overline{S_{\alpha\beta}(t)} = \frac{1}{V}\int d\vec{r}(n_\alpha(\vec{r},t)n_\beta(\vec{r},t) - \frac{1}{3}\delta_{\alpha\beta}) \tag{19}$$

Then time evolution equations for $\overline{S_{\alpha\beta}}$ and the characteristic length scale $a(t)$ are derived. As in the previous argument, the equations consist of two parts, one expresses the effect of the macroscopic flow and the other is the relaxation due to the Frank elasticity. With proper choice of parameters, the constitutive equations can reproduce the experimental behavior shown in Figs. 3 and 4.

4. CONCLUDING REMARKS

In this paper, I have described a class of constitutive equation which is suitable for the system with domains or textures. The characteristic point is that the relaxation time of the system entirely is determined by external conditions. The approach demonstrated here will be useful for other materials such as aggregating suspensions, emulsions and foams. Various modifications are needed for each case, but the essential physics is the same. Detailed discussion will be published in the future.

REFERENCES

1. M. Doi and T. Ohta, J. Chem. Phys. 95 (1991) 1242
2. R. G. Larson and M. Doi, J. Rheology 35 (1991) 539
3. T. Ohta, H. Nozaki and M. Doi, J. Chem. Phys. 93 (1990) 2664
4. R. G. Larson and D.W. Mead, J. Rheology 33 (1989) 1251
5. P. Moldenaers, H. Yanase and J. Mewis, ACS Syposium Miami (1990) chap 26 p370-380
6. G. Marrucci, Proceeding of 9th Rheology Congress, Mexico (1984)p8-13

Theoretical and Applied Rheology, edited by P. Moldenaers and R. Keunings
Proc. XIth Int. Congr. on Rheology, Brussels, Belgium, August 17-21, 1992

OPTICAL RHEOMETRY

Gerald G. Fuller

Department of Chemical Engineering
Stanford University
Stanford, CA 94305-5025

1.0 INTRODUCTION

This paper reviews methods in optical rheometry designed to measure the dynamics and structure of complex liquids. Such applications have a long history in rheology, with the first experiments mainly taking advantage of the stress optical relationship that is valid for many polymeric systems. These were primarily polarimetry measurements, and birefringence was the principal observable. Recent advances in this area have both improved the technology upon which the measurements are based, and made use of other optical interactions such as small angle light scattering and Raman scattering. These have resulted in a number of important advantages that allow the consideration of important, new classes of problems in rheology and nonNewtonian fluid mechanics. In this paper the basic interactions of light with complex, anisotropic liquids are reviewed and a selected number of applications are presented that illustrate these advantages. These applications include (1) separating the dynamics of constituents in multicomponent liquids, and (2) measuring the orientation distribution function in polymeric liquids.

2.0 INTERACTION OF LIGHT WITH COMPLEX LIQUIDS

2.1 Transmission Experiments: The Refractive Index Tensor

The interaction of light with a sample can involve transmission, scattering or reflection. For transmitted light, it is the refractive index tensor that is normally measured (ref. 1). This tensor, $\mathbf{n} = \mathbf{n}' + i\mathbf{n}''$, has both real and imaginary components, causing phase retardation and attenuation of the electric vector, $\mathbf{E}$, respectively. When a material is anisotropic, two optical anisotropies are defined. These are the birefringence, $\Delta n'$, defined as the difference in the principal eigenvalues of $\mathbf{n}'$, and the dichroism, $\Delta n''$, which bears a similar relationship to $\mathbf{n}''$.

Two basic effects contribute to $\mathbf{n}'$ and $\mathbf{n}''$: intrinsic interactions and scattering. Intrinsic causes of birefringence and dichroism are derived from anisotropies that are inherent in the polarizability and absorption cross section of the material, respectively. The polarizability is a measure of how easy it is to distort the electron distribution by an electric field. This property is largely insensitive to wavelength and for that reason intrinsic birefringence is also not usually a strong function of wavelength.

However, intrinsic dichroism is a strong function of wavelength, since it only appears at wavelengths where the material absorbs. This makes intrinsic dichroism an important tool for identifying the orientation dynamics in multicomponent materials. Intrinsic birefringence, however, will generally produce a measure of the bulk dynamics.

For many materials, and especially polymers, there is a simple, linear relationship between the stress tensor, τ_{ij}, and the refractive index tensor. This is the stress optical rule, (ref. 1) which states that,

$$C\tau_{ij} = n'_{ij} + A\delta_{ij}, \tag{1}$$

where C is the stress optical coefficient, and A is a constant associated with the isotropic portions of each tensor. This rule can be derived from many simple molecular models of polymer liquids and is valid whenever the primary contribution to the stress tensor arises from segmental orientation. Equation (1) has been written for the real part of the refractive index tensor, but an analogous equation can be written for the intrinsic dichroism.

Scattering of light also affects the phase and amplitude of the measured electric field. Anisotropic materials scattering light will cause scattering (or form) contributions to the birefringence and dichroism. Whereas the intrinsic effects reflect segmental orientation, scattering effects respond to anisotropies associated with larger length scales, on the order of entire chains, or assemblies of chains (ref. 2).

2.2 Small Angle Light Scattering (SALS): The Structure Factor

Light scattered from an obstacle will radiate at all angles. If these objects are large relative to the wavelength of light, most of the scattered intensity will appear at small angles. In these circumstances, the scattered light can be measured conveniently using two dimensional detectors. The observable in SALS is the structure factor, $S(q)$, defined as

$$S(q) \equiv \frac{1}{c}\int dr \langle \delta c(r)\, \delta c(0) \rangle e^{-iq \bullet r} \qquad (2)$$

where q is the scattering vector and $\langle \delta c(r)\, \delta c(0) \rangle$ is the spatial correlation function for concentration fluctuations, $\delta c(r)$.

Scattering dichroism also arises from light scattering, and Onuki and Doi (ref. 2) have derived the following expression relating the second moment of the structure factor to the imaginary part of the refractive index tensor, $\mathbf{n}''$:

$$n''_{\alpha\beta} = \frac{k^3}{32\pi^2\varepsilon^2}\left(\frac{\partial\varepsilon}{\partial c}\right)^2 \int d\Omega_{\hat{k}_s} S(\mathbf{q})\,(\delta_{\alpha\beta} - \hat{k}_{s_\alpha}\hat{k}_{s_\beta}) \qquad (3)$$

Here, ε is the equilibrium dielectric constant of the suspension, $\mathbf{k}_s$ is the wave vector of the scattered light, $\hat{\mathbf{k}}_s = \mathbf{k}_s/|\mathbf{k}_s|$, $\Omega_{\hat{k}_s}$ is the solid angle over $\hat{\mathbf{k}}_s$, and $k = (2\pi)/\lambda$. The scattering vector is $\mathbf{q} = \mathbf{k}_s - \mathbf{k}_i$, where $\mathbf{k}_i$ is the wave vector of the incident light.

Evidently, the dichroism is proportional to the second moment of the structure factor with respect to the scattering vector.

2.3 Raman Scattering

Rayleigh scattering, where the scattered light occurs at the same wavelength as the incident light, is the dominant contribution to the scattering process. Molecular vibration modes, however, can induce a shift in the frequency of a portion of the scattered light by an amount equal to the vibrational frequency of those modes. This is referred to as Stokes and anti-Stokes Raman scattering. As in the case of infrared absorption, Raman scattering is a vibrational spectroscopy that can be used to measure the orientation of vibrational modes in a system. The intensity of Raman scattered light is:

$$I = I_0 \langle (l_i \alpha'_{ij} l_j)^2 \rangle \qquad (4)$$

where I_0 is the incident light intensity, α'_{ij} is the Raman scattering tensor, and l_i and l_j are unit vectors of the polarization of the incident and scattered light, respectively. The Raman tensor is related to gradients of the polarizability tensor, α_{ij}, with respect to vibrational modes in the system. In other words,

$$\alpha'_{ij} = \left(\frac{\partial \alpha_{ij}}{\partial Q_k}\right)_0 Q_k, \qquad (5)$$

where Q_k is a vibrational mode. (ref. 3)

For a macromolecule characterized by an end-to-end vector, $\mathbf{R}$, the Raman tensor has the following form:

$$\alpha'_{ij} = \gamma(\delta_{ij} + \varepsilon R_i R_j), \qquad (6)$$

where γ is a constant and ε is proportional to the anisotropy in the principal values of the Raman tensor. ε can be expressed in terms of the depolarization ratio of the undeformed polymer (ref. 4). Equations (4) and (6), show that the Raman scattered light intensity provides information on the second moment, $\langle R_i R_j \rangle$, and the fourth moment, $\langle R_i R_j R_k R_l \rangle$, of the orientation distribution function. Intrinsic birefringence and dichroism, however, only provide measures of the second moments.

3.0 APPLICATIONS

3.1 Enhancement of Concentration Fluctuations from a Sheared Polymer Solution

Many polymer liquids (solutions and blends) undergo apparent changes in phase upon the application of flow. This is most often manifested by flow induced turbidity (ref. 5) and for semi-dilute polymer solutions, above the quiescent cloud point temperature, results from the growth of concentration fluctuations. Simple turbidity measurements, however, provide an integrated signal that cannot distinguish between growth, orientation, or distortion of concentration fluctuations. In the present work, time dependent SALS and scattering dichroism have been utilized to investigate the dynamics of growth and orientation of concentration fluctuation enhancement for a solution of polystyrene in dioctyl phthalate. A variety of flow configurations were used, including Couette (ref. 6), parallel plate, and four roll mill flow cells. Experiments were conducted under conditions where the quiescent systεm is optically clear.

In Figure 1, a schematic diagram of the instrument is presented. A HeNe laser is used to illuminate the sample. This light is first polarized and then modulated in its polarization using a half wave plate that rotates at 2 kHz. The light is both scattered and transmitted by the sample. The scattered light forms a pattern on a screen below the sample. This pattern is recorded using a 2D CCD camera interfaced to a computer. Geometric corrections are applied to the data to correct for the distortion of the structure factor by the fact that the camera views the image at an angle. An aperture in the screen allows the unscattered, transmitted light to pass on to a photodiode detector for the measurement of dichroism. This is accomplished by analyzing this signal using a lock-in amplifier that determines its Fourier content.

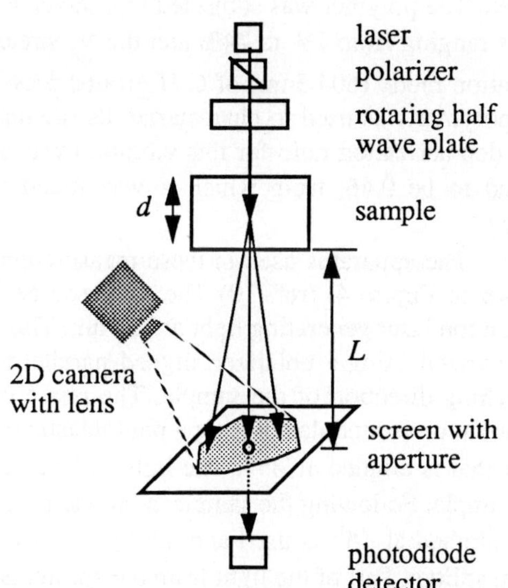

Figure 1. Schematic diagram of the instrument used for SALS and scattering dichroism measurements.

Figure 2 presents the time evolution of $\Delta S(q)$ in a steady shear experiment at a temperature of $16°C$ and a shear rate of 5 s^{-1}. The deviation in the structure factor from quiescence is defined as $\Delta S(q,t) \equiv S(q,t) - S(q,0)$. In general, $\Delta S(q,t)$ is in the form of butterfly patterns consisting of two lobes of high intensity. Inception of steady shear is at t=0 s, while cessation is at t=20 s. The flow direction in Figure 2 is from left to right while the direction of increasing velocity is from bottom to top. On the start of shear, scattering patterns increase in intensity due to shear enhancement of concentration fluctuations. In addition, the pattern is anisotropic and rotates with the vorticity of the flow until steady state is reached (Figure 2b). Comparing scattering patterns just before (Figure 2b) and after flow cessation (Figure 2c), the scattered intensity suddenly increases after cessation due to elastic recoil before $\Delta S(q)$ relaxes to zero. The patterns obtained at steady state are identical in shape to those reported previously on a similar solution by Wu *et al. (ref. 7)* where wide angle light scattering was used to measure the structure factor at steady state conditions.

Results obtained from SALS and scattering dichroism experiments, $\Delta n''_{expt}$, were tested for consistency by using the Onuki-Doi theory to determine dichroism, $\Delta n''_{calc}$. Steady state values of both $\Delta n''_{expt}$ and $\Delta n''_{calc}$ are compared at various shear rates and temperatures. It has been found (ref. 6) that $\Delta n''_{expt}$ and $\Delta n''_{calc}$, are related by a proportionality factor which is essentially a constant over the range of conditions considered. Hence it may be inferred that scattering dichroism and the structure factor are indeed consistent through the Onuki-Doi relation. If the Weissenberg number, $W_i \equiv \dot{\gamma}\tau_r$, is used as a measure of shear, it is seen that data over a range of temperatures and shear rates collapse to a single master curve (Figure 3). Here $\dot{\gamma}$ is the shear rate and τ_r is a characteristic polymeric stress relaxation time.

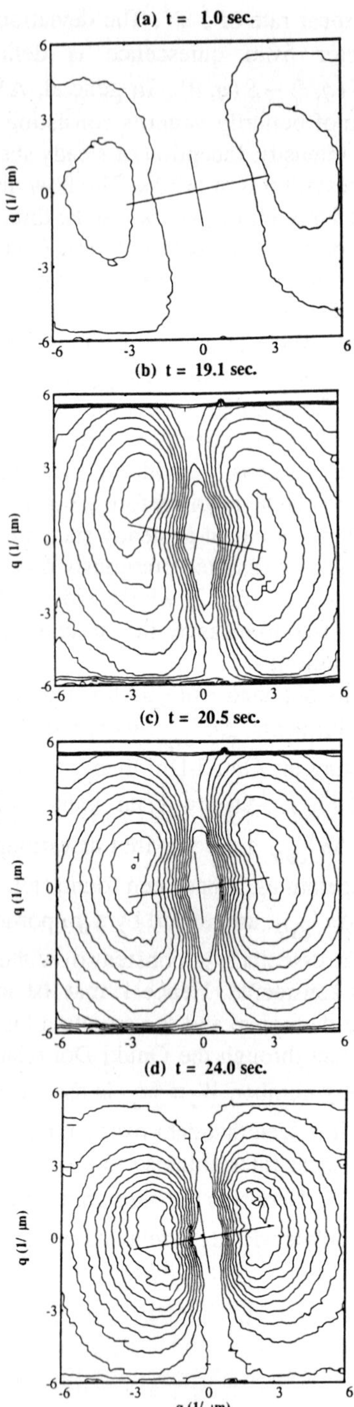

Figure 2. Contour plots of time dependent response to steady shear flow of structure factor in couette geometry; $T = 16^oC$, $\dot{\gamma} = 5s^{-1}$: (a) t = 15 s, (b) 19.1 s, (c) 20.5 s, (d) 24 s

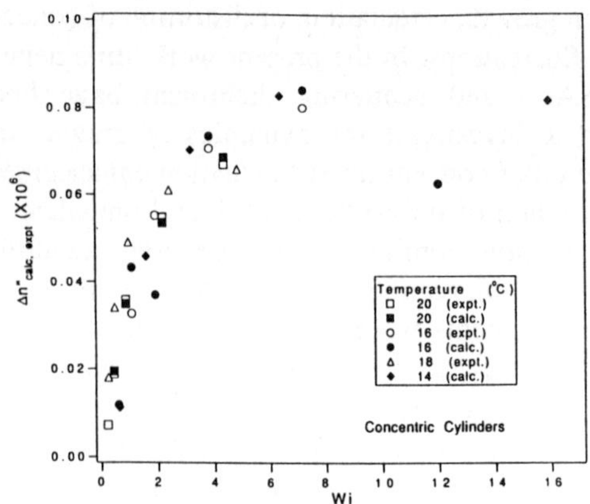

Figure 3. Master curve for steady state $\Delta n''_{calc}$ and $\Delta n''_{expt}$ vs. *Wi*.

3.2 Measurement of the Fourth and Second Moments of the Orientation Distribution in a Polymer Melt

We have used Raman scattering to study the orientation dynamics of a polyisobutylene melt of molecular weight 360,000. In these experiments, the time dependent second and fourth order moments of the orientation distribution function were determined. The polymer was subjected to uniaxial extensions ranging from 2% to 24% and the ν_s stretching vibration mode (604.5nm) of C-H groups present in the polymer was used to characterize its orientation. The depolarization ratio for this vibration was determined to be 0.46, from which ε was found to be -1.73.

The apparatus used in these measurements is shown in Figure 4. (ref. 10) The light source is an Argon ion laser generating light at 514 nm. This light is polarized using a polarizer aligned parallel to the stretching direction of the sample. The phase of the polarization is modulated using a photoelastic modulator that is aligned at 45°. The light is then sent to the sample. Following the sample, a second polarizer, also aligned at 45° is used and the light is sent to a beam splitter. Part of the light from the splitter is sent to a photodiode for the purpose of measuring the birefringence of the sample, which can affect the Raman scattered light. The remaining light is sent to a monochromator/notch filter combination that removes the light at the incident frequency and selects specific vibrational components for study. Finally, a photomultiplier tube is used to measure the intensity of the Raman scattered light.

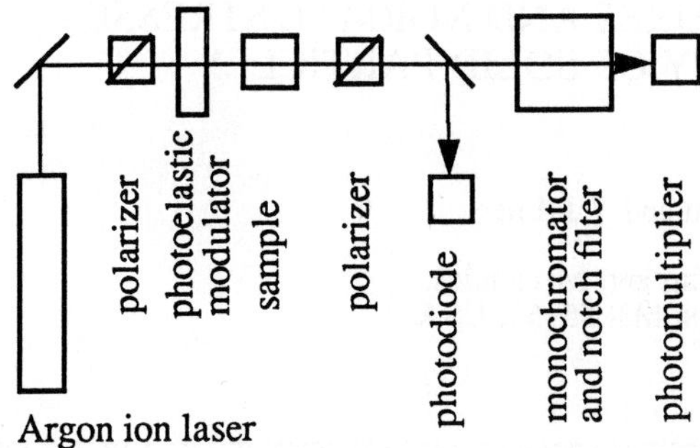

Figure 4. Apparatus for polarization modulated Raman scattering.

The time dependent anisotropies in the second and fourth order moments, are shown in Figure 5. From these data, it is seen that the fourth and second moments are approximately linear in the applied strain and that the fourth moment is slightly larger than the second moment. This results can be used as direct tests of constitutive equations that predict the relative magnitudes of these anisotropies, such as the Doi-Edwards model. The specific magnitudes of the moments reported here are compatible with the estimates offered by Bower for the case where the orientation distribution is a monotonic function of the polar angle of orientation of the Raman tensor. (ref. 11)

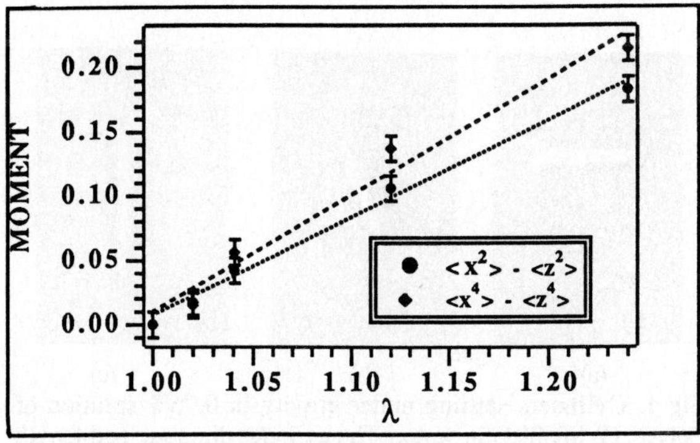

Figure 5. Anisotropies in the second and fourth moments, following a single step reversal extension of 5%.

References

1. Fuller, G.G., Annu. Rev. Fluid Mech., 22, 384 (1990).
2. Onuki, A. and Doi, M., J. Chem. Phys., 85, 1190 (1986).
3. Long, D.A., Raman Spectroscopy, McGraw-Hill, London (1977).
4. Archer, L.A. and Fuller, G.G., "Dynamics of the second and fourth order moments of the orientation distribution function using PMLRS", in preparation.
5. Rangel-Nafaile, C., Metzner, A. B., Wissbrun, K. F., Macromolecules, 17, 1187 (1984).
6. van Egmond, J. W. and Fuller, G. G., J. Chem. Phys., submitted November 1991.
7. Wu. X.-L., Pine, D. J. and Dixon, Phys. Rev. Lett., 66, 2408 (1991).
8. Helfand, E. and Fredrickson, G.H., Phys. Rev. Lett., 62, 2468 (1989).
9. van Egmond, J. W. and Fuller, G. G., "Optical Anisotropy due to Shear Enhanced Concentration Fluctuations in Polymer Solutions", in preparation.
10. Archer, L.A. and Fuller, G.G., "Dynamics of Polymeric liquids using Polarization Modulated Laser Raman Scattering", *Polymer*, accepted (1991).
11. Bower, D. I., *J. Polym. Sci.: Polym. Phys. Ed.* 19, 93 (1981).

Theoretical and Applied Rheology, edited by P. Moldenaers and R. Keunings
Proc. XIth Int. Congr. on Rheology, Brussels, Belgium, August 17-21, 1992

COMPETITION BETWEEN INERTIAL PRESSURES AND NORMAL STRESSES IN THE FLOW INDUCED ANISOTROPY OF SOLID PARTICLES

D.D. Joseph, J. Nelson, H.H. Hu and Y.J. Liu

Department of Aerospace Engineering and Mechanics,
University of Minnesota, Minneapolis, MN 55455, USA

1. INTRODUCTION

It is well known that a long body settling in a viscous liquid will turn its broadside to the stream. The same long body settling in a viscoelastic liquid will turn its broadside parallel to the stream at small speeds [1,2], but heavier long bodies which fall faster again turn broadside. Sedimenting spheres in a fluid filled channel will arrange themselves so that the line of centers between neighboring spheres is across the stream in a viscous liquid and parallel to the stream in a viscoelastic liquid when the fall velocity is small but across the stream again when the fall velocity is large. In both cases the anisotropy is associated with wakes; drafting, kissing and tumbling in a viscous liquid and drafting and kissing but no tumbling in a viscoelastic liquid. Spheres falling close to a wall of the channel rotate as if rolling down the wall in the intuitive way in a viscous liquid, but rotate as if rolling up the wall against intuition in a viscoelastic liquid. Provisional explanations of the peculiar observations can be framed as a competition between inertia and normal stresses with cross stream arrays preferred when inertia dominates and streamwise arrays preferred when normal and extensional stresses dominate.

Certain flows of viscoelastic liquids can be usefully described as a competition between inertia and normal stresses [3] with inertia scaling with U^2/L and normal stresses with U^2/L^2, where U is a typical speed and L a characteristic length for gradients. A similar kind of competition appears to be responsible for the flow induced anisotropies in sedimenting and fluidized suspensions of solid particles which we have observed recently in our laboratory and are reporting here. Elsewhere, it has been noted that small spheres (~70 μm) in an oscillating liquid sheared by the back and forth motion of parallel plates[4] or cone and plates [5,6,7] align in the direction of shear when the liquid is viscoelastic and across the direction of shear when the liquid is Newtonian[7]. It is necessary ultimately to understand why one kind of structure is stable in a flow dominated by inertia and another structure in a flow dominated by viscoelasticity.

2. OBSERVATIONS ABOUT THE TILT ANGLE OF SEDIMENTING CYLINDERS

In this paper we are putting forward the idea that flow induced anisotropy of spherical particles finds its explanation in the flow induced turning couples on long bodies. The streamwise orientation of a settling or fluidized long body is unstable in a viscous liquid, and will always turn its broadside to the stream as in Figs. 1a and 2a. An explanation[8] for this can be found in the couples which are produced by high pressures at the stagnation points on the long body shown in Fig 3. Potential flow is probably a good approximation for viscous flow on the forward side of the body. If the pressures outside a thin boundary layer at the stagnation points were reversed, the long body would not put its broadside into the stream, but instead would put its broadside parallel to the stream, as is in fact the case in the settling of long cylinders in various viscoelastic liquids shown in Figs. 1c and 2c. In practice, the wakes which develop on the back side of bodies give rise to a drag. The turning couples of such wake are self-equilibrating in symmetric cases (α=0 and α=90° in Fig. 3) when the Reynolds number is small and the flow is steady and self-limiting when the Reynolds number is large, due to vortex shedding.[9]

(a) (b) (c)

Fig 1. Cylinders settling under gravity in 0.75% solution of Polyox (WSR-301) in water. The cylinder diameter and length are 0.25 in. and 0.6 in. respectively. The container is 25 in. high, 5 in. wide and 0.44 in. deep. The tilt angle α of the cylinder axis from horizontal is a function of the fall speed. (a) Stainless steel (ρ=0.283 lb/in^3). (b) Aluminum (ρ=0.0975 lb/in^3). (c) Plastic (ρ=0.0476 lb/in^3).

Fig. 2. Aluminum cylinder, $(D,L)=(0.25,0.6)$ in., settling in aqueous Polyox (WSR-301) of different mass concentration ϕ. The tilt angle α is a function of concentration. (a) $\phi=0.5\%$. (b) $\phi=0.75\%$. (c) $\phi=1.25\%$.

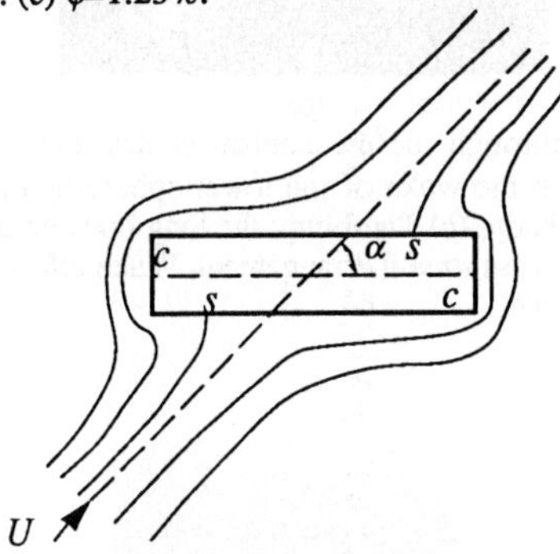

Fig 3. Potential flow past a cylinder. The pressure at stagnation points s will turn the broadside of the body into stream as in Figs. 1a and 2a. If the extensional stress at s were reversed, as may be possible in a viscoelastic liquid, the body would line up with the stream as in Figs. 1c and 2c. The same type of turning with the longside parallel to the stream could be provided by normal stresses caused by strong shears at the corners c. In practice, viscosity will lead to boundary layers and wakes whose effects are not yet understood.

The angle α of tilt between the direction of steady fall and the cylinder axis is evidently determined by a competition between the normal stresses and inertia when neither dominates, as in Figs. 1b and 2b. The normal stresses discussed in the caption of Fig. 3 and in the theory to be discussed in §5 are extensional stresses at points of stagnation. This is clearly at best a partial explanation because forces act all over the body, not just at points of stagnation. The stresses are influenced by the shape of particles, as is seen in Fig. 4.

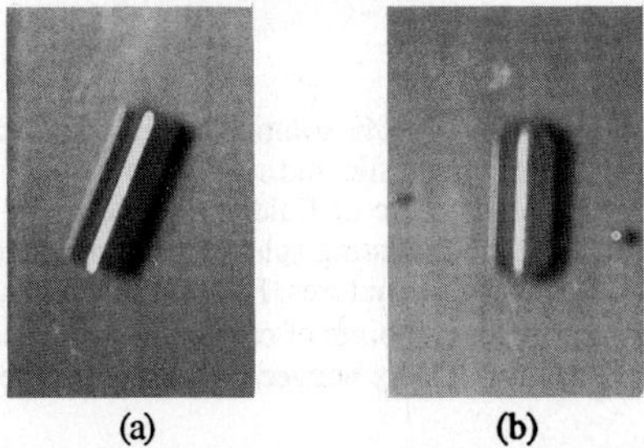

Fig 4. Brass cylinder, $(D,L) = (0.3,0.8)$ in. and $\rho=0.306$ lb/in^3, falling in 1.25% aqueous Polyox (WSR-301). The tilt angle is much smaller when the ends are rounded as in (b). Viscoelastic effects are more pronounced in cylinders with rounded ends.

3. TURNING COUPLES ON ROLLING SPHERES

We have observed a new phenomenon which is evidently associated with turning couples around the point of contact between a sphere and plane wall as the sphere slides and rolls down the plane wall of a liquid filled container. The sphere rotates as if rolling down the wall in a viscous liquid, but will turn as if rolling up the wall in a viscoelastic fluid (Fig 5.) The couple which produces the counter clockwise motion in the glycerin shown in Fig 5a changes sign in the viscoelastic liquids shown in Fig 5b.

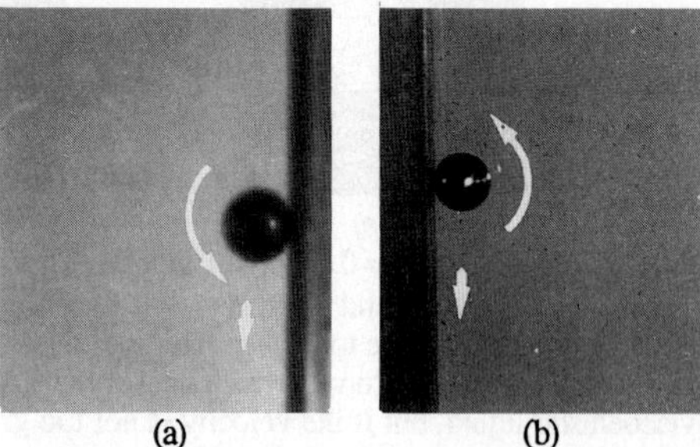

Fig 5. Spheres falling close to a channel wall rotate as if rolling down the wall in viscous fluids, (a) 50/50 glycerin/water solution, and up the wall in viscoelastic fluids, (b) 1.2% aqueous polyacrylamide. The channel in these pictures is tilted slightly to keep the sphere close to the wall.

4. FLOW INDUCED ANISOTROPY OF SPHERICAL PARTICLES

The angle α between the direction of steady fall and the cylinder axis (Fig. 3) is determined by the competition of normal stress and inertia when neither dominate. Broadside on configurations occur in the same Polyox solution for heavier cylinders which fall faster. Inertia dominates in Newtonian liquids, in dilute solutions with small normal stresses, and in more concentrated elastic solutions when the Reynolds number is large enough (Figs. 1a, 2a).

Flow induced anisotropy of sedimenting and fluidized spheres is associated with wakes and surprisingly, with turning couples on long bodies. In viscous liquids this anisotropy is associated with a mechanism called drafting, kissing and tumbling (Fig.6). One sphere is accelerated into the wake of another, they kiss and tumble. The kissing spheres are momentarily a long body which is unstable to the kind of forces shown in Fig.3 which turn long bodies broadside on; so they tumble. This is a local mechanism which implies that globally the only stable configuration is the one in which, on the average, the line of centers between spheres is perpendicular to the stream as seen in Fig. 7a. Stable cross stream arrays of spheres fluidized by water in channels whose small gap confine the motion to two dimensions are a dominant structure which can be observed in nearly every regime of flow.[5,6,7,11,12,13,14] It is apparent from Fig. 7b that a different and even orthogonal anisotropic structure is generated in the viscoelastic liquids in which long bodies settle longside on. The solid spheres in these liquids draft and even kiss, but not tumble. When the fall velocity is large the line of centers between spheres is perpendicular to the stream again as in Fig.7c.

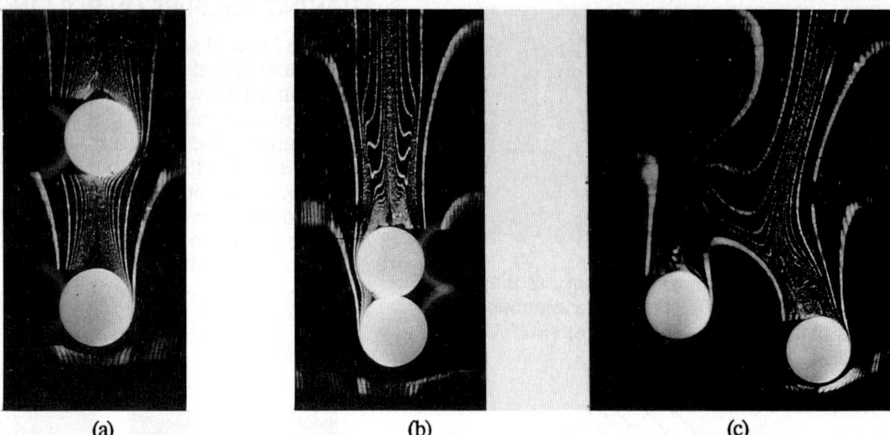

(a) (b) (c)

Fig. 6. Spheres (Delrin, D=0.25in.) in viscous fluid (Glycerin/water) interact through the mechanism of drafting, kissing and tumbling (Joseph, Singh and Fortes[17]). (a) Drafting: the top sphere is caught in the wake of the lower sphere and accelerated downwards. (b) Kissing: the top sphere contacts the lower sphere, creating a long body. (c) Tumbling: the long body arrangement is unstable and experiences a torque which causes the spheres to tumble into a stable cross stream arrangement. Wake effects work also in a viscoelastic liquid, but if the velocity is not too great the spheres will not tumble.

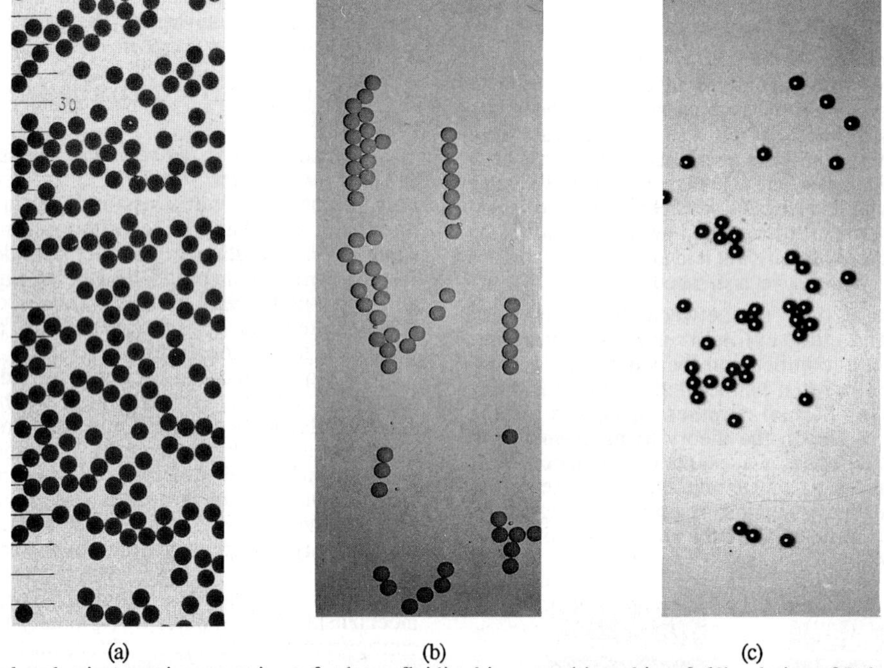

(a) (b) (c)

Fig. 7. Flow induced anisotropy in suspensions of spheres fluidized in water (a), and in a 0.6% solution of Polyox (WSR-301) in water (b) and (c). The plastic spheres in (a) are fluidized by an upward flow of water but the plastic and steel spheres in (b) and (c) are falling under gravity. The cross stream arrangement of particles shown in (a) is characteristic of fluidized suspensions of spheres confined by walls to move in two-dimensions when the suspending fluid is Newtonian. Sedimenting spheres tend to disperse through the dynamics of tumbling; turning couples on long bodies tend to give rise to cross stream structures. Pairs of spheres falling along the line of centers of a tube attract each other in viscoelastic liquids[15]. The normal stress at points of contact between the chains of spheres in (b) must pull the particles together. When the fall velocity is large the line of centers between spheres is perpendicular to the stream again (c).

5. SOME THEORETICAL CONSIDERATIONS

Leal [1] has studied the sedimentation of slender bodies in a second order fluid with inertia neglected. He considers only those non-Newtonian effects resulting from the disturbance velocity field generated by the lowest order geometry independent approximation of the Stokeslet distribution used in slender body theory. He finds that freely translating particles with fore-aft symmetry exhibit a single stable orientation with the axis of revolution vertical. This may suggest that the angle α of tilt observed in experiments may be determined in a competition between inertia and viscoelasticity. The mechanism which aligns the slender body with the stream is not easy to extract from Leal's analysis.

Brunn[16] studied the interaction of two spheres in a second order fluid with inertia neglected and he found an attractive force which draws the spheres together. Riddle, et al [15] discovered that if the initial separation of two spheres settling along their line of centers in a viscoelastic fluid is larger than a certain critical separation the spheres will diverge, whereas it is smaller than this separation they will converge. The analysis of Brunn[16] does not give rise to a critical separation and it cannot treat close approach because it has been assumed that the distance between sphere centers is large compared to the radius.

Joseph[10] has shown that every potential flow is a solution of the equation of motion for second order fluid with stresses given by

$$\sigma_{ij} = -[\, C + \hat{\beta}\, \phi_{,i\ell}\, \phi_{,i\ell} - \rho\phi_{,t} - \rho|\mathbf{u}|^2/2]\, \delta_{ij} + 2[\mu + \alpha_1(\partial_t + \mathbf{u}\cdot\nabla)]\, \phi_{,ij} + 4(\alpha_1+\alpha_2)\phi_{,i\ell}\, \phi_{,\ell j} \tag{1}$$

where

$$\sigma_{ij} = T_{ij} + \rho\mathbf{g}\cdot\mathbf{x}\, \delta_{ij} \tag{2}$$

is the active dynamic stress, $\mathbf{g}$ is gravity, C is a Bernoulli constant, $\hat{\beta}$ is the climbing constant, ρ is the density, μ is the viscosity, α_1 and α_2 are the quadratic constants, $\alpha_1=-n_1/2$ and $\alpha_2=n_1+n_2$ with n_1 and n_2 being the first and second normal stress coefficient. In general, potential flow cannot satisfy no slip conditions at solid walls. In this theory the streamlines are determined say by the prescribed values of the normal component of velocity. You cannot see the effects of changing the values of the material parameters in the values or distribution of velocity which are given by potential flow. In the case of a rod rotating in a sea of second order fluid, the potential flow solution also satisfies the boundary conditions at rod surface and the potential flow is exact. Potential flows of viscous fluids exist outside boundary layer regions and separated regions at the back of bluff bodies. We have learned how to use potential flows in viscous flows and we must learn how to use them in viscoelastic flows.

The case of flow at the stagnation points of a body in steady flow, in an arbitrary direction is of special interest. The steady streaming past a stationary body is equivalent, under a Galilean transformation, to the steady motion of a body in an otherwise quiet fluid. The potential flow of a fluid near a point $(x_1, x_2, x_3) = (0,0,0)$ of stagnation is a purely extensional motion with

$$[\lambda_1, \lambda_2, \lambda_3] = \frac{U}{L}\dot{S}\,[2, -1, -1] \tag{3}$$

where $\dot{S}$ is the dimensionless rate of stretching in the direction x_1, L is the scale of length and

$$[u_1, u_2, u_3] = \frac{U}{L}\dot{S}\,[\,2x_1, -x_2, -x_3]. \tag{4}$$

In this case

$$\begin{bmatrix} \sigma_{11} & 0 & 0 \\ 0 & \sigma_{22} & 0 \\ 0 & 0 & \sigma_{33} \end{bmatrix} = \frac{\rho}{2}U^2\,[\,\dot{S}^2\frac{4x_1^2+x_2^2+x_3^2}{L^2} - 1\,]\begin{bmatrix} 1 & 0 & 0 \\ 0 & 1 & 0 \\ 0 & 0 & 1 \end{bmatrix} + 2\mu\frac{U}{L}\dot{S}\begin{bmatrix} 2 & 0 & 0 \\ 0 & -1 & 0 \\ 0 & 0 & -1 \end{bmatrix} + 2\frac{U^2}{L^2}\dot{S}^2\begin{bmatrix} -\alpha_1+2\alpha_2 & 0 & 0 \\ 0 & -7\alpha_1-4\alpha_2 & 0 \\ 0 & 0 & -7\alpha_1-4\alpha_2 \end{bmatrix} \tag{5}$$

At the stagnation point itself

$$\sigma_{11} = -\frac{\rho}{2}U^2 + 4\mu\frac{U}{L}\dot{S} + 2(2\alpha_2-\alpha_1)\frac{U^2}{L^2}\dot{S}^2. \tag{6}$$

Since $\alpha_1 < 0$, $2\alpha_2-\alpha_1=\frac{5}{2}n_1+2n_2 > 0$, the normal stress term in (6) is positive independent of the sign of $\dot{S}$ but $4\mu\dot{S}$ is negative at the front side of a falling body and is positive at the rear. This is a new manifestation of the competition between inertia and normal stress, which may play a role in the flow induced anisotropy. In practice we would not expect the symmetrical streamlines predicted by potential flow, but the normal stresses that are generated in the non-separated regions of flow around bodies may play an important role in turning long bodies and chaining of spherical bodies.

Dimensionless groups may be formed from the ratios of inertia $\rho U^2/2$, viscosity $4\mu U/L$ and normal-extensional stress $(5n_1+4n_2)U^2/L^2$. We could again speak of an inertial radius for the competition between inertial and normal-extensional stress with inertia dominant when $L>L_c$ and normal stress dominant when $L<L_c$ where

$$L_c^2 \approx \frac{10n_1+8n_2}{\rho} \tag{7}$$

is a material property. This seems agree with the experimental results of Riddle et al [15]. Their experiment should give rise to a potential flow at early times and their critical separation may correspond to our inertial radius.

The concept of an extensional viscosity is a very special one since it is based on the assumption that the flow on which the extensional stress difference $\sigma_{11} - \sigma_{22}$ as defined is exactly the potential flow (4). For a second order fluid

$$\sigma_{11} - \sigma_{22} = 6\mu\frac{U}{L}\dot{S} + 12\,(\alpha_1+\alpha_2)\frac{U^2}{L^2}\dot{S}^2 \tag{8}$$

The extensional viscosity is useless for computing the forces and moments on bodies because the isotropic part of the stress which contains viscoelastic terms in general has been subtracted off.

6. CONCLUSIONS

To our knowledge, the following observations documented in this paper are novel.

(1). It is well known that a long body settling in a viscous liquid will turn its broadside to the stream. The same long body settling in a viscoelastic fluid will turn its broadside parallel to the stream but heavier long bodies which fall faster again turn broadside.

(2). There is a regime in which normal-extensional stresses and inertia compete. This competition evidently decides the tilt angle which the axis of a long body makes with the direction of fall in steady flow.

(3). The tilt angle can be controlled by changing the concentration of the solution using the same long particle or by changing the weight of the particle in the same solution.

(4). The shape of the ends of the particle has an effect on the tilt angle, with rounded ends giving a smaller angle of tilt as in the viscoelastic case.

(5). The natural orientation of a long body in a fluidized suspension is the key to understanding anisotropic structures which develop in sedimenting and fluidized spheres.

(6). Sedimenting spheres in a fluid filled channel will arrange themselves so that the line of centers between neighboring spheres is across the stream in a viscous liquid and parallel to the stream in a viscoelastic liquid when the fall velocity is small but across the stream again when the fall velocity is large.

(7). Spheres falling close to a wall of the channel rotate as if rolling down the wall in the intuitive way in a viscous liquid, but rotate as if rolling up the wall against intuition in a viscoelastic liquid.

(8). Provisional explanation of the peculiar observations can be framed as a competition between inertia and normal stresses with cross stream arrays preferred when inertia dominates and streamwise arrays preferred when normal and extensional stresses dominate.

Quantitative documentation of these conclusions is being prepared for a forthcoming publication.

REFERENCES

1. L.G. Leal, J. Fluid Mech.,69 (1975) 305-337.
2. L.G. Leal, J. of Non-Newtonian Fluid Mech., 5 (1979) 33-78.
3. D.D. Joseph, Fluid Dynamics of Viscoelastic Liquids, Springer-Verlag, New York, 1990, Chapter 17.
4. J. Michele, R. Pätzold and R. Donis, R. Rheol. Acta, 16 (1977) 317-321.
5. D.J. Highgate, Nature, 211 (1966) 1390-1391; D.J. Highgate and R. W. Whorlow (Ed.), Polymer Systems: Deformation and Flow, Macmillan, London, 1968, 251-261.
6. D.J. Highgate and R. W. Whorlow, Rheol. Acta, 8 (1969) 142-151.
7. L. Petit and B. Noetinger, Rheol. Acta, 27 (1988) 437-441.
8. W. Thompson and P.G. Tait, Natural Philosophy, 2nd Ed., Cambridge, 1879.
9. H.H. Hu and D.D. Joseph, Int. Video Journal of Eng. Res., to appear, 1992.
10. D.D. Joseph, J. of Non-Newtonian Fluid Mech., to appear, 42 (1992).
11. D.D. Joseph, A.F. Fortes, T.S. Lundgren and P. Singh, SIAM Advances in Multiphase Flow and Related Problems, (1987) 101-122 (see Fig.17).
12. G. Volpicelli, L. Massimilla and F.S. Zenz, Chem. Engng. Symp. Ser. No. 62, 67 and 42 (1966).
13. A.F. Fortes, D.D. Joseph and T.S. Lundgren, J. Fluid Mech., 177 (1987) 467-483.
14. P. Singh, Ph. Caussignac, A.F. Fortes, D.D. Joseph and T.S. Lundgren, J. Fluid Mech., 205 (1989) 553-571.
15. M.J. Riddle, C. Narvaez, R.B. Bird, J. of Non-Newtonian Fluid Mech., 2 (1977) 23-25.
16. P. Brunn, Rheol. Acta, 16 (1977) 381-475.
17. D.D. Joseph, P. Singh and A. Fortes, in: M. Roco (Ed.), Particulate Two-Phase Flow, Butterworth-Heinnemann Press, 1992.

ACKNOWLEDGEMENTS

This work was supported by the NSF, fluid, particulate and hydraulic systems, by the US Army, Mathematics and AHPCRC, and by the DOE, Department of Basic Energy Sciences.

Theoretical and Applied Rheology, edited by P. Moldenaers and R. Keunings
Proc. XIth Int. Congr. on Rheology, Brussels, Belgium, August 17-21, 1992

TEXTURE OF A LIQUID CRYSTALLINE POLYMER DURING SHEAR

R.G. Larson[1], D.W. Mead[2], and J.T. Gleeson[3]

[1]AT&T Bell Laboratories, Murray Hill, NJ 07974.
[2]Shell Development Co., P.O. Box 1380, Houston Tex., 77251
[3]Dept of Physics, Jadwin Hall, Princeton Univ., Princeton, N.J. 08544

1. *INTRODUCTION*

In transient shearing flows, liquid crystalline polymers (LCP's) show unusual rheological behavior, such as multiple oscillations in stress when the shear rate is suddenly increased, and a relaxation time on cessation of shearing that is inversely proportional to the rate of shearing (refs. 1-3). These phenomena are thought to be caused by shear-induced changes in *texture,* i.e., the spatial pattern of *director* orientation. The director is a unit vector that defines the direction of the local sub-micron-scale average molecular orientation. Inhomogeneities in the director field typically exist in nematic polymers because of *disclinations*, i.e., defect lines at which the director changes orientation nearly discontinuously. Because shearing flow exerts drag on the director and on the disclination lines, it can change both the macroscopically-averaged director orientation, and the length scale, "a," of the texture. Here, we define "a" to be the characteristic distance over which the director is roughly uniform.

Although theories relating the transient textures to rheological properties of LCP's are rudimentary (refs. 4,5), plausible explanations have been given for several of the peculiar features of the transient rheology. In particular, the multiple oscillations in shear and normal stress that can occur during start-up of shearing are believed to be caused by oscillations in the average director orientation that occur because of *director tumbling.* Director tumbling has been both predicted (refs. 6-8) and observed (refs. 9,10) in solutions of rod-like LCP's; it occurs in solutions for which there is no angle between the directions of flow and of the shear gradient at which the viscous torques on the director can balance each other so that the director stops rotating (ref. 11).

While stress oscillations are believed to be caused by shear-induced changes in average director *orientation,* other rheological peculiarities, such as the inverse dependence on shear rate of the stress relaxation time, are thought to be caused by shear-induced changes in the characteristic *length scale* "a" of the director inhomogeneities. Using a scaling argument, Marrucci (ref. 12) proposed that over some range of shear rate "a" decreases as the shear rate increases according to $a^2 \propto \dot{\gamma}^{-1}$. Also, according to this scaling argument, the smaller a is, that is, the finer the texture is, the faster this texture will change when shearing stops. The scaling argument predicts that if τ_{text} is the texture relaxation time, then $\tau_{text} \sim a^2 \sim \dot{\gamma}^{-1}$. As an extension of this argument, it has been proposed that the changes in texture length scale "a" that occur after start-up or cessation of shearing are described by (ref. 8):

$$\frac{dL}{dt} = \alpha\dot{\gamma}L - \beta L^2 \qquad (1)$$

where $L \equiv 1/a^2$ is proportional to the number of disclination lines per unit volume, and α and β are phenomenological parameters. Here

α is dimensionless and β is proportional to K/η, where η is the shear viscosity and K is a typical Frank elastic constant.

From Eqn. (1), we find that after start-up of steady shearing,

$$L^{-1} = a^2 = a_0^2 e^{-\alpha\dot{\gamma}t} - \frac{\beta e^{-\alpha\dot{\gamma}t}}{\alpha\dot{\gamma}} + \frac{\beta}{\alpha\dot{\gamma}} \qquad (2)$$

Here a_0 is the initial value of "a" just before the onset of shearing.

2. TEXTURE OBSERVATIONS

In a previous publication (ref. 13), we described a microscope shearing cell that permits videorecording of textures of LCP's during shearing flow. The sample is confined between two parallel disks; the upper disk is stationary, while the lower one rotates. The light propagates normal to the plates, and passes through an annular glass ring in the lower, rotating, plate, then through the sample, and finally through a glass window in the upper, stationary, plate. Typically, the shearing cell containing the sample resides between crossed polaroids. We examined the liquid crystalline sample designated earlier as PBG118 (ref. 13). This sample contains 21% (by weight) of poly(γ-benzyl-glutamate), or PBG, of average molecular weight 118,000, dissolved in the solvent metacresol. This polymer is a racemic mixture of the enantiomorphs PBLG and PBDG. The rheological properties of this sample were given earlier (ref. 13).

Fig. 1a-e shows videotaped images of PBG118 during start-up of steady shearing; the flow is from right to left. In this experiment, the gap between the upper and lower plates was 175μ, and the shear rate, $\dot{\gamma}$, was 0.51 sec^{-1}. The sample had been sheared at a high rate and allowed to relax for two days to let the texture become very coarse; see Fig. 1a. Because of the residual effects from the previous shearing flow, we expect the initial orientation to be, on average, parallel to the flow direction (ref. 14). Fig. 1b-c shows that when the initially coarse texture is subjected to steady shearing, a *striped texture* forms with stripes parallel to the flow

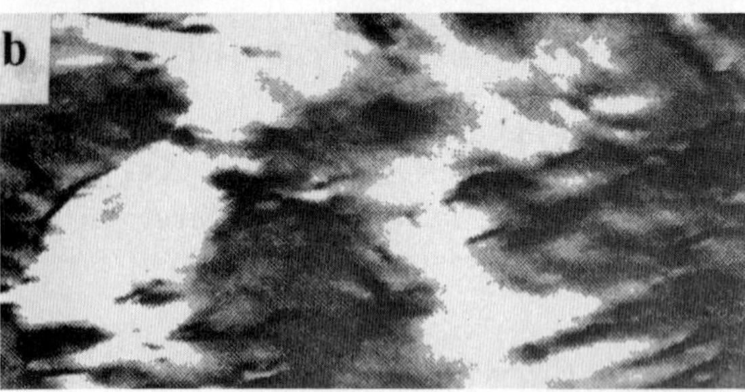

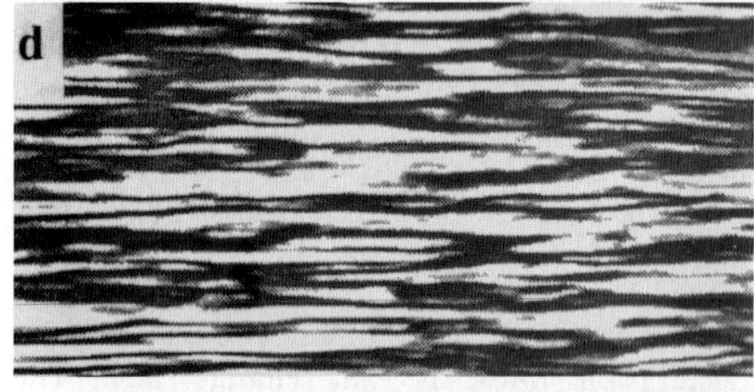

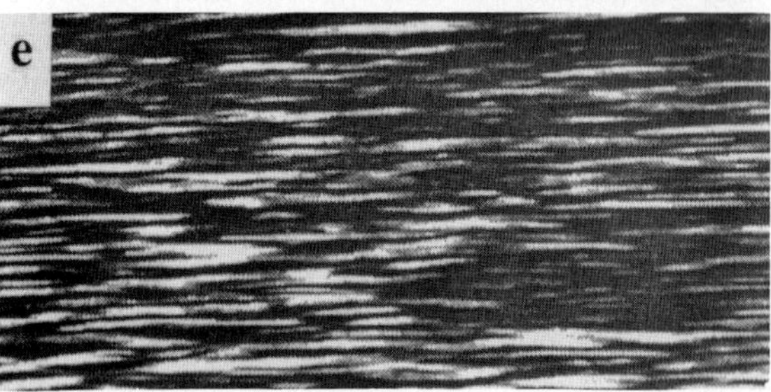

FIGURE 1: Development of the texture of PBG118 sheared at a rate $\dot{\gamma} = 0.51$, and viewed between crossed polaroids. a) before shearing; and after being sheared for b) 5 secs; c) 11 secs; d) 25 secs; and e) 60 secs. The sample gap was 175μ, and the field of view is 890μ wide.

direction. Figs. 1c-e show that as the shearing flow is prolonged, the stripes become finer. Under suitable polarization conditions, disclination lines can also be observed in the fluid (ref. 14); these disclinations are roughly parallel to the flow direction and have a spacing comparable to that of the stripes. Using polarimetry techniques, we have found that the stripes consist of a periodic misalignment of the director with respect to the flow direction (ref.14). Thus in one white stripe, the director resides at an angle of α with respect to the flow direction, while in the neighboring white stripe, the angle is $-\alpha$. The maximum value of α is around 35—45°.

We have analyzed the characteristic spacing of these stripes by taking fast Fourier transforms (FFT's) of the videotaped images. As discussed elsewhere (ref. 13), a characteristic length scale $\lambda_{\perp}^{*}$ of the texture in the direction perpendicular to the flow is obtained by taking an FFT of each vertical line of the image and averaging the FFT's of all vertical lines together to obtain an average Fourier spectrum a(k), where k is the wavenumber, or inverse wavelength. The characteristic wavelength $\lambda_{\perp}^{*}$ is then defined as

$$\frac{1}{2} \equiv \frac{\int_0^{\frac{2\pi}{\lambda_{\perp}^{*}}} a(k)dk}{\int_0^{\infty} a(k)dk} \tag{3}$$

where a(k) is the Fourier amplitude at wavenumber k. Thus, half the integral under the spectrum has wavelengths greater than $\lambda_{\perp}^{*}$; the other half of the integral is in wavelengths less than $\lambda_{\perp}^{*}$.

Fig. 2 plots $\lambda_{\perp}^{*}$ verses time after start-up of shearing for the sequence of videotaped images from which those shown in Fig. 1 were selected. Fig. 2 shows that $\lambda_{\perp}^{*}$ decreases with time after start-up of steady shearing until a steady-state value $\lambda_{\perp}^{*} \approx 25\mu$ is reached after about 30 strain units of shearing. We have found that the spacing between the stripes that one estimates by visual inspection of the original images is typically about half the characteristic value $\lambda_{\perp}^{*}$ given by the FFT's. Also, as the shear rate is increased, the steady-state stripe spacing decreases to a value less than 10μ, which makes it difficult to resolve the spacing. The number of strain units required to reach steady state seems to increase somewhat as the shear rate increases, and is also sensitive to the initial coarseness of the texture, the sample thickness, and to the initial average orientation (ref. 14).

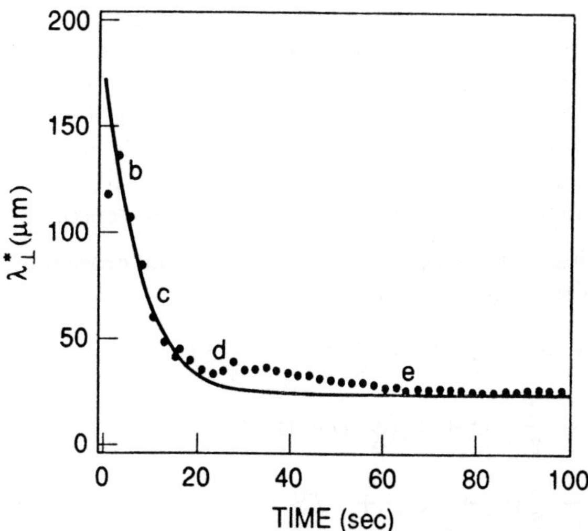

FIGURE 2: Characteristic texture length scale $\lambda_{\perp}^{*}$ verses time obtained by image analysis of videotaped textures. The letters b)—e) correspond to images shown in Fig. 1. The line is the prediction of Eqn. (2).

3. DISCUSSION

Figs. 1 and 2 show that the texture becomes finer as a function of time after the onset of shearing, in qualitative agreement with Eqns. (1) and (2). Eqn. (2) can be fit to the data of Fig. (2); the fit shown is obtained with $\alpha = 0.4$ and $\beta = 121\ \mu^2$/sec, where we have set $a = \lambda_{\perp}^{*}$. The identification we have made between "a" and $\lambda_{\perp}^{*}$ should taken with a grain of salt, since "a" is really the characteristic spacing between disclinations in *three dimensional* space, while $\lambda_{\perp}^{*}$ is obtained from a *two dimensional* projection of all structures within the three dimensional volume that can be focused by the microscope. The shear rate dependence of the texture spacing $\lambda_{\perp}^{*}$ was explored to a limited extent earlier (ref. 13); more work is planned for the future.

Eqn. (1), which was motivated by rheological measurements, is part of a semi-empirical mesoscopic theory for the texture and rheology of liquid crystalline polymers presented in ref. 8. The complete set of equations derived in this theory give the macroscopic stress tensor $\underset{\approx}{\sigma}$ and a tensor $\underset{\approx}{S}$ that is proportional to the macroscopic birefringence tensor:

$$\begin{aligned}\underset{\approx}{\sigma} &= 2\mu \underset{\approx}{D} + 2\mu_1 \underset{\approx}{D} : \underset{\approx}{S}(\underset{\approx}{S} + \frac{1}{3}\underset{\approx}{\delta}) \\ &+ \frac{2}{3}\mu_2 \underset{\approx}{D} + \mu_2(\underset{\approx}{S}\cdot\underset{\approx}{D} + \underset{\approx}{D}\cdot\underset{\approx}{S}) \\ &- \frac{1}{2}(\alpha_2 + \alpha_3)(\epsilon + \phi)\ell \underset{\approx}{S}\end{aligned} \tag{4}$$

$$\begin{aligned}\frac{d}{dt}\underset{\approx}{S} &= \underset{\approx}{\omega}^T \cdot \underset{\approx}{S} + \underset{\approx}{S}\cdot\underset{\approx}{\omega} \\ &+ \lambda\left(\frac{2}{3}\underset{\approx}{D} + \underset{\approx}{D}\cdot\underset{\approx}{S} + \underset{\approx}{S}\cdot\underset{\approx}{D} - 2\underset{\approx}{S} : \underset{\approx}{D}(\underset{\approx}{S} + \frac{1}{3}\underset{\approx}{\delta})\right) \\ &- \epsilon\ell \underset{\approx}{S}\end{aligned} \tag{5}$$

and

$$\frac{d}{dt}\ell = \alpha\dot{\gamma}\ell - \ell^2 \tag{6}$$

Eqn. (6) is the same as Eqn. (1), except that we have rescaled L using $\ell \equiv \beta L$. $\underset{\approx}{D}$ and $\underset{\approx}{\omega}$ in the above are the symmetric and antisymmetric parts of the velocity gradient tensor, and $\underset{\approx}{\delta}$ is the unit tensor. The parameters μ, μ_1, μ_2, and λ can all be obtained from the Doi molecular theory (ref. 4). The phenomenological parameters α, ϕ, and ϵ must all be obtained by fitting data from rheological studies or texture analysis.

In ref. 8, it was shown that the above equations predict qualitatively the shear stress transients in several important shearing flows, including flows involving step-up in the rate of shearing, and reversal of the shearing direction. In these earlier predictions, no attempt was made to adjust the unknown parameter α, which was simply set to unity. In the experiments reported here, however, we have found that to describe the evolution of texture after start-up of steady shearing, α should be somewhat smaller than unity, around 0.4, for PBG118.

A value of α smaller than unity is also indicated by fitting the mesoscopic theory to data from recent interrupted-shear experiments by Moldenaers et al. (ref. 15). In these experiments, prolonged shearing at a given rate of 0.4 sec^{-1} is interrupted for a rest period, after which the shearing flow is resumed at the same rate of shear. The transient shear stresses measured after various rest times t_r are given in Fig. 3(top). These data show a complicated pattern of stress overshoots and undershoots that depend sensitively on the rest period, even when the rest period is as long as hours! Predictions of these transients can be obtained from the mesoscopic theory; these predictions for interrupted shearing prove to be much more sensitive to the values of the parameters in the theory than the predictions for other transient flows in ref. 8. We find that the shapes of the transients are sensitive to all three phenomenological parameters ϵ, ϕ/ϵ, and α. In particular, α must be less than unity, if the predictions are to accord with the measurements.

Fig. 3(bottom) shows that for the sample of high molecular-weight PBDG studied by Moldenaers et al., predictions that accord reasonably well with the measurements can be obtained if $\alpha = 0.2$. Remember that the value of α that we obtained from image analysis of sheared PBG118 was also somewhat less than unity, namely $\alpha = 0.4$. Note several points of agreement between the theoretical predictions and the experimental data: 1.) There are multiple overshoots and undershoots. 2.) The curve of stress verses time continues to change even when the dimensionless rest period $\dot{\gamma}t_r$ becomes very large. 3.) The stress at the first overshoot at first decreases as the rest period t_r increases, and then increases with further increases in t_r. 4.) The stress at the first undershoot decreases continuously as $\dot{\gamma}t_r$ is increased, at least up to very large values of $\dot{\gamma}t_r$. 5.) A significant second undershoot only occurs when $\dot{\gamma}t_r$ exceeds a critical value.

It is obvious, however, that the shapes of the theoretical and experimental curves in Fig. 3 also show some significant differences. Nevertheless, the predictions of mesoscopic theory agree to a significant degree with both texture and rheological data at least for some LCP's; and we have therefore shown that some aspects of both the rheology and the texture of liquid crystalline polymers can be described by a single theoretical framework. Many more combined texture and rheological studies must be done, however, to explore fully the strengths and limitations of this approach.

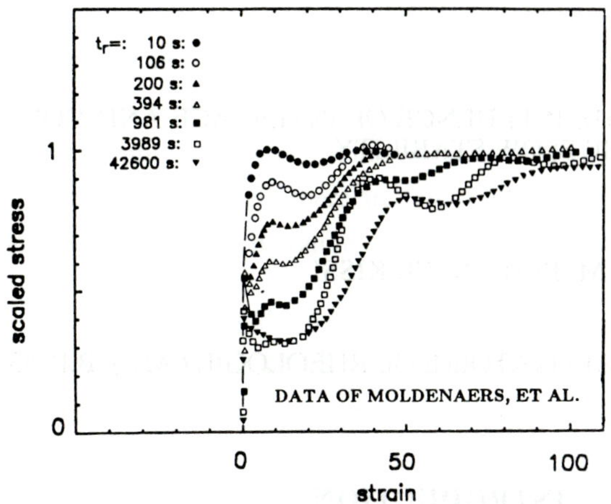

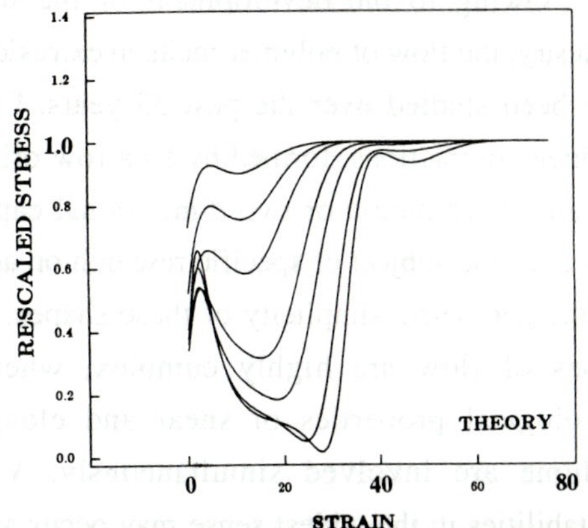

FIGURE 3: The shear stress, normalized by its steady-state value, during steady shearing after a rest period t_r. (top) data of Moldenaers et al. (ref. 15, reproduced from the Journal of Rheology, with permission); (bottom) predictions of the mesoscopic theory in ref. 8, with parameter values $\alpha = 0.2$, $\epsilon = 0.01$, and $\phi/\epsilon = 5$. The lines, from top to bottom, are for $\dot{\gamma} t_r = 3, 10, 25, 100, 250, 1000$, and 2500.

REFERENCES

1. P. Moldenaers, and J. Mewis, J. Rheol., 30 (1986) 567-584.
2. P. Moldenaers, G.G. Fuller, and J. Mewis, Macromolecules, 22, (1989) 960-965.
3. R.G. Larson and D.W. Mead, J. Rheol., 33 (1989) 185-206.
4. N. Kuzuu and M. Doi, J. Phys. Soc., Japan, 52 (1983) 3486-3494; 53 (1984) 1031-1040.
5. G. Marrucci and P.L. Maffettone, Macromolecules, 22 (1989), 4176-4182.
6. R.G. Larson, Macromolecules 23 (1990) 3983-3992.
7. G. Marrucci and P.L. Maffettone, J. Rheol. 34 (1990) 1217-1230,1231-1244; erratum 34 (1991) 313.
8. R.G. Larson and M. Doi, J. Rheol., 35 (1991) 539-563.
9. W.R. Burghardt and G.G. Fuller, Macromolecules, 24 (1991) 2546-2555.
10. M. Srinivasarao and G.C. Berry, J. Rheol., 35 (1991) 379-397.
11. F.M. Leslie, Advances in Liquid Crystals, 4 (1979) 1-81.
12. G. Marrucci, Pure Appl. Chem., 57 (1985) 1545-1552.
13. J.T. Gleeson, R.G. Larson, D.W. Mead, G. Kiss, and P.E. Cladis, Liq. Cryst. (1991) in press.
14. R.G. Larson and D.W. Mead Liq. Cryst.(1992) submitted.
15. P. Moldenaers, H. Yanase, and J. Mewis J. Rheol. 35 (1991) 1681-1699.

flow of a PB in a steel die and in a Teflon die of the same geometry, they demonstrated the following :

- at low flow regimes, the flow curves are the same and the PB has a Newtonian behaviour,
- once a certain flow rate is reached, stress at the wall becomes higher in the steel die that in the Teflon one.

They conclude from this that it is more difficult for the PB to adhere to Teflon than to steel (where there is suitable adhesion). Like fluorous materials in general, Teflon appears to encourage slip.

In addition, it should be noted that the curves obtained by Vinogradov et al. [8] at low flow rates in steel dies are perfectly continuous, which contradicts the results referred to above [4, 6, 7]. In contrast, for higher flow regimes than those considered above for sharkskin, they show that macroscopic slip is likely to occur, even in the steel dies. This is a different type of slip from that discussed previously [4, 6, 7]. Indeed, at these high flow rates, it is no longer surface instabilities that govern the appearance of these phenomena but instabilities in the volume of the entire flow.

3. MELT FRACTURE

Melt fracture is difficult to isolate and study separately. Indeed, in the case of highly entangled products, it is likely to be superimposed on macroscopic slip. In addition, for moderately entangled products, in which there is no macroscopic slip, the phenomenon is superimposed on sharkskin defects. Finally, there remain slightly entangled polymers, where it is possible to observe melt fracture without being hampered by the occurrence of sharkskin or the triggering of slip [13]. Products of this kind, which generally have a low viscosity, have fast response times. The frequency of the phenomena is therefore high, making observations difficult, at least using the techniques and measurement transducers currently available.

In spite of all these difficulties, one major idea is almost unanimously accepted. Indeed, many authors [13, 14, 15, 16, 17, 18] agree in claiming that, above a certain extrusion rate, flow upstream of the contraction becomes unstable. These instabilities occur in the form of sudden pulsations, the frequency of which has been measured in certain cases. This assumption has been confirmed by visualisation [13] and birefringence measurements [19], which show that such instabilities start along the upstream flow axis owing to the high elongation stresses that develop in this area. White's assumption [14] should be pointed out, according to which melt fracture is due to the triggering of a hydrodynamic instability in the form of helicoidal flow upstream of the contraction. With an unconfined jet, these instabilities trigger the phenomenon of melt fracture, which is often seen in the form of a regular helix oscillating at the same frequency as that of the pulsations of the upstream elongational flow [13].

For certain fluid/flow pairs, as soon as the melt fracture regime is reached, macroscopic slip is likely to occur at the die walls. This may be revealed in several ways, depending on the type of installation used :

- Under controlled pressure flow, it is accompanied by a sudden increase in flow and a hysteresis on the flow curves; the more entangled the polymer, the more pronounced these effects are.
- With flow at a controlled average rate, the compressibility of the fluid is an important parameter, and the occurrence of macroscopic slip is accompanied by oscillations in the

instantaneous flow rate and pressure. These oscillations give the extruded rod the characteristic appearance of cork flow.

It should be noted that cork flow has been demonstrated by other authors [9, 4, 6, 7, 10] with various polymers. Friction relations have also been proposed for modelling the phenomenon [7, 11, 12].

Macroscopic slip of flowing polymers is one of the simplest features to study [10]. The results obtained in this field are, however, scanty, the criteria for its appearance poorly established and the triggering mechanisms imperfectly understood. However, it would only seem to occur in the case of highly entangled polymers (i.e. where there is a high Mw/Me). This is all the more true in the case of poorly defined polymers.

In addition, macroscopic slip occurs at high extrusion rates. Flow in such cases is generally unstable and the phenomenon of melt fracture is likely to be superimposed on slip. This complicates determination of the mechanisms involved during the slip, but also of those governing the appearance and development of the melt fracture instability.

4. DISCUSSION

Sharkskin, which has already been discussed at length in this study, is described differently depending on the author concerned. In previous works [10, 13], we have shown that this phenomenon is produced by the fluid cracking just as it leaves the die, under the effect of the high stresses that develop in this area. We have shown that, in the case of highly entangled polymers, the size and spacing of these cracks may be of the same order of magnitude as the diameter of the extruded rod [10]. In the latter case, it is impossible to define sharkskin as a small-amplitude, high-frequency roughness of the unconfined surface. It should also be noted that, even in the case of moderately entangled polymers, this definition concerns only a secondary phenomenon, based on observations carried out well downstream of the outflow section and hence after relaxation of the stresses and of the cracks.

The mechanisms involved in the occurrence of sharkskin are also a matter of controversy. Thus, using similar polymer melts and experimental conditions, Ramamurthy, Kalika and Denn and Hatzikiriakos and Dealy, all of whom associate the appearance of sharkskin with the occurrence of slip, nevertheless obtain different friction relations. This could be explained by adopting assumptions that are questionable (negligible entry pressure effects) or contradictory (pressure effects negligble or not?).

In conclusion, as far as we are concerned, the works referred to above [4, 6, 7] may indicate that adhesion effects in the outflow area are to be incriminated. But under no circumstances do they conclusively prove the existence of macroscopic slip at the wall during the development of sharkskin, though this is indicative of the high slip rates that they have managed to calculate.

5. CONCLUSION

The considerable complexity of the physical mechanisms (including first and foremost the volume rheological properties and interactions with the wall) governing polymer melt flows in sudden contractions is the reason for the difficulties and controversies still existing today. There appear to be three crucial points in examining and attempting to throw light on these types of flow:

- The choice of fluid. Commercially available products are of undeniable interest from the industrial standpoint. However, they are poorly defined and their composition is not often known with precision. The choice of model fluids with a narrow mass distribution and containing no additives may, in contrast, help in understanding the phenomena involved.

- Carefully controlled experimental techniques. Indeed, there are many pitfalls involved in rheometry, visualisation, birefringence and flow curves measurements [13, 20] and these must be carefully avoided. In addition, it is highly profitable to use these various techniques simultaneously.
- Lastly, the results must be interpreted with caution in order to avoid producing further basic divergences between works whose results are in fact similar. By way of example, it would no doubt be worthwhile to establish whether or not pressure changes the viscosity of the polymers studied.

REFERENCES

1. F. N. Cogswell, J. Non-Newt. Fluids Mech., 2 (1977) 33
2. G. V. Vinogradov, A. Ya Malkin, Rheology of Polymers, Mir Moscow, Springer, Berlin, 1980.
3. B. Tremblay, J. Rheol., 35 (6) (1991) 985-998.
4. A. V. Ramamurthy, J. Rheol., 30 (1986) 337-357.
5. C. De Smedt, S. Nam, Plast. Rubber Process. Appl., 8 (1987) 11.
6. D. S. Kalika, M. M. Denn, J. Rheol., 31 (1987) 815-834.
7. S. G. Hatzikiriakos, J. M. Dealy, Submitted in J. Rheol.
8. G. V. Vinogradov, V. P. Protasov, V. E. Dreval, Rheol. Acta, 23 (1984) 46.
9. A. Weill, Rheol. Acta, 19 (1980) 623-632.
10. N. El Kissi, J. M. Piau, J. Non-Newt. Fluids Mech., 37 (1990) 55-94.
11. Y. B. Chernyak, A. I. Leonov, Wear, 108 (1986) 105.
12. N. El Kissi, J. M. Piau, C. R. Acad. Sci. Paris, Sér. II, 309 (1989) 7.
13. J. M. Piau, N. El Kissi, B. Tremblay, J. Non-Newt. Fluids Mech., 34 (1990) 145-180.
14. J. L. White, Appl. Polym. Symp., 20 (1973) 155.
15. C. J. S. Petrie, M. M. Denn, Aiche J., 22 (2) (1976) 209.
16. J. L. den Otter, Plast. Polym., 38 (1970) 155.
17. J. P. Tordella, Trans. Soc. Rheol., 1 (1957) 203.
18. E. B. Bagley, Trans. Soc. Rheol., 5 (1961) 355.
19. J. P. Tordella, F. R. Eirich (Ed.), Rheology, Vol. 5 Academic Press, New-York, NY, 1969.
20. N. El Kissi, J. M. Piau, Submitted in Rheol. Acta.

Contributed papers

MOLECULAR THEORIES

Theoretical and Applied Rheology, edited by P. Moldenaers and R. Keunings
Proc. XIth Int. Congr. on Rheology, Brussels, Belgium, August 17-21, 1992

CAN MOLECULAR THEORY PREDICT POLYMER CHAIN DYNAMICS?

K.H.AHN [1], J.L.SCHRAG [2] AND S.J.LEE [3]

[1]Chemical Engineering Department and Rheology Research Center, University of Wisconsin-Madison, Madison, Wisconsin 53706
[2]Chemistry Department and Rheology Research Center, University of Wisconsin-Madison, Madison, Wisconsin 53706
[3]Chemical Engineering Department, Seoul National University, Seoul, Korea

1. INTRODUCTION

One of the most important goals in polymer rheology is to be able to understand and predict the rheological properties of polymeric materials from their known molecular structural parameters. During the past twenty five years a large body of rheological data has been obtained for polymer solutions, and these data have provided substantial insights into the relation between polymer structure and the unusual properties caused by the presence of these chains. However, for certain properties, experimental studies are not yet complete and the observed phenomena apparently have not been quantitatively explainable by molecular theories; two important examples of such properties for dilute polymer solutions are the 2nd normal stress coefficient and the shear-thickening phenomenon. The ability to predict these properties from known molecular parameters would be important in polymer physics as well as engineering applications. Theoretical treatments for dilute polymer solutions are complicated by the existence of various possible intramolecular interactions. These interactions have been incorporated in such theories by introducing the concepts of hydrodynamic interaction (HI), excluded volume(EV) and nonlinear entropic spring(NS). These three concepts are usually considered to represent the most important intramolecular interaction effects for dilute polymer solutions, to have reasonably well understood physical bases, and to be describable mathematically. However, to date no molecular theory has incorporated all three concepts simultaneously.

Fixman(ref.1,2) solved the diffusion equation for the bead-spring chain model incorporating the concepts of HI and EV. But he did not consider the NS effect, and failed to fit experimental data for the shear rate dependence of the intrinsic viscosity.

Öttinger(ref.3) considered the bead-spring model incorporating HI and NS. He transformed the diffusion equation into a second moment equation and solved it directly without using any assumption except consistent averaging. However, he did not consider the EV effect and his theory required large amounts of computing time and also failed to fit experimental data(ref.4).

Recently, Ahn and Lee(ref.5) developed an algorithm which can consider all three interaction effects simultaneously. Thus the predictions of this theory are expected to be more realistic. In this paper, results of this treatment are compared with experimental data. The comparison suggests that this treatment can explain and predict the rheological properties of dilute polymer solutions in steady shear flows in terms of molecular parameters in a quantitative manner.

2. EXPANSION FACTORS

The EV effect is usually described in terms of expansion factors. Figure 1 shows the radius expansion factor curves for Ahn and Lee, and Fixman theory together with experimental results. Prediction seems to be quite good, while Fixman theory far underestimates the chain expansion. It is also observed that the radius expansion factor is a universal function of the EV parameter z, while the viscosity expansion factor is not(ref.6). This contradicts the two-parameter theory, but coincides with experimental data (ref.7).

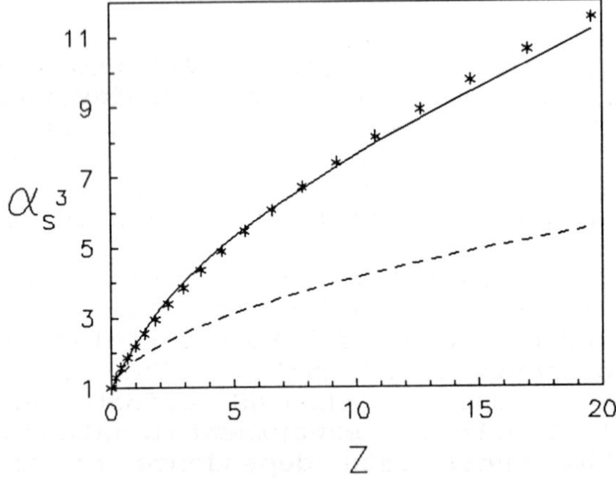

Figure 1. Radius expansion factor vs EV parameter z. Symbols represent experiemtal data, dashed line Fixman's theory, and straight line the prediction of this work.

3. MOLECULAR WEIGHT DEPENDENCE OF THE INTRINSIC VISCOSITY

It is well known that the slope of the Mark-Houwink-Kuhn-Sakurada plot is 0.5 in theta solvents and approaches 0.8 in very good solvents for M>30,000. Theory predicts a slope of 0.5 when HI parameter h^*=0.25, and 1 when h^*=0. It is also observed that h^* determines the slope of the curve and α influences the magnitude. By defining the theta state as h^*=0.25 and α=1, we can determine good solvent condition by varying both h^* and α. Figure 2 shows that we can predict the MW dependence of the intrinsic viscosity quantitatively well if we use both concepts of HI and EV.

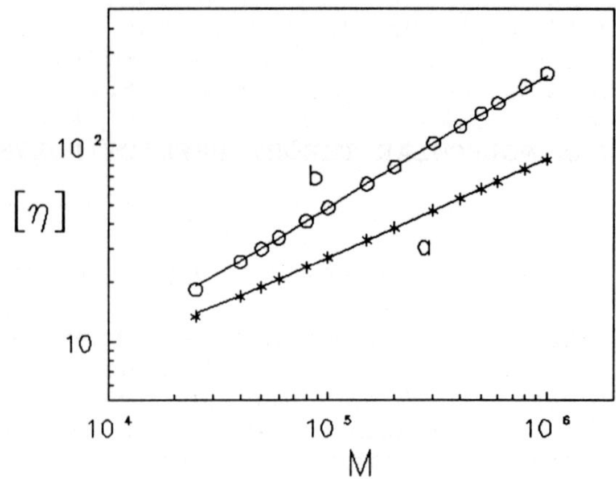

Figure 2. Intrinsic viscosity vs. MW curves for narrow dirtribution linear atactic polystyrenes in both (a)theta (cyclohexane) and (b) good (toluene) solvents. Symbols represent experimental data(ref.8) and lines are calculated results. h* values are 0.25 and 0.095 respectively.

4. FREQUENCY DEPENDENCE OF FLOW BIREFRINGENCE

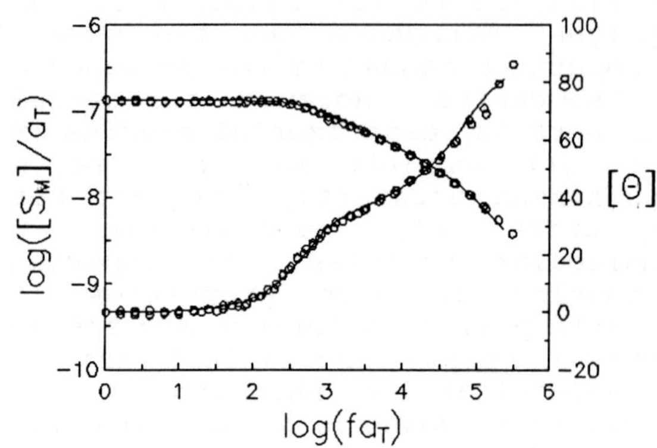

Figure 3. Oscillatory flow birefringence data for narrow distribution linear atactic polystyrene (MW=111,000) in Aroclor 1248. $[S_M]$ is the magnitude of the polymer contribution to the mechano-optic coefficient, θ is the phase angle and a_T is the time/temperature shift factor.

Figure 3 shows the comparison of predictions of this theory with the oscillatory birefringence data(ref. 9). The agreement appears to be excellent. h^* was determined to be 0.15 from the MW dependence of intrinsic viscosity curve, and α was determined to fit experimental data best. NS concept need not be considered here.

5. SHEAR RATE DEPENDENCE OF THE INTRINSIC VISCOSITY

The different interaction parameters affect flow curves differently. The EV parameter determines the onset of shear thinning at low shear rate, the HI parameter affects the curvature, and the NS parameter governs the degree of shear thinning at high shear rate(ref.6). Based on these findings, parameters were adjusted to fit experimental data of polystyrene (ref.10). h^* was first determined from the MW dependence of the intrinsic viscosity, and then α and b were determined to obtain the best fit at low and high shear rates respectively. Figure 4 shows the results, and the comparison seems quite good.

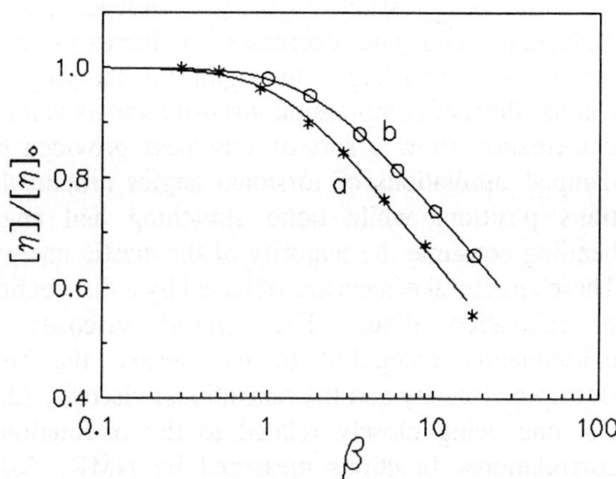

Figure 4. Shear rate dependence of the intrinsic viscosity. Predictions of this work (straight lines) and experimental data (symbols,ref.10) are compared. (a)N=20, h^*=0.25, α=1.28, b=60 and η_s=0.22 (b) N=20, h^*=0.25, α=1.12, b=60 and η_s=5.08

Considering that Wedgewood and Öttinger(ref.4) failed to obtain fits to these data by treating HI and NS, it becomes apparent how important it is to consider the three molecular concepts simultaneously.

6. CONCLUSION

We have shown the possibility that we can explain and predict polymer dynamics if we consider the molecular concepts of HI, EV and NS simultaneously. The success in prediction of the shear rate dependence of the intrinsic viscosity curve seems to be particularly encouraging. However lots of work remains to be done. For example, consistently averaged HI predicts a positive 2nd normal stress coefficient at zero shear rate. Also flow fields other than simple shear need to be examined to see how well this theory works. However we believe that there is enough promise from the present work that we can predict the rheological properties of dilute polymer solutions with the help of molecular theory.

REFERENCES

1. M.Fixman, J.Chem.Phys., 45(1966) 785.
2. M.Fixman, J.Chem.Phys., 45(1966) 793.
3. H.C.Öttinger, J.Non-Newtonian Fluid Mech., 26(1987)207.
4. L.E.Wedgewood and H.C.Öttinger, J.Non-Newtonian Fluid Mech., 27 (1988)245.
5. K.H.Ahn and S.J.Lee, J.Non-Newtonian Fluid Mech., submitted for publication.
6. K.H.Ahn, J.L.Schrag and S.J.Lee, in preparation.
7. H.Fujita, Polymer Solutions, Elsevier, Amsterdam, 1990.
8. J.Brandrup and E.H.Immergut(Ed.), Polymer Handbook, John Wiley & Sons, New York, 1989.
9. R.L.Sammler, C.J.T.Landry, G.R. Woltman and J.L.Schrag, Macromolecules, 23(1990)2388.
10. C.W.Manke and M.C.Williams, J. Non-Newtonian Fluid Mech., 19 (1985)43.

Theoretical and Applied Rheology, edited by P. Moldenaers and R. Keunings
Proc. XIth Int. Congr. on Rheology, Brussels, Belgium, August 17-21, 1992

MOLECULAR INTERPRETATION OF POLYMER MELT RHEOLOGY

J.V. ALEMAN

Instituto de Ciencia y Tecnología de Polímeros
Juan de la Cierva 3, 28006 Madrid, SPAIN

1. INTRODUCTION

The usual molecular interpretation of polymer melt flow is that at equilibrium the polymer coils are spheroids in shape with entanglements between them. Under application of stress some entanglements are broken and the coils reversibly take an elipsoidal shape. Afterwards, a critical stress is reached (melt fracture) at which a sudden change in the polymer physical structure occurs. These changes are the result of two different molecular processes (1).

The intramolecular distortion of small elements of the polymer backbone, also called enthalpy processes, which usually appears as a single relaxation process, although the detailed conformational changes involved may not be identical for every part of the polymer chain.

The distortions of the whole polymer molecule which are independent of chemical structure, and purely a function of molecular mass; they are also known as entropy processes and are observed as a viscoelastic loss which can be described a superposition of a series of relaxation processes.

The complexity of the interplay between the enthalpy (ΔH) and entropy ($T.\Delta S$) processes has made rather empirical the knowledge of the effect of chemical structure on the polymer rheological behaviour. Their combination provides the chain rigidity ($-\Delta F = \Delta H - T.\Delta S$) (2) values of which for different polymers are shown in Table I.

2. SHEAR BEHAVIOUR

Entropy deformation during shear flow is produced by slow long-range intermolecular conformational rearragements which accelerate the relaxation processes destroying entanglements and stretching and aligning the chains. The basic motions taking place are: limited intra- and inter-molecular cooperativity (α-relaxations model), which later extends over short chain segments making longer chain segments to move in Rouse modes and, afterwards, still longer segments to perform snake-like reptation modes. Some computer simulations do not show reptative motion of chains but rather an isotropic movement of segments, making desirable to use a still more microscopic description of polymer liquids.

3. COMPRESSION BEHAVIOUR

Bulk compression viscosity is independent of molecular mass and decreases or increases with temperature according to the rigidity of the polymer chains. Plots of compression viscosity versus volume deformation show a sort of wavyness provided by damped oscillations of torsional angles around the trans position, while bond stretching and angle bending consume the majority of the elastic energy. These structural effects are reflected by a distribution of relaxation times. The volume viscosity is consequently composed of two terms: the bulk isotropic viscosity and the relaxational viscosity (this last one being closely related to the orientational corrrelational functions measured by NMR, ESR, fluorescence resonance, etc.).

4. EXTENSIONAL BEHAVIOUR

Extensional effects usually appear as two normal stress differences. The first is usually positive meaning that an extra tension is produced. The second has been found experimentally to be negative, i.e., the fluid exhibits a compression. A combination of extensional, shear and compression effects provides scaling equations (3) which, in the absence of compressional deformations, reduce to Trouton's law. In rigid polymers (PMMA) the intra-segment barrier to rotation is so high that the conformational change (ΔH) is very unlikely and the dynamic spectrum is reduced to relaxation via whole molecule motion (T.ΔS). In semiflexible chains (PS) the entropy (T.ΔS) and enthalpy (ΔH) processes partially overlap one another. In flexible chains these processes are completly resolved, and their tensile viscosity has a first increasing and then decrasing slope. As result of it a characteristc rapid vortex grow at the corners of a sudden contraction of the flow channel is observed, because the coil configuration-dependent mobility is overcomed by its orientation and continuous stretching, i.e., the second normal stress difference becomes positive and rotational flow is produced.

5. SURFACE BEHAVIOUR

The coil local density, conformation, mobility and relaxation very with distance from the solid surface on a length scale comparable to the chain dimensions. This sensibility to the variation of intermolecular distances is produced mainly by dipolar interactions in the form of electrostatic, Debye, and London forces. Since the distribution of possible configurations adopted by the macromolecules is then less probable than the unperturbed value, a decrease in entropy is produced which causes and increase in stored free energy (elastic energy). When the surface modulus of elasticity is plotted versus the surface deformation rate, a change in slope is observed meaning that the relieve of the surface tension gradient may take place by surface movement (Maragoni) and as surface adsorption.

REFERENCES

1. R.T. Biley, A.M. North, R.A. Petrick, Molecular Motions in High Polymers, Clarendon Press, Oxford (1981).
2. J.V. Alemán, Eur. Polymer J. (1991) 27, 221.
3. J.V. Alemán, Rheol. Acta (1988) 27, 61.

TABLE I

Characteristics of Polymers

Polymer	Formula	Conformation	Dielect. const.	450K (J/g) $-\Delta F$	ΔH	$T.\Delta S$
PMMA	$-CH_2-C(CH_3)(COOCH_3)-$	helicoidal	3.3	561	496	1057
1,2 s-PB	$-CH_2-CH(CH=CH_2)-$	planar (slightly deflected)	2.6	547	603	1147
i-PP	$-CH_2-CH(CH_3)-$	helicoidal	2.3	543	694	1237
PVC	$-CH_2-CH(Cl)-$	planar	3.2	502	316	819
PS	$-CH_2-CH(C_6H_5)-$	helicoidal	2.5	492	448	940
PIB	$-CH_2-C(CH_3)(CH_3)-$	helicoidal	2.5	467	635	1102
EP	$-R'-O-[-R''-O-CH_2-CH(OH)-CH_2-O-]_n-R'-R''-$	planar	3.2	418	438	855
PVAc	$-CH_2-CH(COOC_2H_5)-$	planar	3.2	402	385	788
1,4 cis PB	$-CH_2=CH-CH=CH_2-$	skewed helix	2.5	300	644	945
HDPE	$-CH_2-CH_2-$	planar	2.3	288	747	1050
LDPE	$[(CH_2-CH_2)_n-CH(R)]_{n'}$	planar	2.3	260	797	1057
PEO	$-CH_2-CH_2-O-$	helix/planar	3.7	181	696	877
PBT	$-O-CO-C_6H_4-CO-O-(CH_2)_4-$	planar	3.1	458	588	950
			(at 600K because T_m = 495K)			

R = 3 % branches, mainly of the butyl type and some larger ones.

R' = $-C_6H_4-C(CH_3)_2-C_6H_4-$; R'' = CH_2-CH- (epoxide, bridged by O)

PMMA = poly(methyl methacrylate), 1,2 s-PB = 1,2 syndiotactic polybutadiene, i-PP = isotactic polypropylene, PVC = poly(vinyl chloride), PS = polystyrene, PIB = poly (isobutylene), EP = epoxide prepolymer, PVAc = poly(vinyl acetate), 1,4 cis PB = 1,4 cis polybutadiene, HDPE = high density polyethylene, LDPE = low density polyethylene, PEO = poly(ethylene oxide), PBT = poly(butylene terephthalate).

Theoretical and Applied Rheology, edited by P. Moldenaers and R. Keunings
Proc. XIth Int. Congr. on Rheology, Brussels, Belgium, August 17-21, 1992

MOLECULAR MODEL OF POLYMER MELTS : EFFECTS OF CHAIN DEFORMATION ON NON-LINEAR PROPERTIES

A. BENALLAL and G. MARIN

Laboratoire de Physique des Matériaux Industriels, Université de Pau et des Pays de l'Adour - C.U.R.S., Avenue de l'Université 64000 PAU (France)

1) INTRODUCTION

Transient measurements of shear and normal stresses of high molecular weight polymers exhibit a stress overshoot at high deformation rates : that overshoot is a function of the average relaxation time of the polymer[1,2]. Infrared dichroïsm measurements[3] have shown that the orientational relaxation following a given deformation is in good agreement with the Doï-Edwards model[4]. It appears also that the hypothesis of affine deformation is uncorrect. Some authors have been using the concept of recoverable deformation[6] to explain the experimental observations. That deformation takes into account chain retraction within the tube.

We are giving in that paper a model in which the basic idea is a non-affine deformation of the chain under flow. That model will be compared with a set of experimental results.

2) MODEL

Let us first determine the extension factor of the radius of gyration of a macromolecular chain undergoing a deformation. The calculations of Birshtein and Pryamitsyn[5] allow to derive the extension factor of the radius of gyration.

Let us consider a chain made of N Kuhn segments ; the radius of gyration of the undeformed chain is S_o and S is the radius of gyration of the same deformed chain. The extension factor is defined as[5] :

$$\alpha^2_{av} = \frac{\langle S^2 \rangle}{\langle S_o{}^2 \rangle} \tag{1}$$

Using Flory' theory, the same authors derive the free energy of the gaussian chain, using an elastic force f_{el} and a volumic force f_{int}. Minimization of the free energy leads the equilibrium volume α_{av} :

$$\alpha^2_{av} = \frac{3}{4}\left[f^{-1} - \frac{\coth(2\sqrt{f})}{\sqrt{f}}\right] \tag{2}$$

with $f_{el} = 2\alpha_{av} f$

We assume that the spherical stress tensor σ_m and osmotic pressure π are equal, so we may derive the extension factor.

At equilibrium, $f_{el} = f_{int}$ and $\pi = f_{el} / \langle S^2 \rangle$

so :

$$f = \frac{\alpha_{av}\, \sigma_m \langle S_o^2 \rangle}{2} \tag{3}$$

Let us assume now that :

$$\sigma_m = G(t)\, \varepsilon_m \tag{4}$$

G(t) being the relaxation function and ε_m the spherical strain tensor.

Equations 2, 3, 4 lead to :

$$\alpha_{av} = \left[1 + \left[\frac{G(t)}{G^o_N}(\varepsilon_m - 1)\right]^{1/4}\right] \tag{5}$$

We have used Muller's[6] data in order to test our theoretical assumptions. This author has used small angle neutron scattering on Deuterated labeled chains to analyze chain conformation and determine the radii of gyration. The experimental data have been obtained by Muller on a blend of 10% Deuterated Polystyrene (M_w = 95000 ; P = 1.13) in a matrix of regular polystyrene (M_w = 80000 ; P = 1.12).

The samples have been deformed under uniaxial extension at a constant strain rate $\dot{\varepsilon}_0$ during time t, at 123°C, then annealed.

The maximum relaxation time of the sample is τ_w = 375 s. Zero shear viscosity is 3.8 x 10^7 Pa.s and limiting compliance $J_e^o = 10^{-5}$ Pa^{-1}

Rate (s-1)	Time (s)	Rg // (Å)	Rg ⊥ (Å)
0.023	13	100	63
0.023	30	125	60
0.023	56	164	56+

Table 1 : Variations of the radius of gyration in the direction of flow and in the perpendicular direction as a function of time t.

The equilibrium radius of gyration is 70.5 Å. The full line on figure 1 has been calculated with a relaxation function involving a single relaxation time.

$$G(t) = G^o_N \exp(-t/\tau w) \qquad (6)$$

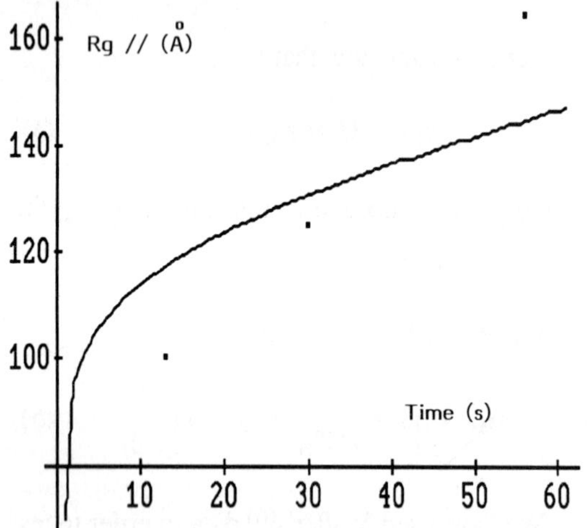

Figure 1 : Variations of the radius of gyration in uniaxial extension.

(• experimental ; full line : theory : Rg = 70.5α_{av})

The good agreement between our model and Muller's data lead us to question the assumption of affine deformation of the chains. Following our hypothesis, the tube diameter of the reptation model and monomer concentration within a blob do change under deformation, but the number of entanglements remains constant. Tube length is given by :

$$l_{(t)} = l_o \alpha^3_{av} \qquad (7)$$

We may conclude that chain deformation is affine only in the limit of infinite chain length. These conclusions lead us to reexamine the non linear constitutive equations, and especially Doï-Edwards[4] approach. It is possible to demonstate that the $Q_{\alpha\beta}$ (λ) "universal" tensor that is a measure of molecular orientation at a given deformation λ is now a function of the extension factor :

$$Q_{\alpha\beta}(\lambda) = Q_{\alpha\beta}^{(IA)}(\lambda)\ \alpha_{av} \qquad (8)$$

That expression allows to understand why the damping function of the BKZ-type non linear constitutive equations depend on the sample molecular weight distribution/polydispersity in the case of polydisperse samples.

REFERENCES

1. K. Osaki, E. Takatori, Y. Tsunshima and M. Kurata Macromolecules, 20, (1987), 525.

2. E.V. Menezes, Ph. D Thesis, Northwestern University, (1980).

3. J.F. Tassin and L. Monnerie, Macromolecules, 21, (1988), 1846.

4. M. Doï and S.F. Edwards, J. Chem. Soc. Faraday Trans. 2, (1978), 74, 1789.

5. T.M. Birshtein and V.A. Pryamitsyn, Macromolecules, 24, (1991), 1554.

6. R. Muller, Thèse d'état, Université Louis Pasteur de Strasbourg, (1988).

Theoretical and Applied Rheology, edited by P. Moldenaers and R. Keunings
Proc. XIth Int. Congr. on Rheology, Brussels, Belgium, August 17-21, 1992

BROWNIAN DYNAMICS STUDY OF CONFORMATIONAL– AND RHEOLOGICAL ASPECTS OF A "REAL" KRAMERS CHAIN UNDER VARIOUS FLOW CONDITIONS

HAN J. M. SLOT

Centre for Computer–Aided–Engineering of Polymers and Composites,
Department of Polymeric Materials Development,
DSM Research, P.O. Box 18, 6160 MD, Geleen, The Netherlands.

1. INTRODUCTION

Polymer microrheology or polymer kinetic theory basically involves two steps, as is clearly stated in the second volume of "Dynamics of Polymeric liquids" by Bird et al (ref. 1). The first step is the establishment of a "diffusion equation" (Fokker–Planck – or Smoluchowski equation) for the configurational distribution function (CDF) and the second step is the development of an expression for the stress tensor, both given some microscopic description of the polymer solution or polymer melt. The stress tensor involves averaging over the CDF, so one needs to solve this diffusion equation for given flow conditions. This can be done only for very special somewhat unrealistic microscopic models, such as Gaussian dumbbells and Gaussian bead–spring chains (ref. 1), and even there one does not really solve the diffusion equation but one only eliminates the CDF in the expression for the stress tensor, using the diffusion equation, resulting in an analytically tractable differential equation for the stress tensor. For slightly more realistic models however, such as for models where the polymer chains have finite extensibility, one needs to resort to numerical means. One way of doing this is to solve the diffusion equation numerically, say by a Galerkin approach (ref. 2), and then using the so obtained CDF to perform the averaging needed for the stress tensor, again numerically. An alternative method, however, is furnished by making use of the statistical equivalence which exists between a Fokker–Planck description and a socalled Langevin description (ref. 3). This Langevin description uses the microscopic degrees of freedom , such as bead positions, directly instead of a CDF.

These microscopic degrees of freedom obey temporal differential equations, consisting of both deterministic – and stochastic terms, and are therefore known as stochastic differential equations (SDE). The alternative method, which is known as Brownian Dynamics (BD), therefore involves numerically integrating the set of SDE, thereby generating subsequent polymer chain configurations in time. The stress tensor or any other macroscopic property of the system is defined as an ensemble–average of the corresponding microscopic property. Thus assuming ergodicity, one can use these numerically generated realisations of the system to calculate any macroscopic property as a time–average. The use of BD in polymer microrheology has seen a lot of activity over the past fifteen years (e.g. refs. 4–9) and still is an active field of research. Most BD studies in polymer microrheology have been limited to bead–spring type models (e.g. ref. 5), i.e. models without constraints. Only in recent years also BD studies, employing models with internal constraints, have appeared in the literature (refs. 4,6,8,9), even though the foundations of this type of simulations were already published at the end of the seventies by Fixman (ref. 4). Although finite extensibility can be incorporated in bead–spring models , it is automatically ensured for bead–rod models. On the other hand the presence of constraints forces one to use generalised coordinates, both in a Fokker–Planck – and a Langevin description, which makes either of the resulting equations unwieldly. However, by introducing socalled constraint forces and enforcing these constraints using a "SHAKE" like algorithm (ref. 10), one can still employ cartesian coordinates. In the present work we will study an extension of one the simplest and oldest bead–rod models around, namely the Kramers freely jointed chain model (ref. 11), describing a dilute polymer solution. The extension consists of two ingredients not present in the original model, namely excluded volume interactions (EVI) and hydrodynamic interactions (HI), making the chain "real" instead of "spooky" The chain is further taken to be osculating, i.e. having spherical beads whose radii are equal to half of the rod-length. The EVI are then introduced by allowing the spherical beads slightly to overlap. These EVI are modelled via a pairwise potential ,which consists of the repulsive part of a shifted Lennard-Jones potential, as was introduced by Weeks et al. (ref. 12). The HI between the beads are described by the Rotne-Prager-Yamakawa (RPY) tensor (ref. 13), which can accomodate for the finite size of the beads, and which ensures a positive definite diffusion tensor for all configurations of the chain. This RPY tensor was introduced in BD by Ermak and McCammon (ref. 14). The present work is to be seen as an extension of the study by Liu (ref. 6) on the BD of a Kramers chain, who neglected both EVI and HI. EVI, although modelled differently, were included in recent BD studies of a Kramers chain by Parnas and Cohen (refs. 9), but with a somewhat different objective than ours. Due to the limited space available, we will describe in the remaining part of this paper, details of the model and of the algorithm which we use. Results will be presented in the oral part of this conference contribution.

2. MODEL AND ALGORITHM

In this section we will describe the equations which form the basis of our BD study. The notation which we use is that of Bird et al. (ref. 1) and Liu (ref. 6). In these equations only scaled quantities will appear, i.e. we will adopt the rod-length a as length-scale, kT as energy-scale and $\lambda \equiv (\zeta a^2/kT)$ as time-scale, where ζ is a scalar bead friction coefficient. The system which we consider is a dilute polymer solution , in which the polymer molecules are idealised by osculating Kramers chains dispersed in an incompressible Newtonian solvent withviscosity η. Each of these chains consists of N spherical (structureless) beads connected by N-1 rigid rods of length 1. Each bead, as mentioned earlier, has a radius equal to 1/2 and a mass equal to m. Now in a dilute polymer solution one can neglect both direct - and indirect (i.e. HI) polymer - polymer interactions. Furthermore on the level of a Langevin description, solvent degrees of freedom are taken into account only implicitly via drag - and stochastic (Brownian) forces acting on the polymer molecules. Therefore only the degrees of freedom of a single chain enter into the description of the system. For a Kramers chain the degrees of freedom, which specify its configuration, can be chosen in a variety of ways (see ref. 1). In our study we need two related choices, namely:

1. N cartesian vectors $\mathbf{r}_\nu$ ($\nu = 1,\ldots,N$) specifying the bead positions with respect to some arbitrary origin in space, or
2. the position of the center of mass $\mathbf{r}_c$ and N bead locations with respect to this center, i.e. $\mathbf{R}_\nu \equiv \mathbf{r}_\nu - \mathbf{r}_c$ ($\nu = 1,\ldots,N$).

The SDE describing the dynamics of the chain can be derived directly from the force-balance for bead ν, i.e.

$$\mathbf{F}_\nu^{(h)} + \mathbf{F}_\nu^{(c)} + \mathbf{F}_\nu^{(ev)} + \mathbf{F}_\nu^{(b)} = \mathbf{0} \tag{1}$$

in which, as is customary, inertial terms have been neglected. From their superscripts it is clear, what the various terms denote: hydrodynamic drag- (including HI), constraint-, excluded volume- and Brownian force. By rewriting the hydrodynamic drag force and using the appropriate expression for the Brownian stochastic force in case of full HI (e.g. ref. 14), (1) leads immidiately to:

$$\dot{\mathbf{r}}_\nu = \kappa \circ \mathbf{r}_\nu + \sum_\mu \Upsilon_{\nu\mu} \circ [\mathbf{F}_\mu^{(c)} + \mathbf{F}_\mu^{(ev)}] + \\ + \sqrt{2} \sum_\mu \Upsilon_{\nu\mu}^{1/2} \circ \mathbf{f}_\mu \tag{2}$$

In this last expression κ is a possibly time-dependent but position-independent traceless velocity gradient tensor, and the tensor $\Upsilon_{\nu\mu}$ denotes the socalled dimensionless diffusion tensor (ref. 1), which is given in the RPY theory by:

$$\Upsilon_{\nu\mu} \equiv \delta_{\nu\mu}\mathbf{I} + (1 - \delta_{\nu\mu}) \frac{h}{r_{\nu\mu}} [(1 + \frac{1}{6r_{\nu\mu}^2}) \mathbf{I} + (1 - \frac{1}{2r_{\nu\mu}^2}) \frac{\mathbf{r}_{\nu\mu}\mathbf{r}_{\nu\mu}}{r_{\nu\mu}^2}] \quad (3)$$

where $\mathbf{r}_{\nu\mu}$ is the vector between the centers of bead ν and μ, and $h \equiv 1/8\pi\eta a$ is the socalled HI parameter. In (2) $\mathbf{f}_\mu$ represents a zero – mean Gaussian white–noise vector. Therefore (2) is a SDE with multiplicative – or colored noise and as it stands is meaningless. We will interpret it in the Stratonovich sense (ref. 3). The constraint force in (1) has the following form:

$$\mathbf{F}_\mu^{(c)} \equiv -2 \sum_i \gamma_i \bar{B}_{i\mu} (\mathbf{r}_{i+1} - \mathbf{r}_i) \quad (4)$$

with $\bar{B}_{i\mu} \equiv \delta_{i+1,\mu} - \delta_{i\mu}$ and where the γ_i (i = 1,...,N–1) are Lagrangian multiplyers, chosen in such a way as to fulfill all N–1 holonomic constraints. Finally the excluded volume force in (1) is given by:

$$\mathbf{F}_\mu^{(ev)} \equiv -24\varepsilon \sum_{\alpha\neq\mu} \Phi\,(r_{\mu\alpha})\, \mathbf{r}_{\mu\alpha} \quad (5)$$

$$\text{with } \Phi(r) \equiv \begin{cases} \frac{2}{r^{14}} - \frac{1}{r^8} & \text{if } r \leq 2^{1/6} \\ 0 & \text{if } r > 2^{1/6} \end{cases} \quad (6)$$

and ε a dimensionless energy parameter, determining the softness of the EVI. As it stands (1) is not directly amendable to numerical evaluation. For that we integrate (1) to first order in Δt. The resulting equation then will contain a term which is a realisation of a vectorial Wiener process with zero mean and a variance–covariance matrix given by $\Upsilon_{\nu\mu}\Delta t$. The algorithm then consists of solving this discretised equation in the absence of the constraint forces, resulting in a unconstrained new configuration. This configuration is then corrected in order to satisfy the constraints. The correction follows directly ounce the Lagrangian multiplyers γ_i are iteratively determined. These chain configurations are then used to evaluate various relevant macroscopic quantities, such as the stress tensor and the radius of gyration auto – correlation function as running (in time) averages. For determining the polymer contribution to the stress tensor we use the Kramers–Kirkwood form (ref. 1) directly, i.e we numerically evaluate:

$$\tau_p \equiv - n_p \sum_\nu \langle \mathbf{R}_\nu \mathbf{F}_\nu^{(h)} \rangle \quad (7)$$

with n_p the polymer density and $\langle\ \rangle$ denoting the average over the realisations of the vectorial Wiener process.

REFERENCES

1 R.B. Bird, C.F. Curtiss, R.C. Armstrong and O. Hassager, Dynamics of Polymeric Liquids, vol.2, Kinetic Theory, Wiley, New York, 1987.
2 X.J. Fan, J. Non–Newtonian Fluid Mech., 17 (1985) 251–265.
3 C.W. Gardiner, Handbook of Stochastic Methods, Springer–Verlag, Berlin, 1990.
4 M. Fixman, J. Chem. Phys., 69 (1978) 1527–1537; 1537–1545.
5 P. Biller and F. Petruccione, J. Non–Newtonian Fluid Mech., 25 (1987) 347–364; J. Rheol., 32 (1988) 1–21.
6 T.W. Liu, J. Chem. Phys., 90 (1989) 5826–5842.
7 K. Kremer and G.S. Grest, J. Chem. Phys., 92 (1990) 5057–5086.
8 H.C. Oettinger, J. Chem. Phys.,91 (1989) 6455–6462; 92 (1990) 4540–4547.
9 R.S. Parnas and Y. Cohen, J. Chem. Phys., 90 (1989) 6680–6690; Macromolecules, 24 (1991) 4646–4656.
10 G. Ciccotti and J.P. Ryckaert, Comp. Phys. Rep., 4 (1986) 345–392.
11 H.A. Kramers, Physica, 11 (1944) 1–19.
12 J.D. Weeks, D. Chandler and H.C. Andersen, J. Chem. Phys., 54 (1971) 5237–5247.
13 J. Rotne and S. Prager, J. Chem. Phys. 50 (1969) 4831–4837; H. Yamakawa, J. Chem. Phys., 53 (1970) 436–443.
14 D.L. Ermak and J.A. McCammon, J. Chem. Phys., 69 (1978) 1352–1360.

Theoretical and Applied Rheology, edited by P. Moldenaers and R. Keunings
Proc. XIth Int. Congr. on Rheology, Brussels, Belgium, August 17-21, 1992

SHEAR FLOW PREDICTIONS OF A HOOKEAN DUMBBELL WITH INTERNAL VISCOSITY USING A GAUSSIAN APPROXIMATION

Jay D. Schieber

Department of Chemical Engineering, University of Houston, Houston, TX 77204-4792, U.S.A.

The simplest model employing internal viscosity is a noninertial Hookean dumbbell utilizing friction points with Stokes' law drag forces, and Brownian forces. The internal viscosity is assumed to be a frictional force with linear dependence on the relative velocity of the two "beads" of the dumbbell projected onto the vector of the dumbbell. That is, changes in the length of the dumbbell experience a friction in addition to that offered by the fluid in a Stokes' law drag, whereas rotations of the dumbbell invoke no additional friction. No rotation vector is employed. If the Brownian forces are then chosen to satisfy the fluctuation-dissipation theorem[1], we may write a stochastic differential equation representing force balance for the dumbbell (neglecting both the inertia of the dumbbell and the inertia of the fluid[2]). The model is essentially the same, aside from some generalizations, as that considered by Kuhn and Kuhn[3], and others. In this work we examine the predictions in shear flow by an approximate scheme[4]. The approximation is accomplished by assuming that all moments appearing in the second moment equation for $\boldsymbol{Q}$ can be approximated by an integral over a Gaussian distribution function. The second moment equation may be found by first writing down the Fokker-Planck equation equivalent to the stochastic differential equation-with proper (Klimontovich[5]) interpretation[1], then multiplying each side by the dyad $\boldsymbol{QQ}$, and integrating over all $\boldsymbol{Q}$ space. This manipulation yields an equation containing higher moments, and, thus, the equation is not closed. However, if we approximate the averages by assuming that the distribution is Gaussian, all averages depend only on $\langle \boldsymbol{QQ} \rangle$, and the equation is closed, although the resulting integrals must be found numerically. On the other hand, for small perturbations to the configuration at equilibrium, we can find exact expressions for $\langle \boldsymbol{QQ} \rangle$ by doing a Taylor expansion on the integrands. The information contained in the resulting expansion is sufficient to provide us with the zero-shear-rate viscosity and the complex viscosity in small-amplitude oscillatory shear flow, if we assume that the Kramer's form for the stress tensor viscosity, first and second normal stress coefficients respectively:

$$\frac{\eta_0 - \eta_s}{nkT\lambda} = \frac{10(1+\epsilon)}{10+3\epsilon}$$

$$\frac{\Psi_{1,0}}{nkT\lambda^2} = \frac{(1+\epsilon)\Psi_{2,0}}{30\epsilon nkT\lambda^2} = \frac{70(1+\epsilon)^2}{245(5+3\epsilon)(10+3\epsilon)}$$

Where ϵ gives the strength of internal viscosity relative to the friction coefficient of the dumbbell beads, n is the number density of dumbbells, η_s is the solvent viscosity, T is the absolute temperature, and λ is the Rouse time constant. We may also use the results from the Taylor expansion to find the predictions for the small amplitude oscillatory shear properties. The complex viscosity is given by

$$\frac{\eta^*}{nkT\lambda} = \frac{(10+6\epsilon)}{10+3\epsilon-10(1+\epsilon)i\omega}$$

where ω is the frequency of the oscillations. By numerical integration of the original coupled, integrodifferential equations, one finds that the steady-state viscosity and first normal stress coefficient are shear-rate dependent. The second normal stress coefficient is also shear rate dependent and changes sign. For nonzero values of the relative internal viscosity parameter ϵ, the transient viscosity and first normal stress coefficient exhibit overshoot. These results are at least in qualitative agreement with experiment.

REFERENCES

(1) J.D. Schieber, J.Non-Newtonian Fluid Mech., in press.
(2) J.D. Schieber and H.C. Öttinger, J.Chem.Phys. **89** (1988) 6972–6981;J.D. Schieber, J.Chem.Phys. **91** (1991) 7526–7533.
(3) W. Kuhn and H. Kuhn, Helvetica Chim.Acta **28** (1945) 1533–1579.
(4) H.C. Öttinger, J.Chem.Phys. **90** (1989) 463; L.E. Wedgewood, J.Non-Newtonian Fluid Mech. **31** (1989) 127.
(5) Yu.L. Klimontovich, Physica A **163** (1990) 515–

Contributed papers

CONSTITUTIVE EQUATIONS

and one.
The retardation function J(t) or its dimensionless form $j(x) = J(t)/J_\infty$ with $J_\infty = G_\infty^{-1}$, $J_0 = G_0^{-1}$ and $\Delta J = J_\infty - J_0$ can be deduced from the inverse Laplace transform of the retardance (= inverse relaxance). A solution for the presented model is possible for c = d because the inverse Laplace transform exists only in this case.

$$j(t) = \frac{J(t)}{J_\infty} = 1 - \frac{\Delta J}{J_\infty} E_{c,1}(-x_{01}{}^{c}) \quad (5)$$

$$\text{with } x_{01} = \frac{t}{\lambda_{01}} \qquad \lambda_{01} = \left(\frac{G_0}{G_\infty}\right)^{\frac{1}{c}} \lambda_0$$

These types of material functions and, consequently, this type of constitutive equation can be used to describe rheological behavior where two straight lines of different slops are connected by a smooth and broad transition in a log-log plot. This behavior will be discussed at the example of three different materials.

3. DISCUSSION

The usefulness of the Eqs. (4) and (5) will be proved subsequently. Two different polymer systems are discussed. The first is a polypropylene at 23°C (ref. 6). This material is in the glassy state and, therefore, it should be especially suitable for the description with a material function showing power law behavior. The data are shown in Fig. 1.

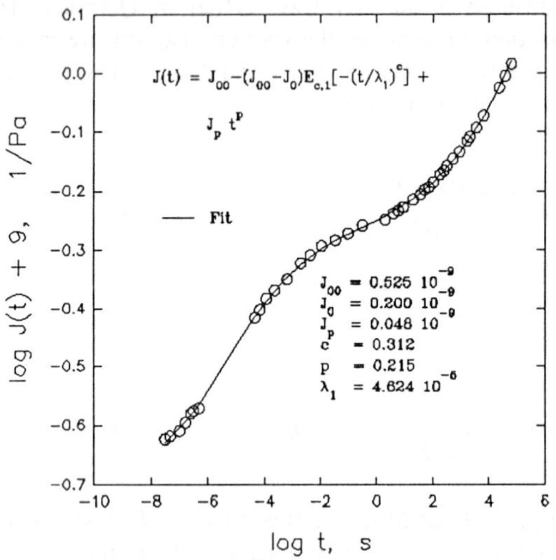

Fig. 1: Creep function of a polypropylene after ref. 6

In order to fit the experimental data, Eq. (5) was used together with a second term representing a simple power law $J_p\, t^p$. This second part of the fit equation can be understood as the short time tail of a second "fractional mode". That figure shows that the fit by 6 parameters is very good over 12 decades in time. That means that this typ of retardation function is suitable to describe relaxation behavior of polymers in the glassy state.
The second example discusses the start up viscosities of a polypropylene/polyamide (70/30 % by weight) blend at a temperature of 240° C and a shear rate of 0.001 s^{-1}. The blends are compatibilized by an appropriate block copolymer (BC). The technical polypropylene used is the matrix material and has a very broad distribution of molecular weights. It shows consequently a very broad transition to the stady state viscosity value. The addition of BSs broadens this transition and for 4 % of BC an overshoot occurs. This overshoot can be avoided in decreasing the shear rate. Of course, overshoot can not described by the here derived formulas. We are restricted to the linear theory of viscoelasticity. Regardless of the humps, one can see from Fig. 2 that the start up viscosities can be described very well. It should be mentioned that the data for the pure PP can not be fited to this model with the same accuracy. Therefore the fit is not presented.

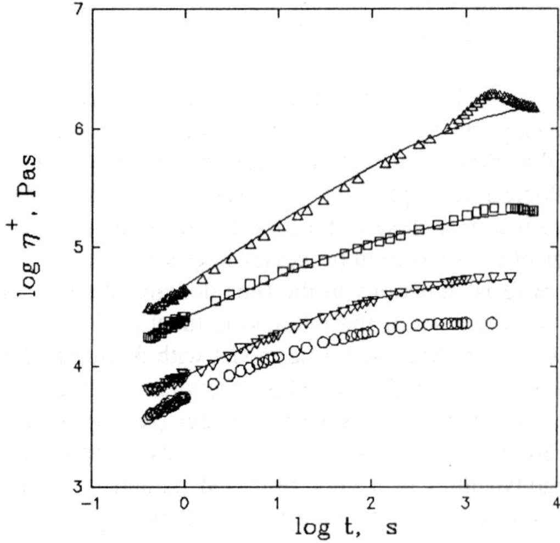

Fig. 2: Start up viscosities for a PP/PA blend and different degrees of compatibilization. The curves are shifted by the factor A = 0.5 Δ, 0.25 □, -0.25 ∇ and -0.5 ○

This material is in the terminal relaxation state and can be described by this type of material functions further. This can be explained by the very broad relaxation time distribution of the matrix and a further enhancement of interactions in the material due to the finer grained structure caused by the BC addition. The occurence of micelles is discussed for BC contents of 4% and higher. Before discussing the next material another type of material behavior must be explained.
There are also materials showing at the long time or

short frequency tails the "classic" exponential or quadratic (for G' e. g.) behavior which is preceded or followed by power law behavior. To describe such a material, further improvement of the fractional differential constitutive equation is necessary.

4. MODEL IMPROVEMENT

In order to improve the rheological constitutive equation with fractional derivatives one can start from Eq. 3. It seems to be reasonable to introduce into the relaxation function a cut-off function coff(x_1) and therefore a second time parameter λ_1. This leads to the following equation:

$$g(t) = ml(x)\ coff(x_1) \qquad (6)$$

$$coff(x_1) = e^{-x_1}, \quad x_1 = t/\lambda_1$$

where ml is the relaxation function of Eq. 3 and coff the cut-off function. The corresponding relaxance and moduli are derived from it and are given in the following formulas:

$$Q(s) = G_0 \frac{s\lambda_0(s\lambda_0 + w)^{d-1}}{1 + (s\lambda_0 + w)^c} \qquad (7)$$

$$0 < c,\ d \leq 1 \quad ; \quad w = \lambda_0/\lambda_1$$

$$G' = y\, z^r \frac{z^c \sin(q\varphi) - \sin(r\varphi)}{1 + 2z^c\cos(c\varphi) + z^{2c}} \qquad (8)$$

$$G'' = y\, z^r \frac{z^c \cos(q\varphi) + \cos(r\varphi)}{1 + 2z^c\cos(c\varphi) + z^{2c}}$$

$$z = \sqrt{y^2 + w^2}, \quad \varphi = \arctan(y/w)$$

$$q = c - r, r = d - 1$$

The derivation of the corresponding constitutive equation with fractional derivatives is under progress.
The usefulness of this type of material function is demonstrated at the third example. This is a polybutadiene modified with urazole providing an associating polymeric network due to the interacting hydrogen bonds (ref. 7). Fig. 3 demonstrates clearly the three different regions in the storage modulus G'. The loss modulus G'' does not show this change from the flow region to the power law region very well. The plateau region follows the power law region. In the most of the cases this zone does not represent a perfect plateau with a slop 0. The here derived type of material function is further able to control the slop of the plateau by the two parameters c and d.
It can be proved that this model gives for c = d = 1 and $\lambda_1 \rightarrow \infty$ the well known results of the Maxwell model. It can also be concluded from Eq. (7) that this model is a generalization of the Cole - Cole relaxation behavior.

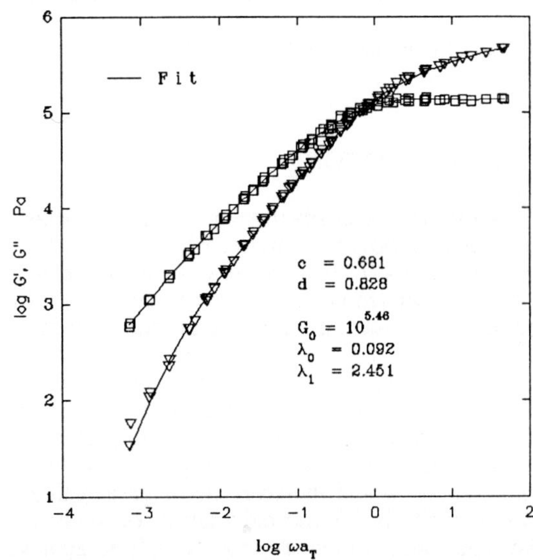

Fig. 3: Dependency of the storage and loss moduli from the reduced frequency for a urazole modified polybutadiene (ref. 6).

5. REFERENCES

1. R.L. Bagley, P.J. Torvik, J. Rheology 27(1983), 201-210
2. W.E. Van Arsdale, J. Rheology, 29(1985), 851-857
3. Chr. Friedrich, in: J. Casas-Vasquez, D. Jou (Eds.), Lecture Notes in Physics 381, Springer, Berlin, Heidelberg, N.Y., 1991
4. Chr. Friedrich, Rheologica Acta, 30 (1991) 151 - 158
5. T.F. Nonnenmacher, W.G. Glöckle, Phil. Mag. Lett., 64 (1991) 89 - 93
6. B.E. Read, G.D. Dean, P.E. Tomlins, Polymer 29(1989), 2159-2169
7. L. de Lucca Freitas, R. Stadler, Macromolecules, (1987) 2478 - 2485

6. ACKNOWLEDGEMENTS

The autor would like to thank Prof. R. Stadler from the University in Mainz and Dr. P.E. Tomlins from the National Physical Laboratories in Teddington for supplying the data.

Theoretical and Applied Rheology, edited by P. Moldenaers and R. Keunings
Proc. XIth Int. Congr. on Rheology, Brussels, Belgium, August 17-21, 1992

APPLICATION OF A WAGNER MODEL FOR THE INTERCONVERSION BETWEEN LINEAR AND NONLINEAR VISCOELASTIC MATERIAL FUNCTIONS

M. GAHLEITNER, R. SOBCZAK and A. SCHAUSBERGER

Institute of Chemistry, J. Kepler University Linz, A-4040 Linz (Austria)

1. INTRODUCTION

While the determination of linear viscoelastic properties of polymer melts, especially of the dynamic moduli, is possible without serious problems on most commercially available rheometers, measuring nonlinear properties remains difficult and mainly restricted to scientific investigations. Nevertheless informations like steady shear viscosity, normal stress coefficients and extensional viscosity are sure to be of importance to the material behaviour in complex processing geometries. It would therefore be of great value to possess methods for the calculation of such data from other ones, which are more simple to achieve. The aim of this study is to investigate the applicability of theoretical concepts from the theory of linear and nonlinear viscoelasticity to practical problems.

2. THEORETICAL BACKGROUND

It is widely accepted, that the linear viscoelastic behaviour of a polymer melt can be described using a discrete relaxation time spectrum. In the constitutive equation [ref. 1]

$$\sigma(t) = \int_{-\infty}^{t} m(t-t')\, C^{-1}(t')\, dt' \qquad (1)$$

(C^{-1} being the finger strain tensor) the memory function m(t-t´) can be expressed by a spectrum of N pairs of relaxation times τ_i and relaxation strenghts g_i:

$$m(t-t') = \sum_{i=1}^{N} g_i \exp\left(-\frac{t-t'}{\tau_i}\right) \qquad (2)$$

The linear material functions, namely the relaxation modulus G(t), the transient viscosity at low shearrates $\eta_0(t)$ and the dynamic moduli can then be calculated from the spectrum. For the storage modulus $G'(\omega)$ and the relaxation modulus $G''(\omega)$ the following formulae are valid:

$$G'(\omega) = \sum_{i=1}^{N} g_i \frac{\omega^2\tau_i^2}{1+\omega^2\tau_i^2} \qquad G''(\omega) = \sum_{i=1}^{N} g_i \frac{\omega\,\tau_i}{1+\omega^2\tau_i^2} \qquad (3)$$

The determination of the spectrum can be carried out using a predefined or variable set of relaxation times (about 1-2 per decade between $1/\omega_{max}$ and $1/\omega_{min}$) and minimizing the sum of deviation squares between experimental and calculated moduli by variation of the g_i - values [refs. 2-6].

In order to describe the nonlinear behaviour of a material, a damping function depending on the deformation has to be included. Using the Wagner model [refs. 7-9], one arrives at the following constitutive equation

$$\sigma(t, \{C^{-1}(t,t'); t'<t\}) = \int_{-\infty}^{t} m(t-t')\, h(C^{-1}(t,t'))\, C^{-1}(t,t')\, dt' \qquad (4)$$

where h is the damping function caused by a deformation-induced disentanglement of the molecules. In the case of shear flow, a simple function of the form

$$h(\gamma) = \exp(-n\,\gamma) \quad \text{with} \quad 0 \le n \le 1 \qquad (5)$$

can be applied. Evaluation of the model for steady shear flow leads to analytical expressions for the steady shear viscosity and the steady first normal stress difference [ref. 7],

$$\eta(\dot\gamma) = \sum_{i=1}^{N} \frac{g_i \tau_i}{(1+n\,\dot\gamma\,\tau_i)^2} \qquad \Psi_1(\dot\gamma) = \sum_{i=1}^{N} \frac{g_i \tau_i^2}{(1+n\,\dot\gamma^2\,\tau_i)^3} \qquad (6)$$

In the case of extensional viscosity, the damping function has to be formulated as a function of the extension rate ε, the actual form depending on the geometry of the extension [refs. 10-12]. For a correlation of the damping functions from shear and extension, the following formula can be derived:

$$h(\dot\varepsilon,t) = e^{-n\sqrt{\alpha(e^{2\dot\varepsilon t}+2e^{-\dot\varepsilon t}) + (1-\alpha)(e^{-2\dot\varepsilon t}+2e^{\dot\varepsilon t}) - 3}} \qquad (7)$$

Here, n is the same damping exponent as in shear flow and α is a factor for the combination of the contributions of first and second invariant of C^{-1}. As no analytical solution for the calculation of the extensional viscosity can be found, a numeric integration procedure has to be carried out yielding

$$\eta_e(\dot\varepsilon,t) = \frac{1}{3\dot\varepsilon} \sum_{i=1}^{N} g_i [\chi_i(\dot\varepsilon,t) + \frac{1}{\tau_i} \int_0^t \chi_i(\dot\varepsilon,t')\,dt'] \qquad (8)$$

with

$$\chi_i(\dot\varepsilon,t) = e^{-t/\tau_i} (e^{2\dot\varepsilon t} - e^{-\dot\varepsilon t})\, h(\dot\varepsilon,t) \qquad (9)$$

3. EXPERIMENTS AND CALCULATIONS

All investigations were carried out using a technical polypropylene, Daplen KS10 (PCD polymers; M_w=510kg/mol, M_w/M_n=5.4). Measurements in the linear region were made in cone/plate or plate/plate geometry on a Rheometrics RDS II, in the nonlinear region partly on an RMS II.

3.1 Linear viscoelastic material functions

The dynamic moduli were measured at three temperatures (200, 230 and 260°C) and combined to a master curve at 200°C using time-temperature - superposition. The following determination of the relaxation spectrum was carried out using the IRIS software [ref. 5]. Table 1 gives the result; a density of about one relaxation time per decade was obtained. The relaxation modulus, which was also measured at two temperatures (200 and 230°C), coincides equally well with the calculated function.

Table 1 - Relaxation spectrum of KS10 at 200°C

i	τ_i / s	g_i / Pa
1	$3.177\ 10^{-3}$	$5.925\ 10^{4}$
2	$1.754\ 10^{-2}$	$2.053\ 10^{4}$
3	$7.955\ 10^{-2}$	$9.876\ 10^{3}$
4	$3.845\ 10^{-1}$	$2.905\ 10^{3}$
5	$2.141\ 10^{0}$	$6.534\ 10^{2}$
6	$1.345\ 10^{1}$	$7.742\ 10^{1}$

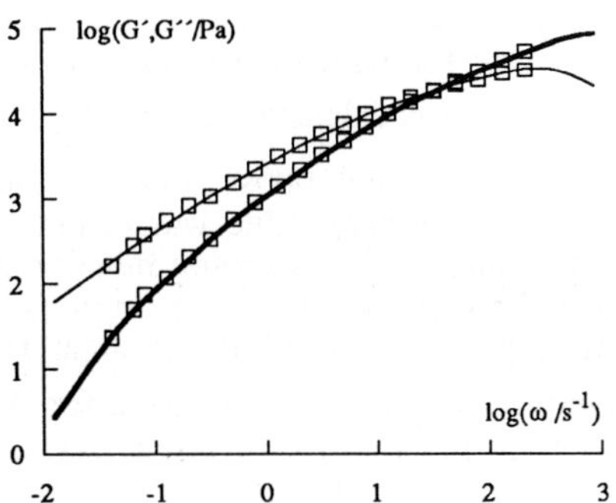

Figure 1 - Dynamic moduli of KS10 at 200°C (squares indicate mesured data, lines calculated functions)

3.2 Nonlinear material functions

The steady shear viscosity, $\eta(\dot\gamma)$, and the steady first normal stress coefficient $\Psi_1(\dot\gamma)$ were measured directly up to shear rates of about 1 s^{-1}. At higher ones, no stationary flow situation could be achieved. So the viscosity there was calculated from dynamic data using the Cox/Merz-rule ($\eta^*(\omega) = \eta(\dot\gamma)$). With

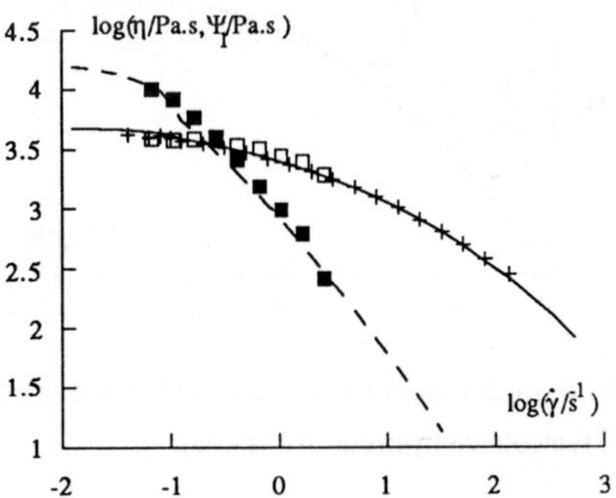

Figure 2 - Viscosity and first normal stress coefficient of KS10 at 200°C (□ - viscosity: measured data , + - viscosity calculated from dynamic moduli using Cox/Merz-rule, ■ - normal stress coefficient: measured data; lines indicate functions calculated according to eqs. (6))

the help of eq.(6) the damping exponent n was determined, again using a least square method. For Daplen KS10, n turned out to be 0.26; the use of a single damping term proved to be sufficient. Figure 2 shows a comparison of the measured and calculated data for steady shear flow.

3.3 Extensional viscosity

The extensional viscosity $\eta_e(\dot{\varepsilon})$ in uniaxial extension was measured on an apparatus of the Mackley type [ref. 13], using a stagnation point flow geometry. This setup allows measurements up to extensional rates of about $500s^{-1}$ for the investigated material. Using the numerical procedure of eqs. (8) and (9), $\eta_e(\dot{\varepsilon},t)$ was calculated for a set of extension rates between 0.01 and $1000s^{-1}$, in each case up to a time where a stationary value was reached. As the direct determination of a damping function from the extensional measurements was impossible, the damping exponent determined in shear flow was used together with a value of 0.032 for α, which was taken from the work of Laun [ref. 10], who determined it for a LDPE melt.

Figure 3 clearly shows the "overshoot" behaviour of the time curves; but as can be seen from figure 4, which also contains the measured data for 200°C, such an "overshoot" also appears in the extension rate dependence. The deviations between measurement and calculation may be attributed to technical problems with the apparatus as well as to the fact, that α was chosen quite arbitrary.

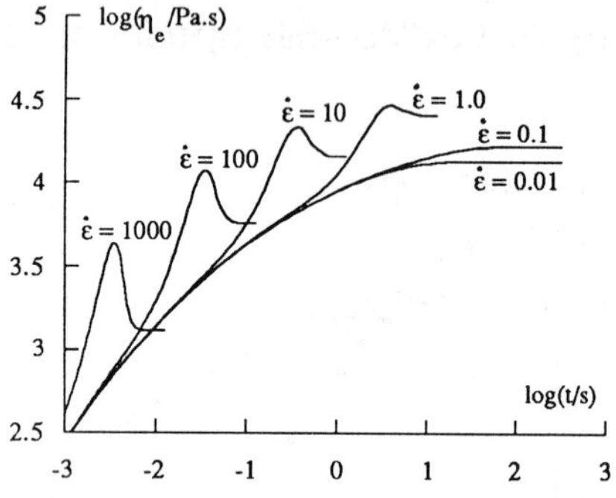

Figure 3 - Startup behaviour of extensional viscosity of KS10 at 200°C (calculated from eqs. (8) and (9))

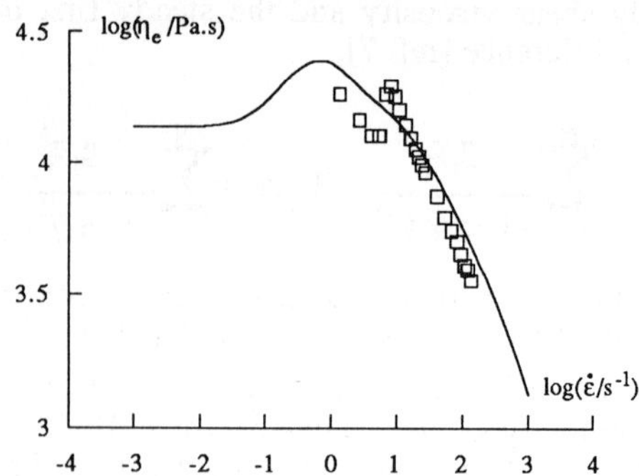

Figure 4 - Stationary extensional viscosity of KS10 at 200°C (squares indicate measured data, line indicates calculated function)

4. CONCLUSIONS

With the method described above, the nonlinear viscoelastic behaviour of polymer melts can be calculated from a linear relaxation spectrum and a single exponential damping function using a Wagner model. These calculations are not restricted to shear flow, but also applicable to extensional flow.

REFERENCES

1. J.D.Ferry, Viscoelastic Properties of Polymers Wiley, New York 1980
2. C.Friedrich, B.Hofmann, Rheol.Acta, 22 (1983) 425-430
3. L.Berger, J.Meissner, Rheol.Acta, in print (1992)
4. J.Honerkamp, J.Weese, Macromolecules, 2 (1989) 4372-4377
5. H.H.Winter, M.Baumgaertel, P.R.Soskey, in:K.T.O´Brien (Ed.), Computer-aided Engineering in Polymer Processing: Applications in Extrusion and other continous processes, Hanser, München, 1990
6. M.Baumgaertel, A.Schausberger, H.H.Winter, Rheol.Acta, 49 (1990) 400-408
7. H.M.Laun, Rheol.Acta, 17 (1978) 1-11
8. M.H.Wagner, Rheol.Acta, 29 (1990) 594-603
9. M.H.Wagner, Angew.Makromol.Chem., 179 (1990) 217-229
10. H.M.Laun, H.Mündstedt, Rheol.Acta, 17 (1978) 415-425
11. M.H.Wagner, J.Meissner, Makromol.Chem., 181 (1980) 1533-1550
12. M.H.Wagner, A.Demarmels, J.Rheol. 34 (1990) 943-958
13. M.R.Mackley, A.Keller, Polymer, 14 (1973) 16-21

Theoretical and Applied Rheology, edited by P. Moldenaers and R. Keunings
Proc. XIth Int. Congr. on Rheology, Brussels, Belgium, August 17-21, 1992

ON THERMODYNAMICS AND STABILITY OF GENERAL MAXWELL-LIKE VISCOELASTIC CONSTITUTIVE EQUATIONS

A.I. LEONOV

Department of Polymer Engineering, The University of Akron
Akron OH 44325-0301 (U.S.A)

1. INTRODUCTION

Viscoelastic constitutive equations (CEs) commonly tested in low strain rate rheometric flows, display various instabilities in flows with high Deborah numbers, seemingly not due to any fundamental reason but because of their poor modeling. This work, started with a new thermodynamic derivation of general Maxwell-like CEs, analyzes the possible origins of the instabilities and demonstrates these in examples of some popular CEs.

2. THERMODYNAMIC DERIVATION OF GENERAL MAXWEL-LIKE CEs

2.1 General formulation

To establish a "canonical" form for the CEs, consistent with thermodynamics, we propose here a new and brief, as compared with recent Poisson-bracket formulation (refs. 1,2), derivation of the CEs based on the methods of irreversible thermodynamics employed earlier for a particular model (refs. 3,4). We suppose that a symmetric second-rank "configuration" tensor **c**, assumed to be positive definite, and temperature T are the state variables for isotropic and generally compresible Maxwell liquids. The stress tensor σ is commonly defined as follows:

$$\sigma = 2\rho \mathbf{c}\cdot \partial F/\partial \mathbf{c} = 2\rho[F_1 \mathbf{c} + F_2(I_1 \mathbf{c} - \mathbf{c}^2) + F_3 I_3 \delta] \quad (1)$$

Here $F(T,I_1,I_2,I_3)$ is the free energy per mass unit, I_k are the basic invariants of tensor **c**, $F_k = \partial F/\partial I_k$, ρ is the density and δ is the unit tensor. Being interested in isothermal flows, we use the Clausius-Duhem dissipative inequality in the form:

$$D = tr(\sigma\cdot \mathbf{e}) - \rho\dot{F}|_T = tr[\sigma\cdot(\mathbf{e} - \mathbf{c}^{-1}\cdot\dot{\mathbf{c}}/2)] \quad (2)$$

Here **e** is the strain rate tensor and the dissipation D is positive for any non-equilibrium processes and vanishs in the equilibrium. Now we should represent the dissipation in the bilinear form: $D = X\cdot Y$ where X is thermodynamic "force" which is equal to σ, and Y is conjugated thermodynamic "flux". To define Y one needs an additional assumtion related to non frame-invariant character of tensor $\mathbf{c}^{-1}\cdot\dot{\mathbf{c}}$ in eq.(2). Instead of this tensor, the tensor $\mathbf{e}_e$, was introduced in (ref. 3)

$$\overset{o}{\mathbf{c}} = \mathbf{c}\cdot\mathbf{e}_e + \mathbf{e}_e\cdot\mathbf{c} \quad (3)$$

to match the local equilibrium hypothesis; $\overset{o}{\mathbf{c}}$ being the corrotational time derivative of tensor **c**. Due to eqs.(1) and (3), the dissipative equation (2) takes the form:

$$D = tr(\sigma\cdot\mathbf{e}_p), \quad \mathbf{e}_p \equiv \mathbf{e} - \mathbf{e}_e \quad (4)$$

where the irreversible strain rate tensor $\mathbf{e}_p$, stands for the thermodynamic flux Y. Then the general phenomenological relation between the thermodynamic flux $\mathbf{e}_p$ and force σ is written as follows:

$$e_{p,ij} = M_{ijst}(T,\mathbf{c})\sigma_{st}(\mathbf{c}) \quad (5)$$

Here the rank-four mobility tensor $\underline{\underline{M}}$ is an isotropic function of tensor **c**. Eq.(5) represents the dissipation as a qudratic form, $D = M_{ijst}(T,\mathbf{c})\sigma_{ij}\sigma_{st}$, which shows that the mobility tensor $\underline{\underline{M}}$ is positive definite. Finally, eqs.(3) and (5) take the form of evolution equation for **c**:

$$\overset{\nabla}{\mathbf{c}} + \psi(\mathbf{c}) = 0, \quad \psi(\mathbf{c}) = 2\mathbf{c}\cdot\mathbf{e}_p(\mathbf{c}) \quad (6)$$

where $\overset{\nabla}{\mathbf{c}}$ is the upper convected time derivative of tensor **c**. Additionally,

$$D = (1/2)\mathrm{tr}(\boldsymbol{\sigma}\cdot\boldsymbol{\phi}\cdot\mathbf{c}^{-1}) \tag{7}$$

Eqs.(1) and (6) form a closed set of CEs for the general Maxwell model under study, written in the "canonical" form. If the new kinetic rank-four tensor $\underline{\underline{\mathbf{L}}}(\mathbf{c})$ is introduced as $L_{ijkl} = c_{im}c_{jn}M_{mnkl}$, the dissipative term $\boldsymbol{\phi}$ in eq.(6) is reduced to the form: $\boldsymbol{\phi} = \underline{\underline{\mathbf{L}}}\cdot\rho\partial F/\partial\mathbf{c}$, which coincides with that proposed in (ref.2). Eq.(6) predicts the mass balance equation, independently from the compressibility condition, if and only if

$$\mathrm{tre}_p(\mathbf{c}) = \mathrm{tr}[\mathbf{c}^{-1}\cdot\boldsymbol{\phi}(\mathbf{c})]/2 = 0, \tag{8}$$

When eq. (8) is satisfied, the density ρ can be expressed through I_3. This means that ρ is a state variable and in the incompressible case when $I_3=1$, $\boldsymbol{\sigma} = -p\boldsymbol{\delta} + \boldsymbol{\sigma}_e$ and $\boldsymbol{\sigma}_e = 2\rho_0\mathbf{c}\cdot\partial F/\partial\mathbf{c}$, p is the equilibrium isotropic pressure. This type (A) of CEs was developed in (refs. 3,4). If eq.(8) is not satisfied (type B of CEs), the density in the compressible case, is not more a state variable and when ρ = const, the pressure is irreversible.

This method was also used to derive the Maxwell-like CEs with mixed upper and lower convected derivatives.

2.2 Examples of general Maxwell-like CEs

Only some incompressible CEs are represented below by specifying $\boldsymbol{\sigma}_e(\mathbf{c})$ and $\boldsymbol{\phi}(\mathbf{c})$ in general CEs (1) and (6).

(i) Giesekus CE (ref. 5):

$$\boldsymbol{\sigma}_e = \mu\mathbf{c}, \quad \boldsymbol{\phi}(\mathbf{c}) = [\boldsymbol{\delta} + a(\mathbf{c} - \boldsymbol{\delta})]\cdot(\mathbf{c} - \boldsymbol{\delta})/\theta$$
$$W=\mu(I_1-3)/2, D=(\mu/2\theta)[I_1-3+a(I_1^2-2I_1-2I_2-3)] \tag{9}$$

(ii) Leonov CE (ref. 3):

$$\boldsymbol{\sigma}_e = \mu\mathbf{c}, \quad \boldsymbol{\phi}(\mathbf{c}) = [\mathbf{c}^2 + \mathbf{c}(I_2 - I_1)/3 - \boldsymbol{\delta}]/(2\theta)$$
$$W = \mu(I_1-3), \; D=\mu(2I_1^2 + I_1I_2 - 6I_2 - 3)/(12\theta) \tag{10}$$

(iii) The Larson CE (ref. 6):

$$\boldsymbol{\sigma}_e = \mu\mathbf{c}/B, \; \boldsymbol{\phi}(\mathbf{c})=B(\mathbf{c} - \boldsymbol{\delta})/\theta, \; B = 1+\xi(I_1-3)/3$$
$$W = (3\mu/2\theta)\ \ln B, \quad D = \mu(I_1-3)/(2\theta) \tag{11}$$

Here $W = \rho_0 F$ is the elastic potential, μ is the Hookean elastic modulus, θ is the relaxation time, a and ξ are numerical parameters, $a,\xi \in [0,1]$.

3. STABILITY OF THE MAXWELL-LIKE CEs

3.1 Hadamard stability of the CEs

The total set consists of the momentum balance, mass conservation and constitutive equations (1) and (6), which satisfy certain initial and boundary conditions; the basic variables being reduced to the tensor **c**, the velocity vector $\underline{v}$ and the pressure p. Now consider infinitesimal disturbances of the basic flow variables, taken in the form:

$$\{\mathbf{c}', \underline{v}', p'\} = \epsilon\{\bar{\mathbf{c}}, \bar{\underline{v}}, \bar{p}\}\exp[i\epsilon^{-1}(\underline{k}\cdot\underline{x} - \omega t)] \tag{12}$$

Here the amplitudes $\bar{\mathbf{c}}, \bar{\underline{v}}, \bar{p}$, wave vector $\underline{k}$ and frequency ω depend on "slow" variables $\epsilon\underline{x}$, ϵt; ϵ being a small amplitude parameter. When $\epsilon \to 0$, eq.(12) corresponds to the method of "frozen coefficients" in studies of Hadamard stability (ref.7), which can be justified by geometro-optical considerations. Using common perturbation technique, one can obtain in the lowest order of ϵ, the dispersion relation:

$$\Omega^2 = k^2 \; B_{ijmn}\xi_i\eta_j\xi_m\eta_n \tag{13}$$

where $\Omega = \omega - \underline{v}\cdot\underline{k}$, $\quad \underline{\xi} = \bar{\underline{v}}/\bar{v}$, $\underline{\eta} = \underline{k}/k$,

$$B_{ijmn} = R_{ijms}c_{sn} + R_{ijsn}c_{sm}\,, \tag{14}$$
$$2R_{ijnl} = \partial\sigma_{ij}/\partial c_{nl} + \partial\sigma_{ij}/\partial c_{ln}$$

and $\boldsymbol{\sigma}$, **c** , $\underline{v}$ are the variables of basic flow. Due to eq.(13), the condition of stability $\mathrm{Im}\omega \geq 0$, is reduced to the positive semi-definiteness of rank-four tensor $\underline{\underline{\mathbf{B}}}$ ($\underline{\underline{\mathbf{B}}} \geq \underline{\underline{\mathbf{0}}}$) expressed in terms of tensor **c**. This is the necessary and sufficient condition for Hadamard stability. If it is satisfied for any flow and for any posible value of positive definite tensor **c**, we call the CEs for elastic liquids <u>globally Hadamard stable.</u>

The stability conditions are proved to be equivalent to the common conditions of thermodynamic stability in elastic solids and expressed in the form of three independent inequalities imposed on elastic potential W and its derivatives, $\partial W \partial I_k$ and $\partial^2 W/\partial I_k \partial I_i$. For the CEs of type A, sufficient conditions for Hadamard stability were also found (ref. 3):

$$W_1 c_i + W_2 c_i^{-1} > 0 \; (i=1,2,3), \; W_{11}W_{22} \geq W_{12}^2 \tag{15}$$

The CEs with mixed upper and lower convected derivatives are proved to be globally Hadamard unstable (e.g.,ref. 7).

Thus it was shown that only those CEs of differential type which have a perfect elastic limit, as eqs.(1) and (6), are globally stable in Hadamard sense, if they are thermodynamically stable.

3.2 On the positive definiteness and boundedness of tensor c

The results obtained above are valid if $\mathbf{c} > \mathbf{0}$, which itself needs to be proved.

Theorem 1 (see also ref. 8):
For any piecewise smooth strain history $\mathbf{c}(t) > \mathbf{0}$, if $\mathbf{c}(t_0) = \boldsymbol{\delta}$, $t \geq t_0$.

Neither the positive definiteness of tensor **c** nor satisfying the conditions of Hadamard stability can guarantee us from "dissipative instability" associated with poor modeling of dissipative term $\psi(c)$ in eq.(6). The common example of this is the behavior of upper convected Maxwell CE in simpe extension.

By definition, a flow is regular if the norm of strain rate tensor is bounded by a positive constant, whether the flow is stationary or not.

Theorem 2:
Let the elastic potential W be a smooth, non-decreasing function of the invariants I_k. If at $|c| \to \infty$, the inequality $D > E \cdot |\sigma|$ holds for any positive number E, then in any regular flow, tensor **c** and stress tensor $\boldsymbol{\sigma}_e$ are bounded.

Both the theorems have a limited character, however, because in most real problems, we usually deal not with a given regular strain history but rather with a mixed strain-stress history, where the theorems do not work.

3.3 Examples of unstable behavior for Maxwell-like CEs (9)-(11)

These CEs are proved to be globally Hadamard stable and satisfy the conditions of theorems 1 and 2 if $\alpha, \xi \in (0,1]$.

In simple steady shearing, Leonov's model (10) is a particular version of the Giesekus model (9) with $\alpha = 1/2$. If $\alpha \leq 1/2$, the flow curves monotonically increase, but when $\alpha > 1/2$ they have a decreasing branch. In both the cases, shear stress is bounded by the value $\mu/(2\alpha)$. In steady simple shearing, the flow curves predicted by Larson's model, have a maximum with the decreasing branches for any $\xi \in (0,1]$.

The 1D stability analyses of the decreasing flow curve branches for both the Giesekus and Larson CEs, showed:

(i) In inertialess approach, these flows are 1D stable, which completely agrees with theorem 2. This result was also confirmed by numerical studies of start up shearing with a given shear rate.

(ii) Common perturbation analyses of 1D stability with allowance for the momentum equation resulted in instabilities for shearing flows related to the decreasing branches of flow curves.

For all three CEs, another kind of 1D inertialess instability was found in start up shearing flow with a given shear stress whose value exceeds the bounding value. Heavy blow up instability with the loss of positive definiteness of tensor **c** was found for all three CEs under study. Also, Larson's CE predicts a non existence of 1D shear flow when the values of given shear stress are high enough.

Only Larson's model displays 1D long wave instability in simple extension under relaxation, and even non-existence of the flow with a given high extensional stress. These features are mostly related to the modeling of the elastic potential.

Abovementioned instabilities happen because of the decreasing branches on the flow curves in simple shear and their boundedness in both types of flow. They can be avoided either by adding a small viscous term or changing elastic or/and dissipative terms in the CEs. The former is valid only for the Giesekus CE ($\alpha \leq 1/2$) and Leonov CEs, whereas the latter is possible only for the general Leonov model (refs. 3,4,9).

REFERENCES

1. M. Grmela, Phys. Lett., A130 (1988) 81-86.
2. A.N. Beris and B.J. Edwards, J. Non-Newtonian Fluid Mech., 34 (1990) 55-78, 503-538.
3. A.I. Leonov, Rheol. Acta, 15 (1976) 85-98.
4. A.I. Leonov, J. Non-Newtonian Fluid Mech., 25 (1987) 1-59.
5. H. Giesekus, J. Non-Newtonian Fluid Mech., 11 (1982) 69-109.
6. R.G. Larson, J. Rheol., 28 (1984) 545-571.
7. D.D. Joseph,Fluid Dynamics of Visco-elastic Liquids, Springer-Verlag, New York, 1990, p.75.
8. M.A. Hulsen, J. Non-Newtonian Fluid Mech., 38 (1990) 93-100.
9. A.N. Prokunin, J. Appl. Math. Mech. (English translation from Russian), 48 (1984) 957-965

Theoretical and Applied Rheology, edited by P. Moldenaers and R. Keunings
Proc. XIth Int. Congr. on Rheology, Brussels, Belgium, August 17-21, 1992

The Effect of Segmental Stretch on Theoretical Predictions of the Doi-Edwards Model

D.W. Mead[1], E.A. Herbolzhiemer[2] and L. Leal[3]
[1]Shell Development Co., P.O. Box 1380, Houston TX 77251
[2]Exxon Research and Engineering Co., Clinton NJ 08801
[3]University of California - Santa Barbara 93106

INTRODUCTION

A fundamental, yet explicit, omission in the original Doi-Edwards tube model was the chain stretching contribution to the stress. Consequently, predictions of the theory for flows where the Deborah number based on the Rouse (retraction) time, De= $\dot{\gamma}\tau_{Rouse}$, was of order unity, were flawed. After several ad hoc attempts to incorporate chain stretching into the DE model, Marrucci and Grizzutti developed a rigorous formulation that accounted for non-uniform segmental stretch along the chain[1]. We have further embellished this chain stretching model by incorporating a finitely extensible freely jointed chain into the description. The force - extension rule for a freely jointed chain is described by a non-linear, inverse Langevin function.The non-linear spring adds another parameter to the model, n_t, the number of Kuhn steps in the freely jointed chain.

This work examines the theoretical predictions of the extended DE model in steady two dimensional flows. More detailed discussion will be provided in subsequent publications. Two dimensional flows can be readily realized experimentally in two and four roll mills. In a suitable coordinate system, the velocity gradient for a general two dimensional flow can be written as;

$$\vec{\nabla}\underset{\sim}{u} = \dot{\gamma}\begin{bmatrix} 0 & 1 \\ \lambda & 0 \end{bmatrix} \qquad \text{Eq.(1)}$$

The degree of extensional character to the flow is parametrized by λ, which is zero for shear flows and unity for planar extension.

The Doi-Edwards Model with Segmental Chain Stretch

We briefly summarize the Doi-Edwards model with segmental stretch. In some respects, the model is identical to the original tube model, and we will not elaborate. Specifically, the equations describing the evolution of the orientational order parameter tensor (2), S(s,t), and the tube survival probability function (3) G(s,t,t') carry over from the original model unchanged. Our notation is identical to the original segmental stretch model publications[1].

$$\underset{\sim}{S}(s,t) = \int_{-\infty}^{t} \frac{\partial G(s,t,t')}{\partial t'} \underset{\sim}{Q}(\underset{\sim}{E}(t,t'))\,dt' \qquad \text{Eq.(2)}$$

$$\frac{\partial G(s,t,t')}{\partial t} = D\frac{\partial^2 G(s,t,t')}{\partial s^2} - v(s,t)\frac{\partial G(s,t,t')}{\partial s} \qquad \text{Eq.(3)}$$

Due to the inclusion of segmental stretch, two additional relations are required. One describes the average relative hydrodynamic velocity between the chain and tube.

$$v(s,t) = \vec{\nabla}\underset{\sim}{u}\int_0^s \underset{\sim}{S}(s',t)\,ds' \qquad \text{Eq.(4)}$$

The last relation describes the one dimensional dynamics of the segmental stretch along the tube coordinate, -L/2 < s <L/2. Equation (5) is a force balance on a differential segment of tube. The net rate of stretch of a chain segment results from an imbalance between the extensional force due to hydrodynamic drag and the retraction force due to non-Gaussian rubber like elasticity. If a random Brownian term were also included, equation (5) would be a one dimensional Langevin equation.

$$\frac{\partial s(s_0,t)}{\partial t} = v(s_0,t) + \left(ND\frac{dL^{-1}(x)}{dx}\right)\frac{\partial^2 s}{\partial s_0^2} \qquad \text{Eq.(5)}$$

The final equation defining the model is the stress calculator, which utilizes an inverse Langevin description for the spring law for a freely jointed chain, $L^{-1}(x)$. Here, x is the fractional extension of the chain, which can be shown to be equal to $\frac{\partial s}{\partial s_0}\sqrt{\frac{N}{n_t}}$. Neither the stress calculator or the birefringence are independent and are completely determined by equations (2-5).

$$\underset{\sim}{\tau}(t) = \frac{2G_N}{3L_0}\int_0^{\frac{L_0}{2}} \underset{\sim}{S}(s_0,t)\,L^{-1}\left(\frac{\partial s}{\partial s_0}\sqrt{\frac{N}{n_t}}\right)\frac{\partial s}{\partial s_0}\,ds_0 \qquad \text{Eq.(6)}$$

$$\Delta\underset{\sim}{n}(t) = K\int_0^{\frac{L_0}{2}} \underset{\sim}{S}(s_0,t)\left(1 - \frac{3\frac{\partial s}{\partial s_0}\sqrt{\frac{N}{n_t}}}{L^{-1}\left(\frac{\partial s}{\partial s_0}\sqrt{\frac{N}{n_t}}\right)}\right)ds_0 \qquad \text{Eq.(7)}$$

Where, $K \equiv \frac{4\pi v n_t}{27}\left(\frac{(\bar{n}^2+2)}{\bar{n}}\right)^2\left(\frac{\alpha_1-\alpha_2}{a}\right)$

The segmental stretch tube model thus has two parallel relaxation processes, chain retraction and reptative diffusion, with widely separated characteristic times. The extent of the separation scales with the 'number' of entanglements, N, as $T_d \sim 3NT_{Rouse}$. Physically, the wide separation of time scales reflects the relative difficulty of stretching free, untethered polymer strands versus orientating them. Hence, a dichotomy exists in the definition of the Deborah number. The two time constants allow us to classify steady flows into three flow regimes, as shown graphically in figure 1. A sketch of the end-to-end distribution function corresponding to each flow regime is also included.

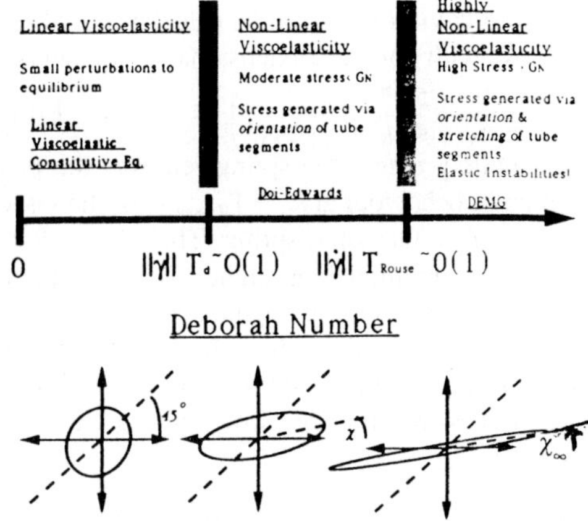

Figure 1) Flow classification scheme and corresponding end-to-end distribution function according to the Marrucci-Grizzutti segmental stretch model.

Reference to (6) and the original Doi-Edwards model shows that whenever chain stretching is significant, the principal stress difference will *exceed* the plateau modulus, G_N. This simple criteria can be readily extended to the birefringence by using the stress optical rule and provides a simple, useful and experimentally observable signature of chain stretch.

Steady State Planar Flow Simulations

We begin by examining the model predictions in steady flow for a specific λ value and light levels of entanglement, N=15 and n_t=1000. Figures 2 and 3 display the behaviour of several key functions versus dimensionless stretch rate, $\lambda^{1/2}\dot{\gamma}T_{Rouse}$. The motivation for scaling the rate of deformation with $\lambda^{1/2}\dot{\gamma}T_{Rouse}$ will become transparent when we examine the high deformation limit of the system.

Steady viscosity predictions show strong departures from the empirical Cox-Merz rule when chain stretching becomes significant. This result is a common feature for all flows we have examined with a finite extensional component, λ>0. In this sense, shear flow is seen to be a special case for the model. The problem of excessive shear thinning in the original model carries over into the segmental stretch version. One plausible means to correct the excessive shear thinning would be to incorporate a structure (orientation) dependent tube diameter into the model. Such a notion is not new, having been explicitly described in the original Doi-Edwards model for rod-like polymers[6]. Counterintuitively, incorporation of this extra relaxation mechanism will actually raise the viscosity level and reduce the degree of shear thinning.

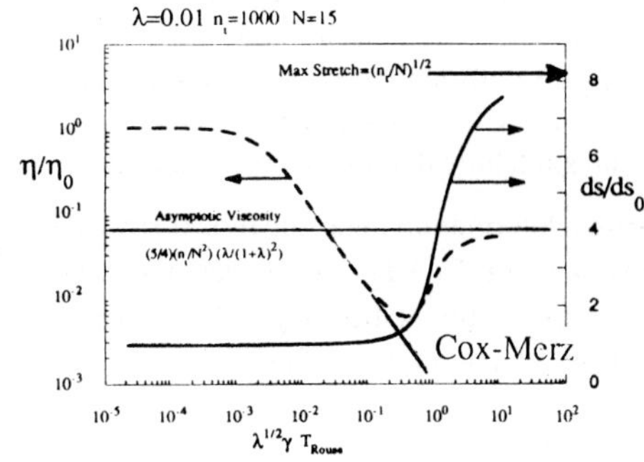

Figure 2) Dimensionless viscosity and chain stretch in steady flow. λ=0.01, nearly simple shear and N=15, n_t=1000.

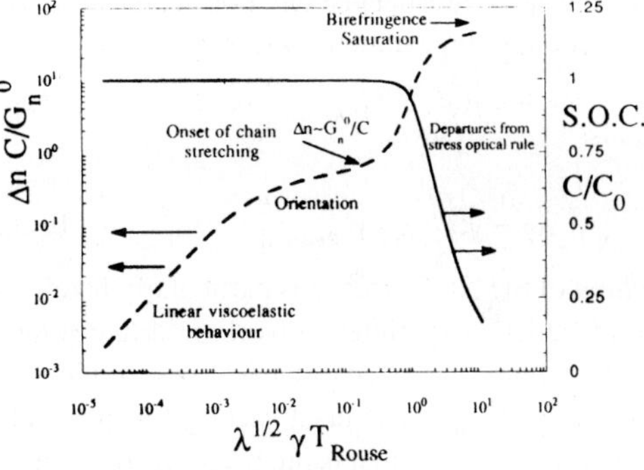

Figure 3) Dimensionless birefringence and stress optical coefficient for the same conditions as figure 2.

Figure 3 also shows the predicted stress optical coefficient for the steady flow and demonstrates that it is possible to have significant chain stretch without producing measurable changes in the stress optical coefficient.To have discernible departures from the stress optical rule, the fractional segmental extension must exceed ~1/3. Of course, any departure of the stress optical coefficient from its Gaussian value is a clear signature of chain stretching.The current model allows for nonuniform stretch through (5). However, the highly nonlinear nature of the inverse Langevin spring law is such that at high deformation rates the chain is nearly *uniformly* stretched, except for a small boundary layer near the end of the tube. This situation does not necessarily persist in transient flow situations.

As the Deborah number is steadily increased, the chain first orients without stretching, ultimately saturating at χ_∞. Only after the chain has completely oriented does it then begin to stretch. This sequential process in steady flow is a natural consequence of the wide separation of time scales for the retraction and reptation processes. The clear separation leads to a distinction between the stress generation mechanisms. In the intermediate flow region, 'orientational stress' is generated via a bias in the distribution function of chain segments that are equilibrated within their tube. At higher deformation rates, the orientation is saturated and defines the directional character of the stress while chain stretching determines the magnitude.

At very high deformation rates, the model simplifies to the extent that analytical results can be generated. For example, the asymptotic orientation angle, χ_∞, and order parameter tensor, $\underset{\sim}{S}^\infty$, can be shown to be.

$$\underset{\sim}{S}^\infty = \frac{1}{1+\lambda}\begin{bmatrix} 1 & \lambda^{1/2} \\ \lambda^{1/2} & \lambda \end{bmatrix} \quad \chi_\infty = \frac{1}{2}\operatorname{atan}\left(\frac{2\sqrt{\lambda}}{1-\lambda}\right) \qquad \text{Eq.(8)}$$

With these asymptotic values, the governing equations (2-5) simplify such that simple scaling rules can be deduced. For example, it can readily be shown that in this limit the chain stretch and birefringence functions collapse to a *universal* function of three dimensionless variables.

$$s = f\left(\frac{s_0}{L_0}, \lambda^{1/2}\dot{\gamma}T_{Rouse}, \frac{N}{n_t}\right) \qquad \text{Eq.(9)}$$

$$\Delta n = g\left(\frac{s_0}{L_0}, \lambda^{1/2}\dot{\gamma}T_{Rouse}, \frac{N}{n_t}\right) \qquad \text{Eq.(10)}$$

The scaling of the birefringence is particularly noteworthy in light of Fuller's experimental work on dilute solutions of flexible polymers in steady planar flows[2]. Fuller found the same $\lambda^{1/2}\dot{\gamma}\,T_{Rouse}$ universal scaling for the birefringence of dilute solutions that is predicted by the segmental stretch model. Thus, the predicted dynamics of highly entangled systems in the limit of high deformation rates are the *same* as that observed in dilute solution. This apparent paradox can be understood in light of the Bueche-Ferry hypothesis[3], which states that for short time scales, i.e. high deformation rates, the effect of the tube constraint is essentially lost and an entangled chain essentially behaves as a free Rouse chain. Thus, the tube model appears to have the correct asymptotic high deformation rate behaviour.

It is instructive to examine the arguments of the universal functions (9-10) in light of the frame indifferent flow classification scheme for planar flows advanced by Drout and Astarita[4]. This scheme essentially classifies planar flows according to the parameters $\lambda^{1/2}\dot{\gamma}\,T_{Rouse}$, an effective stretch rate, and $(1-\lambda)/(1+\lambda)$, a deformation 'efficiency' parameter. The extended tube model can be easily classified with this scheme at high deformation rates. The $\lambda^{1/2}\dot{\gamma}\,T_{Rouse}$ dependence is explicit in (9-10) and the λ dependence is explicit in (8). At more modest deformation rates, the same classification scheme can be used although the functional relationship is less transparent. The steady state model predictions can also be shown to be consistent with the functional form proposed in Larson's general continuum, constant stretch history analysis and classification for planar flows, as of course they must be[5].

Finite asymptotic viscosities can be calculated analytically for all flows with non-zero extensional components.

$$\frac{\eta}{\eta_0} = \frac{5}{4}\left(\frac{n_t}{N^2}\right)\left(\frac{\lambda}{(1+\lambda)^2}\right) \qquad \text{Eq.(11)}$$

Utilizing a finitely extensible spring removes the pathological, singular behaviour at $\lambda^{1/2}\dot{\gamma}\,T_{Rouse} \sim 1$ of the viscosity observed with a Gaussian spring. This allows for the examination of the model at higher Deborah number where significant amounts of segmental stretch are predicted, i.e. the precise region where the model is meant to apply. Without the nonlinear spring, this interesting flow regime is inaccessible for any flow with an extensional component.

Summary

We have briefly examined several properties of the segmental stretch model in steady planar flow. A more detailed account of this work as well as studies of transient planar flows, oscillatory flows and strain recovery will be given in subsequent publications.

References

1) G. Marrucci and N. Grizzutti, Gazzetta Chimica Italiana **118**, 179(1988) ; G. Marrucci and N. Grizzutti, in P.H.T. Uhlherr (Ed.) Xth International Congress on Rheology, Sydney (1988)

2) G.G. Fuller and L.G. Leal, Rheo. Acta **19**, 580 (1980)

3) M. Doi, J.N.N.F.M. **23**, 151(1987)

4) R. Drout, Arch. Mech. Stasow **28**, 189 (1976) ; G. Astarita, J.N.N.F.M. **6**, 69(1979)

5) R.G. Larson, Rheo. Acta **24**, 443(1985)

6) M. Doi and S.F. Edwards, J. Chem. Soc. Faraday trans. II **74**, 568(1978) ; M. Doi and S.F. Edwards, J. Chem. Soc. Faraday trans. II **74**, 918(1978)

Theoretical and Applied Rheology, edited by P. Moldenaers and R. Keunings
Proc. XIth Int. Congr. on Rheology, Brussels, Belgium, August 17-21, 1992

Evaluation of an Upper Convected Maxwell Model for Melts in Large Amplitude Oscillatory Shear

by
J. G. Oakley and A. J. Giacomin
Texas A&M University
Rheology Research Laboratory
Mechanical Engineering Dept.
College Station, TX 77843-3123

INTRODUCTION

The introduction of structure dependent coefficients and a kinetic rate equation into the upper convected Maxwell model yields a versatile model that can be used to predict stress responses to many transient deformations. Different kinetic rate equations have been proposed by several authors [1,2,3,4,5] for the upper convected Maxwell model. Tsang and Dealy [6] found the model by Acierno *et al.* [1] to be inadequate for a polystyrene melt in large amplitude oscillatory shear. Liu *et al.* [5] report good agreement between their model and the behavior of a polymer solution in large amplitude oscillatory shear, but here the stress response contains no higher harmonics. The authors are not aware of any constitutive theory which accurately predicts the higher harmonic content for a polymer melt in large amplitude oscillatory shear.

Recently, Moldenaers developed a kinetic rate equation [4] that successfully predicted the stress response of a melt to tensile stress growth. The transient deformation that is studied here is large amplitude oscillatory shear. Large amplitude oscillatory shear is an excellent deformation to study nonlinear viscoelastic properties since the amplitude of deformation and the time scale of the experiment can be varied independently [7]. Moldenaers' kinetic rate equation was evaluated in large amplitude oscillatory shear and the model predictions showed not enough nonlinearity and too little lost work. Other kinetic rate equations were evaluated [1,2,3,4,5] and the one that best models large amplitude oscillatory shear is a kinetic rate equation developed by Liu *et al.* [5].

THE MODEL

The strain for oscillatory shear is defined as:

$$\gamma(t) = \gamma_0 \sin(\omega t), \tag{1}$$

where γ_0 is the strain amplitude and ω is the radian frequency. The model proposed by Liu *et al.* comprises:

$$\boldsymbol{\tau} = \sum_{i=1}^{n} \boldsymbol{\tau}_i \tag{2}$$

$$\frac{1}{G_i}\boldsymbol{\tau}_i + \lambda_i \frac{\delta}{\delta t}\left(\frac{1}{G_i}\boldsymbol{\tau}_i\right) = 2\lambda_i \mathbf{D} \tag{3}$$

$$G_i = G_{0i} x_i \tag{4}$$

$$\lambda_i = \lambda_{0i} {x_i}^{1.4} \tag{5}$$

$$\frac{dx_i}{dt} = \frac{k_1(1-x_i)}{{\lambda_i}^m} - \frac{k_2 x_i}{{\lambda_i}^m}\left(\frac{\mathrm{tr}\boldsymbol{\tau}_i}{2G_i}\right)^{m/2}. \tag{6}$$

where $\boldsymbol{\tau}$ is the extra stress tensor, $\mathrm{tr}\boldsymbol{\tau}_i$ is the trace of the ith spectral component of the extra stress tensor, G_i is the relaxation modulus at relaxation time λ_i, t is time, $\mathbf{D}$ is $\frac{1}{2}(\nabla v + \nabla v^T)$ ie., the rate of deformation tensor with velocity gradient ∇v, G_{0i} and λ_{0i} are the equilibrium values of G_i and λ_i, x_i is a scalar structural variable ranging from 0 to 1, and $\frac{\delta}{\delta t}$ is the contravariant convected derivative [1]:

$$\frac{\delta \boldsymbol{\tau}_i}{\delta t} = \frac{d\boldsymbol{\tau}_i}{dt} - \nabla v \cdot \boldsymbol{\tau}_i - \boldsymbol{\tau}_i \cdot \nabla v^T. \tag{7}$$

Equation (6) is the kinetic rate equation developed by Liu *et al.* [5] and generalized to this material objective form by Mewis and Denn [8]. The kinetic rate equation contains three parameters, k_1, k_2, and m. k_1 is a rate constant for the creation of entanglements (by thermal diffusion), k_2 is the rate constant for the destruction of entanglements due to shear, and m is an elastic parameter that is typically 0.8-0.9 for polymer melts [7]. The model is evaluated for the branched low density polyethylene IUPAC LDPE X. The relaxation spectrum used is given by Zülle *et al.* [9] and the model constants used are $k_1 = 0.55\ \mathrm{s}^{m-1}$, $k_2 = 0.4\ \mathrm{s}^{m-1}$, and $m = 0.85$. Equations (1)-(6) are solved numerically using a 6th order Runge-Kutta technique [10]. Each solution required approximately 20 seconds of CPU time on a CRAY YMP-2/116 supercomputer.

ANALYSIS

Previously, large amplitude oscillatory shear data have been analyzed with shear stress, σ, vs. shear rate, $\dot{\gamma}$, loops. The extent to which data can be analyzed in this manner is limited to qualitative observations, for example, an ellipse indicates linear behavior while a distorted ellipse indicates a deviation of linear behavior into the nonlinear regime. The shear stress response in the linear region is given by:

$$\sigma(t) = \sigma_0 \sin(\omega t + \delta). \tag{8}$$

There are two material properties in this stress response: σ_0 the stress amplitude and δ the phase angle (the amount the stress leads the strain). A nonlinear stress response is given by a Fourier series. For an isotropic liquid with fading memory, the shear stress is a sum of the odd harmonics:

$$\sigma(t) = \sum_{\substack{j=1 \\ \mathrm{odd}}}^{\infty} \sigma_j \sin(j\omega t + \delta_j). \tag{9}$$

The discrete Fourier transform (DFT) [11] of both the model predictions for the shear stress response and the experimental data is used to determine each σ_j and δ_j. In this manner, a quantitative comparison can be made between the model predictions and the experimental data.

RESULTS

Figures 1,2, and 3 show the model predictions compared with experimental data previously reported by Giacomin *et al.* [12]. Figure 1 is the traditional way of presenting shear stress data in large amplitude oscillatory shear [13]. Both the shape of the predicted loop and area of the predicted loop match well with the experimental loop. Since the predicted loop does not fall exactly on top of the experimental loop, figures 2 and 3 are necessary to clarify the discrepancy. Figure 2 is a frequency spectrum obtained from a DFT on the time domain data. The stress amplitudes of the first (fundamental) and third harmonics (σ_1 and σ_3) agree, while the model predicts a lower stress amplitude for the fifth and seventh harmonics. Two important features of figure 2 are that the model correctly predicts only odd harmonics and that the stress amplitude of each subsequent harmonic is less than the one before it.

Figure 3 is a new way to look at the phase angles. The 45° lines are lines of equivalence, that is, the model's phase angle predictions match the experimental values. The reason the axes range from 0 to 4π is to make clear the fact that angles close to zero are also close to 2π. This would be obscured if the axes only ranged from 0 to 2π. Figure 3 shows that the predicted phase angle of the fundamental (δ_1) matches exactly with the experimental δ_1. The Maxwell model predictions for δ_3, δ_7, and δ_9 all are quite close to the experimental values while δ_5 is the only phase angle that does not show good agreement. Similarly good agreement is found for experiments conducted at different strain amplitudes.

Figure 4 is a standard Pipkin diagram [7] with shear rate amplitude vs. radian frequency. For the construction of figure 4, σ_3 was calculated at frequency intervals of 2π rad/s and shear rate amplitude intervals of 2π rad/s. Altogether 200 model evaluations are represented, requiring 60 minutes of CRAY time. An interesting feature of figure 4 is the rate at which σ_3 increases which is represented by the spacing between contour lines. At frequencies above 18.85 rad/s, σ_3 is small (which means that the stress response is just outside the linear region) and the rate at which σ_3 is increasing is low; however, below 18.85 rad/s the rate at which σ_3 increases is high and fairly constant.

Figure 5 is a three dimensional plot of σ_3 as a function of frequency and strain amplitude. Frequency and strain amplitude were chosen as the independent axes because the features are more pronounced on this type of plot as compared with the traditional Pipkin type axes. Figure 5 shows that for a given frequency, σ_3 increases monotonically with increasing strain amplitude. However, for a given strain amplitude, σ_3 does not increase monotonically with increasing frequency. As the frequency is increased from 1 to 2 Hz, σ_3 increases, but from 3 to 4 Hz, σ_3 decreases as if going back towards the linear regime.

Figure 6 is a contour plot of G_d ($G_d \equiv \sigma_1/\gamma_0$) as a function of frequency and strain amplitude. G_d is the amplitude ratio and has traditionally been used as the only indicator of linearity, where G_d is a function of frequency but independent of strain amplitude [7]. This plot indicates linear viscoelasticity at strain amplitudes of 1 and frequencies greater than 5 Hz with the contour lines being nearly vertical. Another notable feature of figure 6 is the rate at which G_d increases is low in the lower right corner and increases as one moves to the upper left corner.

CONCLUSION

The upper convected Maxwell model with the kinetic rate equation proposed by Liu *et al.* (and generalized by Mewis *et al.*) accurately predicts the shear stress response to large amplitude oscillatory shear for molten IUPAC LDPE X. A quantitative approach to analyzing large amplitude oscillatory shear data can be used to compare model predictions to experimental data with a discrete Fourier transformation of the time domain data. A Pipkin diagram can now be quantitatively produced with the upper convected Maxwell model where in the past only qualitative generalizations could be made about certain regions of the diagram.

REFERENCES

1. D. Acierno, F. P. La Mantia, G. Marrucci and G. Titomanlio, *J. of Non-Newtonian Fluid Mech.*, Vol. **1** (1976) 125-145.
2. G. Marrucci, G. Titomanlio, and G. C. Sarti, *Rheol. Acta*, Vol. **12** (1973) 269-275.
3. J. Mewis and G. De Cleyn, *AIChE Journal*, Vol. **28**, No. **6** (1982) 900-907.
4. P. Moldenaers and J. Mewis, "Integration of Fundamental Polymer Science and Technology,"L. A. Kleintjens and P. J. Lemstra eds., Elsevier Applied Science Publishers, London, 1986, p. 351-355.
5. T. Y. Liu, D. S. Soong, and M. C. Williams, *J. Poly. Sci.*, Vol. **22** (1984) 1561-1587.
6. W. K.-W. Tsang and J. M. Dealy, *J. of Non-Newtonian Fluid Mech.*, Vol. **9** (1981) 203-222.
7. J. M. Dealy and K. F. Wissbrun, "Melt Rheology and its Role in Plastics Processing," Van Nostrand Reinhold, New York, 1990, p. 60,220,221.
8. J. Mewis and M. M. Denn, *J. of Non-Newtonian Fluid Mech.*, Vol. **12** (1983) 69-83.
9. B. Zülle, J. J. Linster, J. Meissner, and H. P. Hürlimann, *Journal of Rheology*, Vol. **31**, No. **7** (1987) 583-598.
10. User's Manual: "IMSL MATH/LIBRARY: FORTRAN Subroutines for Mathematical Applications", IMSL, Houston, TX (1989), p. 633.
11. R. W. Ramirez, "The FFT Fundamentals and Concepts", Prentice-Hall, New Jersey, 1985, p. 69.
12. A. J. Giacomin, T. Samurkas, and J. M. Dealy, *Poly. Eng. and Science*, Vol. **29**, No. **8** (1989) 499-504.
13. T.-T. Tee and J.M. Dealy, *Trans. Soc. Rheol.*, Vol. **19**, No. **4** (1975) 595-615.

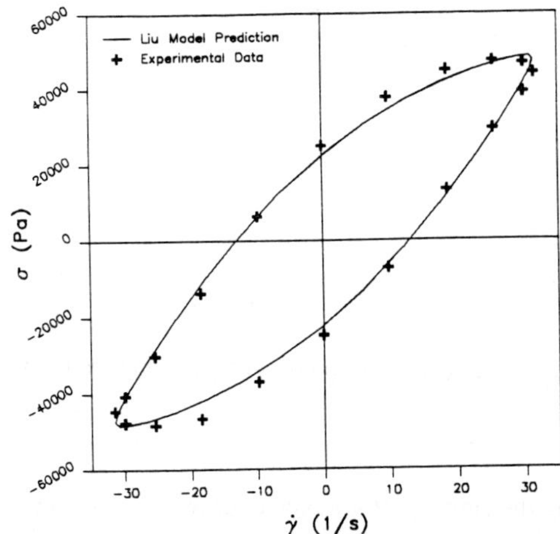

Fig. 1 Shear Stress vs. Shear Rate for IUPAC LDPE X at 150° C, for a Strain Amplitude of 5 and a Frequency of 1 Hz.

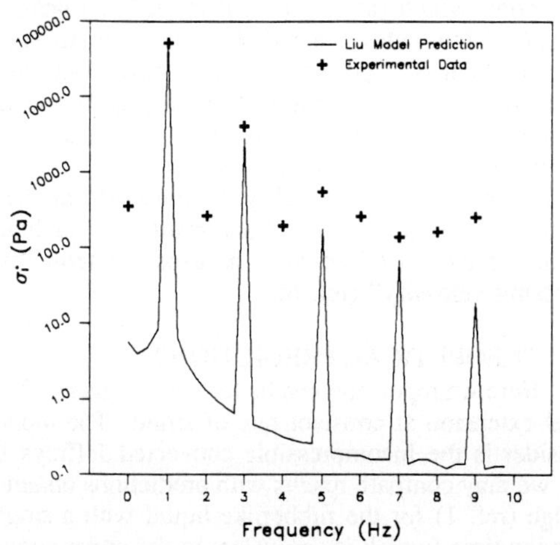

Fig. 2 Frequency Spectrum of Stress Amplitudes for IUPAC LDPE X at 150° C, for a Strain Amplitude of 5 and a Frequency of 1 Hz.

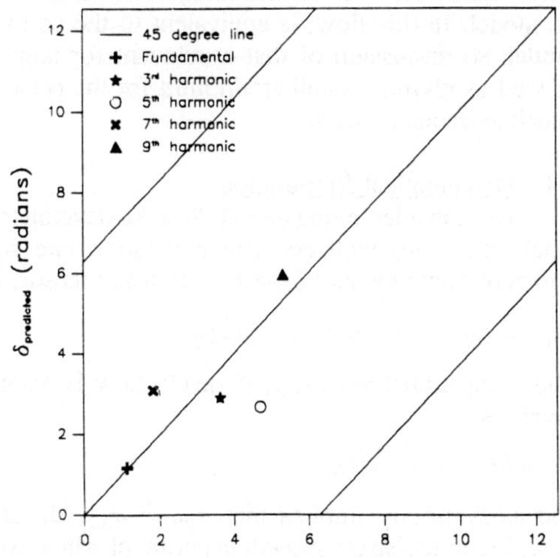

Fig. 3 Phase Angle Plot for IUPAC LDPE X at 150° C, for a Strain Amplitude of 5 and a Frequency of 1 Hz.

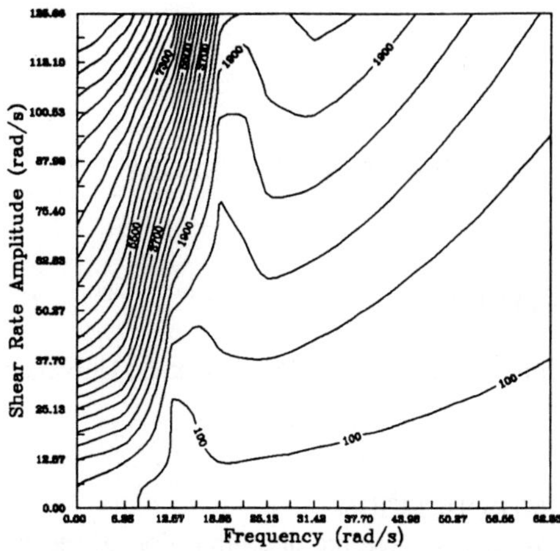

Fig. 4 Pipkin Diagram of the Third Harmonic's Stress Amplitude (in Pa) for IUPAC LDPE X at 150° C.

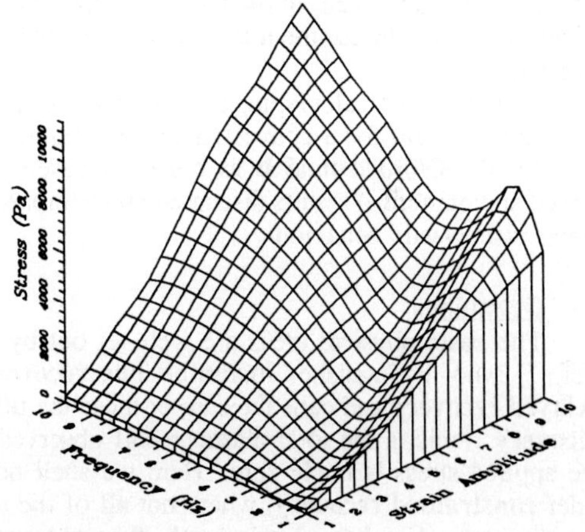

Fig. 5 Three Dimensional Plot of the Third Harmonic's Stress Amplitude as a Function of Strain Amplitude and Frequency for IUPAC LDPE X at 150° C.

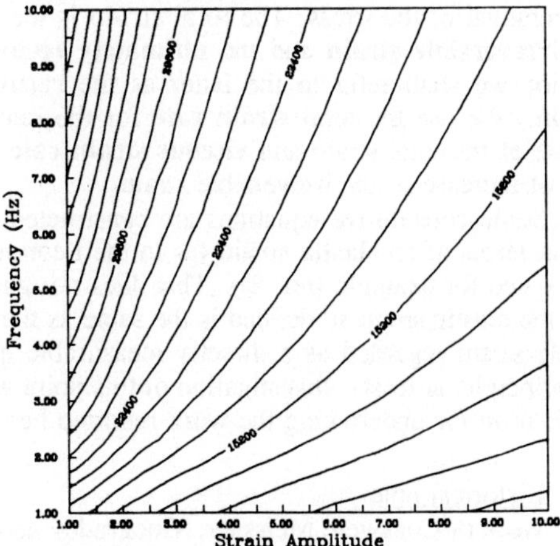

Fig. 6 Contour Plot of G_d (in Pa) as a Function of Strain Amplitude and Frequency for IUPAC LDPE X at 150° C.

Theoretical and Applied Rheology, edited by P. Moldenaers and R. Keunings
Proc. XIth Int. Congr. on Rheology, Brussels, Belgium, August 17-21, 1992

RECOVERABLE STRAIN IN THEORY

C.J.S. PETRIE

Department of Engineering Mathematics, University of Newcastle upon Tyne, Newcastle upon Tyne, NE1 7RU, (U K)

1. INTRODUCTION

We discuss here the decomposition of the strain in an elasticoviscous liquid into recoverable (elastic) and non-recoverable (irreversible or viscous) parts. We discuss use of ratios of transient stress to rate of strain (i.e. stress growth functions), in the context of the question of whether one should use the total strain rate or the viscous strain rate.

Equations for calculating the recoverable and non-recoverable strain for a convected Jeffreys liquid, which includes the Oldroyd fluid **B** as a special case, are presented and numerical results discussed and compared with linear viscoelastic behaviour.

1.1 Definitions

The basic physical ideas are well set out by Lodge (ref. 1) who distinguishes **instantaneous recovery** and **delayed recovery** and calls the sum of these the **ultimate recovery** – this is the total elastic recoil observed when the applied stress is made zero. Here we shall not consider **constrained recovery**, where not all of the applied stresses are reduced to zero but only **free recovery** following extensional flow. We note that while the stresses are zero, the extra-stresses are not (and we cannot control these directly). We define the irreversible part of the strain to be that which remains a very long time after the removal of the stress. The **total strain is** the sum of the **irreversible strain** and the **ultimately recoverable strain**; we shall refer to the latter as the **recoverable strain**. We use the term **strain rate** for the rate of increase of the total strain and **viscous strain rate** for the rate of increase of the irreversible strain.

Some constitutive equations are constructed explicitly in terms of an **elastic strain** (as in the Leonov equation – see for example (ref. 2)). This does not guarantee that the elastic strain so defined is the same as the recoverable strain regarded as a directly measurable quantity (in principle, at least). Investigation of this point was one motivation for undertaking the work reported here.

1.2 Historical note

Work of Cogswell, Meissner, Vinogradov and others in the period from 1968 to 1978 is reviewed in (ref. 3). There was some controversy concerning stress overshoot during the start-up of uniaxial extension at constant strain rate. Vinogradov's results (ref. 4) were originally presented in terms of the ratio of stress to viscous strain rate; Vinogradov called this the **"longitudinal viscosity"** (ref. 4). This led to false claims by others about stress overshoot, which (as is reported in (ref. 3)) were never made by Vinogradov himself. For the materials and conditions which he reported, the *stress* never went through a maximum. It is possible (ref. 5) to present the results of Meissner (ref. 6) in the same way and obtain maxima in the ratio of stress to viscous strain rate. Vinogradov's results can, even more readily, be presented as the ratio of stress to strain rate (ref. 5) and results are qualitatively similar to those of Meissner; he used the term **"tensile stressing viscosity"** (ref. 6).

2. THEORETICAL PREDICTIONS

Here we report the results for free recovery after uniaxial extension at constant rate of strain. The model we consider is the incompressible convected Jeffreys liquid and we may compare results with predictions obtained by Lodge (ref. 1) for the rubberlike liquid with a single relaxation time (which is equivalent to the upper convected Maxwell model, or to our model with zero retardation time – zero "solvent" contribution). The linear viscoelastic model, in this flow, is equivalent to the co-rotational model, so discussion of that is relevant for large strains as well as giving a small strain limit for the behaviour of elasticoviscous models.

2.1 Mathematical formulation

We consider elongational flow (axisymmetric uniaxial extension) with constant elongation rate $\dot{\varepsilon}_0$. The non-zero components of the rate of strain tensor, **D**, are

$$D_{xx} = \dot{\varepsilon}_0 \ , \quad D_{yy} = D_{zz} = -\tfrac{1}{2}\dot{\varepsilon}_0 \ . \tag{1}$$

The elongational viscosity, if steady flow is obtained, is given by

$$\eta_E = (\sigma_{xx} - \sigma_{zz})/\dot{\varepsilon}_0 \tag{2}$$

and axisymmetry implies that $\sigma_{yy} = \sigma_{zz}$. In unsteady flow, i.e. in the stress growth experiment where we apply the rate $\dot{\varepsilon}_0$ for times $t \geq 0$, equation (2) gives the usual stress growth function. We shall calculate the viscous strain rate, $\dot{\varepsilon}_v$, and use that to get a second stress growth function, of the type favoured by Vinogradov (refs. 4-5).

The constitutive equation we study here is

$$\sigma = \sigma^{(p)} + \sigma^{(s)} \ , \tag{3}$$

where $\boldsymbol{\sigma}^{(s)}$ is a "solvent" contribution to the extra-stress tensor, given by $\boldsymbol{\sigma}^{(s)} = 2\eta^{(s)}\mathbf{D}$, and $\boldsymbol{\sigma}^{(p)}$ is a "polymer" or "particle" contribution which is given by a Maxwell model with a generalized convected derivative,

$$\sigma^{(p)} + \lambda \overset{\square}{\sigma}{}^{(p)} = 2\eta^{(p)}\mathbf{D} \ , \tag{5}$$

where λ is the relaxation time and $\eta^{(p)}$ is the "polymer" viscosity. The alternative form of this, written as a convected Jeffreys model, is

$$\sigma + \lambda \overset{\square}{\sigma} = 2\eta \left(\mathbf{D} + \Lambda \overset{\square}{\mathbf{D}}\right) \ . \tag{6}$$

Here the overall viscosity is $\eta = \eta^{(s)} + \eta^{(p)}$ and the retardation time is $\Lambda = \lambda\eta^{(s)}/\eta$.

We define the generalized convected derivative by

$$\overset{\square}{\boldsymbol{\sigma}} = \overset{\nabla}{\boldsymbol{\sigma}} + \xi(\boldsymbol{\sigma}\cdot\mathbf{D} + \mathbf{D}\cdot\boldsymbol{\sigma}) \ ; \tag{7}$$

see (ref. 3), for example. The parameter ξ gives the upper convected derivative for $\xi = 0$ and the co-rotational derivative for $\xi = 1$. If we scale times by λ, extra-stresses by $\eta/((1-\xi)\lambda)$ and strains by $1/(1-\xi)$, we need not concern ourselves with ξ (as long as $\xi < 1$). The ratio of characteristic times, $\beta = \Lambda/\lambda$ is then the only important material parameter.

2.2 Computed results

The differential equations we have to solve are given in (ref. 3) with appropriate initial conditions (there are jumps in strain rate and extra-stress when the stress is removed). We define $\tau_0 = (1-\xi)\lambda)$ and use the notation $x = \tau_0\dot{\varepsilon}$, $y = \tau_0\sigma_{xx}/\eta$, $z = \tau_0\sigma_{yy}/\eta$, $\tau = t/\lambda$ (with dashes to denote differentiation with respect to τ). Then, for extension at constant dimensionless rate x_0, we have

$$y' = (2x_0 - 1)y + 2x_0(1 - 2\beta x_0) \ , \tag{8}$$
$$z' = 1(1 + x_0)z - x_0(1 + \beta x_0) \tag{9}$$

with initial conditions $y(0) = 2\beta x_0$ and $z(0) = -\beta x_0$ for a flow starting suddenly from rest at time zero. This has straightforward exponential solutions, and the total strain is $\varepsilon = x\tau/(1-\xi)$.

For the recovery starting at dimensionless time τ_1 we have $y = z$ (the applied stresses are zero and the extra-stress is isotropic) and

$$x' = x(-1 + \beta x - y)/\beta \ , \tag{10}$$
$$y' = -y - 2\beta x^2 \ , \tag{11}$$
$$\varepsilon' = x/(1-\xi) \ . \tag{12}$$

for $\tau > \tau_1$, with

$$x(\tau_1+) = x(\tau_1-) - \left(y(\tau_1-) - z(\tau_1-)\right)/(3\beta) \ , \tag{13}$$
$$y(\tau_1+) = \left(y(\tau_1-) + 2z(\tau_1-)\right)/3 \ . \tag{14}$$

2.2.1 Calculation of irreversible strain rate. The irreversible strain rate is obtained by a computational technique rather like that used in solving two-point boundary-value problems for ordinary differential equations. The strain during the recovery is a function of the time τ_1 at which the stress is removed and the subsequent time, τ, i.e. $\varepsilon_r(\tau_1,\tau)$ with $\tau > \tau_1$. We obtain a differential equation for $\partial\varepsilon_r(\tau_1,\tau)/\partial\tau_1$ by differentiating the recovery equations (10-12) with respect to τ_1. Then as $\tau \to \infty$ this gives the recoverable part of the strain rate. We subtract this from x_0 to obtain $(1-\xi)\lambda\dot{\varepsilon}_v$. This doubles the order of the system we have to solve, but the equations are not computationally demanding.

2.2.2 Results. Figure 1 shows the results presented as graphs of recoverable strain against total strain for elongation at constant dimensionless rate of strain, x_0, for two values of the ratio of retardation to relaxation time, β. Further computation indicates that the recoverable strain grows monotonically to a steady value which depends on β and x_0. Typically, for $x_0 = 0.4$ we obtain $(1-\xi)\varepsilon_r \to 0.504$ for $\beta = 0.1$ and $(1-\xi)\varepsilon_r \to 0.360$ for $\beta = 0.5$, both reached at a total strain of about $(1-\xi)\varepsilon = 14.0$.

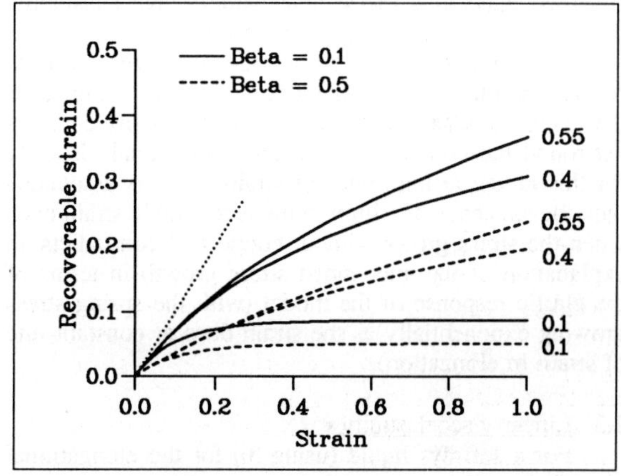

Figure 1: Recoverable strain, effect of parameters. *$\beta = 0.1$ and 0.5; $x_0 = 0.1$, 0.4 and 0.55 as indicated by the numbers next to the graphs. The dotted line shows the maximum recoverable strain (equal to the total strain).*

In Figure 2 we plot the two dimensionless stress growth functions, scaled by the total strain rate and by the viscous strain rate, against time. The smallest value, $x_0 = 0.02$ shows, in effect, the linear viscoelastic response. The strain rates $x_0 = 0.5$ and 0.55 give unbounded stress growth; in these cases, as is well known, there is no steady state and no elongational viscosity (properly defined).

2.2.3 Comments on the results. It was noticed during calculation of the recovery that the rate of strain went to zero and hence the observable recovery ceased much more quickly than the extra-stress components went to zero. We may infer from this, if our model is at all relevant, that allowing a sample of a molten polymer, for example, to recover at the melt temperature until motion ceases

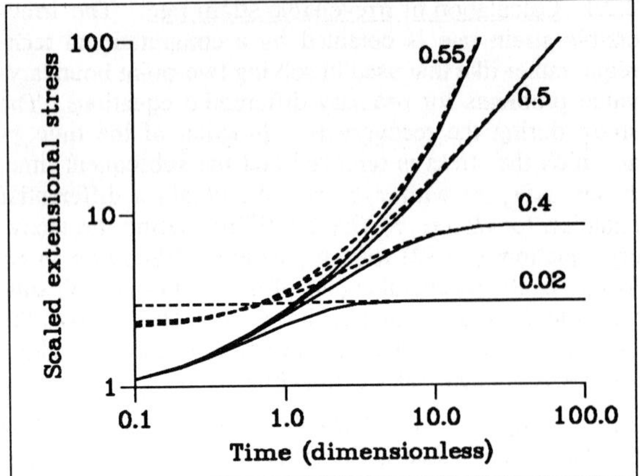

Figure 2: "Transient extensional viscosities".
Full lines: stress divided by the imposed (constant) strain rate. Dashed lines: stress divided by the (time-varying) viscous strain rate. $\beta = 0.1$; $x_0 = 0.02, 0.4, 0.5$ *and* 0.55 *as indicated by the numbers next to the graphs.*

and then quenching the sample would leave substantial frozen-in stresses in the quenched sample.

We note, also, that both viscosities grow without bound when the dimensionless rate of strain, x_0, exceeds 0.5. Thus the idea of the "irreversible" viscosity does not get round the non-existence of an elongational viscosity for this model at high rates of strain. This is associated with the existence of a limit to the recoverable strain even when the stress grows without bound and contradicts an explanation of the unbounded stress growth in terms of the elastic response of the model (with the spring stress growing exponentially as the strain does at constant rate of strain in elongation).

2.3 Linear viscoelasticity

For a Jeffreys liquid (using 3η for the elongational viscosity),

$$\sigma + \lambda\dot{\sigma} = 3\eta(\dot{\varepsilon} + \Lambda\ddot{\varepsilon}) \ , \tag{15}$$

the stress in elongation at constant strain rate $\dot{\varepsilon}_0$, which starts at time $t = 0$ in a stress-free material formerly at rest, is

$$\sigma = 3\eta\dot{\varepsilon}_0\left(1 - e^{-t/\lambda}\right) \ . \tag{16}$$

The recoverable strain is

$$\varepsilon_r = (1-\beta)\lambda\dot{\varepsilon}_0\left(1 - e^{-t/\lambda}\right) \tag{17}$$

and hence from the total strain, $\dot{\varepsilon}_0 t$ we may obtain the viscous strain and finally the viscous strain rate,

$$\dot{\varepsilon}_v = \dot{\varepsilon}_0\left(1 - (1-\beta)e^{-t/\lambda}\right) \ . \tag{18}$$

We note that, in terms of a spring and dashpot model, the strain in the Maxwell spring is $\lambda\dot{\varepsilon}_0(1 - e^{-t/\lambda})$, which is greater than the actual recoverable strain (for $0 < \beta \leq 1$). The extreme values $\beta = 0$ and $\beta = 1$ correctly give the behaviour of the Maxwell model ($\beta = 0$: instantaneous recovery, recoverable strain equal to the strain in the spring) and the Newtonian liquid ($\beta = 1$: no elastic behaviour).

Related results for the nonlinear (convected) model which we are studying, formulated as an integral equation, have been obtained by Lodge, McLeod and Nohel (ref. 7). In particular they show how the system for $\beta \sim 0$ is a singular perturbation of the rubberlike liquid ($\beta = 0$).

3. DISCUSSION

3.1 Interpretation

It has been suggested by Lodge (ref. 8) that use of models with any sort of basis in molecular modelling may lead to inappropriate interpretation of results. We take the view here that, at this stage in our investigations, it is useful to make the calculations we have done and accept the label (from Lodge) of "stress calculator" rather than constitutive equation. It does seem, from much other work, that the simple equations we use (whether we regard them as molecularly based or merely phenomenological) do have difficulty in modelling the behaviour of real liquids in situations where there is flow reversal. It is therefore prudent to take note of Lodge's warning and to be prepared to think much more carefully about elastic recovery predictions from our models.

3.2 Other model liquids

As with the convected Maxwell model, the Leonov model (with one relaxation time) predicts instantaneous recovery on the cessation of elongation at a constant strain rate. It is therefore not surprising that in this case the elastic strain, $\mathbf{C}_e^{-1}$, is identical with the strain which is recovered (ref. 2). If we introduce a second relaxation time, or a "solvent contribution" (retardation time), this is no longer the case. The fact that, in the simplest form of Leonov model, the extra-stress is proportional to the elastic strain is not enough to guarantee that the actual recovery when the stress is removed will be equal to that elastic strain. This may be explained by the fact that any delay in recovery (which arises when there is more than one mode of relaxation) gives the possibility of dissipation during the recovery and hence a loss of some of the elastic energy stored at the instant when the stress is removed.

REFERENCES

1. A S Lodge, Elastic Liquids, Academic Press, New York, 1964, pp. 129-131.
2. R G Larson, Constitutive Equations for Polymer Melts and Solutions, Butterworths, Boston, 1988, p. 159.
3. C J S Petrie, Elongational Flows, Pitman, London, 1979.
4. G V Vinogradov, V D Fikhman, B V Radushkevich and A Ya Malkin, J. Polym. Sci., A2 (1970) 657-678.
5. G V Vinogradov, personal communication, 1977.
6. J Meissner, Rheol. Acta, 10 (1971) 230-242.
7. A S Lodge, J B McLeod and J A Nohel, Proc. Roy. Soc. Edin., 80A (1978) 99-137.
8. A S Lodge, J. Rheol., 32 (1988) 93-95.

Theoretical and Applied Rheology, edited by P. Moldenaers and R. Keunings
Proc. XIth Int. Congr. on Rheology, Brussels, Belgium, August 17-21, 1992

Nonlinear Strain Measures for Extensional and Shearing Flows of Polymer Melts

J. Schaeffer and M.H. Wagner

Institut für Kunststofftechnologie, Universität Stuttgart
Böblinger Str. 70, D-7000 Stuttgart 1, Germany

1. INTRODUCTION

Recent years have seen an increasing activity devoted to modeling rheological behaviour of polymer melts in general flows. Single integral constitutive equations with a memory functional factorized in a time dependent memory function and a nonlinear strain dependent damping functional are able to describe the nonlinear material behavior of polymer melts. By comparing the nonlinear strain measures of the Doi-Edwards theory (ref.1) with data of general biaxial (uniaxial, planar, ellipsoidal, and equibiaxial) extensions of a polyisobutylene (PIB) melt, we derived a constitutive equation containing a molecular stress function (ref.2). The deformation invariant molecular stress function (MSF) models the increasing intensity of the mean field of surrounding chains with increasing deformation, i.e. it represents the ratio of the actual tensil force in the polymer chain to the hypothetical force in equilibrium conformation. Here, we extent this analysis to shear flows.

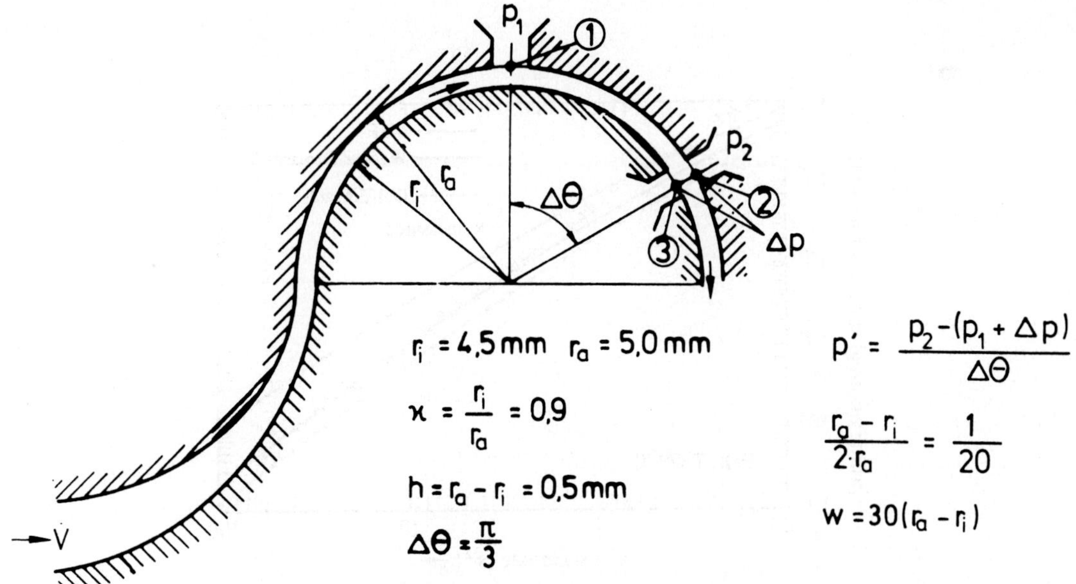

Figure 1: Curved-slit-capillary-rheometer

2. EXPERIMENTAL

We measured steady state shear viscosity and primary normal stress coefficient in the shear range from 0.3 to 50 s^{-1} by means of a curved slit capillary rheometer (ref.3). The state of stress of a viscoelastic material element at steady shear flow with curved streamlines allows for the simultaneous measurement of viscosity and primary normal stress coefficient (Fig. 1). The viscosity can be calculated from the volume flow rate and the pressure drop between the positions 1 and 2, the primary normal stress difference is proportional to the pressure difference between the position 2 and 3. Steady state shear viscosity and primary normal stress coefficient of the same PIB melt as used in the extensional experiments of Demarmels and Meissner (ref.4) are presented in Figs. 2 and 3 respectively.

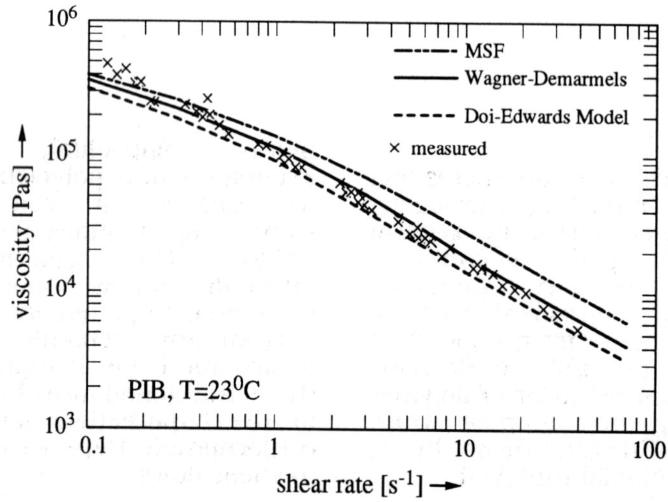

Figure 2: Measured and calculated viscosity as a function of the shear rate

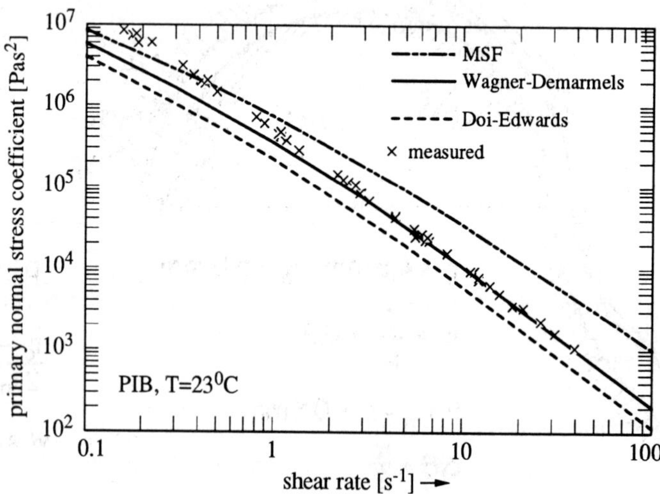

Figure 3: Measured and calculated primary normal stress coefficient as a function of the shear rate

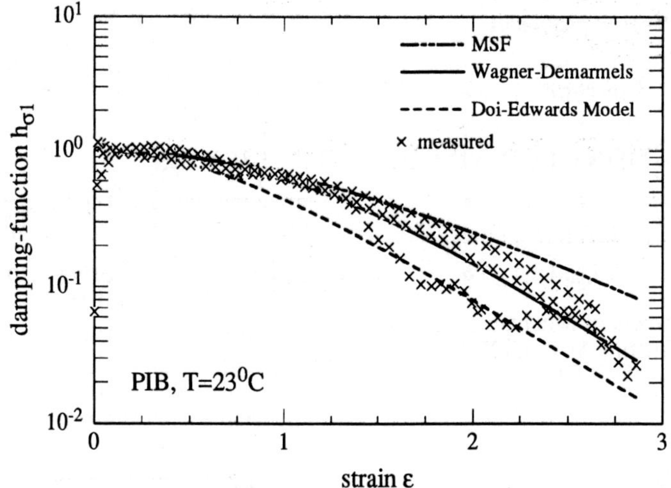

Figure 4: Measured and calculated damping function of planar extension data

3. CONSTITUTIVE DATA ANALYSIS

Predictions of the Doi-Edwards model (including the independent alignment assumption), of the molecular-stress-function (MSF) approach of Wagner and Schaeffer, and of the phenomenological approach of Wagner and Demarmels (ref.5) are compared with measurements of steady state viscosity and primary normal stress coefficient of PIB in Fig. 2 and 3.

It is evident from these figures that the Doi-Edwards model represents the lower limit of the data, while the MSF approach represents the upper limit. The phenomenological approach of Wagner and Demarmels gives a fair description of both steady state shear viscosity and primary normal stress coefficient. This is consistent with the nonlinear strain data in planar extensional flow of PIB, where the effective strain function $h_{\sigma 1}$ of the first normal stress difference σ_1 is eqivalent to the damping function in shear flow.

As seen from Fig. 4, the Doi-Edwards model represents the lower limit of the planar extensional strain data, while the MSF approach represents the upper limit, and the strain function of Wagner and Demarmels is a fair description of the data.

4. CONCLUSIONS

Our data on a well-characterized (linear) PIB melt seem to confirm the equivalence of the strain measures in shear flow (a rotational flow) and planar extensional flow (an irrotational flow). This is in contrast to the results of Dealy, Samurkas, and Larson (ref.6), and Khan and Larson (ref.7) on a branched polyethylene melt.

REFERENCES

1. M. Doi and S.F. Edwards, "Dynamics of concentrated polymer systems. Part 2 Molecular motion under flow", Faraday Trans. II 74, 1978, pp. 1802-1817.
2. M.H. Wagner and J. Schaeffer, Nonlinear strain measures for general biaxial extension of polymer melts", J. Rheol 36(1), 1992.
3. K. Geiger, "Bogenspalt-Kapillarrheometer zur Ermittlung viskoser und elastischer Eigenschaften von Kunststoffschmelzen", dissertation, University Stuttgart,1986.
4. A. Demarmels and J. Meissner, "Multiaxial elongation of polyisobutylene melt with various and changing strain rate ratios", Rheol. Acta 24, 1985, pp. 253-259.
5. M.H. Wagner and A. Demarmels, "A constitutive analysis of extensional flows of polyisobutylene", J. Rheol. 34, 1990, pp. 943-958.
6. T. Samurkas, R.G. Larson and J.M. Dealy, "Strong Extensional and Shearing Flows of a Branched Polyethylene", J. Rheol. 33(4), 1989, pp. 559-578.
7. S.A. Khan and R.G. Larson, "Step planar extension of polymer melts using a lubricated channel", Rheol. Acta 30, 1991, pp. 1-6.

Theoretical and Applied Rheology, edited by P. Moldenaers and R. Keunings
Proc. XIth Int. Congr. on Rheology, Brussels, Belgium, August 17-21, 1992

STRESS-STRAIN RELATIONS FOR VISCOELASTIC LIQUIDS

George B. Thurston

Rheology Laboratory, Department of Mechanical Engineering
The University of Texas at Austin, Austin,TX 78712 U.S.A.

1. INTRODUCTION

The ability of liquids to sustain shear stress over a limited period of time has many technical and industrial applications. Measurements of viscoelasticity under oscillatory shear flow reflect this ability. The viscosity and elasticity of polymer solutions and particle suspensions show a nonlinear dependence on shear rate and when these data are plotted as viscous stress and elastic stress versus amplitude of shear strain, the curves often show large changes in curvature near unit strain. The elastic stress passes through a maximum value which marks a limit to the ability of the liquid structure to sustain shear stress by stored elastic energy (per unit strain). This limit is designated by an elastic yield stress and occurs at a critical value of shear strain. These effects are analyzed using a Maxwell model to represent the multiple relaxation processes which are progressively truncated as the shear strain is increased.

2. MEASUREMENTS OF POLYMER SOLUTIONS

A series of measurements of the viscoelasticity of a solution of hydrolyzed polyacrylamide were made using a viscoelasticity analyzer which operates on the principle of oscillatory flow in a cylindrical tube (Vilastic Scientific, Inc.) (ref. 1). With this instrument a sequence of measurements can be performed over a range of frequencies while holding the shear rate at a constant value: at a low shear rate or low shear strain this provides viscoelastic values due to polymer molecules in their quiescent configurational state. Figure 1 shows measurements extending from 0.03 to 30 Hz. These extended curves are characteristic of a very broad spectrum of relaxation times and the limits of this spectrum are not reached in the three decades of frequency.

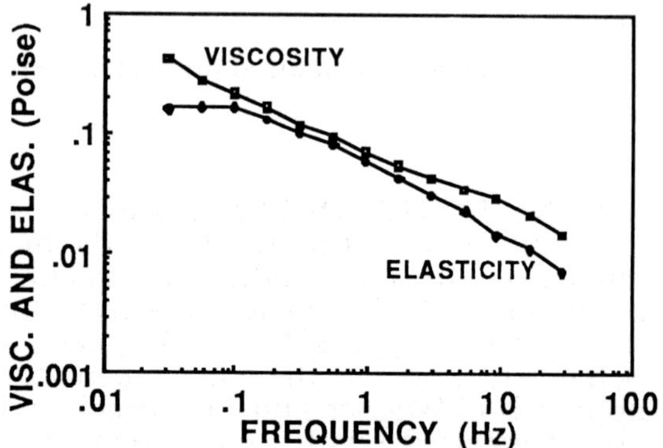

FIGURE 1. Viscosity η' and elasticity η'' for the polyacrylamide ($M_w = 2x10^7$D.) 62 ppm in distilled water. The shear strain was held at 0.5 and the temperature was 22°C.

Further measurements of the viscoelasticity were made at selected fixed frequencies while increasing the shear strain amplitude, forcing new polymer configurations. The measured viscosity η' and elasticity η'' are multiplied by the shear rate $\gamma\,'$ to obtain the viscous stress τ' and elastic stress τ''. The size of the shear strain is $\gamma = \gamma\,' / \omega$, where ω is the radian frequency of the test. Figure 2 shows the rms values of the stress components vs. the shear strain for fixed frequencies in the midrange of figure 1. It is seen that the stresses and strain are linearly related for strain < 1. For strain >1 the elastic stress curves pass through maximum values, marking the frequency dependent elastic yield stresses. The critical strains γ_c, coincident with the maxima, are in the range from 4 to 6.

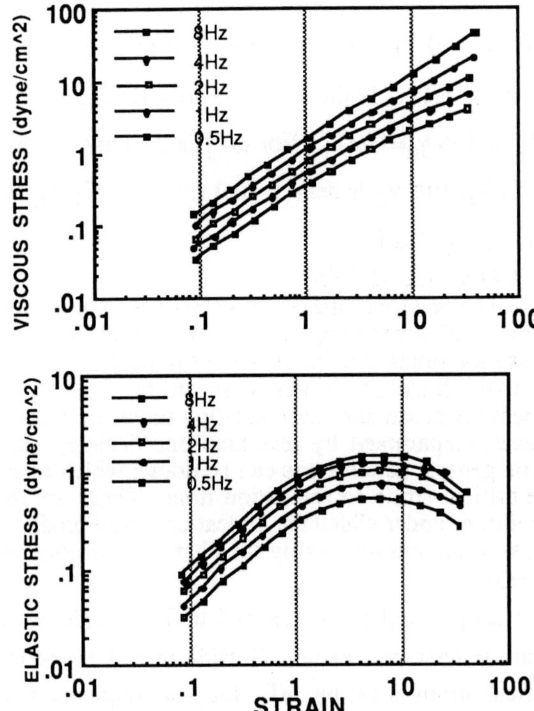

FIGURE 2. Viscous stress and elastic stress vs. oscillatory strain (rms values) for the polyacrylamide solution of figure 1. The curves are for fixed frequencies of 0.5, 1, 2, 4, and 8, Hz.

3. THE TRUNCATION MODEL

A generalized Maxwell model in which the effects of the relaxation processes are progressively truncated with increasing shear rate has been previously described (ref. 2) and has been fitted to measurements of macromolecular solutions (ref. 3) and to human blood (ref. 4). The model can also be used to describe stress-strain behavior like that shown in figure 2. Figure 3 shows the model as a collection of Maxwell elements and a terminating dashpot η_∞. Each element represents a relaxation process and the relaxation time of the p-th element is $T_p = \eta_p / k_p$. The longest (terminal) relaxation time is T_1 and the shortest is T_n.

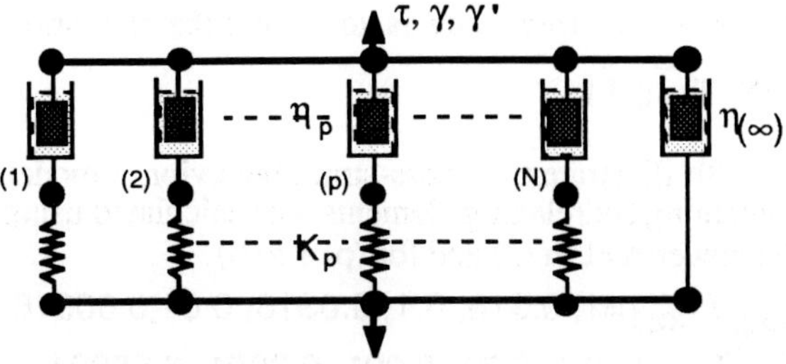

FIGURE 3. Maxwell truncation model for a viscoelastic liquid. The model consists of n-relaxing elements plus a dashpot η_∞. The springs k_p and dashpots η_p serve as energy storage and energy dissipative components.

The complex coefficient of viscoelasticty η^* is given by

$$\eta^* = \eta_\infty + \Sigma(\eta_p/(1 + i\omega T_p), \text{ for } p=1 \text{ to } n. \quad (1)$$

At low shear rates the components have values $\eta_{o,p}$ and $k_{o,p}$ but decrease in size with increasing shear rate to η_p and k_p while T_p remains constant. This truncation can be described using a truncation function $F(\gamma' T_p)$,

$$\eta_p = \eta_{o,p} F(\gamma' T_p)$$

and

$$k_p = k_{o,p} F(\gamma' T_p) \quad (2)$$

where the function may be of the form

$$F(\gamma' T_p) = [1 + (\gamma' T_p)2]^{-A}. \quad (3)$$

This function has the effect of removing the longer relaxation process first as the shear rate increases. To introduce the strain amplitude and to write the stress in dimensionless form, write

$$\gamma' T_p = \gamma (\omega T_1) (T_p/T_1) \quad (4)$$

and

$$\omega T_p = (\omega T_1) (T_p/T_1).$$

The shear stress $\tau^* = \tau' - i \tau'' = \eta^* \gamma'$ can now be written in dimensionless form as

$$[\tau'/(\eta_\infty T_1)] = \gamma (\omega T_1) \times$$
$$\{1 + \Sigma(\eta_p/\eta_\infty)/[1 + ((\omega T_1) (T_p/T_1))^2]^A\} \quad (5)$$

$$[\tau''/(\eta_\infty T_1)] = \gamma (\omega T_1) \times$$
$$\{\Sigma(\eta_p/\eta_\infty)(\omega T_1)/(T_p/T_1)/ [1 + ((\omega T_1)(T_p/T_1))^2]^A\} \quad (6)$$

with

$$(\eta_p/\eta_\infty) = (\eta_{o,p}/\eta_\infty)/[1 + (\gamma (\omega T_1) (T_p/T_1))^2]^A. \quad (7)$$

Using the dimensionless forms of (5) and (6) the stress components can be calculated for various arrays of Maxwell elements while specifying the dimensionless frequency as ωT_1 and the relaxation times as T_p/T_1.

To illustrate the stress-strain behavior, a model containing six relaxing elements was calculated using the power A=1 in (3) and for (p=1 to 6)

$\eta_{o,p}/\eta_{o,1}$=1, 0.316, 0.1, 0.0316, 0.01,0.00316

T_p/T_1 = 1, 0.1, 0.01, 0.001, 0.0001, 0.00001

and $(\eta_{o,1}/\eta_\infty)$ was selected to give a viscosity change $(\Sigma\eta_{o,p}+\eta_\infty)/\eta_\infty = 100/1$.

Figures 4 and 5 show the changes in (dimensionless) viscous stress and elastic stress vs. strain for values of ωT_1 ranging from 0.1 to 10^6. The range is sufficient in this example to show how the critical strain changes for $\omega T_1<<1$ to $\omega T_1>>T_1/T_n$.

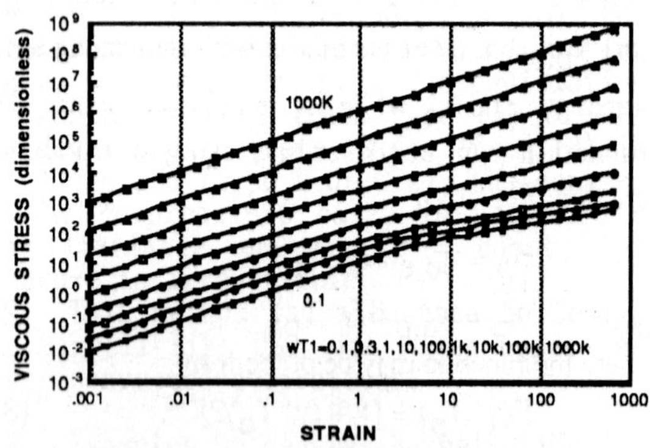

FIGURE 4. Viscous stress vs. strain, six element Maxwell Model. The curves are for values of ωT_1=0.1 to 10^6.

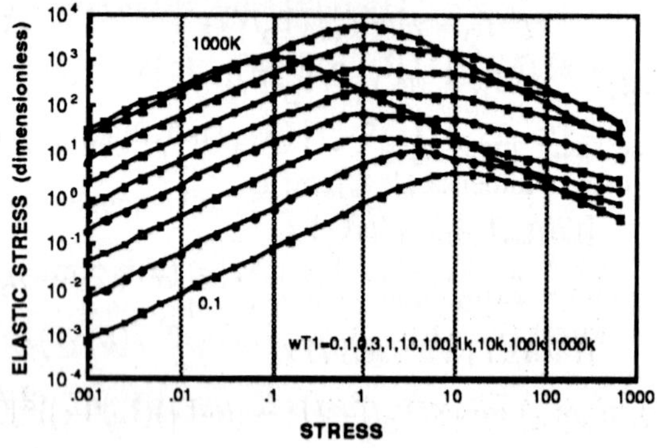

FIGURE 5. Elastic stress vs. strain for the model of figure 4.

The critical strains marking the maxima of elastic stress range form γ_c=0.1 (for $\omega T_1=10^6$) to γ_c=10 (for ωT_1=0.1), and $\gamma_c\approx1$ for most of intermediate values of ωT_1. In this case, the critical strain γ_c for $\omega T_1<<1$ is $\gamma_c=1/\omega T_1$. For ωT_1 in the range from 1 to T_1/T_n the γ_c is near 1, and for $\omega T_1>T_1/T_n$ the strain is $\gamma_c=1/\omega T_n$.

4. CONCLUSIONS

The precise shape of the stress-strain curves and the critical strains depend both on the spectrum of relaxation times and the manner of truncation of their relative influences. When the spectrum of relaxation process is broad and the relaxation times are closely spaced (separataed by less than one decade), then some general conclusions can be drawn which relate the critical strain to relaxation time. These are the conditions under which the truncation model produces stress-strain curves having the character of polymer solutions.

1. If ωT_1 is in the range from 1 to T_1/T_n the critical strain is near unit value. Outside of this range the critical strain is set by ωT_1 for low frequencies or ωT_n for high frequencies.

2. From a single plot of elastic stress vs. strain, the location of the critical strain g_c can be identified with an effective relaxation time $T_{eff}=\gamma_c/\omega$. If $\gamma_c\approx1$ then T_{eff} is in the midrange of the relaxation spectrum. If $\gamma_c>>1$, then T_{eff} is the longest relaxation time T_1. If $\gamma_c<<1$ then T_{eff} is the shortest relaxation time T_n.

REFERENCES

1.G. B. Thurston, Biorheology, 13 (1976) 191-199.
2. G. B. Thurston, J. Non-Newtonian Fluid Mech., 9 (1981)57-68.
3 G. B. Thurston and G. Pope, J. Non-Newtonian Fluid Mech., 9 (1981) 69-78.
4.G. B. Thurston, Biorheology, 16 (1979) 149-162.

Theoretical and Applied Rheology, edited by P. Moldenaers and R. Keunings
Proc. XIth Int. Congr. on Rheology, Brussels, Belgium, August 17-21, 1992

UTILIZATION OF STRESS GROWTH EXPERIMENTS TO DETERMINE STRAIN-DEPENDENT MATERIAL FUNCTIONS

D.C. Venerus[1] and B. Bernstein[2]

[1]Department of Chemical Engineering and [2]Department of Mathematics,
Illinois Institute of Technology, IIT Center, Chicago, IL 60616, USA

INTRODUCTION

Utilization of integral constitutive equations of the K-BKZ type requires the experimental determination of material functions which depend on strain and on time. In principle, once these functions have been determined in a particular flow field from step strain, stress relaxation experiments, the stress tensor in that flow field for any other strain history can be predicted by the constitutive equation. These functions can, with relative ease, be determined in simple shear flows. This is not the case, however, for extensional flows because of the difficulty in making step strain measurements. To overcome this problem, a number of investigators (refs.1-4) have used integrated forms of the K-BKZ model and extensional stress growth data at constant extensional strain rate to determine the strain-dependent, or damping, function. In the present study, the implications and consequences of using this procedure are examined. It is demonstrated, using simple shear flow data on a LDPE melt, that the damping function obtained by this technique is found to depend on strain rate and, furthermore, is not in agreement with the damping function obtained from single-step strain experiments.

THEORETICAL DEVELOPMENT

Several rheological constitutive equations for viscoelastic fluids have the following form:

$$\mathbf{S}(t) = \int_0^\infty \{\phi_1 \mathbf{C}^{-1}(t-s) + \phi_2 \mathbf{C}(t-s)\} ds \qquad (1)$$

In eq.(1), S(t) is the deviatoric stress tensor at the present time, t, C(t-s) is the Cauchy-Green strain tensor relative to t, and s is the backward-running time. The material-dependent functions: $\phi_1 = \phi_1[s, I_1(s), I_2(s)]$ and $\phi_2 = \phi_2[s, I_1(s), I_2(s)]$ depend on elapsed time, s, and on strain through the invariants of C(t-s) which are given by the following:

$$I_1(s) = tr[\mathbf{C}^{-1}(t-s)] \qquad (2)$$

$$I_2(s) = tr[\mathbf{C}(t-s)] \qquad (3)$$

Equation (1) is due to Rivlin and Sawyers (ref.5). If the functions ϕ_1 and ϕ_2 are derived from a strain energy function, then eq.(1) represents the K-BKZ theory (refs.6,7). Bernstein has shown (ref.8) that if the material functions are determined from step strain experiments, then eq.(1) can be used to predict to the stress in any other deformation history.

For many fluids, it is possible to write the material functions as products of time- and strain-dependent functions, i.e.

$$\phi_i[s, I_1, I_2] = \mu(s) h_i[I_1, I_2] \qquad (4)$$

where i=1,2. The time-dependent function $\mu(s)$ is related to the linear viscoelastic shear stress relaxation modulus, G(t), by

$$\mu(s) = -\frac{dG(s)}{ds} \qquad (5)$$

and the strain dependent functions h_i are referred to as damping functions (ref.9). Substitution of eqs.(4) in eq.(1) gives the factored K-BKZ model:

$$\mathbf{S}(t) = \int_0^\infty \mu(s) \{h_1 \mathbf{C}^{-1}(t-s) + h_2 \mathbf{C}(t-s)\} ds \qquad (6)$$

The Wagner model (ref.9) is a special case of eq.(6) where $h_2 = 0$ and is essentially identical to eq.(6) in simple shear flows (if N_2 is not considered) and in uniaxial extension flows. For this investigation, we shall refer to any constitutive equation of the form given by eq.(6) as a factored K-BKZ type model.

It is important to note that the validity of eq.(4),

the time-strain factorability assumption, can be determined only by consideration of experimental data from step strain experiments as is discussed below. Furthermore, if time-strain factorability is shown to be valid for one type of flow (i.e., simple shear) then, according to eq.(6), it must be valid for any other type of flow (i.e. uniaxial extension).

We now consider the predictions of eq.(6) for simple shear flows. For a step strain history with shear strain γ, we have from eq.(6) for the shear stress $\sigma(\gamma,t)$

$$\frac{\sigma(\gamma,t)}{\gamma} = G(\gamma,t) = h(\gamma^2)\,G(t) \tag{7}$$

where

$$h(\gamma^2) = h(I_1,I_2) = h_1(I_1,I_2) = h_2(I_1,I_2) \tag{8}$$

The first equality in eq.(8) is valid since $I_1 = I_2 = 3 + \gamma^2$ in simple shear flow and the second equality can be written if N_2 is not to be considered as is the case here. The time-strain factorability assumption will be valid for a particular fluid if the ratio $G(\gamma,t)/G(t)$ is independent of time. If this is the case, then eq.(7) provides a direct method of determining $h(\gamma^2)$, the strain-dependent material function in simple shear flow, from shear stress data from step strain experiments. Hence, with $h(\gamma^2)$ and $\mu(s)$ determined, the constitutive equation given by eq.(6) can be used to predict the stress, in simple shear flow, for any other strain history.

For the stress growth at constant shear strain rate, $\dot{\gamma}$, history, the shear stress $\sigma^+(\dot{\gamma},t)$ from eq.(6) is given by

$$\frac{\sigma^+(\dot{\gamma},t)}{\dot{\gamma}} = \int_0^t \mu(s)\,h(\dot{\gamma}^2 s^2)\,s\,ds + tG(t)\,h(\dot{\gamma}^2 t^2) \tag{9}$$

From eq.(9), it is possible to derive an expression for the damping function (ref.10):

$$h(\gamma^2) = \frac{1}{\dot{\gamma}t}\left\{\frac{\sigma^+(\dot{\gamma},t)}{G(t)} - \int_0^t \sigma^+(\dot{\gamma},\tau)\,\frac{\mu(\tau)}{G(\tau)^2}\,d\tau\right\} \tag{10}$$

where $\gamma = \dot{\gamma}t$. Equation (10) provides a method for determining $h(\gamma^2)$ which requires shear stress growth experimental data, $\sigma^+(\dot{\gamma},t)$, rather than step strain data. *If eq.(10) and $\sigma^+(\dot{\gamma},t)$ data are to be used to obtain $h(\gamma^2)$, it is important to note that this technique implicitly assumes that constitutive equations of the K-BKZ type (eq.(6)) are able to describe the behavior of the fluid in the stress growth at constant shear strain rate flow.* In reality, eq.(10) represents a consistency test for the factored K-BKZ model for the shear stress in the step strain and stress growth at constant strain rate flows. The function $h(\gamma^2)$ determined from step strain data and eq.(7) can be compared with the function $h(\gamma^2)$ determined from $\sigma^+(\dot{\gamma},t)$ and eq.(10). If the functions $h(\gamma^2)$ determined from eq.(10) over a wide range of strain rates, $\dot{\gamma}$, are in agreement with the $h(\gamma^2)$ from eq.(7), then one could conclude, since the K-BKZ model exactly describes step strain flows, that the factored K-BKZ constitutive equation was able to accurately describe shear stress behavior in the stress growth at constant strain rate flow. If there is a lack of agreement between the damping functions obtained from eq.(7) and eq.(10), then, one could conclude that the K-BKZ model was unable to describe the shear stress growth behavior.

RESULTS AND DISCUSSION

Clearly, if one is interested only in simple shear flows, the simplest and most direct method of obtaining the damping function, $h(\gamma^2)$, is to utilize eq.(7) and shear stress data from step strain tests. Data from the stress growth at constant strain rate experiments could then be collected and compared to the predictions of K-BKZ type models (refs.3,9). However, in the present study, we propose to use $\sigma^+(\dot{\gamma},t)$ data and eq.(10) to obtain $h(\gamma^2)$ and compare these results to the damping function obtained from step strain data and eq.(7). With this procedure, we can determine the suitability of this approach for obtaining strain-dependent material functions, and test K-BKZ type models in stress growth at constant strain rate flows.

A LDPE melt, the same material considered by Samurkas et al. (ref.3), known as IUPACX was used in this study. All experiments were carried out on an RMS-805 at 125°C using the cone and plate geometry. This material was found, in both ref.3 and our lab, to be described by the time-strain factorability assumption. The damping function for this material, measured using a sliding plate apparatus in ref.3, was obtained from step strain experiments and was fit by the expression:

$$h(\gamma^2) = 0.57\exp[-0.29\sqrt{\gamma^2}] + 0.43\exp[-0.095\sqrt{\gamma^2}] \tag{11}$$

The damping function obtained from step strain experiments carried out in our lab is in excellent agreement with the function given in eq.(11). Stress growth experiments to obtain $\sigma^+(\dot{\gamma},t)$ data were also carried out over a wide range of strain rates. Several runs at each strain rate were carried out, in some cases with cone and plate fixtures having different dimensions, and all runs were within 5% of average values.

Equation (10) was evaluated using standard numerical techniques and the stress growth data mentioned above. The damping functions from this method at three strain rates: 0.10, 1.0 and 10 sec^{-1}, are shown in figure 1 along with the damping function given by eq.(11). Clearly, the damping functions at these three strain rates do not coincide and, furthermore, they do not, in general, agree with the damping function obtained from the step strain data. Results similar to those in figure 1 were found for a concentrated polystyrene solution (ref.10). These results indicate that the use of stress growth data and an integrated form of the K-BKZ model, eq.(10), does not provide a unique damping function which is independent of strain rate and which is in agreement with the true damping function obtained from step strain tests.

Similar results have been found in studies were extensional flows were considered and integrated forms of K-BKZ type models, analogous to eq.(10), were used in conjunction with stress growth data from constant strain rate experiments (refs.1-4). Damping functions obtained from stress growth data appeared to depend on the extensional strain rate in a manner similar to that shown in figure 1. In two studies (refs.2,3) where true damping functions (obtained from step strain data) were available, significant departures between the true damping function and the stress-growth determined damping functions were observed.

CONCLUSIONS

As was mentioned earlier, using stress growth data to determine strain-dependent material functions actually represents a consistency relation test of K-BKZ type models. Hence, in the studies where a unique damping function was not obtained using this technique, one might conclude that this represents a failure of K-BKZ type models in stress growth tests. The reliability of this technique to provide an accurate representation of strain-dependent material functions is clearly in question. This is especially true when one considers the fact that this method is used exclusively for extensional flows where true damping functions are not usually measured. Furthermore, extensional flows at constant strain rate have been termed "strong flows" (ref.3), and strong flows are believed to be the most severe tests of *proposed* constitutive equations for viscoelastic fluids.

REFERENCES

1. M.H. Wagner, J. Non-Newt.Fluid Mech., 4 (1978) 39-55.
2. P.J.R. Leblans, J. Sampers and H.C. Booij, J. Non-Newt.Fluid Mech., 19 (1985) 185-207.
3. T. Samurkas, R.G. Larson and J.M. Dealy, J. Rheol., 33 (1989) 559-578.
4. M.H. Wagner and A. Demarmels, J. Rheol., 34 (1990) 943-958.
5. R.S. Rivlin and K.N. Sawyers, Ann.Rev.Fluid Mech., 3 (1971) 117-146.
6. B. Bernstein, E. Kearsley and L.J. Zapas, Trans. Soc. Rheol., 7 (1963) 391-410.
7. A. Kaye, College of Aeronautics, Cranfield, Note No. 134 (1962).
8. B. Bernstein, Acta Mech., 2 (1966) 329-354.
9. M.H. Wagner, Rheol.Acta, 15 (1976) 136-142.
10. E.V. Menezes, J. Non-Newt.Fluid Mech., 7 (1980) 45-62.

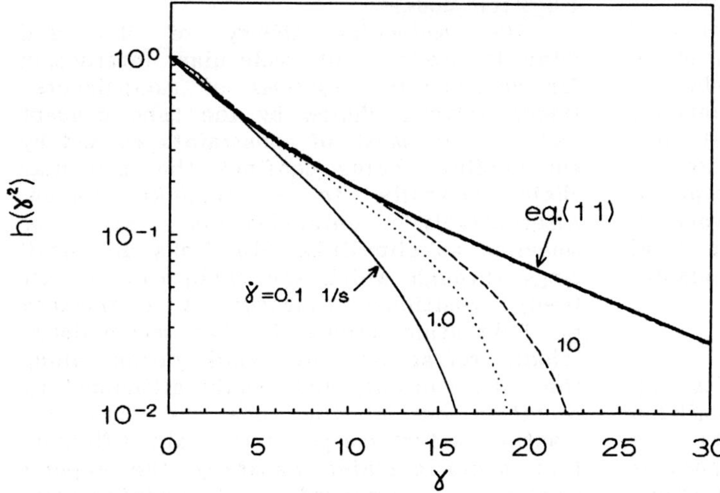

Figure 1. Damping function, $h(\gamma^2)$, vs. shear strain, γ, for LDPE (IUPACX) at 125°C.

Theoretical and Applied Rheology, edited by P. Moldenaers and R. Keunings
Proc. XIth Int. Congr. on Rheology, Brussels, Belgium, August 17-21, 1992

MODELLING NONLINEAR VISCOELASTICITY OF POLYMER MELTS BY CHAIN SLIP AND DISENTANGLEMENT

M.H. WAGNER and J. SCHAEFFER

Institut für Kunststofftechnologie, Universität Stuttgart, Böblingerstr. 70, D-7000 Stuttgart (Germany)

1. INTRODUCTION

1.1 Rubber Elasticity

Large elastic recovery is a unique feature of concentrated systems of flexible polymer chains, and molecular network theories of rubberlike elasticity have enjoyed some success in describing elastic behaviour of crosslinked elastomers above the glass transition temperature (ref. 1). A rubber consists of a single coherent network of flexible macromolecules, which are crosslinked at certain isolated points along the molecular chains. In simplest approximation, the network strands are replaced by Gaussian phantom chains, i.e. each strand is modelled as a linear array of a large number of point masses connected by freely-jointed links, with neglect of mutual interactions between strands and of other than nearest-neighbour interactions along a strand.

Phantom network theories are well suited to explain elasticity of elastomers qualitatively. Quantitatively, however, mechanical behaviour of real systems of crosslinked polymer chains is influenced by intermolecular interactions: The number of possible conformations available to network strands is reduced due to topological constraints. This "entanglement" contribution to the entropic force has been the topic of much theoretical and experimental work in recent years, but no definite and final picture of entanglements in rubber elasticity has emerged so far.

1.2 Temporary Junction Networks

Elastic recovery of polymer melts can be as high as recovery of crosslinked elastomers (ref. 2). Polymer melts are viscoelastic liquids and do not possess permanent crosslinks. Therefore all entropic elasticity has to be attributed to the action of "entanglements". Temporary junction network theories developed by Green and Tobolsky (ref. 3), Yamamoto (ref. 4), and especially Lodge (ref. 5) were the first to explain elasticity of rubberlike liquids. They model entanglements as physical junctions defined by "two points of two molecular chains moving together for at least a certain minimum length of time greater than the fluctuation period which is needed for the molecular chains to pass through almost all of their thermodynamic states". If the further assumptions are made that creation and decay mechanisms of network strands are not affected by flow, the rubberlike-liquid constitutive equation is obtained (ref. 5). Again, many aspects of viscoelasticity are explained qualitatively by the rubberlike-liquid constitutive equation, while its strain measure fails quantitatively at higher strains, indicating that the structure of the temporary junction network is strain dependent.

1.3 Tube Models

The molecular theory of Doi and Edwards models intermolecular interaction for concentrated systems of monodisperse linear polymer chains by the tube concept (ref. 6): The mesh of constraints caused by surrounding chains confines the molecular chain laterally to a tubelike region. Alternatively, entanglements can be modelled as slip-links. Slip links are small rings through which the chain can reptate freely, positioned randomly at a distance r_0. Relaxation occurs by two mechanisms: "chain retraction" by equilibration along the tube contour, and "chain diffusion" by reptation out of the tube. As chain retraction is fast compared to chain diffusion, this model explains naturally the experimentally observed time-deformation separability of the nonlinear relaxation modulus for times greater than the equilibration time. While this molecular theory has thus greatly enhanced our understanding of polymer dynamics for monodisperse linear polymers, it fails quantitatively both in the linear and nonlinear regime if compared to experimental results (refs. 7, 8). Also, its extension to concentrated systems of polydisperse and/or branched polymer chains is not yet clear. Numerous attempts have been made modifying the Doi-Edwards theory by invoking additional ad-hoc assumptions on the molecular level like "tube fluctuations", "chain excursions", "constraint release", or "tube deformation" with the purpose to fit theory to experimental data, and to incorporate the effects of polydispersity. Again, no definite and final picture has emerged so far.

In view of the extraordinary difficulties associated with constructing molecular theories, we propose a temporary slip-link network model as a hybrid theory incorporating both aspects of the temporary junction network models and the Doi-Edwards model. In this way we create a framework containing well defined material functions that can be fitted to experimental data. We hope that this approach will allow us ultimately to derive these functions from purely molecular arguments.

2. TEMPORARY SLIP-LINK NETWORKS

At any fixed time, a "temporary slip-link network" consists of a coherent network of Gaussian strands between slip links. A slip link is produced by the topological constraints of surrounding chains, and restricts lateral motion of the strand, while allowing slip of the chain through the slip link. At equilibrium, strands contain n_0 monomer units of length b, and possess isotropic orientation. We assume affine deformation of slip links. However, due to the slip links, the deformed strand might contain n(u') monomer units instead of n_0 as in equilibrium. We assume (simplistically) that n depends only on the relative length of the strand, $u' = |\underline{u}'|$.

Defining a "slip function" S(u') describing the degree of slip

$$S(u') = n(u') \;/\; n_o \qquad (1)$$

the force $\underline{K}$ in a network strand can be expressed in Gaussian approximation as

$$\underline{K} = \frac{3kT}{n(u')b^2}\,\underline{r}' = \frac{3kT}{S(u')r_o}\,\underline{u}' \qquad (2)$$

We assume that creation and decay mechanisms of network strands are independent of flow as far as diffusion is concerned. However, due to slip in the slip links, we assume that there will be a certain degree of flow-dependent disentanglement, and that this will be independent of time. We define a "disentanglement function" D(u'), which is the probability of a strand created with relative length 1 to survive until the end-to-end vector of that strand has reached a relative length of u'. Again, we assume simplistically that D depends only on the relative length of a strand, u'.

By use of these assumptions, we can write the stress tensor $\underline{\underline{\sigma}}$ as

$$\underline{\underline{\sigma}}(t) = -p\underline{\underline{E}} + 3kT\int_{-\infty}^{t} N(t-t')\langle\frac{D(u')}{S(u')}\,\underline{u}'\underline{u}'\rangle dt' \qquad (3)$$

where $\langle\cdot\cdot\rangle$ denotes an integral over an isotropic distribution of unit vectors. Fitting N(t-t'), the number of strands created at time t' and surviving until time t, to the linear-viscoelastic relaxation modulus G(t) via the memory function m(t-t'), we can rewrite the stress tensor $\underline{\underline{\sigma}}$ as

$$\underline{\underline{\sigma}}(t) = -p\underline{\underline{E}} + \int_{-\infty}^{t} m(t-t')\; g\; \langle\frac{D}{S}\,\underline{u}'\underline{u}'\rangle\; dt' \qquad (4)$$

with a normalization constant g. It is importent to note that slip function S and disentanglement function D are related, because disentanglement and slip must be balanced: on average, surviving network strands of relative length u' must contain n_0 monomer units, i.e.

$$\langle\; D \cdot S \;\rangle = 1 \qquad (5)$$

Eq. (5) represents the mass balance of network strands with regard to slip and disentanglement.

Eqs. (4) and (5) are the central equations of this paper. We call them "the constitutive equations of temporary slip-link networks". It has been shown that they contain as special cases the constitutive equations of Lodge (ref. 5), Wagner (refs. 9-12), Doi-Edwards (ref. 6), Marrucci (ref. 13), and Wagner-Schaeffer (ref. 14).

3. THE MOLECULAR STRESS FUNCTION AND THE DIAMETER OF THE TUBE

Recently, we have analyzed the strain measure of a well-characterized polydisperse polyisobutylene (PIB) melt in uniaxial, planar, ellipsoidal, and equibiaxial extension (ref. 15). We found that

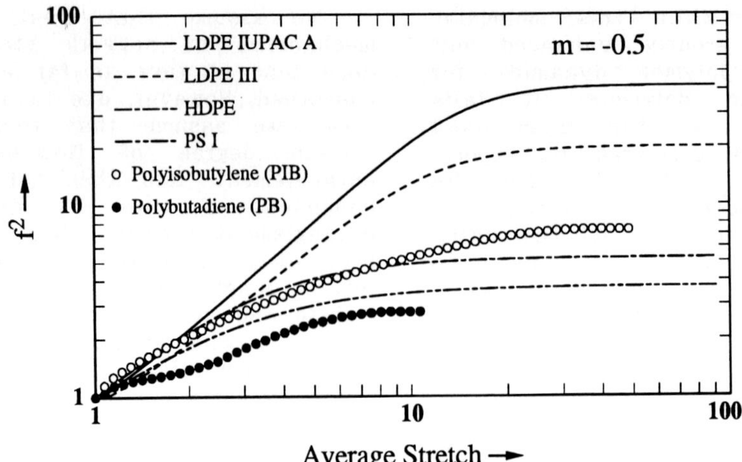

Doi-Edwards' assumption of constant line density of monomer units is not supported by the data, not even in first order in strain. The tension of the average network strand increases with increasing deformation. In contrast to the assumption of Marrucci and de Cindio, however, the line density of monomer units seems to be an average property of the deformed strands. We defined a "molecular stress function" $f(\langle u'\rangle)$ depending on the average relative length of network strands, which describes the increase of force K in the network strands with increasing stretch,

$$K = K_{eq}\, f(\langle u'\rangle) \tag{6}$$

From eq. (2) we find immediately the slip function S corresponding to eq. (6),

$$S(u') = u'\, f^{-1}(\langle u'\rangle) \tag{7}$$

leading (by use of eq. (5) and the so-called "independent-alignment" assumption) to a constitutive equation of the form

$$\underline{\underline{\sigma}}(t) = -p\underline{\underline{E}} + \int_{-\infty}^{t} m(t-t')\, g\, f^2 \left\langle \frac{\underline{u}'\underline{u}'}{u'^2} \right\rangle dt' \tag{8}$$

We call eq. (8) the "molecular stress function constitutive equation". Note that in terms of the tube concept, the molecular stress function f is the inverse of the relative tube diameter. In the figure, the square of the molecular stress function, f^2, is given as a function of the average stretch $\langle u'\rangle$ for several polymer melts tested in uniaxial elongation (m = −0.5) by Laun (LDPE IUPAC A, LDPE III, HDPE, and PS I), Demarmels (PIB), and Schaeffer (PB). At small stretches, f^2 increases linearly with $\langle u'\rangle$, which is consistent with an affine deformation of the total tube. At higher average stretches, the tube deformation is no longer affine, and a maximum molecular stress function corresponding to a minimum tube diameter is reached, the level of which is clearly depending on the degree of long-chain branching.

REFERENCES

1. *A.S. Lodge, R.C. Armstrong, M.H. Wagner, H.H. Winter,* Pure & Appl. Chem. 54, 1349 (1982)
2. *M.H. Wagner, J. Meissner,* Makromol. Chem. 181, 1533 (1980)
3. *M.S. Green, A.V. Tobolsky,* J. Chem. Phys. 14, 80 (1946)
4. *M. Yamamoto,* J. Phys. Soc. Japan, 11, 413 (1956)
5. *A.S. Lodge,* Elastic Liquids, (London-New York 1964)
6. *M. Doi, S.F. Edwards,* Theory of Polymer Dynamics (Oxford 1986)
7. *R.G. Larson,* Constitutive Equations for Polymer Melts and Solutions (London 1988)
8. *M.H. Wagner,* Rheol. Acta 29, 594 (1990)
9. *M.H. Wagner,* Rheol.Acta 15, 136 (1976)
10. *M.H. Wagner,* J.Non-Newtonian Fluid Mech. 4, 39 (1978)
11. *M.H. Wagner, H.M. Laun,* Rheol.Acta 17, 138 (1978)
12. *M.H. Wagner,* Rheol.Acta 18, 33 (1979)
13. *G. Marrucci, G. de Cindio,* Rheol.Acta 19, 68 (1980)
14. *M.H. Wagner, J. Schaeffer,* Rheol. Acta, in print
15. *M.H. Wagner, J. Schaeffer,* J. Rheol., in print

Theoretical and Applied Rheology, edited by P. Moldenaers and R. Keunings
Proc. XIth Int. Congr. on Rheology, Brussels, Belgium, August 17-21, 1992

RHEOLOGICAL BEHAVIOR OF POLYMERS AND A NONLINEAR VISCOELASTIC CONSTITUTIVE RELATIONSHIP*

Ting–Qing Yang and Qinguo Gang

Huazhong University of Science and Technology, Wuhan 430074, China

1. INTRODUCTION

A lot of current literature deals with the inverstigation of rheological behavior for solid polymers (refs.1–2). In this paper, based on study of viscoelastic properties of some polymers (refs.3–5), several sets of experiments have been conducted for a kind of polyurethane and an ethylene–propylene rubber. The nonlinear creep and stress relaxation have been observed. A nonlinear viscoelastic constitutive equation using a generalized strain measure is presented. The material functions and constants can be determined by tests. It has been shown that the theoretical profiles are in good agreement with the experimental results.

2. EXPERIMENTAL STUDY

A set of experiments investigates the creep behavior under different uniaxial stresses, such as $\sigma' = 1.38$ MPa, $\sigma'' = 2.06$ MPa for ethylene–propylene rubber. The second set of tests studies the stress relaxation process under constant strains $\varepsilon_a = 0.095$, $\varepsilon_b = 0.107$ and $\varepsilon_c = 0.176$. It can be shown that the viscoelastic behavior is not linear from the two previous sets of test results.

The third set of experiments investigates the strain response under periodic triangular stress loading. Using the methods and formulae in ref.6 we also find out that the materials are not linear viscoelastic.

From further study, the viscoelastic behavior of the polymers can be described by means of an applied nonlinear constitutive equation which is discussed in section 4.

3. A GENERALIZED STRAIN MEASURE

Deformation gradient tensor is denoted by **F**. The polar decomposition (refs.7–8) is

$$\mathbf{F} = \mathbf{RU} = \mathbf{VR} \tag{1}$$

where **U** and **V** are called the right and left stretch tensor, respectively, and they are symmetric positive definite. The right Cauchy–Green tensor and the left Cauchy–Green tensor are

$$\mathbf{C} = \mathbf{U}^2 = \mathbf{F}^T\mathbf{F}$$
$$\mathbf{B} = \mathbf{V}^2 = \mathbf{F}\mathbf{F}^T \tag{2}$$

Using the Hill's principal axis representation (refs.7–8), it can be shown that

$$\mathbf{F} = \sum_i \lambda_i \, \mathbf{n}_i \otimes \mathbf{N}_i$$
$$\mathbf{U} = \sum_i \lambda_i \, \mathbf{N}_i \otimes \mathbf{N}_i$$
$$\mathbf{V} = \sum_i \lambda_i \, \mathbf{n}_i \otimes \mathbf{n}_i$$
$$\mathbf{C} = \sum_i \lambda_i^2 \, \mathbf{N}_i \otimes \mathbf{N}_i$$
$$\mathbf{B} = \sum_i \lambda_i^2 \, \mathbf{n}_i \otimes \mathbf{n}_i \tag{3}$$

where λ_i (i=1,2,3) are the principal stretches; $\mathbf{N}_i$ and $\mathbf{n}_i$ are the unit eigenvectors in Lagrange and Euler frame, respectively.

* The project was supported by NationalNatural Science Foundation of China.

Hill has given a general representation of generalized strain measure by principal axis representation as follows

$$\mathbf{E}_l = \sum_i f(\lambda_i)\, \mathbf{N}_i \otimes \mathbf{N}_i \qquad (4)$$

where $f(\lambda)$ satisfies: $f(1)=0$; $f'(1)=1$ and $f'(\lambda)>0$. Similarly, a generalized strain measure can be represented as

$$\varepsilon_e = \sum_i g(\lambda_i)\, \mathbf{n}_i \otimes \mathbf{n}_i \qquad (5)$$

where $g(1)=0$, $g'(1)=1$ and $g'(\lambda)>0$. It can be shown that the strain measure ε_e is an objective one, but $\mathbf{E}_l$ is not. In general, $g(\lambda)$ can be expanded as

$$g(\lambda) = (\lambda-1) - c(\lambda-1)^2 + \cdots \qquad (6)$$

where $c = -g''(1)/2!$, and $\lambda_i - 1 = \varepsilon_i$ is sufficiently small. In some cases, ε_e can be simply written as

$$\varepsilon_e = \sum_i [\varepsilon_i - c\varepsilon_i^2]\, \mathbf{n}_i \otimes \mathbf{n}_i \qquad (7)$$

This is the generalized strain measure used in the present paper.

4. CONSTITUTIVE FORMULATION

The general form of constitutive relation for isotropic simple materials can be expressed as

$$\mathbf{R}(t)^T \tilde{\sigma}(t) \mathbf{R}(t) = \pounds[(\mathbf{f}(\mathbf{U}(t-\zeta))] \qquad (8)$$

where $\tilde{\sigma}(t)$ is Cauchy stress. In principal axis representaion,

$$\tilde{\sigma}(t) = \sigma_{kl}\mathbf{n}_k \mathbf{n}_l = \sum_i \sigma_i \mathbf{n}_i \otimes \mathbf{n}_i \qquad (9)$$

and the constitutive relation is

$$\tilde{\sigma}(t) = \pounds[f(\Lambda(t-\zeta))], \quad \text{or} \qquad (10)$$

$$\tilde{\sigma}(t) = \mathbf{K}[f(\Lambda_t(t-\zeta)), f(\Lambda(t)] \qquad (10a)$$

in which

$$\tilde{\sigma}(t) = [\sigma_1(t), \sigma_2(t), \sigma_3(t)]^T,$$

$$\Lambda = (\lambda_1, \lambda_2, \lambda_3)^T,$$

$$f(\Lambda = [f(\lambda_1), f(\lambda_2), f(\lambda_3)]^T,$$

σ_i $(i=1,2,3)$ and λ_i are the principal stresses and the principal stretches, respectively.

For the present purpose, the linear tensor functional $\mathbf{K}$ is assumed to be composed of two parts, i.e.

$$\tilde{\sigma}(t) = \tilde{\sigma}^e(t) + \tilde{\sigma}^a(t) \qquad (11)$$

$$\tilde{\sigma}^e(t) = \mathbf{K}[f(I), f(\Lambda)] = \mathbf{K}_1[f(\Lambda)] \qquad (11a)$$

$$\tilde{\sigma}^a(t) = \mathbf{K}[f(\Lambda_t(t-\zeta)), f(\Lambda)] - \tilde{\sigma}^e(t) = \mathbf{K}_2[\mathbf{J}, f(\Lambda)] \qquad (11b)$$

where $I = (1,1,1)^T$.

$$\mathbf{J} \equiv \mathbf{J}(t,\zeta) = f(\Lambda_t(t-\zeta)) - f(I) \qquad (12)$$

And in static history, $\mathbf{J}=0$,

$$\mathbf{K}_2[0, f(\Lambda)] = 0, \quad \text{and} \quad \tilde{\sigma}(t) = \tilde{\sigma}^e(t)$$

Assume $\mathbf{K}_2[\mathbf{J}, f(\Lambda)]$ is Frechet differentiable. From the principle of fading memory and Riesz representation theorem, we have

$$\mathbf{K}_2[\mathbf{J}, f(\Lambda)] = \int_0^t \mathbf{M}[\zeta, f(\Lambda)]\, \mathbf{J}(t,\zeta)\, d\zeta \qquad (13)$$

where $\mathbf{M}$ is a functional which depends on the fading memory.

According to the conditions of Hill's strain measure, one has

$$f(I) = (0,0,0)^T$$

$$f(\Lambda_t(t-\zeta)) = \left[f\!\left(\frac{\lambda_1(t-\zeta)}{\lambda_1(t)}\right), f\!\left(\frac{\lambda_2(t-\zeta)}{\lambda_2(t)}\right), f\!\left(\frac{\lambda_3(t-\zeta)}{\lambda_3(t)}\right)\right]^T$$

Therefore

$$\mathbf{J}(t,\zeta) = f(\Lambda_t(t-\zeta)) \approx f(\Lambda(t-\zeta)) - f(\Lambda(t))$$

Letting

$$\tilde{\sigma}^e(t) = \mathbf{K}_0 f(\Lambda(t)) \qquad (14)$$

and substituting Eqs.(13) and (14) into Eq.(11) yields

$$\tilde{\sigma}(t) = \mathbf{K}_0 f(\Lambda(t)) + \int_0^t \mathbf{M}[\zeta, f(\Lambda)]\, [f(\Lambda(t-\zeta)) - f(\Lambda(t))]\, d\zeta$$

Denoting

$$\mathbf{G}(\zeta) = -\int_\zeta^t \mathbf{M}[\tau, f(\Lambda(t))]\, d\tau$$

and

$\mathbf{K}(\zeta) = \mathbf{K}_0 + \mathbf{G}(\zeta)$,

the following expression

$$\tilde{\sigma}(t) = \int_{-\infty}^{t} \mathbf{K}(t-\zeta)\frac{\partial}{\partial\zeta} f(\Lambda))\, d\zeta$$

$$= \mathbf{K}(0) f(\Lambda(t)) + \int_{0}^{\infty} \frac{d\mathbf{K}(\zeta)}{d\zeta} f(\Lambda(t-\zeta))\, d\zeta \quad (15)$$

can be finally obtained.

In abstract representation, it is

$$\tilde{\sigma}(t) = \overline{\mathbf{K}}(0) f(\mathbf{V}(t))$$

$$+ \mathbf{R}(t)\int_{0}^{\infty} \frac{d\overline{\mathbf{K}}(\zeta)}{d\zeta}\{f[(\mathbf{U}(t-\zeta)]\}\, d\zeta\, \mathbf{R}^T(t) \quad (16)$$

where $\overline{\mathbf{K}}(\mathbf{t})$ is a fourth rank tensor.

In uniaxial loading, when the strain measure is assumed as Eq.(7), the constitutive equation should be

$$\sigma(t) = \int_{-\infty}^{t} Y(t-\zeta)\frac{\partial}{\partial\zeta}[\varepsilon(\zeta) - c\varepsilon^2(\zeta)]\, d\zeta \quad (17)$$

where Y(t) is relaxation function. For simplicity and applicability, $Y(t)$ is assumed to be a 3–parameter function such as $Y(t) = \alpha + \beta \exp(-\gamma t)$. From the experimental data, the material constants will be $c = 1.52$, $\alpha = 16.7$ MPa, $\beta = 1.2$ MPa and $\gamma = 0.13$.

For 2–dimensional case, the constitutive equations can also be represented by principal axis method from Eq.(15). For instance

$$\sigma_r(t) = \int_{-\infty}^{t} G_1(t-\zeta)\frac{d}{d\zeta} f[\lambda_r(\zeta)]\, d\zeta$$

$$+ \int_{-\infty}^{t} G_2(t-\zeta)\frac{d}{d\zeta} f[\lambda_\theta(\zeta)]\, d\zeta$$

$$\sigma_\theta(t) = \int_{-\infty}^{t} G_1(t-\zeta)\frac{d}{d\zeta} f[\lambda_\theta(\zeta)]\, d\zeta$$

$$+ \int_{-\infty}^{t} G_2(t-\zeta)\frac{d}{d\zeta} f[\lambda_r(\zeta)]\, d\zeta \quad (18)$$

in which $f(\lambda_r(\zeta)) = \varepsilon_r(\zeta) + c_r\varepsilon_r^2(\zeta)$,

$f(\lambda_\theta(\zeta) = \varepsilon_\theta(\zeta) - c_\theta\varepsilon_\theta^2(\zeta)$,

$c_r = -c\,\mathrm{Sign}\varepsilon_r$ and $c_\theta = c\,\mathrm{Sign}\varepsilon_\theta$.

Eqs.(17) and (18) have been used to calculate the bending of beam and to analyze a thick–walled cylinder, respectively, which will be presented in subsequent paper.

REFERENCES

1. J.D. Ferry, Viscoelastic Properties of Polymers, 3rd Ed., New York, John Wiley, 1980.
2. W.V. Chang, R.Bloch and N.W. Tschoegl, On the Theory of the Viscoelastic Behavior of Soft Polymer in Moderately Large Deformations, Rheol. Acta., 15(1976) 367–378.
3. T–Q. Yang and J–Y. Yu, A Constitutive Relationship of Elastic–viscoplastic Materials, in: T.X. Yu and D. Wang (Ed.), Plasticity and Geodynamics, Peking Univ. Press, 1990, pp. 9–16.
4. J–Y. Yu and T–Q. Yang, A Study of Viscoelastic Behavior for Polymers, in: T. Jiang (Ed.), Adv. in Rheol., Shanghai, 1990, pp. 126–128.
5. T–Q. Yang and J–Y. Yu, Investigation on Rheological Behavior and A Constitutive formulation for Polymers, in: L–W. Yuan and H. Odali (Chief Ed.), Proc. of CJICR, Peking Univ. Press, 1991, pp. 103–106.
6. T–Q. Yang and Yu Chen, Stress Response and Energy Dissipation in A Linear Viscoelastic Material Under Periodic Triangular Strain Loading, J. Polym. Sci., Polym. Phys. Ed., 20 (1982) 1437–1442.
7. R. Hill, Aspects of Invariance in Solid Mechanics, in: Adv. Appl. Mech., Vol. 18(1978) 1–75.
8. Z–H. Guo and H–Y. Liang, From Principal Axis representation to Abstract Representation, Adv. in Mech., 20(1990) 303–315.

Theoretical and Applied Rheology, edited by P. Moldenaers and R. Keunings
Proc. XIth Int. Congr. on Rheology, Brussels, Belgium, August 17-21, 1992

THE RECOIL OF RIGID PVC

A.M. ZDILAR and R.I. TANNER

Department of Mechanical Engineering, University of Sydney, N.S.W. 2006, Australia.

1. INTRODUCTION

Polyvinyl chloride is widely used in industry, but it is difficult to find any extensive discussion of rheological relations that describe it. The present paper discusses the behaviour of rigid PVC in extensional deformations at various temperatures. In the tests, a step elongation was applied at an initial time, then after a delay the specimen was cut, permitting recoil. Both linear and nonlinear strain regimes were studied; the linear relaxation properties (relaxation function, storage modulus) were cross-checked with eccentric–disk measurements. In the non–linear strain regime a single-integral constitutive equation of the KBKZ-Wagner type was used. Separability of time and strain effects was demonstrated in our tests and so a "damping function" could be found which was only a function of strain. Video recordings of recoil were made, and detailed predictions of the strain-time behaviour were checked against experiments.

Various constitutive equation proposals were used in the comparison, and the separated kernel integral irreversible model performed reasonably well; some other models of a differential type, and a Doi-Edwards model, were not as accurate.

2. CONSTITUTIVE EQUATIONS

Many constitutive models are currently available on the "market". They are basically divided into two types, the integral and the differential model. The choice of the most suitable model for a particular fluid and flow can be difficult and sometimes not realised. Here some of the "popular" models will be applied to the elastic recoil of PVC.

The starting point for the integral models is the classic KBKZ model (refs. 1,2). The first simplification to this general model is Lodge's equation (ref. 3) which is successful for small strains. For higher strains, where Lodge's equation ceases to be accurate, Wagner (ref. 4) proposed a model using the idea of a strain dependent damping function. The model has been successful for monotonically increasing strains. For reversing flows, Wagner introduced irreversibility (ref. 5) into his formulation. Doi and Edwards (ref. 6) captured the notion of reptation in their integral constitutive model.

The upper convected Maxwell equation (ref. 7) is the equivalent differential version of Lodge's integral constitutive equation. The more complex differential models are divided into separable and non–separable classes. Larson has proposed a separable differential model (ref. 8) which can have irreversibility accounted for (ref. 9).

The non–separable Phan-Thien-Tanner (PTT) differential model (ref. 10) has a trace function which can be expressed in a linear or exponential form. Recoil predictions will be based on all the previous models.

Coding also was divided into integral and differential types. Once both were finalised they could be cross–referenced with Lodge's integral model and Maxwell's differential model. Linear viscoelastic theory was used to validate the coding at small strains.

3. EXPERIMENTAL

The experiments were performed on a modified step rheometer (ref. 11). The single step was achieved by using a stiff spring with a latch for the release mechanism. A knife blade was placed in the frame of the rheometer so a specimen could be step elongated and after a delay, cut, permitting elastic recoil to occur. Video recordings were performed so that both step strains and time dependent recovery could be measured. Testing was performed on two types of rigid PVC at various temperatures in the rubbery regime.

For small step strains the linear relaxation modulus was measured and a discrete relaxation spectrum fitted. These coefficients were then used to predict the storage modulus obtained from eccentric disk rheometry. The comparison was very good.

4. RESULTS

Figure 1 shows the linear relaxation modulus at various temperatures. The master graph for these curves, based on the reference temperature of 120°C, appear in Fig. 2 with the corresponding shift factors plotted in Fig. 3. The master curve is best fitted with a fifth order polynomial, as shown. Note that the data does not follow the classic WLF relationship, but resembles a linear fit, and that the stress has not returned to zero.

The ratio of the non–linear relaxation modulus to the linear modulus yields the damping function. Time strain separability was demonstrated at all times. At higher strains the specimen fractured, thus the damping function was taken at an early time. One such damping function is presented in Figure 4. All models fitted the data reasonably well, except the PTT (linear form) and DE models. Thus, these two models will not be considered for elastic recoil. A linear fit is also able to describe the data.

Strain reversal tests serve as a stringent test for the constitutive equations. Figure 5 shows a typical elastic recoil test, including the various predictions from the models. Lodge's, PTT (exp form), Wagner's and Larson's reversible models do not fit the flow well. Irreversibility does aid in the predictions. The recoil data, in fact, behaves initially close to Wagner's irreversible model and then veers to Larson's irreversible model, lying midway between these two models.

The behaviour could be explained by the morphology of PVC, namely the particulate flow (ref. 12). Possibly another constitutive equation may apply.

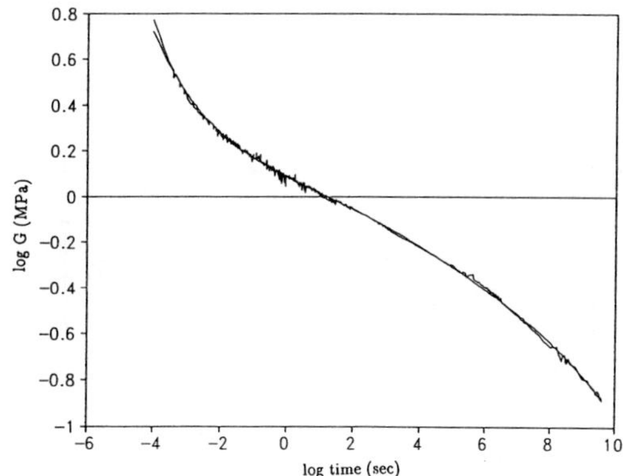

Fig. 2 Master Curve.

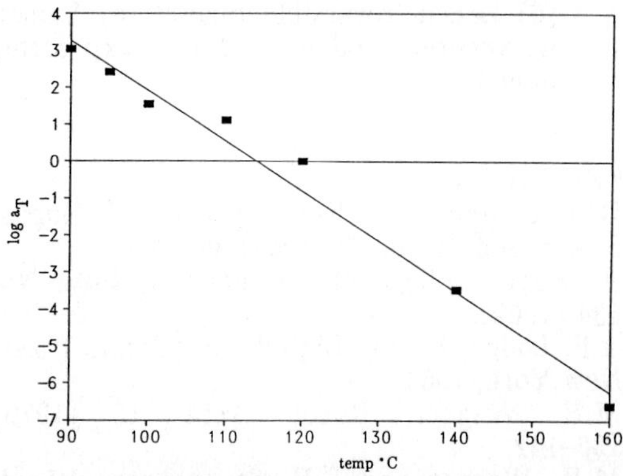

Fig. 3 Shift Factors.

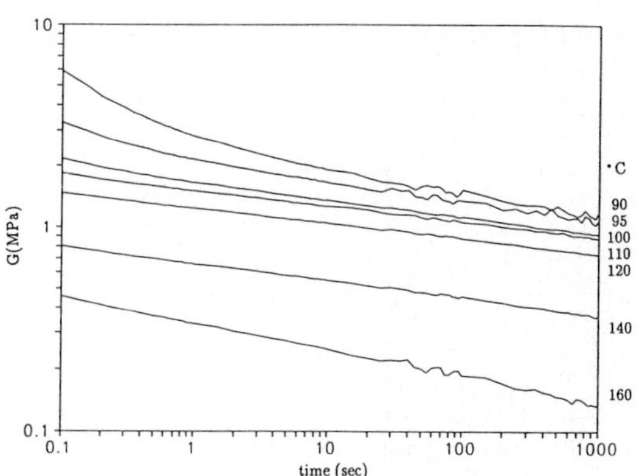

Fig. 1 Linear Relaxation Modulus.

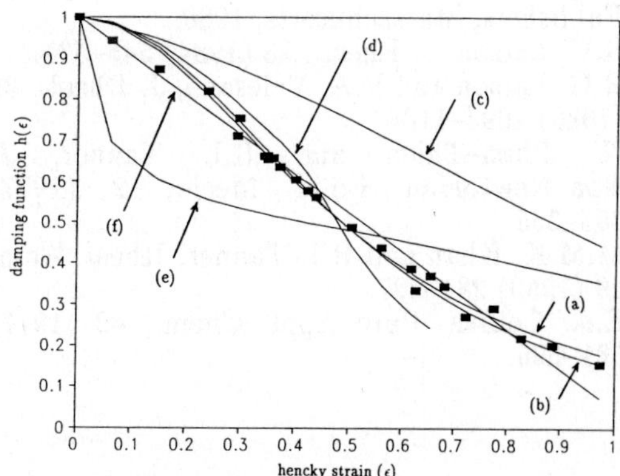

Fig. 4 Damping function at 120°C: ■ data; (a) Wagner fit; (b) Larson fit; (c) DE model; (d) PTT (exp form) fit; (e) PTT (linear form) fit; (f) linear fit.

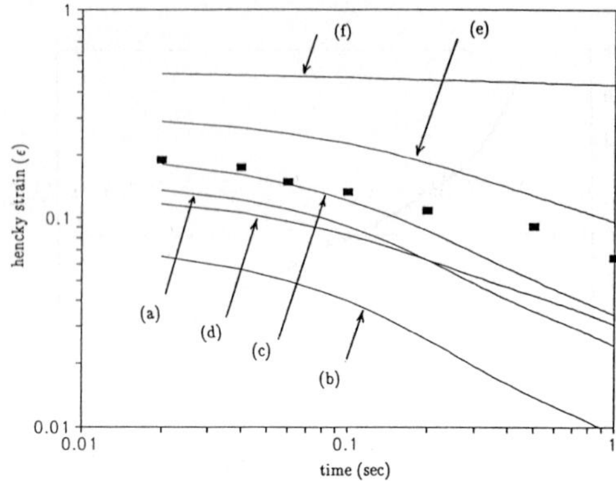

Fig. 5 Elastic recoil at 120°C with $\epsilon_0 = 0.649$ and $t_i = 0.51$ sec: ■ data; (a) Lodge/ Maxwell model; (b) Wagner reversible model; (c) Wagner irreversible model; (d) Larson reversible model; (e) Larson irreversible model; (f) PTT (exp form) model.

REFERENCES

1. B. Bernstein, E.A. Kearsley and L.J. Zapas, Trans. Soc. Rheol., 7 (1963) 391–410.
2. A. Kaye, College of Aeronautics, Note No. 134 (1962).
3. A.S. Lodge, Elastic Liquids, Academic Press, New York, 1964.
4. M.H. Wagner, Rheol. Acta, 15 (1976) 136–142.
5. M.H. Wagner and S.E. Stephenson, Rheol. Acta, 18 (1979) 463–468.
6. M. Doi and S.F. Edwards, Theory of Polymer Dynamics, Oxford University Press, 1986.
7. R.G. Larson, Constitutive Equations for Polymer Melts and Solutions, Butterworths Publishers, Massachusetts, 1988.
8. R.G. Larson, J. Rheol., 28 (1984) 545–571.
9. R.G. Larson and V.A. Valesano, J. Rheol., 30 (1986) 1093–1108.
10 N. Phan–Thien and R.I. Tanner, J. Non–Newtonian Fluid Mech., 2 (1977) 353–365.
11 M.M.K. Khan and R.I. Tanner, Rheol. Acta, 29 (1990) 281–297.
12 E.A. Collins, Pure Appl. Chem., 49 (1977) 581–595.

Theoretical and Applied Rheology, edited by P. Moldenaers and R. Keunings
Proc. XIth Int. Congr. on Rheology, Brussels, Belgium, August 17-21, 1992

DETERMINATION OF THE VISCOELASTIC CONSTITUTIVE DIFFERENTIAL OPERATOR LAW IN TERMS OF THE RELAXATION AND RETARDATION TIMES

R.H. BLANC

Centre National de la Recherche Scientifique Laboratoire de Mécanique et d'Acoustique 31, chemin Joseph Aiguier F-13402 MARSEILLE Cedex 09 France

If one considers the various mathematical methods of specifying the mechanical behaviour of linear viscoelastic media, one can observe that although there exist relationships for deducing from the differential operator form the other stress-strain constitutive equations, the converse is not true. The implicit relationships between the operator law and the mechanical models have generally been mentioned, but not specified by authors dealing with these problems. Most authors have calculated these coefficients only in the case of models comprising few elements. Numerical calculation of these coefficients on the basis of other viscoelastic functions measured within just a fraction of their interval of definition has sometimes led to a scattering of the results which raises doubt as to the physical significance of these parameters. The results presented below explain these discrepancies : these coefficients all depend on all the relaxation or retardation times.

In the present study, it is proposed to establish general, explicit relations, which uniquely specify the coefficients of the stress-strain equation as a function of the relaxation and retardation times. The case of a linear viscoelastic medium is considered. The Carson transforms of the relaxation and retardation functions are expressed in two different ways, taking on the one hand the differential operator form of the constitutive equation, and on the other hand the generalized mechanical models. By identifying the poles of these expressions we deduce the following fundamental expressions for the constant coefficients of the differential operator equation, in terms of the discrete relaxation and retardation spectra.

Let us write the operator equation

$$P\sigma = Q\varepsilon$$

where σ represents the stress, ε the strain and

$$P = \sum_{i=0}^{n} A_i \frac{\partial^i}{\partial t^i},$$

$$Q = \sum_{j=0}^{m} B_j \frac{\partial^j}{\partial t^j}.$$

Let G_k (θ_k) and $J_l(\theta_l)$ be the discrete relaxation and retardation spectra, respectively. We establish

$$A_1 = A_0 \sum_{k=1}^{n} \tau_k$$

$$A_2 = A_0 \sum_{k=1}^{n-1} \sum_{l=k+1}^{n} \tau_k \tau_l$$

$$\dots$$

$$A_i = A_0 \sum_{k=1}^{n-i+1} \dots \sum_{r=q+1}^{n} \underbrace{\tau_k \dots \tau_q \tau_r}_{i}$$

$$\dots$$

$$A_n = A_0 \tau_1 \tau_2 \dots \tau_n .$$

and for the case of a body devoid of instantaneous elasticity

$$A_i = A_0 \sum_{k=1}^{n-i} \dots \sum_{r=q+1}^{n-1} \underbrace{\tau_k \dots \tau_q \tau_r}_{i} .$$

Likewise

$$B_j = B_0 \sum_{l=1}^{m-j+1} \dots \sum_{s=r+1}^{m} \underbrace{\theta_l \dots \theta_r \theta_s}_{j}$$

and for the liquid case

$$B_j = B_1 \sum_{l=1}^{m-j} \dots \sum_{s=r+1}^{m-1} \underbrace{\theta_l \dots \theta_r \theta_s}_{j} .$$

For the constitutive equation to be completely determined, the ratio B_0/A_0 (or B_1/A_0 in the liquid case) remains to be established. This is easily achieved.

REFERENCE

R.H. Blanc, Rheologica Acta, 1988, **27**(5), 482.

Theoretical and Applied Rheology, edited by P. Moldenaers and R. Keunings
Proc. XIth Int. Congr. on Rheology, Brussels, Belgium, August 17-21, 1992

A CONSTITUTIVE EQUATION BASED ON "SUB–CLUSTER THEORY"

LI HANGQUAN, JIN RIGUANG

Dept. of Polym. Sci., Box 61, 100029, Beijing Inst. of Chem. Technol., (P.R.China)

1. INTRODUCTION

Difficulties often appeared when one use classical constitutive equations (ref.1) in very wide range of shear rates. In this work, efforts were made to approach this problem from a new concept based on "Sub–cluster theory" developed by the authors (ref.2,3).

2. DERIVATION OF EQUATION

In "Sub–cluster theory, the deformation of a fluid is considered as the result of two opposite factors: shearing(A) and anti–shearing(B). The interaction of these factors may be described similar to chemical reactions:

$$(A)+(A)\xrightarrow{K_{AA}}(A-A)^{\neq}, \quad (A)+(B)\xrightarrow{K_{AB}}(A-B)^{\neq}$$

$$(B)+(A)\xrightarrow{K_{BA}}(B-A)^{\neq}, \quad (B)+(B)\xrightarrow{K_{BB}}(B-B)^{\neq}$$

Defining $r_1=\frac{K_{AA}}{K_{AB}}$ as shear–rate constant; $r_2=\frac{K_{BB}}{K_{BA}}$ as anti–shear–rate constant. Obviously, the greater the r_1, the greater the probability for appearance of factor A in unit time; the same case is for r_2 and factor B. The average appearing rates of factor A and B can be obtained by statistical calculation (ref.2) as

$$\bar{\lambda}_A = 1+\frac{K_{AA}}{K_{AB}}\frac{\dot{N}_A}{\dot{N}_B} = 1+r_1\frac{\dot{N}_A}{\dot{N}_B} \tag{1}$$

$$\bar{\lambda}_B = 1+\frac{K_{BB}}{K_{BA}}\frac{\dot{N}_B}{\dot{N}_A} = 1+r_2\frac{\dot{N}_B}{\dot{N}_A} \tag{2}$$

where

$$\dot{N}_A=\dot{N}(\dot{\gamma})-\dot{N}_{min},\quad \dot{N}_B=\dot{N}_{max}-\dot{N}(\dot{\gamma}) \tag{3,4}$$

$\dot{N}(\dot{\gamma})$ is appearing rate of factor A at any shear rate; $\dot{N}_{min}$ is appearing rate of factor A at minimum shear rate; $\dot{N}_{max}$ appearing rate of factor A at maximum shear rate.

Since $\dot{N}(\dot{\gamma})\propto\dot{\gamma}$, $\dot{N}_{min}\propto\dot{\gamma}_{min}$, $\dot{N}_{max}\propto\dot{\gamma}_{max}$, then

$$\bar{\lambda}_A = K_1(\tau-\tau_{min});\ \bar{\lambda}_B = K_1(\tau_{max}-\tau) \tag{5,6}$$

Combining Eq.1～6, we obtain:

$$\frac{\tau-\tau_{min}}{\tau_{max}-\tau} = K\frac{1+r_1\dfrac{\dot{\gamma}-\dot{\gamma}_{min}}{\dot{\gamma}_{max}-\dot{\gamma}}}{1+r_2\dfrac{\dot{\gamma}_{max}-\dot{\gamma}}{\dot{\gamma}-\dot{\gamma}_{min}}} \tag{7}$$

where τ is shear stress, $\dot{\gamma}$ is shear rate and K is a parameter denoting flow–strength.

3. DISCUSSION

3.1 When $\dot{\gamma}\to 0$, $\tau\to 0$, Eq.7 become

$$\tau = K\dot{\gamma}\left(\frac{\tau_{max}}{\dot{\gamma}_{max}}\right) = K\eta_{max}\dot{\gamma} = \eta_I\dot{\gamma}_I \tag{8}$$

This is the case in very low shear rate, i.e., in the first Newtonian region.

3.2 When $\dot{\gamma}\to\infty$, Eq.7 become

$$\tau = K\eta_{max}r_1\cdot\dot{\gamma} = \eta_{II}\dot{\gamma}_{II} \tag{9}$$

This is the case in second Newtonian region.

It is clear from Eqs.9 and 10 $\eta_{II}=\eta_I$. For pseudoplastic fluids, $r_1<1$, so that $\eta_I>\eta_{II}$.
For dilatant fluids, $r_1>1$, so that $\eta_I<\eta_{II}$.

3.3 When $\dot{\gamma}_{min}\to 0$, $\tau_{min}\neq 0$, Eq.7 can describe Bingham fluid.

It is seen that Eq.7 can describe all regions of viscosity curve for non–Newtonian fluids. A lot of experimental data were fitted by Eq.7 for polymer melts, such as HDPE, LDPE, PS, PVC, etc. and excellent agreement was obtained.

REFERENCES

1. C.D.Han, Rheology in Polymer Processing, Academic Press, New York, 1976, pp.33–58.
2. Jin Riguang, Vague Sub–Cluster, HEI LONG JIANG Tech. & Sci. Press, Harbin, 1985.
3. Jin Riguang, in Advance in Rheology, Ⅸ th Intl. Cong. on Rheology, 1984, Mexico City, Mexico, Vol.1, p.449.

Theoretical and Applied Rheology, edited by P. Moldenaers and R. Keunings
Proc. XIth Int. Congr. on Rheology, Brussels, Belgium, August 17-21, 1992

RHEOLOGICAL RELATIONS OF THERMOVISCOELASTICITY

B.E.POBEDRIA

Department of Composite Mechanics, Faculty for Mechanics and Mathematics, Moscow University (USSR)

The various rheological relations, describing linear and non–linear mechanical and thermophysical properties of materials as well as it's dependence on temperature, pressure, humidity, irradiation power etc., are considered. The new rheological relations of non–linear viscoelasticity, describing sufficiently broad class of materials (including polymers and composites), are proposed.

The classification of rheological relations by it's structure and possibility to reflect experimentally the observed effects is given. The problem on "interinversity" of the relations describing creep and relaxation as well as the adequacy of stadied models (the possibility to determine the complete set of it's material functions) are discussed.

The method of obtaining of the effective characteristics of viscoelastic composites is expounded. The formulation of boundary thermoviscoelasticity problems, including connected, is given. Here the heat efflux in result of deformation process is taken into account. The conditions on the rheological relations that assure a correctness of boundary problems mathematical formulation are adduced.

The concrete examples are considered.

Theoretical and Applied Rheology, edited by P. Moldenaers and R. Keunings
Proc. XIth Int. Congr. on Rheology, Brussels, Belgium, August 17-21, 1992

RHEOLOGICAL CONSTITUTIVE RELATIONS FOR ASPHALTS

M. G. Sharma and L. Fang
Department of Engineering Science and Mechanics
The Pennsylvania State University
University Park, PA 16802 U.S.A.

1. INTRODUCTION

The paper is concerned with the evaluation of rheological constitutive relations for asphalts using the indentation technique which neither involves elaborate specimen preparation nor require elaborate experimental set up. The rheological data was analyzed using constitutive relations based on a volterra integral equation and KBKZ model.

2. APPARATUS AND TEST PROCEDURE

A mini screw driven universal testing machine provided with a precision load cell of very small capacity (22N) was adapted to perform indentation tests. Asphalt was heated to an elevated temperature of 135°C until it became fluid and then poured into a cylindrical cup (50 mm. diameter and 12.5 mm thick). The exposed plane surface was pressed by a cylindrical indenter with a spherical tip to produce a preselected vertical deformation. The deformation was maintained for a certain duration and the resulting relaxation of load with time was measured. The single step relaxation tests were conducted at three values of deformation and levels of temperature ranging from -30° to 10°C at the intervals of 5°C. Figure 1 shows the typical data obtained from the tests.

3. THEORETICAL INTERPRETATION

In order to determine the constitutive parameters for asphalts a finite element model for the prediction of stress field and contact pressure in a nonlinear semi-infinite medium subjected to indentation by a spherical indentor was developed. This model in conjunction with a parameter identification model was used to back calculate constitutive parameters based upon the following nonlinear volterra integral equation (ref. 1):

$$Sij = \int_o^t m\left(t - t'\right) \hbar(I) C^{-1} dt' \qquad (1)$$

where Sij = deviatoric stress tensor

$$m\,(t - t') = \frac{dG(t - t')}{dt'}$$

G (t) = shear relaxation modulus function

$\hbar(I)$ = damping function

$I = \beta I_1 + (1 - \beta) I_2$

I_1 and I_2 = the first and second strain invariants respectively

C^{-1} = Finger strain tensor

t = time

Figure 2 shows the master stress-relaxation modulus curve for the test material and Figure 3 shows the variation of damping function with I. Figure 1 also shows the comparison of predictions based upon equation (1) and KBKZ against experimental results.

REFERENCE

1. M.M.K. Khan and R.I. Tanner, Rheol. Acta 29:281-297, 1990.

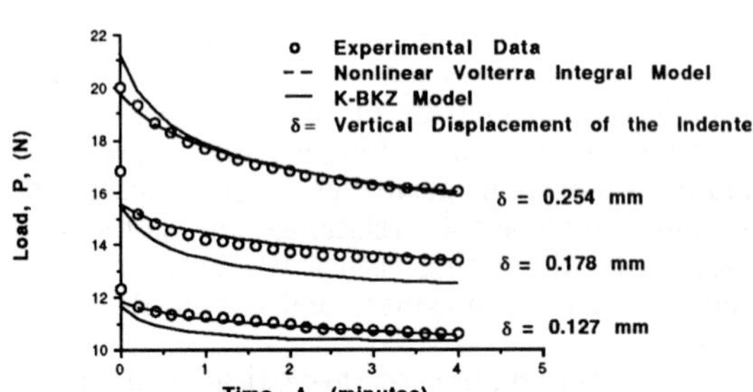

Figure 1. Comparison of Theoretical Predictions with Experimental Data in Single-Step Relaxation Test for Asphalt (AAK-1)

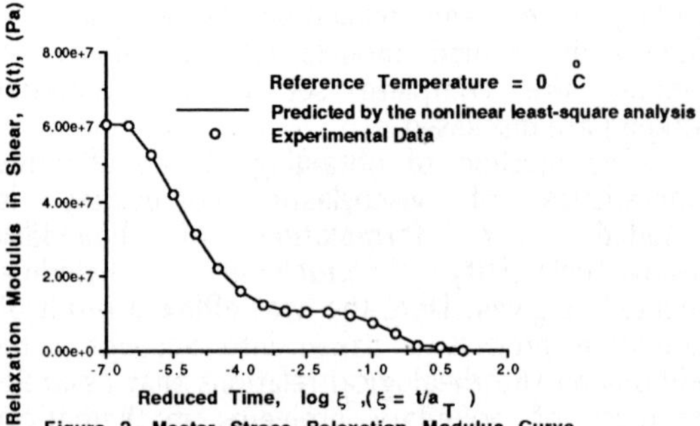

Figure 2. Master Stress Relaxation Modulus Curve for Asphalt (AAK-1)

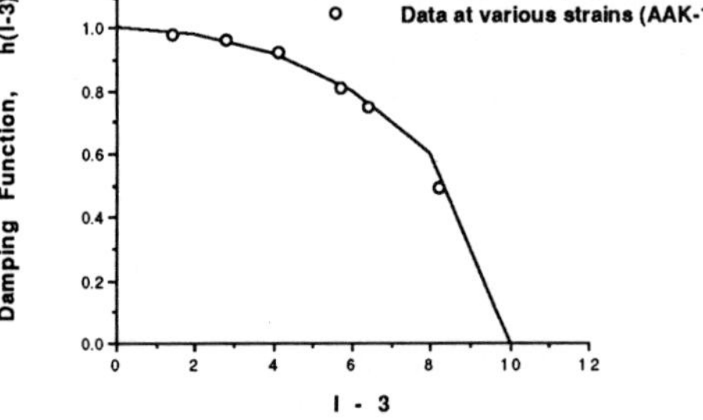

Figure 3. Damping Function for Asphalt (AAK-1)

Theoretical and Applied Rheology, edited by P. Moldenaers and R. Keunings
Proc. XIth Int. Congr. on Rheology, Brussels, Belgium, August 17-21, 1992

CONSTITUTIVE EQUATION FOR THE INK OF BALL-POINT PEN AND PRINTING

PENG WANG AND TI-QIAN JIANG

Chemical Engineering Research Centre, East China University of Chem. Tech., Shanghai 200237, (China)

1. INTRODUCTION

In the past, people like to use the Power law model as the constitutive equation for the ink of ball-point pen, as Tang et al(ref.l) did, but recently, the more inks for ball-point pen and printing exhibit elastic behaviors, so that the development of a nonlinear constitutive equation for them is necessary.

2. DEVEROPMENT OF THE CONSTITUTIVE EQUATION

In order to fine a suitable model which can portray the viscoelastic behavior of the ink for ball-point pen and printing, the ZFD model and Oldroyd-B model have used to do that, the result is that both model were failed in success. Therefore, we put forward the thought of relaxation and retardation time both variate with shear rate. Thus, based on the Oldroyed-B model, the modified three parameters model is developed as follows:

$$\boldsymbol{\tau}+m(\dot{\gamma})\overset{\nabla}{\boldsymbol{\tau}}=\eta_0[\dot{\boldsymbol{\gamma}}+n(\dot{\gamma})\dot{\boldsymbol{\gamma}}_{(2)}] \qquad (1)$$

wherein $m(\gamma)=\lambda_1\gamma^{-p}$, $n(\gamma)=\lambda_2\gamma^{-p}$,

From experimental date we found that for the inks of ball-point pen and printing the value of p can be get indetical number 0.5, i.e., p=0.5, Therefore, all the three rheological properties, such as, viscosity, η, first normal stress diffrence, N_1, and/or second normal stress difference, N_2, can be get.

3. RESULT AND DISCUSSION

Fig. 1 shows the comparison between expari-mental data and prediction from equation (1) for a typical ink named SH-D and made in China. It is evident that the N_1 is absolutly in good agreement but there is somwhate difference for η. We found all the same for other inks which include 602NB made in Germany and H-253 made in Japan.

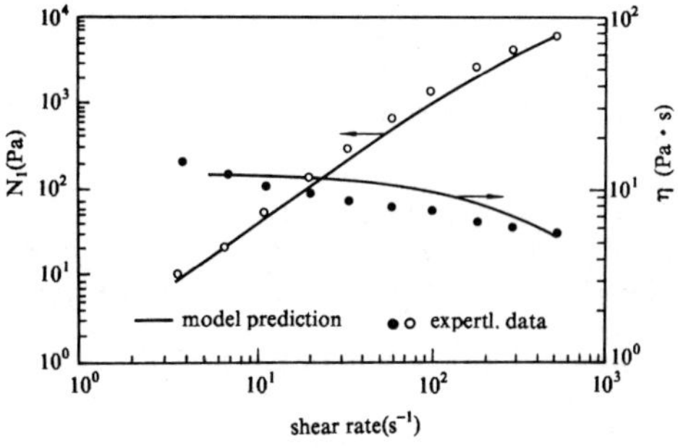

Fig.1 Rheological properties of SH-D ink

Fig.2 shows the comparison between experimental data and prediction of our model for a typical printing ink 1009 made in China. It is clear that both η and N_1 are in good agreement.

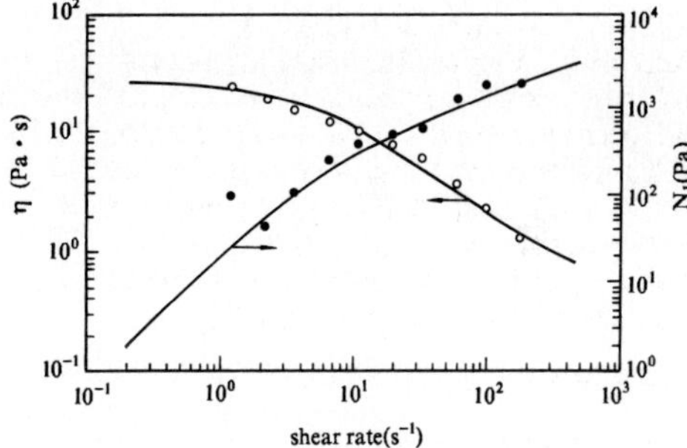

Fig.2 Rheological properties of printing ink

All the experimental data are measured by rheometer RMS- 605. Finally, We came to the conclusion that the modified model(1) can described the rheological properties of the ink for ball-point pen, printing ink and partially hydrolyzed poly-acrylaminde solution.

REFERENCE

Y.N. Tang, J.Y.Zhu and T.Q.Jiang in B.Mena(Ed.), Proc. IX Intl. Cong. on Rheol., Mexico, 1984 Vol. 2, pp. 415-420.

Contributed papers

THEORY

Theoretical and Applied Rheology, edited by P. Moldenaers and R. Keunings
Proc. XIth Int. Congr. on Rheology, Brussels, Belgium, August 17-21, 1992

FRACTAL DYNAMIC THEORY OF VISCOELASTIC RELAXATION IN LINEAR, BRANCHED AND CROSSLINKED GLASSY POLYMERS

T. S. CHOW
Xerox Webster Research Center, 800 Phillips Road, 114-39D, Webster, NY 14580, USA

1. INTRODUCTION

A fractal dynamic theory of viscoelastic relaxation of polymers for temperatures below the glass transition is presented. The theory describes the crossover phenomena from linear, branched, to crosslinked polymers in the glassy state. Fractal is by definition a set made of parts similar to the whole (ref. 1). Self-similarity is the basic notion in fractal structures. In real systems, the self-similarity is a statistical property and the physical properties of materials have to be expressed in terms of the state density (ρ) and correlation function (G) in order to determine their fractal character.

The viscoelastic relaxation of polymeric glasses is a result of the local configurational rearrangements of molecular segments (ref. 2) which is described by the motion of holes (free volumes). In this paper, we shall analyze the hole dynamics and fluctuations on a fractal lattice. The viscoelastic loss modulus is then calculated as a result of energy dissipation caused by the nonequilibrium hole density fluctuations. This leads to a structural interpretation of the relaxation function, time, spectrum.

2. THEORY

Consider polymer is cooled from liquid to glass where the sample is annealed. Since all physical properties of glasses vary slowly in time and space, the motion of holes is treated as an anomalous diffusion process. Let us consider the hole density-density correlation function

$$G(\mathbf{r},t)=\frac{<\delta n(\mathbf{r},t)\,\delta n(0,0)>}{<\delta n^2>} \qquad (1)$$

where δn is the local excess of hole number density, and the angular brackets denote an equilibrium ensemble average. Using Fourier transform in space, the anomalous diffusion equation of a quenched glass can be written as (refs. 3-4)

$$\left(\frac{\partial}{\partial t}-Dq^{2+v}\right)G(\mathbf{q},t)=\delta(t) \qquad (2)$$

where D is the local diffusion constant, **q** is the wave vector of the fluctuation, and δ is Dirac's delta function. When D is a constant, the solution of Eq. (2) displays the well-known Gaussian spreading. The exponent v produces the fractal dimension d_h, which defines a self-similar scaling between wave numbers:

$$q\sim q_v^{\,d_h}\;,\quad with\quad d_h=\frac{2}{2+v}>0 \qquad (3)$$

where q_v is the wave number on the fractal lattice, D_v is a constant and holes exhibit the Gaussian characteristics on the fractal lattice. The self-similarity of the fractal has dilation symmetry shown in Eq.(3). The fractal dimension d_h is introduced in accordance with this spatial scaling. The q^{2+v}-dependence in Eq. (2) is an ansatz because the spatial dependence of D is not known. The consequences of this ansatz are broad and far reaching. Using Eqs (2) and (3), we shall be able to derive the relaxation function (Eq. (9)), the density of states (Eq. (5)), and the stretched exponent (Eq. (11)), and the spatial dependent local diffusivity (Eq. (13)) later. Introducing the Fourier transform in time, we obtain the solution of Eqs. (2-3)

$$G(\omega)=\sum_{q_v}\frac{1}{D_vq_v^{\,2}-i\omega}=\int_0^{q_m}\frac{\rho(q_v)\,dq_v}{D_vq_v^{\,2}-i\omega} \qquad (4)$$

and the density of states

$$\rho(q_v)\,dq_v \sim q_v^{\,d_h-1}\,dq_v \sim \tau_v^{\,-d_h/2}\,d\tau_v \qquad (5)$$

where τ_v is the local relaxation time. Here the hole number spectrum is extended up to the maximum value q_m. Substituting Eq. (5) into Eq. (4) leads to the asymptotic solution:

$$G(\omega) \sim -\frac{2}{d_h}\frac{\tau^{-d_h/2}}{i\omega}, \quad for\ \omega\tau_v >> 1 \qquad (6)$$

where

$$\tau = \frac{1}{D_v q_m^2} \qquad (7)$$

is the macroscopic relaxation time. The change in the state of glass during isothermal annealing is accompanied by absorption of energy, which is related to the density fluctuations of hole. In accordance with the method of the generalized susceptibility (ref. 5), the viscoelastic loss modulus (E"), which measures the energy dissipation, is determined from Eq. (6) as

$$E''(\omega) \sim Im\, G(\omega) \sim \tau^{-d_h/2} \qquad (8)$$

In addition, we have also derived the viscoelastic relaxation modulus in accordance our hole dynamic theory (ref. 3) to be

$$E(t) = E_o exp\left[-\left(\frac{t}{\tau}\right)^{\beta}\right], \quad 0<\beta\leq 1 \qquad (9)$$

where E_o is the unrelaxed modulus. The above equation gives an asymptotic expression of the loss modulus

$$E''(\omega) \sim (\omega\tau)^{-\beta} \qquad for\ \omega\tau>>1 \qquad (10)$$

Comparing Eqs. (8) and (10) yields

$$\beta = \frac{d_h}{2} = \frac{1}{2+v} \qquad (11)$$

The parameter β of stretched exponential should be a function of local interaction, which is related to the non-zero v mentioned in Eqs (2) and (3). When we look at Eqs. (4-6), Eq. (11) confirms the customary way of relating β to the relaxation time spectrum. The glassy state relaxation is dominated by the part of the spectrum having longer relaxation times.

3. DISCUSSIONS

In accordance with Eqs. (5) and (11), one finds the diffusion length

$$<\Delta r^2>^{1/2} \equiv R \sim t^{\beta} \qquad (12)$$

and the local diffusion coefficient

$$D = \frac{R^2}{2t} \sim R^{-v} \qquad (13)$$

For linear polymers, $d_h = 1$, and Eq. (13) shows the diffusion coefficient is spatially independent because the spreading of excess holes is Gaussian (v = 0). In highly crosslinked polymers, the connectivity of hole motions decreases significantly and the diffusion coefficient decreases rapidly with distance. Matching the motion of holes and the chain motions in terms of the segmental mobility (or diffusivity) (ref. 4) of an ideal phantom network, we find that v = 4 and β = 1/6.

The mass (M)-size (L) scaling for chain networks is described by the static Hausdorff dimension (d_f) by (ref. 6)

$$M \sim L^{d_f} \qquad (14)$$

We find that this fractal dimension is related to v by

$$d_f = \frac{2(v+2)}{3} \qquad (15)$$

In the crossover region between the linear and crosslinked polymers, both the structure and property of branched polymers is a strong function of the degree of branching (or crosslinking). As branching increases, so does v, and Eqs. (3) and (15) reveal that the fractal dimension of chain networks increases but that of hole decreases. This results in a decrease in the stretched exponential for viscoelastic relaxation. At the same time, the diffusion in branched polymers becomes localized. Figure 1 shows the relationship between the dynamic stretched exponential and the static fractal dimension, which describes the topological structure of these polymers. The viscoelastic relaxations of linear, branched, and highly crosslinked glassy polymers are compared in Figure 2. The dependence of relaxation time τ on temperature and physical aging below the glass transition is given in ref.3, which shows how the Doolittle equation can be generalized from liquid to solid. Our analysis also reveals that the activation energy controlling the viscoelastic relaxation increases as β decreases, i.e., the system is changed from linear, branched to crosslinked polymers.

REFERENCES

1. B. B. Mandelbrot, The Fractal Geometry of Nature, Freeman, San Francisco, 1982.
2. J. D. Ferry, Viscoelastic properties of Polymers, 3rd Ed. John Wiley & Sons, New York, 1980.
3. T. S. Chow, Macromolecules, 25 (1992) 440-444.
4. T. S. Chow, Phys. Rev., A44 (1991) 6916-6919.
5. L. D. Landau and E. M. Lifshitz, Statistical Physics, 2rd Ed. Pergamon Press: Oxford, 1969, Chap. 12.
6. P. G. de Gennes, Scaling Concepts in Polymer Physics, Cornell Univ. Press, Ithaca, 1979.

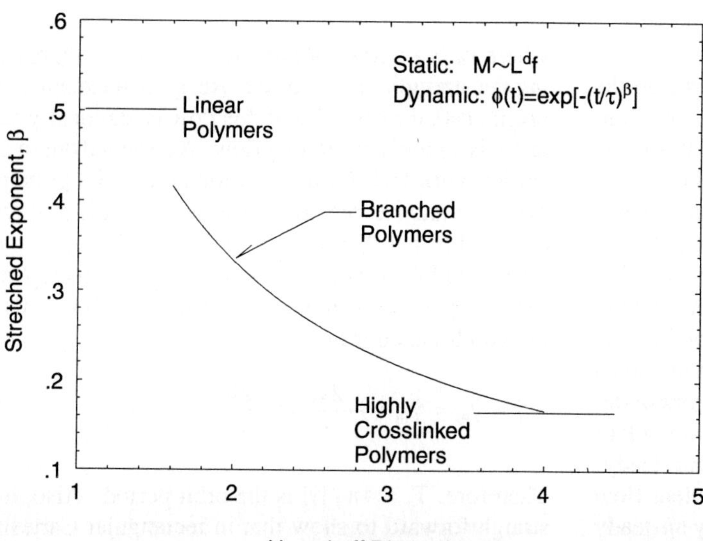

Figure 1 The dynamic stretched exponential versus the static fractal dimension.

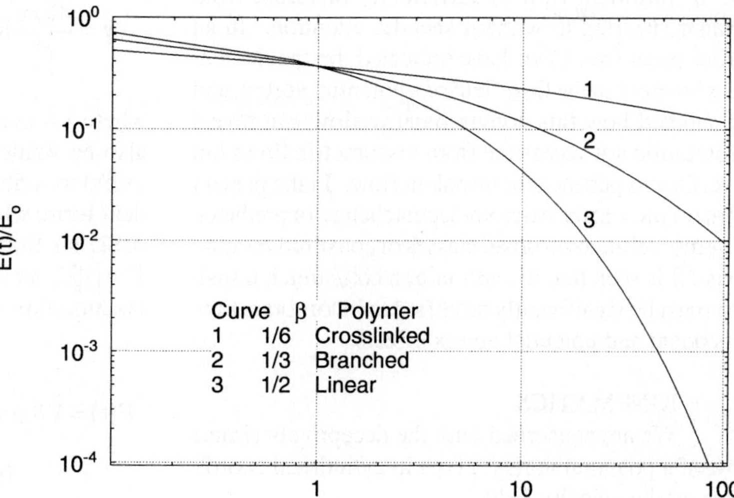

Figure 2 Viscoelastic relaxation of linear, branched and crosslinked glassy polymers.

Theoretical and Applied Rheology, edited by P. Moldenaers and R. Keunings
Proc. XIth Int. Congr. on Rheology, Brussels, Belgium, August 17-21, 1992

THE POTENTIAL VORTEX AS A PROTOTYPE FOR PREDICTIONS OF POLYMER BEHAVIOR IN UNSTEADY AND TURBULENT FLOWS

J.A. DEIBER[1] and W. R. SCHOWALTER[2]

[1]Permanent address: INTEC, Güemes 3450, 3000 Santa Fe, Argentina
[2]Department of Chemical Engineering, University of Illinois at Urbana-Champaign, Urbana, Illinois 61801 U.S.A.

1. INTRODUCTION

There has been a long-standing interest in the effect of macroscopic hydrodynamics on molecular conformations of polymers. One motivation for this interest is a desire to enhance our understanding of turbulence modification by polymers present in minute concentrations in Newtonian solvents. In this regard, conditions for a coil/stretch transition of macromolecular conformation have been cited (refs. 1-4). Most of the work to date has been focused on steady-state behavior; i.e., one considers the average molecular conformation in a given hydrodynamic flow field when that field is time-independent. An important exception is the recent work of Szeri, *et al.* (ref. 5). Association of results from steady-flow calculations or experiments with turbulent flow is, at best, questionable because of the highly unsteady environment experienced by a macromolecule in a turbulent flow field. Furthermore, the highly vortical structure of turbulent flow is sufficiently different from laminar shearing to warrant specific attention. In an earlier paper (ref. 6) we have indicated the special features inherent in the flow field of a potential vortex, and we showed how this nonviscometric flow can reveal information not accessible from viscometric flows but nevertheless pertinent to turbulent flow. In the present work we present the macromolecular behavior predicted by some well-known broad classes of constitutive equations. It is seen that the notion of a coil/stretch transition must be significantly modified to incorporate time dependent and potential vortex effects.

2. KINEMATICS

We are concerned with the deceptively simple flow of a potential vortex, given in cylindrical coordinates by the velocity field

$$\mathbf{v} = \frac{K}{r}\boldsymbol{e}_\theta = v_\theta \boldsymbol{e}_\theta, \tag{1}$$

where $\boldsymbol{e}_\theta$ is a unit vector and K is a constant representing the strength of the vortex. Recall that except at the origin, r=0, the vorticity of this flow is identically zero and it is a purely stretching flow. As pointed out in our earlier work (ref. 6), it is the rotation of the principal axes of stretch as one follows a fluid particle in its circular orbit that distinguishes this flow from the pure stretching flows familiar to polymer rheologists. Non-zero components of the shear rate tensor $\dot{\boldsymbol{\gamma}} = \nabla\mathbf{v} + (\nabla\mathbf{v})^T$ in cylindrical coordinates are

$$\dot{\gamma} = \dot{\gamma}_{\theta r} = \dot{\gamma}_{r\theta} = -\frac{2v_\theta}{r} = -\frac{2K}{r^2}|\dot{\gamma}|. \tag{2}$$

Therefore, $T_0 = 4\pi/|\dot{\gamma}|$ is the orbit period. Also, it is straightforward to show that in rectangular Cartesian coordinates the analog to eqn. (2) is

$$\dot{\boldsymbol{\gamma}} = \frac{4kxy}{r^4}\mathbf{ii} + \frac{2K(y^2 - x^2)}{r^4}(\mathbf{ij} + \mathbf{ji}) - \frac{4Kxy}{r^4}\mathbf{jj}, \tag{3}$$

where $x = r\cos\theta$ and $y = r\sin\theta$. Eqn. (3) can of course also be written in terms of θ and $|\dot{\gamma}|$. It will also be useful to write eqn. (3) in a dimensionless time dependent form, where one follows a material element in its orbit so that, $\theta(t) = 2\pi t/T_0$. Writing, $\hat{t} = t/T_0$, $\Gamma = |\dot{\gamma}|\lambda_H$ and $\boldsymbol{\Gamma} = \dot{\boldsymbol{\gamma}}\lambda_H/2$, where $\lambda_H = \zeta/4H$ is the relaxation time defined for each constitutive model,

$$\boldsymbol{\Gamma}(t) = \Gamma\{(\sin 2\pi\hat{t}\,\cos 2\pi\hat{t})\,\mathbf{ii} + \frac{1}{2}(\sin^2 2\pi\hat{t} - \cos^2 2\pi\hat{t})(\mathbf{ij} + \mathbf{ji}) - (\sin 2\pi\hat{t}\cos 2\pi\hat{t})\,\mathbf{jj}\} \tag{4}$$

Note that ζ is the friction coefficient and H is the spring constant of the connector force.

3. DYNAMICS FOR SELECTED CONSTITUTIVE MODELS

3.1 The FENE-P dumbbell model

This finitely-extensible nonlinear elastic dumbbell model has been used extensively by Bird and his coworkers (ref. 7) to make qualitative inferences about behavior of polymer solutions in various flow fields. Using the Peterlin pre-averaging approximation, one can show that the dynamic equation for the mean distance between beads <**R**> is, in dimensionless form,

$$\frac{\Gamma}{4\pi}\dot{\boldsymbol{\eta}} = \left\{\boldsymbol{\Gamma} - \frac{\left(1-\frac{1}{\eta}\right)}{2\left(1-A^2\eta^2\right)}\boldsymbol{\delta}\right\}\cdot\boldsymbol{\eta}\,, \tag{5}$$

where

$$\eta = \sqrt{\boldsymbol{\eta}\cdot\boldsymbol{\eta}} = 1 \quad \text{at} \quad \hat{t} = 0.$$

Dimensionless quantities are,

$$\boldsymbol{\eta} = \langle\mathbf{R}\rangle / \left(\frac{3}{2}a_v\right), \tag{6}$$

$$A = \frac{3}{2}\frac{a_v}{R_0}, \tag{7}$$

where a_v is the bead radius and η^2 has been used in the denominator of eqn. (5) as a consequence of setting the connector force,

$$\mathbf{F}^c = H\mathbf{R} / \left(1 - \langle\mathbf{R}\rangle\cdot\langle\mathbf{R}\rangle / R_0^2\right). \tag{8}$$

Note that in order to avoid coupling an equation for a vector quantity <**R**> (eqn. (5)) to a tensor quantity <**RR**>, we have imposed an analog to the Peterlin approximation in eqn. (8).

We now solve eqn. (5) by a Runge-Kutta routine to show the evolutionary nature of the conformation as measured by the vector $\eta(t)$. Results indicative of the possible behavior are given in Figs. 1 and 2. The first figure depicts what, in the present context, we call a "weak flow". The interbead distance oscillates between values that include the initial separation η=1. However, at sufficiently large stretch rates, such as Γ=1, a limit cycle behavior is observed, and the interbead distance evolves to an oscillatory state away from the initial separation. Note that because of the finite extensibility of the spring, bead separation will never grow without limit.

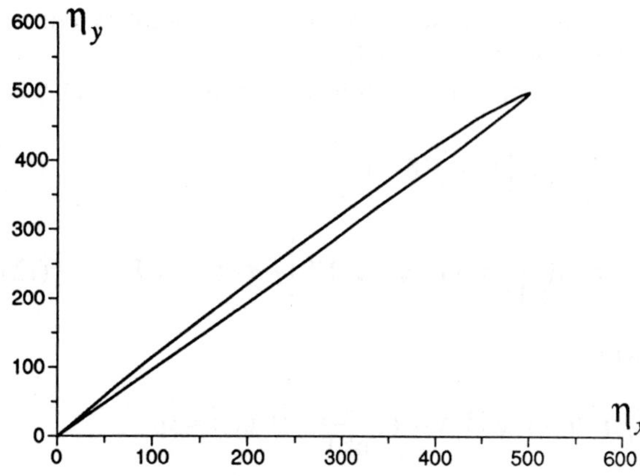

Fig. 1. Components of the dimensionless interbead distance η for the FENE-P dumbbell. Parameters are Γ = 0.01, A = 0.001.

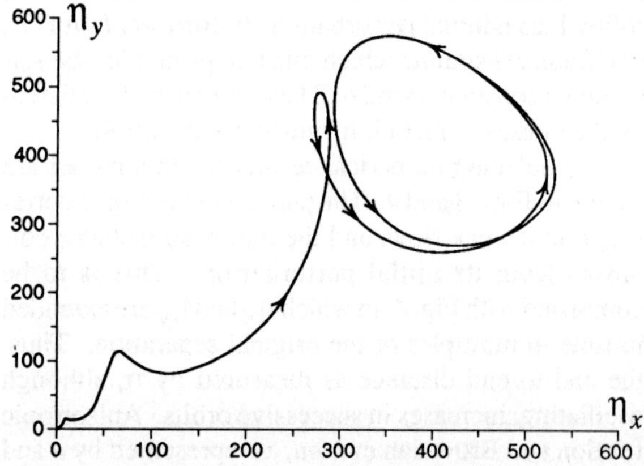

Fig. 2. Components of the dimensionless interbead distance η for the FENE-P dumbell. Parameters are Γ = 1, A = 0.001.

3.2 The Giesekus Model (ref. 8)

A description appropriate for the present work is available in Bird, *et al.* (ref. 7). After much algebra, an expression for the dynamics of the interbead vector can be written as

$$\langle\dot{\mathbf{R}}\rangle = \frac{1}{2}\dot{\boldsymbol{\gamma}}\cdot\langle\mathbf{R}\rangle - \frac{2H}{\zeta}\left(\boldsymbol{\delta} - \frac{a}{nkt}\boldsymbol{\tau}_p\right)\cdot\langle\mathbf{R}\rangle, \tag{9}$$

where $\boldsymbol{\tau}_p$ is related to the separation tensor <**RR**> by

$$\boldsymbol{\tau}_p(1+2b) = -nH\langle\mathbf{RR}\rangle + nkT\boldsymbol{\delta}. \tag{10}$$

The number density is n, and a and b are constants introduced through the anisotropic drag and Brownian forces. The absolute temperature is T and k is the

Boltzmann constant. Also, we have employed a linear connector force $\mathbf{F^c} = \mathbf{HR}$.

A dimensionless form of eqns. (9) and (10) is,

$$\frac{\Gamma}{4\pi}\dot{\boldsymbol{\eta}} = \left\{\boldsymbol{\Gamma} - \frac{1}{2}(\boldsymbol{\delta} - \boldsymbol{\tau})\right\} \cdot \boldsymbol{\eta}, \tag{11}$$

$$\boldsymbol{\tau} + \beta\left[\frac{\Gamma}{4\pi}\dot{\boldsymbol{\tau}} - \boldsymbol{\Gamma}\cdot\boldsymbol{\tau} - \boldsymbol{\tau}\cdot\boldsymbol{\Gamma}^{T}\right] - \boldsymbol{\tau}\cdot\boldsymbol{\tau} = -\alpha\boldsymbol{\Gamma}, \tag{12}$$

with

$$\sqrt{\boldsymbol{\eta}\cdot\boldsymbol{\eta}} = 1 \text{ and } \boldsymbol{\tau} = \boldsymbol{\tau}_p \frac{a}{nkT} = 0 \text{ at } \hat{t} = 0,$$

$$\alpha = \frac{2a}{1+b}, \quad \beta = \frac{(1+2b)}{(1+b)}.$$

We now solve eqns. (11) and (12) subject to the stated initial conditions. Thus, we find the evolution of η following an initial perturbation. In Bird, *et al.* (ref. 7), one finds constraints which must, in principle, be satisfied by constants a and b. These are not fully satisfied by the Giesekus model, in which $b = a$ (ref. 8).

Selected numerical results for this model are shown in Figs. 3 and 4. The parameters in Fig. 3 correspond to a weak flow, and the interbead distance collapses from its initial perturbation. This is to be contrasted with Fig. 4, in which η_x and η_y are extended in time to multiples of the original separation. Thus, the end-to-end distance as measured by η, although oscillating, increases in successive orbits. Anisotropic friction and Brownian motion, as represented by a and b, respectively, have significant effects on the macromolecular conformation.

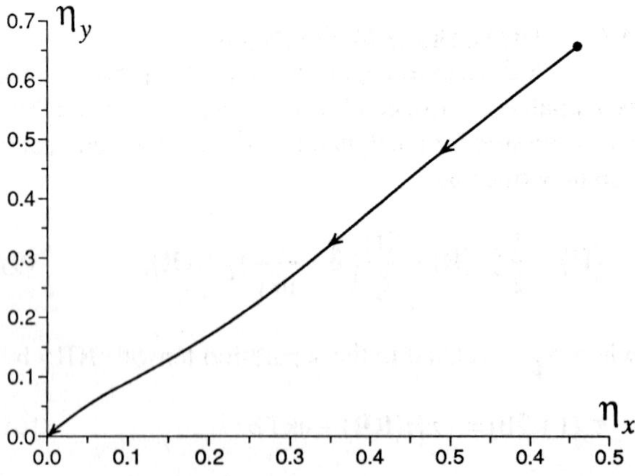

Fig. 3. Components of the dimensionless interbead distance for the Giesekus model. Parameters are $\Gamma = 0.1$, a = 0.3, b = 0.3.

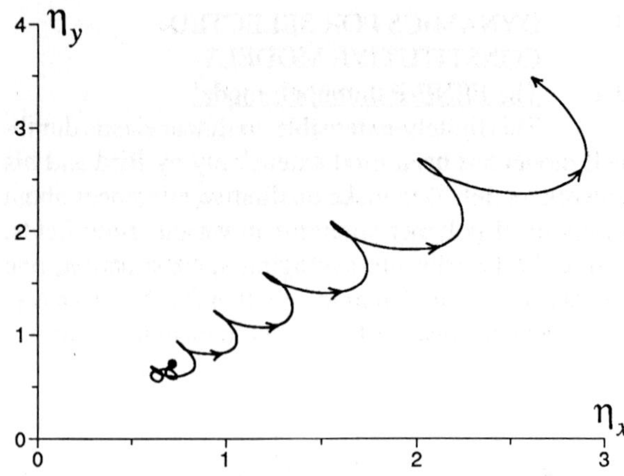

Fig. 4. Components of the dimensionless interbead distance for the Giesekus model. Parameters are $\Gamma = 2$, a = 0.3, b = 0.3.

4. CONCLUSIONS

The potential vortex offers a flow kinematics that provides new information about macromolecular conformations in flows containing important features present in turbulence. Time dependence and vortex strength are shown to be determining factors in assessing the "strength" of a flow in the sense of the influence of that flow on macromolecular stretching.

5. ACKNOWLEDGMENTS

This research was initiated while one of us (W.R.S.) was the recipient of a Guggenheim Fellowship. The hospitality of the Ecole Nationale Supérieure de Techniques Avancées and the Université Pierre et Marie Curie are gratefully acknowledged. J.A.D. acknowledges the financial assistance of the University of Illinois at Urbana-Champaign during a sabbatical leave at that institution.

REFERENCES

1. J.L. Lumley, Ann. Rev. Fluid Mech., 1 (1969) 367.
2. E.J. Hinch, Phys. Fluids, 20 (1977) 522.
3. N.S. Berman, Ann. Rev. Fluid Mech., 10 (1978) 47.
4. N. Phan-Thien and R.I. Tanner, Phys. Fluids, 21 (1978) 311.
5. A.J. Szeri, S. Wiggins and L.G. Leal, J. Fluid Mech., 228 (1991) 207.
6. J.A. Deiber and W.R. Schowalter, presentation at 62nd annual meeting, Society of Rheology, Santa Fe, NM, U.S.A., October, 1990.
7. R.B. Bird, R.C. Armstrong, O. Hassager and C.F. Curtiss, Dynamics of Polymeric Liquids, Vol. 2, (1987) 76-99.
8. H. Giesekus, J. Non-Newtonian Fluid Mech., 11 (1982) 69.

Theoretical and Applied Rheology, edited by P. Moldenaers and R. Keunings
Proc. XIth Int. Congr. on Rheology, Brussels, Belgium, August 17-21, 1992

NEW THEORETICAL ESTIMATES FOR REYNOLDS DILATANCY IN GRANULAR MATERIALS

J.D. Goddard
University of California, San Diego
Dept. Applied Mechanics and Engineering Sciences, La Jolla, CA 92093-0310

1 ABSTRACT

As an extension of the classical analysis of Reynolds (1885), the present paper offers a new theoretical model of shape-coupled volume change (dilatancy) in assemblages of rigid, quasispherical particles of arbitrary size distribution. As in the Reynolds theory, the present work rests on a purely kinematic, "mean-field" treatment of the deformation of a representative minimal particle cluster, which consists of $n = d + 1$ neighboring particles in spatial dimension d. Unlike the Reynolds analysis, the present work employs a well-defined ensemble average over random clusters, and is in principle valid for any packing density (i.e., void ratio).

As one important finding, it is concluded that, because of the inherently non-linear character of dilatancy, certain nearly periodic particle arrays appear anisotropic in dilatancy, even though they might qualify as isotropic in the small-strain limit for elastic particles.

Results from the new theory, which contains an explicit dependence on particle-contact density, are shown to compare more favorably to recent numerical simulations of Bashir and Goddard (1991).

2 EXTENDED ABSTRACT

The yield behavior and plasticity of granular media are strongly influenced by the phenomenon of dilatancy, first revealed by O. Reynolds (1885) and later adopted as the basis for the "stress-dilatancy" theories of Rowe (1962) and subsequent workers. In recent times, it has been shown that the Reynolds-Rowe ideas can be advantageously formulated in terms of the modern theory of internally constrained continua, wherein dilatancy represents the kinematical coupling between shape and volume (Goddard and Bashir 1990).

In Reynolds' conceptual limit of frictionless granules, the above formulation leads in a very direct way to the type of conical yield surface in stress space, often postulated in the literature on granular materials, which represents simple the reactive (work-free) force of constraint. If (Coulomb) friction between particles is now added, and if the resulting "active" or frictional stress is assumed proportional to the reactive, one obtains once more a conical yield surface and, in effect, the above-mentioned stress-dilatancy models.

Because of the importance of granular dilatancy to the above type of theory and in view of the sometimes conflicting or ambiguous experimental data and computer simulations, an attempt has recently been made to carefully compute the dilatancy of idealized two-dimensional granular assemblages, consisting of nearly-rigid and frictional elastic disks or circular rods (Bashir and Goddard 1991). There, it is found that the maximum value of the simple-shearing dilatancy $s = d\epsilon_v/d\gamma$, which generally occurs near states of maximum density, is different from the Reynolds-type estimate $s \cong 0.50$ both for nearly close-packed triangular arrays of equal disks (where $s \cong 0.85$) and more loosely packed polydisperse arrays (where $s \cong 0.35$). These discrepancies motivate in part the present theoretical work.

In the present analysis we imagine a granular assemblage to be replaced by an equivalent graph or network of sites or nodes connected by bonds, with particles representing the former and particle-particle contacts the latter. A fraction f of the bonds are intact, corresponding to active contacts, and a fraction $1 - f$ are broken or inactive. We assume that a set of bonds, both active and inactive, can be defined such as to represent the edges of elementary space-filling volume elements with particles as vertices, which we call simplexes. Each simplex represents the minimal cluster of particles for which volume can be so defined and, in spatial dimension d, a simplex must consist of $n = d + 1$ particles or vertices connected pairwise by $d(d+1)/2$ edges (triangles in two dimensions, tetrahedra in three dimensions, etc.). The effective properties of a granular assemblage (stress, deformation rate, etc.) can now be calculated from appropriate volume or ensemble averages over its underlying simplexes.

In any deformation of a rigid particle assemblage, the active bonds b of the underlying simplexes S obviously cannot be compressed, which leads to the global estimate for volumetric strain rate in terms of an average $<>$ over simplexes:

$$\dot{\epsilon}_v = < \sup_{b \epsilon S} tr(\mathbf{K}_S \mathbf{E}_b) > \qquad (1)$$

where for a given simplex S in space dimension d, $\mathbf{K}$ is d times the compressive deviator,

$$\mathbf{K} := (\dot{\epsilon}_v \mathbf{I} - \mathbf{D}d) \qquad (2)$$

and tr denotes the trace. Here $\mathbf{D}$ is the symmetric stretching or strain-rate tensor for simplex S, and, for any active bond $b \epsilon S$,

$$\mathbf{E}_b := \mathbf{e}_b \otimes \mathbf{e}_b \qquad (3)$$

is the dyad formed from unit bond vector $\mathbf{e}_b$.

As it stands, the above estimate is generally still intractable, since it requires detailed knowledge of the local deformation rate $\mathbf{D} = \mathbf{D}_s$ of each simplex S. If, however, one adopts a mean-field approximation for $\mathbf{D}$, replacing it by the global stretching $< \mathbf{D} >$ and if one further assumes bond activity also to be independent of the simplex S, one obtains simply that

$$\dot{\epsilon}_v = f < \sup_{b \epsilon S} tr(\mathbf{K}\mathbf{E}_b) > \qquad (4)$$

where b refers now to any bond in a simplex, f denotes the fraction of active bonds, and $\mathbf{K}$ is of course given by (2), with $\dot{\epsilon}_v$ and $\mathbf{D}$ referring now to global quantities,

Equation (4) is now amenable to direct evaluation once the simplex statistics (shape, orientation, etc.) are specified. The present work will present some specific calculations, suggesting (a) that the estimate (4) represents an approximate upper-bound on granular dilatancy, and (b) that the Reynolds-type estimates may not be as accurate.

3 REFERENCES

1. O. Reynolds, Phil. Mag. 20 (1885) 469.

2. P.W. Rowe, Proc. Roy. Soc. Lond. A 269 (1962) 500.

3. J.D. Goddard and Y.B. Bashir, in D. DeKee and P.N. Kaloni, (eds.) Rec. Dev. in Structured Continua, Vol. II Longman, 1990.

4. Y.B. Bashir and J.D. Goddard, J. Rheology 35 (1991) 849.

Theoretical and Applied Rheology, edited by P. Moldenaers and R. Keunings
Proc. XIth Int. Congr. on Rheology, Brussels, Belgium, August 17-21, 1992

SOME COMMENTS ON FRACTAL DIMENSION, FLUX AND STRESS TENSOR IN CONTINUUM MECHANICS

R. R. HUILGOL

School of Information Science and Technology, Flinders University of South Australia, G.P.O. Box 2100, Adelaide 5001 (Australia)

1. INTRODUCTION

Currently, there is a great deal of interest in fractals. For example, Feder (ref. 1) argues that viscous fingers are bounded by fractal curves. Regions with fractal boundaries raise some very difficult questions in continuum mechanics and in this paper an attempt is made to answer some of these questions.

Restricting attention to two dimensional regions with fractal curves as boundaries, some of the main problems are : the measurement of the length of the curve; the dimension of the curve; the definition of unit tangent and unit normal to the curve; whether the divergence theorem holds for such regions; and finally does there exist a stress tensor in a continuous body bounded by a fractal curve.

2. LENGTH OF A CURVE

We assume that the fractal curve is continuous. Then, it can be shown that (ref. 2, pp. 16 ff) that the curve can be approximated to any degree of accuracy by a sequence of polygonal lines, i.e., a connected thread of straight lines. Given this, we simply project (ref. 2, pp. 105 ff) each segment onto the x and y axes, measure the lengths of the projections and use the theorem of Pythagoras to arrive at the length of each segemnt. Adding the lengths of all segments and taking the limit as the number of projections goes to infinity, we arrive at the length of the curve. For example, if one examines the construction of the Koch curve (ref. 3), it is obvious that it is made up of straight lines. Also, it is clear that at the jth iteration, the length is

$$L_j = 4^{j+1}3^{-j}, \; j = -1, 0, 1, \ldots \qquad (1)$$

As $j \to \infty$, the length $L_j \to \infty$.

3. DIMENSION OF A CURVE

Since we are interested in a two dimensional region bounded by a curve, the dimension of the curve must be at least 1 and strictly less than 2, for the curve is not by itself an area. Also, since a piecewise smooth curve is of dimension 1, the fractal curve cannot be smooth. In any case, to understand how to define the dimension, consider the real line segemnt $E = [0,1]$. Cover it with $(N+1)$ intervals of length $1/N$. Clearly, the total length $(N+1)/N$ is bigger than the real length, which is 1. Consider the limit

$$\mathcal{H}^s(E) = lim_{N\to\infty}(N+1)(\frac{1}{N})^s. \qquad (2)$$

Obviously, if $0 \leq s < 1$, the limit is infinity; if $s = 1$, then the limit is 1; if $s > 1$, the limit is zero. What has been done is to cover the given line segment (or the curve) by a bunch of overlapping intervals and see what happens to the limit as s is varied. There is, usually, one value of s at which the limit suffers a jump and this value of s is the Hausdorff dimension of the set. In the above case, the Hausdorff dimension is 1.

As a second example, note that for the Koch curve (1)

$$4^{j+1}3^{-js} \to 4 \qquad (3)$$

as $j \to \infty$, if $s = log\ 4/log\ 3$. Hence this fractional value is the Hausdorff dimension of the Koch curve (ref. 3).

The above idea of overlapping intervals is used in determining the upper bound, or the box dimension (ref. 4). Here, the idea is to take the given set (or curve) E and cover it by closed intervals of length δ. Let $\mathcal{N}(E, \delta)$ denote the number of intervals needed. Reduce the size of δ and repeat. Plot the values of $log\ N$ as the ordinate and $log\ (2/\delta)$ as the abscissa. The slope of the line as $\delta \to 0$ is the box dimension of the curve.

So far we have discussed the concepts of length and dimension of a fractal curve. These curves have a curious property (ref. 3) in that every fractal curve whose (Hausdorff) dimension lies between 1 and 2 is of infinite length; moreover, every part of it is of infinite length. Thus, there is no such thing as a fractal curve of finite length with a dimension between 1 and 2.

4. UNIT TANGENT AND NORMAL

If a given curve is of class C^2 then the unit tangent and normal are easy to define. If the curve is not smooth, then we choose a point $\mathbf{x}$ on it and let $\mathbf{t}$ be a unit vector through $\mathbf{x}$. With this point as centre, draw a circle $B(\mathbf{x}, \rho)$ of radius ρ. Draw a double sector S through $\mathbf{x}$ so that it makes an angle $\pm\phi$ with respect to $\mathbf{t}$. If, for every sector angle $\phi > 0$, the length of the curve in $B - S$ goes to 0 as $\rho \to 0$, then $\mathbf{t}$ is the required tangent. This construction of the tangent, called the measure-theoretic tangent, does not use any concepts from calculus. It needs to be modified slightly if the dimension of the curve is greater than 1. Despite the elegant simplicity of construction, in practice it is not very useful because Falconer (ref. 3) has shown that for fractal curves with dimension greater than 1, there is no measure-theoretic tangent almost everywhere.

Let us now consider how to construct a unit external normal to a region Ω bounded by a curve. Again, let $\mathbf{x}$ lie on the boundary. Through this point, draw an external unit vector $\mathbf{n}$ and a line perpendicular to this vector through $\mathbf{x}$. So, we have divided the plane into two regions : P^+ into which the vector $\mathbf{n}$ points and P^- on the opposite side. Let $B(\mathbf{x}, \rho)$ denote the circle with $\mathbf{x}$ as centre and radius ρ. Let us define two other regions : $B^+ = B \cap P^+$, and $B^- = B \cap P^-$. If simultaneously

$$lim_{\rho \to 0} \frac{Area\ of\ [B^+ \cap \Omega]}{Area\ of\ B^+} = 0, \qquad (4)$$

and

$$lim_{\rho \to 0} \frac{Area\ of\ [B^- \cap \Omega]}{Area\ of\ B^-} = 1, \qquad (5)$$

then we say that $\mathbf{n}$ is the measure-theoretic unit external normal. Unfortunately, for fractal curves of dimension between 1 and 2, a measure-theoretic unit normal does not exist almost anywhere.

5. DIVERGENCE THEOREM

The non-existence of a unit normal almost anywhere raises the question : does the divergence theorem hold for a region with a fractal boundary ? Physically speaking, if we imagine a flow of an incompressible fluid into and out of a bounded region with one part of its boundary being a fractal curve, then surely the conservation of mass law must hold. That is, the divergence theorem must hold. The question is how to prove it and what is the best way to approximate the divergence of a vector field in such a region with a fractal boundary. It must be noted that traditional proofs of the divergence theorem (ref. 5) require that a bounded domain have a finite perimeter, i.e., the bounding curve be of finite length. Hence, a new technique is needed to prove the theorem when the boundary has infinite length, as is the case with fractal boundaries. Such a theorem has been proved recently by Harrison and Norton (ref. 6) and the technical details are given there.

Briefly, they calculate the line integral around the region with a smooth approximator to the fractal curve. This approximator must be such that the area between the two goes to zero in the limit. Then the divergence of the vector field over the given domain is equal to that over the approximate domain. The reason the theorem works is because the boundary integral is sensitive to cancellation and approximators to a fractal curve have normal vectors that change directions quite a lot and so create a lot of cancellations.

6. STRESS TENSOR

At this moment, there is no satisfactory answer to the problem of defining a stress tensor in a region bounded by a fractal curve. What is needed is a merger of the work of Harrison and Norton (ref. 6) with the earlier papers by Gurtin and Martins (refs. 7-8) to establish the precise meaning to be given to the relation

$$\mathbf{t}(\mathbf{x},\mathbf{n}) = \mathbf{T}(\mathbf{x})\mathbf{n}, \tag{6}$$

where $\mathbf{t}$ is the traction vector at $\mathbf{x}$ on the boundary through which the unit external normal $\mathbf{n}$ passes and $\mathbf{T}$ is the stress tensor at $\mathbf{x}$.

ACKNOWLEDGEMENTS

I wish to thank my colleague Dr. S. Phadke and Professor A. Norton for their help in the preparation of this paper.

REFERENCES

1. J. Feder, Fractals, Plenum, New York, 1988.

2. A. V. Alexandrov and Yu. G. Reshetnyak, Theory of Irregular Curves, Kluwer Academic, Boston, 1989.

3. K. J. Falconer, Fractal Geometry, Wiley & Sons, New York, 1990.

4. M. F. Barnsley, Fractals Everywhere, Academic, New York, 1988.

5. W. Noll and E. G. Virga, Arch. Rational Mech. Anal., 111 (1990) 1 - 31.

6. J. Harrison and A. Norton, Indiana Univ. Math. J., 40 (1991) 567-594.

7. M. E. Gurtin and L. C. Martins, Arch. Rational Mech. Anal., 60 (1975) 305-324.

8. L. C. Martins, Arch. Rational Mech. Anal., 60 (1975) 325 - 328.

Theoretical and Applied Rheology, edited by P. Moldenaers and R. Keunings
Proc. XIth Int. Congr. on Rheology, Brussels, Belgium, August 17-21, 1992

SECONDARY FLOWS OF NON NEWTONIAN FLUIDS IN RECTILINEAR PIPES

G.MAYNÉ

Département de Mathématique, U.L.B., Campus Plaine, C.P.218/1, 1050 Bruxelles

1. INTRODUCTION

Let us consider the flow of an incompressible Rivlin-Ericksen fluid through a rectilinear tube of non circular cross-section and adopt the constitutive equation valid for viscometric shear flows (refs. 1,2). The extra stress is given by

$$\tau = \eta \overset{(1)}{A} + (N_1 + N_2) \overset{(1)}{A}{}^2 - \frac{N_1}{2} \overset{(2)}{A} \tag{1}$$

where $\overset{(1)}{A}$ and $\overset{(2)}{A}$ are respectively the first and second Rivlin-Ericksen tensor. η, N_1, N_2 are constitutive functions depending on

$$I_2^2 = \frac{1}{2} tr \overset{(2)}{A} \tag{2}$$

Let us suppose that the constitutive functions may be approximated by polynomials. It will be the case if they depend on αI_2^2 where α is a small constitutive constant.

If the cross-section of the pipe is non circular, the flow is generally not rectilinear and a secondary flow appears transversally. We shall assume that N_2 is a small function of αI_2^2:

$$N_2 = \varepsilon \left(\overset{0}{N_2} + \alpha \overset{1}{N_2} I_2^2 + \alpha^2 \overset{2}{N_2} I_2^4 + \ldots \right) \tag{3}$$

where ε is another small constitutive constant. Then, the normal stress difference, measured by N_2, is small with respect to the shear stress and the secondary flow may be considered as a perturbation of the Poiseuille rectilinear basic flow. For simplicity, we shall only consider first order terms in ε and second order terms in α:

$$\eta = \eta_0 + \alpha \eta_1 I_2^2 + \alpha^2 \eta_2 I_2^4 \tag{4}$$

$$N_1 = \overset{0}{N_1} + \alpha \overset{1}{N_1} I_2^2 + \alpha^2 \overset{2}{N_1} I_2^4 \tag{5}$$

2. EQUATIONS OF MOTION

Taking cartesian coordinates x, y, z with axis z parallel to the tube, the velocity field will be of the form:

$$v_x = \varepsilon \psi_{,y} \quad v_y = -\varepsilon \psi_{,x} \quad v_z = w(x, y; \alpha, \varepsilon) \tag{6}$$

where

$$\psi(x, y; \alpha) = \psi_0 + \alpha \psi_1 + \alpha^2 \psi_2 \tag{7}$$

$$w = w_{00} + \alpha w_{01} + \alpha^2 w_{02} + \varepsilon \left(w_{10} + \alpha w_{11} + \alpha^2 w_{12} \right) \tag{8}$$

The first indice indicates the order in ε and the second the order in α. Neglecting the external forces, the equation of motion is

$$\rho\frac{d\bar{v}}{dt} = -grad\ p + div\tau \tag{9}$$

From (9), we obtain six partial differential equations corresponding to each order in ε and α. At order (0,0), (0,1), (0,2), we get respectively

$$\Delta w_{00} = -\frac{G}{\eta_0} \text{ where } G = -\frac{\partial p}{\partial z} \tag{10}$$

$$\Delta w_{01} = \frac{\eta_1}{\eta_0} D_{01}(w_{00}) \tag{11}$$

$$\Delta w_{02} = \frac{\eta_1}{\eta_0} D_{02}(w_{00}, w_{01}) + \frac{\eta_2}{\eta_0} E_{02}(w_{00}) \tag{12}$$

where D and E are definite differential operator acting on w_{00} and w_{01}.
The elimination of p between the x and y-components of (9) gives rise to one equation for the stream function ψ of the secondary flow. At order (1,0), (1,1), (1,2), we obtain respectively:

$$\Delta^2\psi_0 = 0 \tag{13}$$

$$\eta_0\Delta^2\psi_1 = \overset{0}{N_2}\, D_{11}(w_{00}, w_{01}) + \overset{1}{N_2}\, E_{11}(w_{00}) \tag{14}$$

$$\eta_0\Delta^2\psi_2 = \eta_1 D_{12}(w_{00}, \psi_1) +$$

$$\overset{0}{N_2}\, E_{12}(w_{00}, w_{01}, \Delta w_{02}) +$$

$$\overset{1}{N_2}\, F_{12}(w_{00}, w_{01}) + \overset{2}{N_2}\, G_{12}(w_{00}) \tag{15}$$

where D, E, F, G are definite differential operators acting on w_{00}, w_{01}, w_{02} and ψ_1.
The solution of (13) which satisfies the classical adhesion conditions on the boundary of the cross-section is $\psi_0 = 0$.
Of course, (10) is the classical equation of the Poiseuille flow.

3. THE CASE OF ELLIPTICAL CROSS-SECTION

The equation of the cross-section $\mathcal{C}$ is

$$ax^2 + by^2 - 1 = 0 \tag{16}$$

We choose x on the major axis of the ellipse so that $0 < a < b$. The solutions of (10) and (11) which vanish on $\mathcal{C}$ are

$$w_{00} = k(ax^2+by^2-1) \text{ with } k = -\frac{G}{2\eta_0(a+b)} \tag{17}$$

$$w_{01} = \frac{\eta_1}{\eta_0} k^3(ax^2 + by^2 - 1)(qx^2 + ry^2 + s) \tag{18}$$

where q, r, s are definite rational homogeneous functions of a and b. Now, the right hand members of (14) and (15) are known functions of x and y.
The solution of (14) such that $\psi_1 = 0$; $\psi_{1,x} = 0$; $\psi_{1,y} = 0$ on $\mathcal{C}$ is

$$\psi_1 = 4k^4 M_2 \frac{ab(b^2-a^2)}{5a^2+6ab+5b^2} xy(ax^2+by^2-1)^2 \tag{19}$$

The solution of (15) which satisfies similar boundary conditions is

$$\psi_2 = 4\frac{\eta_1}{\eta_0} k^6 M_2 ab(a-b)xy(ax^2 + by^2 - 1)^2$$

$$\times(b\gamma x^2 + b\delta y^2 + h) \tag{20}$$

where

$$\gamma = \gamma_1 - L_2\gamma_2 \qquad \delta = \delta_1 - L_2\delta_2 \qquad h = h_1 - L_2 h_2$$

$$M_2 = \frac{1}{\eta_0}\left(\frac{\eta_1}{\eta_0}\overset{0}{N_2} - \overset{1}{N_2}\right) \qquad L_2 = \frac{\eta_0}{\eta_1}\frac{\eta_2 \overset{0}{N_2} - \eta_0 \overset{2}{N_2}}{\eta_1 \overset{0}{N_2} - \eta_0 \overset{1}{N_2}}$$

$\gamma_1, \gamma_2, \delta_1, \delta_2, h_1, h_2$ are definite rational functions of $c = \frac{a}{b}$.
The stream function of the secondary flow is

$$\psi = 4\frac{\eta_1}{\eta_0} M_2\alpha^2 k^6 ab(a-b)xy(ax^2 + by^2 - 1)^2 \times$$

$$(S + b\gamma x^2 + b\delta y^2 + h) \tag{21}$$

where the dimensionless constant S is defined by

$$S = -\frac{\eta_0}{\alpha k^2 \eta_1} R \text{ with } R = \frac{a+b}{5a^2+6ab+5b^2}$$

4. DISCUSSION

Owing to the symmetry of ψ, we only have to consider this function on the domain $\mathcal{D}$ defined by $x > 0$ $y > 0$ $ax^2 + by^2 - 1 < 0$. The stream lines in this quarter of ellipse are closed curves in the vicinity of the points corresponding to extrema of the surface $z = \psi(x, y)$. This surface will have two extrema if the conic defined by

$$S + b\gamma x^2 + b\delta y^2 + h = 0$$

intersects $\mathcal{D}$. Explicit inequalities can be obtained in order to fulfil this condition. Otherwise, the surface has only one extremum. So, the global look of the secondary flow and specially the number of vortices of the stream lines can be discussed in term of the shape constant c, the dimensionless constitutive constant L_2 and the dimensionless constant S depending on the longitudinal pressure gradient. For each value of L_2 and c, the values of S such that the stream lines have two vortices lie in an interval $\Delta \subset R$. The main conclusion of the discussion are:

1. if $\eta_0\eta_1 < 0$ (pseudo plastic fluids), $S > 0$: Δ goes to the right and increases when L_2 increases from a definite lower bound ℓ'; ℓ' slightly decreases with the eccentricity of the cross-section ($\ell' = 3.94$ when $c = 0.1$, $\ell' = 3.28$ when $c = 0.9$)
2. if $\eta_0\eta_1 > 0$ (dilatant fluids), $S < 0$: Δ goes to the left and increases when L_2 decreases from a definite upper bound ℓ''; ℓ'' slightly decreases with the eccentricity of the cross-section ($\ell'' = 4.94$ when $c = 0.1$, $\ell'' = 3.66$ when $c = 0.9$).

Consequently, in both cases, the stream lines of the secondary flow will have two vortices if the longitudinal pressure gradient G will lie between two definite bounds which decrease and go closer to each other when $|L_2 - \ell'|$ or $|L_2 - \ell''|$ increases. These bounds decrease with the eccentricity of the cross-section if $|L_2 - \ell'|$ or $|L_2 - \ell''|$ are great enough ($L_2 > 5$ or $L_2 < 0$).

REFERENCES

1. G. Böhme, Non Newtonian Fluid Mechanics, North Holland, Amsterdam, 1987, p.66.
2. A.G. Dodson, P. Towsend and K. Walters, Computers and Fluids, 2(1974) 317-338.

Theoretical and Applied Rheology, edited by P. Moldenaers and R. Keunings
Proc. XIth Int. Congr. on Rheology, Brussels, Belgium, August 17-21, 1992

THE MATHEMATICAL SIMULATION OF RELAXATION PHENOMENA IN GAS-LIQUID SYSTEMS

I.SH. AKHATOV, SH.M. FAIZOV

Department of Continuous Media Mechanics, Bashkir State University,
32 Frunze Str., Ufa 450074 (Russia)

The relaxational behaviour of non-Newtonian liquids (clay solutions, polymer solutions) is well known, when thorough compression process takes place (ref.1).For the theoretical description of such phenomena Thomson-Poynting model (ref.2) was usually used. According to this model the pressure of instantly compressed heterogeneous liquid is decreases in time by experimental law. Analogous effect for gas-liquid mixtures was discovered too (ref.3). But our experimental results are show that in this case the systematic divergence from the exponential law is take place. On the initial stage we have more rapid pressure decrease in comparison with exponential law. But on the final stage it is opposite. This effect is takes place when pressure of mixture is considerably larger than saturation pressure. The physical reasons and respective mechanism of this phenomenon are not distinct now. But gas-liquid mixtures for high pressures behave itself as if there are small gas bubbles in it (ref.4).

The new model for description of relaxation processes which occur under the compression of gas-liquid mixtures is suggested in this report. According to this model the pressure relaxation is considered to be the dissolution of compressed gas bubbles. The combine solution of mass balance equation of gas in bubbles and external diffusion problem is considered. As a result we have a "nonlocal" relaxation state equation, which includes derivatives of the fractional order from pressure and density with respect to time (ref.5).

$$a_1 dp'/dt + bD_t^{1/2}p' + p' = a(a_2 dr'/dt + bD_t^{1/2}r' + r')$$

Here pressure p and density r perturbation are noted by dashes; parameters a, a_1, a_2, b are functions of physical properties of mixture.

The method of identification of proposed state equation coefficients on the experimental data is worked out. This method based on some asymptotical qualities of exact solutions of "nonlocal" state equation. Method was tested on liquefied petroleum gas mixtures. It was demonstrated that proposed equation is describes the processes of pressure relaxation more correctly than classical Thomson-Poynting equation.

REFERENCES

1. A.Kh. Mirzajanzade, R.B. Mamedzade, Ya.M. Rasizade, I.A. Shvetsov and N.M. Sherstnev, Isvestiya vusov, Neft i gas, 2 (1976) 53-58.
2. J.H. Poynting and J.J.Thomson, Properties of Matter, London, 1902.
3. I.M. Ametov, A.I. Mamedzade and G.Kh. Melikov, Dokl. AN Azerb. SSR, 41 (1985) 35-41.
4. Yu.A. Buevich, Injenerno-fizicheski journal, 52 (1987) 394-401.
5. I.Sh. Akhatov, Dinamika sploshnoi sredi, 100 (1991) 23-29.

Theoretical and Applied Rheology, edited by P. Moldenaers and R. Keunings
Proc. XIth Int. Congr. on Rheology, Brussels, Belgium, August 17-21, 1992
1992 Elsevier Science Publishers B.V.

THEORETICAL ANALYSIS OF VISCOELASTIC FLUID FLOW IN TUBE

V. M. ARKHIPOV, O. I. SKULSKY, Ye. V. SLAVNOV

Institute of Continuous Media Mechanics, Ural Branch, Russia Academy of Sciences, 614061 Perm (Russia)

1. INTRODUCTION

In this paper the numerical analysis of the flow of a Leonov fluid in straight circular tube has been performed. To solve the problem the finite element method was applied.

2. ALGORITHM OF SOLUTION

To calculate the velocity field of the viscoelastic flow we have developed an algorithm allowing the solution of the following effective viscous problem to be used

$$\nabla \circ \sigma = 0, \quad \nabla \circ v = 0 \tag{1}$$

$$\sigma = -\rho I + 2\eta_{ef} e \tag{2}$$

where σ is the stress tensor, v is the velocity vector, η_{ef} is the effective viscosity, and e is the rate of deformation tensor. After deriving a correspondence condition between the velocity field in the effective viscous problem and that in the viscoelastic flow, it is possible to get the relation for the effective viscosity in such a form

$$\eta_{ef} = D / tr(e^2) \tag{3}$$

where D is the dissipation of viscoelastic fluid. The iterative procedure is based on the equation (3). The velocity field obtained from the solution (1)-(2) is used to calculate equations for the rheological model. This procedure allows to determine the Finger finite elastic strains C, the tensor of irreversible deformation e^p, and the extra-stress tensor τ.

The dissipative function is defined as $\tau : e^p$. According to (3), after determination of a new value of the effective viscosity, the problem (1)-(2) is again solved. Iterations proceed until the effective viscosity and finite strain values will become stable. Further, the pressure value can be found using the dynamic equation.

3. RESULTS

First, it is the solution of a number of test problems that has been performed and the results expected were obtained. Further, we carried out the numerical investigation of the flow of viscoelastic fluids in tube. The analysis of the results shows that in the case of fluids where the elastic strains answer the possibilities of the elastic potential used in this work, the developed numerical scheme is stable. Based on the the results of the numerical solutions we obtained the expressions for the elastic strain components and the volume rate of flow using the similarity method and the dimensional analysis. For the model with one Maxwell relaxator these expressions are

$$C = N + K(El_3 d/l)^a \tag{4}$$

$$Q = K_1/(El_3)^a (d/l)^b De^{-1} \tag{5}$$

where El_3 is the third elastic number, De is the Deborah number, d is the diameter, and l is the channel length. For maximum values of the elastic deformation tensor components (4) we have found the values of constants. The formulae obtained here give a fairly good description of the numerical results. The computational algorithm for viscoelastic flow calculations where components of the stress tensor τ are not used as the unknowns has been developed.

Theoretical and Applied Rheology, edited by P. Moldenaers and R. Keunings
Proc. XIth Int. Congr. on Rheology, Brussels, Belgium, August 17-21, 1992

DETERMINATION OF RELAXATION TIME SPECTRA: DIFFERENT METHODS COMPARED

CHR. FRIEDRICH and H. BRAUN

Materialforschungszentrum der Albert-Ludwigs-Universität, St.-Meier-Str. 31a, 7800 Freiburg, Deutschland

1. INTRODUCTION

From the mathematical point of view, the relaxation time spectrum is a linear integral transform, which is hard to invert. Therefore, it is of interest to find suitable and available procedures to calculate the spectrum.

2. METHOD

Who wants to find the logarithmic density function of the relaxation times (i.e. relaxation time spectrum), those have to solve e.g. the following equation for the searched function $H(\lambda)$, where $G^*(\omega) = G' + iG''$ is the known complex modulus.

$$G^*(\omega) = \int_{-\infty}^{\infty} H(\lambda) \frac{i\omega\lambda}{1 + i\omega\lambda} \, d\ln\lambda \qquad (1)$$

The first class of methods is based on an analytical inversion of relationship (1), where several approximations for the kernel function are used (ref.1). Though these methods are not so exactly, they have been used up to now. Many methods (2. class) have been developed based on the numerical solution of the discretizised Eq.(1). However, only the regularization methods taking into account the ill posed character of this type of equation are exactly enough to find the solution (for details see e.g. ref.2).
A small window, which exists between the forbidden methods of the classical algebra and the very complex regularization methods, was used by Baumgärtel and Winter (ref.3) to create a method for spectrum calculation.
The direct analytical solution of Eq.(1) has had no meaning. This is due to the fact that e.g. the inversion was not found for functions realistic enough to fit the G' data. Recently, a realistic and invertibel function was proposed (ref.4). Eq.2 describes the material function G' of a fractional differential generalization of the Standard solid model, where $y=\omega\lambda_0$. This equation has 5 fit parameters: G_∞ - the equilibrium modulus, G_0 - the plateau modulus, λ_0 a characteristic time and two power law exponents c and d.

$$\frac{G' - G_\infty}{\Delta G} = \frac{y^d\,[\cos(\frac{d}{2}\pi) + y^c\cos(\frac{d-c}{2}\pi)]}{1 + 2y^c\cos(\frac{c}{2}\pi) + y^{2c}} \qquad (2)$$

The corresponding function H was calculated in ref.4 and is given here with $z = \lambda/\lambda_0$.

$$\frac{H(z)}{\Delta G} = \frac{1}{\pi}\,\frac{z^{c-d}\,[\sin((d-c)\pi) + z^c\sin d\pi]}{1 + 2z^c\cos c\pi + z^{2\alpha}} \qquad (3)$$

The efficiency of these equations and this method as well as the comparison with the other mentioned numerical methods will be demonstrated for an oil tar blend. The data were fitted by two modes of Eq.(2), a glassy mode and a terminal relaxation mode. The results are shown in the figure.

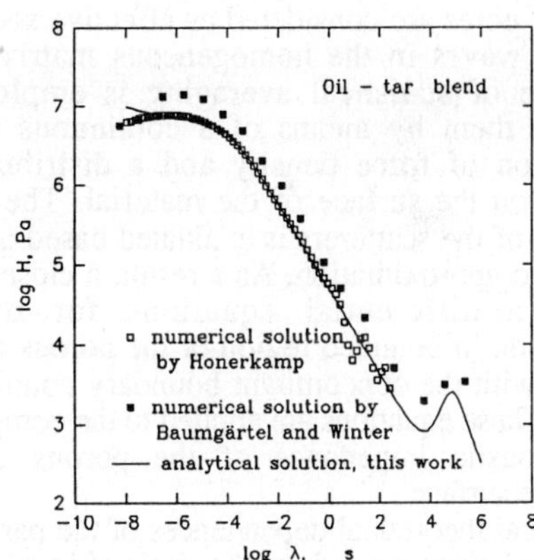

3. REFERENCES

1. J.D. Ferry, Viscoelastic Properties of Polymers, Wiley, N.Y. 1980
2. J. Honerkamp, J. Weese, Macromolecules, 22 (1989), 4372-4377
3. M. Baumgärtel, H.H. Winter, Rheol. Acta, 28 (1989), 511-519
4. Chr. Friedrich, H. Braun, subm. for publication

Theoretical and Applied Rheology, edited by P. Moldenaers and R. Keunings
Proc. XIth Int. Congr. on Rheology, Brussels, Belgium, August 17-21, 1992

PROPAGATION, REFLECTION AND REFRACTION OF ELASTIC WAVES IN POROUS MEDIA

M. GANDELSMAN[1]

[1] Department of Chemical Engineering, Ben-Gurion University, Beer-Sheva 84105 Israel

A new continuous model for multiple scattering of elastic waves in bounded regions of random heterogeneous material with long parallel cylindrical pores is presented. The wavelength is allowed to be comparable with the diameter of the cylinders and the volume concentration of pores is not suggested to be small. The formalism developed is applied to the computation of the characteristics of plane elastic waves reflected from the interface between the porous material under consideration and a homogeneous elastic medium, while the incident wave is directed from the homogeneous medium. The main purpose of the work is the determination of the relationships between the structural parameters of the porous medium and the characteristics of the reflected wave at various angles of incidence and various directions of polarization of incident wave.

In the framework of the theoretical method presented pores are considered as effective sources of scattered waves in the homogeneous matrix, and a procedure of statistical averaging is employed to represent them by means of a continuous volume distribution of force density and a distribution of tractions on the surface of the material. The mutual influence of the scatterers is evaluated based upon the mean-field approximation. As a result, a close system of integro-differential equations for dynamic deformation in bounded region of the porous medium together with the concomitant boundary conditions is derived. These equations are applied to the computation of the acoustic impedance of the porous material with plane surface.

Typical theoretical dependences of the parameters of reflected plane wave from the angle of incidence are shown in Fig. 1. In the present example it is assumed, that the wave vector is perpendicular to the axis of the cylindrical pores and the vector of polarization is directed along the axis of the cylinders. So, the case of anti-plane deformation (SH wave) is considered. The rigidity of the matrix is chosen to be three times more than that of the matched medium, while the densities of the both materials are assumed to be equal. The reflection coefficient (curve 1) and the phase shift between the incident and reflected SH waves (curve 2) are computed as functions of the angle of incidence for the porous medium with volume fraction of pores 0.3 at $K_mR = 1$ (where K_m is the wave number of the shear wave propagating in the homogeneous matrix, R is radius of pores).

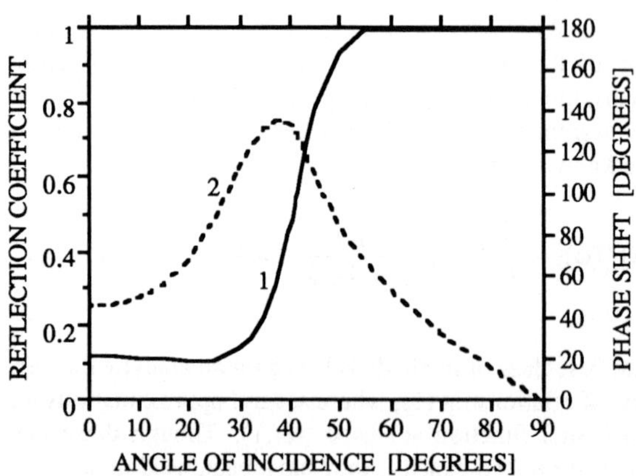

Fig. 1 Calculated characteristics of reflected wave vs. the angle of incidence: solid line (1), reflection coefficient; dashed line (2), phase shift.

The main result of the theoretical study is as follows. At high frequencies the characterization of the reflected wave is not exhausted by either the effective elastic moduli or the effective (in general, complex) wave number of the infinite porous medium. Specific high frequency elastic properties of bounded region of porous material manifest themselves in two effects: First, the reflected wave obtains phase shift at the angles of incidence lesser than the critical angle of total internal reflection (which angle is equal to 41.17 degrees under the chosen conditions). Second, the value of the reflection coefficient can be less than unit at the angles of incidence exceeding the critical angle, due to attenuation of elastic waves in the heterogeneous material. The latter effects can be used for evaluating the fine structure of porous media by means of ultrasound waves.

Theoretical and Applied Rheology, edited by P. Moldenaers and R. Keunings
Proc. XIth Int. Congr. on Rheology, Brussels, Belgium, August 17-21, 1992

GENERALIZED CREEP-RELAXATION FUNCTIONS IN LINEAR VISCOELASTICITY

S. HAZANOV

Department of Materials, Swiss Federal Institute of Technology, EPFL Ecublens MX-G,1015 Lausanne Switzerland

1. Viscoelastic materials of Volterra type are considered,

$$\sigma = E\varepsilon - \int_{-\infty}^{t} G(t-\tau)\varepsilon(\tau)d\tau \tag{1}$$

where G is the heredity kernel. Or, in the convolution form,

$$\sigma = \int_{\sigma}^{t} r(t-u)d\varepsilon(u) \tag{2}$$

where r is the relaxation function.
The problem is to investigate the class of admissible functions G,r. Traditional mechanical models (springs - dashpots), recommend for everyday practice the exponential series. Phenomenological creep-relaxation models of real materials lead to the kernels with weak (integrable) singularity at the initial point - Abel type and their various combinations with exponential ones, such as Mittag-Leffler and exponential-fractional Rabotnov functions. In other words, heredity kernels are usually taken positive and monotone decreasing.

In "(ref. 2)" it was shown that the work positivity during the deformation process can be assured by the positivity of cosine Fourier transformation of the relaxation function (sine Fourier transformation of heredity kernels). The last is satisfied "(ref.1)" by taking the relaxation functions as non-negative, monotone decreasing, limited from below and convex down.

2. Nevertheless the class of thermodynamically admissible kernels (positive cos-F transform), is much more vast and includes as well a lot of non-monotone and non-positive functions, for instance the oscillating ones,

$$K(t) = \exp(-pt)\sin qt \tag{3}$$

$$K^s(\omega) = \frac{2\,p\,q\,\omega}{((p-q)^2 + \omega^2))((p+q)^2 + \omega^2)} \tag{4}$$

(4) shows that its sin-F transform is positive when p,q,ω >0. To satisfy the other, pure thermodynanic criteria, one must make additional assumptions, concerning the free energy expression. For Staverman-Schwarzl bilinear form "(ref.3)",

$$f = 1/2\int_0^t\int_0^t r\,(2\,t - u - v\,)\,d\varepsilon(v)\,d\varepsilon(u) \tag{5}$$

the positivity of the dissipation density is quite assured by the positivity of the relaxation function spectrum r(p), où:

$$r = r_\infty + \int_0^\infty \rho(p)\exp(-\,pt)\,dp \tag{6}$$

Calculations show that the kernel (3) satisfies it as well, at least for the small time values.

Thus the physical admissibility of such "generalized" creep-relaxation functions, locally negative (negative viscosity) can be shown both from the mechanical (work positivity) and. thermodynamic (dissipation positivity) criteria.

3. Moreover, the kernels of this type can be obtained even from classic mechanical spring-dashpot models, inserting in them the inertial terms. For instance, inserting a mass between the spring and the Foigt element, connected in series, we'll obtain for the small viscosity values the creep kernel exactly in the form (3), and for the high values the relaxation kernel is the difference of two exponential terms, otherwise non-monotone.

4. Among the applications of generalized creep-relaxation functions one can give biologic materials with oscillating (falling) stress-strain diagramms. For instance, multitransitional effects in timber "(ref.4)", usually described by a chain of classic viscoelastic models, can be easily explained by means of only one model but with non-traditional kernel. And surely one must mention here the heterogeneous materials (composites), where the homogenization procedure enables to use such kernels in order to describe some specific effects, such as geometric dispersion, oscillating wave profils, etc."(ref.5)".

In other words, a heterogeneous material with relatively simple constitutive laws of the composants can be described quite efficiently by a homogeneous model, but with more complicated "physics" (generalized creep-relaxation functions, negative viscosity). This "equivalence" of geometry and physics can give numerous important applications,for instance opening the new ways in constructing the models of solids.

REFERENCES

1. Y. Rabotnov , Elementi nasledstvennoj mekhaniki tverdykh tel., Nauka, Moscow,1977,p.66.
2. S.Breuer, E.Onat, Quartely of Applied Mathematics XIX (1962), no 4 ,138-145.
3. R.Christensen,Theory of Viscoelasticity.Introduction, Academic Press, New York,1971,p.123.
4. C.Huet, Annales de ITBTP,1988,no 469,35-53.
5. S.Hazanov, Mekhanika Polymerov,no 2 (1976) 364-367

Theoretical and Applied Rheology, edited by P. Moldenaers and R. Keunings
Proc. XIth Int. Congr. on Rheology, Brussels, Belgium, August 17-21, 1992

ON THE CRITERIA OF STABILITY FOR RHEOLOGIC BEHAVIOR OF NONLINEAR VISCOELASTIC MATERIALS

HU JIANYANG[1] and SHI YONGJI[2]

[1]Wuhan University of Technology, 430070, Wuhan (P.R.China).
[2]China Academy of Railway Sciences,100081, Beijing, (P.R.China).

1. INTRODUCTION

For physically nonlinear viscoelastic materials the criteria of stability of isothermal strain processes are studied. The relation between Lyapunov theorem and Drucker's postulate of stability is discussed for both stress relaxation and steady creep situations.

2. CONSTITUTIVE RELATIONS

The constitutive equations of isothermal rheologic processes in physically nonlinear viscoelastic materials are (ref.1)

$$\varepsilon_{ij} = \varepsilon^{e}_{ij} + \varepsilon^{v}_{ij} = S_{ijkl}\sigma_{kl} + \varepsilon^{v}_{ij},$$

$$\dot{\varepsilon}^{v}_{ij} = \dot{\varepsilon}^{v}_{ij}(\sigma_{kl}), \quad i,j,k,l = \overline{1,3} \qquad (1)$$

where S_{ijkl} and $\dot{\varepsilon}^{v}_{ij}$ are tensors of elastic compliance and viscous strain rate, respectively. If the viscous strain rate of the viscoelastic material satisfies the following inequality,i.e.,Drucker's postulate of stability(ref.2)

$$\delta\sigma_{ij}\delta\dot{\varepsilon}^{v}_{ij} > 0 \qquad (2)$$

then the material is stable. The unique solution of the mixed boundary value problem of the isotropic viscoelastic material may be obtained by use of potentialless connection between the stress and strain for creep.

3. STABILITY OF STRESS RELAXATION PROCESS

Suppose that for $t < 0$ the stresses,strains, and displacements are all zero. The description of the stress relaxation process in the nonlinear viscoelastic material can be expressed as

$$S_{ijkl}\dot{\sigma}_{kl} + \dot{\varepsilon}^{v}_{ij} = 0 \qquad (3)$$

Herefrom it follows that the perturbation equation describing the stability of the stress relaxation process under a disturbance $\delta\sigma_{ij}$ are

$$S_{ijkl}\delta\dot{\sigma}_{kl} + \delta\dot{\varepsilon}^{v}_{ij} = 0 \qquad (4)$$

Choosing $v=(1/2)S_{ijkl}\delta\sigma_{ij}\delta\sigma_{kl}$ to be Lyapunov function, thus $\dot{v}= -\delta\sigma_{ij}\delta\dot{\varepsilon}^{v}_{ij} < 0$ due to (2), consequently, all conditions of Lyapunov theorem are holden (ref.3), this means that the stress relaxation process described by Eq.(4) is stable. Thus, the enequality (2) is the sufficient condition of stability of the stress relaxation process in the nonlinear viscoelastic material.

4. STABILITY OF CREEP PROCESS

For an initial state of steady creep, Eq.(4) is transformed into under a disturbance $\delta\sigma_{ij}$

$$S_{ijkl}\delta\dot{\sigma}_{kl}+(\partial\dot{\varepsilon}^{v}_{ij}/\partial\sigma_{kl})\delta\sigma_{kl} = 0 \qquad (5)$$

Here $\partial\dot{\varepsilon}^{v}_{ij}/\partial\sigma_{kl}$ is a function of initial stress. Obviously, if the inequality (2) holds, then the equation (5) will also hold. Inversely, if the condition (2) does not hold, then from (5) have

$$\dot{V} = S_{ijkl}\delta\dot{\sigma}_{ij}\delta\dot{\sigma}_{kl} > 0 \qquad (6)$$

Thus, it follows from **Четаев** theorem that the material is unstable under the disturbance.

5. IN SUMMARY

1) When the viscous strain rate $\dot{\varepsilon}^{v}_{ij}=\dot{\varepsilon}^{v}_{ij}(\sigma_{kl})$ is independent of loading history, the inequality (2) can provide the uniqueness of solution for boundary value problem of viscoelasticity.
2) Drucker's postulate of stability is an agreement with Lyapunov theorem of stability, and they can be regarded as the criteria of stability for physically nonlinear viscoelastic materials.

REFERENCES

1. Yu.N.Rabotnov, Elements of Hereditary Solid Mechanics, Mir, Moscow, 1980, p.387.
2. D.C.Drucker, On the postulate of stability of material in the mechanics of continua,IUTAM-2, Springer, New York, 1964.
3. N.Rouche, P.Habets and M.Laloy, Stability Theory by Lyapunov's Direct Method, Applied Mathematical Sciences, New York,1977.

Theoretical and Applied Rheology, edited by P. Moldenaers and R. Keunings
Proc. XIth Int. Congr. on Rheology, Brussels, Belgium, August 17-21, 1992

ON THE CALCULATION OF DISCRETE RETARDATION AND RELAXATION SPECTRA

J. Kaschta

Chair for Polymers, Institute for Material Sciences, University Erlangen-Nürnberg, Martensstr. 7, 8520 Erlangen (Germany)

Retardation and relaxation spectra, which are real material functions, are not accessible experimentally, but any measurable viscoelastic function may be expressed in terms of the corresponding spectra (ref. 1). In the literature (e.g. ref. 2) a number of approximative solutions for the calculation of either the relaxation spectrum $H(\tau)$ or the retardation spectrum $L(\tau)$ from experimental data may be found. Methods for the calculation of discrete relaxation spectra from dynamic mechanical data specifically in the flow regime have been published recently (refs. 3-4). The object of this paper is to present an easy-to-use method for the calculation of retardation spectra from creep data in all regions of consistency of polymeric materials.

The creep compliance J(t) is related to a discrete retardation spectrum $\{J_i | \tau_i\}$ with N retardation modes $\tau_1 ... \tau_N$ by

$$J(t) = J_0 + \sum_{i=1}^{N} J_i \cdot [1 - e^{t/\tau_i}] + t/\eta_0 \qquad (1)$$

J_0 denotes the glassy compliance and η_0 the shear viscosity. Depending on the material and the regions of consistency over which the data extends, J_0 and/or η_0 might be not determinable. In eq. (1) all 2N+2 adjustable parameters have to be positive.

A least square fit technique with prescribed logarithmically equidistant spacing of retardation times is used in the iterative calculation of the coefficients (ref. 5). Due to the choice of the time spacing and the normally limited experimental time window, the number of modes N is limited and the problem is no longer ill-posed (ref. 4). In figure 1 the mastercurve of a polycarbonate in the glass-rubber transition is plotted together with a optimum discrete spectrum. The recalculated mastercurve is in excellent agreement with the data.

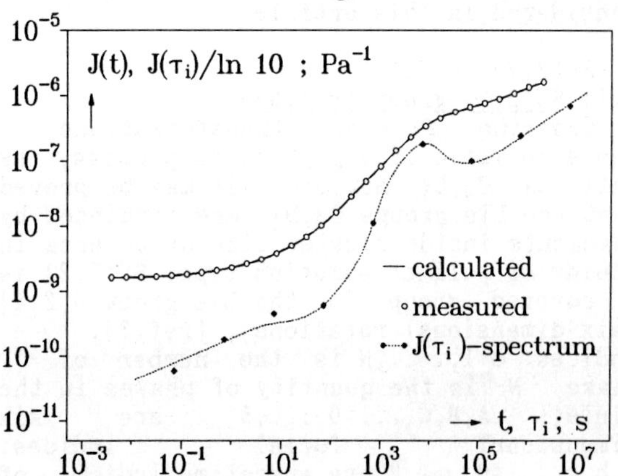

Fig. 1 Creep mastercurve of polycarbonate at 142.5°C; (points=measured, line=calculated from spectrum) and corresponding retardation spectrum

The discrete retardation spectrum shown here can be easily converted to a discrete relaxation spectrum using the method of Gross (ref. 6).

REFERENCES

1. J.D. Ferry, Viscoelastic properties of polymers, 3rd ed., J. Wiley, New York, 1980 p. 60-74
2. N.W. Tschoegl, The Phenomenological Theory of Linear Viscoelastic Behaviour, Springer, Berlin, 1989, ch. 4
3. J. Honerkamp, J. Weese, Macromolecules, 22 (1989) 4372
4. M. Baumgaertel, H.H. Winter, Rheol. Acta 28 (1989) 511-519
5. Kaschta J., Ph. D. Thesis, Erlangen, 1991
6. B. Gross, Mathematical Structure of the Theories of Viscoelasticity, 2nd ed. Hermann, Paris, 1968

Theoretical and Applied Rheology, edited by P. Moldenaers and R. Keunings
Proc. XIth Int. Congr. on Rheology, Brussels, Belgium, August 17-21, 1992

GAUGE MODEL OF HYDRODYNAMICS OF MULTIPHASE VISCO-ELASTIC FLUIDS

Yu.G.NAZMEEV[1], O.F.MOOSSIN[1], T.N.VEZIROGLU[2]

[1]Kazan Branch of Moscow Power Engineering Institute, Krasnoselskaya street, 51, 420066 Kazan, CIS(ex.USSR)

[2]Clean Energy Research Institute University of Maimi, Coral Geables, Florida 33124, USA

1. INTRODUCTION

The fields of gauge models on the basis of the gauge group Lie SU(2,2) have been introduced in (ref.1) for crystals and after it in (ref.2) for fluid. The models of multiphase visco-elastic fluids is considered in this article.

2. SU(2,2) MODEL

2.1. SU(2,2) group (ref.3)

It describe Poincare transformations characteristic of any physical process, as well as C_4,D_1 motions. It may be proved that the Lie groups C_4,D_1 are condioned by movements inside cluster-like structures in fluids of polimer solution type. SU(2,2) is a covered group for the Lie group O(2,4) (six dimensional rotations), (ref.3). Indices: $\Theta=1,\ldots,N$ is the number of phase, N is the quantity of phases in the fluid; $A,B,C,\ldots=0,\ldots,5$ are six dimensional formal indices; $a,b,c,\ldots=0,\ldots,3$ are spacetime indices of objects.

2.2. Lagrangian L

It is the sum of SU(2,2) invariant Lagrangians. It contains: 1) the Lagrangian L_1 which is algebraic function of gauge derivatives $D_a f_{\Theta A}=f_{\Theta A;a}+M_A{}^B{}_a f_{\Theta B}$ of 6 dimensional vector fields-sources $f_{\Theta A}$(";a" is the gauge derivative on the basis of Ricci connection coefficients, $M_A{}^B{}_a$ are the vectors of contorsion which are fields-carriers of interactions) 2) the Lagrangian V(f) is the algebraic function of the $f_{\Theta A}$ 3)visco-elastic Lagrangian L_1:

$$L_1=-\frac{k}{8}R_A{}^B{}_a{}^b R_B{}^A{}_b{}^a,\ k=\mathrm{const} \tag{1}$$

$$R_A{}^B{}_a{}^b=F_A{}^B{}_a{}^b-M_A{}^B{}_a{}^b \tag{2}$$

$$F_A{}^B{}_a{}^b=2M_A{}^B{}_{[m;a]}g^{mb} \tag{3}$$

$$M_A{}^B{}_a{}^b=2M_A{}^D{}_{[a}{}^M{}_{|D|}{}^B{}_{m]}g^{mb} \tag{4}$$

where g^{mb} is spacetime metric tensor, $R_A{}^B{}_a{}^b$ is SU(2,2) curvature tensor 4) the Lagrangian H of elasticity is a polynom of the $M_A{}^B{}_a$ 5) the Lagrangian of other fields. L_0 discribes the density of the mass, kinetic energy, velocities $V_A{}^B{}_a$ of SU(2,2) motions, L_1+ H describes the internal energy and visco-elastic forces, V(f) describes interactions between the phases.

3. VISCO-ELASTIC PROPERTIES OF SU(2,2) FIELDS

Equtions of motion have been obtained from L by means of variational method. They showed that $M_A{}^B{}_a(V_A{}^B{}_a)^{-1} \approx$ const.. Therefore, $F_A{}^B{}_a{}^b$ is the analogue of the tensor of the velocities of the deformations. The forces described by L_1 and linearly dependent of $F_A{}^B{}_a{}^b$ are the analogue of Navier-Stokes viscous forces (and tensor $kM_A{}^B{}_a{}^b$ being the tensor generalization of fluid viscosity scalar); and the forces of L_1 bilinearly dependent of $F_A{}^B{}_a{}^b$ are the analogue of M. Reiner viscous forces.

In our opinion SU(2,2) model has a good properties for the gauge renovation of hydrodynamics.

REFERENCES

1. V.R.Kajgorodov, O.F.Moossin (Ed.), Gravitation and Relativity Theory, Kazan State University, Kazan, 28,(1991), pp. 117-121
2. Yu.G.Nazmeev, O.F.Moossin (Ed.), The works of the Moscow Power Engineering Institute, Moscow, 644, 1991, pp.19-23.
3. A.O.Barut, R.Raczka, Theory of Group Reprezentations and Applications, v. 2, PWN-Polish Scientific Pablishers, Warsaw, 1977, p. 400.

Theoretical and Applied Rheology, edited by P. Moldenaers and R. Keunings
Proc. XIth Int. Congr. on Rheology, Brussels, Belgium, August 17-21, 1992

ON THE DAMPING FUNCTION OF SHEAR RELAXATION MODULUS FOR POLYMERIC LIQUIDS

K. Osaki and E. J. Hwang
Institute for Chemical Research, Kyoto University, Uji, Kyoto 611 (Japan)

The shear relaxation modulus, $G(t,\gamma)$, for entangled polymers can be factorized as

$$G(t,\gamma) = G(t)\ h(\gamma) \qquad (1)$$

at long times. The damping function, $h(\gamma)$, representing the effect of varying magnitude of shear, γ, was found to be a "universal" function for a certain group of polymers (ref. 1). Such a property and the functional form of $h(\gamma)$ was well described with the tube model of de Gennes, Doi, and Edwards. The decrease of $h(\gamma)$ was attributed to the rapid shrink of the extended polymer chain along the tube prior to the relaxation of the overall orientation of the chain resulting from the reptational motion (ref. 2).

In the meantime, the damping function for some commercial polymers were found to decrease more slowly with increasing γ than the "universal" function does. Here we collected the published results for $h(\gamma)$ and compared with each other on a graph where it was plotted against γ with log-log scales. References were strictly limited to the case where $G(t,\gamma)$ was directly measured and the factorization of eq. (1) was confirmed.

Astonishingly many data were close to the theoretical curve if a reasonable range (range A) was alotted to the "close" range. Most of the data can be classified as follows.

<u>Type A (the range A)</u>

polystyrene; polydimethyl siloxane; polyisobutylene; high density polythylene (linear and lightly branched samples);

solutions of polystyrene (linear and star branched), polybutadiene (linear and star branched), and poly-α-methyl styrene.

<u>Type B (well above the range A)</u>

low density polyethylene; linear low density polyethylene; polybutadiene (highly branched); polystyrene solution (no entanglement)

<u>Type C (below the range A)</u>

polybutadiene (linear and star branched); polystyrene solution (high molecular weight: with more than 50 entanglements / chain).

A few data may be classified as A or B depending on the definition of the range (Dow Styron 666; high density polyethylene by Khan et al.; polystyrene solution with bimodal molecular weight distribution).

Obviously, the factor characterizing B, in contrast with A, is not the wide molecular weight distribution but the multiple branching of chain or the lack of entanglement. The results for polyethylene indicate the tendency of deviation from the A type behavior with increasing branching. The result for HDPE by Yoshikawa et al. (ref. 3) includes linear as well as branched (less than one branch /chain) samples with various molecular weight distributions. One branch point/ chain is not a factor to cause the B behavior. Qualitative explanations of the B type deviation of $h(\gamma)$ from the Doi-Edwards prediction may not be so difficult for polymers with no entanglement or with more one branch/chain.

A complete list of references will be published soon.

REFERENCES

1. K. Osaki and M. Kurata, Macromolecules, 13 (1980) 671-676.
2. M. Doi and S. F. Edwards, J. Chem. Soc. Faraday Trans. 2, 74 (1978) 1789-1817.
3. K. Yoshikawa et al., Nihon Reoroji Gakkaishi, 18 (1990) 80-92.

Theoretical and Applied Rheology, edited by P. Moldenaers and R. Keunings
Proc. XIth Int. Congr. on Rheology, Brussels, Belgium, August 17-21, 1992

VARIATIONAL PRINCIPLE FOR LINEAR VISCOELASTICITY, AND APPLICATION TO FAXEN'S THEOREM

J.F. PALIERNE[1] and R. HOCQUART[2]

[1]Laboratoire de rhéologie, IMG BP 53X, 38041 GRENOBLE CEDEX (FRANCE)
[2]Laboratoire de dynamique des fluides complexes et ultrasons, 4 rue Blaise Pascal, 67070 STRASBOURG CEDEX (FRANCE)

The motion equation for small deformations of a viscoelastic fluid

$$\rho\frac{\partial^2\vec{u}}{\partial t^2} = \vec{\nabla}.\sigma \quad \text{where} \quad \sigma(t) = \int \mathbf{M}(t-t'):\gamma(t')dt'$$

involves the memory tensor $\mathbf{M}(t-t')$, vanishing for t<t'.This equation can be derived from the stationarity condition for the action integral

$$S = \int dv \left[\int_{-\infty}^{\infty} dt\, L(t) + \int_{-\infty}^{\infty} dt \int_{-\infty}^{\infty} dt'\, K(t,t') \right]$$

$$\text{where } L = \rho\frac{\partial\vec{u}}{\partial t}.\frac{\partial\vec{u}^+}{\partial t} \quad \text{and} \quad K = -\frac{1}{2}\mathbf{M}(t-t') \vdots \gamma\gamma^+$$

$\vec{u}^+$ being an auxiliary deformation field, and $\mathbf{g}^+$ the associated deformation tensor.This formalism gives as a by-product the motion equation for $\vec{u}^+$:

$$\rho\frac{\partial^2\vec{u}^+}{\partial t^2} = \vec{\nabla}.\sigma^+ \quad \text{where} \quad \sigma^+(t) = \int \mathbf{M}(t'-t):\gamma^+(t')dt'$$

This equation involves the same memory tensor as the constitutive equation for the real medium, with the argument t-t' substituted fort t'-t. A medium obeying such a constitutive equation would exhibit advanced response, i.e. stress setting up before the deformation takes place, and vanishing after the deformation. Though unphysical, the concept of a medium showing advanced response, in association with the real medium with retarded response, permits to generalize to viscoelasticity some theorems pertaining to elasticity and Newtonian fluid mechanics. As an illustration, the Faxèn formulas giving the force and the torque experiences by a rigid obstacle immersed in a flowing Newtonian liquid admit the following generalization. Let a viscoelastic medium be submitted to forces acting at infinity, causing the unperturbed motion $\vec{u}^0$ and stress σ^0. A rigid obstacle inserted into this medium and set in rigid motion $\vec{u}(\vec{x},t) = \vec{U}(t) + \vec{\Omega}(t)\times\vec{x}$ perturbs the motion $\vec{u}^0$, and impresses a force $\vec{F}(t)$ and a torque $\vec{\Gamma}(t)$ on the medium $\vec{F}$ and $\vec{\Gamma}$ can be expressed as functions of $\vec{U}$, $\vec{\Omega}$, and of the *unperturbed* fields $\vec{u}^0$ and σ^0 at the location of the surface of the obstacle :

$$F_k(t) = \int_S dS_j \left[\sigma^0_{jk}(t) + \int_{-\infty}^{t} dt' \left(u_i(t')-u^0_i(t')\right) \chi_{ijk}(t-t') \right]$$

$$\Gamma_k(t) = \int_S dS_j \left[\varepsilon_{kli}\, x_l\, \sigma^0_{ij}(t) + \int_{-\infty}^{t} dt' \left(u_i(t')-u^0_i(t')\right) \psi_{ijk}(t-t') \right]$$

using the tensors $\chi(\vec{x},t-t')$ and $\psi(\vec{x},t-t')$, which are such that the stress in the medium generated by the motion of the obstacle reads

$$\sigma_{ij}(\vec{x},t) = \int_{-\infty}^{t} dt' \left[\chi_{ijk}(\vec{x},t-t')\, U_k(t') + \psi_{ijk}(\vec{x},t-t')\, \Omega_k(t')\right]$$

in absence of unperturbed motion. Consequently, the determination of $\vec{F}(t)$ and $\vec{\Gamma}(t)$.can be made without knowledge of the actual motion of the medium, perturbed by the obstacle.

Theoretical and Applied Rheology, edited by P. Moldenaers and R. Keunings
Proc. XIth Int. Congr. on Rheology, Brussels, Belgium, August 17-21, 1992

A NEW EQUATION FOR ACTUAL FLOW CURVES OF POLYMERIC FLUIDS

JIN RIGUANG, LI HANGQUAN, LUO XIN

Dept. of Polym. Sci., Box 61, 100029, Beijing Inst. of Chem. Technol.. (P.R.China)

1. INTRODUCTION

A typical $\eta-\dot{\gamma}$ curve may be divided into 3 regions: zero-shear-rate region, power-law region and infinite-shear-rate region. Various empirical equations for $\eta(\dot{\gamma})$ have been used to describe the experimental data, such as power law, Ellis'(ref.1), Eyring's(ref.2), Bingham's etc. However, none of those can describe $\eta-\dot{\gamma}$ curves spanning all 3 regions. In this paper, it is considered that the transition from one flow state to another must be accompanied by the change flow-activation-energy (FAE), so the change in FAE is taken as the origin to derive a new viscosity equation which can well be used for $\eta-\dot{\gamma}$ curves in very wide range of shear rates.

2. EQUATION DERIVATION

2.1 PSEUDOPLASTIC FLUIDS

For pseudoplastic fluids, the FAE is linearly decreased with the increase of shear stress:

$$\Delta E_a = \Delta E_o - \alpha\tau = \Delta E_o(1-\beta\tau) \approx \frac{\Delta E_o}{1+\beta\tau} \tag{1}$$

To first approximation, shear stress τ may be expressed by power-law: $\tau = K\dot{\gamma}^n$, then

$$\Delta E_a = \frac{\Delta E_o}{1+m\dot{\gamma}^n} \tag{2}$$

Substituting Eq.2 into "Arrhenius dependence" of the form

$$\eta_a = \eta_o \exp\left(\frac{\Delta E_a}{RT}\right) \tag{3}$$

we have

$$\eta(\dot{\gamma}) = \eta_o \exp\left(\frac{\Delta E_o}{RT}\frac{1}{1+m\dot{\gamma}^n}\right) \tag{4}$$

2.2 DILATANT FLUIDS

For dilatant fluids, the FAE is linearly decreased with the increase of shear stress:

$$\Delta E_a = \Delta E_o + \alpha'\tau = \Delta E_o(1+\beta'\tau) \approx \frac{\Delta E_o}{1-\beta'\tau} \tag{5}$$

The following is similar to that in 2.1 but a minus sign "−" must be added before ΔE_o since the viscosity of dilatant fluids increase with temperature rising. Then we obtain

$$\eta(\dot{\gamma}) = \eta_o \exp\left(-\frac{\Delta E_o}{RT}\frac{1}{1-m\dot{\gamma}^n}\right) \tag{6}$$

Combining Eqs.4 and 6 we have a new generalized Newtonian constitutive equation:

$$\tau = \eta_o \exp\left(\pm\frac{\Delta E_o}{RT}\frac{1}{1\pm m\dot{\gamma}^n}\right)\cdot\dot{\gamma} \tag{7}$$

with "+" for pseudoplastic fluids and "−" for dilatant fluids.

There are 4 parameters in Eq.7. The physical significance of ΔE_o is FAE, that of m is a factor denoting the effect of shear rate on FAE. Obviously, the physical significance of η_o and n are the same as in "Arrhenius dependence" and power-law respectively.

3. DISCUSSION

For the case in first Newtonian region, $\dot{\gamma}\to 0$,

$$\eta_a = \eta_o^o \exp\left(\frac{\pm\Delta E_o}{RT}\right), \quad \tau_{(1)} = \eta_o^o \exp\left(\frac{\pm\Delta E_o}{RT}\right)\dot{\gamma}$$

For the case in second Newtonian region, $\dot{\gamma}\to\infty$,

$$\eta_a = \eta_o^\infty, \quad \tau_{(2)} = \eta_o^\infty\dot{\gamma}$$

It is seen that the sign and magnitude of ΔE_o is responsible for the difference in the rheological behavior of polymer melts. Furthermore, the parameters in Eq.7 have clear physical significances. Theoretical prediction of rheological properties were also made for materials such as PS, PVC, HDPE, PVC/PS blends, computed results agreed with experimental data in high accuracy.

REFERENCES

1. S.B.Ellis et al., AIChE J.,11(1965), pp.588-595.
2. H.Eyring et al., Ind. Eng. Chem., 50(1958), pp.1036-40.

Theoretical and Applied Rheology, edited by P. Moldenaers and R. Keunings
Proc. XIth Int. Congr. on Rheology, Brussels, Belgium, August 17-21, 1992

FUNCTIONAL RELATIONS IN ASYMMETRICAL RHEOLOGY

Z. SOBOTKA

Mathematical Institute of Czechoslovak Academy of Sciences, Žitná 25, 110 00 Praha 1 (Czechoslovakia)

The paper deals with the nonlinear functional relations for anisotropic viscoelastic materials with memory. For anisotropic bodies, the transformed strain tensor and transformed tensors of time derivatives of strain are introduced which have the same principal directions as the stress tensor. Assuming the same anisotropy effects for strain and its derivatives, the transformed tensors are defined by

$$b_{ij} = B_{ijkl}\varepsilon_{kl}, \quad \dot{b}_{ij} = B_{ijkl}\dot{\varepsilon}_{kl}, \dots, \qquad (1)$$

where B_{ijkl} is the dimensionless fourth-rank tensor of moduli of anisotropy.

The stress tensor for an anisotropic body having the memory of the strains and their time derivatives at N values of time $t_0, s_1, s_2, \dots, s_{N-1}$ prior to the present state at the time t is expressed by

$$\sigma_{ij} = f_{ij}[b_{kl}(t_0), b_{kl}(s_1), \dots, b_{kl}(t), \dot{b}_{mn}(t_0), \dot{b}_{mn}(s_1), \dot{b}_{mn}(s_2), \dots, \dot{b}_{mn}(t), \ddot{b}_{pq}(t_0), \ddot{b}_{pq}(s_1), \ddot{b}_{pq}(s_2), \dots, \ddot{b}_{pq}(t), \dots] \qquad (2)$$

Expanding this function according to the rules of tensor algebra, we obtain

$$\begin{aligned}\sigma_{ij} = {} & F_0\delta_{ij} + \sum_{K=0}^{N} F_{K+1} b_{ij}(s_K) + \sum_{K=0}^{N} F_{N+K+2} b_{ip}(s_K) b_{pj}(s_K) + \\ & + \sum_{K=0}^{N-1} F_{2N+K+3}[b_{ip}(t) b_{pj}(s_k) + b_{ip}(s_K) b_{pj}(t)] + \sum_{K=0}^{N} F_{3N+K+3}\dot{b}_{ij}(s_K) + \\ & + \sum_{K=0}^{N} F_{4N+K+3}\dot{b}_{ip}(s_K)\dot{b}_{pj}(s_K) + \sum_{K=0}^{N-1} F_{5N+K+3}[b_{ip}(t)\dot{b}_{pj}(s_K) + \dot{b}_{ip}(s_K) b_{pj}(t)] + \\ & + \dots, \end{aligned} \qquad (3)$$

where F_K are invariant functions.

After determining these functions, carrying out the passage to limit and introducing the modular functions $\tilde{G}[b_I(s),t]$, $\tilde{D}[b_{II}(s),t]$, $\hat{H}\{b_{JJ}[(t,s),t]\}$, $\tilde{P}[\dot{b}_I(s),t]$, $\tilde{Q}[\dot{b}_{II}(s),t]$ and $\tilde{R}[\dot{b}_{JJ}(t,s),t]$, the sums become the Stieltjes integrals and we have the deviatoric relation

$$\begin{aligned}\sigma_{ij} - \sigma_M\delta_{ij} = {} & 2\int_{t_0}^{t}[b_{ij}(s) - b_M(s)\delta_{ij}]\,d\tilde{G}[b_I(s),t] \\ & + 4\int_{t_0}^{t}[b_{ip}(s)b_{pj}(s) - b_S^2(s)\delta_{ij}]\,d\tilde{D}[b_{II}(s),t] + \\ & + 4\int_{t_0}^{t}\{\tfrac{1}{2}[b_{ip}(t)b_{pj}(s) + b_{ip}(s)b_{pj}(t)] - \\ & - b_J^2(s)\,\delta_{ij}\}\,d\tilde{H}[b_{JJ}(t,s),t] + \\ & + 2\int_{t_0}^{t}[\dot{b}_{ij}(s) - \dot{b}_M(s)\,\delta_{ij}]\,d\tilde{P}[\dot{b}_I(s),t] + \\ & + 4\int_{t_0}^{t}[\dot{b}_{ip}(s)\dot{b}_{pj}(s) - \dot{b}_S^2(s)\delta_{ij}]\,d\tilde{Q}[\dot{b}_{II}(s),t] + \\ & + 4\int_{t_0}^{t}\{\tfrac{1}{2}[\dot{b}_{ip}(t)\dot{b}_{pj}(s) + \dot{b}_{ip}(s)\dot{b}_{pj}(t)] - \\ & - \dot{b}_J^2(s)\,\delta_{ij}\}\,d\tilde{R}[\dot{b}_{JJ}(t,s),t] + \dots, \end{aligned} \qquad (4)$$

where b_M, b_I, b_S, b_{II}, b_J, b_{JJ}, $\dot{b}_M$, $\dot{b}_I$, $\dot{b}_S$, $\dot{b}_{II}$, $\dot{b}_J$ and $\dot{b}_{JJ}$ are invariant transformed strains and strain rates.

The asymmetry is expressed by different modular functions and moduli of anisotropy in tension and compression and in the positive and negative shear.

Theoretical and Applied Rheology, edited by P. Moldenaers and R. Keunings
Proc. XIth Int. Congr. on Rheology, Brussels, Belgium, August 17-21, 1992

THE GENERALIZATION OF EQUATIONS OF STATE IN VOLUME RHEOLOGY

A.Ye. TERENTYEV

Riga Technical University, 1 Kalku St. 226355 Riga (Latvia)

1. INTRODUCTION

In order to describe the behaviour of the material in the general case of mechanical action it is usually necessary to test it under shear and also under three-dimensional, one-dimensional or uniaxial uniform loading. The elasticity theory states the interrelation of six values: four elasticity moduli corresponding to various types of uniform loading and two ratios — Poisson's ratio and that of Lamee. Two test values are sufficient for calculating the other four.

The aim of the present work is to establish a similar relationship of 6 sets of values — instead of 6 values — for linear isotropic elastoviscous media under four kinds of uniform loading.

2. PROBLEM FORMULATION IN THE THEORY OF VISCOELASTICITY

Compared with the elasticity theory the problem under discussion becomes considerably wider and more complicated and is expressed in the following way. By using test data from two types of uniform loading of the material, determine for the other two types of loading: 1) equations of state (integral or differential); 2) bilinear viscoelastic operators; 3) compliance curves; 4) types of rheological models and model parameters; 5) a specific expression of Hooke's law obtained from each of the above.

The solution of each of the 5 problems requiring an independent approach results in the solution of the whole task. A complete solution is achieved by solving all the above formulated problems that are interrelated and nevertheless are characterized by different levels of generalization, problems 3 even requiring an individual approach.

3. METHODS OF PROBLEM SOLVING

It has been found that the problem can be solved completely in two ways:

3.1 by using elastoviscous analogy i.e., by substituting the corresponding bilinear visco-elastic operators, characterizing various types of uniform loading of media, for elasticity moduli in the familiar relationships of elasticity theory, and this proved to be more convenient in solving problem 2 and consequently also problems 1, 3 and 5 [1];

3.2 by proposing a new structural elastoviscous analogy which means the representation of the familiar elastic moduli relationships in the form of structural models of series and parallel connections between elastic elements, and their further replacement by the respective rheological models; this leads to an unexpectedly easy solution of problem 4 and consequently of problems 1, 2 and 5.

4. CONCLUSIONS

1) A new structural elastoviscous analogy has been found.

2) The problem of interrelationship of 6-set rheological values has been solved in 5 formulations for linear isotropic elastoviscous media under for types of uniform loading, and this permits the determination of the whole 6-set value complex by two (or even one) of the simplest tests and also permits the finding of coefficients of generalized Hooke's law, i.e., arriving at the law itself in a substantiated way and using it for practical three-dimensional problem solutions in material processing.

REFERENCES

1. A.Terentyev, G.Kunnos. Method of Determining Rheological Parameters for Linear Viscoelastic Medium Under Four Kinds of Uniform Loading, in: P.F.G. Banfill (Ed.), International Conference on the Rheology of Fresh Cement and Concrete, Liverpool, U.K., March 26-29, 1990, E. & F.N. Spon, Cambridge, 1991, pp. 93-102.

Contributed papers

FLUID MECHANICS

Theoretical and Applied Rheology, edited by P. Moldenaers and R. Keunings
Proc. XIth Int. Congr. on Rheology, Brussels, Belgium, August 17-21, 1992

ABOUT THE INFLUENCE OF FLUID INERTIA IN UNSTEADY COUETTE FLOW

D. Aschoff and P. Schümmer

Institut für Verfahrenstechnik, RWTH Aachen, Turmstr. 46, D–5100 Aachen

1. INTRODUCTION

Investigations of structural properties of dilute polymer solutions are usually carried out by rheometrical experiments. Besides the measurement of zero–shear–rate viscosity an instantaneous change of shear rate or a shear oscillation are realized. To evaluate the measurements for moderately concentrated solutions the inertia of the fluid can be neglected; and the methods developed, indeed, ignore inertia absolutely. But for dilute polymer solutions this causes an erroneous interpretation of the results. Restricting the motion to the linear viscoelastic regîme it is possible to take into account inertia and to calculate its influence. Discussing these results one gets rules for evaluating experiments correctly.

For reasons of measuring accuracy it is neccessary to use cylindrical Couette systems which means that there has to be considered the problem of curvature of the streamlines. But from comparison between the solutions for cylindrical and rectilinear flow one can estimate if it is allowed to evaluate the rectilinear flow solution instead of the Couette flow. We will discuss the special boundary condition of a motion of the outer cylinder (or one plate) while the inner cylinder (or the other plate) will be kept at rest. The shear stress will be measured at this place, and its temporal development shall be compared with the theoretical result and such be evaluated.

2. LINEAR VISCOELASTICITY

For the relation between local shear stress S and shear rate $\dot{\gamma}$ linear viscoelasticity yields the following integral law

$$S = \int_{-\infty}^{t} G(t-t')\,\dot{\gamma}(t')\,dt' + \eta_s\,\dot{\gamma}(t) \qquad (1)$$

By introducing a "solvent"–viscosity η_s it is warranted that there exists a Laplace transform of the relaxation modulus G(t).

We call it $\hat{G}(s)$. The complex viscosity in oscillation flow then is:

$$\eta^* = \hat{G}(i\omega) + \eta_s.$$

3. THE FLOW FIELD

We have to solve cylindrical and rectilinear flow.

A. Cylindrical flow

The shear rate is given by

$$\dot{\gamma} = r\,\frac{\partial}{\partial r}\left[\frac{v}{r}\right] \qquad (2)$$

and the equation of motion reads

$$\rho\,\frac{\partial v}{\partial t} = \frac{1}{r^2}\,\frac{\partial (r^2\,S)}{\partial r} \qquad (3)$$

With the constitutive equation we get

$$\rho\,\frac{\partial v}{\partial t} = \int_{-\infty}^{t} G(t-t')\left[\frac{\partial^2 v(t')}{\partial r^2} + \frac{1}{r}\frac{\partial v(t')}{\partial r} - \frac{v(t')}{r^2}\right]dt' + \eta_s\left[\frac{\partial^2 v}{\partial r^2} + \frac{1}{r}\frac{\partial v}{\partial r} - \frac{v}{r^2}\right] \qquad (4)$$

We will distinguish the two cases of instantaneous change of steady flow and oscillatory flow. The first case is solved by using a Laplace transformation (which we mark by a circumflex):

$$s\,\hat{v} - v_0(r) = \frac{\hat{G}(s)+\eta_s}{\rho}\left[\frac{\partial^2 \hat{v}}{\partial r^2} + \frac{1}{r}\frac{\partial \hat{v}}{\partial r} - \frac{\hat{v}}{r^2}\right] \qquad (5)$$

for $t' \geq 0$

$v_0(r)$ is the steady velocity–field for $t' \lesssim 0$.

We call $\hat{G}^*(s) = \hat{G}(s) + \eta_s$.

With the boundary conditions mentioned above we get the solution

$$\hat{v} = \frac{v_0(r)}{s} \pm \frac{v_a}{s} \cdot \frac{N_1(i\beta R_i)\, J_1(i\beta r) - J_1(i\beta R_i)\, N_1(i\beta r)}{N_1(i\beta R_i)\, J_1(i\beta R_A) - J_1(i\beta R_i)\, N_1(i\beta R_A)} \tag{6}$$

$$\text{with } \beta = i \left[\frac{\rho\, s}{\hat{G}^*} \right]^{1/2}$$

J_1 and N_1 are Bessel functions of the first and second kind. "+" means start and "–" stop flow realisation of the steady shear flow. The shear stress at the inner cylinder which has to be measured for the evaluation of the experiments has the Laplace form (from (1)):

$$\hat{S}(R_i) = \dot{\gamma}_0(R_i)\,\frac{\eta}{s} \pm \frac{\frac{v_a}{s}\,\hat{G}^*(s) \cdot \frac{2}{\pi}\frac{1}{R_i}}{J_1(i\beta R_i)\, N_1(i\beta R_a) - N_1(i\beta R_i)\, J_1(i\beta R_a)} \tag{7}$$

For oscillatory flow we introduce the usual formulation $v(r,t) = \hat{\hat{v}}(r)\, e^{i\omega t}$ into the equation of motion which then reads:

$$\frac{\partial^2 \hat{\hat{v}}}{\partial r^2} + \frac{1}{r}\frac{\partial \hat{\hat{v}}}{\partial r} - \frac{\hat{\hat{v}}}{r^2} = \alpha^2\, \hat{\hat{v}}$$

$$\text{with } \alpha = \left[\frac{i\,\omega\,\rho}{\eta^*} \right]^{1/2}$$

which is very similar to the previous differential equation. The solution is

$$\hat{\hat{v}} = \hat{\hat{v}}_a \cdot \frac{N_1(i\alpha R_i)\, J_1(i\alpha r) - J_1(i\alpha R_i)\, N_1(i\alpha r)}{N_1(i\alpha R_i)\, J_1(i\alpha R_A) - J_1(i\alpha R_i)\, N_1(i\alpha R_A)} \tag{8}$$

The complex amplitude of the shear stress at the inner cylinder is

$$\hat{\hat{S}}(R_i) = \hat{\hat{v}}_a \cdot \frac{\eta^* \frac{2}{\pi}\frac{1}{R_i}}{J_1(i\alpha R_i)\, N_1(i\alpha R_a) - N_1(i\alpha R_i)\, J_1(i\alpha R_a)} \tag{9}$$

B. <u>Rectilinear flow</u>

We write down the respective solutions: The Laplace form of the velocity for instantaneous change is:

$$\hat{v} = \frac{v_0(h)}{s}\frac{x}{h} \pm \frac{v_h}{s}\frac{\sinh(\beta x)}{\sinh(\beta h)} \tag{10}$$

$v_0(h)$ is the upper plate velocity for $t \lesssim 0$, "+" for start flow, "–" for stop flow (then one must put $v_h = v_0(h)$)

The Laplace form of the shear stress at the lower plate reads:

$$\hat{S}(0) = \eta\,\frac{v_0(h)}{h}\frac{1}{s} \pm \frac{v_h}{s}\frac{\hat{G}^*\,\beta}{\sinh(\beta h)} \tag{11}$$

The complex amplitude of the velocity is:

$$\hat{\hat{v}} = \hat{\hat{v}}_h \frac{\sinh(\alpha x)}{\sinh(\alpha h)} \tag{12}$$

The complex amplitude of the shear stress at the lower plate is:

$$\hat{\hat{S}} = \eta^*\, \hat{\hat{v}}_h \frac{\alpha}{\sinh(\alpha h)} \tag{13}$$

C. <u>Comparison of cylindrical and rectilinear flow</u>

For cylindrical flow we may introduce $r = R_i + x$ and $R_a = R_i + h$. There exist asymptotic forms for the Bessel functions which for large values of the argument will lead to the results for rectilinear flow. By analytical and numerical methods we have estimated the possibility of describing Couette flow by a rectilinear flow approximation. We found out that for a sufficiently small gap it is allowed to evaluate the Couette flow as rectilinear flow, perhaps taking into account a correction factor.

4. THE SHEAR STRESS AT THE LOWER PLATE

To get the viscoelastic properties of a dilute polymer solution we have measured the shear stress at the lower plate, held at rest. The evaluation must by all means take into account the influence of inertia. After it has been shown, that rectilinear flow evaluation works well, for oscillation the amplitude and phase-shift measurement gives the material functions η' and η'' by the aid of equation (13).

The instantaneous change experiment has to be evaluated by comparison with the inversion of the Laplace form, eq. (11), which is available via the well known inversion integral of the Laplace transformation. We will write it down only for the relaxation experiment of Jeffreys type of the rheological equation of state for which we get

$$\hat{G}^*(s) = \eta \frac{1+t_2 s}{1+t_1 s} .$$

The poles of the equation (11) are simple poles so that we may write down the shear stress as sum of the residues of eq. (11):

$$S^*(0) = \left[1 - \frac{t_2}{t_1}\right] \cdot e^{-\frac{t^*}{d_1}} + 2 \sum_{n=1}^{\infty} (-1)^{n+1} \cdot e^{-\frac{1}{2}\left[\frac{1}{d_1} + \frac{t_2}{t_1} n^2\right] t^*} \left(\left[\frac{1}{d_1} - \frac{1}{2}\frac{t_2}{t_1}\left[\frac{1}{d_1} + \frac{t_2}{t_1} n^2\right]\right] \cdot \frac{\sinh \alpha_n t^*}{\alpha_n} + \frac{t_2}{t_1} \cosh \alpha_n t^* \right) \tag{14}$$

with the abbreviations

$$S^* = \frac{S}{\eta} \frac{h}{v_0(h)}, \quad t^* = \pi^2 \frac{\eta}{\rho h^2} t, \quad d_1 = \pi^2 \frac{\eta}{\rho h^2} t_1$$

$$\alpha_n = \left[\frac{1}{4}\left[\frac{1}{d_1} + \frac{t_2}{t_1} n^2\right]^2 - \frac{n^2}{d_1}\right]^{1/2}$$

It may happen that α_n gets imaginary; in this case the functions sinh and cosh will change to sin and cos. The series gives the deviation by inertia. The figure shows the difference for typical liquids. The time dependence will be of the same type for other rheological models with simple poles, even in cylindrical flow.

For the latter the poles are given by the zeros $(i\beta R_i)$ of the denominator of the second RH–term of eq. (7) which are all real. For Newtonian liquids with $t_1 = t_2 = 0$ one has to calculate a limiting value which gives

$$S_N^*(0) = 2 \sum_{n=1}^{\infty} (-1)^{n+1} e^{-n^2 t^*} \tag{15}$$

This is in some analogy to the relaxation curve of a Rouse–fluid without influence of inertia.

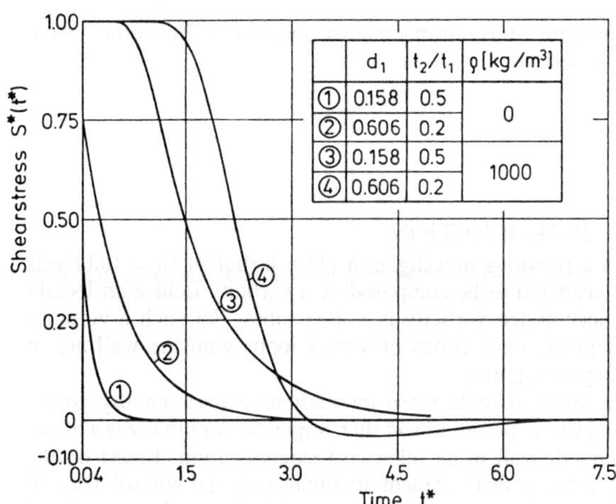

Fig. 1: Shear stress as a function of time at the lower plate (relaxation experiment)

5. CONCLUSION

The influence of inertia within typical rheometry of dilute polymer solutions is not neglectable. The correct evaluation of the measurements is possible by the aid of the results given in this paper.

SYMBOLS

S shear stress
$\dot{\gamma}$ shear rate
v velocity field
v_0 steady velocity field
v_h velocity at x=h
v_a velocity at $r=R_a$
h extension of the gap
R_i inner radius
R_a outer radius

REFERENCES

1 J. G. Oldroyd: Q. J. Mech. Appl. Math. 4, 271 (1951)
2 H. Markovitz: J. Appl. Phys. 23, 1070 (1952)
3 J. Pawlowski: Kolloid–Zeitschrift 138, 20 (1954)
4 G. Böhme and M. Stenger: J. Rheol. 34, 415 (1990)

redistribution of vorticity occurs which leads to a complex problem, especially when the fluid in the vortex ring exhibits non-Newtonian rheological properties.
Nevertheless, at the point where the interaction with the wall starts, the state of the vortex ring can be evaluated. This calculation is important for an understanding of the mechanism of interaction.
When the coordinate system is turned by an angle α in the new coordinate system (ξ, η) an induction velocity acts on the vortex element (1), and (2) respectively, see Fig. 3.
The induced velocities $v_i(1)$ and $v_i(2)$ are marked in Fig. 3.

$$v_{Ti}(1) = \frac{\Gamma}{4\pi}$$

$$(\frac{1}{4R^2}\sin\alpha + \frac{1}{4\eta_1^2} - \frac{1}{(\xi_1-\xi_2)^2+(\eta_1+\eta_2)^2}\cos\beta;$$

$$-\frac{1}{4R^2}\cos\alpha + \frac{1}{(\xi_1-\xi_2)^2+(\eta_1+\eta_2)^2}\sin\beta)$$

$$v_{Ti}(2) = \frac{\Gamma}{4\pi}$$

$$(\frac{1}{4R^2}\sin\alpha - \frac{1}{4\eta_2^2} - \frac{1}{(\xi_1-\xi_2)^2+(\eta_1+\eta_2)^2}\cos\beta;$$

$$-\frac{1}{4R^2}\cos\alpha - \frac{1}{(\xi_1-\xi_2)^2+(\eta_1+\eta_2)^2}\sin\beta)$$

$$\mathrm{tg}\beta = \frac{(\xi_1-\xi_2)}{(\eta_1+\eta_2)}$$

$$\xi = \cos\alpha x - \sin\alpha y; \quad \eta = -\sin\alpha x + \cos\alpha y$$

The branch (1) of the vortex is rotated around the centre of mass of the vortex while branch (2) is elongated. In this state an azimuthal transport of fluid along the core region of the vortex starts which could be obtained by integration.

$$v_{Hi} = \frac{\Gamma}{4\pi}\left(\frac{1}{\rho_4^2}\cos(\beta_4-\alpha) - \frac{1}{\rho_3^2}\cos(\alpha+\beta_3)\right)$$

$$\mathrm{tg}\beta_3 = \frac{\xi_1-\xi_2}{3\eta_2+\eta_1}, \quad \mathrm{tg}\beta_4 = \frac{\xi_1-\xi_2}{3\eta_1+\eta_2}$$

$$\rho_4^2 = \left(\frac{3\eta_1+\eta_2}{2}\right)^2 + \left(\frac{\xi_1-\xi_2}{2}\right)^2$$

$$\rho_3^2 = \left(\frac{3\eta_2+\eta_1}{2}\right)^2 + \left(\frac{\xi_1-\xi_2}{2}\right)^2$$

In order to calculate the influence of the inclination angle α on this process it is sufficient to calculate the horizontal velocity v_H of the centre of mass at the point where the start of the interaction with the wall was observed. The determination of these properties is coupled with many problems, e.g. the visualization of the vortex by dye, that at the moment only significant trends can be reported.

In order to compare the data obtained for different test fluids the dimensionless variables

$$L^* = \frac{Lv_T}{\Gamma} \quad \text{and} \quad T^* = \frac{Tv_T^2}{\Gamma}$$

are introduced.
The most important result is that the dimensionless time T^* which is needed to redistribute the vorticity, is significantly reduced in polymer solutions. No differences were found between water and the viscoelastic surfactant solution. The dimensionless spread L^* did not depend on the test fluid; a small decrease was found in surfactant solutions which can be related to the enhanced reflection at the wall in these fluids.
In conclusion, it can be stated that the non-Newtonian rheology of the polymer solutions strongly enhances the rate of redistribution of vorticity. The mechanism for this process is not understood till now and in contrast to the expectation that this rate is reduced due to an increase of the elongational viscosity.

4. REFERENCES

[1] A. Gyr, H.-W. Bewersdorff, J. Bühler, and D. Papantoniou, J. Non-Newtonian Fluid Mech. (1992) (accepted for publication)

[2] J.D.A. Walker, C.R. Smith, A.W. Cerra, and T.L. Doligalski, J. Fluid. Mech. **181** (1987) 99

[3] K. Vissmann and H.W. Bewersdorff, J. Non-Newtonian Fluid Mech. **34** (1990) 289

[4] R.E. Falco and D.C. Wiggert, in: G.R. Hough (Ed.) Progr. in Astronautics and Aeronautics **72** (1980) 275

Theoretical and Applied Rheology, edited by P. Moldenaers and R. Keunings
Proc. XIth Int. Congr. on Rheology, Brussels, Belgium, August 17-21, 1992

MODELLING OF PULLOUT IN SOLUTION SPINNING; ANALOGY WITH THE VORTEX SIZE IN CONTRACTION FLOWS

M.J.H. BULTERS[1] and H.E.H. MEIJER[2]

1 DSM Research Department of Material Development, P.O. Box 18, 6160 MD Geleen (The Netherlands)
2 Eindhoven University of Technology, Faculty of Mechanical Engineering, Polymer Technology, P.O. Box 513, 5600 MB Eindhoven (The Netherlands)

1. INTRODUCTION

In the search for ultimate properties of polymers, solutions of high molecular weight materials are spun using conventional spinning techniques. A striking example is the so-called "gelspinning" process, named after the intermediate spun yarn which forms a gel during the crystallisation of the polymer. In the gelspinning process the solutions are pulled out of the capillary at usual values of the drawratio. In the upstream contraction, in the inflow of the capillary, vortices occur which at higher throughputs tend to become unstable. This is reflected in a poor extrudate quality and hence imposes a limitation on the process.

In this analysis pullout and specifically the position of the detachment point are modelled.

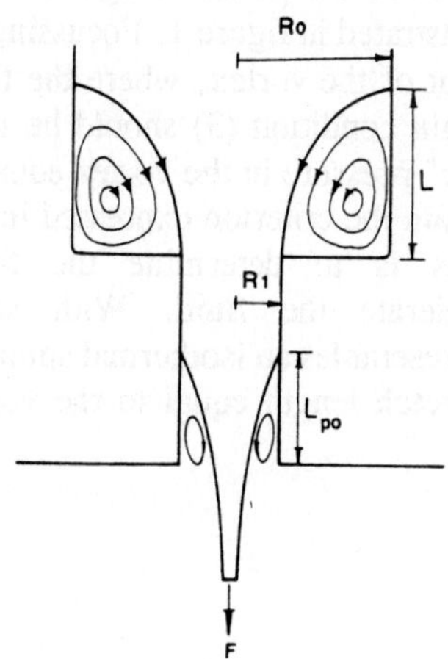

Fig. 1. Analogy between the pullout length and the vortex size.

The same analysis can be used to predict the vortex size in the upstream contraction (figure 1). The aim is a model which is as simple as possible, while still accounting for the dominant viscoelastic effects (ref. 1).

2. PULLOUT

The mechanism that determines the position of the detachment point is a balancing of the force necessary to stretch the filament (spinning force) and the stress that the fluid exerts normal to the wall due to its deformation history. The total normal stress due to the spinning force σ_{zz} and the normal stress at the wall σ_{rr} are connected through the first normal stress difference N_1, generated by the upstream flow:

$$N_1=\sigma_{zz}-\sigma_{rr} \tag{1}$$

In a first approximation the spinning tension at the detachment point equals the average normal stress in z-direction:

$$\sigma_{zz}=\frac{F}{\pi R^2} \tag{2}$$

Furthermore, at the detachment point the total normal stress exerted on the wall should equal zero in order to allow the fluid to leave the wall:

$$\sigma_{rr}=0 \tag{3}$$

Combining eqs. (1,2 and 3) gives the relation between the spinning force F and the normal stress difference N_1 at the detachment point:

$$F_c = N_1 \pi R^2 \tag{4}$$

Or, more accurately, from an integral momentum balance (ref. 1):

$$F_c = 2\pi \int_0^R (N_1 + \frac{N_2}{2}) r dr \tag{5}$$

Above the critical spinning force F_c, the solution is pulled out of the die.

Also the position of the detachment point (i.e. pullout length) can now be modelled, see figure 2.

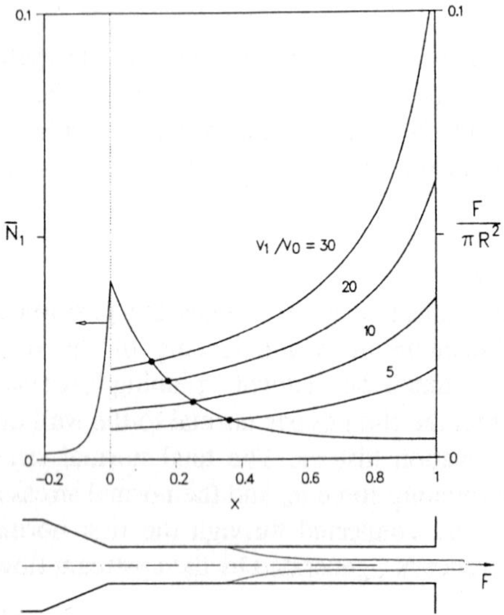

Fig. 2. The averaged normal stress N1 versus the distance z from the contraction (qualitatively) and the spinning force as function of the isothermal stretch length (1-z). The intersection of both lines gives the position of the detachment point according to eq. 5. Parameter: draw ratio.

Outside the die strongly non-isothermal conditions exist due to the evaporation of solvent, resulting in a steep stress-strain curve. Whereas inside the die more or less isothermal stretching conditions exist. As a consequence, the spinning force necessary to accelerate the fluid from v_0 to v_1 strongly depends on the pullout length (i.e. misothermal stretch length). A longer isothermal stretch length would result a lower spinning force. On the other hand, the normal stress on the wall depends on the shearrate as well as on the distance from the contraction. The intersection of both lines gives the position of the detachment point. In this position the spinning tension just balances the stress at the wall according to eq. 5.

The model is compared with experimental data of Sridhar and Gupta (ref.2) on the onset of pullout using a model liquid (PAA in 50% aqueous glycerol solutions), see figure 3.

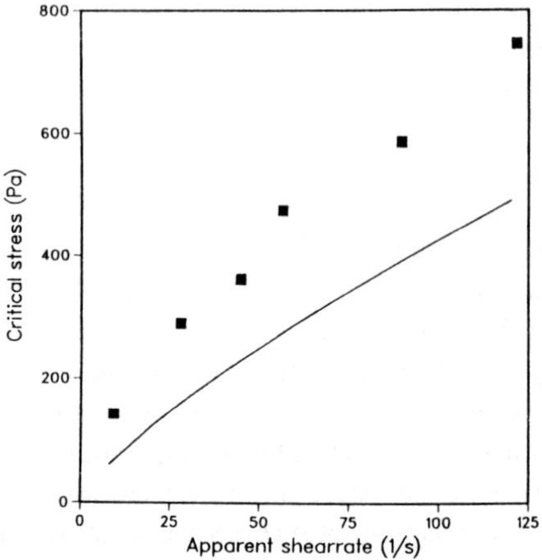

Fig. 3. Critical stress at which pullout starts versus apparent shearrate. Experimental data from ref. 2 compared with model predictions according to eq.(5).

3. VORTEX SIZE IN CONTRACTION FLOW

The analogy between the pullout length and the vortex size is demonstrated in figure 1. Focussing on the detachment point of the vortex, where the fluid leaves the wall, again condition (3) should be met, taking the "isotropic" pressure in the vortex equal to zero. This yields again the criterion expressed in eq. (5). What remains is to determine the force necessary to accelerate the fluid. With some simplifications this resembles an isothermal spinning process with the stretch length equal to the vortex size and with:

$$v_0 = \frac{Q}{\pi R_0^2} \tag{6}$$

$$\frac{v_1}{v_0}=(\frac{R_0}{R_1})^2 \tag{7}$$

The vortex size can be found by combining the expression for the critical force (eq. (5)) with an equation for the spinning force. For example, for the generalised Maxwell fluid the spinning force is given in approximation (ref. 3) by:

$$F=\frac{3Q\eta\ln(\frac{v_1}{v_0})}{L+\lambda(1-2a)(v_1-v_0)} \tag{8}$$

Combined with eq. (5), this yields the vortex length:

$$L=\frac{3Q\eta\ln(\frac{v_1}{v_o})}{2\pi\int_0^R(N_1+\frac{N_2}{2})rdr}-\lambda(1-2a)(v_1-v_0) \tag{9}$$

Inspection of eq. (9) shows that the vortex size is determined by both: the (transient) elongational viscosity aswell as the shearrate dependent N_1.
The results from the present model compare reasonably well with experimental data. In figure 4 results are compared with data on aqueous solutions of PAA from Binding and Walters (ref. 4).

4. CONCLUSIONS

- N_1 determines the force at which pullout starts and, combined with the
transient elongational viscosity, the position of the detachment point.

- The vortex size is determined by essentially the same mechanism as pullout.

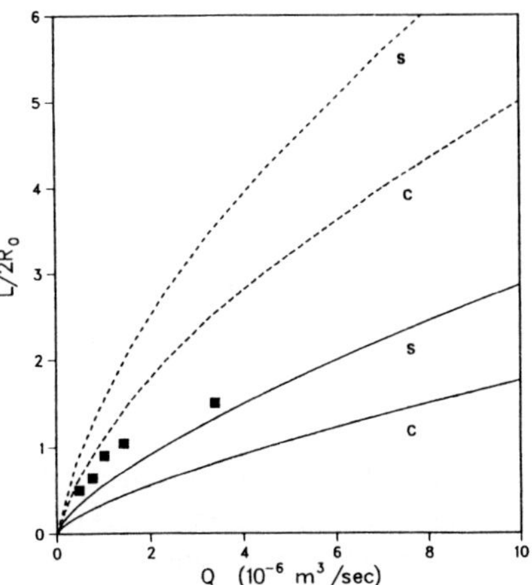

Fig. 4. Vortex size versus throughput. Experimental data (▪) and predictions (dashed lines) from ref. 4 compared with predictions according to the present model (draw lines). The elongational viscosity is determined from spinline data (S) and from contraction flow data (C).

REFERENCES

1. M. J. H. Bulters, H.E.H. Meijer, J. Non-Newtonian Fluid Mech., 38 (1990) 43
2. T. Sridhar, R.K. Gupta, J. Non-Newtonian Fluid Mech., 30 (1988) 285
3. G.R. Zeichner, M.Ch.E. Thesis, University of Delaware, (1973)
4. D.M. Binding, K. Walters, J. Non-Newtonian Fluid Mech., 30 (1988) 233

Theoretical and Applied Rheology, edited by P. Moldenaers and R. Keunings
Proc. XIth Int. Congr. on Rheology, Brussels, Belgium, August 17-21, 1992

RETENTION OF COLLOIDS IN POROUS MEDIA FLOWS

G. CHAUVETEAU

Institut Français du Pétrole, Division Gisements, B.P. 311 - 92506 Rueil-Malmaison Cedex, France

The retention of colloids in porous media plays an important role in various academic and industrial areas, such as chromatography, biology, filtration processes and oil recovery. Our analysis is restricted to the case of dispersion of particles or solutes which are stable (i.e. non aggregating under Brownian motion) and concerns both rigid and deformable colloids having a small size a (a < a few microns) in such a way that both gravity and inertia effects can be neglected. When colloid-to-pore size ratio is very small the only retention mechanism is adsorption. However, as soon as colloid radius a becomes non negligible compared to that of pore r_p, the problem is more complex and retention may due to various mechanisms including hydrodynamical, physico-chemical and sterical effects.

This paper aims at describing these mechanisms, showing that they have their own characteristics and sensitivity to various parameters and thus affect transport properties specifically. Are examined also the consequences of these different types of retention on both the rheology of colloidal suspension in porous-media flow and the permeability variations consecutive to colloid retention. Retention R occurs when the mean concentration of colloid inside the porous medium C_p is not equal to that of injected suspension C_o. Thus retention may be positive ($C_p > Co$) or negative ($C_p < Co$), reversible or irreversible. In particular, any difference between colloid velocity v_c and that of the solvent v_s at the center of mass of the solvent in absence of colloid induces a retention (R > 0 if $v_c < v_s$ and R < 0 if $v_c > v_s$), showing that retention is not restricted to the trivial case of colloid immobilization either at the pore wall surface or in pore constrictions.

The various mechanisms possibly responsible for colloid retention are classified below according to the location where retention occurs 1) in bulk flow i.e. at distances from the pore wall d larger than a, 2) at interfaces, i.e. for $d = a$, 3) near pore constrictions having size $r_p < a$. For each mechanism described is also given the dominant phenomenon : sterical, hydrodynamical or physico-chemical (see table 1 at the end of this preprint). An encircled number has been attributed to each mechanism.

1. RETENTIONS IN BULK FLOWS

1.1. Convection-induced concentration gradients ①

When the local grain Peclet number Peg becomes higher than 1 concentration gradients may be induced in converging and diverging pores if pore constriction is smaller than that of colloid (ref. 1). At equilibrium between hydrodynamic convection and Brownian diffusion, mean colloid velocity v_c is zero, and colloid concentrations are higher and lower than injected concentration upstream and downstream respectively. Overall retention at equilibrium is positive, increases continuously with Pe for rigid particles, but may decrease at high deformation rates for deformable or orientable colloids when hydrodynamic forces becomes high enough to deform colloid and thus to induce passage through pore constrictions (ref. 2). It should be noticed that this retention occurs at distances d up to grain diameter, i.e. much larger than a. This retention is expected to be reversible with flow rate and to have minor effect on pressure drop since it occurs inside small pores which are not plugged and participate to a small extent to the overall flow rate.

1.2. Velocity drifts ②

A velocity drift, i.e. a difference between colloid velocity and that of the solvent at the colloid center in the absence of this colloid, has been shown to occur in any flow situation where velocity gradient are non homogeneous at colloid length scale (ref. 3). Colloid lags the bulk solvent in a rectilinear Poiseuille flow, thus causing a positive retention, while a cross streamline drift is expected in non rectilinear flows when $\nabla\nabla v$ normal to the streamlines is non-zero, i.e. both upstream to grains or inside pore bodies. The overall effect of cross-streamline drift on retention is probably small since porous media flow are statistically symetrical but this point must be studied further. Of course, velocity drift being hydrodynamically driven will give a effect increasing with flowrate for Pe > 1, proportional to the square of colloid-to-pore size ratio. Velocity drifts increase pressure drop and thus decrease suspension mobility only for dense colloids like solids or coiled polymers.

1.3. Hydrodynamic interactions (3)

When a colloid moving on its streamline approaches at a distance of pore wall close to their radius either upstream to grains constituting the porous medium or in the converging zone towards a pore throat entrance, hydrodynamic interactions between colloid and pore wall colloid trajectory separates from streamlines (ref. 4) and a nonuniform concentration profile develops with an increase in concentration near the wall. This overconcentration is increased if there is inertia effects (non-negligible Stokes number).

A second type of effect arises when colloid concentration increases beyond dilute regime. Under these conditions, hydrodynamic interations between particles cause colloid trajectory to deviate from streamlines, thus creating higher concentrations in quiet zones and increase in viscous frictions in strong flow zones, particularly at pore entrance.

A third type of hydrodynamic interactions occurs, along the axis of a pore throat (ref. 5), when a dense colloid having a size non-negligible compared to pore throat arrive near the pore entrance. Its velocity is strongly reduced compared to that of solvent in the absence of colloid giving both local overconcentration and reduction in flow rate in small pores higher than in large pores, thus causing higher hydrodynamic dispersion.

1.4. Entropy-driven concentration gradients (4)

When oriented or deformed in strong flow zones, non spherical or deformable colloids losses entropy, which must be compensated by differences in colloid concentrations. This induces higher concentrations in weak flow or pseudo-stagnant zones, and thus a positive retention (ref. 6).

1.5. Convective exclusion volume (5)

Since in pore throats colloids excluded by either hydrodynamic interactions (see Section 1.3) or by steric hindrances (see Section 2.4) from streamlines close to the wall, they move on trajectory deviating from wall vicinity at high Peclet number and by this way does not experience all pore space, giving a negative contribution to overall retention (ref. 2).

2. RETENTION AT INTERFACES

2.1. Flow induced deposition and release (6)

In purely shearing zones, hydrodynamic forces tends to release deposited or adsorbed colloids (ref. 7), or in other terms to decrease adsorption or deposition energy. Same effect is expected in extensional zones located downstream from grains constituting the porous medium (ref. 2). On the contrary, in compressional zone upstream from grains, hydrodynamic forces increase adsorption or deposition energy (ref. 2).

2.2. Irreversible deposition or adsorption (7)

When colloid/surface interactions are attractive at short distances (Van der Waals forces, hydrogen bonding, hydrophobic interactions or more generally entropic effects coming from removal of adsorbed small molecules) colloids are deposited at the pore wall surface, thus giving positive retention and a pore size reduction which decreases porous medium permeability (ref. 1). For flexible or orientable colloids such as polymers, an increase in concentration enhances surface accessibility by increasing osmotic pressure, leading to an increased adsorption (see ref. 2 and Section 3.4).

2.3. Reversible deposition or adsorption (8)

When static adsorption isotherm curve is of low energy type, showing a linear part at low colloid concentrations, adsorption is reversible. In these conditions, adsorption and depletion may occurs simultaneously and the overall retention and its effects on permeability results from the competition between these two effects (ref. 8). After displacement by water desorption and thus full restoration of permeability is expected.

2.4. Steric repulsions from the pore wall (9)

When colloid/surface interactions are strictly repulsives at short distances (i.e. with strong hydration or high electrostatic energy barrier >> kT), there is a depletion near the interface (ref. 8) which decreases apparent vicosity and increases colloid velocity, thus leading to an overall negative retention, despite the substantial slip ($u/v < 1$ where u is colloid velocity) which develops at small gaps mainly for low internal permeability colloids (ref. 9).

3. RETENTIONS IN PORE CONSTRICTIONS

3.1. Hydrodynamic immobilization (10)

Such immobilization in close contact with the pore walls upstream from a pore constriction ($a > r_p$) requires much higher velocity, by a factor r_g/a than for convection-induced retention (see Section 1.1) and the corresponding retention level is very small (ref. 2).

3.2. Flow-induced aggregation (11)

If we except the case of strong hydration forces, colloid-colloid interactions are always attractive at short distance. Repulsive electrostatic energy barrier which is responsible for Brownian dispersion or solution stability may be overcome by hydrodynamic forces at high deformation rates. Under these conditions, aggregates or precipitates may be formed which may be adsorbed or retained by different mechanisms described above, leading to porous medium plugging.

3.3. Steric entrapment (12)

Such a retention which corresponds to the trapping inside a pore body surrounded by pore throats smaller than the colloid can occur only with deformable colloids which may have been deformed by hydrodynamics or by high colloid concentration to enter inside this pore body (ref. 1).

3.4. Deformation-induced squeezing (13)

Squeezing, i.e. penetration of colloids in regions of pore space having dimensions smaller than the initial colloid size may be induced by hydrodynamic forces for highly deformable colloids like polymers. This causes positive retention increasing with flow rate (ref. 2).

3.5. Deformation-induced bridging (14)

This type of retention may occur for deformable colloids (flexible polymers) flowing through adsorbing porous media having pore throat not too large compared to polymer size (ref. 2). At low strain rates, macromolecules move through the pores without difficulty, but as soon as strain rates are high enough to deform substantially the macromolecule, a pore bridging can occur by adsorption since stretched macromolecules have no time enough to relax both in length and orientation between two successive pore throats. After this first step, newly arrived macromolecules become entangled around bridging molecules thus leading progressively to a complete plugging.

3.6. Inaccessible pore volume (15)

The fraction of pore volume which is inaccessible to colloid due to steric exclusion increases as porous medium permeability decreases and leads to a colloid velocity higher than that of solvent, i.e. a negative retention (ref. 1).

Conclusion

Most of the retention mechanisms analyzed above are governed by a few adimensional numbers, such as colloid-to-particle ratio, convection-to-diffusion velocity ratio (Peclet number), and two numbers characterizing the ratio between hydrodynamical and interactional energy between colloids and colloid/pore wall respectively. The evalution of these adimensional numbers is a prerequesite to identify the type of retention possibly acting in a given system and thus to put forward any prediction concerning the amount and the consequences of this retention.

References

1. G. Chauveteau and K.S. Sorbie, in: Basic Concepts in Oil Recovery Processes, M. Bavière Ed. published for SCI by Elsevier Applied Science (1991) 43-87.
2. G. Chauveteau, Ph. Delaplace, J.F. Argillier, M. Bagassi and L. Léger, in: Physical Chemistry of Colloids and Interfaces in Oil Production, H. Toulhouat and J. Lecourtier Ed., Editions Technip, Paris, 1992, 309-316.
3. J.H. Aubert and M. Tirrell, J. Chem. Phys., 72, 4 (1980), 2694-2701.
4. J.N. Kao, Y. Wang, R. Pfeffer and S. Weinbaum, J. Colloid Interface Sc. 121, 2 (1988) 543-557.
5. Z. Dagan, S. Weinbaum and R. Pfeffer, J. Fluid Mech., 117 (1982) 143-170.
6. A.B. Metzner, Y Cohen and C. Rangel-Nafaile, J. Non-Newtonian Fluid Mech., 5 (1979) 449-462.
7. K.C. Khilar and H.S. Fogler, Reviews in Chemical Engineering, N.R. Amundson and D. Luss (Ed.), 4, 1 and 2 (1987), Freund Publishing House L.T.D., London.
8. G. Chauveteau, M. Tirrell and A. Omari, J. Colloid Interface Sc. 100, 1 (1984) 41-54.
9. A.J. Goldman, R.G. Cox and H. Brenner, Chem. Eng. Sc. 22 (1967) 653-660.

Location / Dominant Phenomenon	IN BULK FLOW ($d > a$)	AT INTERFACES ($d \approx a$)	NEAR PORE CONSTRICTIONS ($a < r_p$)
HYDRODYNAMIC	(1) Convection-induced concentration gradients (2) Velocity drifts (3) Hydrodynamic interactions	(6) Flow-induced deposition and release	(10) Hydrodynamic immobilisation (11) Flow-induced aggregation (12) Steric entrapment
PHYSICO-CHEMICAL	(4) Entropy-driven concentration gradients	(7) Irreversible deposition and release (8) Reversible deposition or adsorption	(13) Deformation-induced squeezing (14) Deformation-induced bridging
STERIC	(5) Consecutive volume exclusion	(9) Steric repulsions from the wall	(15) Inaccessible pore volume

Table 1: Classification of the Retention Mechanisms

Theoretical and Applied Rheology, edited by P. Moldenaers and R. Keunings
Proc. XIth Int. Congr. on Rheology, Brussels, Belgium, August 17-21, 1992

NON-NEWTONIAN FLUID FLOW THROUGH POROUS MEDIA

R.P. CHHABRA

Department of Chemical Engineering, Indian Institute of Technology, Kanpur, India 208016

1. INTRODUCTION

The flow of non - Newtonian polymer solutions and melts through porous media occurs in many industrially important processes including enhanced oil recovery using polymer solutions and the filtration of molten polymers prior to processing, etc. It is readily recognized that the volumetric flow rate-pressure drop relationship is an important design parameter in all such applications. While a vast body of knowledge is now available on the estimation of the frictional pressure loss associated with the flow of purely viscous non-Newtonian fluids through model porous media consisting of spherical particles (ref.1), much less is known about the analogous situation involving beds of non-spherical particles, as encountered in actual field applications. In this work, it is demonstrated that how an existing method originally developed for the flow of Newtonian fluids through beds of spherical particles can be extended to encompass non-Newtonian flow in beds of non-spherical particles. The paper is concluded by presenting detailed comparisons between predictions and experiments.

2. EXPERIMENTAL

A glass column of 100mm diameter and 1800mm long packed with Raschig rings (two sizes) or gravel chips (as shown in Figure 1 a & b) was used to measure pressure drop for flow through packed beds of non-spherical particles. The three different types of packings used were characterised by assigning a volume equivalent diameter, d_e, (6.25, 8.02 and 12.02mm) and a sphericity shape factor ψ (0.222, 0.318 and 0.604); the latter was back calculated from the well known Ergun equation (ref.2) using experimental values of pressure drop for Newtonian fluids. The mean bed voidage, ε, varied between 0.41 and 0.71. Twelve aqueous solutions of different concentrations of a medium viscosity grade carboxymethyl cellulose were used as model non-Newtonian test liquids. Though these were not checked for possible viscoelastic effects but in view of the low molecular weight and low concentrations used, these are likely to be negligible. Thus, the usual two parameter power law model $\{\tau = m\ (\dot{\gamma})^n\}$ was found to yield satisfactory representation of shear stress-shear rate data; the resulting ranges of the model parameters are: $1 \geq n \geq 0.49$ and $0.23 \leq m \leq 3.77$ Pa.s^n. Thus, pressure drop was measured as a function of particle diameter, sphericity factor, bed voidage and non-Newtonian characterstics.

3. RESULTS AND DISCUSSION

It is customary to use dimensionless groups to represent presssure drop results as a function of the flow rate, and properties of the bed (d_e or d, ψ, ε) and the fluid (m, n, ρ). For the flow of Newtonian fluids through a bed of spherical particles, the following definitions of the Reynolds number and friction factor have gained wide acceptance:

$$Re = \rho V_o d/\mu(1-\varepsilon) \quad (1)$$

$$f = (d/\rho V_o^2)(-\Delta p/L)(\varepsilon^3/(1-\varepsilon)) \quad (2)$$

where V_o is the superficial velocity.

Based on extensive experimental results ($Re \le 1000$), Ergun (ref.2) proposed the following expression:

$$f = (150/Re) + 1.75 \quad (3)$$

Eq.(3) has also been found to be applicable for beds of non-spherical particles provided d is replaced by ($d_e\psi$) in Eqs.(1) and (2) (refs.3,4).

For power law fluids, while one can still define the friction factor via Eq.(1) but considerable confusion exists regarding the definition of Reynolds number (ref.5). The following modified definition of Reynolds number proposed by Kemblowski and Michniewicz (ref.6) will be used here:

$$Re^* = (\rho V_o^{2-n}(d_e\psi)^n/m(1-\varepsilon)^n)(4n/3n+1)^n$$

$$(15 \quad 2/\varepsilon^2)^{1-n} \quad (4)$$

Note that for n=1 and ψ = 1, Eq.(4) reduces to Eq.(2).

The present experimental results are plotted in Figure 2 along with the predictions of Eq.(3) with Re replaced by Re^* ($0.001 \le Re^* \le 20$); the agreement between the two is seen to be excellent. The resulting average and maximum deviations for about 425 data points are 13% and 29% respectively.

Similar type of comparisons result when the other data available in the literature for cylinders and pellets (refs.7 and8) are analyzed using the method outlined herein.

4. CONCLUSIONS

It is established that the well known Ergun equation can also be used to predict pressure loss for the flow of power law type fluids through packed bedsof non-spherical particles simply by using a volume equivalent diameter times a sphericity factor, i.e., $d_e\psi$ in lieu of particle diameter and a modified definition of Reynolds number. However, this observation is restricted primarily to the so called creeping flow region only.

REFERENCES

1. J.G. Savins, Ind. Eng. Chem. **61**(1969) 18-47.
2. S.Ergun, Chem. Eng. Prog., 48 (1952) 89-94.
3. R.P.Chhabra and B.K. Srinivas, Powder Technol.,**61**(1991) 15-19.
4. M.K.Sharma and R.P. Chhabra, Can. J. Chem. Eng., 70 (1992) in press.
5. Z.Kemblowski M. Michniewicz, and J.Sek, Adv. Transport Processes, 5(1987) 117-175.
6. Z. Kemblowski and M. Dziubinski, Rheo. Acta 18(1979) 730-739.
7. S. Kumar and S.N. Upadhyay, Ind.Eng.Chem.Fundam., 20 (1981).
8. Yu, Y.H., C.Y. Wen and R.C. Bailie, Can. J. Chem. Eng.,46 (1968) 149-154.

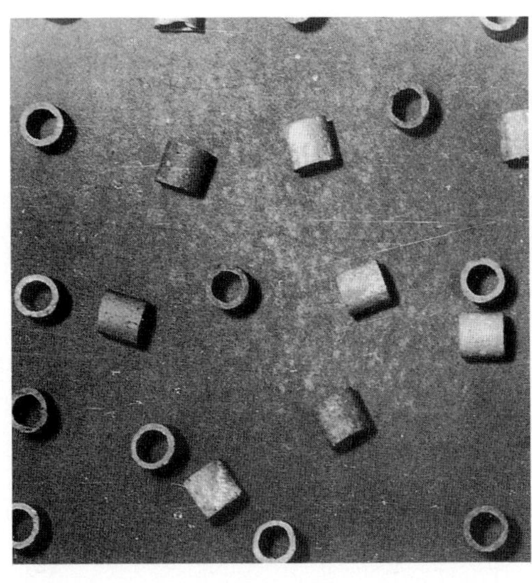

FIGURE 1a : Photograph of 12 mm Raschig rings.

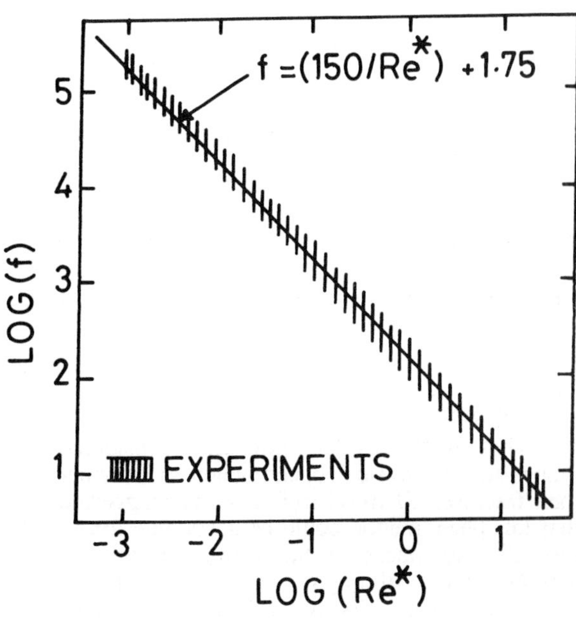

FIGURE 2 : Comparison between experiments and predictions of Eq. (3) with $Re = Re^*$ and $d = d_e\psi$.

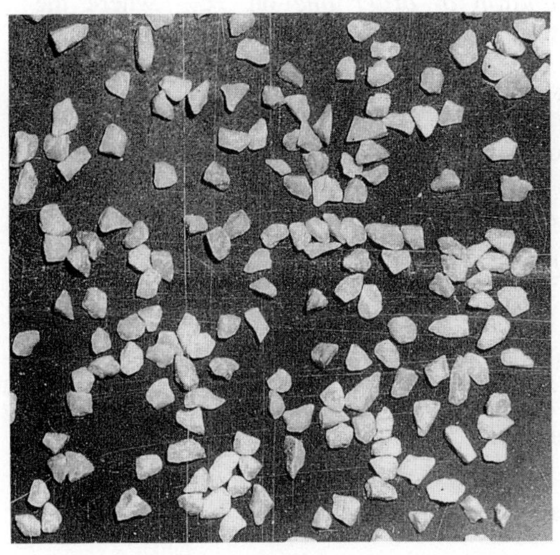

FIGURE 1b : Photograph of 6.25 mm gravel chips.

Theoretical and Applied Rheology, edited by P. Moldenaers and R. Keunings
Proc. XIth Int. Congr. on Rheology, Brussels, Belgium, August 17-21, 1992

STABILITY OF RECTILINEAL FLOW OF VISCOELASTIC FLUIDS

J. DUNWOODY

Department of Engineering Mathematics, The Queen's University of Belfast, Belfast BT9 5AH.

1. INTRODUCTION

The mathematical problem considered relates to the physical phenomenon of polymer melt fracture. Slemrod (ref. 1–2) has suggested that this phenomenon could be accounted for by loss of smoothness in the flow, and this he demonstrated through anlaysis of a one–dimensional velocity field perturbing plane Couette flow, using a particular model, when the stress–strain rate curve has the required properties which he related to the experimental data of Tordella (ref. 3), and/or certain derivatives of initial data are large. In all instances the initial data itself is small. However, standard eigenvalue analysis shows this flow to be linearly stable.

In its most general form, the problem is posed for inline perturbations of a steady plane lineal flow of an incompressible fluid which may be driven by a pressure gradient or not. The appropriate non–linear equations are

$$\left.\begin{aligned} \rho\partial_t u(x,t) &= \partial_x \mathcal{F}(\gamma(.;t);\kappa) \\ \gamma(s;t) &= -\int_0^s \partial_x u(x,t-r)dr, \quad 0 \le s \le \infty, \\ u(0,t) &= u(L,t) = 0, \quad -\infty < t < \infty \end{aligned}\right\} \quad (1)$$

where $u(x, t)$ is the perturbation velocity, x the cross–flow co–ordinate and $\mathcal{F}(\gamma(.;t);\kappa)$ the perturbation shear stress with $\mathcal{F}(0;\kappa) = 0$, a non–linear functional of the perturbation shear strain history $\gamma(s;t)$ and dependent on the basic shear rate $\kappa = \kappa(x)$.

With attention restricted to this particular problem, the model proposed by Slemrod (ref. 1) is not as special as might first appear.

2. DOMAIN OF $\mathcal{F}$

Firstly, the set of continuous functions on $[0, L]xR$

$$C_{\mathcal{F}} = \left\{\xi(x,t-s): \int_0^\infty e^{-\alpha s}\xi(x,t-s)ds < \infty, \ Re(\alpha) > 0\right\} \quad (2)$$

with each $\xi(x,t) = 0$ and $\xi(0,t-s) = \xi(L,t-s) = 0$ is a *proper subspace* of $C([0, L]xR)$ and in the usual way

$$<D_s\xi,\phi> = -<\xi,\phi'> = -\int_{-\infty}^{\infty} \xi(x,t-s)\phi'(s)ds, \quad (3)$$

valid for all *test functions* ϕ on R, defines a *distribution of order one* on ξ, where the function ξ has been extended by setting

$$\xi(x,t-s) = 0 \ \forall \ s \in (-\infty,0). \quad (4)$$

A companion distribution of order one, $D_x\xi$, may be similarly defined:

$$<D_x\xi,h> = -<\xi,h'> = -\int_0^L \xi(x,t-s)h'(x)dx \quad (5)$$

for all test functions h on $[0, L]$. Also, since $\xi(x,t-s)$ is continuous in both its arguments, generalized derivatives of all orders $D^n_{xx...ss}\xi$ may be defined to exist in which the order of differentiation is immaterial. For example,

$$<D^2_{xs}\xi;\ \phi,\ h> = \int_0^L\int_{-\infty}^{\infty} \xi(x,t-s)h'(x)\phi'(s)dxds$$

$$= <D^2_{sx}\xi;\ h,\ \phi>. \quad (6)$$

so that the *perturbation velocity* and *relative strain* in a fluid flow with *relative perturbation displacement* ξ, which are

$$u(x,t-s) = D_s\xi, \quad \gamma(x,t-s) = D_x\xi \tag{7}$$

respectively, satisfy a generalized compatibility relation.

From standard theorems quoted by Doetsch (ref. 4), it is known that the function

$$F^k(x,t;\alpha) = \int_0^\infty e^{-\alpha s} D_s^k \xi(x,t-s)ds \tag{8}$$

is $0(|\alpha|^k)$, $Re(\alpha) > \sigma > 0$, and that the mapping from the set of all such $F^k(x,t;\alpha)$ to the set of distributions defined on $C_{\mathcal{F}}$ is one–to–one to within a *null function.* Also, it is evident from the property in (6) and $(1)_2$ that

$$\int_0^\infty e^{-\alpha s} D_x u(x,t-s)ds = \alpha \int_0^\infty e^{-\alpha s}\gamma(x,t-s)ds, \tag{9}$$

and so the argument of the functional $\mathcal{F}$ with *domain* in $C_{\mathcal{F}}$ may be transformed to give

$$\mathcal{F}(\gamma(.;t);K) = \mathcal{G}\left[\int_0^\infty e^{-\alpha s}\partial_x u(x,t-s)ds;\kappa\right], \tag{10}$$

where $\partial_x u(x,t-s)$ is a *second order distribution* defined on $C_{\mathcal{F}}$. In terms of (10) equations (1) reduce to

$$\left.\begin{aligned} \partial_t u(x,t) &= \partial_x \mathcal{G}\left[\int_0^\infty e^{-\alpha s}\partial_x u(x,t-s)ds;\kappa\right] \\ u(0,t) &= u(L,t) = 0 \end{aligned}\right\}. \tag{11}$$

where the derivatives are open to interpretation in the generalized sense.

3. TRANSFORMED HYPERBOLIC SYSTEM

For convenience the parameter α in (11) is restricted to be real and positive definite. The transforms

$$\left.\begin{aligned} y(x,t) &= \int_0^\infty e^{-\alpha s}\partial_t u(x,t-s)ds \\ w(x,t) &= \int_0^\infty e^{-\alpha s}\partial_x u(x,t-s)ds, \end{aligned}\right\}, \tag{12}$$

the first implying $\partial_t y = \partial_t u - \alpha y$, if substituted in (11) then yield

$$\left.\begin{aligned} &\partial_t y = \partial_x g(w;\kappa) - \alpha y \\ &\partial_t w = \partial_x y, \quad y(0,t) = y(L,t) = 0 \ \forall\, 0 \le t \end{aligned}\right\}, \tag{13}$$

and when appropriate initial conditions

$$y(x,0) = y_0(x), \ w(x,0) = w_0(x) \tag{14}$$

are added (13) and (14) form an initial boundary value problem (cf. ref. 1). The differential system in (13) is non–linear and hyperbolic if $\partial_w g > 0$, which is a reasonable physical assumption to make in general, and it takes its most simple form when κ is constant, the only case considered by Slemrod (refs. 1–2). By extending y and w to be odd and even periodic functions in x respectively, with periods 2L, (13) and (14) may be reformulated as a pure initial value problem and estimates for the solution sought in terms of the Riemann invariants

$$\begin{matrix} r \\ s \end{matrix} = y \begin{matrix} + \\ - \end{matrix} \int_0^w \sqrt{\partial_w g}\, dw, \quad \begin{matrix} r_0 \\ s_0 \end{matrix} = \begin{matrix} r(0,x) \\ s(0,x) \end{matrix}. \tag{15}$$

The main results quoted by Slemrod (refs. 1–2) for κ *constant* are these:

Suppose $\sup_{x\in r}|r_0|$, $\sup_{x\in r}|s_0|$ *are sufficiently small and* $\partial_w g(0;\kappa) > 0$, $\partial_w^2 g(0;\kappa) > 0$. *If* $\partial_x r_0$ *or* $\partial_x s_0$ *is positive and sufficiently large at any point* $x \in R$, *then* (13), (14) *has a solution* $(y,w) \in C^1(RxR)$ *for only a finite time. For* $\partial_w^2 g(0;\kappa) < 0$ *and* $\partial_x r_0$ *or* $\partial_x s_0$ *sufficiently large and negative the same conclusion applies. In addition if* $(y,w) \notin C^1(RxR)$ *the transforms* (12) *imply* $(\partial_t u, \partial_x u) \notin C^1(RxR)$. But lack of smoothness and decay are not antagonistic properties of solutions to viscoelasticity problems.

4. SYMMETRIC FORM

The system (13) and (14) may be transformed to a *symmetric system* by the transformation $w^* = \sqrt{\partial_w g}\, w$. Since interest is focused on small initial data (y_0,w_0^*), terms involving y and w* are neglected in comparison to unity so as to obtain the linear *symmetric hyperbolic system* within the definition of Friedrichs:

$$\left.\begin{aligned}\partial_t y &= \mu\partial_x w^* - \alpha y + d_x\mu w^* \\ \partial_t w^* &= \mu\partial_x y, \qquad \mu = \sqrt{\partial_w g(0;\ \kappa)}\end{aligned}\right\} \quad (16)$$

For perturbed plane *Couette flow* $d_x\kappa \equiv 0$, so that the coefficients in (16) are constant. Hence, for this case the initial data may be extended to $-\infty < x < \infty$ so that $y_0(x)$ is an *odd periodic* function, while $w_0^*(x)$ is an *even periodic* function, both with period 2L. With this initial data (16) becomes an initial value problem. But, in order to put the solution of this problem within the range of theorems quoted by Mizohata (ref. 5) initial data $u_0 = (y_0, w_0^*)^T$ is prescribed for each half-period $[nL,(n+1)L]$ so that $u_0 \in \mathscr{D}^1_{L^2}(0,L)$, the *closure* of $C_0^\infty(0,L)$, with norm

$$\|f\| = \|f\|_{L^2} + \|f_x\|_{L^2}, \quad f \in \mathscr{D}^1_{L^2}(0,L).$$

Taking each half-period in turn and setting $u_0(x) = 0$ for $x \notin [nL,(n+1)L]$, so that $u_0 \in \mathscr{D}^1_{L^2}(R)$, a countable set of initial value problems is posed. It is evident from the properties of $\mathscr{D}^1_{L^2}(0,L)$ and the theory of characteristics that the *domain of influence* for each solution is bounded by $x + ct = nL$ and $x - ct = (n+1)L$. Hence for $0 < t \le T$ only a finite number of the solutions contribute when $nL \le x \le (n+1)L$. The relevant existence and decay theorem for the solutions is this (cf. ref. 5): *For* $u_0 \in \mathscr{D}^1_{L^2}(R)$ *there exists a unique solution* u(t) *with values in* $\mathscr{D}^1_{L^2}(R)$, $t \ge 0$, *and derivative* $d_t u$ *with values in* $L^2(R)$, $t > 0$, *both sets of values continuous in* t. *Further, the solution satisfies the energy equation*

$$d_t\|u\|^2_{L^2(R)} = -\alpha\|y\|^2_{L^2(R)}. \quad (17)$$

It is readily argued from the initial conditions and (17) that $\|u\|^2_{L^2(R)}$ decays *exponentially.*

For plane *Poiseuille flow* $d_x\kappa \not\equiv 0$ and the above arguments may not be used because in order to extend the solution as a periodic function in x the coefficients in (16) are at best piecewise smooth and piecewise continuous in extension. However, equations (13) when linearized are equivalent to

$$\left.\begin{aligned}\partial_t^2 y + \alpha\partial_t y &= \partial_x(\partial_w g(0;\kappa)\partial_x y) \\ y(0,t) &= y(L,t) = 0\ \forall\ 0 \le t\end{aligned}\right\}. \quad (18)$$

Standard separation of variables and Sturm–Liouville analysis lead to a formal solution for prescribed initial conditions

$$y(x,0) = y_0(x),\ \partial_t y(x,0) = \partial_w g(0;\kappa)\partial_x w_0 - \alpha y_0(x) \quad (19)$$

which is *exponentially stable* if $\partial_w g(0;\kappa) > 0\ \forall\ x \in [0,L]$, but the classification of *the space to which* (y_0, w_0) *must belong for a solution to exist is not known in this case.*

It is tempting to consider the effects of *change of type* (cf. ref. 2) on solutions to (18), (19), but on circumspection elementary analysis reveals that

$$\partial_w g(0;\kappa) \le 0 \iff d_\kappa\tau(\kappa) \le 0, \quad (20)$$

where $\tau(\kappa)$ is the shear stress in the basic flow and

$$\tau(\kappa) = a_0(x - L/2) = -\tau(-\kappa),\ a_0 \text{ constant}, \quad (21)$$

because the particle histories $\xi_0(x,t-s) \in C_{\mathscr{F}}$ for steady Poiseuille flow. *If there exists a set* $[\kappa_1,\kappa_2] \in (0,\infty)$ *and* $d_\kappa\tau(\kappa) < 0\ \forall\ \kappa \in [\kappa_1,\kappa_2]$ *but is positive definite elsewhere, a smooth solution of* (21) *with* $\kappa = d_x u_0(x)$ on $0 \le x \le L$ *and* $u_0(0) = u_0(L) = 0$ *does not exist for* $2\tau(\kappa_1) \le a_0 L$. If this latter condition is not satisfied the solution (or flow) is linearly stable in the commonly accepted sense.

REFERENCES

1. M. Slemrod, Arch. Rational Mech. Anal., 68, 1978, p. 211.
2. M. Slemrod, Unstable Viscoelastic Fluid Flows (private communication), c. 1980.
3. J.P. Tordella, J. App. Phys., 27, 1959, p. 454.
4. G. Doetsch, Laplace Transformation, Springer–Verlag, Berlin–Heidelberg–New York, 1974.
5. S. Mizohata, Partial Differential Equations, Cambridge University Press, 1973.

Theoretical and Applied Rheology, edited by P. Moldenaers and R. Keunings
Proc. XIth Int. Congr. on Rheology, Brussels, Belgium, August 17-21, 1992

STUDIES ON THE MECHANISM OF HETEROGENEOUS DRAG REDUCTION

K. HOYER[1], H.-W. BEWERSDORFF[1], A. GYR[1], and A. TSINOBER[2]

[1] Institute of Hydromechanics and Water Resources Management, Swiss Federal Institute of Technology, CH-8093 Zürich, Switzerland

[2] Faculty of Engineering, Tel-Aviv University, Tel-Aviv 69978, Israel

1. INTRODUCTION

Heterogeneous drag reduction is of atmost importance for the physical mechanisms of turbulence. It is achieved by injecting a concentrated polymer solution, e.g. a 0.5 - 1 % polyacrylamide solution, into the core-region of a turbulent pipe or channel flow. The injected polymer solution forms a thread, and a number of observations have shown that the thread preserves its identity for very long distances. Surprisingly, drag reduction occurs. This leads to a controversy: according to one opinion the drag reduction is due to an interaction of the viscoelastic thread with the turbulence of a Newtonian fluid, whereas according to the other opinion the drag reduction is due to minute amounts of polymer removed from the thread and dispersed into the bulk of the fluid (see for example the discussion in [1]).

The first experiment on heterogeneous drag reduction was reported by *Vleggaar and Tels* [2]. These authors found that the drag reduction was two to three times higher than that of a homogeneous pre-mixed dilute polymer solution with the same effective polymer concentration and identical flow-rates. In another contribution [3] they correlated the reduction in heat transfer with the drag reduction and concluded that the polymer thread damps the radial momentum transfer. Later on *Bewersdorff* [4 - 6] found that depending on the concentration of the injected polymer solution the resulting non-homogeneous mixture downstream of the injection point may consist of a single thread, several long threads, or many small polymer strings polymer solution. Although the concentration of polymers in the near-wall region was less than 1 ppm , drag reduction as high as 24 % could be observed. Laser-Doppler-anemometry measurements of the turbulent structure showed a different velocity profile than that found for homogeneous polymer solutions: Up to a dimensionless wall distance of approximately $y^+= 30$ the dimensionless velocity profile was the same as that for a Newtonian fluid, but further towards the core the slope of the velocity profile increased to give relatively larger velocity values. In the core region the slope of the profile was also increased. The Reynolds shear stresses were drastically reduced. These results were supported by those of *Berman* [7] who additionally found by analyzing the skewness and flatness curves of the axial velocity fluctuations that the polymer thread alters the large scale structure near the wall implying that there is a different type of drag reduction than that found in dilute homogeneous polymer solutions.

Usui et al. [8, 9] suggested that the main interaction between the polymer thread and the turbulence which results in drag reduction takes place in the central part of the flow and suppresses the large scale motions of turbulence. By varying the ratio of the polymer injection velocity to the bulk velocity, and by mounting contractions in the pipe, *Bewersdorff* [10] showed that the drag reduction increases with the elongation of the polymer thread as long as the thread stays intact and is not broken into pieces.

Smith and Tiederman [11] concluded from dye diffusion experiments that the drag reduction occurred mainly due to diffused polymer from the polymer thread.

In a recent study *Hoyt and Sellin* [12] reported that the effect is due to dissolved molecules removed from the thread at high Reynolds numbers, whereas at low Reynolds numbers an interaction of the thread with the large-scale eddies should be responsible for the observed drag reduction.

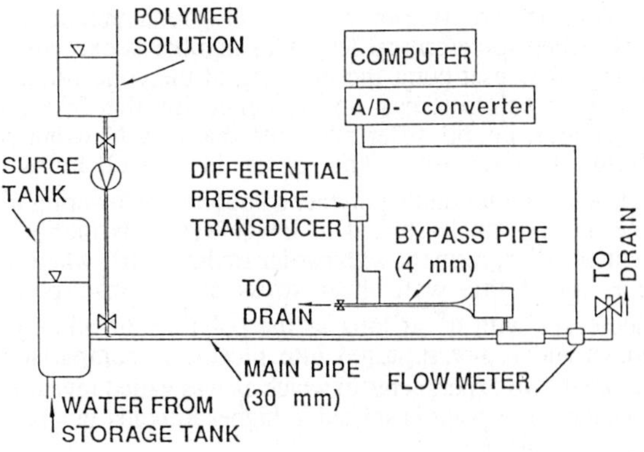

Figure 1: Experimental arrangement

The aim of the present study is an attempt to resolve the above mentioned controversion whether this type of drag reduction is caused by minute amounts of polymer molecules removed from the thread into the bulk fluid, by to an interaction of the polymer thread with the ambient fluid, or due to a combination of both effects. This studyis based on investigation which is based only on the data about the frictional behaviour of the flow and some visualization experiments, and is part of a larger research program on heterogeneous drag reduction.

2. EXPERIMENTAL ARRANGEMENT

The experimental arrangement is described in detail in Bewersdorff et al. [13] and shown in Figure 1. The main goal was to check whether the fluid surrounding the thread is drag reducing in the usual sence. For this purpose the thread was sucked out and the drag reduction downstream of the withdrawl point was measured. In a second series of experiments near-wall fluid was withdrawn instead and the drag reduction of that fluid in a by-pass was determined.

3. RESULTS

The wall shear stress τ_w, the friction factor f and the drag reduction DR at a constant flow-rate can be calculated as

$$\tau_w = \frac{\Delta p\, d}{4L}, \quad f = \frac{2\tau_w}{\rho U^2}, \quad DR = 1 - \frac{f_p}{f_s}$$

where $\Delta p/L$ is the pressure drop per unit length, d the hydraulic diameter, U the bulk velocity, and ρ the density of the fluid. The subscripts p and s denote experiments with polymer injection and pure solvent, i.e. water.

Heterogeneous drag reduction reaches its constant value at about l/d = 150 - 200, [5-7] with l the distance from the injection point. Bewersdorff [10] found that an important parameter for this kind of drag reduction is u^*, the ratio of the injection velocity to the bulk velocity. In the following, c_p will be used for the concentration of the thread material, and c_R for the effective polymer concentration averaged over the cross-section.

The friction behaviour of the injection experiments exhibits an onset-phenomenon, i.e. drag reduction occurs only when some critical Reynolds number is exceeded. Below the onset-point the injection of the concentrated polymer solution has no influence on the friction behaviour, i.e. no difference from that of a Newtonian fluid is detectable. The onset is a well-known phenomenon for dilute polymer solutions. By lowering u^* the onset-point is shifted to lower Reynolds numbers which is in agreement with earlier findings [10] where it was found that wall shear stress at the onset-point decreases with u^* as long as the polymer thread stays intact and is not ruptured into pieces. A comparison between two experiments in which c_p was varied revealed that the onset point is shifted to higher Renolds numbers with increasing c_p.

As was mentioned in Section 2 a series of experiments was carried out in which a suction device was mounted in the pipe in order to trap the polymer thread. In this way the friction behaviour with and without trapping but with the suction device in place can be compared. This is done in Figure 2. At high Reynolds numbers the friction behaviour downstream from the suction device appears to be identical for the two cases. However, the trapping results in a shift of the onset-point to higher Reynolds numbers. Therefore the question remains whether the movement of the polymer thread is responsible at least for a part of the observed drag reduction at low Reynolds nimbers [12].

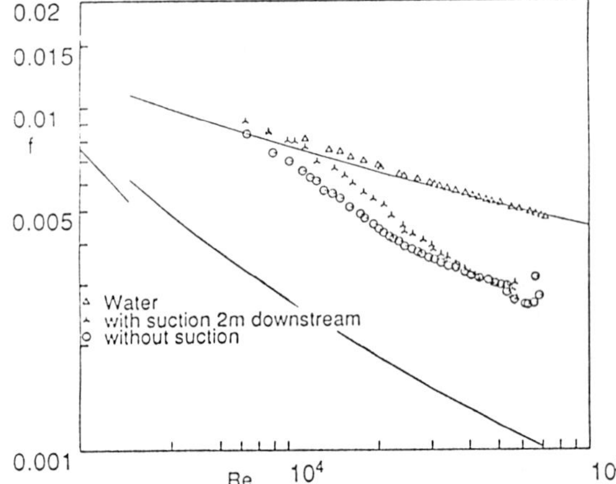

Figure 2: Friction behaviour with and without suction in a round pipe with c_P= 0.8%, c_R= 10 ppm and u^*= 1.13

In Figure 2 the Prandtl-Kármán law

$$f^{-1/2} = 4 \log (\mathrm{Re}\, f^{1/2}) - 0.4$$

is the upper straight line. Furthermore, the lower line represents the empirical maximum drag reduction asymptote [14]

$$f^{-1/2} = 19 \log (\mathrm{Re}\, f^{1/2}) - 32.4 .$$

Figure 3 shows the friction behaviour of the fluid leaving the main flow through a porous pipe wall into a bypass pipe of 4 mm in diameter. In all cases, drag reduction occurred in the bypass, even when no drag reduction in the main flow was observed. The level of drag reduction in the bypass increased with the Reynolds number of the main flow which indicates that the concentration of drag reducing material or the length of its agglomerates in the near-wall region of the main flow also increases with the Reynolds number of the main flow. Furthermore, it is interesting to note that for the lower Reynolds numbers in the main flow (Re = 10,000 and 20,000) the drag reduction does not increase with the Reynolds number in the bypass, although one should expect to get a higher concentration of drag reducing material in the bypass with increasing offtake due to a concentration gradient in near-wall region of the main flow. This may be due to the fact that during the porous media flow through the porous wall polymer associates or filaments may have been destroyed [15] or single molecules may have been degraded. (Degradation is known to be more severe at extremely low polymer concentrations.)

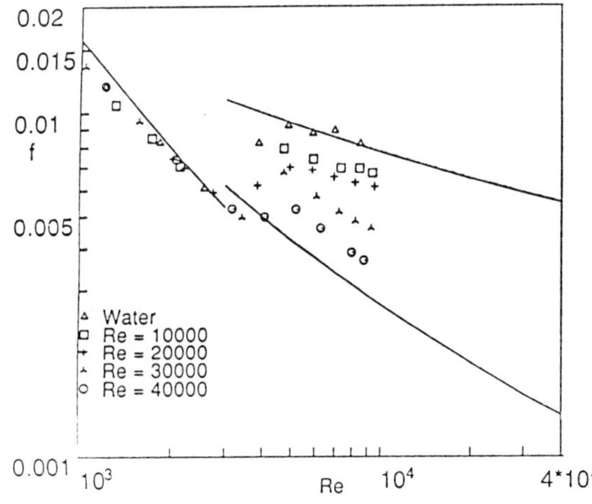

Figure 3 Friction behaviour in the bypass in terms of the Reynolds number in the main flow,.c_R= 10 ppm, c_p= 0.8 %, , and u^* = 0.12

First of all the experiments described above confirm unequivocally that heterogeneous drag reduction is a strong effect.

The thread trapping and the bypass experiments indicate that the heterogeneous drag reduction is at least partly due to minute amounts of polymers removed from the thread and dispersed in the bulk of the fluid. This, however, does not resolve the controversy on the mechanism since nothing is known about how much (if any) of drag reduction is produced by the direct interaction of the thread and the flow.

An incredibly small amount of polymer present in disperse form is, therefore, responsible for part of the effect. It was found that: (a) The effect decreases if the thread ruptures into pieces instead of being enhanced as expected for a diffusion process. (b) With decreasing u^* the effect increases and the onset point is lowered. This suggests that diffusion is not the mechanism by which the polymer enters the bulk of the fluid

We propose the following explanation of these findings: (a) the polymer in the bulk of the flow is present as agglomerates much longer than single molecules and (b) the dispersion mechanism is due to rheological properties of the thread. This is suggested by the fact that the thread is most efficient for values of u* just below the load necessary to rupture the thread.

In addition, video observations show that drag reduction occurs whenever the thread starts forming loops. In these loops the thread often unravels locally into various strands, which unify again when the thread is stretched. In the unravelled state, small disturbances are

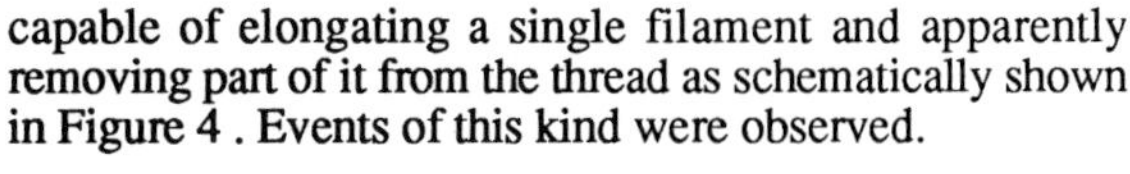
capable of elongating a single filament and apparently removing part of it from the thread as schematically shown in Figure 4 . Events of this kind were observed.

Figure 4. Scheme of thread splitting

REFERENCES

1. A. Gyr (ed.), Structure of Turbulence and Drag Reduction, Springer-Verlag, Berlin (1990)
2. J. Vleggaar and M. Tels, Chem. Eng. Sci. **28,** (1973), 965
3. J. Vleggaar and M. Tels, Int. J. Heat and Mass Transfer **16,** (1973), 1629
4. H.-W. Bewersdorff in: Drag Reduction (eds.) R.H.J. Sellin, R.T.Moses, University of Bristol (1984), B4
5. H.-W. Bewersdorff, Rheol. Acta **23,** (1984), 522
6. H.-W. Bewersdorff in: The Influence of PolymerAdditives on Velocity Fields B. Gampert, Springer-Verlag, Berlin (1985), 337
7. N.S. Berman, Chem. Eng. Commun. **42,** (1986), 37
8. H. Usui , K. Maeguchi and Y. Sano, Phys. Fluids **31**, (1988), 2518
9. H. Usui, (1990) in: [1], 257
10. H.-W. Bewersdorff in: Drag Reduction in Fluid Flows (eds.) R.H.J.Sellin, R.T. Moses, Ellis Horwood, Chichester (1989), 279
11. R.E. Smith and W.G. Tiederman, Rheol. Acta **30**, (1991), 103
12. J.W. Hoyt and R.H.J. Sellin, Rheol. Acta **30**, (1991), 307
13. H.-W. Bewersdorff, A. Gyr, K. Hoyer and A. Tsinober (to be published)
14. P.S. Virk, AIChE J. **21**, (1975), 625
15. E.J. Hinch and C. Elata, J. Non-Newtonian Fluid Mech. **5**, (1979), 411

ACKNOWLEDGMENT

The authors wishes to thank the Swiss National Science Foundation for providing the Research Grant for the present study.

Theoretical and Applied Rheology, edited by P. Moldenaers and R. Keunings
Proc. XIth Int. Congr. on Rheology, Brussels, Belgium, August 17-21, 1992

A MONTE CARLO SIMULATION OF UNSTABLE VISCOUS FINGERING

TI-QIAN JIANG and XIANG-FANG LI

Chemical Engineering Research Centre, East China University of Chem. Tech., Shanghai 200237 (China)

1. INTRODUCTION

When a fluid is forced into porous media to displace more viscous fluid, the interface between the two fluids is unstable. As time proceeds the interface develops a highly complex fingered pattern, termed Viscous Fingering Phenomenon (ref.1). It is report that the Viscous fingering, specially that in the Hale-show cell, have the fractal nature developed by mandelbort (ref.2) and can have serious economic consequence. Therefore, it has led to a number of experimental and theoretic methods to characterize the phenomena. For simulation, the first model is DLA (Diffusion Limited Aggregation) of Witten and Sander (ref.3), whose main principle is that when a particle is add, one at a time, to a growing cluster or aggragate of particles via random walk trajectories. A particle moves a long way from the cluster when it reaches the killing circle, it will be killed. The DLA model is self-similar object with fractal dimension $d_f = 1.7 \pm 0.5$ for two dimension. Kadanoff (ref.4) has shown that regular DLA model correspounds to viscous fingering at zero-surface tension and infinite-viscous-ratio limit. Sherwood and Nittman (ref.5) has proposed a GGG (Gradient Governed Growth) model, in which the walk distance of driving fluid is proportional to local pressure gredient for thier view. They, therefore, get the result under infinite viscous ratio with the fractal dimension $d_f = 1.9$ that is different with DLA model.

As shown in above, all the authors look the unstable viscous fingering phenomena as static state and mainly focus on Newtonian fluids. However, in this paper, we should like to consider the dynamic characterzations of a miscible system in which at least one is power-law Non-Newtonian fluid.

2. DEVELOPMENT OF A NEW MODEL MGGM

Based on the above models, a new model so called MGGM (Monmentum Governed Growth Model) in which a momentum modified term is introduced, was developed as follows. For a miscible system, the mixture can be regraded as a continual medium. Therefore, under low Re number, Darcy law becomes:

$$\nabla \overline{P} + \frac{\mu}{K} |\overline{U}|^{n-1} \overline{U} = 0 \tag{1}$$

wherin n denotes the power-law index of the fluid; $m = 1/n$; for Newtonian fluids $n = 1$, $m = 1$. For horizontal flow of the imcompressible fluids, the pressure distribution is satisfied by the following equation:

$$\nabla (|\nabla P|^{m-1} \nabla P) = 0 \tag{2}$$

The boundary conditions are:

$P = 1$ at $r = 0$;
$P = 0$ at $r = R$.

The solution of equation (2) with the boundary conditions is given by

$$P(r) = \frac{R^{(m-1)/m} - r^{(m-1)}}{R^{(m-1)/m} - 1} \qquad (m \neq 1)$$

and (3)

$$P(r) = 1 - \frac{lgr}{lgR} \qquad (m = 1)$$

Equation (3) is the expression of the pressure distribution in the whole flow field. The growth at the fop of fingering parttern is depended on the momentum which provied by the pressure difference between a particle and another being apart one particle distance from former. The possibility of the growing direction i of a particle may be evaluated by the percent:

$$\beta_i = \frac{\nabla P_i^m}{\sum \nabla P_i^m} (\%) \tag{4}$$

and $\sum \beta_i = 1.$

In our view, particle may deviates the direction in which the particle has maximal momentum, if the particle is influenced by the force came from the more viscous fluid. In this case, it hence changes the moving direction as shown in Fig.1, where the momentums are

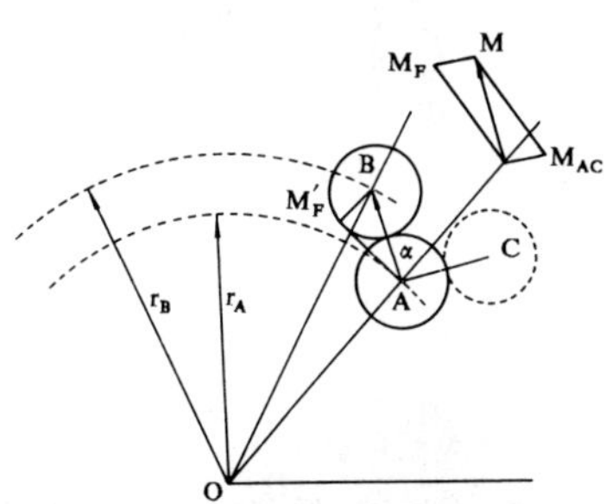

Fig.1 Production of momentum modified term

$$M_{AB} = m_S U = m_S \left[-\frac{k}{\mu} \triangle P_{AB} \right]^m \quad (5)$$

$$M_F = M_{AB} \sin\alpha \quad (6)$$

M dirction is already changed direction of the particle.

In our model, we have introduced a speciel parameter K as the function of viscous ratio and Power-law index n as follows:

$$K = K(\mu_1 / \mu_2, m) \quad (7)$$

In the monte carlo simulation process, the probabilistic seeds are take from the probabilitic number generator in computer.

3. RESULT AND DISCUSSION

3.1 Model's parameter K

Our model's parameter K absolutely can propory the influences of variant viscous ratio on the fingering pattern. The simulation results for n= 1, 0.4, 0.1, K=0.3, 0.1, 0.015, 0.01 were shown in Fig.2. comparing Fig.2 with the experimental results (see Fig.3) it should be found that the fingering pattern are similar and based on the sweep efficience a relation between simulation paramter K and viscous ratio is obtaind:

$$V_r = 0.204K^{-4.21} \quad (8)$$

3.2 Comparing the fractal dimension with DLA

When n=0.1, say, for Non-Newtonian fluid, comparing the simulate results under K=0.01, i.e., the viscous ratio approaches to infinit, the fingering patterns have fractal structure. If we plot lgN(r)~lgr, from the slope of the line the fractal dimension d_f should be obtained, as shown in Fig.4.

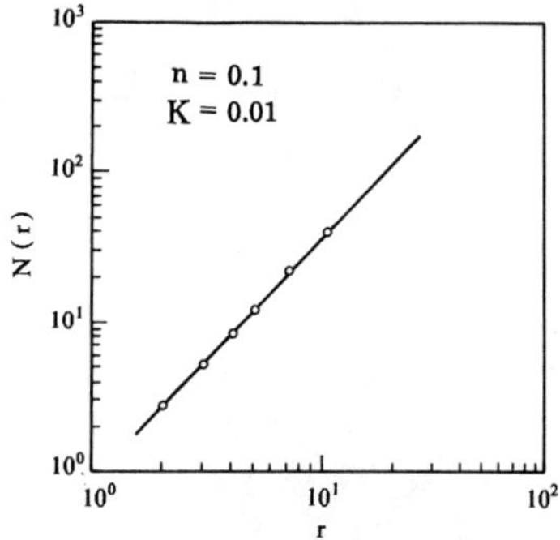

Fig.2 lg(total numb. of parts.)~lg(radius)

Our Computation showed the fractal dimension d_f variating with the value of K but around 1.70, consistent with the result of DLA.

3.3 Difference between Non-Newtonian and Newtonian fluids

It may be proved that the more shear thinning the more growing probability alone radial direction for fingering pattern. If one compare the simulation result for n= 1, 0.4, 0.1 with the experimental result (refs.1,6) it should be found that the more small for the value of n, the more straight for the trunk of fingering pattern. Therefore, why the fingering pattern is thick for Newtonian displacement and is thin for Power-law Non-Newtonian displacement can be explained. It is evident that the preceding supposition, that is, the growing direction of particles is depended on the momentum, in our new model is correct.

REFERENCES

1. E.Allen, D.V.Boger, 10th Internl. Cong. on Rheol., Vol.2, pp.146-148. 1988,
2. B.B.Mandelbrot, Fractals Form, Chance and Dimension, Freeman 1977.
3. T.A.Witten, L.M.Sander, Phys. Rev. Latt., **47**, (1981) 1499.
4. L.J.Kadanoff, J.Statist.Phys., **39**, (1985) 217.
5. J.D.Sherwood, J. Nittmann, J. Phys (Paris) **47**, (1986) 15.
6. H.K.He, L.X.Hui and T.Q.Jiang, J. Chem. Eng. China Univ. **4**(2), (1990) 101.

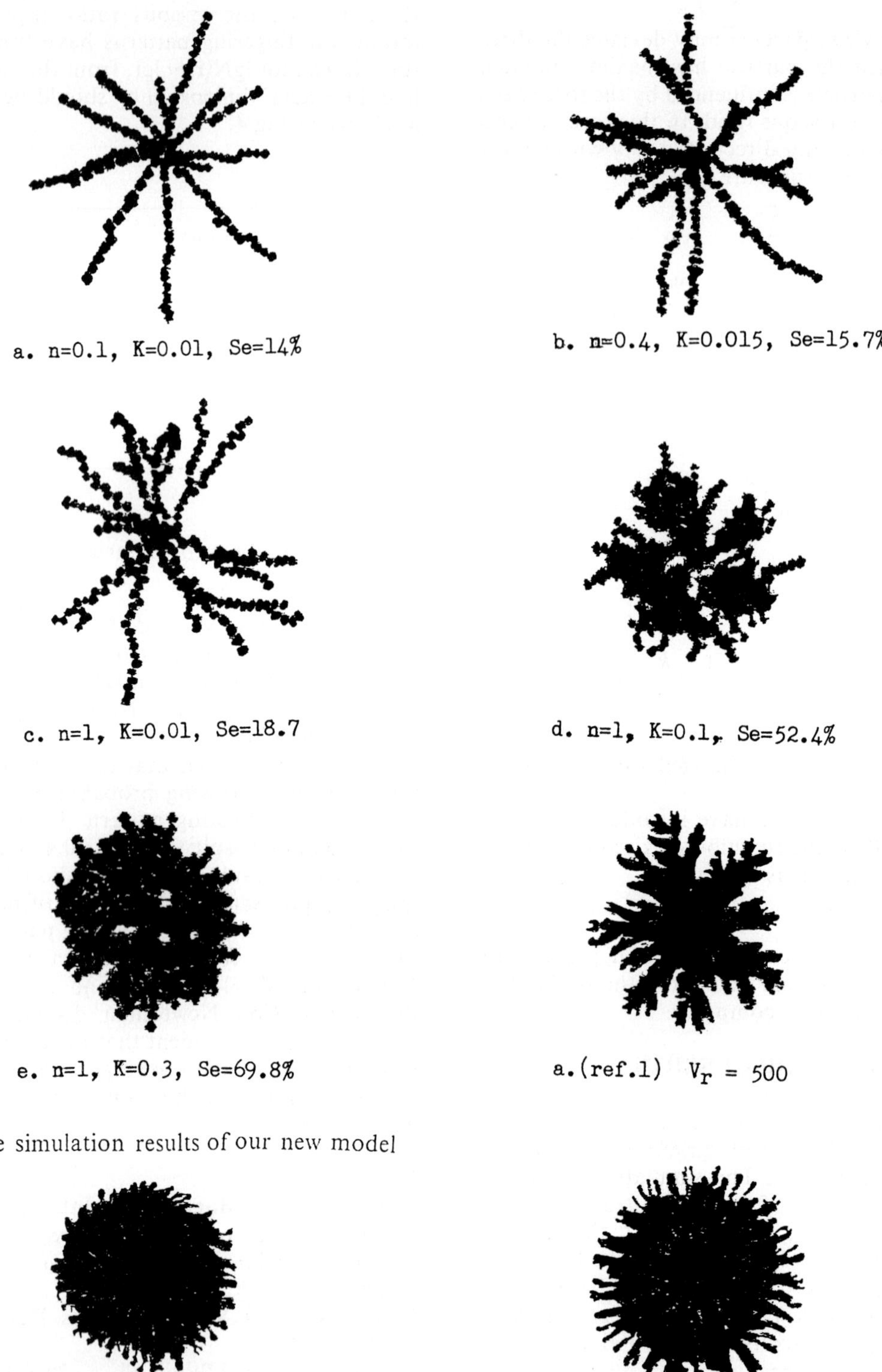

a. n=0.1, K=0.01, Se=14%

b. n=0.4, K=0.015, Se=15.7%

c. n=1, K=0.01, Se=18.7

d. n=1, K=0.1, Se=52.4%

e. n=1, K=0.3, Se=69.8%

Fig.2 The simulation results of our new model

a.(ref.1) V_r = 500

b.(ref.6) V_r = 12

c.(ref.1) V_r = 10

Fig.3 The experimental results

Theoretical and Applied Rheology, edited by P. Moldenaers and R. Keunings
Proc. XIth Int. Congr. on Rheology, Brussels, Belgium, August 17-21, 1992

Overstability of Viscoelastic Fluid Heated from Below

Chongyoup KIM and Gookhyun BAEK

Department of Polymer Engineering, Chungnam National University
220 Goong-dong, Yoosong-goo
Taejeon 305-764, Korea

INTRODUCTION

In this study, we present an experimental study on the instability of a viscoelastic fluid layer heated from below(Bénard convection), noting that the theoretical research on the system has been revitalized recently(refs. 1,2). The governing equations for the present system are the momentum, mass, energy and constitutive equations under the Boussinesque approximation. The constitutive equation does not have to be fixed for this research. But a proper parameter representing elasticity is required to include the case of polymer solutions. If the linear viscoelastic Maxwell model is chosen as the constitutive equation for this purpose, the relevant dimensionless groups are found to be Ra, Pr and De defined as follows:

Ra : Rayleigh number ($= g\alpha\Delta T d^3/\kappa\nu$)
Pr : Prandtl number ($= \nu/\kappa$)
De : Deborah number ($= \lambda\kappa/d^2$)

where g is gravitational acceleration, α is thermal expansion coefficient, ΔT is temperature difference, d is gap distance, κ is themal diffusivity, ν is kinematic viscosity and λ is the relaxation time of fluid.

In the linear stability analysis, the disturbances are expressed as $e^{\sigma t}f(\mathbf{x})$, where t is time and $\mathbf{x}$ is the coordinate of position. In the case of exchange of stability, the imaginary part of σ vanishes when the system crosses the marginal state. In the case of overstability, on the other hand, the imaginary part does not vanish when it passes the marginal state so that oscillations are present when the system becomes unstable. Also, for the case of overstability, the Ra at the onset of instability(crtical Ra; Ra_c) is smaller than Ra_c for the case of exchange of stability.

Stability analyses showed that the principle of exchange of stability did not hold for the case of viscoelastic fluid and overstability could take place due to the destabilizing effect of elasticity(refs. 3-6). It has been argued that, however, overstability could not be attained in normal laboratory conditions. This is because an unreasonably large ΔT is required for normal polymer solutions to attain the required Ra. Overstability was predicted only for very dilute solutions of high molecular weight polymers(refs. 3,4).

Despite the several theoretical predictions, no experimental observation on overstability has been reported to date to the author's knowledge. The only experimental report on the instability in a horizontal layer of viscoelastic fluid heated from below was done by Liang and Acrivos(ref. 7). But they assumed that no overstability existed, and argued that the experimental technique could be used for determining the zero shear rate viscosity which was otherwise very difficult to measure.

The principal objective of this research is to experimenatally examine the stability characteristics and especially the existence of overstability. The result shows that the fluid elasticity has destabilizing effects. Also, the experimental observations supported the theoretical analyses reported in the literature.

EXPERIMENT

The experimental apparatus was set up by following ref. 7 as closely as possible except that no optical observation unit was installed. Therefore the detailed description of the apparatus will not be given in the present paper and can be found in ref. 8. For more accurate measurements, the temperature difference between the upper and lower surfaces was measured by means of a T-type differential thermocouple with eight junctions. The measurement was performed using an HP3852s data acquisition system and the accuracy of measurement in the temperature difference was 0.001°C.

Reagent grade ethylene glycol(EG) was used as the solvent for this study. Polymers tested in this study were xanthan gum(XG) and polyacrylamide(PAAM) made by Aldrich Chem. Co. The weight average molecular weight of PAAM was about 6,000,000 (supplied by the manufacturer). Polymer solutions were prepared separately for each test while stirring slow to prevent from possible mechanical degradations. It took about a week to prepare the test fluid. For XG, 50, 100 and 125 ppm solutions were tested and for PAAM, 50, 100, 200 and 300 ppm solutions were tested.

The zero shear rate viscosity was measured by means of a Zimm-Crothers viscometer(ref. 9). The zero shear rate viscosities of XG and PAAM solutions were found to follow the following formulae at the mean temperature of the fluid in the convection cell(25°C):

$$\text{XG} \quad : \eta = 18.0 + 0.346852c + 0.0021265c^2 \quad (0 \le c \le 125) \tag{1}$$

$$\text{PAAM} : \eta = 18.0 + 0.05718c \quad (0 \le c \le 300) \tag{2}$$

where c is in ppm and η is in cp. The zero shear rate intrinsic viscosities $[\eta]_0$ of XG and PAAM were found to be 0.02 and 0.0026 /ppm at 25°C, respectively. Since the polymer concentration was very dilute, the other physical properties such as density, thermal expansion coefficient, thermal conductivity, etc were assumed to be the same as those of pure ethylene glycol as was done by Liang and Acrivos(ref. 7)

To check the accuracy of the apparatus, experiments were performed twice for pure EG, first by increasing Ra(ΔT) then decreasing Ra and the results are shown in Fig. 1. The onset point of instability(convection) was determined from the plot of Nu vs. Ra. In Fig. 1, Ra_c is 1680 ± 30, which is close enough to the theoretical value of 1708. Also, at the conduction regime, Nu is equal to 1 ± 0.02 and the slope in the convection regime coincides with the result collected in ref. 9. In view of the good agreement between the present results and those of the previous studies given in ref. 9 for Newtonian fluids, the apparatus appears to work satisfactorily.

RESULTS AND DISCUSSION

A typical set of experimental results for XG solution is plotted in Fig. 2 when $c = 100$ppm. As in the case of pure EG, experiments were performed first by increasing Ra then decreasing Ra. We first observe that Ra_c is less than 1700, therefore confirm the existence of overstability. But the result obtained by decreasing Ra did not follow that by increasing Ra and the Ra_c for the case of decreasing Ra was smaller than for the case of increasing Ra. The same phenomenon was always present for other XG solutions.

The difference in Ra_c was caused by the degradation of polymers. This was confirmed by the viscosity decrease of the solution (The viscosities of the test fluid were measured before, during and after each experimental run). Therefore, only the critical Ra obtained by increasing Ra was regarded meaningful. It was because before the onset of convection, the fluid remained in the state of rest and there should be no degradation.

The degradation of XG in the convection regime also could be expected from the theoretical stability analysis. If the exchange of stability is satisfied, the rate of shear is given by Re(σ), and should be small in the convection regime if Ra is not too large. But, if the system is overstable, the rate of shear is given by Im(σ) which is not zero and could be very large. Using the equation in ref. 4 with experimentally determined Ra_c and Pr, we calculated Im(σ) for the XG solution tested here and found that it had an order of 10/sec. Even though XG is not very susceptible to shear degradation, this value appears to be large enough to induce shear degradation considering the viscosity of EG is about 20 times larger than that of water.

In Fig 3., Ra_c vs. c is plotted. For XG, we note that as the concentration increases, Ra_c decreases. This means that the fluid elasticity has a destabilizing effect. Since the coil overlap concentration (c^*) of XG in EG is estimated to

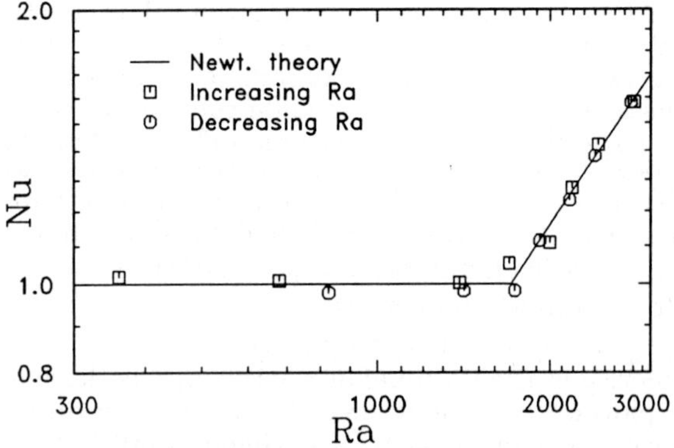

Fig. 1. Nu vs. Ra for pure ethylene glycol.

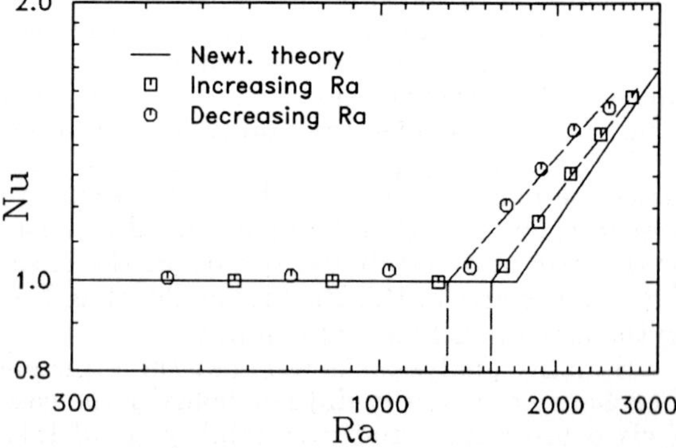

Fig. 2. Nu vs. Ra for a 100ppm xanthan gum solution.

be 50 ppm($1/[\eta]_0$), the relaxation time of the solution for the concentration range considered here increases as the concentration increases. For XG in EG, the relaxation time in the dilute limit, $\tau_0 = [\eta]_0\eta_s M/N_A kT$, is 0.87sec, where η_s is the solvent viscosity, M is the molecular weight, N_A is the Avogadro's number, k is the Boltzmann constant and T is temperature. If this value is inserted into the equation in ref. 4, no overstability is predicted. But in the case of 100 ppm solution, $\lambda = 6.7$sec is obtained if the experimentally determined Ra_c and Pr are inserted into the equation in ref. 4. This result is in agreement at least qualitiatively with the theoretical prediction that λ increases sharply as the concentration increases for $c \geq c^*$.

In Fig. 4., a set of results for a PAAM solution is plotted when $c = 200$ppm. Even though the same procedure was followed as in the case of XG, the result was almost the same as in the Newtonian case and no degradation was observed. Considering c^* for PAAM in EG is 390ppm, the concentration is within the dilute limit and the relaxation time λ of this solution is then given by $\tau_0(= 0.13\text{sec})$. For this value, no overstability,

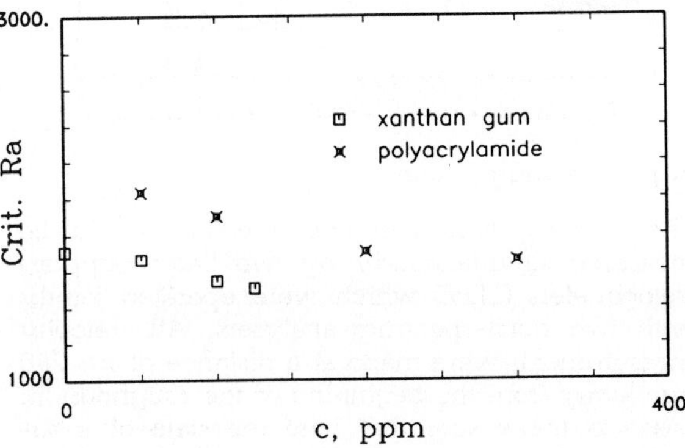

Fig. 3. Critical Ra vs. concentration for xanthan gum and polyacrylamide.

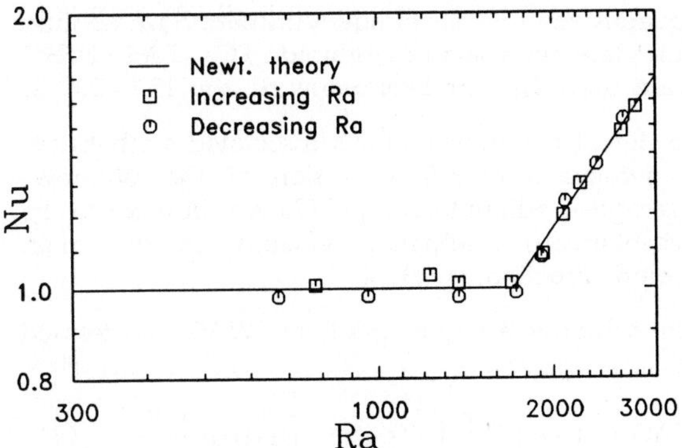

Fig. 4. Nu vs. Ra for a 200ppm polyacrylamide solution.

hence no oscillation is predicted by the theoretical analysis as observed in the experiment. The fact that no oscillation is present also could be confirmed by the experimental fact that no degradation was observed.

For PAAM, no overstability was observed in the concentration range investigated here. But solutions of higher concentration were not investigated since sufficient elasticity could not be attained. An unexpected observation was that for very dilute solutions of PAAM, convection was delayed to large Ra as shown in Fig. 3. The same result was also observed in preliminary experiments using acquous solutions of PAAM. But we have not been able to find out any reasonable explanation for the delay yet.

CONCLUDING REMARKS

In this research, we report on the experimental observation of overstability of a horizontal viscoelastic fluid layer for the first time. The result was in agreement with the linear stability analysis reported in the literature.

The result of this study implies that the stability experiment be utilized in measuring the relaxation time of a polymer solution in semi-dilute regions. Since this method is independent of the conventional rheometry, this will be another criterion in choosing the proper constitutive equation if theoretical analyses on various constitutive equations are performed in parallel.

REFERENCES

1. S. Rosenblat, *J. non-Newtonian Fluid Mech.*, **21**(1986) 201-223.
2. J. Stastna, *J. non-Newtonian Fluid Mech.*, **18** (1985) 61-69.
3. M. Sokolov and R. I. Tanner, *Phys. Fluids*, **15**(1972) 534-539.
4. C. M. Vest and V. S. Arpaci, *J. Fluid Mech.*, **36**(1969) 613-623.
5. T. Green III, *Phys. Fluids*, **11**(1968) 1410-1412.
6. I. A. Eltayeb, *Proc. Roy. Soc. Lond.*, **A356** (1977) 161-176.
7. S. F. Liang and A. Acrivos, *Rheol. Acta*, **9** (1970) 447-455.
8. G. Baek, MSE thesis, Chungnam Nat'l Univ., 1992, pp. 19-23.
9. S. Chandrasekhar, Hydrodynamic and hydromagnetic stability, Clarendon Press, Oxford, 1961, pp. 59-71.
10. B. H. Zimm and D. M. Crothers, *Proc. Nat'l. Acad. Sci.*, **48**(1962) 905-911.

Acknowlegement: Support for this research from the Korea Science and Engineering Foundation is gratefully acknowledged (Contract No. 911-1001-001-2).

Theoretical and Applied Rheology, edited by P. Moldenaers and R. Keunings
Proc. XIth Int. Congr. on Rheology, Brussels, Belgium, August 17-21, 1992

THE INFLUENCE OF A DRAG REDUCING SURFACTANT ON THE COHERENT MOTIONS IN ROUGH AND SMOOTH WALL TURBULENT BOUNDARY LAYERS

H. Lyko and K. Strauß

Universität Dortmund, Fachbereich Chemietechnik, Lehrstuhl Energieprozeßtechnik, Postfach 500 500, 4600 Dortmund 50, Germany

1. ABSTRACT

It is assumed that the drag reducing effect of certain cationic surfactants is due to their influence on the coherent structures in turbulent boundary layers.

The purpose of the present article was to study the effect of a surfactant additive on the frequency and duration of coherent motions by analysing recorded tracks of the instantaneous velocity signal. Some remarkable results of two-component measurements in the rough wall boundary layer are presented additionally.

2. INTRODUCTION

Some aqueous solutions of cationic surfactants show a strong drag reducing behaviour in turbulent channel and pipe flows [1]. One important feature of the solution considered here is a kind of 'anisotropic' turbulence, i.e. an almost complete reduction of the turbulence intensities in the lateral directions while fluctuations remain in the main stream direction. This effect could be observed in the wake flow of a cylinder and the plane mixing layer ([2],[3],[4]), beyond the rough wall boundary layer at moderate Reynolds numbers. There is a strong influence on the occurrence and shape of vortices in those flow geometries, e.g. the critical Reynolds number for the Karman vortex street is shifted to much higher values compared to the Newtonian flow [2].

These effects are strongly related to the complex rheological behaviour of those solutions. In laminar shear flow they reveal a so-called shearinduced state (SIS) which is indicated by a steep increase of viscosity and the occurrence of elasticity when a critical shear rate is exceeded [5].

The main feature of turbulence structure in wall-bounded flows is the so-called burst-sweep-mechanism. During previous examinations a reduced detection frequency of sweep-like events was observed for the flow of surfactant solution along the rough wall [6].

3. EXPERIMENTAL SETUP AND PROCEDURE

The experiments were performed in a closed water channel with a removable rough plate at the upper wall. The size and spacing of the roughness elements (in mm) are given in figure 1.

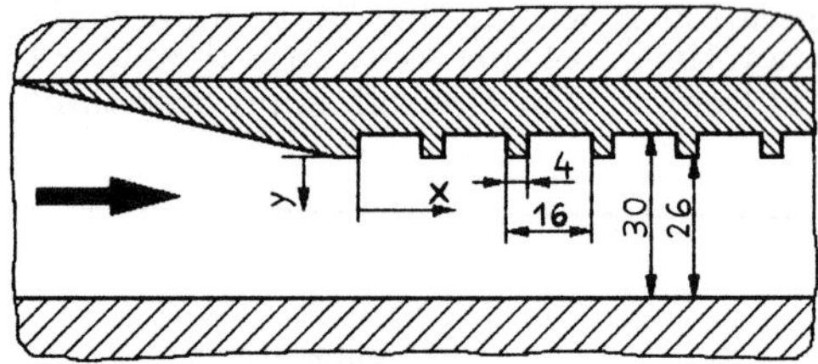

Fig.1: The rough plate.

The velocity signals in two directions could be measured simultaneously by two Laser-Doppler-velocimeters (LDV) which were operated jointly with two burst-spectrum-analysers. All velocity measurements were made at a distance of $x > 280$ mm away from the beginning of the roughness in order to make sure that here the state of a so-called self-preserving boundary layer is reached with regard to the chosen range of Reynolds numbers.

As test fluids desalinated water and a 2.4mM aqueous solution of tetradecyltrimethylammonium salicylate with sodium bromide ($C_{14}TAS + NaBr$) were used. The fluid temperature was 19.7 ± 0.1° C.

To detect coherent motions associated with bursts or sweeps a modified version of the window-average-gradient technique [7] was applied to the instantaneous fluctuation velocity in the main stream direction, u'(t).

The window-average-gradient (WAG) is defined as

$$WAG\,(t,T) = 1/T \left[\int_{t}^{t+T/2} u'(t)\,dt - \int_{t-T/2}^{t} u'(t)\,dt \right]. \qquad (1)$$

This value serves as an expression for the acceleration of a fluid element. It is compared with a positive (sweep) or negative (burst) threshold value of k× urms. The sign of u'(t) must change within the period, in which WAG(t,T) exceeds the threshold level, as a further condition for a sucessful detection. This modification was invented by Komori et al. [8] who stated that really vigorous bursts and sweeps (in smooth wall layers) are always accompanied by a change of sign of the fluctuation velocity.

The detection frequency was determined as a function of the distance from the wall, y, and the time window, T. The product $k \times u_{rms}$ was kept constant for a given Reynolds number and fluid. $k \times u_{rms}$ was set equal to $k_O \times u_{rms,max}$, which was calculated at $k_O = 0.8$ and the maximum RMS-value of the fluctation velocity within the boundary layer for each Reynolds number and fluid.

4. EXPERIMENTAL RESULTS

4.1 Smooth wall

The figures 2 and 3 show the maximum detection frequencies and the accompanying time windows for a positive threshold level (detection of sweeps) as functions of the dimensionless wall distance pertaining to different Reynolds numbers Re_Θ. Re_Θ is defined by the momentum thickness Θ, the velocity U_l outside the boundary layer and (for all cases) the viscosity of water. The time window can be understood as a measure for the mean duration of the most frequent events. The scaling of the water data with the friction velocity u_τ and υ corresponds to the results of Shah and Antonia [9] who discovered the same scaling for bursts in the Newtonian boundary layer at low Reynolds numbers.

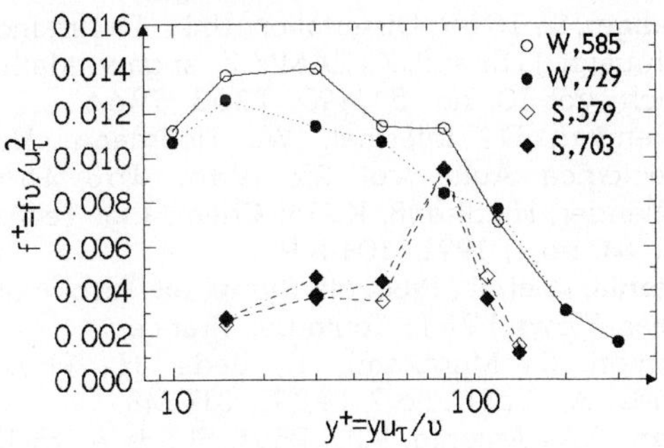

Fig.2: Maximum detection frequencies as functions of y^+ and Re_Θ at the smooth wall (letters stand for water and surfactant solution, respectively).

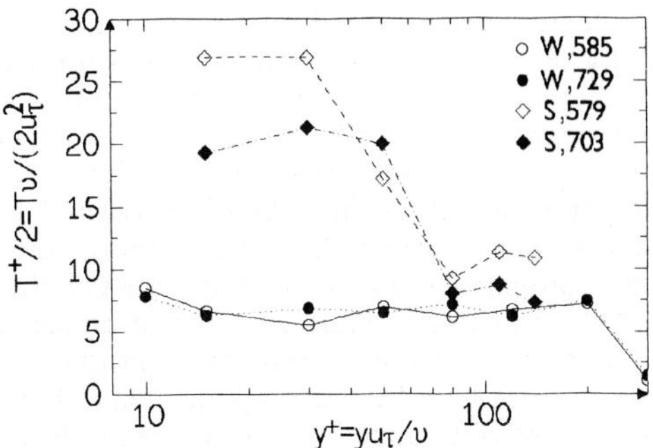

Fig.3: Time windows as functions of the wall distance y^+ and Re_Θ at the smooth wall.

The region with an almost constant time $T^+/2$ includes the buffer layer and the logarithmic region of the turbulent boundary layer.

The surfactant data reveal an abrupt increase in the detection frequency and decrease in the corresponding time value at a certain wall distance y^+ which is beyond the place of maximum turbulence intensity within the boundary layer. At these places mean velocity gradients were found which are quite similar to the shear rates indicating the transition to the SIS, when the testfluid was examined in a couette-type viscometer after it had been pumped around in the test channel for a few days.

4.2 Rough wall

The detection frequencies obtained in the rough wall boundary layer are given in fig. 4. Here the scaling with the variables δ (boundary layer thickness) and U_l was chosen because for the flow of water it is more adequate than the scaling with the wall variables u_τ and υ. Negative values for y/δ were realized in the groove between two roughness elements.

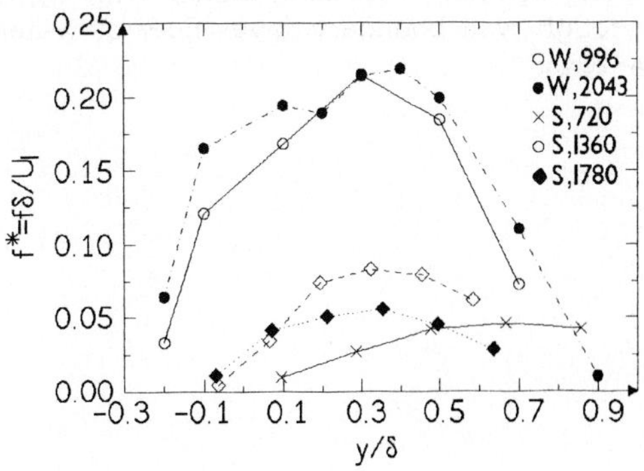

Fig.4: Maximum detection frequencies for the rough wall boundary layer

The detection frequencies obtained for the flow of surfactant solution do not show such a 'structural' change as could be observed at the smooth wall. Beyond the fact that the frequencies are strongly reduced for the surfactant solution, the values for both fluids show quite similar tendencies for Re_Θ above about 1000. When visualizing the flow and measuring the velocity in the direction normal to the wall one notes a considerable fluctuation velocity in this direction only at the higher Reynolds number. This indicates a break-up of the anisotropic flow mentioned beforehand. Here two-component measurements lead to the assumption, that there must be a great difference in the coherent structures in the water flow and the flow of the surfactant solution.

For illustration, the cross correlations of the fluctuation velocities u'(t) and v'(t), which were measured near the leading edge of a roughness element, are given in figs. 5 and 6. For small wall distances the values of the Reynolds shear stress (obtained from R_{uv} at zero time delay) have different signs for the two fluids. Beyond that τ_{xy} changes its sign within the boundary layer of surfactant solution.

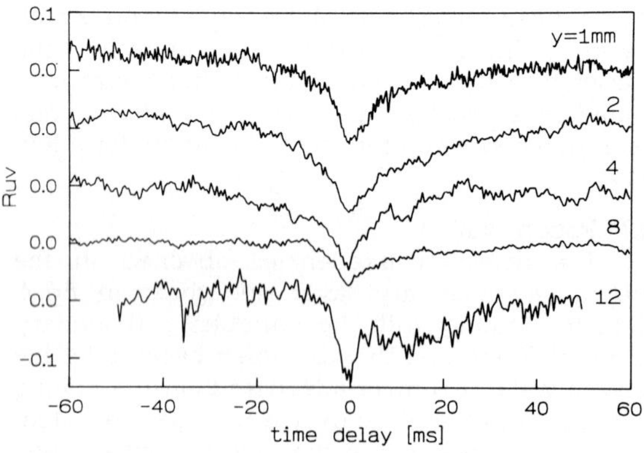

Fig.5: Cross correlation function scaled with u_{rms} for the rough wall boundary layer flow of water at Re_Θ=2043.

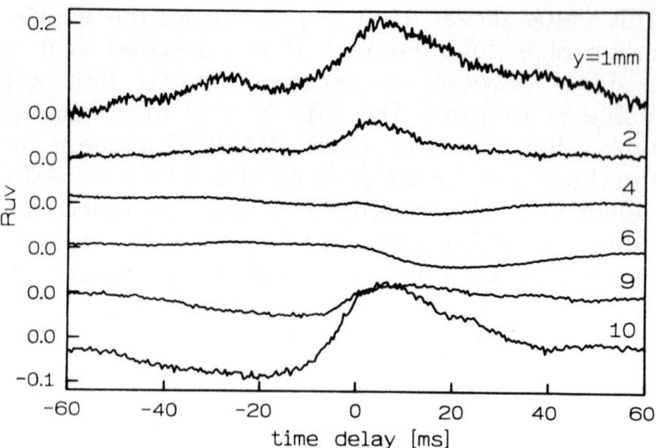

Fig.6: Cross correlation function scaled with u_{rms} for the rough wall boundary layer flow of surfactant solution at Re_Θ=1360.

5. CONCLUSIONS

In the smooth wall boundary layer of a surfactant solution a structural change from longer coherent motions of low frequencies towards shorter events of higher frequencies was found at a wall distance where the mean shearrate may correspond to the onset of the SIS. At the rough wall measurements of one component do not lead to similar results, but a strong deviation from tubulence structures in the Newtonian flow can be deduced from the cross correlation function and the Reynolds shear stress as functions of the wall distance.

6. ACKNOWLEDGEMENTS

This research project is supported by the Deutsche Forschungsgemeinschaft under Str 192/6 - 1.

7. REFERENCES

[1] Thiel, H.; 1990, Dissertation, Univ. Dortmund

[2] Dohmann, J.; Proc. 8th Symp. on Turbulent Shear Flows, 1991, Munich, Germany

[3] Riediger, S.; 1989, Dissertation, Univ. Dortmund

[4] Dohmann, J.; Strauß, K.; ZAMM Z. angew. Math. Mech. vol 70, no. 5, 1990, T363—T366

[5] Ohlendorf, D.; Interthal, W.; Hoffmann, H.; Rheologica Acta, vol 25, 1986, 468-486

[6] Holländer, H.; Strauß, K., in: Chem. Ing. Tech., vol 64, no 1, 1992, 104-109

[7] Antonia, R. et al., Proc. 6th Symp. on Turbulent Shear Flows, 1987, Toulouse, France

[8] Komori, S.; Murakami, Y.; Ueda, H.; Phys. Fluids A, vol 1, no 2, 1989, 339-348

[9] Shah, D.A.; Antonia, R.A.; Phys. Fluids A, vol 1, no 2, 1989, 318-325

Theoretical and Applied Rheology, edited by P. Moldenaers and R. Keunings
Proc. XIth Int. Congr. on Rheology, Brussels, Belgium, August 17-21, 1992

FLOW OF YIELD STRESS FLUIDS THROUGH A SUDDEN CHANGE OF SECTION

A. MAGNIN and J.M. PIAU

Laboratoire de Rhéologie (IMG), Domaine Universitaire, B.P. 53 X - 38041 GRENOBLE CEDEX (France)

1. INTRODUCTION

Yield stress fluids are of considerable interest in many sectors of industry. However, they are distinctly less studied than viscoelastic polymers. In order to improve processing and transportation procedures, it is first necessary to deepen fundamental knowledge of these materials and to develop predictive models. Experimental results and numerical simulations of flow of yield stress fluids following the Herschel-Bulkley law through sudden circular contractions and expansions are presented. A specific test loop and a test yield stress fluid are presented. In particular, the influence of yield stress on static zones, vortex, pressure and energy loss is studied and the effects of inertia and shear thinning are also considered.

2. EXPERIMENTAL METHODS

An original experimental loop adapted to the study of yield stress fluid flowing through pipes and singularities was built [1,2]. The presence of yield stress required the development of technological solutions and original procedures, to :

- obtain a homogeneous yield fluid in the installation by gelation of the fluid in the test loop
- determine pressure without disturbing the flow by measuring pipe deformation under the effect of pressure,
- have access to the kinematic field by chronophotography and laser anemometry through transparent geometries.

3. TEST FLUIDS

A transparent aqueous polymer gel was chosen as test yield stress fluid [1, 2].

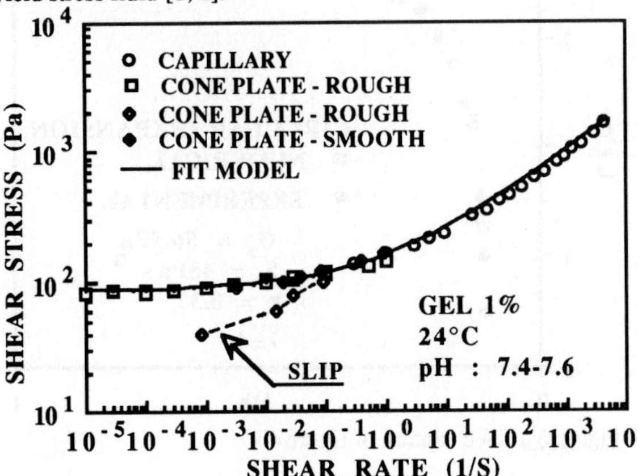

Figure 1 : Rheological measurements on the polymer gel used as a test yield stress fluid

By using recently developed rheometric techniques, the study was able to cover a wide range of shear rates in both steady and unsteady states, in cone-plate and capillary testing facilities [1, 2]. Special attention was paid to slip at the wall. The shear stress as a function of shear rate may be represented in steady state conditions by a Herschel-Bulkley relation (figure 1). The value of yield stress can be fixed by adjusting the polymer concentration [1, 2].

4. THEORY

We studied the isothermal flow of an incompressible fluid through a circular die of which the radius ratio λ of the large to the small tubes is taken equal to 4 in this study. The Eulerian description of this problem is written :

$$\nabla \cdot \mathbf{u} = 0 \qquad (1)$$

$$Re(\mathbf{u}.\nabla)\mathbf{u} = -\nabla p + \nabla \mathbf{T} \qquad (2)$$

p, isotropic pressure; **u**, velocity; **T**,shear stress tensor
- The Herschel-Bulkley law:

$$\mathbf{T} = 2\left[\frac{Hb}{\dot\gamma} + \dot\gamma^{(n-1)}\right].\mathbf{D} \qquad \text{for } |\sigma| \geq \sigma_s \qquad (3)$$

$$\dot\gamma = 0 \qquad \text{for } |\sigma| < \sigma_s \qquad (4)$$

$$H_b = \frac{\sigma_s}{k\left(\frac{U_2}{R_2}\right)^n}, \quad R_e = \frac{\rho R_2^n U_2^{2-n}}{k} \qquad (5)$$

σ_s, the yield stress; σ, second invariant of the stress tensor; $\dot\gamma$, second invariant of the strain rate tensor **D**; n, power-law index, k, consistency index.

The variables were made dimensionless by radius R_2, and average velocity U_2 in the small tube and by the viscous stress $k(U_2/R_2)^n$.

Three dimensionless parameters thus govern the problem:
Hb : Herschel-Bulkley number (yield stress/viscous stress),
Re : Reynolds number (inertial stress/viscous stress,
n : magnitude of shear thinning.

Numerical simulations were run by the finite elements program Polyflow. A biviscosity model was used to simulate the yield stress behavior [5]. The following set of boundary conditions is assumed :

- no-slip at the wall is imposed.

- entry and exit tubes are long enough so that fully developed conditions are imposed.
- the axial axis is one symmetry.

The pressure change Δp_s and mechanical energy loss ΔE_s due to the change of section. are represented under the form of the equivalent length :

$$Leq = \frac{\Delta p_s}{2\sigma_{w2}}, \quad L'eq = \frac{\Delta E_s}{2\sigma_{w2}} \qquad (6)$$

where σ_{w2} is the wall shear stress in the small tube.

5. CIRCULAR EXPANSION

5.1 Vortex and static zone

Figures 2 show the streamline pattern observed in the case of flow of the yield stress fluid through a sudden circular expansion.

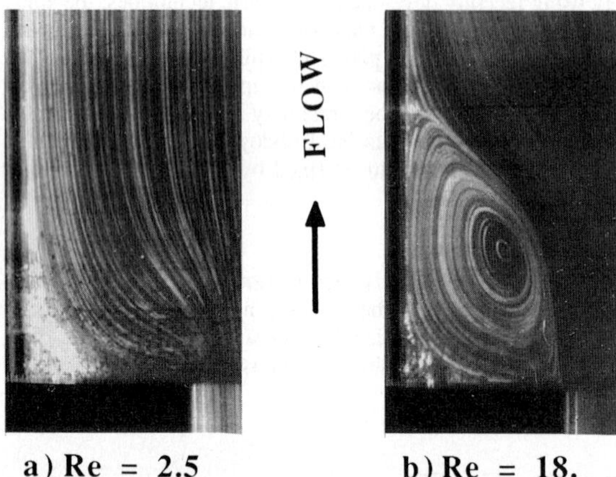

a) Re = 2.5 Hb = 0.66 b) Re = 18. Hb = 0.098

Figure 2 : Photographs of streamlines in sudden expansion of ratio λ= 4 for gel (n = 0.37).

A static zone can be seen in the corner of the expansion where the fluid does not flow owing to the presence of the yield stress (fig. 2a). When inertia increases a vortex appears, together with the presence of a static zone in the corner of the expansion (fig. 2b). A static zone can be observed in the zone of re-attachment of the vortex.

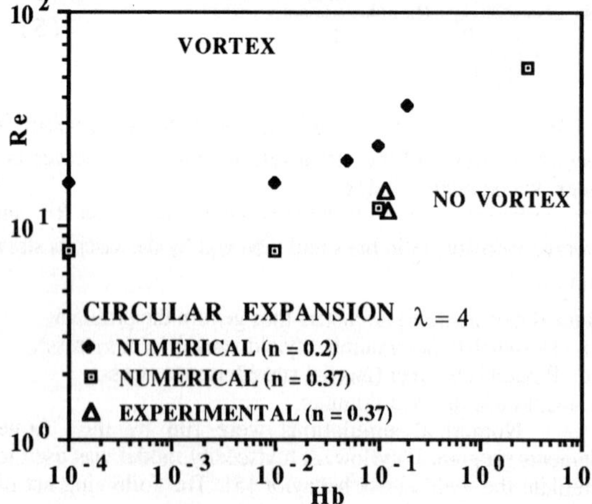

Figure 3 : Re-Hb map as function of n showing the boundaries of inception of vortex. No vortex below boundary,vortex above.

Figure 3 shows the Re-Hb map as a function of n showing the boundaries of inception of the vortex. The yield stress prevents and delays the appearance of the vortex created by the inertia but checked by the shear thinning. The influence of yield stress is notable at low values of Hb (0.01 about). The vortex does not appear in the corner of the expansion (see X*1 in fig. 4). It is jammed between the static zone in the corner of the expansion and the central flow running from the small tube as shown in fig. 4. The vortex grows with the inertia.

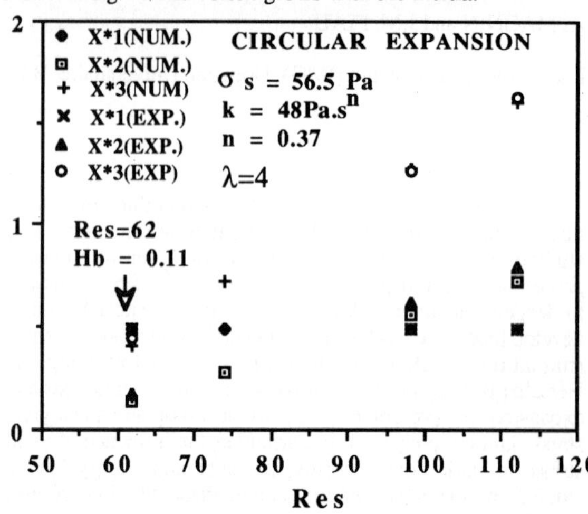

Figure 4 : X*1 radial distance of the center of the vortex from the symmetry axis; X*2 axial distance of the center of the vortex from the plane of expansion; X*3 axial distance of the re-attachment point of the vortex from the plane of expansion as a function of Reynolds number. Arrow shows the inception of the vortex. X*1, X*2, X*3 are made dimensionless by the radius of large tube

5.2 Drop of pressure and energy loss

Figure 5 shows the correction of pressure as a function of Hb and Re number during an experiment when the flow rate is increased (Re increases and Hb decreases) Numerical results agree well with the experimental results.

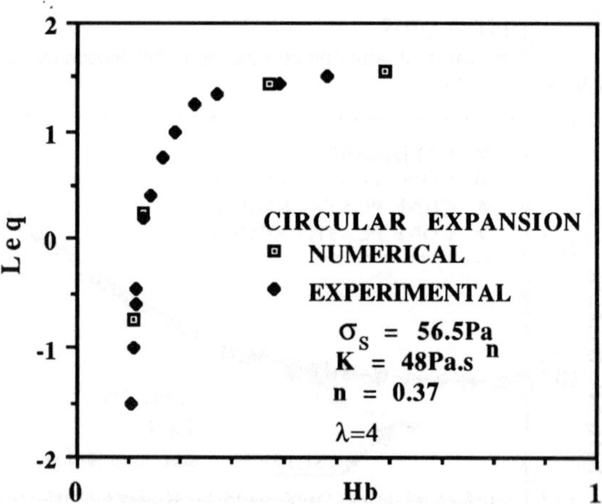

Figure 6 a : See caption of figure 6b.

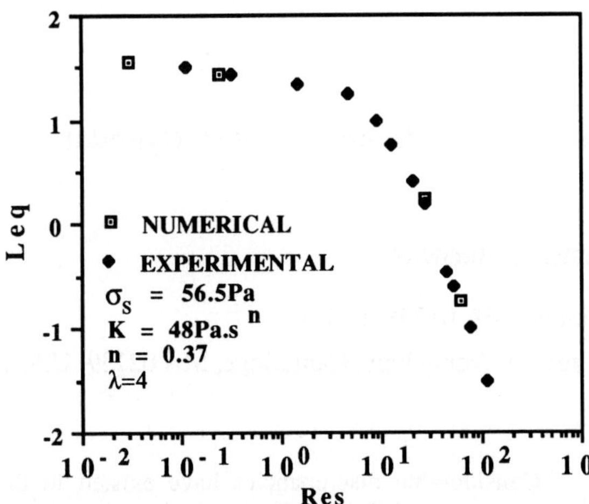

Figure 6b : Pressure correction during an experiment when the flow rate is changed. a) as a function of Reynolds number, b) as a function of Hb number.

Figure 7 shows that yield stress increases energy loss when inertia is low. But when inertia effects are high, yield stress, by reducing vortex, reduces energy loss.

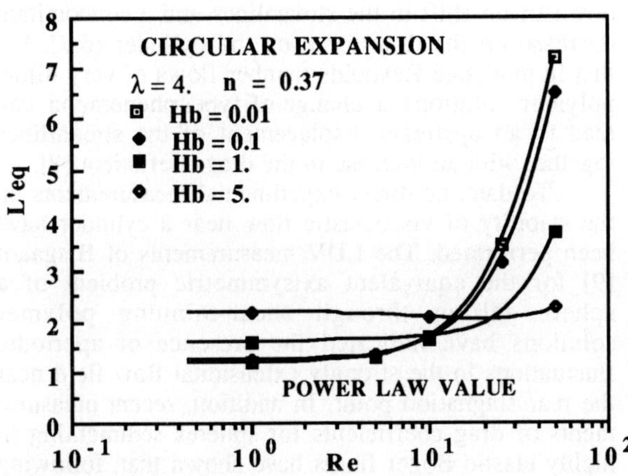

Figure 7 : Energy loss L'eq versus Hb number as a function of shear thinning index n.

6. CIRCULAR CONTRACTION

Experimental and numerical studies were also performed. The results are only summarized in this short paper.

6.1 Vortex and static zone

For flow without inertia, vizualization and numerical simulation show that yield stress creates a static zone in the corner of the contraction. The Newtonian vortex (Hb=0, n= 1) disappears for low values of Hb= 0.01 about (fig. 8). In the same manner, when shear thinning and inertia increase, the vortex zone diminishes and may disappear but a static zone remains in the corner of the contraction.

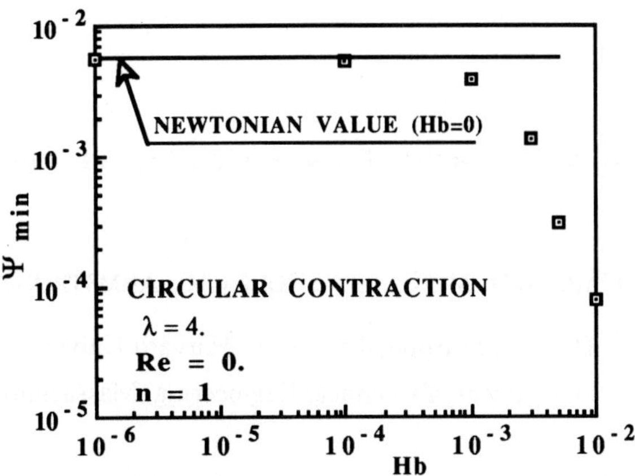

Figure 8 : Minimum value of stream function as a measure of vortex as a function of parameter Hb.

6.2 Drop of pressure and energy loss

For flow without inertia, yield stress increases the drop in pressure (fig. 9) and the energy loss The influence of yield stress becomes high for Hb>0.1 approx.. In the same manner, inertia and shear thinning increase the drop in pressure and energy loss.

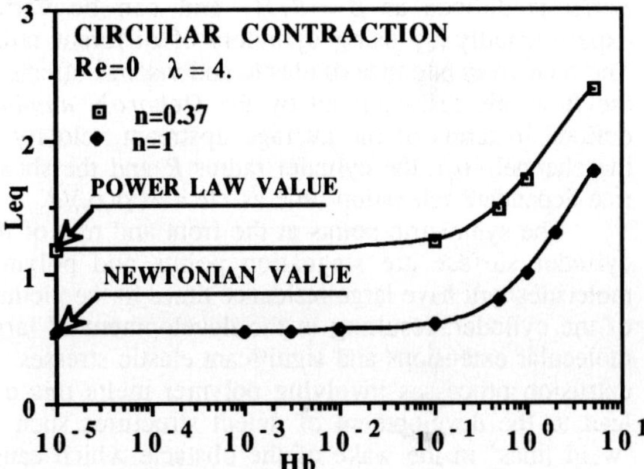

Figure 9 : Equivalent length Leq versus Hb number as a function of shear thinning index n.

7. CONCLUSIONS

Experimental study and numerical simulations of Herschel-Bulkley flow through sudden circular expansions and contractions have demonstrated the influence of yield stress. With low values of Hb number, flow is appreciably modified. Interaction with inertia and shear thinning effects has been characterized. Experimental results compare well with numerical simulations by finite elements.

REFERENCES

1. Kouamela , D., Thesis, (1991), Grenoble, France
2. Magnin, A. and J.M. Piau, (submitted)
3. Magnin, A. and J.M. Piau, J. Non New. Fluid Mech., 36 (1990) 85
4. Magnin, A. and J.M. Piau, J. Non Newtonian Fluid Mech., 23 (1987) 91
5. Magnin, A. and J.M. Piau, EUROTHERM Seminar n°20, Nancy, 1991

Theoretical and Applied Rheology, edited by P. Moldenaers and R. Keunings
Proc. XIth Int. Congr. on Rheology, Brussels, Belgium, August 17-21, 1992

A Viscoelastic Flow Instability in the Wake of a Confined Circular Cylinder

GARETH H. MCKINLEY[1], ROBERT C. ARMSTRONG[2], ROBERT A. BROWN[2]

[1] Division of Applied Sciences, Harvard University, Cambridge, MA 01238, U.S.A.

[2] Department of Chemical Engineering, Massachusetts Institute of Technology, Cambridge, MA 02139, U.S.A

1. INTRODUCTION

The measurements presented in this paper document in detail the kinematic effects of viscoelasticity on the stability of strong planar extensional flows in the wake of a circular cylinder. The experimental test geometry shown in Fig. 1 consists of a cylinder of radius R that is centrally mounted in a long planar channel with a half-width H. The *cylinder-channel ratio* is defined as $\beta = R/H$, and can be varied experimentally by using cylinders of different radii. The relative importance of elastic and viscous effects in the flow are characterized by the *Deborah number,* defined in terms of the average upstream velocity in the channel $\langle v_z \rangle$, the cylinder radius R and the shear-rate-dependent relaxation time as $De = \lambda(\dot{\gamma})\langle v_z \rangle / R$.

The symmetric points at the front and rear of the cylinder surface are stagnation points and polymer molecules will have large residence times in the vicinity of the cylinder, resulting in the development of large molecular extensions and significant elastic stresses. In extrusion processes involving polymer melts this can lead to the development of defect structures such as 'weld lines' in the wake of the obstacle which cause considerable degradation in the ultimate material properties of the product [1].

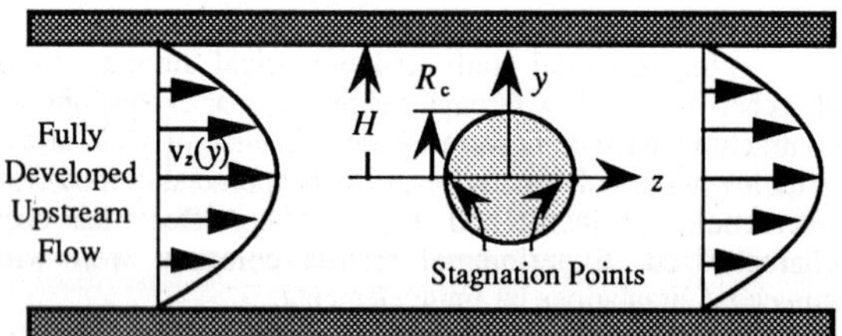

Figure 1 Viscoelastic flow past a circular cylinder constrained in a planar channel. The cylinder of radius R is mounted centrally in the channel which has a width $2H$.

This test geometry, together with the analogous axisymmetric problem of viscoelastic flow around a sphere constrained in a cylindrical tube has also been adopted as a benchmark problem for numerical computations of viscoelastic flows [2,3].

Considerable discrepancies have existed in the literature concerning the effects of elasticity on the shape of the fluid streamlines near a cylinder. Some experimental flow visualization studies and theoretical analyses have indicated a progressive downstream displacement in the streamlines as De is increased [4], whereas others have shown a pronounced upstream shift [5]. It now appears clear that for creeping flows of highly viscous fluids, elasticity leads to a gradual downstream shift in the streamlines and a concomitant decrease on the drag force on the cylinder [6,7], but that in moderate Reynolds number flows of very dilute polymer solutions a change-of-type phenomena can lead to an upstream displacement of the streamlines together with an increase in the drag coefficient [8].

To date, no direct experimental measurements on the stability of viscoelastic flow near a cylinder have been performed. The LDV measurements of Bisgaard [9] for the equivalent axisymmetric problem of a sphere falling through shear-thinning polymer solutions have indicated the presence of aperiodic fluctuations in the strongly extensional flow field near the rear stagnation point. In addition, recent measurements of drag coefficients for spheres sedimenting in highly elastic Boger fluids have shown that, following the initial drag decrease expected at low De, the drag coefficient at higher De increases substantially [10] possibly due to the onset of such a flow instability. The LDV measurements presented in this work show that, for flows above a critical Deborah number, the planar extensional flow in the wake of a cylinder is unstable and undergoes a transition to a complex three-dimensional motion.

2. EXPERIMENTAL METHODS

2.1 Test Fluid and Geometry

The viscoelastic test fluid used in this study is a 'Boger fluid' composed of 0.31 wt% PIB dissolved in a PB/C14 solvent. The rheological properties of the fluid have been thoroughly characterized over a range of temperatures in steady and transient shear flows [11]. The fluid is highly elastic with a characteristic relaxation time of $\lambda_0 \sim 0.79$ s, and a zero-shear-rate viscosity η_0 = 13.76 Pa·s which remains almost constant across four decades of shear rate. It is thus possible to study elastic effects in the flow without additional complicating factors such as shear-thinning or fluid inertia.

A Cartesian coordinate system $\{x,y,z\}$ is defined as shown in Fig. 1 with the origin at the center of the cylinder, the z-axis aligned along the flow direction, the y-axis in the 'transverse' direction and the x-axis pointing in the 'neutral' direction along the cylinder's axis of symmetry. In order to ensure that the flow in the rectangular channel far upstream of the cylinder closely approximates idealized planar Poiseuille flow, the geometry has been designed with a large aspect ratio $\Delta x/\Delta y = 6$ [see ref. 12 for further details]. The results presented in this paper consider flow past a cylinder of radius R = 3.18 mm equivalent to a cylinder-channel ratio of $\beta = R/H = 0.50$.

2.2 Laser-Doppler Velocimetry

Non-invasive measurements of the velocity field near the cylinder were obtained using a three-color, six-beam Laser Doppler Velocimetry (LDV) system shown schematically in Fig. 2.

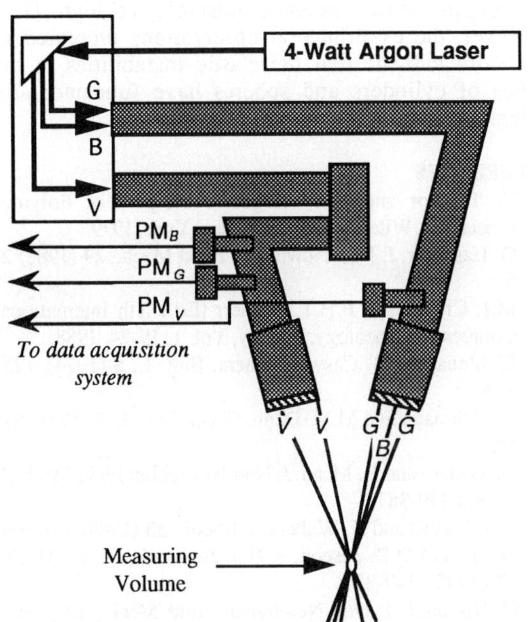

Figure 2 Schematic diagram showing the optical and geometric arrangement of the three-color, six-beam laser Doppler velocimetry system.

The apparatus is an extension of a 2-color system developed by Raiford [13] and enables point-wise coincident measurement of all three velocity components. The dimensions of the ellipsoidal measuring volume are determined from the focal lengths of the lenses, the beam diameters, the initial separation of the beam pairs and the intersection angle of the beams to be approximately 50 μm × 50 μm × 400 μm. The Doppler-shifted light that is scattered by particles in the flow field is measured in an off-axis, side-scatter mode by three photo-multipliers (PM) as shown in Fig. 2. Three frequency trackers plus a dual-channel Spectrum Analyzer are employed to determine the Doppler frequency of the velocity component measured by each beam-pair. Steady and time-dependent velocity measurements in the range $0.1 \leq \upsilon \leq 100$ cm/s can be measured with accuracies of ± 1%. The entire optical train is mounted on a computer-controlled, three-dimensional translating table which can be positioned with an accuracy of ± 4μm throughout the test geometry. The global dynamics of the flow are also recorded directly by using a CCD video camera and macro lens assembly.

3. RESULTS

3.1 Steady Two-dimensional Flow, $De < 1$.

At very low flow rates ($De \ll 1$) the velocity field around the cylinder is symmetric and the LDV measurements compare extremely well with finite element simulations of creeping flow past a cylinder.

The effects of increasing the Deborah number of the flow past the cylinder have been investigated in a series of LDV measurements along the channel center-plane (y = 0). Upstream of the cylinder, velocity profiles for increasing De superpose when scaled with the average velocity $\langle \upsilon_z \rangle$ and the cylinder radius R. However, measurements in the wake of the cylinder show a progressive *downstream* shift in the streamlines as the flow rate and Deborah number increases.

3.2 Transition to Three-Dimensional Flow, $De \geq 1.3$.

Further increases in De result in progressively larger downstream shifts in the velocity profiles until the onset of a flow instability occurs at De = 1.30. The flow bifurcates from a steady two-dimensional planar extensional flow to a steady, three-dimensional cellular structure. LDV measurements in the wake close to the cylinder reveal large spatial fluctuations in the axial velocity that extend in the neutral x-direction along the length of the cylinder. A sample axial velocity profile $\upsilon_z(x)$ at z/R = 1.75 is shown overleaf in Fig. 3. Fourier analysis of these periodic oscillations shows that the characteristic wavelength λ_x of the perturbations is almost equal to the radius of the cylinder, $\lambda_x = 0.95R \pm 0.05R$. Video-imaging results of the flow also clearly show that these velocity oscillations are associated with the development of a regularly spaced 'banded structure' in the wake.

The complex flow field in the cylinder wake has been carefully explored using the three-color LDV system. Velocity profiles show that periodic oscillations in the wake of the cylinder extend a long distance downstream, but that the cellular three-dimensional structure is confined to the narrow region of strongly extensional flow near the center-plane of the channel. Further away from the centerline the flow between the cylinder and the channel walls is primarily a steady shearing flow that remains two-dimensional for all values of *De*.

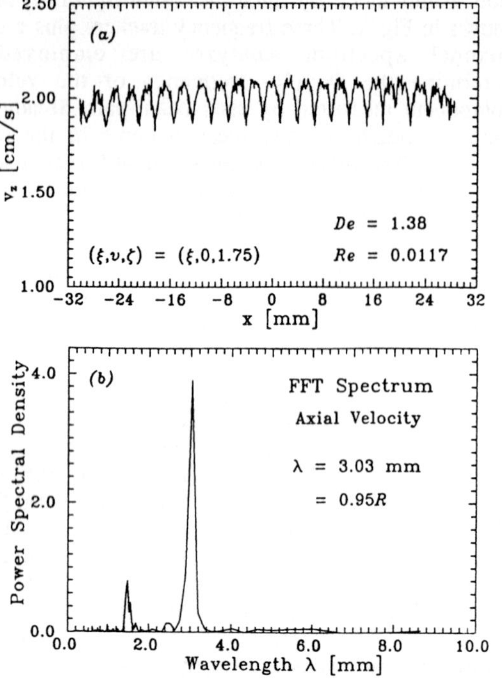

Figure 3 *(a)* Spatial fluctuations in the axial velocity $v_z(x)$ measured in the downstream wake of the cylinder; *(b)* FFT spectrum showing spatial wavelength of the fluctuations.

LDV measurements of the axial velocity in the wake are shown in Figure 4. In contrast to the time-dependent cellular structure that develops across the length of the cylinder, the stagnation flow profiles vary smoothly and monotonically in the downstream direction. Upstream of the cylinder the profiles superpose when normalized with the average velocity $\langle v_z \rangle$ and elasticity does not affect the upstream stagnation flow even at Deborah numbers of $De > 3$.

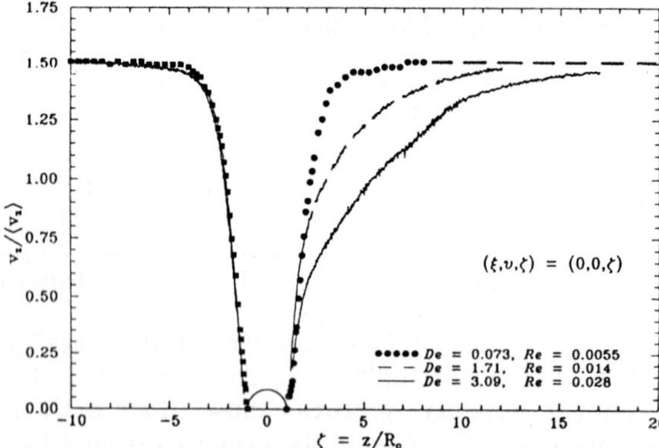

Figure 4 Progressive downstream shift of the centerline axial velocity profiles in the three-dimensional cylinder wake as *De* is increased from 0.07 (●) to $De = 3.08$ (—).

In the wake of the cylinder, however, elastic effects result in a progressive downstream shift in the position of the streamlines around the cylinder. The symmetric LDV measurements for Newtonian flow at $De = 0.07$ are also shown in Fig. 4 to emphasize the magnitude of this downstream displacement; at $De = 3.08$ the flow does not recover a fully developed parabolic profile for distances of over 15*R* downstream of the cylinder. These LDV measurements show no indication of the 'negative wake' observed in shear-thinning fluids, and are in agreement with earlier visualization studies using spheres and Boger fluids [14].

4. CONCLUSIONS

These LDV measurements are the first experimental observations of a three-dimensional elastic instability that occurs within the planar extensional flow in the wake of a circular cylinder. At a critical Deborah number the steady two-dimensional flow bifurcates to a steady, three-dimensional motion consisting of a spatially-periodic cellular structure that extends in the neutral direction along the length of the cylinder. The instability develops at low *Re* and is not associated with change-of-type. A similar elastic instability has previously been documented in the wake of spheres [9], and it thus appears that both uniaxial and planar extensional flows near stagnation points of submerged bodies become unstable at high *De*.. However, the experimental observations presented in this work indicate that the elastic instabilities in the wakes of cylinders and spheres have fundamentally different spatial and temporal characteristics.

REFERENCES

1. Z. Tadmor and C.G. Gogos, Principles of Polymer Processing, Wiley-Interscience, New York, 1979.
2. O. Hassager, J. Non-Newtonian Fluid Mech., **29** (1988) 2-5.
3. M.J. Crochet, in P.H.T. Uhlherr (Ed.) Xth International Congress on Rheology, Sydney, Vol. 1, 19-24, 1988.
4. B. Mena and B. Caswell, Chem. Eng. J., **8** (1974) 125-134.
5. J.S. Ultmann and M.M. Denn, Chem. Eng. J., **2** (1971) 81-89.
6. O. Manero and B. Mena, J. Non-Newtonian Fluid Mech., **9** (1981) 379-387.
7. S.A. Dhahir and K. Walters, J. Rheol., **33** (1989) 781-804.
8. H. Hu and D.D. Joseph, J. Non-Newtonian Fluid Mech., **37** (1990) 347-377.
9. C. Bisgaard, J. Non-Newtonian Fluid Mech., **12** (1983) 283-302.
10. C. Chmielewski, K.L. Nichols and K. Jayaraman, J. Non-Newtonian Fluid Mech., **35** (1990) 37-49.
11. L.M. Quinzani, G.H. McKinley, R.A. Brown and R.C. Armstrong, J. Rheol, **34** (1990) 705-748.
12. G.H. McKinley, Ph.D. Thesis, M.I.T., 1991.
13. G.H. McKinley, W.P. Raiford, R.A. Brown and R.C. Armstrong, J. Fluid Mech., **223** (1991) 411-456.
14. A. Maalouf and D. Sigli, Rheol. Acta, **23** (1984) 497-507.

ACKNOWLEDGEMENTS: The authors are grateful to the National Science Foundation for support of this work.

Theoretical and Applied Rheology, edited by P. Moldenaers and R. Keunings
Proc. XIth Int. Congr. on Rheology, Brussels, Belgium, August 17-21, 1992

DIRECT SECOND KIND BOUNDARY INTEGRAL FORMULATION FOR STOKES FLOW PROBLEMS

L. A. MONDY[1] and M. S. INGBER[2]

[1]Sandia National Laboratories, Albuquerque, NM 87185 (USA)
[2]Dept. Mech. Eng., University of New Mexico, Albuquerque, NM 87131 (USA)

1. INTRODUCTION

The boundary element method (BEM) has proven to be a valuable tool in analyzing Stokes flow problems (refs. 1-4). For these problems, the classical direct boundary element method (DBEM) is based on a boundary integral equation for the components of velocity (ref. 1). In most applications, the velocities on the bounding walls or on the surface of particles are prescribed, while the corresponding surface tractions are unknown. In these situations the velocity boundary integral equation (VBIE) represents a Fredholm integral equation of the first kind. In many cases discretized Fredholm integral equations of the first kind lead to poorly conditioned matrix equations (refs. 2,3). Power and Miranda (ref. 4), recognizing the numerical problems associated with Fredholm integral equations of the first kind, developed a novel boundary integral equation based on the double-layer potential for studying flows about immersed particles, which represents a Fredholm integral equation of the second kind. This is an indirect boundary element method (IBEM) in contrast to Youngren and Acrivos' (ref. 1) DBEM. In the DBEM the source densities are given by velocity and traction components, whereas the source densities for the IBEM cannot be interpreted in terms of physical field quantities.

In this paper we describe a DBEM for the Stokes problem which results in a Fedholm integral equation of the second kind. This is accomplished by deriving a boundary integral equation for the components of traction (TBIE). We examine the applicability and compare the accuracy of boundary element methods based on the VBIE and the TBIE for two example problems. These examples demonstrate the efficacy of the method of regularization of the hypersingular integrals and the method of constraining out the rigid body eigenmodes for the TBIE. This work also demonstrates the robust nature of the VBIE for these problems that require a moderate number of boundary elements.

2. THE TBIE

The derivation of the TBIE is accomplished by formulating a boundary integral equation for the surface tractions as opposed to the velocities. We start with the classical boundary element formulation for the Stokes flow problem (ref.1), which recasts the creeping flow equations into integral form by considering a weighted residual statement of the differential equations:

$$c_{ij}(\mathbf{x})u_{ij}(\mathbf{x}) + \int_\Gamma q^*_{ijk}(\mathbf{x},y)u_k(y)n_j(y)d\Gamma = -\int_\Gamma u^*_{ik}(\mathbf{x},\mathbf{y})f_k(y)d\Gamma \qquad (1)$$

$$\eta(\mathbf{x})p(\mathbf{x}) = -\int_\Gamma p^*_k(\mathbf{x},\mathbf{y})f_k(\mathbf{y})d\Gamma - 2\frac{\partial}{\partial x_j}\int_\Gamma p^*_k(\mathbf{x},\mathbf{y})u_k(\mathbf{y})n_j(\mathbf{y})d\Gamma \qquad (2)$$

where the f_k's are the components of the traction along the surface Γ, the n_j's are the components of the unit outward-normal vector to the boundary, and $\partial/\partial x_j$ is the derivative in the jth coordinate direction at the field point $\mathbf{x}$. Also, $u^*_{ij}(\mathbf{x},\mathbf{y})$, $p^*_j(\mathbf{x},\mathbf{y})$ and $q^*_{ijk}(\mathbf{x},\mathbf{y})$ are the fundamental solutions for the velocity field, the pressure field, and the associated stress field, respectively. The coefficients c_{ij} and η can be determined from the problem geometry (ref. 5).

If the tractions along the boundary are unknown (and, hence, Eq. 1 represents a Fredholm integral equation of the first kind), it may be desirable to reformulate the problem by writing a boundary integral equation for the components of traction (representing a Fredholm integral equation of the second kind). We note that the components of traction are given by

$$f_i = \left[-\delta_{ij}p + \frac{\partial u_i}{\partial x_j} + \frac{\partial u_j}{\partial x_i}\right] n_j \,. \qquad (3)$$

Hence, using Eq. 2 and the appropriate derivatives of Eq. 1, we can write a TBIE.

The order of differentiation and integration can be interchanged if the resulting hypersingular integrals are interpreted in the finite-part sense. The final form of the TBIE is:

$$f_i(\mathbf{x}) = \int_\Gamma \pi^*_{ikl}(\mathbf{x},\mathbf{y})u_k(\mathbf{y})n_l(\mathbf{y}) - \int_\Gamma W^*_{ik}(\mathbf{x},\mathbf{y})f_k(\mathbf{y})d\Gamma(\mathbf{y}), \quad (4)$$

where

$$\pi^*_{ikl}(\mathbf{x},\mathbf{y}) = \left\{ -\frac{\delta_{ij}\delta_{lk}}{2\pi \mid \mathbf{x}-\mathbf{y} \mid^3} + \frac{15 d_i d_j d_k d_l}{2\pi \mid \mathbf{x}-\mathbf{y} \mid^7} - \frac{3[\delta_{jl}d_i d_k + \delta_{jk}d_i d_l + \delta_{il}d_j d_k + \delta_{ik}d_j d_l]}{4\pi \mid \mathbf{x}-\mathbf{y} \mid^5} \right\} n_j(\mathbf{x}) \quad (5)$$

and

$$W^*_{ik}(\mathbf{x},\mathbf{y}) = \frac{3 d_i d_j d_k}{4\pi \mid \mathbf{x}-\mathbf{y} \mid^5} n_j(\mathbf{x}), \quad (6)$$

where $d_i = (x_i - y_i)$ and x_i and y_i are the components of $\mathbf{x}$ and $\mathbf{y}$, and δ_{ij} is the Kronecher-delta function.

We note that the kernal function π^*_{ikl}(x,y) contains a singularity of order $1/r^3$ (where r is the distance between $\mathbf{x}$ and $\mathbf{y}$) as the field point $\mathbf{x}$ approaches the source point $\mathbf{y}$, and, hence, the TBIE is hypersingular. A major difficulty associated with the TBIE formulation (and perhaps one reason why this formulation has not been developed previously) is the evaluation of the hypersingular integral in Eq. 4. We proposed a method of regularization similar to the one proposed by Stallybrass (ref. 6) for the derivative of the Helmholtz integral equation. The method is discussed in detail in ref. 7. The regularized form of the TBIE is given by:

$$f_i(\mathbf{x}) = \int_\Gamma \pi^*_{ikl}(\mathbf{x},\mathbf{y})[u_k(\mathbf{y}) - u_k(\mathbf{x})]n_l(\mathbf{y})d\Gamma(\mathbf{y}) - \int_\Gamma W^*_{ik}(\mathbf{x},\mathbf{y})f_k(\mathbf{y})d\Gamma(\mathbf{y}). \quad (7)$$

We note that the kernal function W^*_{ik}(**x**,**y**) represents a double-layer potential, and, therefore, the integral representation for the tractions given by Eq. 7 is not complete because of the existence of six rigid body eigenmodes (ref. 4). The problem of the rigid body eigenmodes associated with the double-layer potential in the TBIE becomes immediately apparent by considering uniform flow about a rigid particle for which the perturbed velocity is constant on the surface of the particle. In this case, the first integral in Eq. 7 vanishes identically, which indicates that the classical exterior solution for the stress field is not unique. The rigid body modes only pose a problem for exterior domains or domains whose complements are not simply connected, and only for problems in which the boundary conditions themselves are compatible with rigid body motions.

For exterior flows about particles, the discretized boundary element equations are generated by collocating the TBIE at each collocation node on the surface of the particles. In order to preclude rigid body motions, we supplement this system of equations by collocating the VBIE at points interior to the particles. For general three-dimensional flows about multiparticle systems, we choose three non-colinear interior collocation nodes which preclude all possible rigid body modes. The incorporation of the additional collocation points overspecifies the system of equations which we solve in the least square sense.

3. NUMERICAL EXAMPLES

In this paper, we wish to compare the boundary elements methods based on the VBIE and TBIE. In order to perform a fair comparison between the two formulations, we use the same number and class of boundary elements and the same quadrature routine for both. Further, the particular class of boundary elements (quadrilateral superparametric) and the quadrature routine (a variable order Gauss) have been selected in order to minimize the contribution of the geometric and quadrature errors on the overall accuracy of the numerical solutions (refs. 8,9). We consider two example problems in this section to show the relative efficiency and accuracy of the VBIE and TBIE.

The first example problem considers flow between counter-rotating parallel plates. Since this is an interior problem, there are no problems associated with the rigid body modes and a direct comparison between the VBIE and TBIE can be performed. The diameter of the counter rotating parallel plates is 1.0 and the gap width is also 1.0. The upper plate rotates with an angular velocity of 1.0 and the lower plate with angular velocity of -1.0. For this problem we prescribe the velocity boundary conditions everywhere.

In this example both the TBIE and VBIE yield excellent results (less than 1% different from the exact solution for points away from the boundaries and about 3% nearer the boundaries) with little to distinguish between the two sets of computations. Although no systematic study was performed, it appears in general that the VBIE solution is slightly more accurate away from the boundary, whereas the TBIE solution is slightly more accurate closer to the boundary.

For the finest grid examined (having 272 boundary elements), the condition number associated with the discretized TBIE equations was 2526 and the condition number associated with the discretized VBIE equations was 4290. Thus, there is little difference in the matrix condition numbers between the two methods. Furthermore, these condition numbers are relatively low considering that 816 simultaneous linear equations are solved. The CPU times, measured on an HP 720 workstation for this finest grid, was 80 CPU seconds for the VBIE and 91 CPU seconds for the TBIE.

The second example problem considers a pair of identical spheres falling with their line of center making a 45° angle with the horizontal. We consider three separation distances given by $\epsilon = 1.0a$, $\epsilon = 0.1a$, and $\epsilon = 0.01a$, where a is the radius of each sphere. Here, neither the velocities nor the tractions are specified on the surfaces of the particles. Since the particles are freely falling, we assume that the particles are in equilibrium, which is consistent with neglecting the fluid accelerations in the Stokes flow approximations. To close the system of equations, the equilibrium equations and the kinematic equations (which relate the velocities at the centroids of the particles to the surface velocities) supplement the boundary element equations (ref. 8). It is important to note that, for this problem, the TBIE and VBIE do not represent either a Fredholm integral equation of the first or second kind, but rather some type of mixed integral equations. Also, here, three additional interior collocation nodes for each sphere are chosen for the TBIE to constrain out the rigid body modes.

We compare the results generated by the VBIE and TBIE to the exact solution (ref. 10). For the case with $\epsilon = 1.0$, the maximum error using a grid containing 216 elements was less than 1% in all velocities. In general, the results for sedimentation velocities were most accurate while the results for the angular velocities were the least accurate. The results provided by the VBIE were more accurate than the results provide by the TBIE. Similar observations were made for the case with $\epsilon = 0.1$. In this case the maximum error using the grid containing 216 elements was 1.32%, which occurred in the computation of the angular velocity using the TBIE.

The case in which the separation distance $\epsilon = 0.01$ is the most difficult one from a computational point of view. Quadrature errors could severely degrade the solutions. A systematic study showed that quadrature errors had a more pronounced effect on the TBIE, with its singularities in the kernel functions being one order higher than the VBIE's.

The solutions provided by the VBIE were again more accurate that those provided by the TBIE. The maximum error for the VBIE was less than 1% in the angular velocity for the finest grid (here 294 elements). The solutions provided by the TBIE were also good when enough Gauss quadrature points were used. With this method, for the finest grid, the maximum percent error occurred in the horizontal drift velocity, where the errror was 4.7%. The TBIE equations were, however, better conditioned. For the finest grid, the condition number for the VBIE was about 250,000 and for the TBIE, about 2900.

4. CONCLUSIONS

We have developed a new direct second kind boundary integral formulation for Stokes flow problems based on the traction boundary integral equations. We developed a method of regularization for the hypersingular integral contained in the TBIE and a method for constraining out the rigid body modes associated with the double-layer representation of the tractions. Example problems show the capabilities of the TBIE. Results were compared to analytic solutions and to results generated by the more traditional VBIE. In general, we found that the VBIE provided slightly more accurate solutions than the TBIE, but that both methods performed well. The TBIE may provide an attractive alternative to the VBIE for problems requiring very fine discretizations because of geometric complexities or large gradients in velocity or traction.

REFERENCES

1. G. K. Youngren and A. Acrivos, J. Fluid Mech. 69 (1975) 377-403.
2. M. A. Goldberg, Solution methods for integral equations: Theory and applications, Plenum Press, New York, 1978.
3. H. Niessner, in: C. A. Brebbia, W. L. Wendland, and G. Kuhn (Eds.), Boundary Elements IX, Vol. 1. 1987, pp. 213-227.
4. H. Power and G. Miranda, SIAM J. Appl. Math. 47 (1987) 689-698.
5. M. S. Ingber and J. Li, Comm. Appl. Num. Meths. 7 (1991) 367-376.
6. M. P. Stallybrass, J. Math. Mech. 16 (1967) 1247-1289.
7. M. S. Ingber and L. A. Mondy, submitted for publication (1992).
8. M. S. Ingber, Int. J. Num. Meths. Fluids 10 (1990) 791-809.
9. M. Rezayat, D. J. Shippy, and F. J. Rizzo, Comp. Meths. Appl. Mech. Engrg. 55 (1986) 349-367.
10. A. J. Goldman, R. G. Cox, and H. Brenner, Chem. Eng. Sci. 21 (1966) 1151-1170.

Theoretical and Applied Rheology, edited by P. Moldenaers and R. Keunings
Proc. XIth Int. Congr. on Rheology, Brussels, Belgium, August 17-21, 1992

LEVELING OF THIN LIQUID FILMS

TASOS C. PAPANASTASIOU, K. ELLWOOD, N. MALAMATERIS, S. M. ALAIE AND J.O. WILKES

Chemical Engineering Department, University of Michigan, Ann Arbor, MI 48109

1. INTRODUCTION

Dynamic, nonlinear leveling of Newtonian and non-Newtonian liquid thin films on semi- and unbounded domains are analyzed by means of a free boundary condition at the artificial outflow boundary of the truncated domain. It is shown that traveling and solidary waves are predicted accurately on short truncated domains. Predicted leveling rates agree with reported experiments. Reported behavior of shear-thinning and viscoelastic films are also predicted. Further application of the free boundary condition include three-dimensional jets falling under gravity, and stability of multilayer extrusion.

2. GOVERNING EQUATIONS

Some of the computational domains used are shown in Fig. 1.

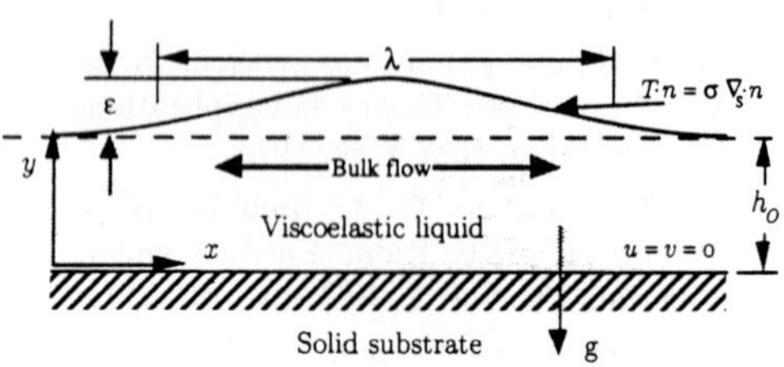

(a) Spreading on a horizontal substrate.

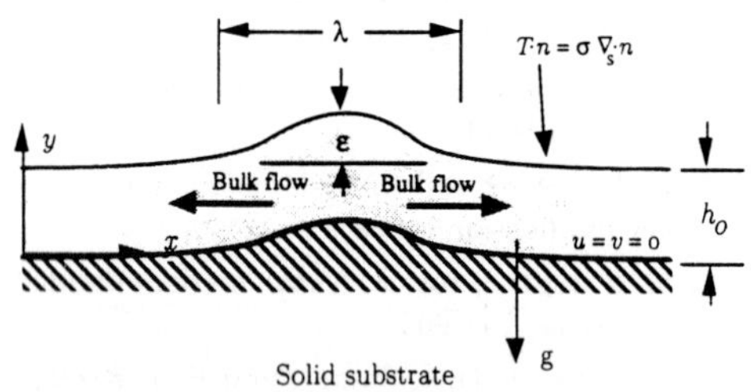

Fig.1 Leveling of liquid over horizontal substrate

The governing equations are:

$$\nabla \cdot \mathbf{u} = 0 \tag{1}$$

$$\mathrm{Re}(\mathbf{u}_t + \mathbf{u} \cdot \nabla \mathbf{u}) = \nabla \cdot (-\mathrm{p} + \tau) + \mathrm{Stg} \tag{2}$$

$$\mathbf{n} \cdot (\mathbf{u} - \dot{\mathrm{x}}) = 0 \tag{3}$$

The stress tensor τ in the momentum equation differentiates between Newtonian, shear-thinning and viscoelastic films. The Oldroyd-B model is used here:

$$\tau = \tau^1 + 2\Gamma\mathrm{D} \tag{4}$$

$$\tau^1 + \mathrm{De}[\tau^t + \mathbf{u} \cdot \nabla \tau - (\tau \cdot \nabla \mathbf{u} + \nabla^{\mathrm{T}} \mathbf{u} \cdot \tau) = 2(1-\Gamma)\mathrm{D} \tag{5}$$

In the dimensionless Eqns. (1) to (5) the definitions used are:

$$\mathrm{St} = \frac{\rho g h_0^2}{\mu_0 \mathrm{V}}, \ \mathrm{Re} = \frac{\rho \mathrm{V} h_0}{\mu_0}, \ \mathrm{De} = \frac{\lambda_1 \mathrm{V}}{h_0}, \ \Gamma = \frac{\lambda_2}{\lambda_1} \tag{6}$$

To represent the shear-thinning behavior, the Carreau viscosity model

$$\frac{\mu - \mu_\infty}{\mu_0 - \mu_\infty} = [1 + (\lambda\, \mathrm{II_D})^2]^{\frac{n-1}{2}} \tag{7}$$

has been used in Eqns. (4) and (5).

The initial conditions are those of rest under the initial configuration of the film of dimensions shown in Fig. 1. The boundary conditions, also shown in Fig. 1, are those of no-slip at the solid substrate and stress balance under the surface tension at the free surface, where any surface tension gradients are neglected. At the midplane of symmetry, the appropriate symmetry conditions are applied. *At the synthetic outflow boundary, the free boundary condition is implemented as described below.*

3. THE FREE BOUNDARY CONDITION

The momentum weighted residual is brought to the form,

$$\mathrm{R}_{\mathrm{m}}^{\mathrm{i}} = \int_V \left[\mathrm{Re}(\mathbf{u}_t + \mathbf{u} \cdot \nabla \mathbf{u})\phi^i + \mathbf{T} \cdot \nabla \phi^i - \mathrm{Stg}\phi^L\right] - \int_S \mathbf{n} \cdot \mathbf{T}\phi^L \mathrm{ds} \tag{8}$$

The surface integral is replaced by the well-defined essential and/or natural boundary conditions mentioned earlier. At the synthetic outflow such conditions do not exist, even for steady-state processes. However, as explained and demonstrated in a series of earlier publications (ref. 1, 2), such synthetic outflows, without any influence from flows on semi-infinite domains become hyperbolic atdownstream, much the same way as a boundary layer flow is modeled (ref. 3). For these kinds of flows, the boundary term of Eqn. (8) is evaluate as is, in terms of the as of yet unknown modal values along the synthetic outflow. *This is the free boundary condition.* Figs. (2) and (3) show two characteristic cases of application. Fig. (2), shows that the free boundary condition can be applied even where the synthetic outflow cuts a developing eddy. Fig. (3) shows that traveling waves are allowed to exit the artificial outflow without backreflection or distortion.

The applicability of the free boundary condition has also been demonstrated with jets falling under gravity (Ref. 4, 5), and with linear and nonlinear stability analysis of multilayer coating (ref. 6) and extrusion (ref. 7).

More selected results obtained with the free boundary condition are shown in Figs. 4 to 9 and summarized below.

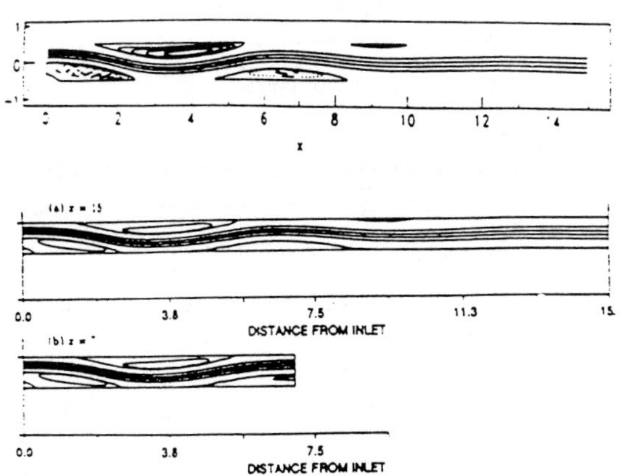

Fig. 2 Comparison of streamline patterns obtained with the free boundary condition applied at x=15 and x=7, for a stratified backward facing step (Taken from ref. 1).

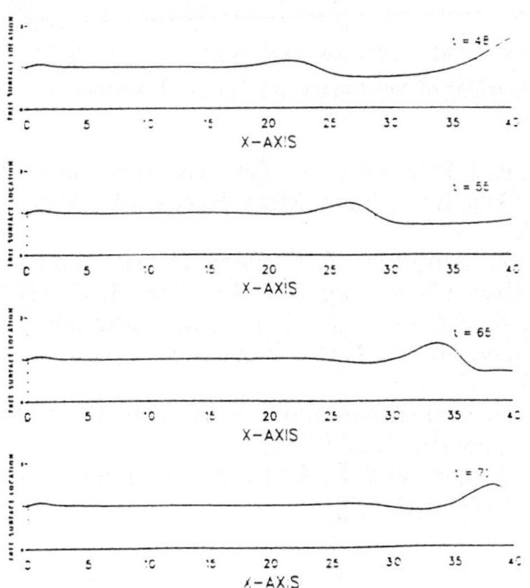

Fig. 3 Development and traveling of surge in start-up flow along a vertical plate, and exit through free boundary condition (Taken from ref. 2).

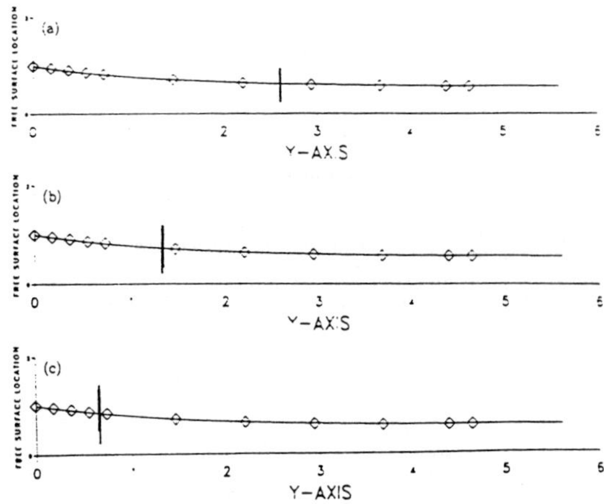

Fig. 4 Extrusion and leveling of film on vertical plate, in comparison with data (ref. 8).

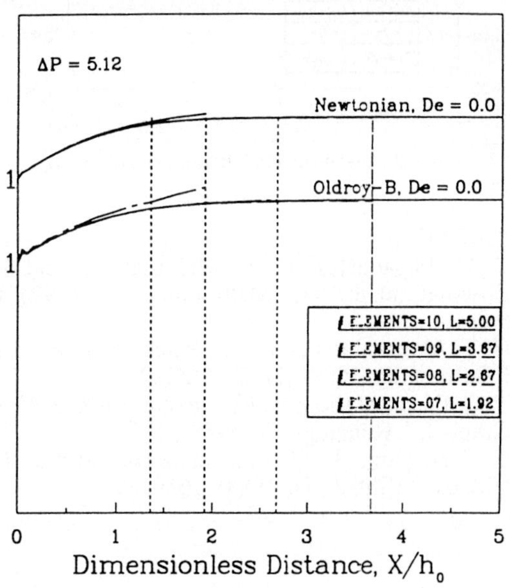

Fig. 5 Extrusion of Newtonian and Oldroyd-B viscoelastic liquids.

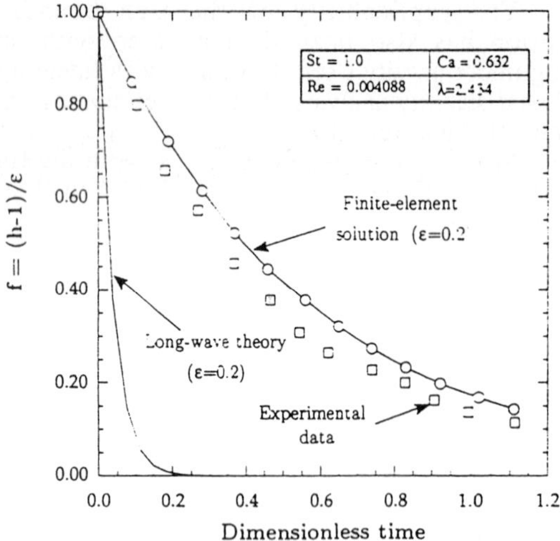

Fig. 6 Leveling of peak of sinusoidal disturbance, in comparison with data (ref. 9).

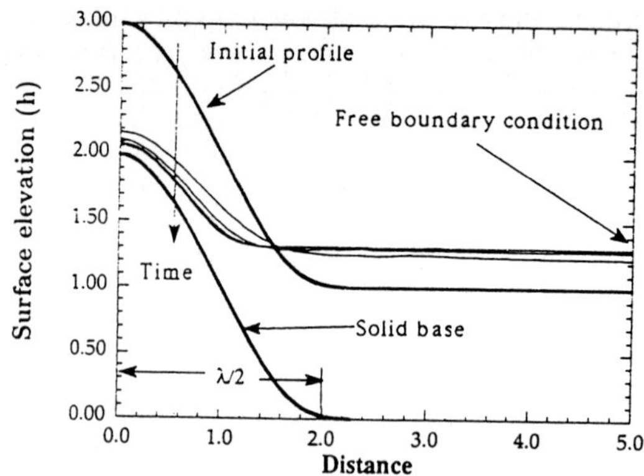

Fig. 7 Leveling of Newtonian liquid over substrate with bump.

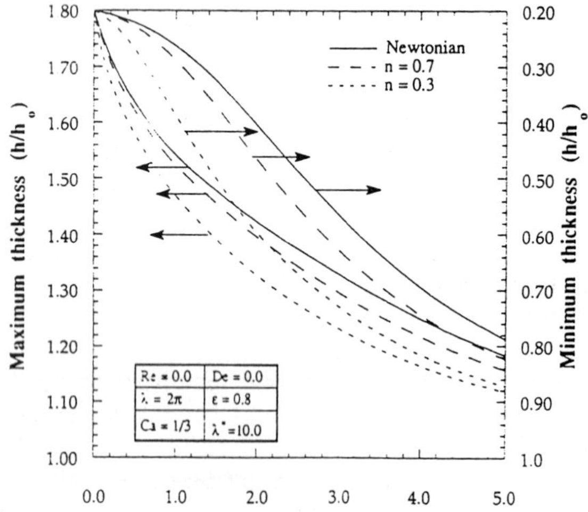

Fig. 8 Leveling of Newtonian and Carreau shear-thinning liquids.

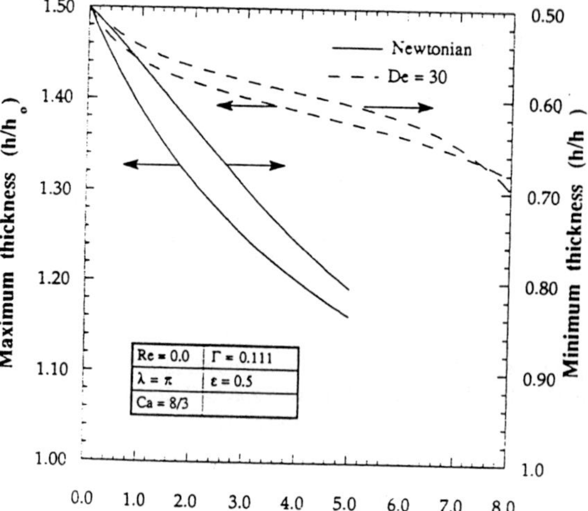

Fig. 9 Leveling of Newtonian and Oldroyd-B viscoelastic liquids.

REFERENCES

1. T. C. Papanastasiou, N. Malamataris and K. Ellwood, Int. J. Num. Meth. Fluids, 14(1992), in press
2. N. T. Malamataris and T. C. Papanastasiou, I & EC Research, 30(1991), 2211-2220.
3. T. C. Papanastasiou, A. N. Alexandrou and W. P. Graebel, J. Rheology, 32(1988), 485-510.
4. G. Georgiou, T. C. Papanastasiou and J. O. Wilkes, AIChE J., 34(1988), 1559-1562.
5. K. R. J. Ellwood, T. C. Papanastasiou and J. O. Wilkes, Int. J. Num. Meth. Fluids, 14(1992), 13-24.
6. N. R. Anturkar, T. C. Papanastasiou and J. O Wilkes, Chem. Eng. Sci., 45(1990), 3271-3281.
7. N. R. Anturkar, T. C. Papanastasiou and J. O Wilkes, J. non-Newt. Fluid Mech., 41(1991), 1 26.
8. R. N. Nedderman and J. O. Wilkes, Chem Eng Sci., 38(1983), 525-534.
9. D. Degani and C. Gutfinger, Comput. Fluids 4(1976), 149-161.

Theoretical and Applied Rheology, edited by P. Moldenaers and R. Keunings
Proc. XIth Int. Congr. on Rheology, Brussels, Belgium, August 17-21, 1992

COMPLETED DOUBLE-LAYER BOUNDARY INTEGRAL EQUATION METHOD: A NUMERICAL IMPLEMENTATION AND SOME EXPERIMENTAL RESULTS

N. PHAN-THIEN[1], D. TULLOCK[1], V. ILIC[1] and S. KIM[2]

[1]Department of Mechanical Engineering, The University of Sydney, NSW 2006, Australia
[2]Department of Chemical Engineering, University of Wisconsin-Madison, WI 53706, USA

Abstract

This paper reports on the Completed Double Layer Boundary Integral Equation Method for the solution of Stokes flows of a system of rigid particles in a container. It uses an iterative solver and therefore can handle a large number of particles with complex geometries. Particles' trajectories for a few typical problems are presented to illustrate the feasibility of the method. Experimental data were obtained with a two-camera system for the cases of two and eight spheres sedimenting in a container. A comparison with the numerical prediction in these cases showed a good agreement, to within the bounds of the experimental error.

1. INTRODUCTION

The determination of the bulk properties of suspensions and particulate solids requires the knowledge of the motion of the individual particles in the composite fluid or solid. We are mainly interested, however, in the simulation of suspensions of rigid particles of arbitrary shape in a viscous fluid. In most cases of practical interest, the fluid inertia is negligible (Reynolds numbers of the order $10^{-3} - 10^{-4}$), and the flow is predominantly isothermal. From the knowledge of the forces and torques acting on each particle, it is possible numerically to deduce sedimenting particles' linear and rotational velocities in a number ways from the Stokes' equations of motion,

$$-\nabla P + \mu\nabla^2\mathbf{u} = 0, \;\; \nabla\cdot\mathbf{u} = 0, \;\; \mathbf{x}\in\mathcal{D}, \qquad (1)$$

where $\mathbf{u}$ is the velocity field, P is the hydrostatic pressure, μ is the constant viscosity, and $\mathcal{D}$ is the domain of interest, ie., the fluid volume between the container and the particles surfaces.

The Boundary Element Method (BEM), (refs. 1,2), has proved ideally suited to the solution of such problems. However, with increasing number of elements required to model the geometry accurately, a limiting number of elements is reached such that it is either beyond the capacity of the available computer, or the problem becomes ill-conditioned. As an extension of the BEM, the Completed Double Layer Boundary Integral Equation Method (CDL-BIEM), (ref. 3), based on an alternative formulation of the Stokes' equations which are then solved with a fixed point iteration scheme, (ref. 4), overcame this difficulty as well as enabling efficient use of parallel computing architecture. This paper reports a comparison of the CDL-BIEM benchmark solution against the standard BEM results for a sphere sedimenting in a square cylinder. In addition, trajectories, in a square tank, of two and eight sedimenting spheres were calculated and compared with experimental data.

2. BACKGROUND

Following refs. 3,4, the integral representation of the Stokes equations is achieved with a distribution of a double-layer, the surface density φ of which satisfies a Fredholm integral equation of the second kind. After completing and deflating the spectral radius of the double-layer integral operator $\mathcal{K}$, the CDL-BIEM numerical technique is then applied to a fixed point iterative solution of the resulting boundary integral equations,

$$\varphi_j(\zeta) + (\mathcal{K}\varphi)_j(\zeta) + \varphi_j^{(p,l)}(\zeta) < \varphi^{(p,l)}, \varphi >$$
$$= \sum_{p=1}^{M}\left(F_i^{(p)} - \frac{1}{2}(\mathbf{T}^{(p)}\times\nabla)_i\right)G_{ji}(\zeta, \mathbf{x}_c^{(p)}), \qquad (2)$$
$$\zeta \in S.$$

where $\mathbf{F}^{(p)}$ and $\mathbf{T}^{(p)}$ are the force and torque acting on particle p, respectively, and $\mathbf{G}(\mathbf{X}, \mathbf{x}_c^{(p)})$ is the singularity solution that corresponds to a point force being placed at $\mathbf{x}_c^{(p)}$, any point inside the particle p. In this paper, $\mathbf{x}_c^{(p)}$ is chosen to be the center of mass of the particle.

3. IMPLEMENTATION

The numerical implementation of the CDL-BIEM proceeds in several stages. First, from the geometries of the particles the normalised null functions $\varphi^{(p,l)}$, for degrees of freedom $l = 1, \ldots, 6$ of particles $p = 1, \ldots, M$, the normalised "right" and "left" eigenvectors are constructed.

Next, following the discretisation of S, the main iterative equation to solve for φ_t is

$$\varphi_t' = \mathbf{b}_t - (\mathcal{H} - \mathcal{P}_L - \mathcal{P}_R)\varphi_t, \qquad (3)$$

where φ' is the next corrected solution; $\mathcal{P}_L$ and $\mathcal{P}_R$ are first-rank operators, while $\mathcal{H}$ is the surface integral operator such that $(1+\mathcal{H})\varphi = \mathbf{b}$, being the LHS of (2). This step is repeated until the norm

$$\max_{\zeta \in S} ||\varphi'(\zeta) - \varphi(\zeta)|| < \epsilon,$$

where ϵ is a set tolerance. Our experience indicates that a tolerance of $O(10^{-3})$ is adequate.

The boundary elements used in this implementation are super-parametric elements, described in (ref. 5). They are functionally constant elements, ie., over which φ_t is considered to be constant. Integrals on the right hand side of (3) as well as those involving the double-layer kernel when $\zeta \notin \Delta$ can be calculated by standard Gaussian quadrature. We use an adaptive integration scheme (up to 64-point quadrature) on the double-layer operator. The order of the quadrature for the adaptive scheme depends on the relative distance from the collocation point to the centroid of the element containing the field point to the maximum side of the element.

The surface velocity of a particle can then obtained directly from the original representation of the velocity fields of point forces and torques.

In the calculation of the particles' trajectories, the particles' inertia is neglected and their positions are updated in time using the numerically predicted values of the translational and rotational velocities of each particle.

4. EXPERIMENTS

The suspended liquid, dimethylpolysiloxane (silicone oil) (density 0.974 g/cm^3 and the viscosity 14.5 kPa s at 20.0^o C) was contained in a 200 x 200 x 1000 mm glass square tank. The tank lid housed a vacuum operated particle releasing device for two sphere (12.7±0.05 mm diameter, brass/PMM) tests or the eight eight sphere (4.75±0.05 mm diameter, ruby) drop fixture. Two video cameras were used to obtain particle dynamics measurements following the method of Dingman et al.(ref. 6). Combined video image was played back on a video tape recorder in single frame mode at chosen time intervals enabling manual recording of particles' image co-ordinates from the two camera views. An algorithmic transformation into space co-ordinates of the two sets of image co-ordinates then obtained the particles' trajectory with an accuracy of about 3 %.

5. RESULTS AND CONCLUSIONS

The sedimentation of a sphere at the centre of a cubic container was chosen as the numerical benchmark problem because of the ease of generating the boundary mesh. The agreement between the CDL-BIEM and BEM techniques for a sedimenting sphere was excellent, Figure 1.

For experimental data, the simulation of the trajectory of the collision of an eccentric fast sphere with a slow sphere sedimenting axially was also very good, Figure 2. Experimental trajectories were also in excellent agreement with calculated data for the eight sphere experiment, in which the spheres were initially in a planar cross formation, Figure 3.

Although we report only a modest number of particles, we believe the CDL-BIEM techniques will prove to be more powerful with the emergance of massively parallel machines.

Acknowledgements

This research was funded by an Australian Research Council Grant and a University of Sydney Special Grant (to NP-T), and an NFS Grant CTS-9015377 and an ACS Petroleum Research Fund Grant (to SK).

REFERENCES

1. P.K. Banerjee and R. Butterfield, Boundary Element Methods in Engineering Science, McGraw Hill: London.
2. C.A. Brebbia, J.C.F. Telles and L.C. Wrobel, Boundary Element Techniques: Theory and Applications in Engineering, Springer-Verlag: Berlin (1984).
3. H. Power and G. Miranda, SIAM J. Appl. Math., **47** (1987) 689–698.
4. S. Kim and S.J. Karrila, Microhydrodynamics: Principles and Selected Applications, Butterworth-Heinemann: Boston (1991).
5. N. Phan-Thien, D.Tullock and S. Kim, Computational Mech. (in press).
6. S. Dingman, M.S. Ingber, L. Mondy and J. Abbott, preprint (1991).

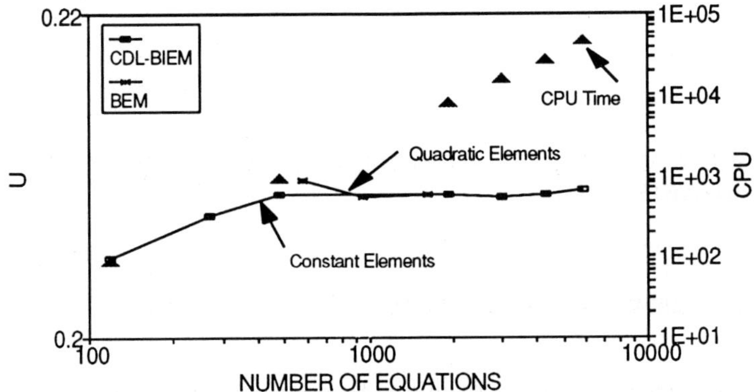

Figure 1. The normalised sedimentation velocity and the CPU time (on an IBM RS6000 computer) versus the number of equations.

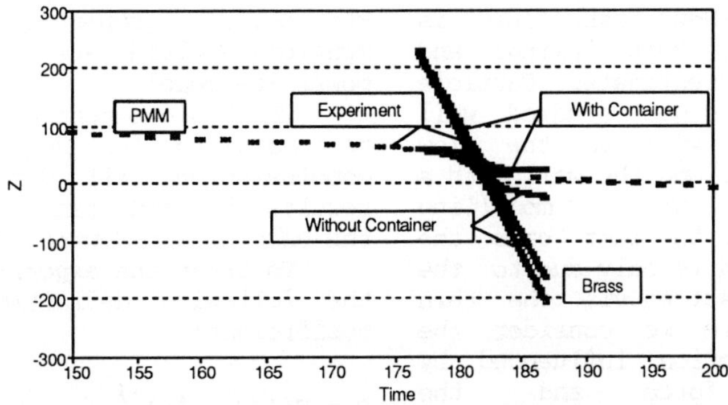

Figure 2. The vertical positions of the spheres' centres as functions of time.

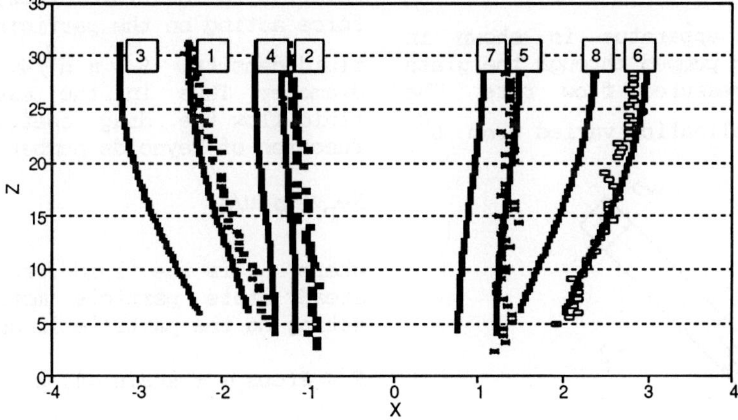

Figure 3. The trajectories of eight spheres in $z - x$ plane.

Theoretical and Applied Rheology, edited by P. Moldenaers and R. Keunings
Proc. XIth Int. Congr. on Rheology, Brussels, Belgium, August 17-21, 1992

ON THE MOTION OF SPHERICAL PARTICLES ALONG THE WALL IN THE SHEAR FLOW OF NEWTONIAN AND NON-NEWTONIAN FLUID

A.N. PROKUNIN, YU.A. BUKMAN and YU.P. GUPALO

Institute for Problems in Mechanics, Russian Academy of Sciences, pr. Vernadskogo 101, 117526 Moscow (Russia)

1. INTRODUCTION

The motion of particles in the flow of Newtonian and non-Newtonian fluid is studied experimentally. Both laminar and turbulent flows are investigated. Particle motion in uniform flows was studied well enough elsewhere, as well as the slow motion of a particle along the wall with a lubrication layer while modelling lubricated bearing. In the last case the influence of the fluid is only due to the particle - wall interaction via the thin lubrication layer. Here we consider the case of a particle motion influenced by the gravitation force and the particle-fluid interaction force when both forces can be of the same order.

2. METHOD

Experimental apparatus is shown in Fig. 1. Fluid was pumped through the glass tube 1 with a measured flow rate. The angle of tube inclination varied from 0^0

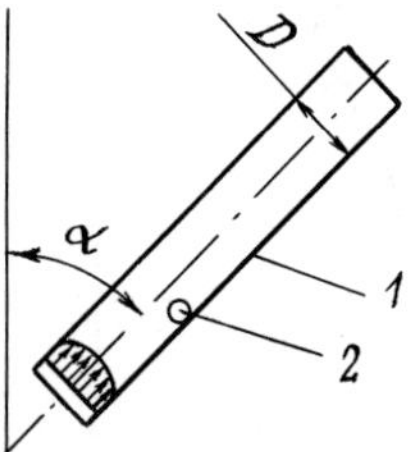

Fig. 1. Diagram of experimental apparatus

(vertical position) to 90^0(horizontal position). Spherical particles 2 were made of lead, steel or glass. Newtonian fluids (water and glycerol solutions) and non-Newtonian "milling yellow" were used. Rheological properties of the last mentioned fluid are well described by power-law model.

The basic experimental results will be considered for Newtonian fluids. In conclusion we will demonstrate how the results obtained can be generalized to the case of non-Newtonian fluids.

To treat the experimental data we use the following definition of the drag coefficient:

$$C = F(1/2\ \rho u^2 S)^{-1} \tag{1}$$

where u is the particle velocity relative to the fluid flow, F is the force acting on the particle, ρ is the fluid density, $S = \pi\ d^2/4$, d is particle diameter. Thus in the case of uniform fluid flow the drag coefficient is the function of Reynolds number

$$Re_p = \rho du/\eta \tag{2}$$

where η is the fluid viscosity. For the steady state particle motion the force acting on the particle is as follows:

$$F = P(\cos\alpha \pm k \sin\alpha),\ k = F_{fr}/P \sin\alpha \tag{3}$$

where k is the traditional friction coefficient. Friction force F_{fr} depends upon the weight of the particle P in fluid and the hydrodynamic force, which tends to press it towards the wall. Plus or minus sign depends on whether the particle moves up or down the tube (co- or contra- the fluid flow). For the spherical particle from (1) and (3) we have

$$C = 4/3gd(\rho_s/\rho - 1)u^{-2}(\cos\alpha \pm k\sin\alpha) \quad (4)$$

where g is the acceleration of the gravitation field, ρ_s is the particle density. In the case of the particle motion along the wall in the horizontal or inclined tube we put in (4):

$$u = \langle v_d \rangle \mp w \quad (5)$$

where $\langle v \rangle_d$ is the mean (averaged over the distance equal to a particle diameter) local undisturbed flow-rate, w is the particle velocity along the tube wall.

3. EXPERIMENTAL RESULTS FOR NEWTONIAN FLUIDS

Consider the motion of a spherical particle in the upward fluid flow in horizontal and inclined tubes. In this case it was found that

$$\langle v \rangle_d - w = u \approx const \quad (6)$$

for any Reynolds number Re based on the tube diameter. The constant in (6) depends upon the tube inclination α, particle size and density, viscosity and density of a fluid, properties of the particle and the tube material. The constant value might change when laminar flow transforms to a turbulent one.

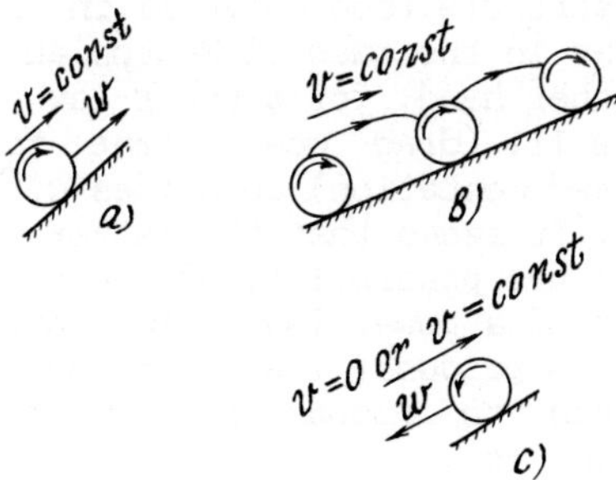

Fig. 2. Types of spherical particle motions along the wall of the tube

In experimental data treatment (see (6)) the motion of a particle rolling up along the tube wall was considered (Fig. 2a), rotation rate being increased together with flow rate. When the particle transport in tubes with slight inclination (i.e. in almost vertical ones) was studied, the other types of particle motion appeared. For instance, rather light particles (glass ballotini) demonstrated saltation when fluid viscosity was increased (see Fig. 2b).

The drag coefficient for the sphere rolling up along the tube wall was calculated from the formula (4) (the values of u being given in (6)) for small inclinations (small values of α), when the friction is of minor importance. These results are given by the curve II in Fig. 3 and it corresponds to the classical diagram I if the drag coefficient according to this diagram is multiplied by the factor 1.5.

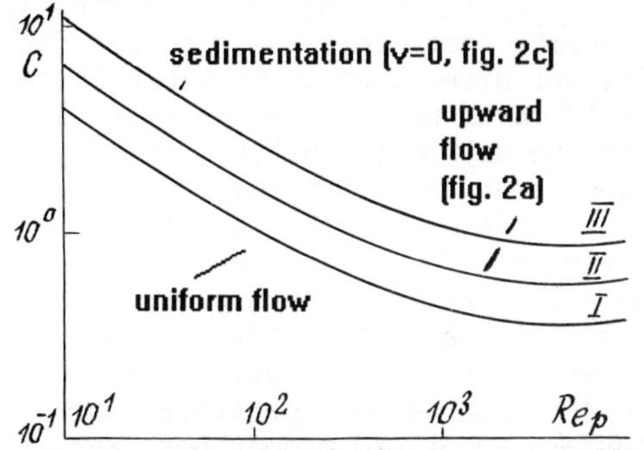

Fig. 3. Dependences of the drag coefficient C on Re_p.

After the friction coefficient k was found from the curve II (it appeared to be almost constant as it was in the case of dry friction), the validity of the given assumption of small friction effect at small α was checked.

The curve II represents the experimental data for various particle and fluid parameters and various particle and fluid parameters and various tube diameters (for the case $d/D \ll 1$).

While the spherical particle is settling in an inclined tube it is rolling down along the tube wall and is rotating as shown in **Fig. 2c**. The drag coefficient for a settling sphere was calculated in a similar way (as for a sphere moving up the tube) by means of formula (4) for small α when the friction effect was negligible. The result is shown in Fig. 3 by the curve III, corresponding to values given by the curve I if multiplied by the factor 2.5.

If rates are small enough to keep the downward motion of a particle along the tube wall, then

$$\langle v \rangle_d + w = u \approx const \tag{7}$$

The constant value of u was equal to the sedimentation velocity (within the ±10% error).

It is worth noting that in the case of a turbulent flow the particle could be detached from the tube wall. This effect was pronounced in particular for light particle moving in tubes with slight **inclinations (v=0, sedimentation).**

4. DISCUSSION OF THE RESULTS OF EXPERIMENTS

From Fig. 3 it follows that the drag coefficient for a particle rolling up the tube wall (curve II) is smaller than that for a particle rolling down (curve III). It seems that this effect is due to different directions of particle rotation when it moves up or down the tube as well as to the asymmetric nature of the flow pattern near the particle.

Consider now the effect of the friction of a particle moving along the tube wall. In the first approximation the value of the coefficient $h = kd/2$ like the case of a dry friction can be put constant for given particle and tube materials and fluid properties. For fixed particle and tube materials the value of h is greater for more viscous fluids.

This effect is sharpy pronounced for light glass ballotini, when the friction coefficient h could change by the factor 25 if the fluid viscosity increased or decreased only by the factor 3.

When the particle moves up the tube the friction is higher than in case of downward movement, particularly for light particles and high fluid viscosity. It seems to be due to the additional hydrodynamic forces acting on the particle and directed to the wall for upward motion or from the wall for downward motion. The latter effect causes the "floating" of a particle when it moves down the tube of small inclination.

Consider now the influence of a turbulence on the friction force acting on the particle. If rather light particles move up the tube the turbulence tends to increase the friction coefficient h, i.e. to increase the hydrodynamic force, which presses the particle against the wall. The opposite effect takes place if the particle moves down the tube.

Due to the greater friction force for more viscous fluids it can become almost equal to the gravitation force (for light particles). Therefore the existence (confirmed in experimens) of the maximum in flow velocity (vs α diagram), corresponding to the stopping or starting the motion of a particle, is possible.

It is worth noting that for $\alpha > \alpha_{max}$ the particle usually rolls along the tube wall, while for $\alpha < \alpha_{max}$ some small saltations could be observed (see Fig. 2b).

5. THE MOTION OF SPHERICAL PARTICLES IN NON-NEWTONIAN POWER-LAW FLUID.

The power- law fluid is governed by the equation

$$\sigma = m\dot{\gamma}^n \tag{8}$$

where σ is the stress, $\dot{\gamma}$ is the deformation rate, m and n are the rheological constants. We use the generalized Reynolds number introduced by Metzner [1]:

$$Re_p = 8^{1-n}\left(\frac{4n}{3n+1}\right)^n \frac{u^{2-n}d^n\rho}{m} \tag{9}$$

It is well known that in terms of Re_p many formulas for Newtonian fluids can be applied to describe the non-Newtonian behaviour.

Our experiments have demonstrated that most of the above described properties of particle motions in the flow in tubes are sustained. For instance, dependences (6) for small α and the drag coefficient as function of Re_p for a rolling particle (see curve II in Fig. 3) is same as in the case of Newtonian fluid. On the other hand, for a power-law fluid the curve III (drag coefficient for the case of sedimentation) coincides with the curve II. It seems that the latter effect is due to the possibility of the change in viscosity of a power-law fluid under the deformation process. The viscosity change also leads to inconstancy of friction coefficient k .

The experiments with non- Newtonian fluids were carried out only in laminar flow of fluid.

REFERENCES

1. A.B. Metzner, I.C. Reed. AIChE Journal, 1 (1955) 434-445.

Theoretical and Applied Rheology, edited by P. Moldenaers and R. Keunings
Proc. XIth Int. Congr. on Rheology, Brussels, Belgium, August 17-21, 1992

MODIFICATION OF INERTIAL FILM INSTABILITY BY VISCOELASTICITY

T. R. Salamon[1], R. C. Armstrong[1] and R. A. Brown[1]

[1]Department of Chemical Engineering, Massachusetts Institute of Technology, Cambridge, MA 02139 U.S.A.

1. INTRODUCTION

The unperturbed flow of a viscous liquid flowing down an inclined plane is a flat film with a semi-parabolic velocity profile. Although an exact solution to the equations of motion, the flat film is not stable for all flow rates and inclination angles and goes unstable to wave-like motions which do not have any regular character. If the fluid film is artificially perturbed two characteristic types of waves are observed (ref. 1). The first is a periodic wave which arises in synchronization with the perturbing pulses. The second is a single wave which appears when the perturbing pulses are less frequent and very strong.

The challenge to analytical and computational analysis is to describe the multiple flow states seen in these experiments. Due to the complexity of treating the full Navier-Stokes equations and boundary conditions, long-wave approximations have been used to simplify the full equation set (refs. 2-4). The long-wave approximation assumes that the length of a characteristic disturbance l_0 is much greater than the film thickness h_0, so that the long-wave parameter $\mu \equiv h_0/l_0$ is small. This gives rise to a single partial differential equation for the free surface height which is valid to some order of μ. Various forms of this equation for the free surface height have been studied both analytically and numerically. Periodic and solitary waves are solutions to the equations, in agreement with the types of waves which have been observed experimentally (refs. 3-6). However, little is known about the validity of these approximate solutions, because extensive numerical solutions have not been reported.

The effect of viscoelasticity on the inertial film instability has not been investigated in detail. The linear stability analysis of the flat film flow of an Oldroyd-B model have been studied over a wide range of Deborah number (ref. 7). These results indicate that the effect of viscoelasticity is primarily destabilizing at small Reynolds numbers whereas at moderate Reynolds numbers viscoelastic effects stabilize the film. Nothing is known theoretically about the effect of elasticity on the finite amplitude waves which evolve after linear stability.

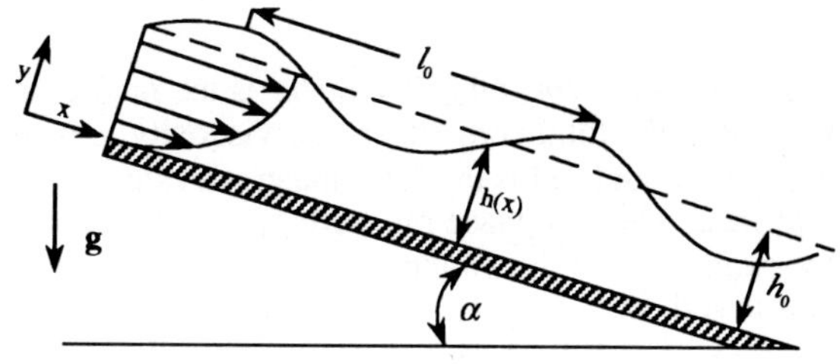

Figure 1. Diagram of flow.

Our goal is to present full numerical solution of the traveling wave states and document the effect of elasticity on the nonlinear states. The influence of viscoelasticity on traveling waves for the flow of an Oldroyd-B fluid is investigated numerically by using the finite-element method. The formulation is based on the Elastic-Viscous Split-Stress Method developed by Rajagopalan *et al.* (ref. 8) for free surface flows.

2. GOVERNING EQUATIONS

The flow is depicted in Figure 1. Fluid of density ρ flows down a plane which is inclined at an angle α to the horizontal. Periodic traveling wave solutions of period l_0 in the x-direction are determined as stationary states in a coordinate system that travels at the wavespeed c. The following scalings are used:

$$length \approx h_0 , \tag{1}$$

$$velocity \approx U_0 = \rho g\, h_0^2 \sin \alpha / 2\, \eta_0 , \tag{2}$$

$$stress \approx \rho g\, h_0 \sin \alpha. \tag{3}$$

These scalings give the dimensionless equations for the flow of an Oldroyd-B model

$$\frac{Re}{2}\,\mathbf{v}\cdot\nabla\mathbf{v} - \mathbf{g} + \nabla p - \beta\nabla^2\mathbf{v} + \nabla\cdot\tau_p = \mathbf{0}, \quad (4)$$

$$\nabla\cdot\mathbf{v} = 0, \quad (5)$$

$$\tau_p + De\,\tau_{p(1)} = -(1/2)(1-\beta)\dot{\gamma}, \quad (6)$$

with the following boundary conditions,

$$\mathbf{v} = -c\,\mathbf{e}_x \qquad \text{at } y = 0, \quad (7)$$

$$\mathbf{tn}{:}\sigma = 0 \qquad \text{at } y = h(x), \quad (8)$$

$$\mathbf{nn}{:}\sigma + p_{gas} - 2H\mathrm{We} = 0 \qquad \text{at } y = h(x), \quad (9)$$

$$\mathbf{v}\cdot\mathbf{n} = 0 \qquad \text{at } y = h(x), \quad (10)$$

where $\mathbf{g} = \mathbf{e}_x + \cot\alpha\,\mathbf{e}_y$, $\beta = \eta_s/\eta_0$ is the ratio of the solvent to the total zero-shear-rate viscosity, c is the wavespeed and $2H$ is the local mean curvature of the interface. The gas phase pressure p_{gas} is determined by requiring that the volume of the traveling wave is conserved, i.e.

$$\int_0^{l_o} h(x)\,dx = V, \quad (11)$$

and the wavespeed c is determined by requiring that the free surface satisfy a phase constraint

$$\int_0^{l_o} h_x(x)\sin(2\pi x/l_o)\,dx = 0. \quad (12)$$

The dimensionless groups which arise from these scalings are

$$Re \equiv \rho U_0 h_0/\eta_0 \qquad \text{Reynolds number,} \quad (13)$$

$$De \equiv \lambda\,U_0/h_0 \qquad \text{Deborah number,} \quad (14)$$

$$We \equiv T/\rho g h_0^2 \qquad \text{Weber number,} \quad (15)$$

where λ is a characteristic relaxation time for the polymer and T is the surface tension.

3. NUMERICAL METHOD

The governing equations and boundary conditions are approximated by the finite element method. The method used is the Elastic-Viscous Split-Stress formulation of Rajagopalan *et al.* (ref. 8) which is based on splitting the stress into purely elastic and viscous contributions. This is done by defining the elastic stress tensor as

$$\Sigma \equiv \tau_p + (1/2)(1-\beta)\dot{\gamma}, \quad (16)$$

and rewriting the momentum (4) and constitutive (6) equations in terms of the new dependent variables $(\mathbf{v}, \Sigma, p)$

$$\frac{Re}{2}\,\mathbf{v}\cdot\nabla\mathbf{v} - \mathbf{g} + \nabla p - \nabla^2\mathbf{v} + (1-\beta)\nabla\cdot\Sigma = \mathbf{0}, \quad (17)$$

$$\Sigma + De\,\Sigma_{(1)} - De\,\dot{\gamma}_{(1)} = \mathbf{0}. \quad (18)$$

The reformulated equations are discretized by applying Galerkin's method to the momentum and continuity equations and the Streamline-Upwind/Petrov-Galerkin method to the constitutive equation. The velocity and elastic stress are approximated by biquadratic Lagrangian basis functions and the pressure is approximated by bilinear Lagrangian basis functions. Special treatment of the term $\dot{\gamma}_{(1)}$, which has second derivatives in velocity, is done by interpolating the discontinuous velocity gradients onto a continuous representation by the least squares method. The free surface is represented by one-dimensional quadratic basis functions with the kinematic condition taken as the distinguished condition for the interface. The normal and shear stress balances are incorporated naturally into the weak form of the governing equations. Newton's method is used to solve the resulting set of nonlinear algebraic equations.

4. RESULTS

All of the results reported here were obtained for $We = 1.0$, $\mu = h_0/l_0 = 0.02$ and $\alpha = 90°$. The finite element mesh used had 80 elements in the x-direction and 20 elements in the y-direction and corresponded to 38,002 unknowns.

Contours of the streamfunction (ψ) and pressure (p) for the Newtonian flow, $De = 0.0$, $\beta = 1.0$, and $Re = 0.035$, 0.235 and 0.535 are presented in Figure 2. The contours are equispaced between the maximum (●) and minimum (o)values of the solution field. In Figure 2a with $Re = 0.035$ the traveling wave is near the onset of instability for a Newtonain fluid and the disturbance is small. As the Reynolds number increases, both c and the maximum free surface height (h_{max}) increase.

Contours of ψ, p, τ_{pxx}, τ_{pyy} and τ_{pyx} for the case $De = 0.25$, $\beta = 0.5$ and $Re = 0.035$ are presented in Figure 3. The magnitude of the viscoelastic traveling wave for this case is substantial compared to the Newtonian case depicted in Figure 2a. The wave speed has also increased.

The wavespeed (c) and h_{max} are plotted versus the Reynolds number (Re) in Figure 4 for the Newtonian flow (●) and an Oldroyd-B model with $De = 0.25$ and $\beta = 0.5$ (o). Both c and h_{max} increase substantially as the Reynolds number increases for both fluids.

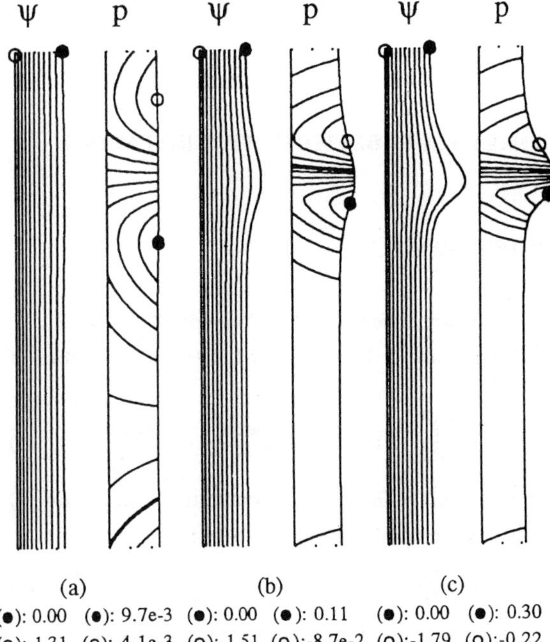

Figure 2. Contours of the streamfunction (ψ) and pressure (p) for Newtonian flow with $We = 1.0$, $\mu = 0.02$ and $\alpha = 90°$ on an 80x20 mesh: (a) $Re = 0.035$, $c = 1.98$ and $h_{max} = 1.03$; (b) $Re = 0.235$, $c = 2.19$ and $h_{max} = 1.27$; and (c) $Re = 0.535$, $c = 2.51$ and $h_{max} = 1.61$. Maximum (●) and minimum (o) values of the contours are denoted.

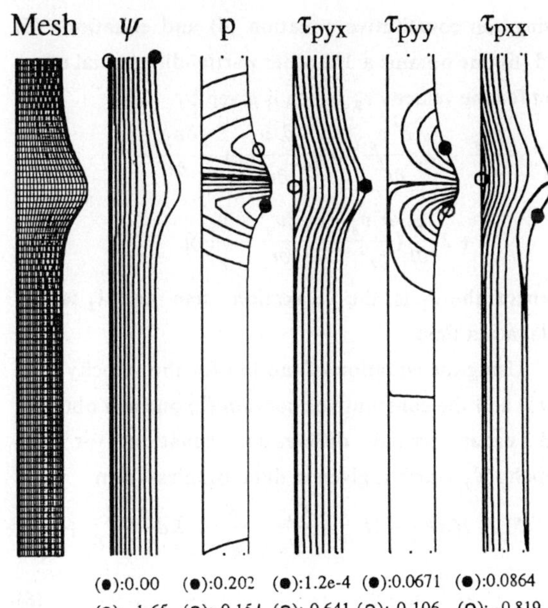

Figure 3. Contours of ψ, p, τ_{pxx}, τ_{pyy} and τ_{pyx} for an Oldroyd-B model with $De = 0.25$, $\beta = 0.5$, $We = 1.0$, $\mu = 0.02$ and $\alpha = 90°$ on an 80x20 mesh with Re = 0.035, $c = 2.35$ and $h_{max} = 1.44$. Maximum (●) and minimum (o) values of the contours are denoted.

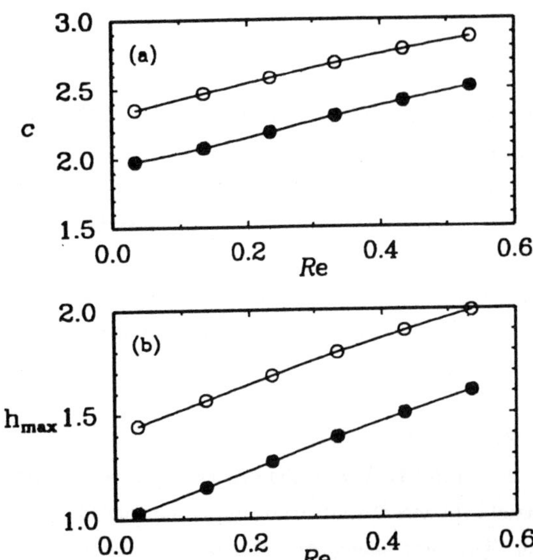

Figure 4. (a) Wavespeed c versus Reynolds number (Re) and (b) h_{max} versus Reynolds number (Re) with $We = 1.0$, $\mu = 0.02$, $\alpha = 90°$ for Newtonian flow (●) and the flow of an Oldroyd-B model with $De = 0.25$ and $\beta = 0.5$ (o) on an 80x20 mesh.

5. SUMMARY

The finite element method has been used to compute the traveling wave state for an Oldroyd-B fluid flowing down a vertical plane. For small values of the elasticity the traveling waves have both a larger amplitude and wavespeed than their Newtonian counterparts, indicating that viscoelasticity magnifies the nonlinear state and makes the use of numerical solutions essential for studying the evolution of the waves.

REFERENCES

1. P. L. Kapitza, Wave Flow of Thin Layers of a Viscous Fluid, in: D. Ter Haar, ed., Collected Papers of P. L. Kapitza, Pergamon, Oxford, 1965, pp. 662-709.
2. D. J. Benney, J. Math. Phys., 45 (1966) 150-155.
3. C. Nakaya, Phys. Fluids, 18(11) (1975) 1407-1412.
4. H.-C. Chang, Phys. Fluids A, 1(8) (1989) 1314-1327.
5. C. Nakaya, Phys. Fluids A, 1(7) (1989) 1143-1154.
6. A. Pumir, P. Manneville and Y. Pomeau, J. Fluid Mech., 135 (1983) 27-50.
7. E. S. G. Shaqfeh, R. G. Larson and G. H. Frederickson, J. Non-Newtonian Fluid Mech., 31 (1989) 87-113.
8. D. Rajagopalan, R. J. Phillips, R. C. Armstrong, R. A. Brown, and A. Bose, J. of Fluid Mech., 235 (1992) 611-642.

Theoretical and Applied Rheology, edited by P. Moldenaers and R. Keunings
Proc. XIth Int. Congr. on Rheology, Brussels, Belgium, August 17-21, 1992

TIME-DEPENDENT FLOW OF UPPER-CONVECTED JEFFREY FLUID BETWEEN TWO ROTATING CYLINDERS

HAN SHIFANG[+] and ROESNER K.G.[++]

[+]Res. Lab. of non-Newtonuian Fluid Mech., Chengdu Branch, Academia Sinica, 610015, Chengdu, P.R.China
[++]Inst. fur Mechanik, Arbeitsgruppe dynamik der Fluide, TH Darmstadt, Hochschustr.1, D-6100, Darmstadt F.R.Germany

INTRODUCTION

An unsteady flow in the gap between two rotating coaxial cylinders is studied for upper-convected Jeffrey fluid. The time-dependent character of the transient flow starting from rest or from a stationary motion is of special interest. The improved approach due to Kantorovich is used. The constitutive equation of upper-convected Jeffrey fluid is introduced. Using the Macsyma-software based on LISP of ARTIFICIAL INTELLIGANCE the solution is found computationally and anallytically by solving an eigenvalue problem of ordinary differential equations for different approximations. The approach developed in the article is called the COMPUTATIONAL ANALYTICAL method due to ARTIFICAIL INTELLIGENCE.

GOVERNING EQUATUINS

We consider the unsteady flow of a viscoelastic fluid in the gap between two coaxial cylinders. The fluid is assumed to be incompressible. The cylindrical coordinate system (r, θ, z) is used. The velocity field is assumed to be of

$$v_r = 0, \quad v_r = v(r,t), \quad v_z = 0. \tag{1}$$

The constitutive equation of the Jeffrey fluid has the form of

$$S^{ik} + \lambda_1 \overset{\nabla}{S}{}^{ik} = \eta_0 (A^{ik} + \lambda_2 \overset{\nabla}{A}{}^{ik}) \tag{2}$$

where

$$\overset{\nabla}{S}_{ik} = \frac{\partial S^{ik}}{\partial t} + v^m \frac{\partial S^{ik}}{\partial x^m} - S^{im}\frac{\partial v^k}{\partial x^m} - S^{mk}\frac{\partial v^i}{\partial x^m} \tag{3}$$

For the velocity field (1) the constitutive equation (2) can be reduced in the following form :

$$S_{r\theta} + \lambda_1 \frac{\partial S_{r\theta}}{\partial t} = \eta_0 [\frac{\partial v_\theta}{\partial r} - \frac{v_\theta}{r} + \lambda_2 \frac{\partial}{\partial t}(\frac{\partial v_\theta}{\partial r} - \frac{v_\theta}{r})] \tag{4}$$

Then the equation of motion gets a simple form :

$$\rho \frac{\partial v_\theta}{\partial t} = \frac{\partial S_{r\theta}}{\partial r} + \frac{2}{r} S_{r\theta}. \tag{5}$$

Differentiation of equation (5) with respect to t yields an equation of the following form :

$$\rho \frac{\partial^2 v_\theta}{\partial t^2} = \frac{\partial^2 S_{r\theta}}{\partial t \partial r} + \frac{2}{r}\frac{\partial S_{r\theta}}{\partial t}. \tag{6}$$

Using the constitutive equation (4) and equations (5) and (6) one obtains a 3rd order partial differential equation for the velocity v_θ which is given by

$$\rho(\frac{\partial v_\theta}{\partial t} + \lambda_1 \frac{\partial^2 v_\theta}{\partial t^2}) = \eta_0 [\frac{\partial^2 v_\theta}{\partial r^2} + \frac{1}{r}\frac{\partial v_\theta}{\partial r} - \frac{v_\theta}{r^2}$$

$$+ \lambda_2 \frac{\partial}{\partial t}(\frac{\partial^2 v_\theta}{\partial r^2} + \frac{1}{r}\frac{\partial v_\theta}{\partial r} - \frac{v_\theta}{r^2})]. \tag{7}$$

where the λ_1 is the relaxation time and λ_2 is the redaxation time.

Using the equation of motion for the velocity field of (1) and the constitutive equation (2) one can obtain a 3rd order partial differential equation for the velocity v_θ which is given in dimensionless form

$$F_1 + HaF_{tt} - [F_{yy} + \frac{F_y}{y} - \frac{F}{y^2} + Ka\frac{\partial}{\partial t}(F_{yy} + \frac{F_y}{y}$$

$$- \frac{F}{y^2})] = 0 \tag{8}$$

$$Ha := \frac{We}{Re}, \; Ka := \frac{\lambda_2}{\lambda_1} Ha. \tag{9}$$

$$F := \frac{v_0}{\omega_1 R_1}, \; \hat{t} := \frac{\omega_1 t}{R_1}, \; y := \frac{r}{R_1} \tag{10}$$

APPROACH

An improved method of Katorovich type is developed. The general solution is assumed in the special form, a sum of f_n. For the N–th approximation we shall use the f_1, f_2, ···, f_{N-1}, obtained frm the (N−1)–th approximations and the f_N is to find for the N–th approximation.

If the equation of the problem is given in the general form, then we can write as

$$H(v) = 0. \tag{11}$$

The shape functions are assumed to be:

$$(y - \frac{1}{y}), \; (y - \frac{1}{y})y^2, \cdots \tag{12}$$

The general solution of the time dependent equation (8) truncated to a finite number of terms is assumed in the special form

$$F_N = \sum_{n=1}^{N} f_n, \tag{13}$$

The variational principle of the problem is described as

$$\iint H[\sum_{n=1}^{n} f_n](y - \frac{1}{y})y^{2(n-1)} \delta G_n(t) dt dy = 0 \tag{14}$$

SOLUTION

According to the approach proposed the solutions of different approximations are obtained by the Macsyma–software.

1st Approximation

$$F(y,t) = F_0 + a\left(y - \frac{1}{y}\right)G_1(t) \tag{15}$$

$$a_1 \frac{d^2 G_1(t)}{dt^2} + a_2 \frac{dG_1(t)}{dt} = 0. \tag{16}$$

2nd Approximation

$$F(y,t) = F_0 + a\left(y - \frac{1}{y}\right)[G_1(t) + y^2 G_2(t)] \tag{17}$$

$$\iint \{H[f_1](y - \frac{1}{y})y^2 + H[f_2](y - \frac{1}{y})y^2\} \delta G_2 \cdot$$

$$dt dy = 0, \tag{18}$$

$$b_1 \frac{d^2 G_2(t)}{dt^2} + b_2 \frac{dG_2(t)}{dt} + b_0 G_2(t) = 0, \tag{19}$$

$$b_1 := HaP_1(m), \; m := \frac{R_2}{R_1},$$

$$b_2 := HaP_1(m) + KaP_2(m), \; b_0 := P_2(m),$$

$$P_1 := 8m^7 - 35m^4 + 42m^2 - 15,$$

$$P_2 := 8m^2(14m^5 - 35m^2 + 21).$$

The roots of equation (19) are:

$$K_{1,2} = -\frac{-1}{2HaP_1(m)}\{P_1(m) + KaP_2(m)$$
$$\pm [P_1^2(m) + 2(Ka - 2Ha)P_1(m)P_2(m)$$
$$+ Ka^2 P_2^2(m)]^{1/2}\} \tag{20}$$

It should be noted that the expression $P_1(m)$ has 5 real and 2 complex zeros which are given by

$$m_1 = 1, (threefold),$$

$$m_{2,3} = \pm i\frac{\sqrt{21 - 2\sqrt{105}}}{4} - \frac{3}{4}, \tag{21}$$

$$m_{4,5} = \pm i\frac{\sqrt{21 + 2\sqrt{105}}}{4} - \frac{3}{4}.$$

The following cases are discussed:

Case 1:The roots are real ($K_1 < 0$, $K_2 < 0$ and $K_1 \neq K_2$).

The solution of the problem has the following form:

$$F(y,t) = F_0 + a(y - \frac{1}{y})\{[1 - e^{-\frac{t}{Ha}}] + y^2[e^{K_2 t}$$
$$- e^{K_2 t}]\}, \tag{22}$$

where $K_1$and $K_2$are given by(20)

Case 2:$K_1 = K_2$

Case 3:K_1, $K_2$are complex.

CONCLUSIONS

Using Macsyma–software and improved approach due to Kantorovich the problem of time dependent flow of upper–convected Jeffrey fluid between two rotating cylinders is reduced to problem of system of ordinary defferential equations. The unsteady behaviour of the flow is Studied by the present method.

The computational analytical Approach due to Macsyma is developed.

Reference

1. Han Shifang, Continuum Mechanics of Non–Newtonian Fluids, Sichuan Press of Science and Technology (1988), Chengdu, China
2. Han Shifang et al, Acta Mechanica Sinica, vol6. No.3 p221–p226 (1990)

Acknowledgments : The author Han Shifang would like to a knowledge the Alexander Humboldt–Stiftung for support of this research.

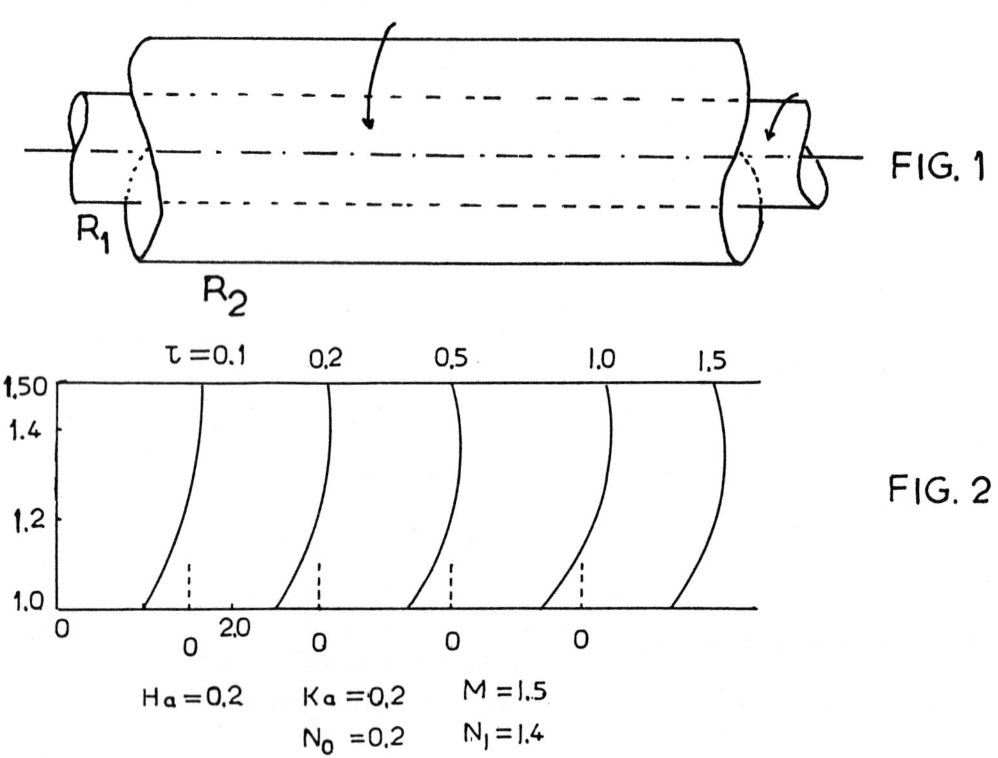

FIG. 1

FIG. 2

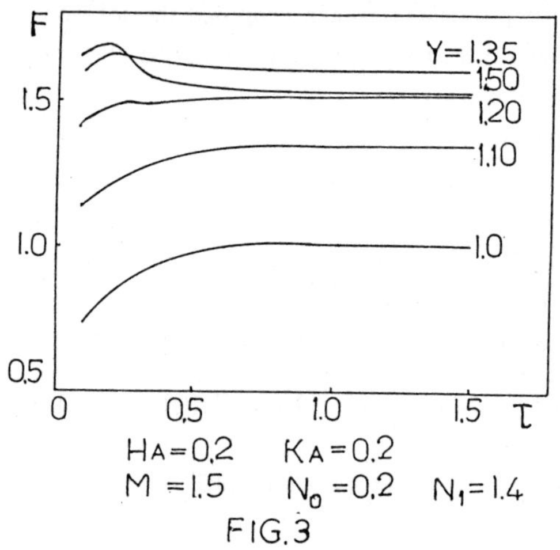

FIG.3

Theoretical and Applied Rheology, edited by P. Moldenaers and R. Keunings
Proc. XIth Int. Congr. on Rheology, Brussels, Belgium, August 17-21, 1992

Flow of viscoplastic fluids in eccentric annular geometries

Peter Szabo Ole Hassager

Department of Chemical Engineering
The Technical University of Denmark
DK-2800 Lyngby

1 Introduction

In some practical flow situations such as transportation of drilling-fluids in deviation drillings it is often a good approximation to regard the fluid as a viscoplastic liquid. This assumption involves some difficulties in the prediction of flow fields. The computational problem is a consequence of the existence of a yield stress in the fluid. This yield stress will cause the liquid to stiffen if it is not loaded with a stress in excess of the yield stress.

In simple geometries it is usually possible to determine the location of plug zones by simple arguments. For example for axial flow in a pipe the shear stress varies linearly from zero at the centerline to the maximum value at the wall. Hence it is simple to determine the plug zone around the centerline. Another example is unidirectional flow in a wide horizontal slot where the velocity field just varies with the vertical position. In that situation the fluid will be like a solid in the middle of the slot. Besides these simple examples there are some more complicated ones. The flow of a viscoplastic fluid in an eccentric annulus is a relevant example in connection with the transportation of drilling-fluids mentioned above. In this geometry it is not simple to predict the actual shape of the possible plug zones.

It has been argued that plug zones cannot exist in complex geometries [1], but later analysis have revealed plug zones in complex geometries [2, 3, 4, 5]. Also in the analysis in the following sections we show that there will exist plug zones in the general eccentric annulus. Furthermore the shape of the plug zones is in some way surprising.

2 Definition of model and geometry

Viscoplastic fluids behaves like *Generalized Newtonian* fluids when a given stress invariant exceeds the yield stress. Below the yield stress one can regard the fluid as a solid. According to this we write the constitutive equation as

$$\boldsymbol{\tau} = -\eta\dot{\boldsymbol{\gamma}} \quad \text{for } \tau > \tau_o \tag{1}$$

$$\dot{\boldsymbol{\gamma}} = \mathbf{0} \quad \text{for } \tau \leq \tau_o \tag{2}$$

We here analyse stationary axial annular flow with a constant pressure gradient. Therefore the momentum equation reduces to

$$[\boldsymbol{\nabla} \cdot \boldsymbol{\tau}]_z = \frac{\mathrm{d}\mathcal{P}}{\mathrm{d}z} \tag{3}$$

where $\mathcal{P}$ is a modified pressure, and the annulus is oriented in the z-direction.

Based on the constant pressure gradient it is convenient to define a characteristic velocity V_o as

$$V_o = -\frac{\mathrm{d}\mathcal{P}}{\mathrm{d}z}\frac{R_2^2}{\eta_o} \tag{4}$$

where R_2 represents the radius of the outer cylinder in the annular geometry.

After nondimensionalysing equations (1–3) and the *Bingham* viscosity [6] we get

$$\boldsymbol{\nabla}^* \cdot [\eta^* \boldsymbol{\nabla}^* v_z^*] = -1 \tag{5}$$

and

$$\eta^* = \frac{\eta}{\eta_o} = 1 + \frac{\text{Bn}}{\dot{\gamma}^*} \quad (6)$$

where

$$\text{Bn} = \frac{\tau_o R_2}{\eta_o V_o} \quad (7)$$

is the *Bingham*-number.

The computational domain is shown in figure 1, where we have defined the parameters $\kappa(= R_1/R_2)$ and $\delta^*(= \delta/R_2)$ respectively as the inner radius and the center difference in dimensionless variables.

Hence the axial flow in the eccentric annulus will be characterized by the three nondimensional numbers κ, δ^* and Bn.

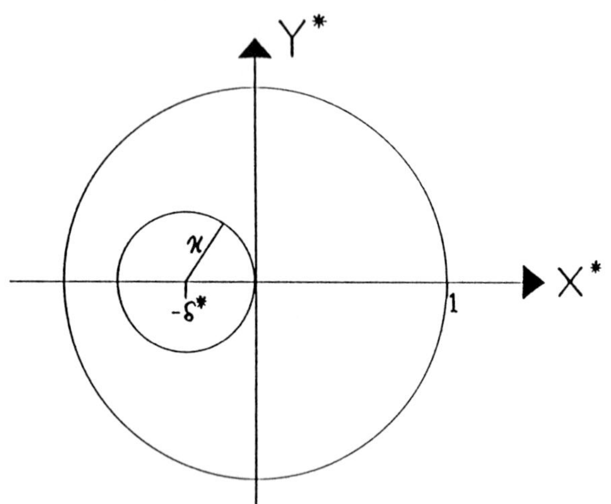

Figure 1

3 Results

We believe to have revealed some surprising new properties of a yield-stress fluid, examplified by the *Bingham* fluid.

First, we have found a new analytical solution describing the flowfield in the annulus for small eccentricities.

Second, we have classified the flow fields for all eccentricities by use of simple forcebalances. The result of this analysis is shown in figure 2 for a particular choice of the κ-parameter (=0.4).

In area (I) there is no flow at all since the *Bingham* number is too high. In area (II) at smaller *Bingham* numbers but high eccentricity there is flow, but only in the wide part of the annulus. Thus in this region there is a plug attached to the walls in the narrow part of the annulus, and a moving plug surrounded by shearing fluid in the wide part of the annulus. In area (III) there are two separate moving plugs, one in the narrow part of the annulus, and one in the wide part. This situation is promoted by a lower eccentricity. Finally in the near-concentric domain (IV) the inner cylinder is surrounded by a single moving plug. In this area the velocity profile is expressed by the analytical solution in mentioned above.

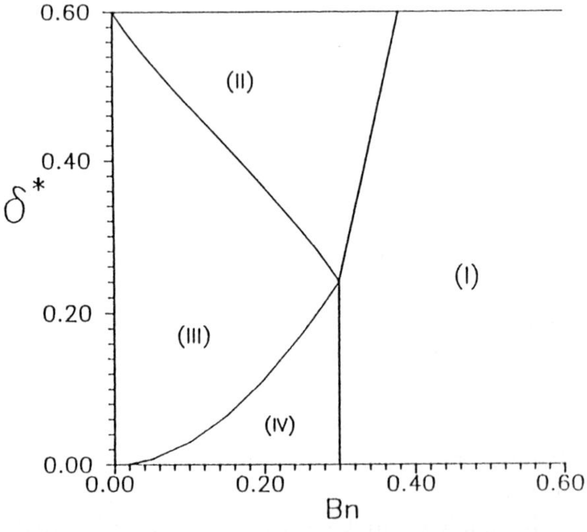

Figure 2

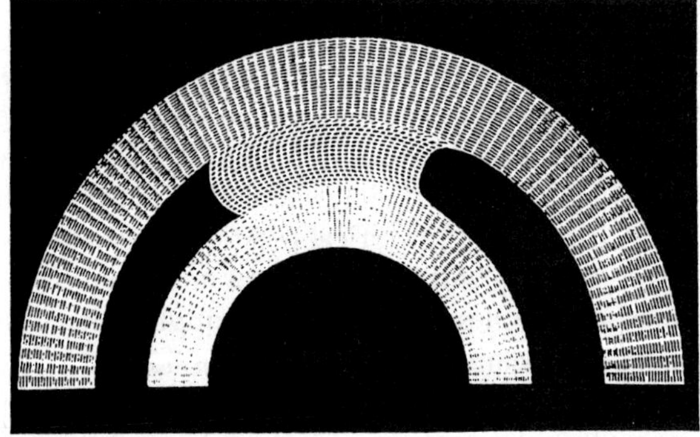

Figure 3

Finally, the plug in the narrow part of the annulus is interesting inasmuch as there will be a condition $\tau_{nz} = +\tau_o$ on part of the surface, and $\tau_{nz} = -\tau_o$ on another part of the surface. The reason for this is that fluid will be pulling the plug near the opening to the wider part of the annulus ($\tau_{nz} = +\tau_o$) while the plug will be pulling the fluid near the annular walls ($\tau_{nz} = -\tau_o$). The transition from the condition $\tau_{nz} = +\tau_o$ to $\tau_{nz} = -\tau_o$ must necessarily occur at a point, i.e. with a singularity in the stress gradient. We see from figure 3 that this transition takes place at a sharp corner. We believe the demonstration of such transitions in yield stress fluids to be novel.

References

[1] Lipscomb, G. G; Denn, M. M: *Flow of Bingham fluids in complex geometries.* J. Non-Newtonian Fluid Mech. **14**, 337–346, 1984.

[2] Beris, A. N; Tsamopoulos, J. A; Armstrong, R. C; Brown, R. A: *Creeping motion of a sphere through a Bingham plastic.* J. Fluid Mech. **158**, 219–244, 1985.

[3] Gartling, D. K; Phan-Thien, N: *A numerical simulation of a plastic fluid in a parallel-plate plastometer.* J. Non-Newtonian Fluid Mech. **14**, 347–360, 1984.

[4] O'Donovan, E. J; Tanner, R. I: *Numerical study of the Bingham squeeze film problem.* J. Non-Newtonian Fluid Mech. **15**, 75–83, 1984.

[5] Atkinson, C; El-Ali, K: *Some boundary value problems for the Bingham model.* J. Non-Newtonian Fluid Mech. **41**, 339–363, 1992.

[6] Bird, R. B; Armstrong, R. C; Hassager, O: *Dynamics of Polymeric Liquids.* 2 edition. John Wiley & Sons. New York. 1987.

[7] Crochet, M. J; Davies, A. R; Walters, K: *Numerical Simulation of Non-Newtonian Flow.* Rheology Series. Vol.1. Elsevier. Amsterdam. 1984.

[8] Walton, I. C; Bittleston, S. H: *The axial flow of a Bingham plastic in a narrow eccentric annulus.* J. Fluid Mech. **222**, 39–60, 1991.

is defined by the next equation;

$$v_{da} = v_{dr}(z) - V_j(z) \quad (1)$$

where $V_j(z)$ and $v_{dr}(z)$ denote the fall velocity of the jet and the real transmission velocity of deformation at a position on the jet, z, respectively. The apparent transmission velocity is given by $v_{da} = f \cdot \lambda$. Hence, its value in this condition is calculated from the empirical formula shown in Fig.6, that is, v_{da}=1.6 m/s. This value is affected by the nozzle diameter and the fluid property.

In generally, the transmission velocity of wave on the string is obtained by the next equation;

$$v = \sqrt{\frac{T}{\sigma}} \quad (2)$$

where T and σ denote tension of the string and linear density, respectively. In our experiment, T and σ are function of the z-coordinate. From eqn.(1) and (2), the elongational stress distribution along the jet length is given by

$$P_{zz}(z) = \rho v_{dr}^2 = \rho\left(V_j(z) + v_{da}\right)^2. \quad (3)$$

Here, ρ denotes density. Hence, the elongational stress distribution is obtained when the velocity distribution or the jet diameter along z-coordinate is measured. As the result, the elongational viscosity is estimated by the elongational stress and the local elongation rate calculated from the velocity distribution on the jet. One of the results which were obtained in the way described above is shown in Fig.7.

4. CONCLUSIONS

We studied the behavior of a viscoelastic fluid jet subjected to the periodic deformation. As a result, the following conclusions were obtained.

1) When a viscoelastic fluid jet is subjected to the periodic deformation, a standing wave appears at resonance frequency.

2) The wavelength of the standing wave is affected by the infliction frequency, the nozzle diameter and the fluid property, not by the infliction position and the flow rate. The apparent transmission velocity is evaluated from the relationship between the wave length and the infliction frequency.

3) A new method to estimate the elongation stress distribution along the jet length and the elongation viscosity was presented.

REFERENCES

1. Y.Tomita and T.Takahashi, JSME International Journal, SereisII, 35-1 (1992) 10-15.

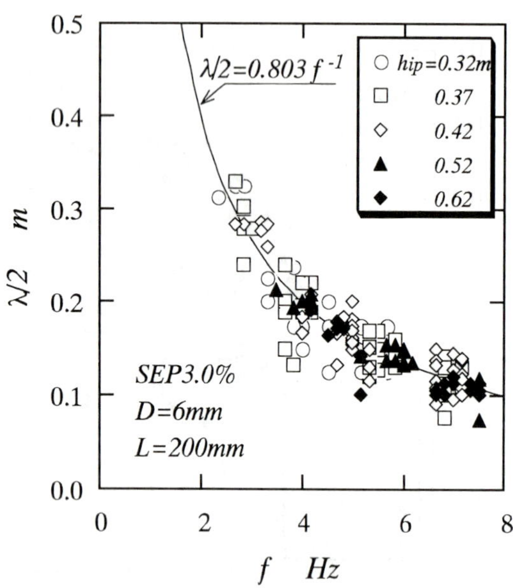

Fig.6 Relationship between the infliction frequency and the wavelength of the standing wave

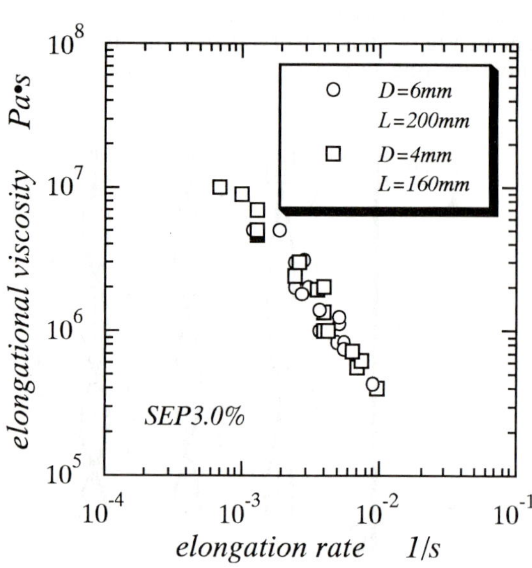

Fig.7 Relationship between the elongation rate and the elongational viscosity

Theoretical and Applied Rheology, edited by P. Moldenaers and R. Keunings
Proc. XIth Int. Congr. on Rheology, Brussels, Belgium, August 17-21, 1992

DRAG REDUCTION BY THE POLYMER INJECTION INTO A PIPE FLOW

H. USUI and Y.SAKUMA

Department of Chemical Engineering Yamaguchi University Tokiwadai, Ube 755, (Japan)

1. INTRODUCTION

Recent investigations have revealed that a large amount of drag reduction was obtained by injecting a highly concentrated polymer solution into a turbulent pipe flow. The injected polymer solution exists as a polymer thread, and the flow condition becomes a heterogeneous drag-reducing system. Although many injection methods have been considered, recent papers have mainly dealt with the centerline injection scheme. If the viscoelasticity of the injected polymer solution is sufficiently high, the single polymer thread is observed even at a very long downstream distance from the injection point. Drag reduction due to the centerline injection of polymer was first reported by Vleggaar and Tels[1]. From visual observation of the injected polymer thread, they concluded that the majority of the polymer thread existed only in the central portion of the pipe flow, and therefore, the thread must modify the turbulence in the central core of the pipe to produce the drag reduction. This hypothesis has been confirmed by Bewersdorff[2], Berman and Sinha[3], and Usui[4].

On the other hand, McComb and Rabie[5] measured the concentration profiles of the salt which was dissolved from the polymer thread. They concluded that the polymer molecules must be presented in the near-wall region to produce the drag reduction. Recently, Smith and Tiederman[6] have confirmed McComb and Rabie's observations. They concluded that low concentration of polymer in the near-wall region are sufficient to cause the drag reduction obtained in the polymer thread injection experiments.

Another unsolved problem in the polymer thread drag-reducing system is the defect of Reynolds stress. Both Bewersdorff[2] and Willmarth et al.[7] showed that the Reynolds stress defect was observed in the polymer thread drag-reducing system. The origin of the Reynolds stress defect seems to be a key point to answer the dispute mentioned above.

2. EXPERIMENTAL APPARATUS AND PROCEDURE

A flow visualization technique was employed to obtain the quantitative data for the movement of injected polymer thread. The flow loop shown in Fig. 1 is almost the same as described in the previous paper[8]. The test section was an 8 m long × 51.3 mm i.d. vertical acrylic pipe which formed a part of the once through flow system. Polyethylene oxide (grade Alcox E-160 supplied by Meisei Chemical Corp.) was used as the polymer additive. The aqueous solution of the polymer additive containing 8,000 ppm Alcox E-160 was pressurized to be injected into the water flow through the injection nozzle. The centerline injection mode was accomplished by a single nozzle located at the center of the test tube. The injection flow rate was adjusted so that the bulk averaged polymer concentration, C_{av}, becomes 30 ppm. Pressure drop at the flow

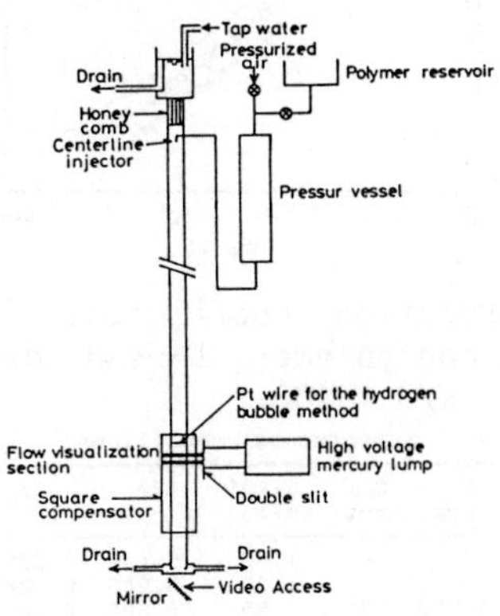

Fig. 1 Experimental apparatus

visualization section was measured by means of U tube manometer with CCl_4.

The light from a high voltage mercury lump passed through the single or double slit to illuminate the coloured polymer thread or hydrogen bubbles at x/D = 100. The fluorescent dye was added to the injected polymer solution. Also the circular platinum wire (100 μm dia.) was set at y = 1 mm to produce the hydrogen bubbles. The cross sectional view illuminated by the single slit was observed through the video access window at the bottom end of the vertical test pipe. The video image was digitized by a transparent digitizer placed in front of the video monitor, and the digitized data were processed by means of a micro computer.

3. EXPERIMENTAL RESULTS AND DISCUSSION

3.1 Drag reduction caused by polymer injection

In Fig.2. the experimental results of the friction factor measurements are shown. The Reynolds number, Re, was based on the viscosity of water. The experimental conditions for the flow visualization run are shown in Table 1. DR is the drag reduction rate defined by,

$$DR = (f_N - f_P)/f_N \times 100\ (\%) \qquad (1)$$

where f_N and f_P are the friction factors in a Newtonian fluid flow and a polymer induced drag-reducing flow, respectively. Present experimental results are in good agreement with those obtained by Frings[9]. Both experimental data show the effective

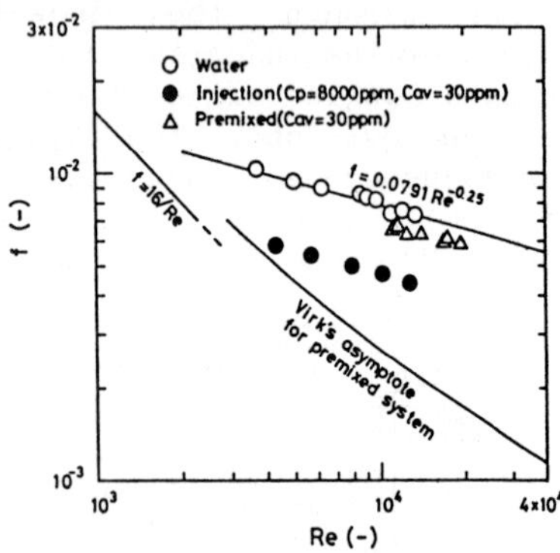

Fig. 2 Friction coefficient for water, premixed and polymer thread drag-reducing flows at x/D = 100.

Table I Experimental conditions

	C_P (ppm)	C_{av} (ppm)	$\nu \times 10^6$ (m^2/s)	Re (–)	f (–)	$u_\tau \times 10^2$ (m/s)	DR (%)
water	–	–	1.35	5,420	0.00922	0.99	–
water	–	–	1.35	9,040	0.00811	1.55	–
injection	8,000	30	1.35	5,390	0.00545	0.76	41
injection	8,000	30	1.35	9,010	0.00479	1.19	41

drag reduction by the polymer injection. while the premixed drag reduction is less effective at the same averaged polymer concentration. These results confirm the result obtained Bewersdorff[2]. On the other hand, recent centally-injected polymer drag reduction experiments by Hoyt and Sellin[10] show that the drag reduction data obtained for very dilute (several ppm) premixed polymer solutions are almost the same as the drag reduction observed for the polymer injection experiments with the same averaged polymer concentration. Their results partially confirm the results obtained by Smith and Tiederman[6]. The capability of drag reduction under the very dilute condition is still open to question.

3.2 Low speed streak spacing

The cross sectional photograph illuminated by the single slit is shown in Fig.3. This photograph contains the single polymer thread and the hydrogen bubbles near the wall. As the platinum wire to produce the hydrogen bubble was located at $y^+ \simeq 10$, the peaks of the hydrogen bubbles (indicated by the arrows A and B) are thought to indicate the lift up of the low speed streaks. It was found that the clear lift up phenomenon was observed when the polymer thread approached to near wall region. Thus, the low speed streak spacing was conditionally determined when the polymer thread came to the near wall region across the region ; r/R= 0.5 ∼ 0.6. The averaged values of the experimentally determined low speed streak spacing, λ^+, are shown in Fig.4. The standard deviation of the individual low speed streak spacing data was 0.006. It is evident that the enlargment of the low speed streak spacing occurs in polymer thread drag-reducing flow, and the λ^+ values obtained in this work are similar to the data measured by Donohue, Tiederman and Reischman[11] for premixed drag reducing system.

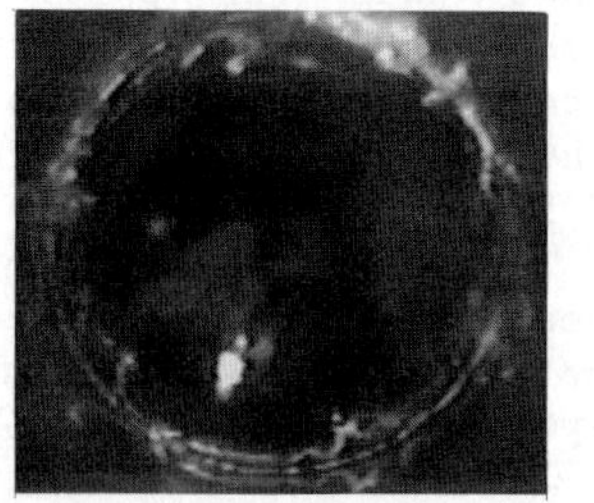

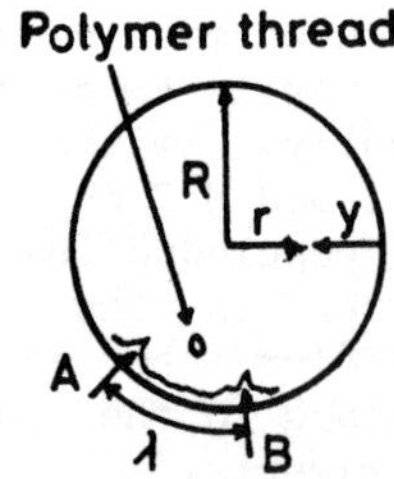

Fig. 3. The cross sectional photograph showing the single polymer thread and the hydrogen bubbles illuminated by the single slit light.

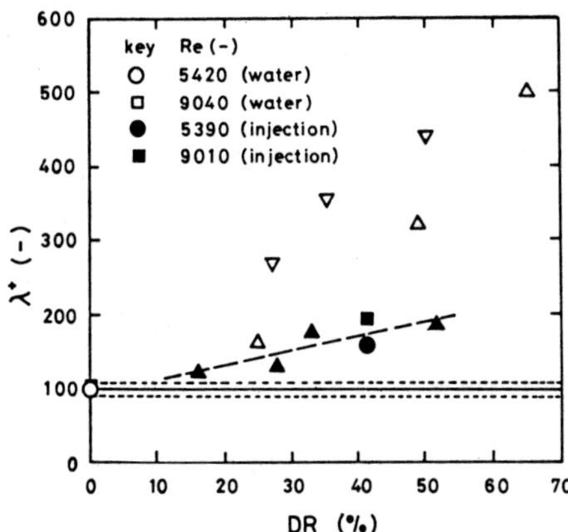

Fig. 4. Averaged values of the low speed streak spacing, λ^+, vs the drag reduction rate. ▲:Donohue et al.(1972), △:Eckelman et al.(1972), ▽:Fortuna and Hanratty(1972)

3.3 Velocity measurements of the injected polymer

The video image of the single tracer particle contained in the polymer thread was used to determine the velocity of the polymer thread. The polystylene sphere (350 ~ 550 μm dia.) mixed in the injecting polymer solution with very low concentration was illuminated by the coloured double slit light. The tracer particle appeared successively twice on the video image, and it was easy to obtain the position data of the both tracer images and the time interval of their appearance. Then we can measure the instantaneous velocity vector of the tracer particle which is contained in the polymer thread. Both the time averaged velocity distribution and the fluctuation of injected polymer thread were determined experimentally. However these results are not shown here because of the space limitation of this paper. The experimental observations are summarized as follows.

1. A significant suppression of turbulence level of the single polymer thread was observed. In particular, the radial fluctuation of polymer threads was more anomalously suppressed, and the suppression of the radial eddy motion appeared to be essential in the drag-reducing flow.
2. The circumferential component of the polymer thread fluctuation did not show any significant suppression. We can conclude that the movement of the polymer thread is not restricted in the circumferential direction, but the radial and the longitudinal movement is strongly damped because of the wall effect and of the viscoelasticity of polymer string.

4. CONCLUDING REMARKS

The flow visualization experiments developed in this study may give the following concluding remarks on the interaction between the polymer thread and turbulent eddies.

The polymer threads were relatively stable in the pipe flow, and the highly viscoelastic polymer thread flowing mainly at the center portion of the pipe flow has non-turbulent but a wave like motion. A significant suppression of the longitudinal and radial fluctuation of the polymer thread was observed. However, the circumferential movement of the polymer thread seemed to be rather free. The enlargment of the low speed streak spacing occurred in polymer thread drag-reducing flow, and the λ^+ values obtained in this work were similar to the previous data measured for the premixed drag reducing system.

REFERENCES

1. J. Vleggaar and M. Tels, Chem. Eng. Sci., 28 (1973)965-968
2. H. W. Bewersdorff, Rheol. Acta., 21, (1982)587-589 or Drag Reduction-84, ed. by R.H.L.Sellin and R.T.Moses, Univ. of Bristol,(1984)B.4-1
3. N. S. Berman and P. K. Sinha, Drag Reduction-84, ed. by R.H.J.Sellin and R.T.Moses, Univ. of Bristol (1984) B.3-1
4. H. Usui, Structure of Turbulence and Drag Reduction, ed. by A. Gyr, Springer-Verlag (1990)257-274
5. W. D. McComb and L. H. Rabie, AIChE. J., 28,(1982)547-565
6. R. E. Smith and W. G. Tiederman, Rheol. Acta, 30 (1991)103-113
7. W. W. Willmarth, T. Wei and C. O. Lee, Phys. Fluids, 30,,(1987)933-935
8. H. Usui, K. Maeguchi and Y. Sano, Phys. Fluids, 31,(1988)2518-2523
9. B. Frings, Rheol. Acta,27(1988)92-110
10. J. W. Hoyt and R. H. J. Sellin, Rheol. Acta, 27(1988)518-522
11. G. L. Donohue, W. G. Tiederman and M. M. Reischman, J. Fluid Mech.,56 (1972) 559-575
12. L. D. Eckelman, G. Fortuna and T. J Hanrraty, Nature, 236 (1972)94
13. G. Fortuna and T. J Hanrraty, J. Fluid Mech., 53(1972)575

Theoretical and Applied Rheology, edited by P. Moldenaers and R. Keunings
Proc. XIth Int. Congr. on Rheology, Brussels, Belgium, August 17-21, 1992

FLOWS WITH DOMINATING EXTENSION AS APPLIED TO VISCOELASTIC FLUIDS

Stefan ZAHORSKI

Institute of Fundamental Technological Research, Polish Academy of Sciences
Swietokrzyska 21, 00-049 Warszawa, Poland

1. INTRODUCTION

In many practical situations flows of viscoelastic fluids are neither extensional nor shearing. In this paper we present an attemp to define a class of flows in which extensional effects are dominating but shearing effects are also taken into account.

2. GOVERNING EQUATIONS

Consider the velocity gradient:

$$[\nabla v^*] = \begin{bmatrix} -\frac{1}{2} & 0 & 0 \\ 0 & -\frac{1}{2} & 0 \\ 0 & 0 & 1 \end{bmatrix} q =$$

$$= \begin{bmatrix} \frac{\partial u}{\partial r} & 0 & \frac{\partial u}{\partial z} \\ 0 & \frac{u}{r} & 0 \\ \frac{\partial w}{\partial r} & 0 & \frac{\partial w}{\partial z} \end{bmatrix} \quad (1)$$

where the first term describes a purely extensional flow, while the second one corresponds to an additional velocity field. If we assume, moreover, that the ratio of the dimension h in the r-direction to the dimension l in the z- direction is a small quality $\epsilon = h/l \ll 1$, we can use simplifications applied in thin-layer flows (lubrication approximation). Introducing dimensionless quantities for coordinates and additional velocities (e.g. $U = qh$ denotes the characteristic velocity for additional flows), we can prove that, for flows with relatively small vorticity components, the first matrix components may be notably more meaningful than the second matrix components. In our previous papers (ref.1,2) we defined the flows with dominating extensions (FDEs) as such thin-layer flows for which the constitutive equations describing purely extensional flows of an incompressible simple fluid can be used in a form linearly perturbed with respect to the additional velocity gradients, viz.

$$\begin{aligned} \mathrm{T}^* &= -p\mathrm{I} + \beta_1 \mathrm{A}_1 + \beta_1 \mathrm{A}_1' + \beta_2 \left(\mathrm{A}_1^2\right)' \\ &\quad + \frac{\partial \beta_1}{\partial q} q' \mathrm{A}_1 + \frac{\partial \beta_2}{\partial q} q' \mathrm{A}_1^2 + \cdots, \\ tr\mathrm{A}_1 &= 0, \end{aligned} \quad (2)$$

where T^* is the stress tensor, p - the hydrostatic pressure, A_1 - the first Rivlin-Ericksen kinematic tensor, and $\beta_i (i = 1, 2)$ denote the material functions depending on the invariants of A_1 . Primes denote the corresponding increments related to the additional gradients.

Bearing in mind the inertialess equation of equilibrium and retaining only terms of the highest order of magnitude with respect to ϵ, we arrive at the governing equations of the type:

$$\begin{aligned} \frac{\partial p^*}{\partial r} &= 0, \\ \frac{\partial p^*}{\partial z} &= \beta \frac{1}{r} \frac{\partial}{\partial r} \left(r \frac{\partial w}{\partial r} \right) + \\ &\quad + \frac{1}{2} \frac{\partial}{\partial z} \left[\left(\frac{\partial \beta_1}{\partial q} + \frac{\partial \beta_2}{\partial q} q \right) \left(\frac{\partial w}{\partial r} \right)^2 \right] = C(z) \end{aligned} \quad (3)$$

where $p^* = p - \mathrm{T}^{*11}_E$ and $\beta = \beta_1 + \beta_2 q$ can be called the modified pressure and the extensional viscosity, respectively.

The above nonlinear differential equations with appropriate boundary conditions can be solved either numerically or analytically for slightly non-Newtonian fluids, if the quantity

$$k = \frac{1}{\beta} \left(\frac{\partial \beta_1}{\partial q} + \frac{\partial \beta_2}{\partial q} q \right) q \quad (4)$$

is small enough and a perturbation method is applied.

For fibre spinning processes under the so-called quasi-elongational approximation, such a procedure leads to the additional velocity profile:

$$\begin{aligned} w &= -\frac{3}{4\beta}\frac{v'^2}{v}\left(r^2 - \frac{h^2}{2}\right) \\ &\quad -\frac{9}{32}\frac{v'}{v}\frac{h^2}{\beta}\,\Omega\left(\frac{v'^2}{v}\right)^2\left(r^2 - \frac{h^2}{2}\right) \\ &\quad -\frac{9}{128\beta}\frac{\partial}{\partial z}\,\Psi\,, \end{aligned} \tag{5}$$

where

$$\Omega = \frac{\partial \beta_1}{\partial q} + \frac{\partial \beta_2}{\partial q}q,$$

$$\Psi = \left[\Omega\left(\frac{v'^2}{v}\right)^2\left(r^4 - \frac{h^4}{3}\right) - 2h^2\left(r^2 - \frac{h^2}{2}\right)\right]$$

and where v denotes the velocity along the spin-line, v' - its derivative with respect to z, and h is the variable radius of the filament.

3. CERTAIN RESULTS

The concept of FDEs was used in many cases of interest showing a good qualitative agreement with available experimental data. These cases are: squeezing flows, flows between rotating cylinders, flows past blunt profiles and bodies, fibre spinning processes, opposing jets flows, flows in converging slits and dies, etc. All the obtained results essentially depend on the extensional viscosity function as well as on its variability with respect to the extension rate.

In particular, for plane and axisymmetric continuous squeezing flows definitive load-enhancement effects are observed, if the extensional viscosity is an increasing function of the extension rate. Similarly, in constant-load squeezing flows the velocity of approaching plates is always less for viscoelastic fluids as compared with Newtonian ones (ref.1).

In flows between two rotating cylinders (rolling or calendering) viscoelastic properties of a fluid with increasing extensional viscosity reduce significantly the distances from the nip to the exit and to that position where the thrust distribution reaches its maximum. In the case of frequently observed S-shaped variability of the extensional viscosity function, the maximum thrust and the total loading force increase with increasing velocity of the rotating cylinders. The total friction force also increases monotonically, on the contrary to some experimental evidence. The flow rate and the friction coefficient usually decrease for higher velocities; the latter quantity may show a minimum value (cf.ref.3).

It is also shown that the thickness of the elastic type boundary layer accompanying viscoelastic flows past blunt profiles or bodies may be much greater than in the case of classical Prandtl-type boundary layers. The corresponding velocity profiles depend on the ratio of the extensional viscosity to the Weissenberg number.

For the above mentioned non-isothermal fibre spinning processes, it can also be proved that the velocity profiles across the filament are always convex and any "skin effects" do not appear at all (ref.2). Other interesting and important results can be presented for plane and axisymmetric converging flows, opposing jets flows, etc.(cf.ref.3). For increasing extensional viscosity the velocity profiles in the converging regions of slits and pipes are remarkably "flattened" as compared with those for purely viscous fluids. It is also shown that for decreasing extension rates the planar extensional viscosity takes values less than those resulting from measurements in the lubricating-die Converging Flow Rheometer. Similarly, for increasing extension rates higher values of the planar extensional viscosity are obtained. Such a behaviour of the viscosity function is in full qualitative agreement with the results determined from the Spin-Line Rheometer flow (cf.ref.5).

REFERENCES

1. S.Zahorski, Arch.Mech., 38 (1986) 191-207.
2. S.Zahorski, J.Non-Newtonian Fluid Mech., 36 (1990) 71-83.
3. P.Doremus and J.M.Piau, J.Non- Newtonian Fluid Mech., 13 (1983) 79-91.
4. S.Zahorski, J.Non-Newtonian Fluid Mech., in print.
5. D.M.Jones, K.Walters and P.R. Williams, Rheol.Acta, 26 (1987) 20-30.

Theoretical and Applied Rheology, edited by P. Moldenaers and R. Keunings
Proc. XIth Int. Congr. on Rheology, Brussels, Belgium, August 17-21, 1992

THE THERMOCONVECTIVE INSTABILITY IN HYDRODYNAMICS OF RELAXATIONAL LIQUIDS

I.SH. AKHATOV, R.G. CHEMBARISOVA

Department of Continuous Media Mechanics, Bashkir State University,
32 Frunze Str., Ufa 450074 (Russia)

The main purpose of this report is to show the influence of rheological characteristics on the beginning and development of thermal convection in non-Newtonian liquids which possesses relaxational properties. There are three different problems considered in this report.

1. Thermoconvection in horizontal plane layer of viscoelastic liquid. For the description of rheological behaviour of liquid the Oldroyd's linearized model (ref.1) is used. It shows that in this case the convective motion is described by Lorenz system (ref.2) with modified Prandtl number, which depends on relaxation times of liquid.

2. Heat convection of relaxation liquids in porous media. When studying the motion of liquids in porous media it is usually supposed that the equilibrium between the pressure gradient and filtration velocity is settled instantly (Darcy's law). But for many rheologically anomalous liquids this equilibrium may be settled after some delay. The non-equilibrium effects of these systems lead to the necessity of using modified Darcy's law (ref.3) as

$$(T_1 d/dt+1)\mathbf{v}= -k/m(T_2 d/dt+1)\,\mathbf{grad}\,p$$

where T_1,T_2 - relaxation times, k - permeability of porous medium, m - viscosity of liquid, v , p - filtration velocity and pressure. It shows that the convective motion in horizontal plane porous layer saturated by relaxational liquid is described by new nonlinear dynamical system

$$D_1 dX/dt= -D_2XZ-(1-rD_2)X+(1-D_2)Y$$

$$dY/dt= -XZ+rX-Y$$

$$dZ/dt=XY-bZ$$

where parameters r and b are calculated samely as for Lorenz system (ref.2); D_1 , D_2 - dedimensionalized relaxation times. The analytical and numerical research of this new system chaotic solutions is realized. In particular is noted that stability disturbance for Rayleigh number less than classical Darcy-modified Rayleigh critical value is possible in certain cases.

3. Heat convection of relaxational liquids in toroidal shaped porous medium. It shows that the solution of thermoconvective problem in this case is approximated by above-stated new nonlinear dynamical system exactly, so that the higher garmonics of Galerkin's expansion decreases exponentially in time and their influence on lowest garmonics is equal to zero.

REFERENCES

1. J.G. Oldroyd, Second order effects in elasticity, plasticity and fluid dynamics. Proc. Intern. Symp. 1962, Pergamon Press, 1964, p.p.520-529.
2. E.N. Lorenz, Journal of the Atmospheric Sciences, 20 (1963) 130-141.
3. M.G. Alishaev and A.Kh. Mirzajanzade, Izvestiya vuzov. Neft i gaz, 6 (1975) 71-74.

Theoretical and Applied Rheology, edited by P. Moldenaers and R. Keunings
Proc. XIth Int. Congr. on Rheology, Brussels, Belgium, August 17-21, 1992

THE SHEAR INSTABILITY IN TWO-LAYER VISCO-ELASTIC LIQUIDS

I.SH. AKHATOV, I.U. SUBAEV

Department of Continuous Media Mechanics, Bashkir State University,
32 Frunze Str., Ufa 450074 (Russia)

It is known that the dissolution of small quantities of polymers lead to the lowering of hydraulic resistance for turbulent rate of liquid's motion (Toms effect). However the using of polymer make-weights can be accompanied by additional hydrodynamic peculiarities, for example, it can be for two layer condition of mixture motion in pipe. Our experimental results led us to the inference that the small polymer make-weights are intensify the interior waves in two-layer liquid's motion in pipe and therefore the hydraulic resistance is increase.

The stability problem for interior waves in two-layer liquids with tangential shock of velocity was firstly formulated by Helmholtz and Kelvin. At that time this problem is studied in linear approximation. For short-wave disturbances the strong instability takes place (Kelvin-Helmholtz instability), but for long-wave disturbances the wave motion is stabilized by stratification action (neutral stability). However in last case the disturbance becomes to be unstable because of viscous dissipation. In hydrodynamics such waves ("negative energy waves") were firstly investigated by Benjamin (ref.1).

In this report the influence of visco-elastic qualities on waves stability in two-layer non-Newtonian liquid is discussed. For the description of relaxational qualities of liquids the following form of rheology law is used

$$\hat{R}T_{ij}=\hat{Q}E_{ij}$$

Here T_{ij} , E_{ij} - components of a stress deviator tensor and of a rate of strain tensor; $\hat{R}$, $\hat{Q}$ - differential operators based on Oldroyd derivatives (ref.2), which reduce by linear approximation to partial derivatives on time.

This problem was previously analyzed under the assumption of depth infinity (ref.3). Now we are take into account the arbitrary depth of liquids layers. The dispersion relation of appropriate problem is derived in this report. The dependence of growth index of unstable perturbations on liquid's rheological parameters is analyzed. The analysis allowed us to come to the conclusion that in some cases visco-elastic properties can lead to intensification of interior waves in two-layer liquids. These theoretical results are qualitatively coordinated with our experimental data.

Further long nonlinear waves in two-layer visco-elastic liquids of finite depth is discussed. The modified Korteweg-de Vrize equation with the additional term which includes the fractional derivatives with respect to time is suggested. The instability of modulated waves and solitones of envelopes in two-layer visco-elastic liquids is analyzed by Whitham's method of averaged lagrangians.

REFERENCES

1. T.B. Benjamin, J.Fluid Mech., 16 (1963) 436-450.
2. J.G. Oldroyd, Second order effects in elasticity, plasticity and fluid dynsmics. Proc. Intern. Symp. 1962, Pergamon Press, 1964, p.p.520-529.
3. I.Sh. Akhatov, Prikladnaya matematika i mehanika, 53 (1989) 630-635.

Theoretical and Applied Rheology, edited by P. Moldenaers and R. Keunings
Proc. XIth Int. Congr. on Rheology, Brussels, Belgium, August 17-21, 1992

NUMERICAL AND EXPERIMENTAL STUDY OF VISCOELASTIC FLOW IN CAVITY

S.I. BAKHTIYAROV

Department of Theoretical Mechanics, Azerbaijan Industrial University, P.O. Box 370601, Baku (Azerbaijan Republic)

The results of computations and experiments on the flow of a viscoelastic liquid in cavities are reported.

1. COMPUTATIONAL STUDY

Finite-difference solutions of the Navier-Stokes equations are given in stream function-vorticity formulation for a wide range of Reynolds numbers from 0 to 500 for varied configurations of the cavity in the channel at We=0. The solutions show that only at high Reynolds numbers does the flow in a square cavity completely corresponds to that assumed by Batchelor's constant vorticity model for separated flows. At Reynolds numbers from 100 to 250 does the flow in the cavities corresponds to that assumed by Lavrentyev's model for separated flows.

The second order Oldroyd's model was used for the viscoelastic fluid. The solutions show that at all Reynolds numbers the flows of viscous and viscoelastic fluids (We= 0.1,0.2,0.5) in the cavities are different.

The dynamic interaction between the main flow and the captive eddy is analysed for both fluids.

2. EXPERIMENTAL STUDY

The flow of an 0.8 percent PAA solution and of a viscoelastic separator used in petroleum technology was investigated. Flow visualization was accomplished in a channel wuth rectangular cavities whose length-to-depth ratios were 2:1 and 1:1. The strong difference between viscoelastic and Newtonian flows at low Reynolds numbers and the absence of such a difference at large Reynolds is attributable to the marked differences in the Deborah number in these cases.

A short series of tests was run to visualize the eddy zone in a channel downstream of a step.The tests were run with 65% aqueous glycerol and PAA solutions.

Theoretical and Applied Rheology, edited by P. Moldenaers and R. Keunings
Proc. XIth Int. Congr. on Rheology, Brussels, Belgium, August 17-21, 1992

FLOW OF GENERALIZED NEWTONIAN LIQUIDS THROUGH FIXED BEDS OF PARTICLES

J.CAKL and I.MACHAČ

Chemical Engineering Department, Institute of Chemical Technology, 532 10 Pardubice (Czechoslovakia)

1. PRESSURE DROP RELATIONSHIP

Making use of a modified capillary bed model, the relationships were suggested predicting volume flow rate - pressure drop dependence for creeping flow of a generalized Newtonian liquid through fixed beds of spherical and non-spherical particles (ref.1). For a power law liquid, we have

$$f = \frac{72.\varphi^{n+1}}{Re_n} . \left[1 + \frac{2.d_p}{3\varphi.(1-\varepsilon).D_h}\right], \qquad (1)$$

where

$$f = \Delta p.d_p.\varepsilon^3/[\rho.u^2.L.(1-\varepsilon)] \qquad (2)$$

is the bed friction factor,

$$Re_n = \frac{\rho.u^{2-n}.d_p^n}{K} . \frac{\varepsilon^{2(n-1)}.12^{1-n}}{(1-\varepsilon)^n} . \left(\frac{4n}{3n+1}\right)^n \qquad (3)$$

is the modified Reynolds number,

$$\varphi = \omega.(1 + \alpha.\psi) \qquad (4)$$

is the dynamic characteristic of the bed (ref.2).

The validity of eq.(1) can be extended to transient flow region as well if dependence of the resistance number ψ of spherical particle bed on Reynolds number is considered.

For Newtonian fluid flow (n=1), the dependence $\psi = \psi(Re)$ and the wetted surface correction coefficients ω and the shape factors α were determined (ref.2).

2. EXPERIMENTS AND RESULTS

The proposed method of pressure drop prediction in transient flow region was verified experimentally. In experiments, the pressure drop of water solutions of glycerol, hydroxyethylcellulose Natrosol 250 MR and polysacharide Xanthane CX 12 was measured in the flow through fixed beds of balls, cylinders, and polyhedrons (ref.3).

From experimental data, the values of friction factor were calculated and the dependences $f.Re_n$ vs. Re_n were plotted. The dependences obtained for non-spherical particles are given in the following Fig.

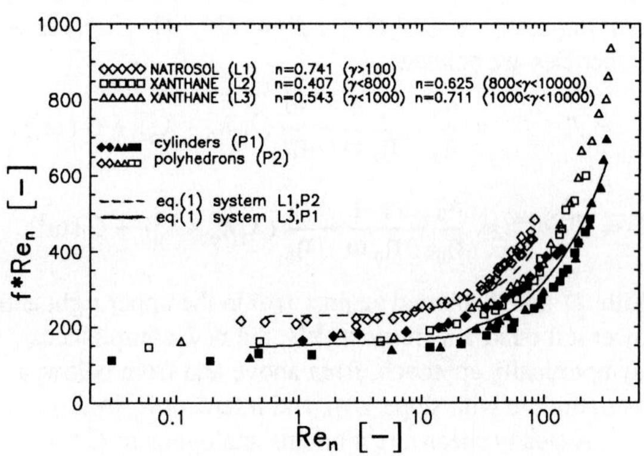

It was confirmed that the bed factor φ can be considered to be independent on rheological properties of the liquid, and the quantities $\psi(Re)$, ω, and α obtained for Newtonian liquid flow can be used for pressure drop prediction in the fixed bed flow of a power law liquid. The mean relative deviation of the experimental pressure drop data and those calculated according to the equations suggested was 12%.

REFERENCES

1. I. Machač and V. Dolejš, Chem.Eng.Sci. 36 (1981) 1679-1686
2. V.Dolejš and I.Machač, Inter.Chem.Eng. 27 (1987) 730-736
3. J.Slabá, Diploma Thesis, Inst.of Chem. Technol. Pardubice, 1988

Theoretical and Applied Rheology, edited by P. Moldenaers and R. Keunings
Proc. XIth Int. Congr. on Rheology, Brussels, Belgium, August 17-21, 1992

COMBINED LOW FREQUENCY & LOW SHEAR RATE ASYMPTOTICS

Bruce Caswell and Charles V. Callahan, Division of Engineering, Brown University, Providence, RI 02912 USA
Support for this work under NSF Grant CTS-8921668 is gratefully acknowledged.

The moment expansion of the complex modulus, $G^*(i\omega)$ at low frequency, ω, is well known to be

$$G' + iG'' = \alpha_2\omega^2 + i(\alpha_1\omega - \alpha_3\omega^3) + O(\omega^4).$$

Here i is $\sqrt{-1}$ and α_1, α_2, α_3 are the first three moments of the relaxation modulus $G(t)$, α_1 is the zero shear rate viscosity η_0, α_2/α_1 is the mean relaxation time λ_0, and $\lambda_1 = \alpha_3/\alpha_2 \geqslant \lambda_0$. These data were not measured with a plan for their analysis with the scheme presented here. Our asymptotic forms for J^+ and J^- suggest that for optimal extrapolation to zero ω and $\dot\gamma$, the data should be taken at equal intervals of $1/\omega$ and $\dot\gamma$. From the real and imaginary compliances $J' = G'/|G^*|^2$, $J'' = G''/|G^*|^2$ two new clemencies are defined:

$$J^+ = J' + J'' = \frac{\lambda_0}{\eta_0} + \frac{1}{\eta_0}\frac{1}{\omega} + \frac{\omega}{\eta_0}(\lambda_1\lambda_0 - \lambda_0^2) + O(\omega^2),$$

$$J^- = J' - J'' = \frac{\lambda_0}{\eta_0} - \frac{1}{\eta_0}\frac{1}{\omega} - \frac{\omega}{\eta_0}(\lambda_1\lambda_0 - \lambda_0^2) + O(\omega^2).$$

With J^+ and J^- plotted against $1/\omega$ in the upper right and lower left quadrants respectively, the new compliances asymptotically approach, from above and from below, a common line with slope $1/\eta_0$ and intercept λ_0/η_0.

A steady shear-rate modulus analogous to G^* is obtained in terms of the shear rate, $\dot\gamma$, by replacement of G' with half the normal stress difference $N_1(\dot\gamma)$, and G'' with the shear stress $\tau(\dot\gamma)$. The compliances corresponding to J^+, J^- asymptotically approach the same line as the frequency functions. Hence frequency data and steady shear-rate data can be combined in a single plot which simultaneously yields the constants η_0, λ_0 and clearly demonstrates the Coleman-Markovitz relation. This is illustrated in Figures 1. and 2. in which plotted are J^+ and J^- derived from the Rouse Theory and values calculated from the data of references 1 and 2.

Figure 1: 19% PIB-Cetane

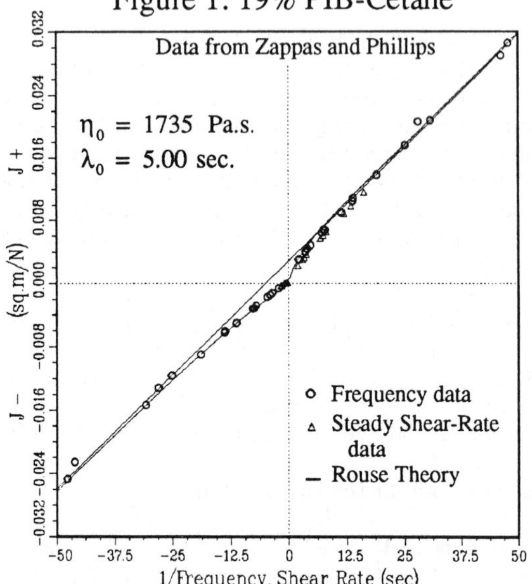

Figure 2: 15% PS in DiBPhth

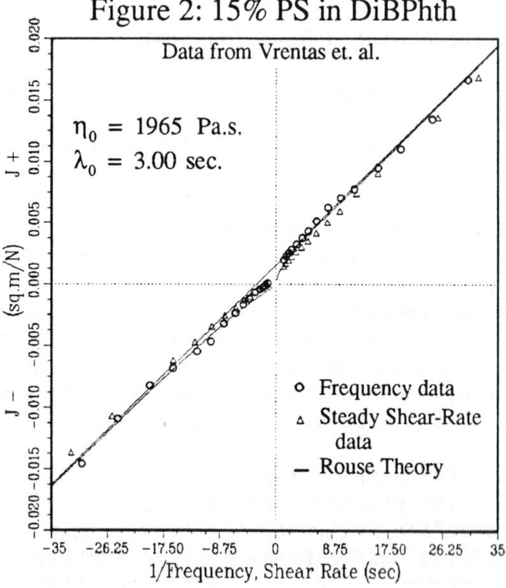

REFERENCES

1. Zappas, L. J. and J. C. Phillips, J. Rheol., 25 (1981) pp. 405-420.
2. Venerus, D. C., C. M. Vrentas, and J. S. Vrentas, J. Rheol., 34 (1990) pp. 405-420.

Theoretical and Applied Rheology, edited by P. Moldenaers and R. Keunings
Proc. XIth Int. Congr. on Rheology, Brussels, Belgium, August 17-21, 1992

FLOW OF A VISCOELASTIC FLUID IN A CYLINDRICAL TUBE OF SLOWLY VARYING CROSS-SECTION

C.F. CHAN MAN FONG

Department of Chemical Engineering, University of Sherbrooke, Sherbrooke, Quebec, J1K 2R1 (Canada)

1. INTRODUCTION

The flow of a fluid in a tube of varying cross-section is of importance in industry and in biomechanics, where the fluid is usually non-Newtonian. Several authors (refs. 1, 2) have considered the flow of a power-law fluid and a 4-constant Oldroyd fluid in tubes of varying cross-section. These fluids are special cases of so-called simple fluids.

It has been shown that in a viscometric flow, the simple fluid reduces to the Criminale-Ericksen-Filbey fluid (ref. 3). Since we assume the radius of the tube to vary slowly along the axis, we may assume the flow to be nearly viscometric. Thus for the flow under consideration, the CEF equation may be adequate.

2. MATHEMATICAL FORMULATION

The constitutive equation of a CEF fluid may be written as:

$$\underline{\tau} = \eta \underline{A}_1 + (\psi_1 + \psi_2) \underline{A}_1^2 - \frac{1}{2} \psi_1 A_2 \quad (1)$$

where $\underline{\tau}$ is the deviatoric stress tensor, η, ψ_1 and ψ_2 are the viscosity function and the primary and secondary normal stress coefficients respectively. These are functions of II, the second invariant of $\underline{A}_1$. $\underline{A}_1$ and $\underline{A}_2$ are the first and second Rivlin-Ericksen tensors respectively.

Assuming the flow to be symmetrical, the physical components of the velocity, in cylindrical polar coordinates (r, θ, z), are:

$$V_{(r)} = -\frac{1}{r}\frac{\partial \Psi}{\partial z}, \quad V_{(\theta)} = 0, \quad V_{(z)} = \frac{1}{r}\frac{\partial \Psi}{\partial r} \quad (2)$$

where $\Psi(r, z)$ is the stream function.
The radius of the tube is assumed to be given by:

$$r = a_o + \varepsilon\, a_1(z) \quad (3)$$

where a_o is a constant and ε is a small parameter. Van Dyke (ref. 4) has shown that it is possible to obtain a convergent series solution in ε by scaling z. Thus we choose a new independent variable $z_1(=\varepsilon z)$ instead of z.

All quantities are now expanded in powers of ε. For example η is expanded as:

$$\eta(II) = \eta(II_o) + \varepsilon\, II_1\, \eta'(II_o) + \ldots \quad (4)$$

where ' denotes differentiation with respect to $II_o(r, z_1)$. The zeroth approximation corresponds to viscometric flow.

The stress tensor $\underline{\tau}$ can be calculated and then substituted into the equation of motion. By comparing powers of ε and assuming some forms for η, ψ_1 and ψ_2 it is then possible to determine the pressure drop in powers of ε for a given volumetric flow rate.

ACKNOWLEDGMENT

The author is grateful to Dr. D. De Kee for his comments and support.

REFERENCES

1. A.B. Jarzebski and W.L. Wilkinson, J. Non-Newtonian Fluid Mech., 8 (1981) 239.
2. S.R. Kasivisvanathan, P.N. Kaloni and K.R. Rajagopal, Int. J. Non-Linear Mech., 26 (1991) 777.
3. W.O. Criminale, J.L. Ericksen and G.L. Filbey, Arch. Rat. Mech. Anal., 1 (1958) 410.
4. M. Van Dyke, Adv. Applied Mech., 25 (1987) 1.

Theoretical and Applied Rheology, edited by P. Moldenaers and R. Keunings
Proc. XIth Int. Congr. on Rheology, Brussels, Belgium, August 17-21, 1992

HELICAL FLOW OF POWER-LAW FLUIDS IN CONCENTRIC ANNULI WITH A ROTATING INNER CYLINDER

J.DAVID, P.FILIP and Z.BERAN

Institute of Hydrodynamics, Czech.Acad.Sci., Podbabská 13, 166 12 Prague 6, Czechoslovakia

Steady laminar helical flow of incompressible non-Newtonian fluids in a concentric annulus is analyzed for the case of a stationary outer cylinder and an inner one rotating under a constant torque. A constant pressure drop acts on the fluid in the axial direction. The solution is derived for both pseudoplastic and dilatant power-law fluids.

Numerical computation - based on relations obtained from the balance equations describing this problem and transformed into a suitably chosen non-dimensional form - provides a complete solution with sufficiently high accuracy. This solution covers a broad region of all relevant parameters including limiting cases from kinematical, rheological, dynamical, and geometrical aspects, viz.

(a) axial Poiseuille flow and Couette-Poiseuille flow resulting for a dimensionless rotation of the inner cylinder approaching to 0 and ∞, respectively;

(b) extremal values of both pseudoplasticity and dilatancy;

(c) axial pressure drop and torsional torque applied over several orders of magnitude (within the laminar region of flow);

(d) narrow and wide clearances of annulus.

For transforming the above stated helical flow to the non-dimensional form it is necessary:

(1) to determine accurately the volumetric flow rate of the corresponding axial Poiseuille flow. In (ref.1) there is derived the analytical explicit relation between flow rate vs pressure drop, $Q \sim (\Delta P)^{1/n}$. The coefficient of the proportionality substantially depends on the precise determination of a radial location of maximum axial velocity. The present procedure simplifies the computation of this coefficient, and in comparison to previous works covers continuously a large region of rheological and geometrical parameters.

(2) to determine a functional dependence of axial flow rate of the corresponding Couette-Poiseuille flow on the frequency of rotation of the inner cylinder ω for the limiting case $\tau_{rz} \ll \tau_{r\phi}$. This dependence is of the form $Q \sim \omega^{1-n}$.

The solution of helical flow reduces for n=1 to the obvious Newtonian result $Q \equiv$Const independently on ω.

For the extreme case of pseudoplasticity n=0 the result for ideally plastic material (ref.2) - non-existence of helical flow - is confirmed.

The approach to a helical flow substantially depends on a choice of given parameters. The presented functional dependences of obtained dimensionless characteristics enable to solve this flow for various possible choices.

In comparison with the hitherto published papers the present analysis of helical flow provides a more accurate solution in significantly broadened ranges of parameters, therefore it ensures better applicability in practice.

REFERENCES

1. R.W.Hanks, K.M.Larsen, Ind.Eng.Chem. Fundam.,18(1979) 33-35.
2. D.Durban, M.R.Sitzer, Acta Mechanica 50(1983) 135-140.

Theoretical and Applied Rheology, edited by P. Moldenaers and R. Keunings
Proc. XIth Int. Congr. on Rheology, Brussels, Belgium, August 17-21, 1992

UNSTEADY POWER FLOW OF THE PLASTIC VISCOELASTIC FLUIDS IN LONG CANALS

O.R.DORNJAK

Department of Applied Mathematics and Mechanics, Voronezh State University, Universitetskaja Sq.1,Voronezh,394693 (Russia)

1. INTRODUCTION

A tendency to form gels often exshibits liophobic colloidal suspension of high concentration, if the suspension is at rest. The particles can form a spatial network and the suspension develops an internal structure with elastic properties. If the suspension is sheared, the network is broken and flows as plastic. The paper presents the results of the numerical research of shear flow transition regimes of plastic viscoelastic fluids in long canals taking into account elastic properties of structure network.

2. A HYDRODYNAMIC MODEL

For the research of the dynamic behavior of the fluids like clayey suspensions we have chosen the rheological equation in the following form:

$$(T-Y\,\mathrm{sign}(dv/dr)/n\ 1(|T|-Y)+$$
$$+(dT/dt)/G\ 1(Y-|T|)=dv/dr \qquad (1)$$

where 1(t) is Heaviside function, T is the shear stress, dv/dr is the shear rate, Y is the yeld stress, n is the plastic viscosity of the medium after the destruction of the network, G is the shear modulus in elastic deformation of the network. Equation (1) can be received from the different tensor correlations, for example,from White model for the plastic viscoelastic fluids (ref.1). Model (1) doesn't take into account the relaxation processes, but allows to make clear the principal peculiarities of the transition regimes in plastic viscoelastic systems.

A change of a velocity profil v(r,t) across the canal in cylindrical coordinates is given by the equations

$$j\,dv/dt=-dp/dz+1/r\,d(rT)/dr \qquad (2)$$
$$dp/dz=(P+a\sin kt)\ 1(t) \qquad (3)$$

where j is the density, P is the average pressure gradient, a is the amplitude of the pulsatile pressure gradient and k is the angular frequency of the pressure fluctuation.The cases of the flat and circular canals are considered.

The boundary conditions at the external surface are v(R1,t)=v(R2,t)=0.

At the boundary between areas with the destructed and undestructed network l(r,t) the value of the velocity and the tangential stress are

$$T(t,l-0)=T(t,l+0)=Y,v(t,l-0)=v(t,l+0) \qquad (4)$$

Location of the boundary l(r,t) is unknown beforehand and has to be found from the solution of the above formulated problem.

3. CONCLUSIONS

Due to wave processes in elastic structure network under the influence of an axial pressure gradient different hydrodynamic types of transition regimes are possible (ref 2).

In case Lagrange number (La=H(P/Y), H is the width of the canal) has large values the structure is not destroyed and oscillaties elastically. In case Lagrange number has relatively small values the transition process leads to a steady flow of destructed network near walls that takes away the elastic shear kernel.

The specific transient regime exhibits itself in case of moderate Lagrange numbers (for the flat canal and a=0, - 1<La<2). In this case the structure is destroyed only at the beginning. Then after repeated passages of the waves across the canal, the structure is restored everywhere and oscillates elastically.

REFERENCES

1. J.L.White, D.S.Huang, J.Non-Newtonian Fluid Mech., 9 (1981) 223 - 233.
2. O.P. Дорняк, З.П.Шульман, Б.М.Хусид, Э.А.Зальц гендлер, ДАН СССР, 4 (1988) c.802 - 805.

Theoretical and Applied Rheology, edited by P. Moldenaers and R. Keunings
Proc. XIth Int. Congr. on Rheology, Brussels, Belgium, August 17-21, 1992

SHEAR FLOWS OF GEOMATERIALS WITH COMPLEX RHEOLOGY : NON-STATIONARITY AND STABILITY

D.V.GEORGIEVSKII

Chair of Composite Mechanics, Faculty of Mechanics & Mathematics,
Moscow State University, Moscow, 119899, Russia

As is generally known, linear and non-linear viscoplastic models are of great importance among many rheological one's that were proposed in geophysical and hydrodynamic problems for the earth's crust behaviour description. The number of independent material constants in the corresponding constitutive laws may be vary within the broad limits. The most general relation connecting the maximum shear stress T and maximum strain rate U in viscoplastic models is the following

$$T = t + f(mU) , \qquad (1)$$

where t - shear yield stress, m - effective viscosity, f(V) - arbitrary hardening function. On the basis of variational principles a plastic flow is put into effect only where $T > t$, so called rigid zones occupy the rest of the field. The problem of rigid zones distribution and it's boundaries motion in non-stationary flows is of great interst. A shear flow with broken linear velocity disturbance by thickness is a classical example simulating the horizontal motion of pliable salt strata under tangent loading (ref.1). With the certain class of boundary disturbances as well as kinematical and dynamical imperfections within the field, instable regimes resulting in vertical mass transfer may appear. An amplitude growth speed becomes maximum under the certain wave length (dominating length). The critical difference of shear rates on the boundaries is obtained and the corresponding dispersion-wave relations connecting complex oscillation frequency with disturbance wave length are analysed. The constitutive law in form (1) allows to apply the received results to viscous liquids (t=0), linear viscoplastic geomaterials (f is linear) and other models that have been dealt with in geotectonics. It should be noted that many results are corroborated both quantitatively and qualitatively by facts known from solutions of the classical problems (H.Lamb, O.Reynolds, H.Hencky, A.A.Il'yushin, B.E.Pobedria).
Side by side with stated above, the problems on gravitational instability of stratificated structures in gravity field are of interest in geophysics. The phenomena taking place by density inversion in the system consisting of two viscous incompressible liquids are studied minutely enough in literature. On the basis of low disturbances method for such models the initial stage of saltdomes break through sedimentary rocks as well as the sinking of heavy platforms are investigated. The experimental verification of the analytical results is obtained by means of scale modelling on centrifuges (ref.1). Taking the road of the subsequent generalization of geomaterials rheology, the gravitational instability of heavy viscoelastic structures is investigated (refs.2,3). It was found that the choose of viscoelastic constitutive law corroborates well the hypothesis on block structure of lithosphere.

REFERENCES

1. H.Ramberg, Gravity, Deformation and the Earth's Crust, Academic Press, London - N.-Y., 1981, p.400.
2. A.A.Il'yushin and B.E.Pobedria,The Foundations of Mathematical Theory of Thermoviscoelasticity, Nauka, Moscow, 1970, p.280.
3. V.A.Dubrovskii, Proc. USSR Acad. Sci., Earth Physics, 1985, No.1, 29-34.

Theoretical and Applied Rheology, edited by P. Moldenaers and R. Keunings
Proc. XIth Int. Congr. on Rheology, Brussels, Belgium, August 17-21, 1992

RHEOLOGY IN SELF-PROPAGATING HIGH-TEMPERATURE SYNTHESIS

B.M.KHUSID

Heat and Mass Transfer Institute, Byelorussian Academy of Sciences, 15, P.Brovka St., 220728, Minsk (Republic of Belarus)

The self-propagating high-temperature synthesis (SHS) is the kind of heterogeneous combustion aimed at producing various materials. To synthesize TiC, for example, Ti and C powders are mixed together and press-formed into a pellet. The pellet is ignited by an electrically heated wire coil which provides the heat impulse. The exothermal gasless reaction Ti + C $\longrightarrow$ TiC forms a combustion wave which propagates along the sample. Typical parameters of the SHS wave for inorganic systems are: front temperature, 1500 to 3000 ^{o}C; wave velocity, 0.1 to 10 cm/s; and heating rate, 10^{3} to 10^{5} ^{o}C/s. The SHS is a cost-effective method for manufacturing a variety of ceramic materials. The high temperature of the final SHS products enables one to deform them, and therefore, finished articles of refractory composite materials may be produced by die-casting, extrusion, and pressing immediately after combustion. The SHS method and continuous metal-forming technology based upon it have raised new fundamental problems of micro- and macrorheology.

The main features of structure transformations in the SHS wave have been examined for the reactions in which the materials before, during, and after the synthesis are solid or liquid. The appearance of transient liquid caused by heating plays the main role in the mechanism of structural transformations in the combustion wave. The difference between the adhesion and the cohesion energies drives melt wetting. Wetting dynamics is governed by the transient melt rheological properties. The hot metal-forming technology depends upon the macrorheology of sponge-like combustion products. They are very sensitive to the rate of deformation.

Mathematical models for micro- and macrorheological phenomena have been constructed and examined. It is found that the rheological properties of two-phase compositions with co-existing solid and liquid components determines the process parameters.

Theoretical and Applied Rheology, edited by P. Moldenaers and R. Keunings
Proc. XIth Int. Congr. on Rheology, Brussels, Belgium, August 17-21, 1992

BOUNDARY LAYER FLOWS OF INELASTIC NON-POWER-LAW FLUIDS

P. MITSCHKA
Institute of Chemical Process Fundamentals, Czechoslovak Academy of Sciences
Rozvojová 135, 165 02 Praha 6 - Suchdol (Czechoslovakia)

Considered are boundary - layer flows of Ree-Eyring fluids past cylinders governed by the following equations

$$u_x + v_y = 0 \quad (1)$$

$$uu_x + vu_y = UU_x + \tau_y/\rho \quad (2)$$

$$\tau = \mu_E u_y + \operatorname{arsinh}(u_y/C)/B \quad (3)$$

$$U/U_\infty = 2\sin(x/R) \quad (4)$$

together with usual boundary conditions. Introducing into (1) and (2) the so called Falkner-Skan transformation modified for the class of fluids considered by using the zero shear-rate viscosity $\nu_0 = (\mu_E + 1/BC)/\rho$ instead of ν (see ref.1) we obtain

$$(b\,f'')' + 0.5(m+1)ff'' + m(1-f'^2) = x(f'f'_x - f''f_x) \quad (5)$$

where the dimensionless non-linear effective viscosity for Ree-Eyring fluids is

$$b = \{[1+Ey\,\operatorname{arsinh}(\alpha f'')]/\alpha f''\}/(1+Ey) \quad (6)$$

The meaning of f, m and the apostrophe in (5) and (6) is the same as for the Newtonian fluids; $Ey=1/\mu_E BC$ is a dimensionless material constant of the fluid considered and α a second x-dependent parameter

$$\alpha = \alpha(x) = \beta\,\Phi(x/R) \quad (7)$$

where $\beta = \rho U_\infty^3/[\mu_E(1+Ey)RC^2]$ (8)

and $\Phi(x/R) = [(U/U_\infty)^3/(x/R)]^{\frac{1}{2}}$ (9)

The problem considered is thus fully described by two parameters, Ey and β. For Ey = 0, it reduces to the Newtonian case, for Ey $\to \infty$ to the case with Eyring's two-parameter model $\tau = \operatorname{arsinh}(u_y/C)/B$.

The transformed momentum equations (5) and (6) have been solved by Keller's Box Method (ref.1), using a full linearization of b, as suggested by Andersson and Toften (ref.2). In Fig. 1, first results, the dependencies of the friction parameter along the cylinder surface, are shown.

REFERENCES

1. T. Cebeci and P. Bradshaw, Physical and Computational Aspects of Convective Heat Transfer, Springer, New York, 1984.
2. H.I. Andersson and T.H. Toften, J. Non--Newtonian Fluid Mech., 32(1989) 175-195.

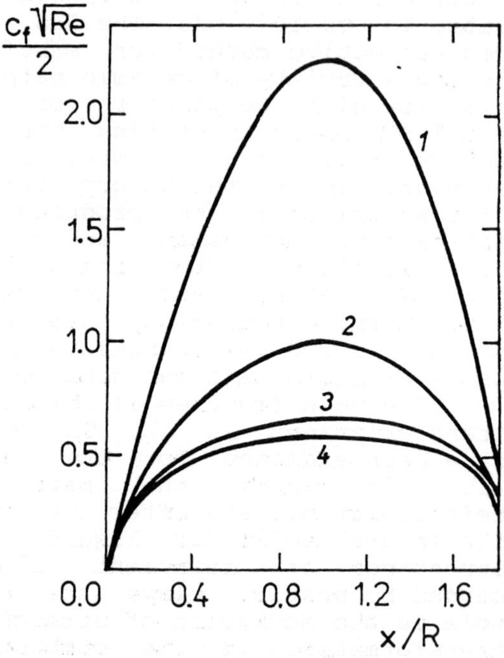

Fig.1 Calculated wall friction variation along cylinder surface for β=10, 1 Ey=0, 2 Ey=10, 3 Ey=100, 4 Ey≥1000

Theoretical and Applied Rheology, edited by P. Moldenaers and R. Keunings
Proc. XIth Int. Congr. on Rheology, Brussels, Belgium, August 17-21, 1992

AN EXPERIMENTAL/THEORETICAL INVESTIGATION OF INTERFACIAL STABILITY IN MULTILAYER COEXTRUSION PROCESSES

G. M. Wilson, Y. Y. Su and B. Khomami

Department of Chemical Engineering and the Materials Research Laboratory, Washington University, Campus Box 1198, 1 Brookings Drive, St. Louis, MO 63130 (USA)

1. INTRODUCTION

A major problem in the manufacture of multilayer polymeric films is the formation of interfacial waves which can result in a significant deterioration of product properties. Hence, to establish processing windows for the stable operation of coextrusion processes, a better understanding of the problems associated with the multiphase flow of viscoelastic fluids, particularly with respect to interfacial stability is required. Therefore, our objective in this study is to examine the nature of interfacial deformation and the stability of multilayer viscoelastic fluids. In particular, we have carried out both theoretical and experimental studies to examine the stability of the interface as a function of viscosity, depth and elasticity ratios as well as disturbance wavelength in both parallel and converging channel geometries.

2. METHODS

In the theoretical portion of the study, we have used both asymptotic and spectral based numerical techniques (ref. 1) to solve the set of equations governing the linear stability of superposed modified Oldroyd-B fluids in converging and parallel plate die geometries. The results of our analysis are presented in terms of neutral stability diagrams in the parametric space of viscosity, depth and elasticity ratio.

In our experiments, the stability of the interface is examined by introducing disturbances of a desired wavelength and amplitude into the system and subsequently measuring their growth or decay using state-of-the-art digital image processing techniques (ref. 2). A typical unstable interface as viewed in the fourth window of our coextrusion test die is shown in Figure 1. This approach has been successfully used to investigate the stability of both compatible and incompatible polymer systems as a function of viscosity, elasticity and layer depth in both parallel channel and converging geometries. The results of these investigations are contained in a variety of experimental stability contours which show stability as a function of layer depth ratio and disturbance wavenumber.

3. CONCLUSIONS

The theoretical and experimental results for various polymer systems have been compared and preliminary results suggest that the linear stability analysis based on the modified Oldroyd-B constitutive equation is qualitatively correct in its predictions of disturbance growth rate as well as the shape of the neutral stability contour. However, before any final conclusion is drawn, more vigorous comparisons of theoretical and experimental results should be carried out. This is currently underway in our research group.

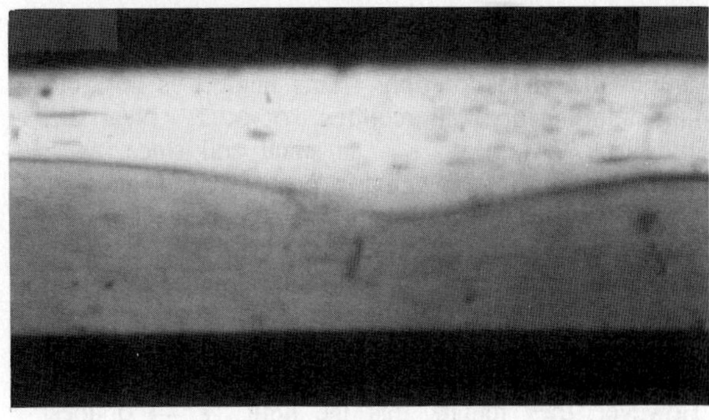

Figure 1 -Unstable interface as viewed in window #4.

REFERENCES

1. Su, Y. Y. and B. Khomami, J. Rheol., 36 (2) 357-387 (1992).
2. Wilson, G. M. and B. Khomami, J. Non-Newtonian Fluid Mech., to appear (1992).

Theoretical and Applied Rheology, edited by P. Moldenaers and R. Keunings
Proc. XIth Int. Congr. on Rheology, Brussels, Belgium, August 17-21, 1992

SQUEEZING FLOW OF A BINGHAM PLASTIC

S.D.R. WILSON

Mathematics Department, University of Manchester, Manchester M13 9PL, England

Suppose the narrow gap, of width h, between two parallel circular discs, of radius a, is filled with a material which obeys a constitutive law of the Bingham viscoplastic type, and that the discs are squeezed together. There is an immediate and obvious difficulty in calculating the flow. For the material in the gap has to get out of the way; on the other hand the shear stress falls to zero on the central plane and is small on either side, so that the material there is solid.

Here we treat the Bingham model as a limiting case of the "biviscosity" model (ref.1) for which

$$\underline{\tau} = \begin{cases} \eta_1 \underline{d} & \frac{1}{2}\underline{\tau}:\underline{\tau}<\tau_1^2 \\ \eta_2 \underline{d} + \dfrac{\tau_0}{(\frac{1}{2}\underline{d}:\underline{d})^{\frac{1}{2}}}\underline{d} & \frac{1}{2}\underline{\tau}:\underline{\tau}>\tau_1^2 \end{cases}$$

The limit under discussion is $\eta_2/\eta_1 \longrightarrow 0$.

The theory contains 2 parameters, $\delta = h/a$ and $\epsilon = \eta_2/\eta_1$, and the paradoxes arise because letting both go to zero involves a clash of limits.

The difficulty stems from the neglect of the normal stresses, τ_{rr} $\tau_{\theta\theta}$ and τ_{zz} compared with the shear stress τ_{rz}, in the equations of motion and in the yield criterion. This neglect is part of lubrication theory and requires $\delta \ll 1$. Near the mid-plane, however, τ_{rz} is small and since, as has been shown, the velocity profile corresponds to plug flow plus a small variation, the neglect of the other components may not be valid if η_1 is large. It is the ability to withstand these stresses (so to speak) which prevents a solid from deforming in the required manner, as argued convincingly by Lipscombe and Denn (ref.1). However, they remark that the limit $\epsilon \longrightarrow 0$ does not correspond to a Bingham material because for η_1 arbitrarily large the material can still flow, while a solid cannot. This seems unsatisfactory. There is no obvious mathematical objection to letting $\epsilon \longrightarrow 0$ in the equations and one feels that $\epsilon = 0$ cannot be utterly different from (say) $\epsilon = 10^{-3}$.

A careful analysis shows that the shear stress τ_{rz} is of order Gh (where -G is the pressure gradient) and that the normal stresses are of order $Gh^2\eta_1/a\eta_2$. The neglect of these components therefore requires $\delta/\epsilon \ll 1$. If δ/ϵ is of order one, the normal stresses must be included in the yield criterion; but they may be neglected in the equations of motion, which contain stress gradients, if $\delta^2/\epsilon \ll 1$, and on this basis a corrected theory is feasible. This shows that there is a yield surface located more or less as indicated in ref.1, and that the material inside the "unyielded" plug is (in the limit $\epsilon \longrightarrow 0$, $k = \delta/\epsilon$ fixed) at the yield point. It is also possible to show that when k is small the stand-off distance of the yield surface, on the axis, is 3kh, and that there is no change to the pressure gradient to leading order.

REFERENCE

1. G.G. Lipscomb and M.M. Denn, J. Non-Newtonian Fluid Mech, 14 (1984) 337-346.

Theoretical and Applied Rheology, edited by P. Moldenaers and R. Keunings
Proc. XIth Int. Congr. on Rheology, Brussels, Belgium, August 17-21, 1992

VIBRATION OF A SPHERE IN A YIELD STRESS FLUID

O. Wünsch

Institut für Strömungslehre und Strömungsmaschinen, Universität der Bundeswehr, Holstenhofweg 85, W–2000 Hamburg 70 (Germany)

1 Introduction

In a viscoplastic fluid it is possible, that suspended particles do not move under the action of gravity although the density of particles is greater than the density of the fluid. The reason for this behaviour is the yield stress. On the other hand, if the fluid is set in sinusoidal vibration, the particle undergo forced oscillations. In this case the particles begin to sediment, if the vibration reaches certain values of amplitude and frequency.

2 Model

This effect is investigated for single sphere. The motion of the sphere can be described by a non–linear mechanical model. A mass is fixed over linear and non–linear springs and dashpots with a frame which is set in sinusoidal vibration (fig. 1).

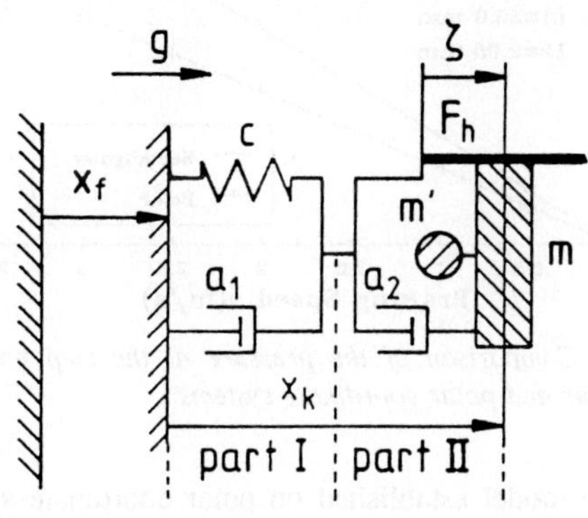

Figure 1: Non–linear mechanical model

If the values for amplitude and frequency of the oscillation of the frame are small, the forces on the mass do not reach the Coulomb–force F_h. In this case only the linear part (I) of the model works and the mass oscillate about the normal position. Otherwise the second part (II) of the model comes in action additionally and the mass changes the position in time–average.

By a field theoretical investigation it is possible to connect the free parameters of the model with the rheological parameters of the fluid above the yield stress and with the parameters of the sphere (ref.1,2). The model leads to a criterion for the beginning of sedimentation and allows to determine the average velocity of the falling sphere.

3 Experiment

The results of the theoretical investigation are verified by experiments. The motion of glass spheres in an oscillating cylindrical tube filled with aqueous carbopol–solutions are detected. One main result is shown in figure 2. A borderline seperates two regions. Below the line the amplitude and frequency of the oscillation of the frame are small so that spheres oscillate about the normal position only. Above the line a drift–motion is overlaid the vibration and spheres sediment in the solution.

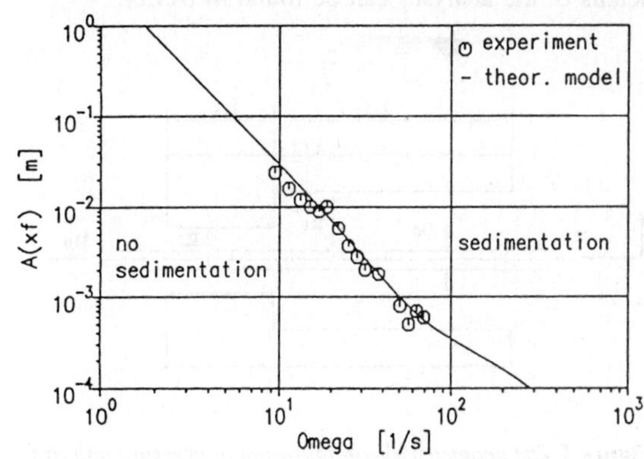

Figure 2: Beginning of sedimentation

4 References

1. Beris, A.N, et.al., J. Fluid Mech. **158** (1985), 219–244
2. Wünsch, O., Z. angew. Math. Mech. **72** (1992) 5, T349–T352

Theoretical and Applied Rheology, edited by P. Moldenaers and R. Keunings
Proc. XIth Int. Congr. on Rheology, Brussels, Belgium, August 17-21, 1992

A Closed Form Analytical Solution in Polar Coordinate for the Hydrodynamic Pressure in a Stepped Bore Unit for Polymer Coating of Wires

J. YU and M. S. J. HASHMI

School of Mechanical & Manufacturing Engineering, Dublin City University, Ireland

1. INTRODUCTION

In wire drawing process, a wire is pulled through a reduction die and the material deforms plastically whilst passing through the die. In industrial wire drawing practice, lubrication is used to reduce the drawing load and die wear and hence improve the machine life and surface finish of the products.

Christopherson and Naylor(ref.1) presented a paper which showed a method of reducing friction in wire drawing by hydrodynamic lubrication. It had been assumed that friction in conventional wire drawing was of a boundary nature and that a change of mode to hydrodynamic lubrication should greatly reduce friction.

In this paper models of steady uniform laminar Newtonian and non-Newtonian fluid in the process of plasto-hydrodynamic wire drawing in a stepped bore unit for polymer coating have been developed. The closed form analytical solution is established in polar (cylindrical) coordinate since the practical case mathematically belongs to axisymmetric problem. The details of the analyses can be found in (ref.2).

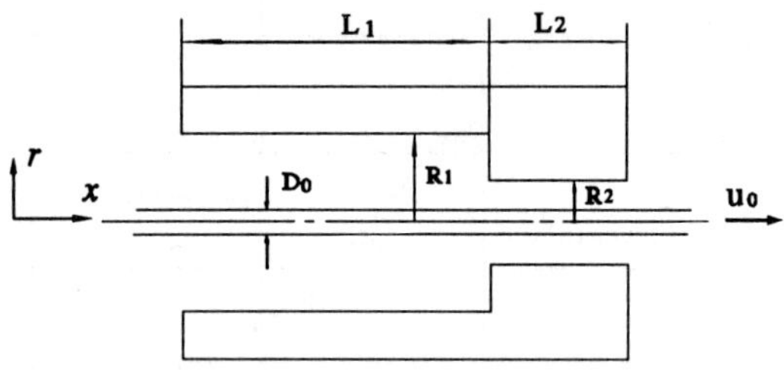

Figure 1 *The geometrical configuration of pressure unit and the wire drawing*

2. ANALYSES

The analysis is based on the geometrical configuration, shown in Fig.1, during wire drawing using a unit filled with the fluid (polymer melt). In order to analyze the drawing process, the following assumptions were made: (1) The flow of the fluid is incompressible, steady and uniform; (2) The flow of the polymer is laminar; (3) The material of the wire is rigid.

3. RESULTS

The results were calculated on the basis of equation derived from the theoretical analyses.

The pressure at the step under the polar coordinate system are compared with that calculated under the rectangular coordinate system corresponding to the drawing speed in the case of Newtonian fluid. From Fig. 2, the pressure from the first method is smaller than that from the second one and the discrepancy increases with the increase of the drawing speed.

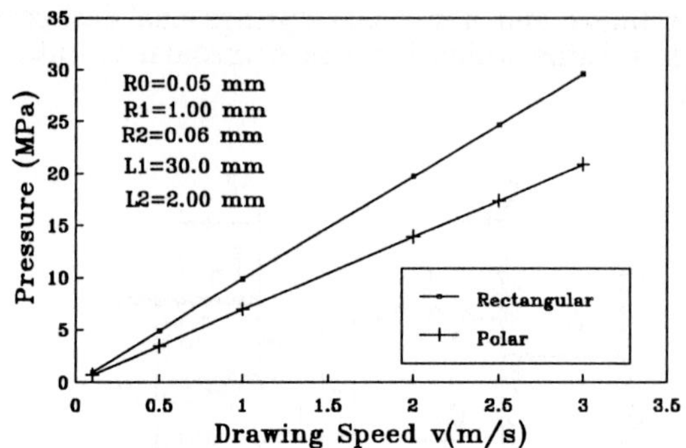

Figure 2 *Comparison of the pressure at the step under rectangular and polar coordinate systems*

The model established on polar coordinate system should be most appropriate since the real unit is axisymmetrical. However, experimental results are necessary to validate the theoretical results.

REFERENCES

1. D.G. Christopherson, H. Naylor, Instn. Mech. Engrs., 169 (1955) pp. 643-653
2. J. Yu and M.S.J. Hashmi,"submitted for publication".

CONTRIBUTED PAPERS

NUMERICAL SIMULATION

Theoretical and Applied Rheology, edited by P. Moldenaers and R. Keunings
Proc. XIth Int. Congr. on Rheology, Brussels, Belgium, August 17-21, 1992

COMPARISON OF NUMERICAL SIMULATIONS OF POLYMER FLOW WITH EXPERIMENTAL DATA

THOMAS A. BAER AND BRUCE A. FINLAYSON[1]

[1]Department of Chemical Engineering, BF-10, University of Washington, Seattle, WA 98195 USA

1. INTRODUCTION

The ultimate goal of modeling viscoelastic flows is to represent physical processing problems with sufficient accuracy that design decisions can be made and operating parameters extracted from the computer model. There are two major problems that must be solved before that goal is reached. The first is that many of the standard computational methods do not converge for shear rates in the region observed experimentally, and the special ones that do are limited in their ability to model real fluids. The second problem is how well a given constitutive equation, developed and tested in one-dimensional experiments, can be applied to a fully two-dimensional case; the concomitant concern is a comparison of the model and experimental data. This paper employs a streamline integration method to try to overcome problem one and provides answers to problem two, at least over the range of shear rates where both experimental and computational data exists for two-dimensional flows. Full reviews of the literature are provided in the references cited here.

2. PREDICTION OF ELONGATIONAL STRESS GROWTH VISCOSITY

The 8-mode Phan-Thien-Tanner model is:

$$\tau = \sum_{i=1}^{8} \tau_i - \eta \dot{\gamma}, \quad Z_i = \exp(-\varepsilon \lambda_i \operatorname{tr} \tau_i / \eta_i)$$

$$Z_i \tau_i + \lambda_i \tau_{i(1)} + \frac{\xi}{2} \lambda_i (\tau_i \cdot \dot{\gamma} + \dot{\gamma} \cdot \tau_i) = -\eta_i \dot{\gamma}$$

Bird, *et al.* (ref. 1, p. 382) fit the 8-mode model to this data as follows. First eight relaxation times $\{\lambda_i\}$ are chosen. Then the values of viscosity coefficients $\{\eta_i\}$ were obtained by fitting the data for linear viscoelastic storage and loss moduli (ref. 2, p. 273). The parameters ξ and ε were then chosen to give a best fit of shear data. The fit to the shear viscosity and first normal stress data (ref. 3-4) is illustrated in Figure 1. The polymer is low-density polyethylene, IUPAC Melt I; the agreement is excellent, although there is a slight departure at very high shear rates.

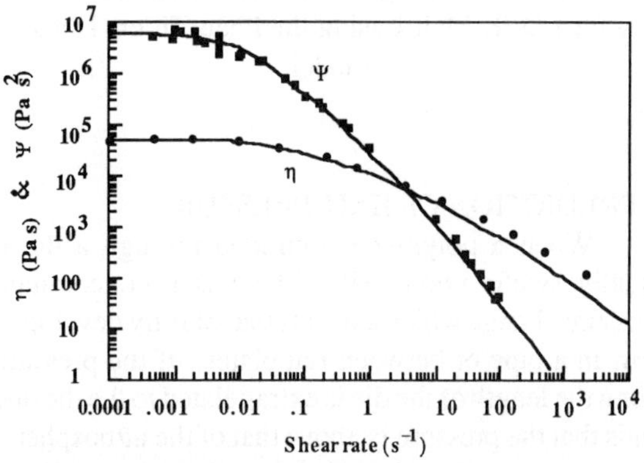

Figure 1. Fit of Shear Vicosity and First Normal Stress Data for LDPE-Melt I using the Phan-Thien-Tanner Model

These parameters were then used in a simulation of the elongational stress growth viscosity to compare with data provided by Meissner (ref.3). A fiber is stretched at a fixed rate ($\dot{\varepsilon}$), the velocity field is known, and the stress equations are integrated in time. The agreement between data and simulations is excellent, as shown in Figure 2. These results are comparable to the comparison made by Bird, *et al.* (ref. 2, p. 413) with a Giesekus model. These results are especially powerful because the experiment is an elongational one, yet the experiments used to provide the parameters are linear viscoelasticity and shear experiments; the elongational data was not used to determine the param-

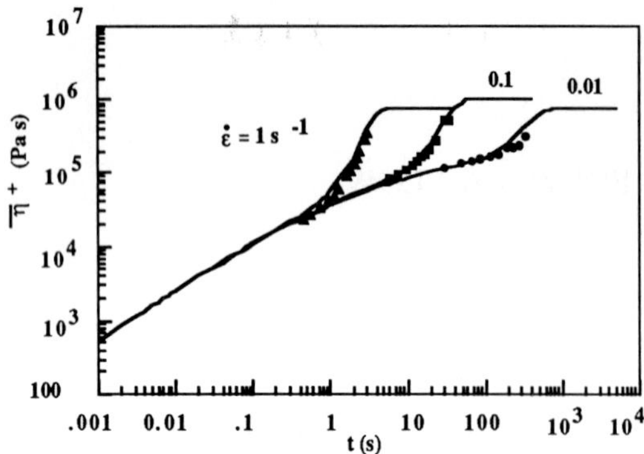

Figure 2. Fit of Elongational Stress Growth Viscosity for LDPE-Melt I using the Phan-Thien-Tanner model

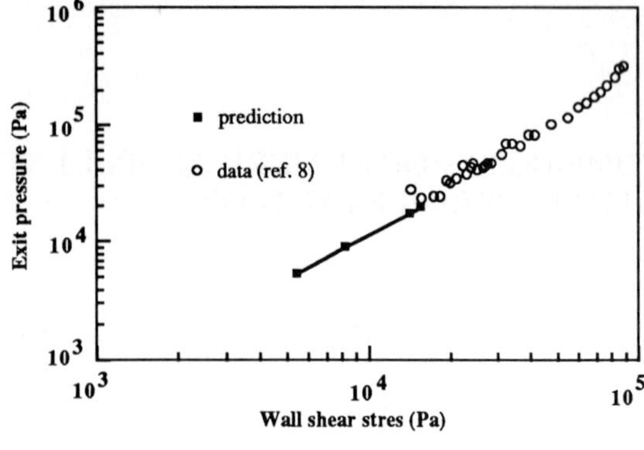

Figure 3. Comparison of Exit Pressure Predictions with Experiments for LDPE-Melt X using the Phan-Thien-Tanner model

3. PREDICTION OF EXIT PRESSURE

When a polymer is extruded through a die it usually swells upon exit. There is a concomitant pressure change which exceeds that of fully developed flow in a pipe or between flat plates. If the pressure along the length of the die is extrapolated to the die one finds that the pressure is above that of the atmosphere. That pressure excess is called the exit pressure. Some authors believe that the exit pressure can be used to obtain the normal stress (ref. 5-6); others disagree (ref. 7). A definitive experimental answer to this question was provided by Tuna and Finlayson (ref. 8), who measured the exit pressure for low density polyethylene in a planar slit die. They found that this data could be analyzed to obtain a first normal stress difference which was in excellent agreement with that measured by Meissner (ref. 3). However, simulations of the same case could not be done then because the iterative method failed at a lower shear rate than was used experimentally; the experiment was too insensitive to be run at lower shear rates.

Calculations reported here were performed with an 8-mode Phan-Thien-Tanner model using a streamline integration technique. The complete details are reported elsewhere (ref. 9); space limitations prevent a full treatment here. The velocity is solved from the momentum equation with a known elastic stress, and the stress equation is solved from the constitutitive equation with a known velocity. The iterative technique then alternates between solving for velocity and stress. The velocity is solved using a standard Galerkin finite element methods with biquadratic velocity elements. The stress equation is solved by integrating along streamlines using an ODE solver, LSODAR. Since the finite element mesh does not conform to the streamlines, it is necessary to track the streamline within an element. The stream function is first solved on the same mesh. Also needed is a routine that will take a given x-y point and determine which element it is in, so that the correct velocity and stream function can be obtained within an element. At the conclusion of the integration along one streamline one has a list of x, y, stress values at somewhat arbitrary, but known, points. This procedure is done along several streamlines so that the region of flow is covered well. Then the stress values are interpolated onto the same finite element mesh as used for the velocity solution. The process is repeated, sometimes using a relaxation parameter for velocity. This process does not converge for all shear rates, but does converge for high enough shear rates to include a small portion of the experimental range. Since there are 32 stress components (8 modes with 4 stress components each) a direct solution of the whole problem would be prohibitive, especially when one requires sub-meshes or cubic trial functions for stresses.

The finite element mesh used 162 elements extending from z = -10 to z = 11. The refinement of the mesh at the singularity was about 5 percent radially and axially. For the stick slip problem the mesh was refined to have elements as small as 0.00625 of the radius, with consistent results. The stress values are obtained at locations that depend on the streamlines. We used values which started out 0.0153 of the radius away from the wall; near the singularity they get closer to the wall. Thus the mesh is reasonably well refined.

Calculations of exit pressure are compared with Tuna and Finlayson's data (ref. 8) in Figure 3. There is only a small region of overlap, but the agreement there is excellent. Furthermore the trend of the calculations and experiment are in agreement with each other. Our other simultions with one-mode models have not gotten to this high a shear rate, nor has the trend of the calculations been in agreement with the trend of the data.

4. CONCLUSION

These results show agreement between simulations with an 8-mode Phan-Thien-Tanner model and experimental data for unsteady stress growth viscosity and exit pressure. The parameters of the model are determined from other experiments, indeed ones emphasizing shear behavior. Thus this model is sufficiently representative of real phenomena that data obtained from linear viscoelasticity and shear experiments can be used to predict experimental results with a significant elongational component.

REFERENCES

1. R. B. Bird, C. F. Curtiss, R. C. Armstrong, O. Hassager, Dynamics of Polymeric Liquids, Vol. 2: Kinetic Theory, 2nd edition,Wiley-Interscience, Wiley, New York, 1987.
2. R. B. Bird, R. C. Armstrong, O.Hassager, Dynamics of Polymeric Liquids, Vol. 1: Fluid Mechanics, 2nd edition, Wiley-Interscience, Wiley, New York, 1987.
3. Meussner, J., Kunststoffe, 61 (1971) 576-582 .
4. Laun, H. M., Rheol. Acta., 17 (1978) 1-15.
5. C. D. Han, J. Appl. Polym. Sci., 15 (1971)2567-2577.
6. C. D. Han, K. U. Kim, N. Siskovis, C. R. Huang, J. Appl. Polym. Sci., 17 (1973) 95-103.
7. D. V. Boger and M. M. Denn, J. Non-Newt. Fluid Mech., 6 (1980) 163-185.
8. N. Y. Tuna and B. A. Finlayson, J. Rheology, 32 (1988) 285-308.
9. T. A. Baer, Ph. D. Thesis, Univ. Washington (1991).

ACKNOWLEDGEMENT

This research was supported by the Petroleum Research Fund, administered by the American Chemical Society.

Theoretical and Applied Rheology, edited by P. Moldenaers and R. Keunings
Proc. XIth Int. Congr. on Rheology, Brussels, Belgium, August 17-21, 1992

SOME TRANSIENT STUDIES OF NON-NEWTONIAN NON-ISOTHERMAL FLOWS IN TWO AND THREE DIMENSIONS

A BALOCH, D. DING, P. TOWNSEND and M.F. WEBSTER

University of Wales Institute of non-Newtonian Fluid Mechanics, Department of Mathematics and Computer Science, University College, Swansea, SA2 8PP, U.K.

1. INTRODUCTION

The work presented here constitutes part of a programme to develop a general purpose code for predicting flows of non-Newtonian materials. The code, which is based on a Taylor-Galerkin/pressure correction algorithm has been described in detail elsewhere [1,2]. It has been validated in a series of two-dimensional model problems for thermal non-Newtonian conditions [3,4,5,6], and some three-dimensional problems have also been investigated [7]. Here the performance of the algorithm for two new flows is investigated. In the first we consider the effect of non-Newtonian behaviour and of non-isothermal conditions on the flow described in [7] in which fluid passes through a curved channel of square cross section, and in the second problem we study the flow of Newtonian and viscoelastic liquids through sudden expansions.

2. MATHEMATICAL MODEL

For inelastic, non-Newtonian materials, incompressible flow under non-isothermal conditions is governed by the equations of motion and energy. In conventional notation these are as follows:

$$\rho \frac{\delta \underline{v}}{\delta t} = - \nabla p + \nabla(\underline{T}) - \rho \mathbf{v}\nabla \mathbf{v} + \rho\, \mathbf{g},$$

$$\rho\, c_p \frac{\delta \Theta}{\delta t} = \nabla(\kappa \nabla \Theta) - \rho c_p \mathbf{v}\nabla(\Theta) + \mu \Phi,$$

where $\mathbf{v}$ is the velocity vector, p is the pressure, Θ is the temperature, $\underline{T}$ is the viscous stress tensor, $\mathbf{g}$ is the body force vector, ρ is the density, c_p is the thermal heat capacity at constant pressure, κ is the thermal conductivity, μ is the fluid viscosity, and Φ is a dissipation function dependent on shear rate.

Appropriate boundary conditions on the velocity components or their derivatives, will need to be applied, according to the flow conditions. The fluid viscosity, μ, is a function of shear rate, γ, and temperature. Various forms for this are adopted in the literature. For the first problem described below we choose to use a simple power-law shear dependence together with a temperature dependence of the form

$$\mu = 10^{(a + b/(\Theta + c))} \gamma^{m-1}$$

where a, b, c are material constants and m is the power-law index.

For viscoelastic behaviour we choose to make use of a simplified form of the model due to Phan Thien and Tanner given by

$$\{1 + \frac{\varepsilon\lambda_1}{\mu_1} \mathrm{Trace}(\underline{\tau})\}\, \underline{\tau} + \lambda_1 D\underline{\tau} = 2\mu_1 \underline{D}$$

$$\underline{T} = \underline{\tau} + 2\mu_2 \underline{D}$$

where $\underline{\tau}$ is an extra-stress tensor, λ_1 is the relaxation time, and μ the fluid viscosity may be decomposed into the sum of a solvent viscosity μ_2 and a solute viscosity μ_1. From these one can compute the retardation time λ_2, normally associated with the Oldroyd-B model from $\lambda_2/\lambda_1=\mu_2/\mu$. The convected derivative is a combination of both the upper and lower convected derivatives, given by

$$D\underline{\tau} = (1-\xi/2)\overset{\nabla}{\underline{\tau}} + (\xi/2)\overset{\Delta}{\underline{\tau}}\ .$$

3. THE SOLUTION ALGORITHM

The basic methodology has already been described in detail in the literature [2 - 5] and only a brief outline is presented here. A temporal discretisation in a Taylor series is employed prior to a spatial Galerkin discretisation. In the most basic form of the algorithm, an initial non-divergence-free velocity field is initially computed using a half time-step scheme and this leads a Poisson equation to solve for the pressure difference field. This field is then used to produce the final divergence-free velocity field. Piecewise continuous quadratic polynomials are used to approximate the velocity and stress components and the temperature, whilst piecewise linear functions are employed for the pressure, either over a triangular subdivision of the spatial domain for two-dimensional problems or a tetrahedral subdivion for three-dimensional problems. The algorithm is carefully coded using conditional compilation to provide efficient versions for whatever combinations of spatial dimensions (2 or 3), fluid properties(Newtonian, generalised Newtonian or viscoelastic) and thermal or isothermal conditions. Viscoelastic properties give rise to the most demanding conditions and only the isothermal two-dimensional version has been fully tested todate. It is hoped to present some three-dimensional results at the conference.

4. THE GLASS FLOW PROBLEM.

The first problem concerns the flow under gravity of a highly viscous liquid (molten glass) through an axisymmetric contraction. Figure 1 shows the schematic diagram of the problem, in which z is the vertical direction (i.e. the flow direction) and r is the radial direction. The cylinder undergoes a contraction some distance along its length. The length of the cylinder is 0.6 metres and the radius is R=0.1 metres. A solution mesh of 128 triangular elements is employed.

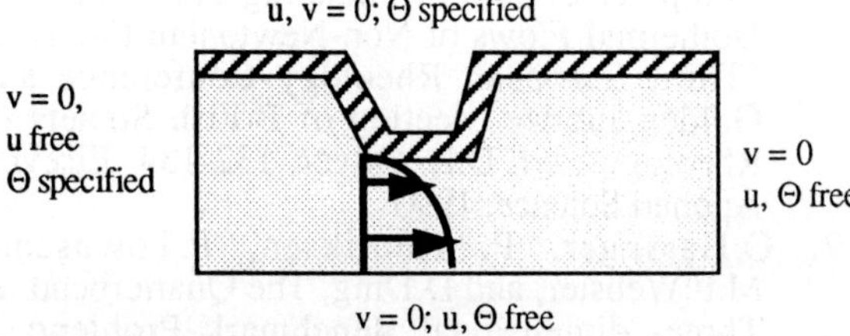

FIGURE 1. Schematic diagram for glass flow problem

An appropriate non-dimensionalisation of the relevant equations here gives rise to three parameters, the Reynolds number Re, the Graetz number or thermal Péclet number Pé and the Griffith number G_f. Full definitions are given elsewhere in [8]. Adopting typical values for flow parameters gives rise to values of Re=4.31×10^{-11}, Pé=100, and G_f=24.9. With these values it becomes clear that the momentum equation is almost completely dominated by the diffusion terms and the energy equation is strongly dominated by the thermal transient and convection terms.

Initially the fluid is assumed to start from rest with all boundaries and the fluid itself at 500° C. When a steady state has been reached, the boundary wall temperature is suddenly increased to 900°C and the transient development of the flow tracked.

If we concentrate first on thermal effects and assume no shear dependence in the viscosity function (m=1), the important parameter is the Péclet number. At high values, heat is convected down the tube with little diffusion towards the centre of the tube. For low Péclet numbers, diffusion is more important so that the effect of the hotter wall is felt in a much greater portion of the tube. The significantly different temperature distributions in the two cases gives rise to very different viscosity fields and consequently, ver different velocity fields. At early stages in the transient development, temperature boundary layers recirculation regions upstream and downstream of the contraction. When shear thinning behaviour (m=0.2) is included, one sees the characteristic flattening of the axial velocity profiles.

5. THE EXPANSION FLOW PROBLEM.

A schematic diagram for this problem is shown in figure 2. Fluid enters through a small channel of square cross-section (shown at the bottom of the diagram), and flows out into a second channel of rectangular cross-section, and subsequently into a third channel, of square cross section. The net effect is that the fluid experiences two separate expansion flows, the first a clearly three-dimensional flow and the second, provided one is concerned with effects away from the side walls, a two-dimensional flow. This geometry represents a region where one is able experimentally to compare two and three-dimensional expansion flow in one experiment.

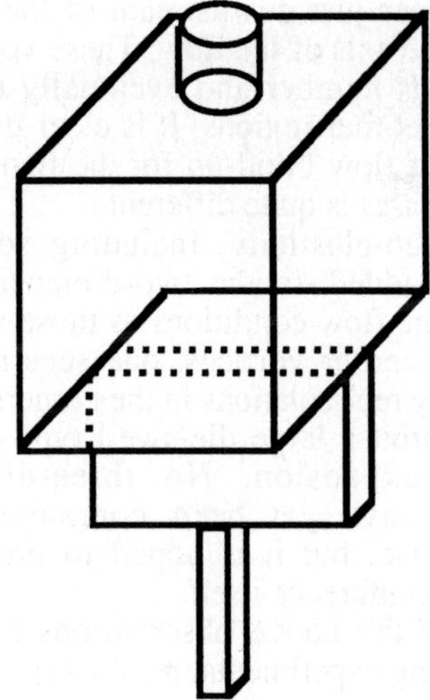

Figure 2 Schematic diagram of the expansion flow.

We will refer to such experiments later. For the purposes of our computations it is convenient to consider the two expansion portions of the geometry separately .

We have therefore investigated flow in both two-dimensional and three-dimensional expansions. Attention has been initially focussed on Newtonian results, for some recent experiments [9] have indicated that inertial effects in such fluids give rise to some interesting observations. We also, however, have made some preliminary viscoelastic computations, and here the results are also significant.

If we concentrate on Newtonian flow simulation results for the moment, for low Reynolds numbers, near unity, the fluid enters the two-dimensional expansion and fans out in a jet, leaving quite significant dead regions in each of the expansion corners. Here, Reynolds number is defined in terms of the inflow dimension and the maximum inflow velocity. These regions stretch right up to the inflow channel and some distance down the expansion. As the Reynolds number is increased to order ten, the size of these regions does not change greatly, initially, but a vortex which fills the regions develops and strengthens. If the computation is continued to much higher Reynolds numbers, in excess of fifty, the vortices grow very large and project some considerable distance down the expansion. For the comparable three-dimensional flow, the build up is quite different.We have some difficulty, now, observing the flow but are able to define streamlines in the symmetric mid-plane. At low Reynolds numbers no dead regions are apparent, and streamlines entering the expansion project right into the corners without any recirculation. As the inertial effects are increased, a pair of so-called 'lip-vortices' appear just downstream of the expansion and near to the axis of the flow. These vortices grow with Reynolds number and eventually occupy the whole of the corner regions. It is clear then that the mechanism of flow build-up for the two and three-dimensional cases is quite different.

If visco-elasticity, including some shear-thinning, is 'added' to the two-dimensional flow, under the same flow conditions as those which gave rise to large corner vortices, one sees no vortices other than tiny recirculations in the corners. The fluid seems to exhibit a large die-swell type effect as it enters the expansion. No three-dimensional simulations have yet been completed for the viscoelatic case, but it is hoped to present some results in the conference itself.

All of the above observations follow very closely existing experimental results [9].

6. CONCLUSIONS.

The performance of a simulation code for the prediction of non-Newtonian flows, has been investigated for a number of different flow conditions and material properties. Where comparison with other work was possible, the code performed well.

7. REFERENCES

1. P. Townsend and M.F. Webster, An Algorithm for the Three-Dimensional Transient Simulation of non-Newtonian Fluid Flows, in "Proceedings of Numeta Conference, Numerical Methods in Engineering: theory and applications", NUMETA 87: Vol 2, T12/1-11, Nijhoff, 1987.
2. D.M. Hawken, H.R. Tamaddon-Jahromi, P. Townsend and M.F. Webster, A Taylor-Galerkin based Algorithm for Viscous Incompressible Flow; Int. J. Num. Meth. Fluids, 10, 327-351, 1990.
3. D. Ding, P. Townsend and M.F. Webster, The development of flow simulation software for polymer processing applications; Revista Portguesa De Hemorreologia, 4 (suppl. 1/Pt A), 31-38, 1990.
4. M.F. Webster and P. Townsend, A Taylor-Galerkin Finite Element Method for Non-Newtonian Flows; presented at II World Congress on Computational Mechanics, Stuttgart, August, 1990.
5. D. Ding, P. Townsend and M.F. Webster, A Study of Some Model Problems for Validating Polymer Processing Computer Codes; presented at XII ICHMT Symposium on Manufacturing and Material Processing, Dubrovnik, August 1990.
6. D. Ding, P. Townsend and M.F. Webster, A Computer Code for Simulating Transient Non-Isothermal Flows of Non-Newtonian Fluids; in "Third European Rheology Conference and Golden Jubilee Meeting of British Society of Rheology", ed. D.R. Oliver, 132-134, Elsevier Applied Science, 1990.
7. O.Hassager, P. Henriksen, P.Townsend, M.F.Webster, and D.Ding, The Quarterbend: A Three- dimensional Benchmark Problem; J. Comput. Fluids, 20, **4**, 373,1990.
8. D. Ding, P. Townsend and M.F. Webster, Computer Science Report CSR 1-92, University of Wales Swansea.
9. K.Walters, private communication.

Theoretical and Applied Rheology, edited by P. Moldenaers and R. Keunings
Proc. XIth Int. Congr. on Rheology, Brussels, Belgium, August 17-21, 1992

NUMERICAL ANALYSIS OF SOME FINITE ELEMENT METHOD FOR THE APPROXIMATION OF DIFFERENTIAL MODEL FOR VISCOELASTIC FLOW

J. BARANGER, D. SANDRI

U.A. CNRS-740 LAN - bât. 101 - Université de Lyon 1 - 69622 VILLEURBANNE CEDEX, FRANCE.

1. INTRODUCTION

The Oldroyd B model in differential form is frequently used in numerical simulation with finite element (FE) method. The unknows are σ the viscoelastic part of the extra stress tensor, u the velocity and p the pressure. We show that for a suitable choice of finite element space coupled with a good upwinding technique for the state equation in σ, it is possible to prove that the method is convergent and we give error bounds.

For example a P_2/P_1 continous finite element approximation of (u,p) can be associated to either P_1 a discontinuous approximation of σ with upwinding by the Galerkin discontinuous method, or a P_1 continuous approximation of σ with upwinding Petrov Galerkin method (SUPG). In both case the errors are in $h^{3/2}$ for the three unknowns in the «energy» norm.

The result has been proved only in case the continuous problem has sufficiently small and smooth solution (see [1], [2], [7]) excluding situation where the geometry has a rentrant corner. The fact that the newtonian viscosity is non vanishing in the Oldroyd B model is also crucial and the Maxwell model is not included in this study.

Some other differential models can also be considered.

We begin by describing the Oldroyd B model for viscoelastic flow. In §3 we set the FE approximation to be studied, and existence of an approximate solution and error bounds are given.

2. CONSTITUTIVE LAWS OF DIFFERENTIAL TYPE

We consider a fluid flowing in a bounded, connected open set Ω in $\mathbb{R}^N$ with Lipschitzian boundary Γ ; n is the outward unit normal to Γ.

We use the following notations : $\mathbb{R}^N$ is equiped with cartesian coordinates x_i, $i = 1$ to N ; for a function u, $\frac{\partial u}{\partial x_i}$ is written $u_{,i}$ and $\frac{\partial u}{\partial t}$ is written u_t ; Einstein's convention of summation is used.

For a scalar function p, gradient of p is a vector ∇p with $(\nabla p)_i = p_{,i}$; for a vector function u, gradient of u is a tensor ∇u with $(\nabla u)_{ij} = u_{i,j}$.

For a vector function u, divergence of u is a scalar, $\nabla.u = u_{i,i}$ and $u.\nabla = u_i \frac{\partial}{\partial x_i}$; for a tensor function σ, divergence of σ is a vector, $\nabla.\sigma$ with $(\nabla.\sigma)_i = \sigma_{ij,j}$; if τ is an another tensor function $\sigma : \tau = \sigma_{ij}\tau_{ij}$.

In order to describe the flow we use the pressure p (scalar function) ; the velocity vector u and the (total) stress tensor σ_{tot}.

∇u is the velocity gradient tensor, $d(u) = 1/2(\nabla u + \nabla u^\top)$ the rate of strain tensor, $\omega(u) = 1/2(\nabla u - \nabla u^\top)$ the vorticity tensor. For $a \in [-1,1]$ one defines an objective derivative of a tensor σ by

$$\frac{\partial_a \sigma}{\partial t} = \sigma_t + (u.\nabla)\sigma + g_a(\sigma, \nabla u), \text{with } g_a \text{ given by}$$

$$g_a(\sigma,\nabla u) = \tfrac{1-a}{2}(\sigma\nabla u + \nabla u^\top \sigma) - \tfrac{1+a}{2}(\nabla u\sigma + \sigma\nabla u^\top).$$

This work has been supported in part by the GDR CNRS 901 "Rhéologie des polymères fondus".

Re is the Reynolds number, λ the Weissenberg number. We use a third dimensionless number α, $0 < \alpha \leq 1$ which may be considered as the fraction of viscoelastic viscosity ($\alpha = 1$ for Maxwell's Model).

The fluid is submitted to a density of forces f. Then the momentum equation is written :

$$\mathrm{Re}\Big(u_t + (u.\nabla)u\Big) - \nabla.\sigma_{\mathrm{tot}} = f.$$

The Oldroyd B model is described by $\sigma_{\mathrm{tot}} = -p\mathrm{I} + \sigma_N + \sigma$ where σ_N and σ are respectively the newtonian and viscoelastic part of the extra stress tensor $\sigma_{\mathrm{tot}} + p\mathrm{I}$; σ_N is defined by $\sigma_N = 2(1-\alpha)d(u)$. Substituting this expression in the momentum equation gives

$$\mathrm{Re}\Big(u_t+(u.\nabla)u\Big)-2(1-\alpha)\nabla.d(u)-\nabla.\sigma+\nabla p = f.$$

The viscoelastic stress tensor σ satisfies the constitutive equation

$$\sigma + \lambda\frac{\partial_a\sigma}{\partial t} - 2\alpha d(u) = 0. \tag{1.1}$$

We add the incompressibility condition $\nabla.u = 0$.

In the following we consider the case of a stationnary creeping flow (inertial terms are neglected).

We need also some boundary conditions. A possible one for u is $u = u_0$ given on Γ. Regarding σ and the hyperbolic character of (1.1) for u fixed one would try $\sigma = \sigma_0$ given on $\Gamma^- = \{x \in \Gamma, u.n(x) < 0\}$.

In order to make some theoretical numerical analysis of the problem we make the assumption that $u_0 = 0$; this implies that $\Gamma^- = \emptyset$ and there is no boundary condition for σ.We then obtain the following Oldroyd B problem :

Problem (O) Find (σ, u, p) such that

$$\begin{aligned}&\sigma + \lambda(u.\nabla)\sigma + \lambda g_a(\sigma, \nabla u)\\ &\qquad -2\alpha d(u) = 0 \quad in \quad \Omega, \qquad (1.2)\\ &-\nabla.\sigma - 2(1-\alpha)\nabla.d(u) + \nabla p = f \quad in \quad \Omega,\\ &\nabla.u = 0 \quad in \quad \Omega,\\ &u = 0 \quad on \quad \Gamma.\end{aligned}$$

The hyperbolic character of (1.2) implies that some upwinding is needed. It was considered first in the pioneering works [6] where streamline upwinding methods are used, and [3] using upwinding by discontinuous finite element. The choice of an upwinding technique depends on the choice of the FE space used to approximate σ as explained below.

3. FE APPROXIMATION

Let us introduce some notations. We denote by $(p,q) = \int_\Omega pq\,dx$, respectively $(\sigma,\tau) = \int_\Omega \sigma{:}\tau\,dx$ the scalar product in $L^2(\Omega)$, resp. $L^2(\Omega)^{N^2}$. The corresponding norm is denoted $|\ |$. We set

$$T = \{\tau,\ \tau_{ij} = \tau_{ji};\ \tau_{ij} \in L^2(\Omega);\ i,j = 1\ to\ N\},$$

$$X = H_0^1(\Omega)^N, \text{ and}$$

$$Q = L_0^2(\Omega) = \{q \in L^2(\Omega);\ \textstyle\int_\Omega q = 0\}.$$

$\Omega \subset \mathbb{R}^2$ is supposed to be polygonal and is equiped with a uniformly regular family of triangulations $\mathcal{T}_h$ made of triangles K. h denote the size of the mesh.

Let $P_k(K)$ denotes the space of polynomials of degree less or equal to k on $K \in \mathcal{T}_h$. We use for the approximation of (u,p) finite elements spaces satisfying

$$C^0(\overline{\Omega})^2 \supset X_h \supset \{v \in X \cap C^0(\overline{\Omega})^2;\quad v_{|K} \in P_\ell(K)^2,\ \forall K \in \mathcal{T}_h\},$$

$$Q_h \supset \{q \in Q;\ q_{|K} \in P_r(K),\ \forall K \in \mathcal{T}_h\}, \text{or}$$

$$Q_h \supset \{q \in Q \cap C^0(\overline{\Omega});\ q_{|K} \in P_r(K),\ \forall K \in \mathcal{T}_h\},$$

with $r \geq 1$ in the case of continuous approximation of the pressure.

and we assume that (X_h, Q_h) satisfies the inf sup condition (see [4]) :

$$\inf_{q\in Q_h}\ \sup_{v\in X_h} \frac{(q,\nabla.v)}{|q||d(v)|} \geq \beta > 0.$$

σ is approximated either by P_k discontinuous FE :

$$T_h = \{\tau \in T;\ \tau_{|K} \in P_k(K)^4,\ \forall K \in \mathcal{T}_h\},$$

or by P_k continuous FE :

$$T_h = \{\tau \in T\cap C^0(\overline{\Omega})^4;\ \tau_{|K} \in P_k(K)^4,\ \forall K \in \mathcal{T}_h\}.$$

We describe here FE approximation with continuous constraints. The SUPG method [5] used for the first equation of problem (O) is described by the following operator B defined for $(u, \sigma, \tau) \in X \times H^1(\Omega)^4 \times H^1(\Omega)^4$ by :

$$B(u,\sigma,\tau) = \Big((u.\nabla)\sigma, \tau + h(u.\nabla)\tau\Big) + \tfrac{1}{2}(\nabla.u\,\sigma,\tau).$$

We shall denote : $\tau_u = \tau + \lambda h(u.\nabla)\tau$.

Then problem (O) is approximated by

<u>Problem (O)</u>

Problem $(\mathrm{O})_\mathrm{h}$: *Find* $(\sigma_h, u_h, p_h) \in T_h \times X_h \times Q_h$ *such that*

$$\begin{aligned}
&(\sigma_h, \tau_{u_h}) + B(\lambda u_h, \sigma_h, \tau) + \lambda(g_a(\sigma_h, \nabla u_h), \tau_{u_h}) \\
&\qquad -2\alpha(d(u_h), \tau_{u_h}) = 0 \qquad \forall \tau_h \in T_h \\
&(\sigma_h, d(v_h)) + 2(1-\alpha)(d(u_h), d(v_h)) \\
&\qquad -(p_h, \nabla.v_h) = \langle f, v_h\rangle \quad \forall v_h \in X_h \\
&(q_h, \nabla.u_h) = 0 \quad \forall q_h \in Q_h.
\end{aligned}$$

The main result is the following :

Theorem. *Assume* $0 < \alpha < 1$, $k \geq 1$, $\ell \geq 1$ *and* $r \geq 0$. *Then there exists* C_0 , h_0 *such that if problem (O) admits a solution* $(\sigma, u, p) \in H^{k+1}(\Omega)^4 \times H^{\max\{3,\ell+1\}}(\Omega)^2 \times (H^{r+1}(\Omega) \cap L^2_0(\Omega))$ *satisfying*

$$\max\{||\sigma||_{k+1,2}, ||u||_{\max\{3,\ell+1\},2}, ||p||_{r+1,2}\} \leq C_0,$$

and if λ *is sufficiently small when* $\ell = 1$ *or* $r = 0$, *then for all* $h \leq h_0$, *problem* $(\mathrm{O})_\mathrm{h}$ *admits a solution* $(\sigma_h, u_h, p_h) \in T_h \times X_h \times Q_h$ *and there exists a* C *independant of* h *such that :*

$$|\sigma - \sigma_h| + |d(u-u_h)| + |p - p_h| \leq C h^{\min\{\ell, r+1, k+\frac{1}{2}\}}.$$

Sketch of the proof. For this purpose we define an application $\Phi : T_h \times V_h \longrightarrow T_h \times V_h$, wich to (σ_1, u_1) associates $(\sigma_2, u_2) = \Phi(\sigma_1, u_1)$ where $(\sigma_2, u_2) \in T_h \times V_h$ satisfies

$$\begin{aligned}
&(\sigma_2, \tau_{u_1}) + B(\lambda u_1, \sigma_2, \tau) + \lambda(g_a(\sigma_1, \nabla u_1), \tau_{u_1}) \\
&\qquad -2\alpha(d(u_2), \tau_{u_1}) = 0 \qquad \forall \tau_h \in T_h,
\end{aligned}$$

$$\begin{aligned}
&(\sigma_2, d(v_h)) + 2(1-\alpha)(d(u_2), d(v_h)) \\
&\qquad = \langle f, v_h\rangle \quad \forall v_h \in V_h,
\end{aligned}$$

with

$V_h = \{v \in X_h;\ (q, \nabla.v) = 0,\ \forall q \in Q_h.\}$

Then proof is decomposed in four parts : we show that

1) Φ is well defined and bounded on bounded sets.

2) Φ is continuous on $T_h \times V_h$.

3) There exists a ball B_h in $T_h \times V_h$ with center (σ, u) solution of problem (O) such that B_h is non empty and $\Phi(B_h) \subset B_h$ if h, $||\sigma||_{k+1,2}$, $||u||_{\max\{3,\ell+1\},2}$, and $||p||_{r+1,2}$ are sufficiently small. B_h being defines as follows :

$$B_h = \{(\tau, v) \in T_h \times V_h;$$
$$|\tau - \sigma|_{0,2} + |d(v-u)|_{0,2} \leq C^* h^{\min\{\ell, r+1, k+1/2\}}\}.$$

4) Brouwer's theorem gives then the existence of a fixed point (σ_h, u_h) of Φ solution of problem $(\mathrm{O})_\mathrm{h}$ and satisfying the error bounds of the theorem. Existence of p_h and error bound comes from the inf sup condition on (X_h, Q_h).

Remark. The result above is also valid, when Galerkin discontinuous method is used for the approximation of the first equation of the problem (O).

REFERENCES

1. J. Baranger and D. Sandri, C.R. Acad. Sci. Paris, Sér. I., t. 312, (1991) 541-544.

2. J. Baranger and D. Sandri, 10ème Colloque international sur les méthodes de Calcul Scientifique et Technique., INRIA, Paris, France.

3. M. Fortin and A. Fortin, J. Non-Newtonian Fluid Mech. 32, (1989) 295-310.

4. V.Girault and P.A. Raviart, Finite element method for Navier-Stokes equations,Theory and Algorithms, Springer, Berlin Heidelberg New-York 1986.

5. C. Johnson, U. Nävert and J. Pitkäranta, Comput. Methods Appl. Mech. Engrg. 45, (1984) 285-312.

6. J.M. Marchal and M.J. Crochet, J. Non-Newtonian Fluid Mech. 26, (1987) 77-114.

7. D. Sandri, C.R. Acad. Sci. Paris, Sér. I., t. 313, (1991) 111-114.

Theoretical and Applied Rheology, edited by P. Moldenaers and R. Keunings
Proc. XIth Int. Congr. on Rheology, Brussels, Belgium, August 17-21, 1992

NUMERICAL STUDY OF THE STABILITY OF VISCOELASTIC FLOWS

Ch. BODART, M.J. CROCHET
Unité de Mécanique Appliquée, Université Catholique de Louvain, 2 Place du Levant, 1348 Louvain-la-Neuve, Belgium

1. INTRODUCTION

A number of efficient and accurate finite element algorithms have been developed over the last ten years for calculating flows of viscoelastic fluids (see e.g. Crochet [1]). Most available solutions refer to steady-state flows ; in particular, flows through abrupt contractions have been the object of several papers, motivated by experimental observations with Boger fluid (see e.g. Boger [2] ; Walters [3]).

Experiments show that viscoelastic flow can bifurcate from steady-state to time-dependent behaviour. This was already noted by Nguyen and Boger [4] who observed transition from steady-state axisymmetric to swirling and periodic flow through an abrupt contraction. The transition between steady-state and oscillatory regimes was confirmed by Lawler et al [5] for the flow of organic Boger fluids.

The purpose of our paper is to examine whether time-dependent finite element simulations might generate observed instabilities and above all transitions to other types of flows. Our present paper is limited to plane flow through a four to one contraction, while the analysis of flows through a circular contraction is in progress. We have been motivated by the work of Fortin at al. [6] whose numerical results exhibited a non-symmetric flow regime.

After a brief reminder of the numerical method and of the time marching algorithm, we explain in more detail the type of boundary conditions adopted for the present work. Since our purpose is to generate time-dependent and possibly oscillatory flows while axial symmetry might be lost, it is essential to select boundary conditions which do not constraint the flow at a finite distance form the plane of contraction. We then perform our analysis with three finite element meshes and two different algorithms in order to verify the invariance of our results with respect to the finite element discretization.

2. FINITE ELEMENT FORMULATION

We calculate the flow of an Oldroyd-B fluid for which the extra-stres tensor is decomposed as the sum of a viscoelastic component (Maxwell-B fluid) and a purely viscous one. The ratio of the retardation to the relaxation time λ is selected as 1/9 throughout our calculations. The flow of an Oldroyd-B fluid through an abrupt contraction is characterized by the Reynolds number (Re) and the Deborah number (De) defined as follows,

$$Re = \rho V D_d / \eta , \qquad De = \lambda \dot{\gamma}_w ;$$

in these expressions ρ is the density, V is the average velocity in the downstream tube of diameter D_d, η is the total shear viscosity, λ is the relaxation time and $\dot{\gamma}_w$ is the wall shear rate in the downstream tube.

For our calculations, we use the finite element representation of Marchal and Crochet [7]. The velocity field is biquadratic and the pressure bilinear over quadrilateral elements while the stress tensor is bilinear over 4x4 quadrilateral sub-elements. For discretizing the constitutive equations we either use the Galerkin or the streamline-upwind (SU) formulation introduced by Marchal and Crochet [7]. The streamline-upwind / Petrov Galerkin formulation (SUPG) [8] is inefficient in the presence of stress singularities at sharp corners but can be used for smooth contractions.

The time integration consists of a fully implicit coupled scheme. We use a classical predictor-corrector scheme with automatic monitoring of the time-step for reducing the discretization error [9]. The predictor step is an explicit Adams-Bashforth scheme. The corrector step is a Cranck-Nicholson scheme solved with Newton's method ; the predicted values are used as first estimates in the iterative process while convergence is required within one iteration. Lack of convergence results in a decrease of the time step. In our examples below, the relative convergence criterion in 10^{-3}.

The fully coupled implicit calculation is expensive in computer time. In order to alleviate its cost, we have considered several decoupled schemes which result is an upper limit for the time step in view of conditional stability. The saving in computer time was found irrelevant for fluids with a single relaxation time but it might be significant for fluids endowed with a spectrum of relaxation times. A similar observation was made by Northey et al [10] in a recent paper.

The algorithm has been fully tested on start-up simple shear and Poiseuille flows for which an excellent agreement has been found between numerical and analytical results over a wide range of Weissenberg and Reynolds numbers.

3. BOUNDARY CONDITIONS

We wish to verify the stability of the flow of an Oldroyd-B fluid through an abrupt contraction. For that purpose, we first obtain a steady-state solution for an assigned value of the Weissenberg and the Reynolds number. The steady-state solution, which is obtained by means of a steady-state algorithm, allows us to calculate the pressure loss between the entry and the exit sections.

With the steady-state as an initial solution, we suddenly increase the pressure in the entry section with a step of 10 % of its initial value. We wish to find whether the flow tends to another steady-state configuration, corresponding to the higher pressure. We are however confronted with the difficulty of imposing correct boundary conditions. For a generalized Newtonian fluid, the absence of normal stress differences allows us to impose natural boundary conditions in the entry and in the exit sections corresponding to the applied pressure. For viscoelastic fluids, one usually imposes velocity and stress profiles in the entry sections and a fully developed velocity profile in the exit section. Such boundary conditions cannot be applied to our time-dependent problem since the resulting velocity profile is a priori unknown.

In the calculation below, we have introduced in the upstream channel a "slip-stick" section where the slip coefficient on the wall evolves smoothly from a vanishing value (full slip) to an infinite value (no slip). We are thus entitled to impose in the entry section a uniform non-vanishing pressure with vanishing components of the extra-stress tensor. The smooth slip-stick transition in the upstream section reduces the high stress level associated with an abrupt slip-stick transition.

In the exit section, we impose vanishing tangential velocity components. The normal contact force t_z is decomposed as the sum of a (vanishing) pressure and the extra-stress component T_{zz}. The latter, which is a priori unknown in the exit section, is expressed in terms of nodal values and transferred to the left hand-side of the discretized equations of motion. The above combination of inflow and outflow boundary conditions allows us to impose arbitrary pressure differences without constraints on the development of the velocity field which can possibly evolve freely towards a non-symmetric configuration.

4. FLOW THROUGH A CONTRACTION

Let us first consider the flow of an Oldroyd-B fluid through a four to one abrupt contraction. The flow domain contains an upstream section with a length of 45 D_d while the downstream length measures 25 D_d. The smooth slip-stick section ends at a distance of 15 D_d before the contraction.

We have calculated the flow on three finite element meshes (denoted I, II and III) which contain respectively 264, 504 and 760 elements. The configuration of the meshes near the plane of the contraction is shown in Fig. 1.

4.1. Galerkin method

Since our results are limited to relatively low values of the Deborah number, we have first used the Galerkin method for discretizing the constitutive equations, in view of its higher order of accuracy in terms of the finite element size h. The numerical tests are summarized in Table 1.

Table 1. Numerical tests for the flow through an abrupt contraction

Deborah number		1.0	2.0	3.0	4.0	5.0
Galerkin	MESH I			Stab.	Stab.	Stab.
Galerkin	MESH II	Stab.	Stab.	± Unst.	Unst.	
Galerkin	MESH III		Stab.	Unst.		
SU	MESH I			Stab.		
SU	MESH II		Stab.	Stab.	Stab.	Stab.
SU	MESH III		Stab.	Stab.		

With the coarse mesh I, we have not encountered stability problems up to De=6. The flow solution exhibits a smooth transition towards a new steady state while the variable time-step reaches its maximum allowable value once the new solution has been reached.

The flow behavior is quite different with mesh II. Here, a stable solution is reached up to De=2. At De=3, the behavior is uncertain : the time-step is very low but it does not collapse, while the velocity field exhibits an oscillatory behavior with a period of order 2 λ ; the velocity field remains fully symmetric. Fig. 2 shows the time-dependence of the axial velocity component at various points of the contraction plane ; the pressure jump occurs at time t=0. The upper curve corresponds to a point on the axis of symmetry. The other curves are double, since we have superposed the value of the velocity on both sides of the plane of symmetry. At De=4, the time-step collapses to such small values that it is impossible to pursue the calculations. The corresponding graph is shown on Fig. 3 ; again, we do not observe a lack of symmetry.

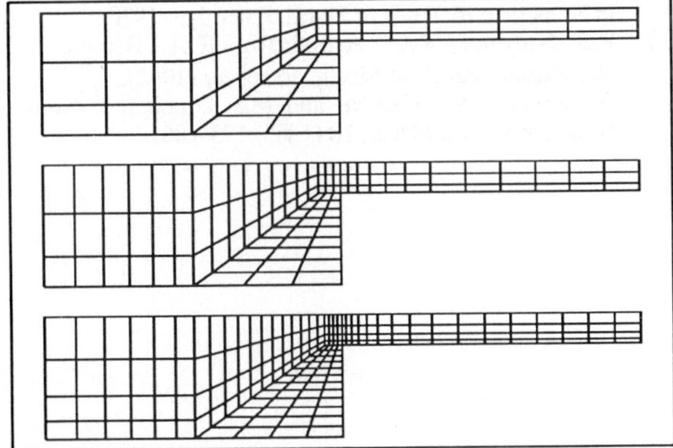

Fig. 1. Enlarged view of the finite element meshes near the contraction ; the flow is calculated over both sides of the plane of symmetry.

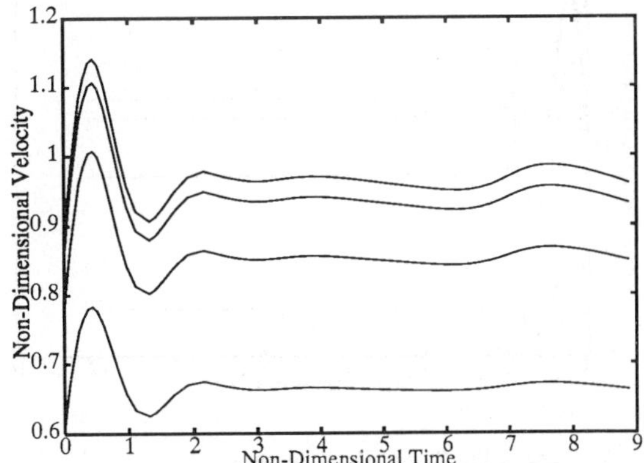

Fig. 2. Axial velocity component at various points of the contraction plane as a function of the non-dimensional time t/λ ; Galerkin, De=3, Mesh II

With mesh III, we have also found a stable behavior at De=2 and an unstable one at De=3.

We cannot however assert that the loss of stability is not a consequence of numerical error. In all cases of instability, we found that the loss of stability was preceded by the loss of positive-definiteness of the tensor $\mathbf{T_A} = \mathbf{T} + \eta/\lambda\, \mathbf{I}$ (see eg Dupret et al. [11]) which signals an artificial change of type.

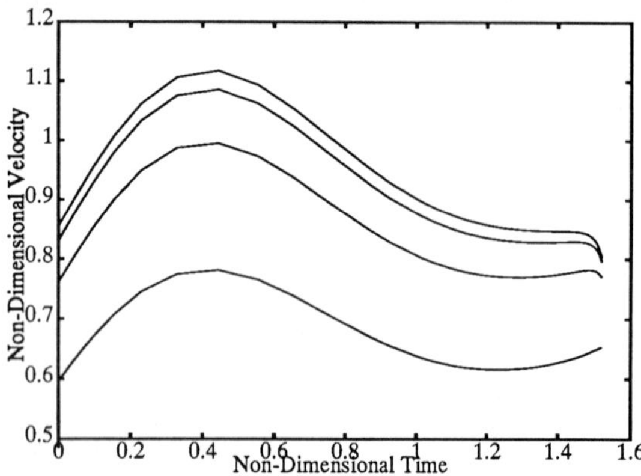

Fig. 3. Same as Fig. 2 ; Galerkin, De=4, Mesh II

4.2. Streamline-upwind method

It was shown in [7] that the use of SU for integrating the constitutive equations delays the loss of positive-definiteness of $\mathbf{T_A}$ and allows one to pursue the calculations up to large values of De. We have thus run the same tests with SU and we have obtained a stable behavior in all cases, as indicated in Table 1. Fig. 4 shows the evolution of the velocity field towards a stable behavior while the time-step reaches its allowable maximum. The tensor $\mathbf{T_A}$ is positive-definite in almost all cases. The velocity field remains symmetric.

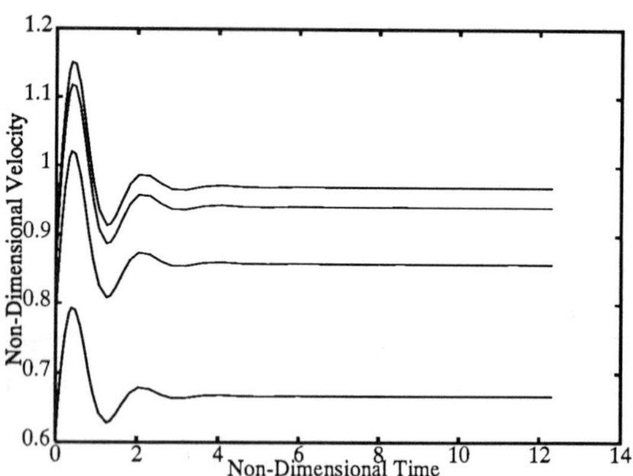

Fig. 4. Same as Fig. 2 ; SU, De=4, Mesh II

4.3. Flow through a smooth contraction

Since the loss of positive-definiteness of $\mathbf{T_A}$ is essentially due to the corner singularity, we have designed finite element meshes with rounded edges with radii respectively equal to 0.5, 0.25 and 0.1 D_d. The mesh refinement was similar to that of mesh II. Pressure jumps have been analyzed with Galerkin's method at De=3 and 4 for the three geometries. No lack of stability was observed.

5. CONCLUSIONS

The results of the present paper reveal the difficulty of discriminating between a true instability and the consequences of discritization errors. Instabilities have been observed for the flow through an abrupt contraction, but the loss of positive-definiteness of $\mathbf{T_A}$ signals a loss of accuracy. The SU method produces a stabler behavior. It is unfortunate that the collapse of the time increment associated with an instability prevents further study of the numerical solution.

ACKNOWLEDGEMENTS

The results presented in this paper have been obtained within the framework of Interuniversity Attraction Poles initiated by the Belgian State, Prime Minister's Office, Science Policy Programming.

REFERENCES

1. M.J. Crochet, Rubber Chemistry and Technology, 62/3 (1989) 426-455
2. D.V. Boger, Annual Review of Fluid Mechanics, 19 (1987) 157-182.
3. R.E. Evans and K. Walters, J. Non-Newtonian Fluid Mech., 20 (1986) 11-29.
4. H. Nguyen, D.V. Boger, J. Non-Newtonian Fluid Mech., 5 (1979) 353-368.
5. J.V. Lawler, S.J. Muller, R.A. Brown and R.C. Armstrong, J. Non-Newtonian Fluid Mech., 20 (1986) 51-92
6. M. Fortin, D. Esselaoui, Intern. J. for Numerical Methods in Fluids, 7 (1987) 1035-1052
7. J.M. Marchal, M.J. Crochet, J. Non-Newtonian Fluid Mech., 26 (1987) 77-114.
8. M.J. Crochet , V. Legat, J. Non-Newtonian Fluid Mech., 42 (1992) 283-300.
9. P.M. Gresho, R.L. Lee and R.L. Sani, Recent Advances in Numerical Methods in Fluids, Pineridge (1980).
10. P.J. Northey, R.C. Armstrong, R.A. Brown, J. Non-Newtonian Fluid Mech., to appear (1992).
11. F. Dupret, J.M. Marchal and M.J. Crochet, J. Non-Newtonian Fluid Mech., 18 (1985) 173-186.

Theoretical and Applied Rheology, edited by P. Moldenaers and R. Keunings
Proc. XIth Int. Congr. on Rheology, Brussels, Belgium, August 17-21, 1992

ON THE APPLICATION OF A MULTIMODE DIFFERENTIAL MODEL

M.B. BUSH

Department of Mechanical Engineering, The University of Western Australia, Nedlands 6009, Australia.

1. INTRODUCTION

There now exist a variety of sophisticated techniques for the simulation of complex viscoelastic fluid flow. The success of these methods lies in their exploitation of specific features of the equations to be solved. Multiple relaxation mode constitutive models, for example, may be solved efficiently by various streamline integration strategies (refs. 1-2), which can be used in uncoupled solution procedures. On the other hand, high Weissenberg number solutions can be obtained within coupled procedures using spectral methods (ref. 3) and finite element schemes which in some way exploit the 'type' of the coupled set of momentum, continuity and constitutive equations (refs. 4-6). Thus, use of the 'explicitly elliptic momentum equation' (EEME) formulation and streamline upwinding applied to the constitutive equations has gone a long way to relieving the so called 'Weissenberg number limit' restriction on calculations in complex flows.

Armed with such successful techniques for simulating viscoelastic flows, the natural next step is to apply the methods to more and more complex problems. An obvious way in which the complexity should be increased is to include multiple mode viscoelastic models, which represent to varying degrees the complex relaxation spectrum exhibited by most real viscoelastic fluids.

Treatment of multiple relaxation mode models has been relatively scarce to date. Luo and Tanner (ref. 1) applied a multimode KBKZ integral model (8 modes) to the simulation of the extrusion of low density polyethylene LDPE, while Bush (ref. 2) applied a multimode Leonov differential model (7 modes) to the same problem, with comparable results. More recently, Rajagopalan et al (ref. 5) incorporated a two-mode Maxwell model into their EEME formulation.

The introduction of additional relaxation times normally increases the computation time in a nonlinear fashion. There is strong motivation, therefore, to utilize the minimum number of relaxation times to achieve a sufficiently accurate solution. The usual procedure for simulation of the flow of a real polymer material will start with reliable characterisation data for that material, frequently in the form of a relaxation spectrum. This can then be used to provide the material properties corresponding to each of the discrete relaxation times chosen to represent the spectrum. Any other parameters in the numerical model can then be adjusted to produce the correct viscous, elongational and normal stress behaviour if necessary. The question then arises as to how well this experimental data needs to be represented in order to produce adequate results when pressed into service to simulate the flow of the material in question.

The answer to this question no doubt depends on the particular model being used and the type of flow problem to be solved. In this paper we study this question for the case of a Phan-Thien-Tanner (PTT) model applied to the extrusion flow of LDPE.

2. GOVERNING EQUATIONS

The momentum and continuity equations are given as

$$\nabla.\sigma = \mathbf{0} \quad \text{and} \quad \nabla.\mathbf{u} = 0 \tag{1}$$

where $\mathbf{u}$ is the velocity vector and σ the stress tensor. For N relaxation modes we have

$$\sigma = -p\mathbf{I} + \sum_{i=1}^{N} \tau_i \tag{2}$$

where p is the pressure and τ_i is the extra stress corresponding to the i-th relaxation mode, which is given by the PTT model as

$$\lambda_i \frac{\Delta\tau_i}{\Delta t} + \tau_i \exp\left(\frac{\varepsilon\lambda_i}{\eta_i}\right) + \lambda_i\zeta(\mathbf{D}\tau_i + \tau_i\mathbf{D}) = 2\eta_i\mathbf{D} \tag{3}$$

where **D** is the strain rate tensor, λ_i is the relaxation time of mode i and ε, ζ are the usual parameters. In the current work we have chosen $\varepsilon = 0.02$ and $\zeta = 0.075$.

3. SOLUTION METHOD

The boundary element method was used to solve the momentum and continuity equations while a streamline integration scheme was used to solve the constitutive equations (ref. 2). The streamlines in the flow coincide with characteristic curves of the steady form of (3). This then permits the stress field along a streamline to be computed by integration of a set of ordinary differential equations corresponding to (3), starting at an upstream location where the stress tensor is known. Integration has been performed by either Gear or Runge Kutta algorithms, depending on the stiffness of the corresponding set of differential equations. When relaxation times ranging over several decades are to be employed, it is likely that one or more of the equation sets are stiff. The particular relaxation times at which this occurs depend on the kinematics of the flow, which is itself a function of the model flow rate.

The boundary element formulation uses quadratic isoparametric elements and domain cells. The entire solution is obtained by an iterative scheme involving alternate computation of the kinematic and kinetic solution fields. Details can be found in ref. 2.

4. RESULTS

The problem treated in this study is that of extrusion of LDPE from a long circular die. This is the geometry utilized in refs. 1-2, and was also considered experimentally by Meissner (ref. 7). The inlet and outlet lengths treated in the calculation are each 4R, where R is the radius of the tube. Boundary conditions consist of fully developed flow upstream, no slip on the die wall, no stress on the jet surface and plug flow downstream. The mesh consists of 43 boundary elements (87 nodes) and 224 domain cells.

The relaxation spectrum has been divided into N regions of approximately equal width on the $\log_{10}$ scale. Relaxation times ranging between 0.01 seconds and 3000 seconds were adopted.

Figure 1 shows the corresponding steady shear, normal stress and steady elongational response of the resulting PTT model for N = 3, 5 and 7 relaxation times. These are compared with measured values (refs. 7-8). In general, all the models produce an apparently reasonable prediction of the data, although the effect of the poor representation of the relaxation spectrum produced by the 3 mode model is obvious. The accuracy of the predicted shear and elongational data clearly improves as the number of modes is increased.

In Figure 2 we see the computed swell ratio as a function of the mode number, for various values of the apparent shear rate in the tube ($D = 4\bar{u}/R$, where $\bar{u}$ is the average velocity). It is clear that the behaviour becomes somewhat unpredictable as the number of relaxation modes is reduced. For $N > 5$, however, the swell ratio is approximately steady.

In Figure 3 the computed results are shown as a function of apparent shear rate. Once again the convergence for $N > 5$ is apparent. We note that despite this convergence, the comparison with the experimental data is not good, particularly at higher values of flow rate. The reason for this is not clear (refs. 1-2), but it is likely that a different choice of model or parameters may improve the picture.

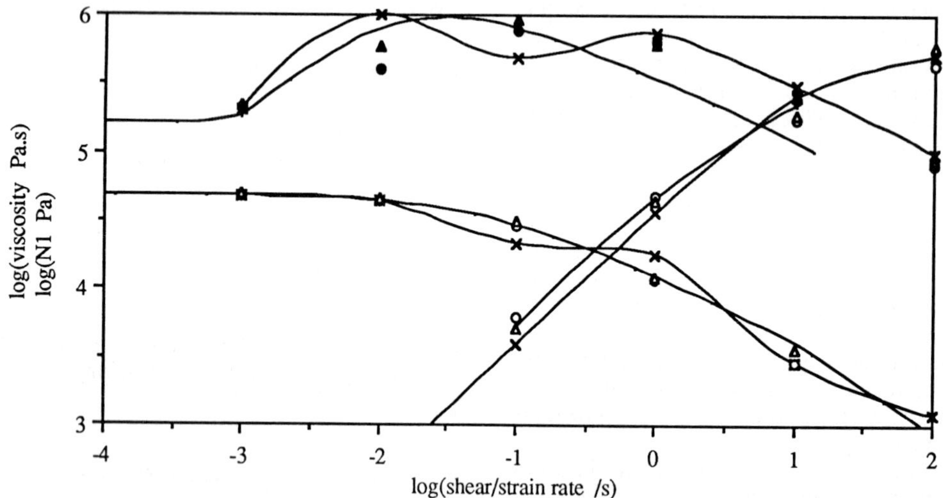

Figure 1. The predicted steady shear and elongational response compared with measured values: ———, measured values (refs. 7-8); —X—, 3 mode; O, 5 mode; Δ, 7 mode. Filled symbols correspond to elongational properties.

5. CONCLUSIONS

This study indicates that at least 5 modes should be used to successfully model this problem. Although the result will be somewhat fluid and problem dependent, it clearly sheds doubt on the wisdom of applying single mode models to the simulation of the flow of complex fluids.

As pointed out above, the relaxation spectrum was divided into approximately equal increments in relaxation time. The effect of a different scheme of subdivision has not been investigated, but will form part of future work.

REFERENCES

1. X.L. Luo and R.I. Tanner, J. Non-Newt. Fluid Mech., 22(1886)61.
2. M.B. Bush, J. Non-Newt. Fluid Mech., 31(1989)179-191.
3. S. Pilitsis and A.N. Beris, J. Non-Newt. Fluid Mech., 31(1989)231-287.
4. J.M. Marchal and M.J. Crochet, J. Non-Newt. Fluid Mech., 26(1987)77-114.
5. D. Rajagopalan, R.C. Armstrong and R.A. Brown, J. Non-Newt. Fluid Mech., 36(1990)135-157.
6. X-L. Luo and R.I. Tanner, J. Non-Newt. Fluid Mech., 31(1989)143-162.
7. J. Meissner, Pure Appl. Chem., 42(1975)553.
8. H.M. Laun and H. Munstedt, Rheol. Acta., 17(1978)415.

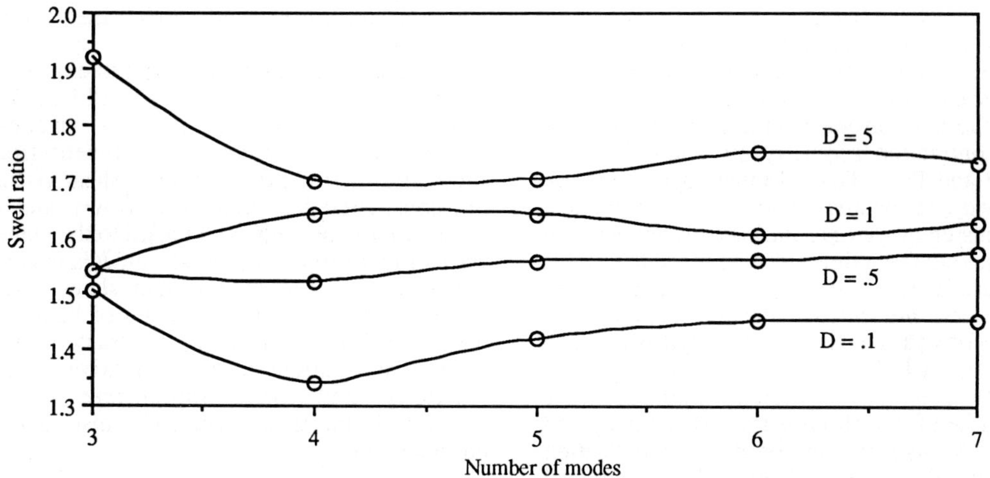

Figure 2. The computed swell ratio for a variety of values of apparent shear rate, D.

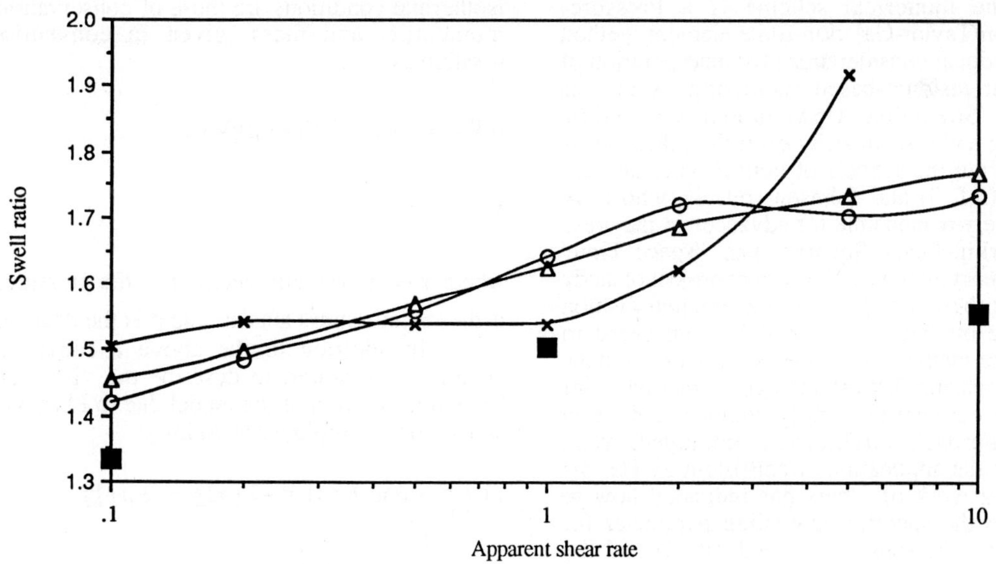

Figure 3. The predicted swell ratio as a function of apparent shear rate: ■, measured values (ref. 7); —X—, 3 mode; —O—, 5 mode; —Δ—, 7 mode.

Theoretical and Applied Rheology, edited by P. Moldenaers and R. Keunings
Proc. XIth Int. Congr. on Rheology, Brussels, Belgium, August 17-21, 1992

TRANSIENT VISCOELASTIC FLOW CALCULATIONS FOR HIGHLY ELASTIC FLUIDS

E.O. CAREW, P. TOWNSEND and M.F.WEBSTER

University of Wales Institute of non-Newtonian Fluid Mechanics, Department of Mathematics and Computer Science, University College, Swansea, SA2 8PP, U.K.

1. INTRODUCTION

This paper specifically addresses the issue of attaining consistent accurate solutions for viscoelastic flows of differential constitutive model fluids. The main problem investigated is the 4:1 planar contraction geometry on which flows of Oldroyd and Phan Thien-Tanner type fluids are considered (for details see ref. 1). The numerical algorithms employed are intrinsically transient in nature and involve a time stepping procedure. In an earlier article (ref. 2) much necessary background material was presented and some results were reported for transient flows of Oldroyd-B fluids in a plane channel. There analytical solutions were available and the numerical methods were benchmarked for such properties as accuracy and stability. An appropriate non-dimensionalisation was employed and the availability of this analytical solution provides the means by which a consistent transient flow may be analysed for contraction type flows.

The numerical scheme is a Pressure-Correction/Taylor-Galerkin finite element method with additional considerations for incorporation of consistent residual-based stabilizing extensions such as, 'Streamline Upwind Petrov-Galerkin' weighting and discontinuity capturing, these ideas coming from the schools of both Hughes and co-workers (ref. 3) and Johnson (ref. 4), who have been largely responsible for advances in the areas of 'Galerkin-Least Squares' and 'Space-Time' finite element methods. The current project of study addresses these aspects in an incremental fashion and some of these issues will be considered in detail. The method used here is a fractional-stage transient scheme that offers novel attractions over the more conventional fully-coupled and often steady approaches frequently employed. With respect to the application of consistent SUPG, the pioneering work of others has indicated how to generalise the specific upwinding parameter for unstructured meshes (ref. 5 and ref. 6) and the effectiveness of its application is studied herein for various flows. The often conflicting issues of time-stepping instability and model instability are examined.

Particular attention is paid to exploring the differences between the Oldroyd and Phan Thien-Tanner models, for a range of viscoelastic behaviour. The physical representation of properties such as extensional and shear viscosities are considered for contraction flows, and some observations are made with regard to the numerical tractability of some viscoelastic models in complex flows. In this regard, constant shear viscosity models of Oldroyd-B and a class of Phan Thien-Tanner (PTT) models are contrasted in their numerical solutions at representative values of elasticity, and where appropriate additional variation in the shear viscosity is also taken into consideration.

2. MATHEMATICAL EQUATIONS

The basic equations relevant to the incompressible flow of viscoelastic materials under isothermal conditions are those of conservation of momentum and mass, given in conventional notation as

$$\rho \frac{\delta \underline{\mathbf{v}}}{\delta t} = - \nabla p + \nabla(\underline{T}) - \rho \mathbf{v} \nabla \mathbf{v} ,$$

$$\nabla . \mathbf{v} = 0 ,$$

where $\mathbf{v}$ is the velocity vector, p is the pressure, $\underline{T}$ is the viscous stress tensor, and ρ is the density.

In addition to the above a constitutive equation is required to describe the viscoelastic behaviour. A form of the model due to Phan Thien and Tanner is employed as follows:

$$\{1 + \frac{\varepsilon \lambda_1}{\mu_1} \mathrm{trace}(\underline{\tau})\} \underline{\tau} + \lambda_1 \mathrm{D}\underline{\tau} = 2 \mu_1 \underline{D}$$

$$\underline{T} = \underline{\tau} + 2 \mu_2 \underline{D}$$

where $\underline{\tau}$ is an extra-stress tensor, λ_1 is the relaxation time, and μ the fluid viscosity may be decomposed into the sum of a solvent viscosity μ_2 and a solute viscosity μ_1. From these one can compute the retardation time λ_2, normally associated with the Oldroyd-B model from $\lambda_2/\lambda_1=\mu_2/\mu$. The convected derivative is a combination of both the upper and lower convected derivatives, given by

$$D\underline{\tau} = (1-\xi/2)\overset{\nabla}{\underline{\tau}} + (\xi/2)\overset{\Delta}{\underline{\tau}} .$$

The material parameters ε and ξ are defined such that $\varepsilon\geq0$ and $0\leq\xi\leq2$ providing a class of such fluids. For the particular case of the Oldroyd-B model $\varepsilon=\xi=0$, and then only the upper convected derivative is present with no additional contribution of nonlinearity from the trace($\underline{\tau}$) term.

The statement of the problem mathematically is completed with the provision of appropriate boundary conditions on the velocity components or their derivatives, and the stress components which are dictated by the flow conditions. A schematic diagram of the problem is presented in fig.1 below.

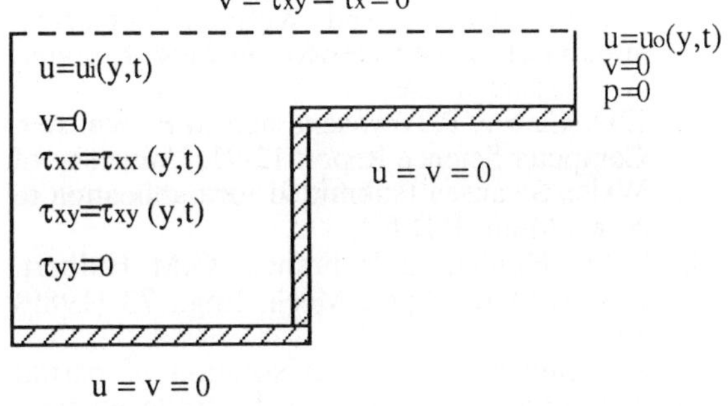

Figure 1. The 4:1 planar contraction

3. DISCUSSION

The Oldroyd-B model is considered first. A nondimensional relaxation time λ_1 is quoted (ref. 7) to characterise each instance of fluid elasticity studied, though it is clearly understood that the retardation time is taken as $\lambda_1/9$ in all instances. To derive a Deborah number based on a constant downstream shear rate a multiplicative factor of three must be applied to the above. A unit value of Reynolds number is generally assumed and interest is focused on steady state solutions. On adopting the usual continuation approach through increasing λ_1 solutions, the conventional Taylor-Galerkin (TG) algorithm applied to this problem provided solutions up to a limit point of around $\lambda_1=1.5$. Beyond such a point the transient development of the solution encounters a sudden and sharp exponential explosion that renders the scheme divergent. Decreasing the time-step does not surmount this instability. The form of the algorithm used here is one such that in a first stage a nondivergence-free velocity field is generated over a predictor-corrector pair of equations for both velocity and stress. A second stage then solves for a pressure difference over the time-step as a result of the degradation in incompressibility, and a third stage then completes the computation on the time-step cycle to recover a divergence-free velocity field. The variational weighting functions are taken as the standard (mixed) Galerkin forms, with velocity and stress governed by quadratic functions and pressure by linears over triangles subdividing the flow domain. The momentum equations are discretised in time through a semi-implicit scheme (ref. 8), whilst the constitutive equations adopt the explicit form (ref. 9). This factorization of the system of equations gives rise to much smaller problems to solve at each stage, hence making it possible to solve larger problems in their entirety (such as three-dimensional instances). By selecting the subcritical value of $\lambda_1=1$ and a time-step of 0.01, it is possible to compare the quality of solution obtainable from the TG scheme as quoted and a Taylor-Streamline Upwind Petrov Galerkin (TSUPG) form, where only the weighting function that acts on the constitutive equation is modified. Focusing attention upon the shear stress field one observes that the TG solution displays a nonsmooth form and that the TSUPG solution corrects this situation to present a high quality result, recapturing the smoothness globally, as associated with the TG scheme at a level of λ_1 one order of magnitude lower.

Pursuing this success to $\lambda_1=2$ illustrates the point that the TG scheme fails to converge, but so does the TSUPG scheme with a time-step of 0.01. Here break-up of the shear stress field occurs and is most prominent beyond the contraction plane being associated with the channel wall. Discontinuities then arise in shear stress gradients in the cross-stream direction. Also the nature of the lack of convergence for TSUPG has now changed substantially from that observed for TG. One now sees in place of the sudden exponential growth, high amplitude and frequency oscillations in the associated error norms as time evolves, that do not decay. This behaviour is associated with a time-stepping instability and may be removed by decreasing the time-step to 0.005, at which point a convergent solution is then attained. It is instructive to also note what effect introducing locally discontinuity capturing terms has on the numerical scheme and the solution here. Indeed at a time-step of 0.01, such locally smoothing factors do smooth out the stress field and are only brought into effect within the neighbourhood of sharp changes in the solution gradient as required. However such factors cannot in themselves cure the presence of instabilities such as that due to time-stepping here. For TSUPG and a semi-implicit approach, a limit point of λ_1=5 emerges for the Oldroyd-B model. This situation cannot be remedied by reducing the time-step, but is modified by further introduction of implicitness into the scheme.

However shifting attention to the PTT class of models, it is possible to progress further with this same scheme. This class of models is well categorised in terms of a pair of constant material parameters (ε, ξ), and suitable values of these are proposed by Tanner (ref. 9). Accordingly utilising values of (ε, ξ)=(0.02, 0) introduces a constraint on the extensional behaviour whilst retaining the upper convected derivative of the Oldroyd-B fluid. For such a fluid solutions are derived at λ_1=5, 10 and 15 and there appears to be no upper limit imposed on λ_1. Such a switch in model has incorporated two new factors, one being a limit on the extensional viscosity (only implicit to the numerics) and a second being a variable shear viscosity. To provide a check on the influence of the second factor two further PTT instances were considered those of (ε, ξ)=(0, 0.00005) and (0, 0.01). For such values these PTT models possess unconstrained extensional characteristics (for ε=0), whilst simultaneously displaying shear-thinning behaviour that bounds, from above and below, the shear-dependent viscosity function of the prior converged result. Both such PTT choices again diverged, and hence the natural conclusion to reach is that the constraint on the extensional behaviour is the critical factor here and not the effects of shear-thinning.

It can be substantiated that the critical factor in extension rate that governs the singularity in extensional viscosity for the Oldroyd-B fluid is certainly exceeded for λ_1=5, and that one should therefore naturally expect the factors closely associated with extension in the computation (such as the first normal-stress difference) to behave accordingly. Indeed along the centre line of the channel, where no shearing occurs, discontinuities in first normal-stress difference are encountered for the Oldroyd-B fluid as λ_1=5 is approached. This is not found to be the case for the PTT model with (ε, ξ)=(0.02, 0), further justifying the above conclusions.

4. CONCLUSIONS

Instabilities of both numerical and fluid model origin have been investigated here and it is shown how such effects may be overcome. The TSUPG scheme offers clear advantages above and beyond the TG scheme whenever solution gradients primarily in the flow direction require control. When solution gradient tangential to the flow become significant and interior or boundary layers are encountered, then some form of discontinuity capturing is advocated. The form adopted here is localised in its action being governed by the residual of the equations themselves, and the solution gradient. Such strategies are consistent and their properties are well understood.

The limitations of the Oldroyd-B fluid have been explored and by contrasting the behaviour of the solution for this model under identical working conditions with that of the class of PTT models, the critical limiting factors involved have been established. It is in this manner that by providing control to the extensional behaviour of these models solutions may be reached without apparent limit in elasticity. Though shear-thinning has a role to play here its influence is shown to be superficial to this central issue.

REFERENCES

1. Crochet, Davies and Walters, Numerical simulation of non-Newtonian flow, Elsevier, Amsterdam, 1984.
2. E.O. Carew, P.Townsend and M.F. Webster, Computer Science Report 12-91, University of Wales Swansea (submitted for publication to Num. Meth. P.D.E.).
3. T.J.R Hughes, L.P Franca, G.M Hulbert, Comp. Meth. Appl. Mech. Eng., 73 (1989) 173.
4. C. Johnson, Numerical solution of partial differential equations by the finite element method, Cambridge University Press, 1987.
5. F. Shakib, Ph.D. diss., Stanford University (1987).
6. F. T. P. Baaijens, private communication, 1990.
7. N.D. Waters and M.J. King, Rheol. Acta, 9 (1970) 245-355.
8. D.M. Hawken, H.R. Tamaddon-Jahromi, P. Townsend and M.F. Webster, Int. J. Num. Meth. Fluids, 10 (1990) 327-351.
9. P. Townsend and M.F. Webster, in NUMETA 87: Vol 2, Nijhoff, 1987.
10. R.I. Tanner, Definition of the PTT model for VIIth Workshop, private communication, 1989.

Theoretical and Applied Rheology, edited by P. Moldenaers and R. Keunings
Proc. XIth Int. Congr. on Rheology, Brussels, Belgium, August 17-21, 1992

INITIAL CONDITIONS FOR MULTIMODE FLUIDS

Bruce Caswell, Division of Engineering, Brown University, Providence, RI 02912 USA.
Support for this work under NSFGrant CTS-8921668 is gratefully acknowledged.

Initial value problems for linear viscoelastic fluids are most simply formulated by Laplace transformation of time (ref 1., 2., 3.). However, the inversion of the transforms presents certain difficulties in the case of fluids described by a spectrum of relaxation times (multimode fluids). For shear flows solutions of the governing equations are most easily evaluated for those obtained by direct methods, such as the method of separation of variables used in this work. In the Laplace transform formulation the initial conditions on the time derivatives of the velocity V are imposed implicitly in the case of startup from rest. Direct methods require explicit initial conditions on $\partial^n V/\partial t^n$ for n=1,..,N. Shear flows with shear planes defined by y=constant and with non-zero component of velocity u(y,t) is in the x-direction are the motions to studied here. The fluids are taken to be linearly viscoelastic with relaxation moduli G(t) represented by multimode Prony series,

$$G(t) = \sum_{i=1}^{N} g_i e^{-t/\lambda_i} \ ,$$

where the set $\lambda_i, g_i \ ; i=1, N$ is the N-mode spectrum of relaxation times and moduli. The constitutive law for the shear stress **T** can be written as

$$L_1(\partial /\partial t)\mathbf{T}=L_2(\partial /\partial t)u_y \ , \tag{1}$$

where the the subscript y stands for the spatial derivative, and the operators L_1, and L_2 have the structure

$$L_1(s)=(1+\lambda_1 s)(1+\lambda_2 s)...(1+\lambda_N s) \ , \tag{2a}$$

$$L_2(s) = \eta + ... + \lambda_1\lambda_2...\lambda_N G_0 s^{N-1} \ . \tag{2b}$$

Here η is the zero shear-rate viscosity, G_0 is the zero-time value of G(t), and the argument s is the Laplace transform image of the time derivatives. The transform of G(t),$\overline{G}(s)$, is related to these operators by by

$$\overline{G}(s) = L_2(s)/L_1(s) \ . \tag{3}$$

In the absence of a pressure gradient the equation of motion is

$$\rho\partial u/\partial t = \mathbf{T}_y \ . \tag{4}$$

A single governing equation is obtained by elimination of the stress between equations (1) and (4),

$$\rho L_1(\partial /\partial t)(\partial u/\partial t) = L_2(\partial /\partial t)u_{yy} \ . \tag{5}$$

In order to examine the initial behavior of the solution of (5) it is operated upon with

$$L_0(s) = 1 + \lambda'' s \ . \tag{6}$$

The transform of equation (5) then becomes

$$\rho L_0(s).L_1(s)s\overline{u} = \lambda'' G_0 L_1(s)\overline{u}_{yy} + L_3(s)\overline{u}_{yy} \ , \tag{7}$$

where $L_3 = L_0.L_2 - \lambda'' G_0 L_1$. At very short times(large s) it follows from the structure of the L_1 and L_2 operators that the governing equation (7) reduces to

$$\rho L_0(s)s\overline{u} = \lambda'' G_0\overline{u}_{yy} + O(s^{-2}) \ , \tag{8}$$

The order of the error in (8) follows directly as s^{-1} from equations (2a,b), (6), and the definition of the L_3 operator; the additional accuracy is obtained by chosing λ'' to be $-G_0/\dot{G}_0$, where the denominator is the zero time value of dG(t)/dt. This follows from the terms in the L_2 operator not given explicitly in (2b). Equation (8) is then recognizable as the wave equation governing the motion of shear waves propagating with speed $c = (G_0/\rho)^{1/2}$ and whose amplitude attenuates as $e^{-t/2\lambda''}$; it is also the exact governing equation (5) for N=1, the Maxwell fluid.

It follows from (8) that for the multimode approximation initial values of the first N partial time derivatives must be set to ensure satisfaction of the wave equation. This condition has been deduced from the transform of the governing equation (5) together with the assumption of startup from rest. Other initial conditions such as the decay from an initially steady flow require a reexamination of the short time behavior of the governing equation. In all cases initial conditions must satisfy a necessary condition of degeneracy derived from the observation that an N-mode fluid degenerates to N-1 modes when any two relaxation times are equal. This means the solution of the governing equation must also reduce to that of the (N-1)-mode case. This principle is best illustrated with the initial conditions for the startup of shear flow of multimode fluids with N=0, 1, 2, 3. With subscript t to denote partial time derivatives these conditions are:

$$N = 0\ (Newtonian):\ u(y,0) = 0\ . \tag{9}$$

$$N = 1\ (Maxwell):\ u(y,0) = 0\ ,\ u_t(y,0) = 0\ . \tag{10}$$

For 2-modes a condition on u_{tt} is obtained from the equation of motion(8) for the Maxwell fluid and the second of conditions (10); hence

$$N = 2:\ \ u(y,0) = 0\ ,\ u_t(y,0) = 0\ ,$$

$$u_{tt}(y,0) = c^2 u_{yy}(y,0)\ . \tag{11}$$

The governing equation(5) for the 2-mode fluid is

$$\lambda_1\lambda_2 u_{ttt} + (\lambda_1+\lambda_2)u_{tt} + u_t = (\eta/\rho)u_{yy} + c^2 u_{tyy}\ . \tag{12}$$

When the initial conditions (11) for the 2-mode fluid are applied to (12) it reduces to the time derivative of the wave equation (8)

$$u_{ttt} + (1/\lambda'')u_{tt} = c^2 u_{tyy}\ . \tag{13}$$

The right hand side of (13) vanishes in view of the second condition of (11), and consequently the appropriate initial conditions for the 3-mode fluid are

$$N = 3:\ conditions\ (11),\ \text{and}$$

$$u_{ttt}(y,0) = -\,(c^2/\lambda'')u_{yy}\ . \tag{14}$$

This procedure can be generalized for arbitrary N, the additional initial condition required for the N-mode fluid is determined from the governing equation of the N-1 case. Initial conditions derived in this way satisfy the degeneracy condition because of the way they incorporate the initial conditions of the N-1, N-2, ...-mode fluids, and because the coefficients appearing in these initial conditions are properties of G(t) which can be represented in terms of the modal parameters for any N.

The initial conditions derived above apply to instantaneously elastic fluids. For fluids exhibiting instantaneous viscosity the viscoelastic stress as given by (1) is augmented by a purely viscous term characterized by the 'solvent' or instantaneous viscosity η_s. An analysis at short times, similar to that given above, leads to the viscous analog of the wave equation,

$$u_{tt} + (1/\lambda'')u_t = (c^2+\nu_s/\lambda'')u_{yy} + \nu_s u_{tyy}\ , \tag{15}$$

where the positive coefficient $\nu_s = \eta_s/\rho$ of the last term guarantees ellipticity. For small values of this parameter the jumps predicted by the wave equation($\nu_s = 0$) become steep but smooth fronts. With (15) in hand the initial conditions equivalent to (11) and (14) can be written down by inspection.

With the above initial conditions solutions were obtained by the method of separation of variables for the run-up problem (ref 1.) between plane walls, and are shown in Figures. 1, 2, 3. In this problem the planes at y = + and -1 are set in motion with unit velocity at t = 0; the fluid, initially at rest, adheres to the planes and is dragged along by the boundary motion. These Figures show the velocity profiles at six successive, equally spaced times for 1, 2, and 3-mode instantaneously elastic fluids. The largest relaxation times are unity and the others are successively smaller by 1/10. The moduli are: 1.0 for N = 1, 0.04, 0.96 for N = 2, and 0.04, 0.24, 0.72 for N = 3; hence with unit density the wave speeds are equal in each case. The feet of the arrows are located at the wave fronts. Figure 1. is typical of the Maxwell fluid, see (ref 2). Figure 2 shows that with two modes the wavefronts are initially discontinuities of the velocity gradient, but are smoothed after a single encounter with the centerline. For three modes, Figure 3, the initial wavefronts are dicontinuities of higher order gradients.

The initial conditions given above are examples which satisfy the degeneracy condition. An example for which this condition does not hold is obtained with a generalization of conditions (10) for the Maxwell fluid by setting to zero all of the $\partial^i u/\partial t^i\ ,\ i = 1, N$. Figure 4 shows velocity profiles obtained when such initial conditions are used in the run-up problem for a 3-mode fluid having the same properties as the fluid of Figure 3. The wall spacing has been reduced by 1/10 to enhance the apparent slip which is evident in the solution even though the adherence condition was imposed at the walls.

REFERENCES

1. Rivlin, R. S., Rheol. Acta 21(1982), 213-222

2. Kazakia, J. Y. and R. S. Rivlin,
Rheol. Acta 20(1981), 111-127

3. Preziosi, L. and D. D. Joseph,
J.Non-Newtonian Fluid Mechanics 33(1987), 239-259

Figure 1: 1-Mode Fluid

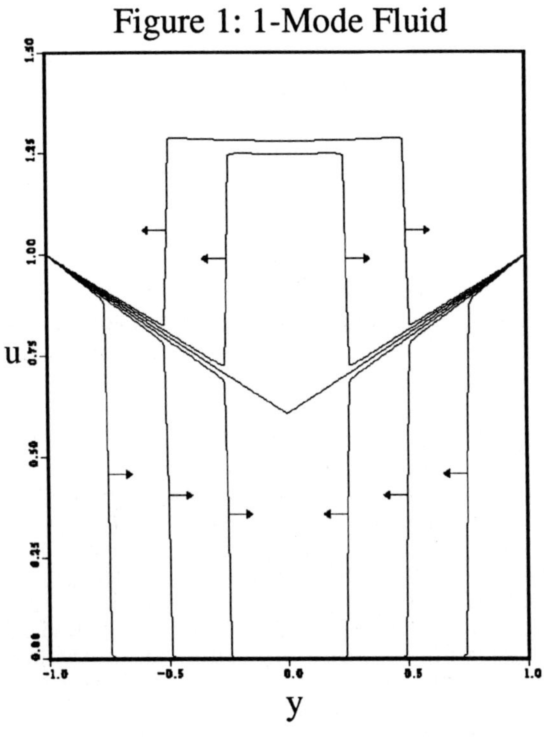

Figure 3: 3-Mode Fluid

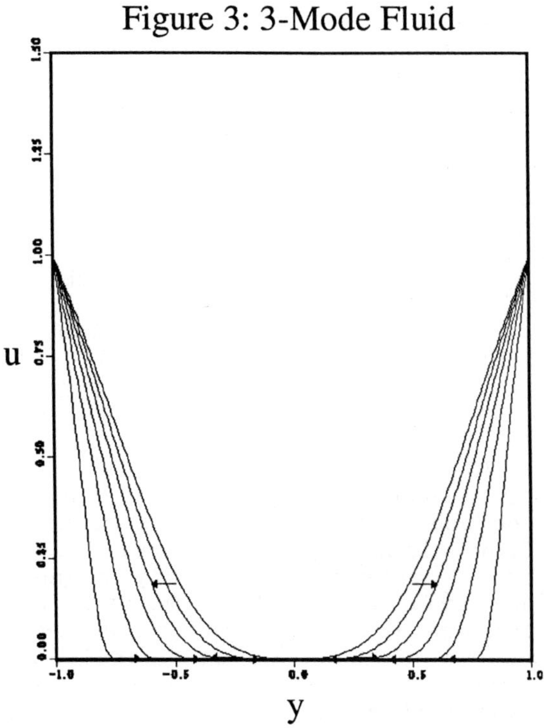

Figure 2: 2-Mode Fluid

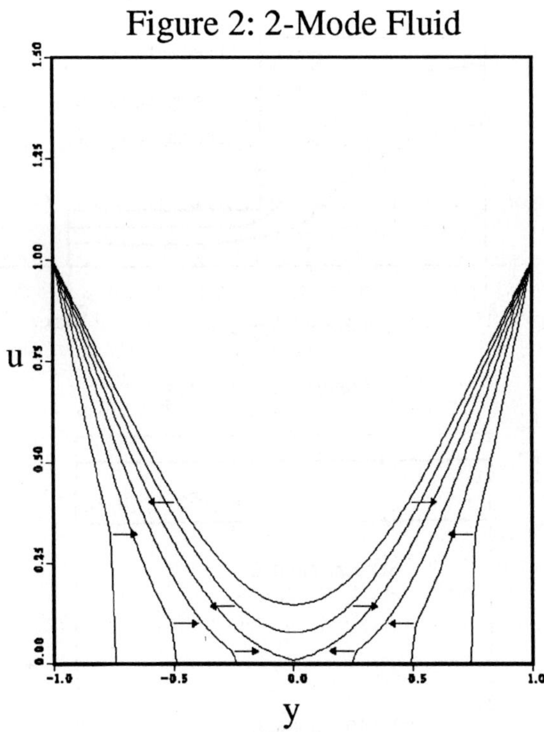

Figure 4: 3-Mode Fluid

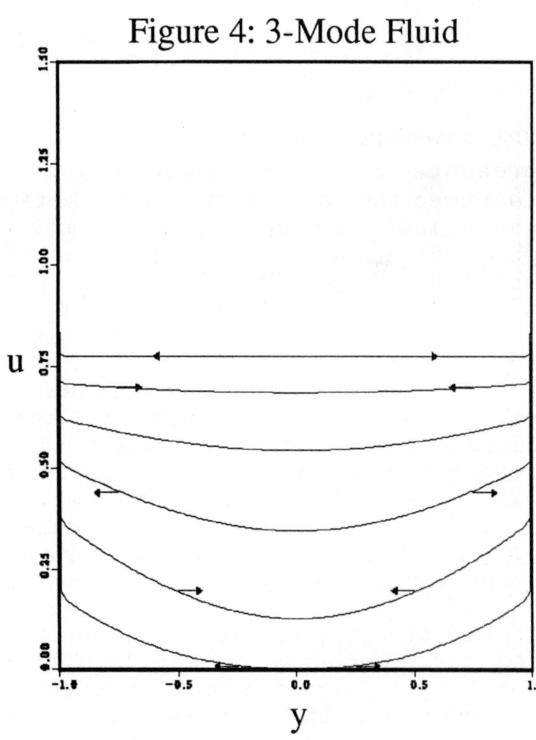

Theoretical and Applied Rheology, edited by P. Moldenaers and R. Keunings
Proc. XIth Int. Congr. on Rheology, Brussels, Belgium, August 17-21, 1992

THE MAIN FLOW OF A MEMORY INTEGRAL FLUID IN AN AXISYMMETRIC CONTRACTION AT HIGH WEISSENBERG NUMBERS

J.R. CLERMONT, M.E. DE LA LANDE

Laboratoire de Rhéologie, I.M.G, Domaine Universitaire, B.P. 53X, 38041 Grenoble Cedex (France)

1. INTRODUCTION

Previous papers have concerned the use of the stream-tube method in flow simulation (e.g. ref.1-2). In this context, a simply-connected physical domain D is mapped into a computational domain D' where the mapped streamlines are parallel and straight. Various constitutive equations were considered including Newtonian, non-Newtonian purely viscous and memory-integral fluids. We now investigate the complex flow of a memory-integral fluid (modified Goddard-Miller) in a 4/1 axisymmetric contraction.

The present formulation leads to the numerical determination (with the pressure p) of a transformation $\mathcal{T}$ (assumed to be one-to-one) between the computational domain D' and the physical domain D, using an unknown streamline mapping function f. The distinguishing features of the stream-tube method may be summarized as follows:

1° The mapped streamlines are rectilinear in domain D' and a rectangular mesh can be built on the straight streamlines related to the main flow of the physical domain.

2° The particle-tracking problem is avoided since the mapped streamlines are straight.

3° The incompressibility condition is automatically verified and the computation may be carried out by considering successive sub-domains, the stream tubes, from the outer boundary to the central flow region.

Possibilities for application of the stream-tube analysis are related to the existence of a one-to-one transformation. So, recirculating flows which may occur in complex flow geometries cannot be taken into account. The present work considers the main flow in the axisymmetric contraction involving determination of open streamlines.

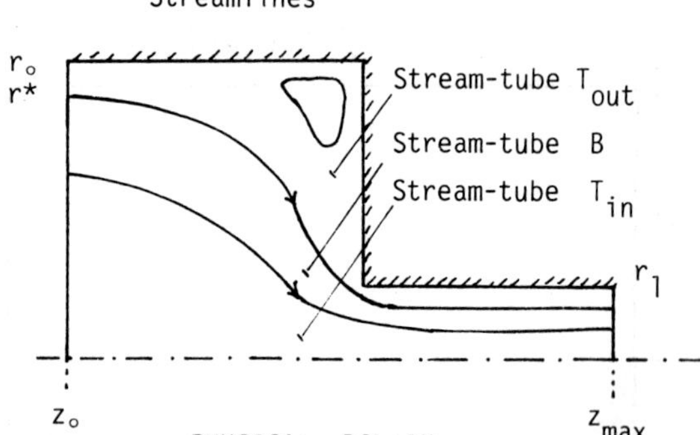

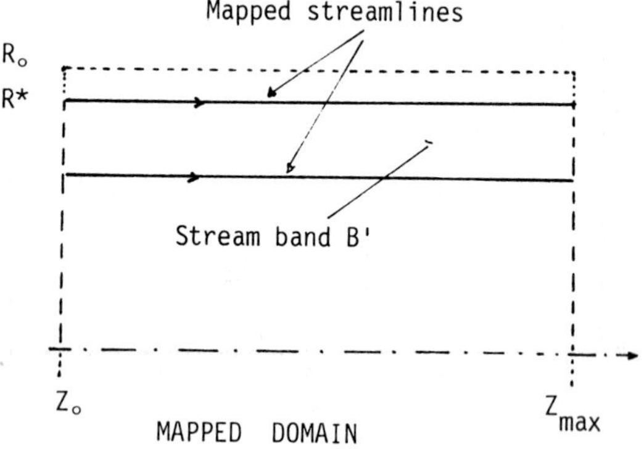

Fig 1 Mapping of the contraction flow geometry

2. BASIC EQUATIONS

A cylindrical coordinate system $x^1=r$, $x^2=\theta, x^3=z$ is associated to the axisymmetric flow domain of the 4/1 contraction (Fig 1) limited by Poiseuille flow sections z_o and z_{max}. We now consider the following transformation $\mathbb{T}$ defined by

$$r = f(R,Z)\ ,\ \theta = \Theta,\ z = Z \qquad (1)$$

verifying the initial conditions:

$$r_o = f(R,Z_o) = R_o,\ \theta = \Theta,\ z_o = Z_o \qquad (2)$$

The stream function Ψ at the upstream section is given by:

$$\Psi(r,z_o) = -\int_0^r w(s,z_o)s\ ds \qquad (3)$$

We may define a transformed stream function as $\Psi^*(R)=\Psi(r,z_o)$, $r=R$, at section $z_o=Z_o$. The velocity components u and w are:

$$u = -f'_Z\Psi^{*\prime}(R)/(ff'_R),$$
$$w = -\Psi^{*\prime}(R)/(ff'_R) \qquad (4)$$

The Jacobian of the transformation is given by:

$$J=\partial(r,\theta,z)/\partial(R,\Theta,Z) = f'_R. \qquad (5)$$

In absence of secondary flows, the streamlines define a simply-connected region and the Jacobian is non-singular.

Equation (4) verifies the incompressibility condition. In the isothermal case, we must only verify the momentum conservation equations:

$$\partial\sigma^{rr}/\partial r + \partial\sigma^{rz}/\partial z + (\sigma^{rr} - \sigma^{\theta\theta})/r = 0 \quad (6)$$

$$\partial\sigma^{rz}/\partial r + \partial\sigma^{zz}/\partial z + \sigma^{rz}/r = 0 \qquad (7)$$

where σ^{ij} are the components of the stress tensor $\boldsymbol{\sigma}$

$$\boldsymbol{\sigma} = -p\ \mathbb{I} + \mathbb{T} \qquad (8)$$

p denotes the pressure and $\mathbb{T}$ the extra-stress tensor.

The memory-integral equation has been used in a previous study on swelling (ref. 3) and is given in a corotational frame at time t by:

$$\mathbb{T}^*(t) = \int_{-\infty}^{t} \phi(II_D)\exp[-\lambda(t-\tau)]\mathbb{D}^*(\tau)d\tau \quad (9)$$

The symbol (*) indicates that this equation is written in a co-rotational frame at time t, λ is related to an average characteristic time, $\mathbb{D}$ is the strain-rate tensor and ϕ is a kinematic function of the second invariant II_D of D. The data were provided from experiments on a Gedex polystyrene at 200°C (ref. 3). The non-zero stress components of $\mathbb{T}$ have been given elsewhere.

The time evolution of a particle is given by the following equation:

$$\tau-\tau_o = \int_{Z_o}^{Z} d\zeta/w(R,\zeta)\ ,\ r \le R^* \qquad (10)$$

where τ_o and τ are the corresponding times to the particle positions Z_o and Z, respectively. This equation is considered in the computational domain D*' where the streamlines are rectilinear (Fig 1).

3. GOVERNING EQUATIONS

For the numerical computation of the flow in a stream tube, using as unknowns the function f, the pressure p and the non-zero stress components, we consider a mixed system of the following equations:

1° The dynamical equations (6-7)

2° The constitutive equations (9)

3° Simple boundary conditions for the unknowns and non-linear boundary condition equations, due to the action of complementary domains of the stream tube under consideration:

The first stream tube B in the main flow is shown in Fig 1. The vortices are embedded in the stream tube T_{out} and, in the inside, T_{in} denotes the complementary flow domain.

The action of the domains T_{out} and T_{in} of the tube B is involved by the following integral form of the momentum conservation law, expressed in the axisymmetric case by (ref.2):

$$\left[\int_{S}[-p\ \mathbb{I} + \mathbb{T}].\mathbf{n}\ ds\right].\mathbf{e_z} = 0 \quad (11)$$

where $\mathbf{e_z}$ denotes the unit vector along the z-axis, $\mathbb{S}$ the closed surface limiting the stream tubes T_{out} or T_{in}, $\mathbf{n}$ the unit outward normal vector to the surface $\mathbb{S}$.

4. APPROXIMATION OF UNKNOWNS - COMPUTATIONAL PROCEDURE

The mapping function f is assumed is expressed with 6 nodal values and basic functions. The pressure and stress components are given at the nodal points in the rectangular stream band built on the streamlines (ref.4).

The closed system of the equations is associated with the 2 constraints given by equ. 11 (2 complementary stream tubes). This optimization problem is solved by building an objective function Φ of the vector $\mathbf{X}$ of unknowns, from the original equations such that:

$$\Phi(\mathbf{X}) = 0 \quad (12)$$

The equations are solved by the Levenberg-Marquardt algorithm (ref. 4)

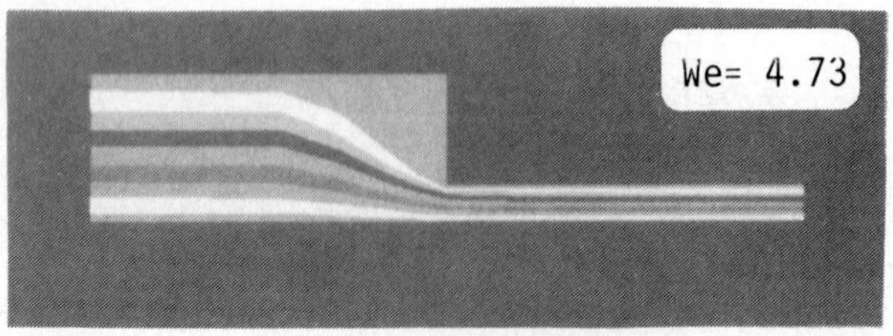

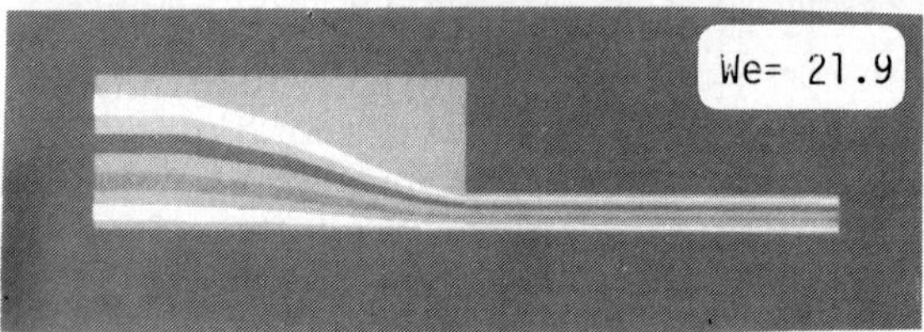

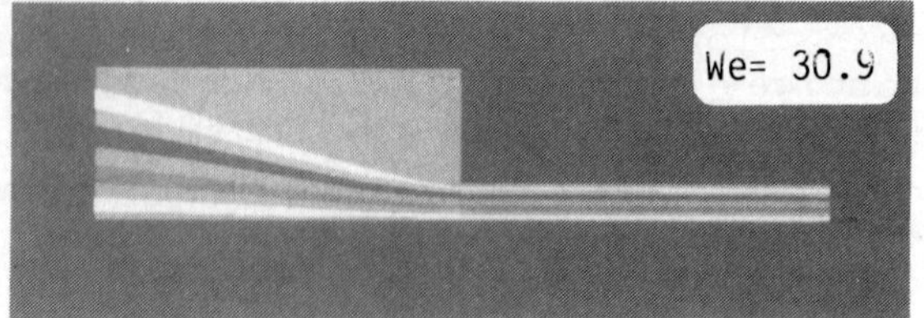

Fig 2 Evolution of streamlines for different We

5. NUMERICAL RESULTS

The numerical results were obtained at high Weissenberg numbers. The algorithm was found to converge up to We=31. We give in the following some results related to the main flow.

Fig 2 shows the evolution of the streamlines and the importance of vortex zones at different We. Fig 3 illustrates the evolution of the non-dimensional stress component T^{33} at different We, on the first computed streamline. The growing importance of the stresses versus the Weissenberg number is to be underlined.

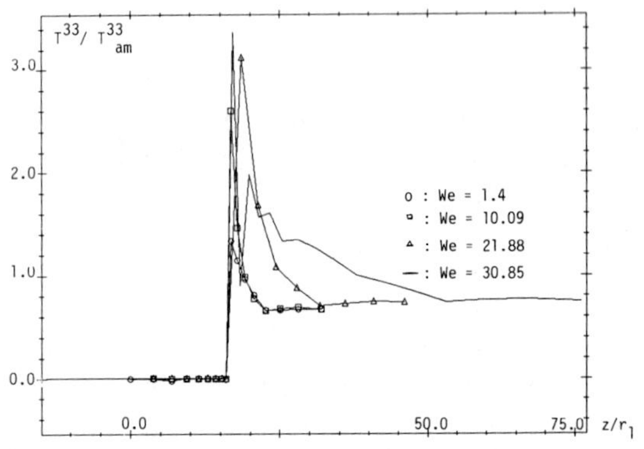

Fig 3 Evolution of the stress component T^{33} at different We

A robust algorithm, which permits computation of main flows has been presented in this work. We are presently investigating the flows of a K-BKZ fluid in complex geometries.

REFERENCES

1. J.R. Clermont, M.E. de la Lande, Engineering Comput. Vol 3, 4 (1986) 339

2. J.R. Clermont, M.E. de la Lande, T. Pham Dinh and A. Yassine, Int. Journal Num. Meth. Fluids, Vol 13 (1991) 391

3. P. André, J.R. Clermont, Journal Non-Newt. Fluid Mech., 38 (1990) 1

4. J.R. Clermont, M.E. de la Lande, Mech. Res. Commun. , Vol 18, 5 (1991) 303

Theoretical and Applied Rheology, edited by P. Moldenaers and R. Keunings
Proc. XIth Int. Congr. on Rheology, Brussels, Belgium, August 17-21, 1992

NUMERICAL SIMULATION OF THE WEISSENBERG EFFECT OF A GIESEKUS FLUID

Benoît DEBBAUT

POLY*FLOW* s.a., 16 Place de l'Université, B - 1348 Louvain-la-Neuve, Belgium.

1. INTRODUCTION

The first experimental observation on rod climbing was done by Garner and Nissan (ref. 1) with a solution of rubber in benzene. It led Weissenberg to introduce his hypothesis of pull or normal stress acting along streamlines (ref. 2). Since then, several papers were devoted to experimental measurements and theoretical developments on rod climbing (see e.g. refs. 3-6). Besides, it was shown that the climbing is a quadratic function of the rod angular velocity.

A first numerical work on rod climbing of viscoelastic fluids of the differential type was presented by Debbaut and Hocq (ref. 7). In particular, using a Johnson-Segalman model, they were able to predict the experiments with STP (a 26.6 % polyisobutylene solution in petroleum oil) done by Beavers and Joseph (ref. 4).

Our purpose in this paper is to extend the results by predicting the rod climbing of a Giesekus fluid (ref. 8). Numerical results are displayed, while a comparison with experimental measurements and analytical predictions on rod climbing is presented. Furthermore, we focus on the free surface shape and on the development of secondary motions.

2. EQUATIONS AND DISCRETIZATION

For flows of incompressible viscoelastic fluids, the momentum and continuity equations are given by :

$$-\nabla p + \nabla . \boldsymbol{T} + \rho \boldsymbol{g} = \rho \boldsymbol{v} . \nabla \boldsymbol{v} \,,$$
$$\nabla . \boldsymbol{v} = 0 \,. \qquad (1)$$

Here $\boldsymbol{T}$, $\boldsymbol{v}$ and p are the viscoelastic extra-stress tensor, the velocity vector and the pressure, respectively. In (1), $\boldsymbol{g}$ is the gravity vector and ρ the fluid density. For closing the system (1), we add a constitutive equation which relates the extra-stress to the flow kinematics. We select the Giesekus model, the constitutive equation of which is written as follows (ref. 8) :

$$\boldsymbol{T}\left[\boldsymbol{I} + \frac{\alpha\lambda}{\eta}\boldsymbol{T}\right] + \lambda\overset{\nabla}{\boldsymbol{T}} = 2\eta\boldsymbol{D}, \qquad (2)$$

where $\boldsymbol{D}$ is the rate of deformation tensor, and is given by $\boldsymbol{D}=\frac{1}{2}(\nabla \boldsymbol{v} + \nabla \boldsymbol{v}^{\mathrm{T}})$, while $\boldsymbol{I}$ is the

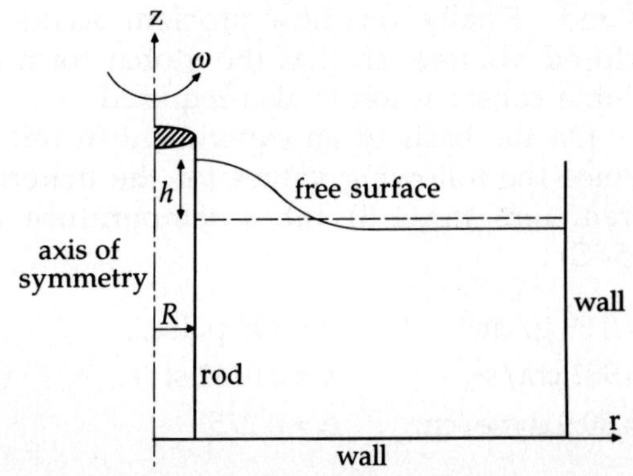

Fig. 1. Rod climbing : flow description and boundary conditions ; *h* is the climbing.

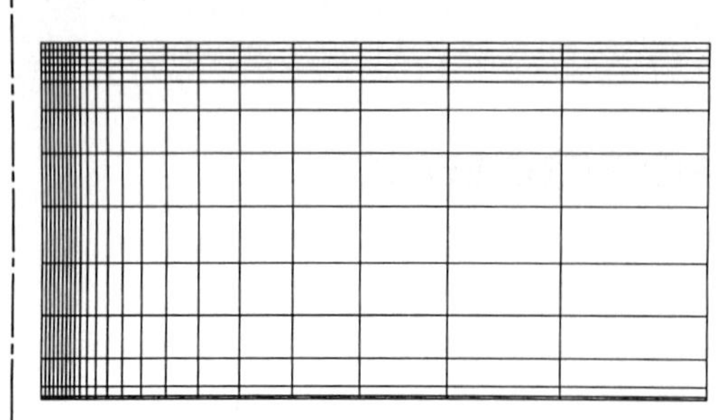

Fig. 2. Finite element mesh.

unit tensor. In (2), η is the zero shear rate viscosity, λ the relaxation time, and the symbol ∇ denotes the upper-convected derivative operator. The material parameter α controls the ratio of the second to the first normal stress coefficients ν_2 and ν_1 in a viscometric flow. In particular, for low shear rates, we have $\alpha = -2\,\nu_2/\nu_1$.

The rod climbing flow is sketched in Fig. 1 with the boundary conditions. It involves a free surface which obeys the kinematic equation $\boldsymbol{v}.\boldsymbol{n} = 0$ where $\boldsymbol{n}$ is the normal vector to the free surface. Surface tension σ_s is also taken into account and is given by :

$$\sigma_s = \Gamma\left(\frac{1}{R_1} + \frac{1}{R_2}\right)\boldsymbol{n}\,, \tag{3}$$

where Γ is the surface tension coefficient, while R_1 and R_2 are the principal curvature radii of the free surface. Together with (3), we impose a perpendicular contact of the fluid at the rod. Finally, this flow problem occurs in a closed volume, so that the global form of volume conservation is also required.

On the basis of an experiment in ref. 4, we use the following values for the material parameters in (1-3) (at a temperature of 25.5 °C) :

$$\begin{aligned} &\rho = 0.89\ \mathrm{g/cm^3}, && \eta = 146\ \text{poise}, \\ &g = 981\ \mathrm{cm/s^2}, && \lambda = 0.0162\ \mathrm{s}, \\ &\Gamma = 30.9\ \mathrm{dyne/cm}, && \alpha = 0.275. \end{aligned} \tag{4}$$

The diameter of the cylindrical vessel is 30.5 cm, the height of fluid at rest is 7.7 cm, and we have selected the radius R = 0.635 cm for the rotating rod.

Equations (1-3) are solved by means of the finite element algorithm developed by Marchal and Crochet (refs. 9,10). This algorithm combines a 4x4 sub-element interpolation for T with a streamline upwinding technique applied to the convective term appearing in $\overset{\nabla}{T}$. The numerical simulations are performed with the finite element mesh displayed in Fig. 2. It is characterized by 34698 degrees of freedom.

For low rotation speed ω of the rod, when viscoelastic effects, inertia forces and surface tension are taken into account, the climbing h, specified in Fig. 1, is predicted analytically as follows (refs. 4-6) :

$$h = \frac{R}{2\Gamma\Lambda}\left[\frac{4\beta}{4 + R\Lambda} - \frac{\rho R^2}{2 + R\Lambda}\right]4\pi^2\omega^2, \tag{5}$$

where $\Lambda = \sqrt{\rho g/\Gamma}$, and $\beta = \frac{\nu_1}{2} + 2\nu_2 = \eta\lambda - 2\alpha\eta\lambda$ is the so-called climbing constant and is a linear combination of the first and second normal stress coefficients. Equation (5) predicts a quadratic dependence of h upon ω.

3. RESULTS

Numerical simulations have been achieved for various rotation speeds ω of the rod. In Fig. 3 we display the climbing h as a

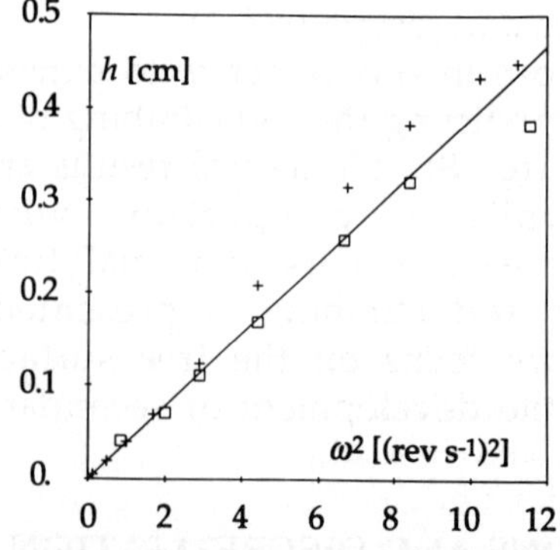

Fig. 3. Climbing h as a function of ω^2. (+) : numerical prediction, (—) : analytical prediction given by (5), (o) : experimental measurements of ref. 4.

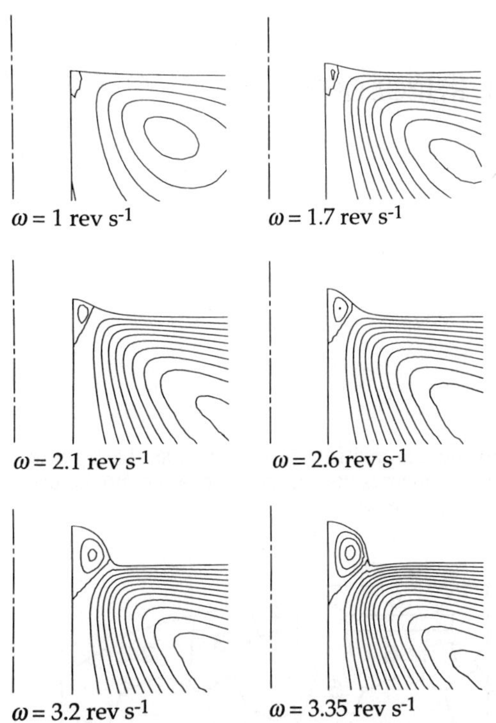

Fig. 4. Secondary motions under the bulge and free surface shape for various ω.

function of ω^2, together with the experimental measurements of Beavers and Joseph (ref. 4). We also compare these results with the analytical prediction given by (5). Good agreement is obtained among all these results. In particular, the quadratic dependence of h upon ω is clearly evidenced for low ω.

Flow patterns show a complex development as a function of ω. Fig. 4 displays the secondary motions under the bulge together with the free surface shape, for various rotation speeds ω. We observe that the upper vortex under the bulge grows with increasing ω, in a fashion similar to that sketched by Yoo et al. (ref. 5).

In this upper vortex, fluid particles have a ckockwise motion, while they rotate counterclockwise in the lower one. Both vortices originate from the interaction between normal stresses, gravity and free surface boundary conditions. Indeed, they disappear when the free surface is replaced by a slip wall. It is worth noting that the intensity of this secondary activity remains several orders of magnitude lower than the main azimuthal flow.

4. CONCLUSIONS

A finite element technique has been used for simulating axisymmetric swirling flows of viscoelastic liquids of the differential type. It has been successfully applied to the Weissenberg effect ; numerical results on rod climbing and surface displacement agree with experimental measurements and analytical predictions. Moreover, we have also been able to display the secondary motions which remain extremely weak.

REFERENCES

1. F.H. Garner and A.H. Nissan, Rheological properties of high viscosity solutions of long molecules, Nature, 158 (1946) 634-635.
2. K. Weissenberg, A continuum theory of rheological phenomena, Nature, 159 (1947) 310-311.
3. H. Giesekus, Einige Bemerkungen zum Fließverhalten elasto-viskoser Flüssigkeiten in stationären Schichtströmungen, Rheol. Acta, 1 (1961) 404-413.
4. G.S. Beavers and D.D. Joseph, The rotating rod viscometer, J. Fluid Mech., 69 (1975) 475-511.
5. J. Yoo, D.D. Joseph and G.S. Beavers, Higher-order theory of the Weissenberg effect, J. Fluid Mech., 92 (1979) 529-590.
6. D.D. Joseph, Fluid Dynamics of Polymeric Liquids, Springer-Verlag, New-York, 1990, chap. 17.
7. B. Debbaut and B. Hocq, On the numerical simulation of axisymmetric swirling flows of differential viscoelastic liquids : the rod climbing and the *Quelleffekt*, J. non-Newtonian Fluid Mech., in press.
8. H. Giesekus, A simple constitutive equation for polymer fluids based on the concept of deformation dependent tensorial mobility, J. non-Newtonian Fluid Mech., 11 (1982) 69-109.
9. J.M. Marchal and M.J. Crochet, A new mixed finite element for calculating viscoelastic flow, J. non-Newtonian Fluid Mech., 26 (1987) 77-114.
10. B. Debbaut, J.M. Marchal and M.J. Crochet, Numerical simulation of highly viscoelastic flows through an abrupt contraction, J. non-Newtonian Fluid Mech., 29 (1988) 119-146.

Theoretical and Applied Rheology, edited by P. Moldenaers and R. Keunings
Proc. XIth Int. Congr. on Rheology, Brussels, Belgium, August 17-21, 1992

PARALLEL ALGORITHMS IN COMPUTATIONAL RHEOLOGY

R. KEUNINGS, O. ZONE and R. AGGARWAL

Unité de Mécanique Appliquée, Université Catholique de Louvain,
B-1348 Louvain-la-Neuve (Belgium)

1. INTRODUCTION

The present paper gives a brief account of our current work towards the development of efficient parallel algorithms for the simulation of rheologically-complex flows on distributed memory parallel computers. Two main topics are currently being investigated : the development of a parallel frontal solver for solving generic finite element equations, and the use of a parallel strategy for simulating the flow of integral viscoelastic fluids. The proposed frontal solver has been implemented within a research code ported on the Intel iPSC hypercube. On the other hand, the parallel implementation of the integral viscoelastic flow solver is based upon the commercial code POLYFLOW developed in Louvain-la-Neuve.

2. PARALLEL FRONTAL SOLVER

Our goal is to develop a parallel frontal algorithm for computing solutions to finite element flow equations. Details on the relevant mathematical models and numerical techniques are given in recent reviews (Refs.1-2). The programming model used in our work is that of a set of P concurrent processors with local memory and message-passing capabilities. We assume that P is an integral power of 2. The proposed parallel algorithm (Refs.3-4) is based on a *topologically* linear domain decomposition approach illustrated in Fig.1 : each subdomain has an interface with only two neighbor subdomains.

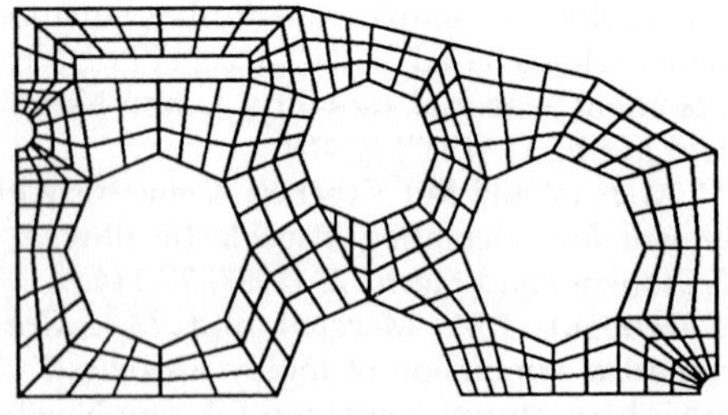

Fig.1a. Non-structured finite element mesh.

This decomposition procedure can be applied to complex geometries discretized by means of non-structured finite element meshes.

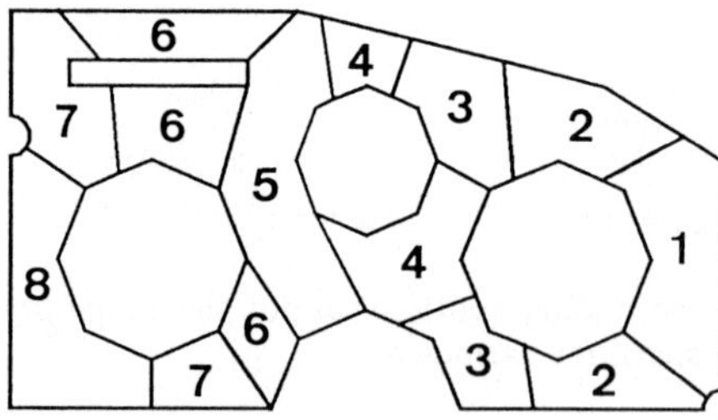

Fig.1b. Topologically-linear decomposition (P=8).

At the beginning of the calculations, each processor is allocated the elements of a subdomain. All processors then perform the frontal method on their subdomain in parallel. At the end of this process, each processor holds an active system which corresponds to the nodes located at the interface with neighbor processors. A communication step is then needed to assemble the contributions of those interface variables. Once the set of equations for the interface variables has been solved, the processors compute their internal nodal values independently in a backsubstitution step.

The way the interface system is communicated, assembled, and solved is of course the critical issue addressed in this work. Briefly put, the assembly and solution of the interface system are handled through a binary tree whose leaves are the processors (Fig.2). The algorithm involves two phases. A *first phase* involves traversal of the tree from the leaves to the root; it requires a number of communication steps equal to $log_2 P$. At each step, half of the active processors send in parallel their interface system to their neighbor processor, after which they become idle; the receiving processors assemble in parallel the contribution of their neighbor to their own, and eliminate the common variables by means of Gaussian elimination. The process is repeated until the root processor is reached. The *second phase* involves traversal of the tree from the root to the leaves. It consists in log_2P steps involving communication of interface nodal values computed at the previous tree level together with a backsubstitution process in half of the processors at each tree level. A processor having performed the elimination of an interface variable in the first phase computes this variable through a backsubstitution step.

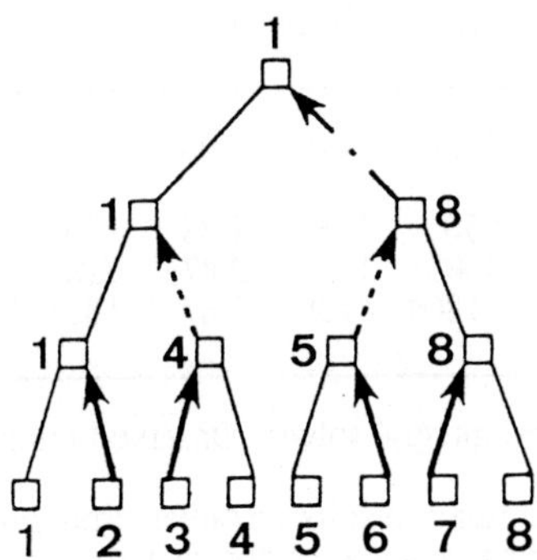

Fig.2. Communication scheme for 8 processors.

For one-dimensional finite element problems, we have shown (Ref.3) that this algorithm enjoys close-to-optimal efficiencies as the problem size (i.e. the number of one-dimensional finite elements) increases. Indeed, the problem size must only grow like $P \cdot log_2P$ as the number of processors P increases in order to keep the efficiency constant.

For multi-dimensional problems decomposed as in Fig.1b, new issues arise however relative to the one-dimensional case. Indeed, the size of the interface systems is problem-dependent and not necessarily homogeneous among processors; the same remark holds for the frontal width within the subdomains. As a result, *load balancing* becomes an important issue. As far as efficiency is concerned, the decomposition must be such that the overhead due to the communication and computation of interface variables be negligible with respect to the workload within the subdomains. The latter is not only a function of the number of elements within each subdomain, but also of the local elements numbering.

For a class of *structured* two-dimensional test problems, we have developed (Ref.4) a performance model that predicts the algorithm efficiency as a function of problem size and number of available processors. The model predictions are in good agreement with the experimental efficiencies we have observed on a 16 processor Intel iPSC/2 hypercube for the solution of Poisson's equation. High efficiencies are reached for large size problems such that the workload within each processor dominates that for the interface system.

By the time of this writing, results for *non-structured* finite element meshes have been obtained for Poisson and Navier-Stokes problems, using a 16 processor Intel iPSC/2 hypercube. We have conducted experiments on a finite element mesh with 918 elements and 3993 nodes. Table 1 gives wall-clock times and efficiencies as a function of the number of available processors. These data are for Poisson's equation. Two different decompositions were used in each case. The first decomposition only aims at balancing the number of elements within the subdomains, without taking into account the relative imbalance of the local *front widths*. The second decomposition is more sophisticated in that it allocates more elements to the subdomains with lower frontal width. It also improves on the elements numbering in the sequential case. It should be noted that both decompositions have been defined manually in these preliminary experiments.

Total Time (sec)	P=1	2	4	8	16
Decomp.#1	266	113	85	37	23
Decomp.#2	183	89	56	27	16
Efficiency	**P=1**	**2**	**4**	**8**	**16**
Decomp.#1	1.	1.18	0.78	0.90	0.72
Decomp.#2	1.	1.03	0.82	0.85	0.71

Table 1. Parallel frontal solver : Algorithm's performance as a function of number of processors (Poisson's equation).

Very good levels of efficiency are obtained with both Decompositions 1 and 2. It is worth noting that the efficiency does not behave monotonically as the number of processors increases (efficiencies greater than 1 can even be obtained!). This results from the complex interplay between number of elements, elements numbering, and number of interface variables in the definition of the workload within the subdomains and the parallel overheads. Results for the full steady-state Navier-Stokes equations are given in Table 2. In this case, the problem involves 9069 nodal values and the code does not fit into the local memories of less than 4 processors. Efficiencies are thus measured relative to the 4 processor timings.

Total Time (sec)	P=4	8	16
	402	205	138
Efficiency	**P=4**	**8**	**16**
	1.	0.98	0.73

Table 2. Parallel frontal solver : Algorithm's performance as a function of number of processors (Navier-Stokes).

The code is currently being extended to treat the case of differential viscoelastic fluids. Large-scale viscoelastic flow results obtained on a 128 processor iPSC/860 will be available by the time of the conference. Work is also under way towards the automatic optimal decomposition of non-structured finite element meshes, by means of non-deterministic heuristics. These recent techniques of combinatorial optimization are able to compute sub-optimal decompositions that minimize the interface sizes and balance the number of elements per subdomain.

3. INTEGRAL VISCOELASTIC MODELS

The basic algorithm for solving integral viscoelastic equations consists of a two-step decoupled iterative scheme (Refs.1-2) At iteration *n*, the flow kinematics is first updated on the basis of the current values of viscoelastic stresses. This amounts to solve a pseudo-Navier-Stokes problem with the viscoelastic stresses taken as body-force terms. A standard finite element velocity-pressure formulation is used to that purpose. The finite element equations are solved by a Newton scheme and a direct frontal solver is used to compute the Newton increments. The second step consists of computing new viscoelastic stresses on the basis of the updated kinematics. We need to compute those stresses at all Gauss points of the finite element mesh used to update the kinematics. This is a large computational task, even in two-dimensional flow problems. Indeed, viscoelastic stress computation implies the tracking of each Gauss point along its upstream trajectory, the computation of the strain history along that path, and finally the computation of the memory integral.

For a typical two-dimensional sequential simulation, we find that more than 95% of the cpu cycles are spent in the computation of viscoelastic stresses on the basis of known kinematics. The direct Navier-Stokes solver is thus very marginal as far as computing time is concerned. Integral viscoelastic calculations are ideal candidates for parallel computing since the stress computations can be done for each Gauss point independently of the others.

The basic parallel algorithm for integral models goes as follows. We assume that we have access to *P* processors with message-passing capabilities and local memory. The same copy of the code is loaded on all processors, together with the input data on finite element mesh, material properties and boundary conditions. A specific processor is in charge of the frontal assembly and elimination of the pseudo-Navier-Stokes problem. The remaining *(P-1)* "slave" processors compute in parallel the viscoelastic stresses at their allocated Gauss points. In addition, they compute the resulting local finite element matrices which are communicated to the "frontal solver" processor. In view of the non-local character of tracking and memory integral operations, each of the "slave" processor must have at its disposal the whole velocity field computed at the previous iteration. The velocity field is broadcast by the "frontal solver" processor to the other "slave" processors at the end of each iteration.

The issue of load-balancing is a rather peculiar one in the present application. It would seem that equipartition of the mesh into *(P-1)* subdomains would guaranty perfect load-balance. This is not true for viscoelastic integral equations. Indeed, the computation load involved in the determination of the viscoelastic stress at a given Gauss point depends on the material parameters, on the current kinematics, and on the location of the Gauss point within the overall flow domain. Indeed, it is impossible to predict a priori the computational load required by the stress calculation, and this load can vary significantly from one Gauss point to the other. Clearly, we can think of an adaptive allocation procedure which would take measured timings of the previous iterations into account in order to dynamically re-allocate the Gauss points to the "slave" processors. The preliminary results described below are based upon a static allocation.

Results have been obtained for the stick-slip flow of a K.B.K.Z. fluid, by means of the Intel iPSC/860 hypercube with 8 processors. We have used three finite element discretizations of increasing refinement. All three problems can be solved within the local memory of one node of the hypercube. Table 3 gives the observed timings *T(P)* for the three meshes with *P*=1, 4, and 8. Parallel speedup *Sp(P)* is defined as *T(1)/T(P)*. It should be noted that the optimal speedup is *(P-1)*, since only the slave processors participate to the parallel tasks.

#Elements	T(1)	T(4)	Sp(4)	T(8)	Sp(8)
16	650	273	2.38	183	3.55
80	4305	1497	2.87	728	5.91
180	11804	4109	2.87	1818	6.49

Table 3. Parallel integral solver : Observed timings (sec)

These preliminary results on rather small problems are very encouraging in view of the fact that no particular measure has been taken to do more than just equipartitioning of the mesh to approach load-balance. In particular, the results with 8 processors tend to the optimal speedup of *(P-1)=7* when the size of the problem increases. Large-scale simulation results obtained with the present code on a 128 processor iPSC/860 machine will be available by the time of the conference.

ACKNOWLEDGMENTS

The doctoral work of O. Zone is supported financially by the Programme *FIRST* of the Région Wallonne (Belgium). The postdoctoral work of R. Aggarwal is supported financially by the BRITE/EURAM Programme. We wish to thank A. Goublomme, B. Hocq, V. Legat, J.M. Marchal, and D. Vanderstraeten for their helpful comments. We also wish to thank Prof. M. Cosnard (ENSL, France), Dr. P. Leca (ONERA, France), and Dr. D. Roose (KUL, Belgium) for providing us with ample computer time on the iPSC/860 hypercubes located at their respective institutions. The results presented in this paper have been obtained within the framework of Interuniversity Attraction Poles initiated by the Belgian State, Prime Minister's Office, Science Policy Programming. The scientific responsibility rests with its authors.

REFERENCES

1. R. Keunings, in: C.L Tucker III (Ed.), Fundamentals of Computer Modeling for Polymer Processing, Carl Hanser Verlag, 1989, pp. 402-470
2. M.J. Crochet, Amer. Chem. Soc., 62 (1989) 426-455
3. O. Zone, R. Keunings, and D. Roose, in : Distributed Memory Computing, A. Bode (Ed.), Lecture Notes in Computer Science, Vol. 487, Springer Verlag, 1991, pp. 294-303
4. O. Zone and R. Keunings, in: High Performance Computing II, M. Durand and F. El Dabaghi (Eds.) North-Holland , 1991, pp. 333-344

Theoretical and Applied Rheology, edited by P. Moldenaers and R. Keunings
Proc. XIth Int. Congr. on Rheology, Brussels, Belgium, August 17-21, 1992

NUMERICAL SIMULATION OF THREE-DIMENSIONAL EXTRUDATE SWELL OF SEMICONCENTRATED FIBER SUSPENSIONS

Seong Jae LEE and Seung Jong LEE

Department of Chemical Engineering, Seoul National University,
San 56-1 Shinlim-dong, Kwanak-ku, Seoul 151-742 (Korea)

1. INTRODUCTION

Mechanical properties of short fiber reinforced composite materials strongly depend on the orientation distribution of fibers in the final products, which in turn is determined by the history of the flow field experienced during the processing steps. And the flow field itself is also influenced by the orientation distribution of fibers. Therefore, it is important to understand the interaction between the complicated flow patterns of fiber suspensions and the orientation distribution of fibers.

Most of analytical studies on the fiber motion within the fluid flow have been based on early Jeffery's model (ref. 1), which described the motion of an ellipsoid immersed in a Newtonian suspending medium. Givler et al. (ref. 2) developed a numerical simulation scheme to predict the fiber orientation in several flow geometries by solving the steady state velocity field and then integrating Jeffery's equation along the streamlines. Folgar and Tucker (ref. 3) proposed a phenomenological model which could represent the interaction among fibers in concentrated fiber suspensions by adding a diffusion term into Jeffery's equation.

In an attempt to consider the interaction between fluid and fiber, Batchelor (ref. 4) derived the bulk stress tensor in a dilute suspension of rigid ellipsoids as an ensemble average when the suspension is statistically homogeneous. Lipscomb et al. (ref. 5) considered the fluid/fiber interaction in the numerical simulation of the flow of fiber suspensions through axisymmetric contraction geometries, but they assumed fibers were fully aligned along their own streamline directions.

In this work we have developed a three-dimensional finite element code to solve the flow problems through arbitrary geometries, which may include free surfaces, and solved simultaneously the coupled system of equations for the flow and the orientation distribution of fibers in the three-dimensional extrudate swell flow of semiconcentrated fiber suspensions.

2. MATHEMATICAL MODELING

The flow and fiber orientation distribution in the steady isothermal flow of fiber suspensions are governed by the continuity, momentum, constitutive, and fiber orientation equations described in the followings. The continuity equation is

$$\nabla\cdot\mathbf{v} = 0 \tag{1}$$

And the Cauchy momentum equation is

$$-\nabla p + \nabla\cdot\mathbf{T} + \mathbf{f} = \rho\mathbf{v}\cdot\nabla\mathbf{v} \tag{2}$$

where $\mathbf{T}$ is the extra stress tensor given by the constitutive equation described below.

The general continuum model approach for formulating the constitutive equation is based on the theories of the stress field around the ellipsoid given by Hinch and Leal (ref. 6). The model for dilute and semiconcentrated fiber suspensions leads to

$$\mathbf{T} = 2\mu\mathbf{D} + 2\mu\phi\{C_1\mathbf{D}:\langle\mathbf{pppp}\rangle + C_2(\mathbf{D}\cdot\langle\mathbf{pp}\rangle + \langle\mathbf{pp}\rangle\cdot\mathbf{D}) + C_3\mathbf{D}\} \tag{3}$$

Here, μ is the viscosity of the Newtonian suspending medium, ϕ is the fiber volume fraction and $\mathbf{p}$ is the unit vector along the axis of the ellipsoid.

By the order of magnitude analysis on the stress shape coefficients C_1, C_2, C_3 for large fiber aspect ratio (r), the continuum constitutive model expressed as Eq.(3) may reduce to

$$\mathbf{T} = 2\mu\mathbf{D} + \mu\alpha\mathbf{D}:\langle\mathbf{pppp}\rangle \tag{4a}$$

$$\alpha = \phi\,\frac{r^2}{\ln 2r - 1.5} \tag{4b}$$

The equation for the fiber orientaion is also required as follows,

$$\mathbf{v}\cdot\nabla\mathbf{a}_2 = -\mathbf{W}\cdot\mathbf{a}_2 + \mathbf{a}_2\cdot\mathbf{W} + \lambda(\mathbf{D}\cdot\mathbf{a}_2 + \mathbf{a}_2\cdot\mathbf{D} - 2\mathbf{D}:\mathbf{a}_4) + 2C_I\dot{\gamma}(\mathbf{I} - \nu\mathbf{a}_2) \tag{5}$$

Here, $\mathbf{a}_2$ and $\mathbf{a}_4$ are the orientation tensors proposed by Advani and Tucker (ref. 7) defined as

$$\mathbf{a}_2 = \langle\mathbf{pp}\rangle = \int \mathbf{pp}\Psi(\mathbf{p})\,d\mathbf{p} \tag{6a}$$

$$\mathbf{a}_4 = \langle\mathbf{pppp}\rangle = \int \mathbf{pppp}\Psi(\mathbf{p})\,d\mathbf{p} \tag{6b}$$

And C_I is a phenomenological interaction coefficient which models the effects of interaction among fibers.

3. NUMERICAL METHODS

In this work, the standard Galerkin finite element method has been used to solve the three-dimensional flow and orientation equations. The velocity components and the orientation tensor components are approximated in terms of the 27-node triquadratic shape function, and the pressure is approximated in terms of the 8-node trilinear shape function in the hexahedron elements. The integration scheme used here is 27-point Gaussian quadrature.

The resulting nonlinear algebraic system of equations must be solved by an iterative technique. After every nonlinear iteration based on the latest velocity information, the free surface is calculated separately by using both the pathline approach method and the mapping method. This procedure is repeated until the convergence is obtained at a given value of α and then the value of α is increased gradually to obtain the solutions at higher α of interest.

4. RESULTS AND DISCUSSION

4.1 Axisymmetric extrudate swell

In Fig.1, the swell ratio is shown as a function of suspension intensity (α) for the various C_I values in the axisymmetric extrudate swell with the random fiber input at the die inlet. The swell ratio always increases as α increases. Figure 2 as the case of aligned fiber input at the die inlet shows rather different results. As α increases, the swell ratio increases in case of a relatively high C_I value (0.1) but decreases in case of a relatively low C_I value (0.001). At a moderate C_I value (0.01), the swell ratio decreases at first then increases with the increase of the α value. It is easily seen that the random fiber input case yields larger extrudate swell than the aligned input case. This phenomenon can be explained by considering that the swell ratio depends on the elongational viscosity difference between the outer surface part and the inner core part.

Two major and minor axes at each node in Fig.3 give an orientation ellipsoid which indicates the degree of orientation along every direction. The maximum elongational viscosity is obtained when fibers are aligned to the streamline direction, and on the contrary, the minimum case is obtained when aligned perpendicularly to the streamline direction. If the outer part has very large elongational viscosity compared with that of the inner part, higher swells are produced, but if the outer part has a little large elongational viscosity, lower swells are produced.

4.2 Three-dimensional extrudate swell

Finite element mesh, which is comprised of 160 hexahedron elements, is used for the extrudate swell of fiber suspensions through a square die. The geometry is chosen such that the total length of the system is 8cm, the die exit is at x=3cm, and the die half-width and half-height is 1cm, respectively. The convergence is obtained in about 10 iterations when $\alpha=0.1$, $C_I=0.1$ and the error allowance is set to 0.001, but it requires more iterations as α becomes higher value or C_I becomes lower value.

In Fig.4, the predicted extrudate swell and the orientation distributions of fibers with the random fiber input are shown at the outer surface and cross section of the far downstream, and at the center symmetric plane. Figure 5 represents the similar to Fig.4 except using the aligned fiber input condition. As C_I decreases in both random and aligned fiber input cases, fibers are more easily aligned to the streamline direction. As shown in Fig.5, the initial aligned fibers along the centerline pass through and exit the die and move to the downstream direction of extrudate producing the perpendicular orientation distribution gradually in spite of the initial zero distribution in the perpendicular direction. As a result, the perpendicular orientation distributions to the streamline direction occur along the centerline because of the effects of the fiber/fiber interactions and the expansional flow.

5. CONCLUSIONS

In the simulations of the extrudate swell of fiber suspensions through a two-dimensional circular die and a three-dimensional square die, the die swell ratio decreases as the suspension intensity increases when the fibers at the die inlet are aligned. But in the case of random orientation states of fibers at the die inlet, roughly thought as the case of short dies, the swell ratio increases, however. Different trends in the swell ratio caused by the different fiber orientation states at the die inlet can be explained in terms of the elongational viscosity difference between the outer surface part and the inner core part.

REFERENCES

1. G.B. Jeffery, Proc. Roy. Soc., A102 (1923) 161.
2. R.C. Givler, M.J. Crochet and R.B. Pipes, J. Compos. Mat., 17 (1983) 330.
3. F. Folgar and C.L. Tucker, J. Reinf. Plast. Compos., 3 (1984) 98.
4. G.K. Batchelor, J. Fluid Mech., 41 (1970) 545.
5. G.G. Lipscomb, M.M. Denn, D.U. Hur and D.V. Boger, J. Non-Newtonian Fluid Mech., 26 (1988) 297.
6. E.J. Hinch and L.G. Leal, J. Fluid Mech., 52 (1972) 683.
7. S.G. Advani and C.L. Tucker, J. Rheol., 31 (1987) 751.
8. J.S. Yoon, MS thesis, Dept. of Chemical Engineering, Seoul National University, 1992.

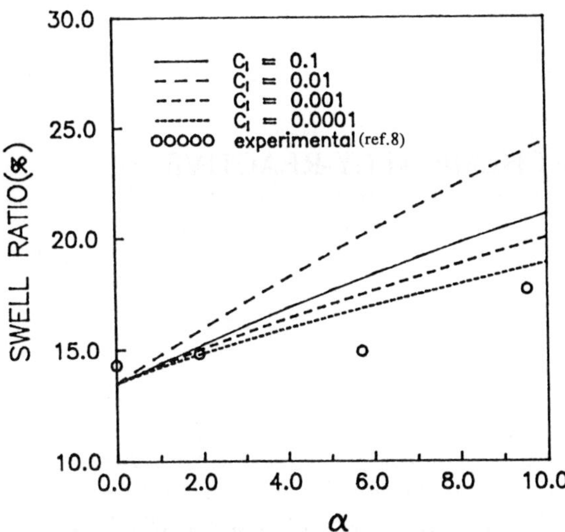

Fig.1. Swell ratio vs. α in the axisymmetric extrudate swell with random fiber input.

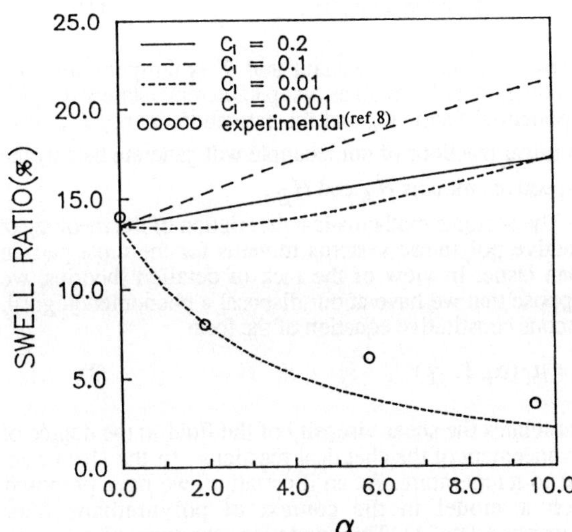

Fig.2. Swell ratio vs. α in the axisymmetric extrudate swell with aligned fiber input.

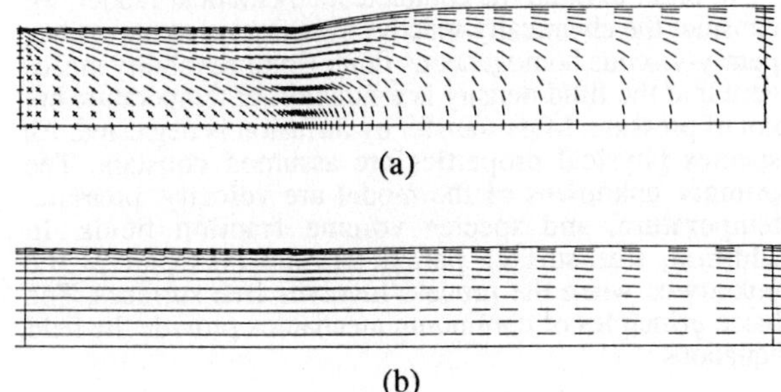

Fig.3. Fiber orientation distribution in the axisymmetric extrudate swell at $C_I = 0.001$ and $\alpha = 10$: (a) random fiber input, (b) aligned fiber input.

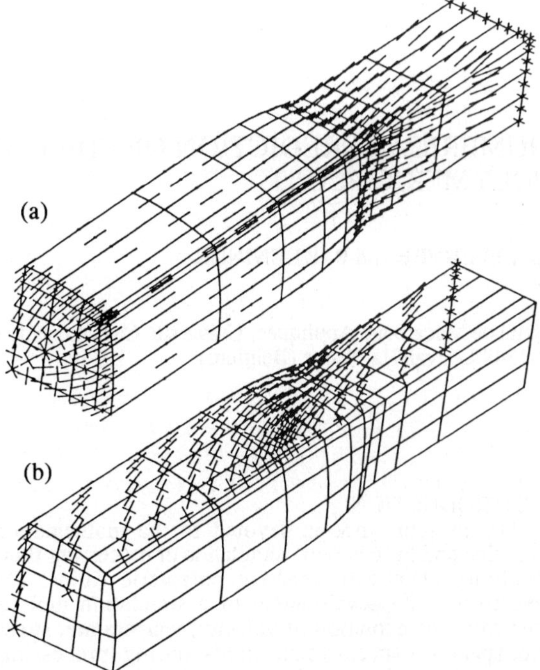

Fig.4. Converged finite element mesh and fiber orientation distributions at $\alpha = 1$, $C_I = 0.01$ with random fiber input: (a) at surface and cross section, (b) at center symmetric plane.

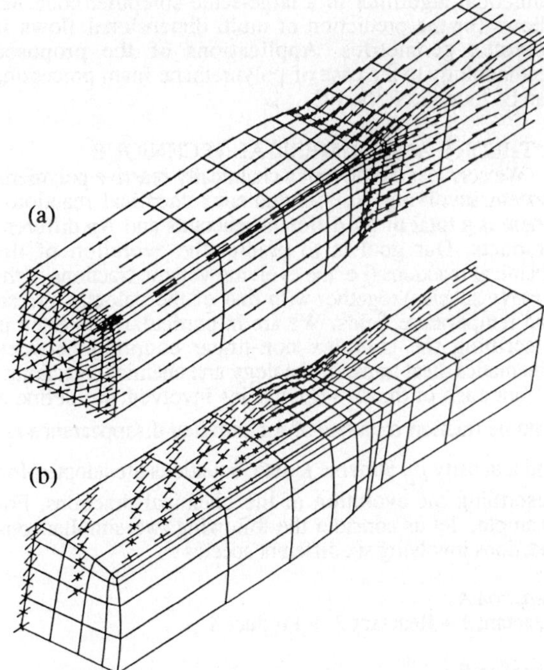

Fig.5. Converged finite element mesh and fiber orientation distributions at $\alpha = 1$, $C_I = 0.01$ with aligned fiber input: (a) at surface and cross section, (b) at center symmetric plane.

Theoretical and Applied Rheology, edited by P. Moldenaers and R. Keunings
Proc. XIth Int. Congr. on Rheology, Brussels, Belgium, August 17-21, 1992

NUMERICAL PREDICTION OF THE FLOW OF CHEMICALLY-REACTIVE POLYMERIC FLUIDS

L. LEFEBVRE and R. KEUNINGS

Unité de Mécanique Appliquée, Université Catholique de Louvain,
B-1348 Louvain-la-Neuve (Belgium)

1. INTRODUCTION

The present work is devoted to the mathematical modeling and the computer simulation of processing flows involving chemically-reactive polymeric fluids. The objective is the development of a simulation tool for predicting the evolution of velocity, temperature, stress, and species concentration fields during processing. Governing equations are derived on the basis of the first principles of continuum mechanics and phenomenological relations for the chemical kinetics and the evolving fluid rheology. The set of coupled, non-linear partial differential equations is solved numerically by means of a specialized finite element technique. We have implemented the numerical algorithm in a large-scale computer code that allows for the prediction of multi-dimensional flows in complex geometries. Applications of the proposed methodology to the case of polyurethane foam processing can be found in Refs. 1-2.

2. THEORY AND NUMERICAL TECHNIQUE

We consider the flow of a chemically-reactive polymeric system involving *Nc* simultaneous chemical reactions. There is a total of *Nr* different reactants and *Np* different products. Our goal is to predict the evolution of the chemical reactions (i.e. the evolving volume fractions of the *Nr+Np* species) together with that of the velocity, stress, and temperature fields. We are in particular interested in describing the complex non-linear coupling between kinematics, heat transfer, rheology and chemical reactions.

For each of the *Nr+Np* species involved, we define a volume fraction α_i, a rate of formation or disappearance r_i, and a density ρ_i. Suitable kinetic equations are adopted for describing the evolution of the chemical reactions. For example, let us consider the following two simultaneous reactions involving six different species :

Reaction A :
Reactant 1 + Reactant 2 -> Product 3 ,

Reaction B :
Reactant 1 + Reactant 4 -> Product 5 + Product 6 .

We have here *Nc=2*, *Nr=3* and *Np=3*. If second order kinetics are assumed to hold, we have for example

$$r_3 = K_A \exp(-E_A/RT)\, \alpha_1 \alpha_2 , \qquad (1)$$

where *R* is the gas constant and *T* is temperature. The kinetic model (1) involves two parameters, namely the pre-exponential factor K_A and the activation energy E_A. The chemical reactions of our example will generate heat by the respective amounts H_A and H_B.

The accurate mathematical description of the rheology of reactive polymeric systems remains for the most part an open issue. In view of the lack of detailed theories, we suppose that we have at our disposal a phenomenological, viscous constitutive equation of the form

$$\mu_f = \mu_f(\alpha_j, T, \dot{\gamma}) \quad , \qquad (2)$$

that relates the shear viscosity of the fluid to the degree of advancement of the chemical reactions, to the shear rate, and to temperature. As an illustration, we have proposed such a model in the context of polyurethane foam processing (Ref.1). This model has the form of a mixing rule that includes as control variables the concentrations in reactants, shear rate, temperature, and gas phase volume fraction.

In order to build the complete mathematical model, we consider the chemically-reactive system as a compressible, purely-viscous homogeneous fluid. Compressibility means here that the fluid density is a function of temperature but not of pressure. Mass transfer by diffusion is neglected. All species physical properties are assumed constant. The primary unknowns of the model are velocity, pressure, temperature, and species volume fraction fields. In addition, free surface parameters are also part of the unknowns when the process involves free surfaces. The basic principles of continuum mechanics provide the field equations :

Momentum equation

$$\rho_f \left[\frac{\partial \mathbf{v}}{\partial t} + \mathbf{v} \cdot \nabla \mathbf{v}\right] = -\nabla p + \nabla \cdot [2\mu_f(\alpha_j, T, \dot{\gamma})\, \mathbf{D}] \; (3)$$

Energy equation

$$\rho_f C_f \left[\frac{\partial T}{\partial t} + \mathbf{v} \cdot \nabla T\right] = \nabla \cdot [k_f \nabla T] + H(\alpha_j) , \tag{4}$$

Mass conservation for i^{th} species $(i = 1, 2, ..., Nr+Np-1)$

$$\frac{\partial(\rho_i \alpha_i)}{\partial t} + \nabla \cdot [\rho_i \alpha_i \mathbf{v}] = r_i(\alpha_j, T) , \tag{5}$$

Constraint on volume fractions

$$\sum_{i=1}^{Nr+Np} \alpha_i = 1 , \tag{6}$$

Global mass conservation

$$\frac{\partial \rho_f}{\partial t} + \nabla \cdot [\rho_f \mathbf{v}] = 0 \quad . \tag{7}$$

In these equations, $\boldsymbol{v}$ stands for the velocity vector, p is pressure, μ_f is the fluid viscosity, $\boldsymbol{D}$ is the rate of strain tensor, k_f is the fluid heat conductivity, C_f is the fluid heat capacity, $H(\alpha_j)$ is the total heat generated by the Nc reactions, and $r_i(\alpha_j,T)$ is the rate of formation or disappearance of the ith species. Finally, ρ_f is the fluid density given by

$$\rho_f = \sum_{i=1}^{Nr+Np} [\alpha_i \rho_i] \ . \tag{8}$$

Note that viscous dissipation has been neglected in the formulation of (4).

Equations (3-7) form a set of coupled, non-linear partial differential equations to be solved with suitable initial and boundary conditions. We solve these equations approximately by means of a mixed finite element discretization of the primary unknown fields. Galerkin's principle is used to discretize the momentum, volume fraction constraint, and global mass conservation equations. A consistent streamline-upwind Petrov-Galerkin method is implemented for the energy and species conservation equations, in view of their hyperbolic nature. In the case of steady-state flows, we obtain a large set of non-linear algebraic equations for the nodal values of the unknown fields. We solve this set of equations by means of a full Newton scheme. Finally, an adaptive continuation procedure on material properties and operating conditions is used to provide initial estimates for the iterative procedure. Further details are given in Ref. 2, where the case of time-dependent flows is also considered.

We have used the above numerical model to simulate several steady-state and time-dependent flows of reactive polymeric systems, including flows with free surfaces. An illustrative flow problem is discussed in the present paper. The corresponding simulations have been performed on a CRAY X-MP supercomputer with typical CPU times of the order of one hour.

3. TYPICAL SIMULATION RESULTS

The physical process under consideration is shown in Fig.1. The computational domain is divided into three parts: a heater, a solid die, and the chemically-reactive fluid. The fluid enters on the left with a specified temperature and a flat velocity profile. It is assumed to slip at the die wall, and fully-developed conditions are imposed at the exit section. We assume here that the viscosity of the fluid remains constant.

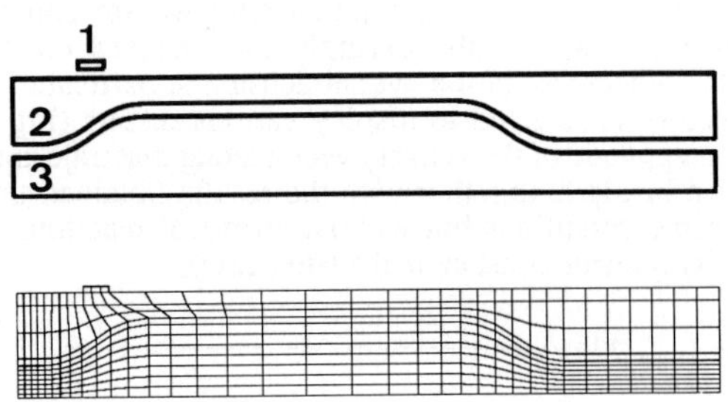

Fig.1 Exploded view of computational domain and coarse finite element discretization. Domain 1: heater; Domain 2: die; Domain 3: fluid.

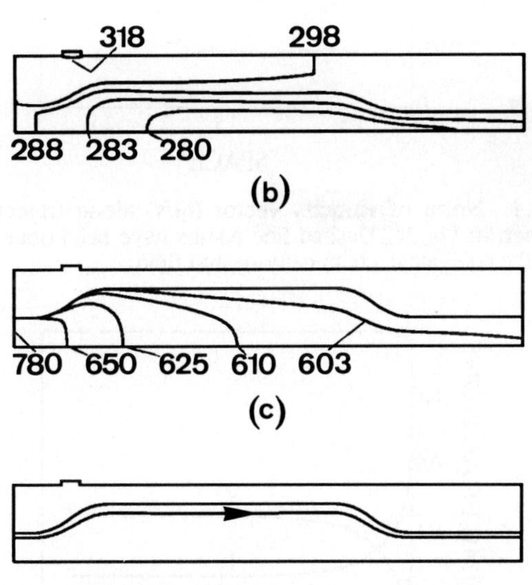

Fig.2 (a) Temperature contours (K); (b) Fluid density contours (kg/m^3); (c) Flow trajectory selected for display in subsequent figures.

This example involves the two chemical reactions discussed in the previous section. One assumes that reaction A is exothermic, while reaction B is endothermic. Vanishing values for the volume fractions of products are specified at the entry section, while non-stoechiometric values are imposed for the reactants. Conditions of perfect thermal insulation are imposed at the outer boundary of the "heater" and "die" computational domains. We solve the heat conduction equations in those two subdomains (with a specified, uniform heat source in the heater), together with the governing equations (3-7) in the fluid phase. Continuity of temperature and heat flux is obtained automatically in the finite element sense at the interface between the three subdomains.

We have used two finite element meshes to discretize the computational domain. The coarse mesh is shown in Fig.1; it involves a total of 6782 unknown nodal values. The refined mesh has 25970 nodal values. There is good agreement between the results obtained with these two meshes.

Figure 2 gives the computed temperature and fluid density contours. In this example, the chemical reactions lead to a decrease of the overall density. A particular fluid trajectory is selected to display various results (Fig.2c). The magnitude of the velocity vector along that trajectory is shown in Fig.3, together with the results obtained under the same conditions but without chemical reactions (the density remains constant in the latter case).

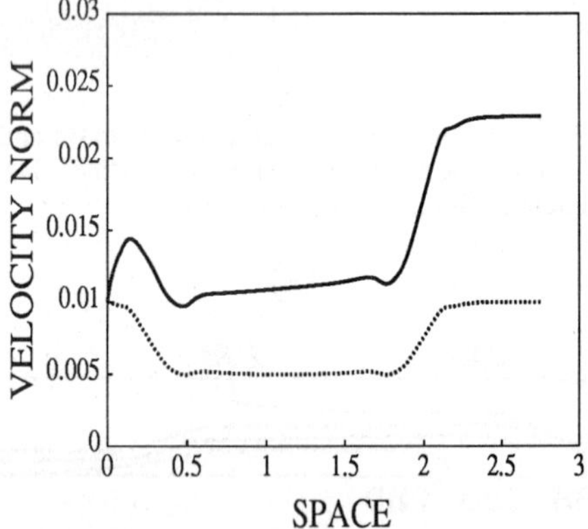

Fig.3 Norm of velocity vector (m/s) along trajectory shown in Fig.2c. Dashed line results have been obtained for the equivalent, chemically-neutral fluid.

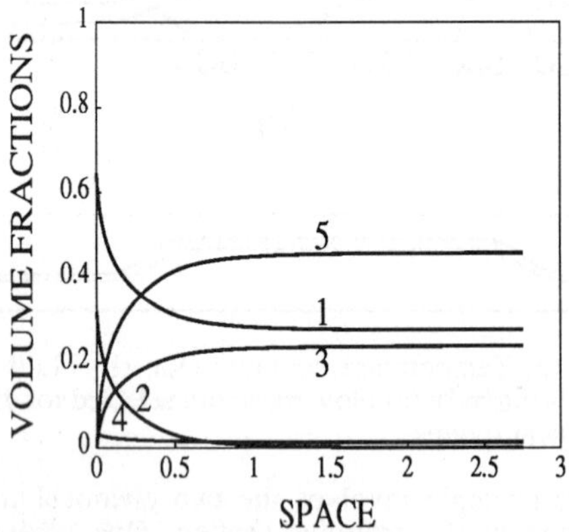

Fig.4 Evolution of species volume fractions along trajectory shown in Fig.2c.

There is a marked difference between the two cases, which is due to the coupling between kinematics and chemically-induced density variations.

Volume fractions predicted along the selected trajectory are shown in Fig.4. In view of the non-stoechiometric data specified at the inlet section, Reactant 1 is not totally consumed when the reactions are completed.

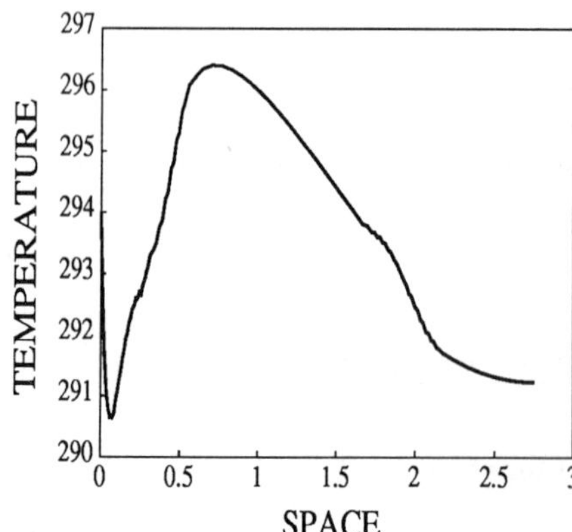

Fig.5 Evolution of temperature (K) along trajectory shown in Fig.2c.

Finally, we show in Fig.5 the computed temperature along the selected trajectory. One observes four main regimes. First, a rapid decrease due to the dominating endothermic character of Reaction *B*; this is followed by an increase in temperature caused by heat conduction from the heater; sufficiently far away from the heater, the endothermic character of Reaction *B* prevails again; finally, when the chemical reactions are completed, the temperature reaches a constant value.

Additional details on this and other simulations can be found in Refs.1-2.

ACKNOWLEDGEMENTS

The doctoral work of L. Lefebvre is supported financially by a grant from Shell Research Louvain-la-Neuve. This text presents research results obtained in the framework of the Belgian Programme on Interuniversity Poles of Attraction initiated by the Belgian State, Prime Minister's Office, Science Policy Programming. The scientific responsibility is assumed by its authors.

REFERENCES

1. L. Lefebvre and R. Keunings, in: M. Cross et al. (Eds.), Mathematical Modelling in Materials Processing, Oxford University Press, 1992, in press
2. L. Lefebvre, Ph.D. Thesis, Université Catholique de Louvain, Belgium, in preparation

Theoretical and Applied Rheology, edited by P. Moldenaers and R. Keunings
Proc. XIth Int. Congr. on Rheology, Brussels, Belgium, August 17-21, 1992

THREE-DIMENSIONAL EXTRUSION : AN IMPLICIT FORMULATION FOR GENERALIZED NEWTONIAN FLUIDS

Vincent Legat[1], and Jean-Marie Marchal[2]

[1]Applied Mechanics, Université Catholique de Louvain, Place du Levant 2, 1348 Louvain-la-Neuve (Belgium)

[2]Polyflow SA, Place de l'Université 16, 1348 Louvain-la-Neuve (Belgium)

1. INTRODUCTION

Implicit Finite Element techniques for solving three-dimensional free surface problems in complex geometries have recently been proposed by Legat and Marchal. A first publication (ref. 1) addresses the Direct Extrusion Problem in the absence of surface tension. A method is proposed to calculate the shape of the extrudate on the basis of a known die geometry.

A second paper (ref. 2) is relative to the Inverse Problem which consists of finding the die geometry that produces an extrudate of prescribed shape. We have developed an original method which is able to solve the Inverse Extrusion Problem while avoiding "trial and error" iterations on the die geometry. The method is geometrically general. In particular, the die section shapes may be complex, while very few topological restrictions apply on the mesh. The remeshing of the interior nodes is well adapted to a large class of extrusion problems. In (ref. 2), the method is fully implicit and does not include surface tension, the iterative scheme being of Newton-Raphson's type. Power law as well as Newtonian fluids have been considered.

In this paper, we use the method presented in (ref. 2) to solve Inverse Extrusion Problems for power law fluids. Two geometries are considered, each of those including abrupt corners. Surface tension has been introduced in order to analyze its effect upon the extrudate shape for the Direct and Inverse Problems. In particular, the effect of surface tension upon the presence of corners is considered.

2. DIRECT AND INVERSE PROBLEMS

Let us consider the numerical techniques presented in (ref. 1,2) for the Direct and Inverse Problems respectively. We refer to Fig. 1. for a schematic two-dimensional description of the free surface problem.

We solve the (Generalized) Navier-Stokes equations together with kinematic conditions on free surfaces. Two geometrial degrees of freedom as well as two kinematic conditions have been introduced along corners, in order to handle their displacement in planes othognal to the direction of extrusion. For the Direct Problem a vanishing displacement at the die lip C_{LIP} is imposed as a boundary condition for the kinematic condition. For the Inverse Problem, this vanishing displacement can no longer be introduced along the line C_{LIP}. We introduce an additional variable, which stands for the displacement of the die lip. Due to the introduction of this Lagrange multiplier, one is allowed to impose a zero displacement at the end C_{EXIT} of the extrudate where the shape is prescribed.

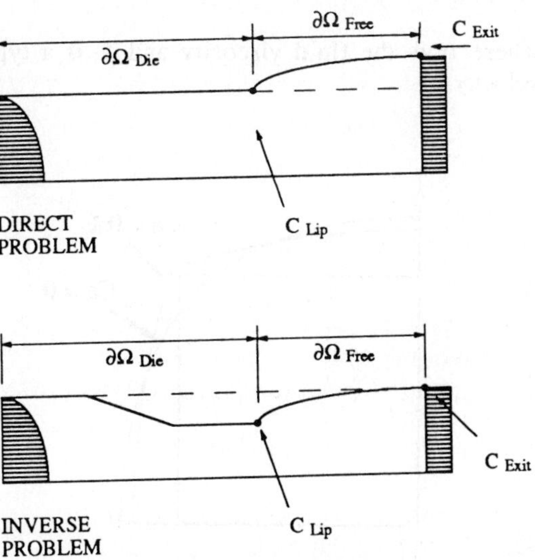

Fig. 1. Direct and Inverse Problems

We look for solutions of the Inverse Extrusion Problem within a peculiar class of dies. The shape of the die cross section is continuously adapted from a constant inlet section to the unknown die lip section (which is result of the simulation). In all sections where the positions of the boundary nodes are updated, we remesh the interior nodes according to an Euclidian distance rule which is described in (ref. 1). A full Newton-Raphson scheme is implemented in order to allow for quadratic convergence rate.

3. SURFACE TENSION

The classical Euler-Lagrange formulation of surface tension correlates normal force on the free surface f_N to the mean radius of the curvature R of the surface :

$$f_N = \sigma R^{-1}, \quad (1)$$

where σ is the surface tension coefficient.

However form (1) of the surface tension force cannot be used as such in a Finite Element formulation in view of the presence of second order derivative of the nodal positions, and in view of absence of boundary conditions. A variational formulation of (1) can however be introduced on the basis of Green's theorem applied on Rieman's surfaces (ref. 3). The "angle" conditions along the boundary of the free surface are automatically introduced. Solutions are presented as a function of the capillary number Ca defined as :

$$Ca = \sigma (\eta V)^{-1}, \quad (2)$$

where η is the fluid viscosity and V is a typical velocity.

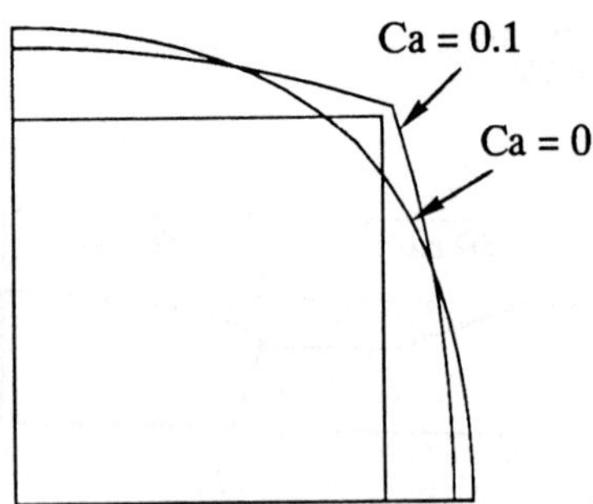

Fig. 2. Extrudate shape for a square die.

3. RESULTS

Let us first consider the Direct Problem with a square die. In Fig. 2, we analyze the extrudate shape of a Newtonian fluid as a function of Ca. Fig. 3. shows a perpective view of the converged mesh at Ca = 0.1. The final shape of the extrudate becomes circular in view of the capillary tensions.

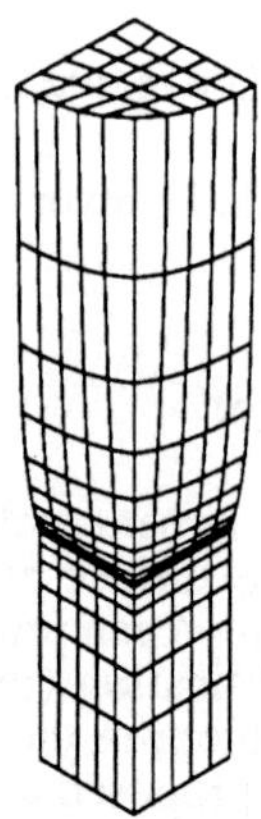

Fig. 3. Mesh used for the Direct Problem.

In order to validate our method for the Inverse Problem, we have also calculated the die geometry which produces a square profile in the case of a Newtonian fluid. This benchmark problem has been considered by Trang-Cong Phan-Thien (ref. 4), Wambersie and Crochet (ref. 5), Yokoi and Scriven (ref. 6). Our solutions agree with those results. Quadratic convergence of the iterative scheme and convergence with mesh refinement have been observed.

In this paper, we have calculated the die section for a Generalized Newtonian fluid. Most polymers present a shear-thinning behaviour, which within a range of shear rates can be described by a dependency of the viscosity upon the shear rate $\dot{\gamma}$ of the power law type :

$$\eta(\dot{\gamma}) = K \eta \dot{\gamma}^{m-1}. \quad (3)$$

The power law index lies for many polymers between 0.6 and 0.2. The velocity distribution in the die for a power law fluid will in general reduce swelling effects.

Let us consider the rectangular die section for a fluid with a power law index equal to 0.6. Solutions of the Inverse Problem are presented in Fig. 4. For different values of Ca, we present die sections that produce rectangular extrudates. It must be noted that due to the presence of surface tension, the length of the extrudate modifies solution. For the rectangular die of dimensions 1 / 1.5, we have used a extrudate length equal to 4. For an infinitely long product and an isothermal model, all dies produce a circular extrudate in the presence of surface tension.

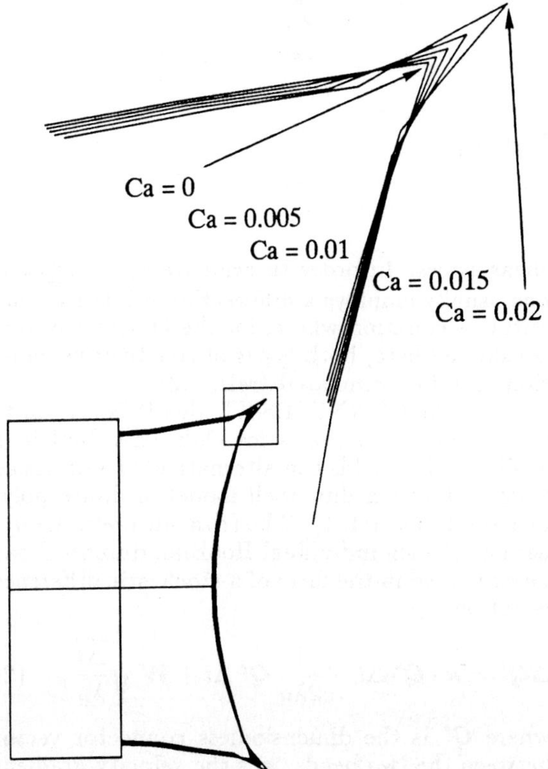

Fig. 4. Die sections as function of Ca in order to produce rectangular extrudate.

As a final example, we have considered a star-like extrudate. The final extrudate is presented as a function of Ca, in Fig. 5. The presence of surface tension increases the difficulty of the problem. It is impossible to converge for a value of Ca greater than 0.01, in view of the deformation of the meshes.

4. CONCLUSIONS

We have considered the combined effects of shear thinning and surface tension upon extrudate swelling. Both the direct and inverse problem have been addressed for rectangular and star-like sections. In all cases, a shear thinning behaviour reduces the swelling and modifies the deformation, whereas surface tension tends to smooth all angles and therefore has a strong influence upon shape predictions.

For the inverse problem with surface tension, angles in the die section might become so accute that no solution exists for long extrudates or large values of the capilary number. However viscous effects are generally the most important for polymers. As all extrudates become circular in the presence of surface tension for infinitely long free surfaces, other effects such as cooling and freezing of the polymer must be considered if one wishes to simulate long output sections.

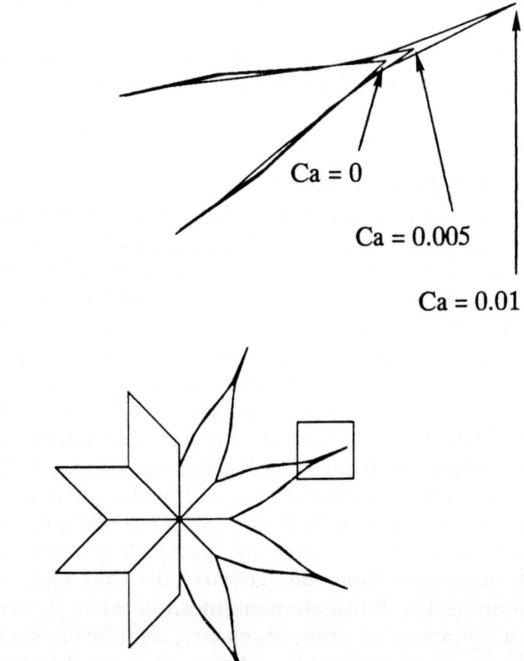

Fig. 5. Die sections as function of Ca in order to produce star-like extrudate.

REFERENCES

1. V. Legat, J.M. Marchal, Int. J. Numerical Methods in Fluids. (1992) (in press).

2. V. Legat, J.M. Marchal, Int. J. Numerical Methods in Fluids. (submitted for publication).

3. K.J. Ruschak, Computers & Fluids, Vol. 11, 4, (1983) 391-401.

4. T. Tran-Cong, N. Phan-Thien, J. Non-Newtonian Fluid Mech., 30, (1988) 37-48.

5. O. Wambersie, M.J. Crochet, Int. J. Numerical Methods in Fluids., 14, (1992) 343-360.

6. T. Yokoi, L.E. Scriven, J. Non-Newtonian Fluid Mech. (submitted for publication)

Theoretical and Applied Rheology, edited by P. Moldenaers and R. Keunings
Proc. XIth Int. Congr. on Rheology, Brussels, Belgium, August 17-21, 1992

"SMART" POLYMERS IN FINITE-ELEMENT CALCULATIONS

HANS CHRISTIAN ÖTTINGER and MANUEL LASO

Institut für Polymere, Eidgenössische Technische Hochschule,
ETH-Zentrum, CH-8092 Zürich (Switzerland)

1. INTRODUCTION

In plastics processing, polymeric liquids suffer complex deformation histories, and they show a stress response which is a functional of the entire deformation history. *In real liquids*, the full information about the stresses is contained in the current configuration of the polymer molecules which results from the deformation history. *In numerical calculations of flows*, the strain-history-dependent stresses are traditionally obtained from constitutive equations of the differential or integral type. It is the purpose of this paper to show *how one can let model polymer molecules do the work of "computing" stresses in numerical flow calculations*. To that end, we combine the finite-element method with Brownian dynamics or other stochastic simulation techniques (in homogeneous flows, it is well-known how stresses can be evaluated very efficiently by simulating typical polymer configurations). We here present the first steps of an extensive project, which we refer to as the CONNFFESSIT-project (Calculation Of Non-Newtonian Flow: Finite Elements and Stochastic SImulation Technique).

2. MODEL AND PROBLEM

In order to illustrate the basic ideas of the CONNFFESSIT-procedure we consider the start-up of steady shear flow for Oldroyd's fluid B. Initially, the fluid rests between two fixed plates separated by a distance L (the plates are located at $y = 0$ and $y = L$). At time $t = 0$, the lower plate (at $y = 0$) starts to move with a constant velocity U (in the x-direction). The velocity profile $v_x = v_x(y,t)$ is to be calculated from the momentum balance equation

$$\rho \frac{\partial}{\partial t} v_x = \eta_s \frac{\partial^2}{\partial y^2} v_x - \frac{\partial}{\partial y} \tau^p_{xy} , \qquad (1)$$

where ρ is the fluid density, η_s is the solvent viscosity, and τ^p_{xy} is the polymer contribution to the shear stress. In order to evaluate $\tau^p_{xy} = \tau^p_{xy}(y,t)$ one usually employs a differential or integral constitutive equation where, for the Oldroyd-B fluid considered here, both types of constitutive equations can be formulated (refs. 1,2).

For the CONNFFESSIT-idea it is crucial to realize that the stresses for Oldroyd's fluid B in arbitrary flow fields can alternatively be obtained from a Hookean dumbbell model of dilute polymer solutions (ref. 1). The (time-discretized) dynamics of each individual Hookean dumbbell can be expressed in the form of a stochastic difference equation,

$$\Delta \boldsymbol{Q}^* = \boldsymbol{\kappa} \cdot \boldsymbol{Q}^* \, \Delta t - \frac{1}{2\lambda_H} \boldsymbol{Q}^* \, \Delta t + \boldsymbol{W} \sqrt{\frac{\Delta t}{\lambda_H}} , \qquad (2)$$

where $\boldsymbol{Q}^*$ is the dimensionless connector vector between the two beads, $\boldsymbol{\kappa}$ is the velocity-gradient tensor (for our shear-flow problem, its only non-vanishing component is $\kappa_{xy} = \partial v_x / \partial y$), λ_H is the relaxation time of the dumbbells, and Δt is the time step. The column vector $\boldsymbol{W}$ contains three independent standard Gaussian random numbers with mean 0 and variance 1. Furthermore, the Gaussian vectors $\boldsymbol{W}$ for different time steps are independent. The stochastic difference equation (2), which is equivalent to a diffusion equation in polymer configuration space, is the starting point for Brownian dynamics simulations (ref. 3). When a large number of dumbbells is simulated, the polymer contribution to the shear stress is given as the ensemble average

$$\tau^p_{xy} = -nkT \left\langle Q^*_x Q^*_y \right\rangle , \qquad (3)$$

where nkT is the product of the dumbbell number density, Boltzmann's constant, and the absolute temperature.

3. METHOD

Our basic idea for solving start-up of steady shear flow within the CONNFFESSIT-project is as follows: We throw a large number of Hookean dumbbells into the Newtonian solvent between the plates (the larger the number of dumbbells is, the smaller the statistical error bars are). The velocity of the center of mass of each dumbbell is equal to the local fluid velocity (for our simple shear-flow problem, the position y of each dumbbell does not change in time), and the dynamics of the internal degrees of freedom of the dumbbells are given in terms of the velocity gradients in (2). From the dumbbell configuration(s) at time t and position y we obtain the polymer contribution to the shear stress according to (3). Rather than v_x and τ^p_{xy}, we use v_x and the dumbbell configurations $\boldsymbol{Q}^*$ as the primary variables in solving the momentum balance equation (1), and τ^p_{xy} is considered here as an auxiliary variable only. The dumbbell configurations contain all the relevant information about the polymeric liquid, in particular, about the shear stress. This is why, in the title, we referred to the dumbbells, which represent polymer molecules, as "smart".

In the numerical calculations, the y-range is divided into M finite elements of equal length L/M, and on each element we approximate v_x by a linear function of y (approximation of the type P^1–C^0). We distribute the total number of N dumbbells evenly over the M elements. At $t = 0$, v_x decreases linearly from U to 0 in the first element and is identically zero in all other elements; all dumbbell configurations are chosen according to the known equilibrium distribution. In each further time step, we have to evolve the velocity function and the dumbbell configurations. From the dumbbell configurations we calculate the shear stress τ^p_{xy} as an auxiliary quantity from (3); since all dumbbells in any given finite element feel the same velocity gradients it is natural to simply average τ^p_{xy} over all dumbbells in the given element (approximation of the type P^0–C^{-1}). The velocity profile at the end of the time step is then obtained from (1). The new dumbbell configurations are obtained by simulating (2). The time-integration proceeds until the linear steady-state velocity profile is reached within a specified precision. For the time-integration, we use an explicit Euler scheme.

In all numerical calculations, we fix the units of length, time, and mass by setting the distance between the plates, the velocity of the lower plate, and the total zero-shear-rate viscosity equal to one ($L = U = \eta_0 = 1$).

4. RESULTS

In order to assess the validity and efficiency of the CONNFFESSIT-idea we here discuss results for the special case of the upper-convected Maxwell model ($\eta_s = 0$). This is the most crucial test because (i) the shear stress stems entirely from the Brownian dynamics simulation and the effect of fluctuating stress data is hence most obvious, and (ii) the Maxwell model is a particularly delicate model in numerical calculations of flow (our explicit calculations for Oldroyd's fluid B confirmed that the method works better for $\eta_s \neq 0$). We did not try to refine the finite-element method in order to account for the hyperbolic character of the Maxwell model which leads to a finite propagation speed $c = \sqrt{nkT/\rho}$ for velocity perturbations (ref. 4).

Figure 1 shows the time-dependent velocity profile for a combination of parameters for which the linear steady-state profile is reached monotonically. The jump in velocity propagates with a finite speed from the lower to the upper plate; there exists a sharp wave front. In front of the velocity jump, the exact velocity is zero (ref. 4); in our numerical solution, we find fluctuations of the order of 0.1 which are suppressed in the figure.

The results for a combination of parameters for which elasticity is more important are shown in Figure 2. The velocity at the wave front is now much less damped and, due to multiple reflections at the walls, the wave front moves back and forth between the walls (fig. 2a). The time-dependent velocity at a fixed position, e.g. at the center of the channel, displays oscillations resulting from the weakly damped wave front repeatedly passing through the position under consideration (fig. 2b). All important features in start-up of shear flow for the Maxwell model are captured by the method developed here, and within the statistical error bars perfect agreement with the analytical solution is found.

5. DISCUSSION

The above discussion for the Maxwell model shows that the CONNFFESSIT-idea can be successfully employed in order to calculate nontrivial flows. The method has many important advantages compared to the usual calculation of stresses from differential or integral constitutive equations:

- One can easily switch to another constitutive equation by simply exchanging the subroutine for the stochastic simulation.
- Typically, even those kinetic-theory models which are commonly considered to be very complicated can be simulated without much effort and can thus be treated with the method suggested here. It is not even necessary to have closed-form constitutive equations in order to make kinetic-theory models tractable.

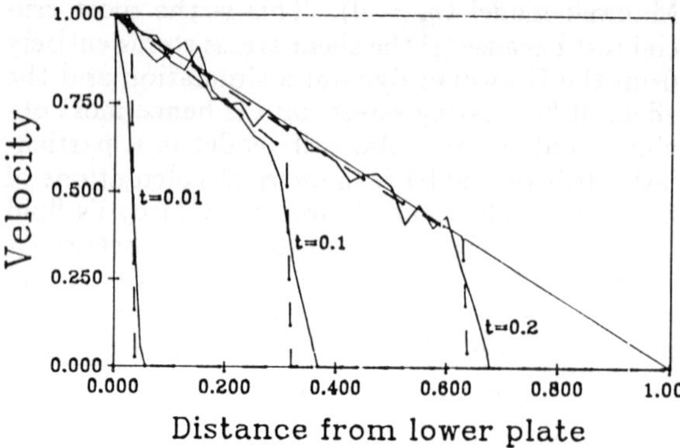

Figure 1: Velocity as a function of position for different times. Parameters: $\lambda_H = 0.1$, $c^2 = nkT/\rho = 10$; $N = 750,000$, $M = 40$, $\Delta t = 10^{-3}\lambda_H = 10^{-4}$. In all figures, the continuous (dashed) lines represent the simulation (analytical) results. The diagonal line indicates the linear steady-state profile.

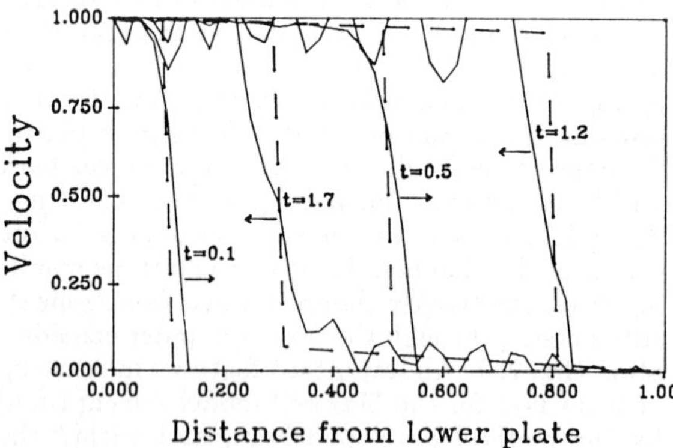

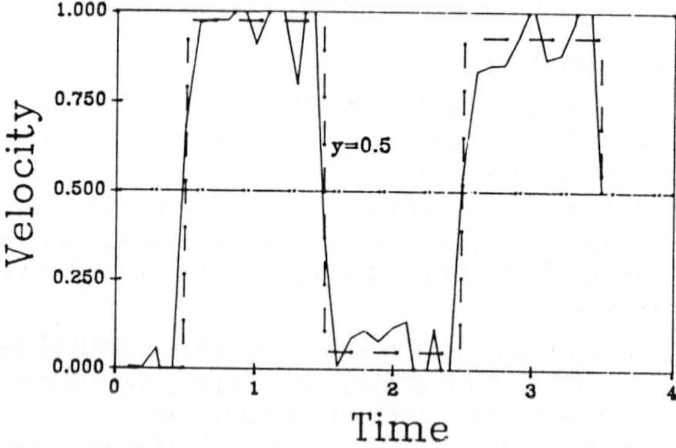

Figure 2: (a) Velocity as a function of position for different times. (b) Velocity at the center of the channel as a function of time. The dash-dotted line represents the steady-state value of the velocity. Parameters: $\lambda_H = 10$, $c^2 = nkT/\rho = 1$; $N = 1,750,000$, $M = 40$, $\Delta t = 10^{-3}\lambda_H = 10^{-2}$.

- Complete information about the polymer configurations is directly available, particularly, the molecular stretching and orientation responsible for the mechanical properties of finished plastic products (this information being stored in the "smart" molecules simulated).
- Since the actual motion of individual molecules is simulated it is possible to take effects of polydispersity, polymer diffusivity, polymer migration, or walls into account in a straightforward way without any artificial assumptions or mathematical complications.
- The method is well-suited for time-dependent problems.

A drawback of the CONNFFESSIT-idea is that a large amount of memory is required to store all the polymer configurations, and the simulations may be very time-consuming. (However, it is always possible to simulate a number of molecules which is much larger than the number of finite elements, $N \gg M$.) Another disadvantage is that the result for the flow field is, of course, subject to statistical fluctuations.

Results for the start-up of steady shear flow for the Oldroyd-B, FENE, Doi-Edwards, and Curtiss-Bird constitutive equations, together with further details about the method, will be published elsewhere. Currently we are working on several improvements and generalizations of the implementation of the CONNFFESSIT-idea, including implicit time-integration schemes, more general methods for fitting finite-element functions to simulation data, generalization to flow problems in two space dimensions, and formulation of rules for simulating constitutive equations not derived from kinetic theory.

Generous allocations of CPU-time on the Swiss Federal Institute of Technology's CRAY Y-MP computer are gratefully acknowledged. H.C.Ö. thanks the Swiss National Foundation for Scientific Research (Grant No. 21-29571.90) for financial support.

REFERENCES

1. R. B. Bird, C. F. Curtiss, R. C. Armstrong and O. Hassager, Dynamics of Polymeric Liquids, Vol. 2, Kinetic Theory, 2nd Ed., Wiley-Interscience, New York, 1987, §13.4, pp. 71–76.
2. M. J. Crochet, A. R. Davies and K. Walters, Numerical Simulation of Non-Newtonian Flow, Elsevier, Amsterdam, 1984.
3. H. C. Öttinger, J. Non-Newtonian Fluid Mech., 19 (1986) 357–363.
4. A. Narain and D. D. Joseph, Rheol. Acta, 21 (1982) 228-250.

Theoretical and Applied Rheology, edited by P. Moldenaers and R. Keunings
Proc. XIth Int. Congr. on Rheology, Brussels, Belgium, August 17-21, 1992

HIGHER ORDER FINITE ELEMENTS FOR FLOW OF VISCOELASTIC FLUIDS

K. K. Talwar and B. Khomami

Department of Chemical Engineering and the Materials Research Laboratory, Washington University, Campus Box 1198, 1 Brookings Drive, St. Louis, MO 63130 (USA)

1. INTRODUCTION

The governing equations describing the flow of Upper Convected Maxwell (UCM) and Oldroyd-B (OLD-B) fluids have been known to provide unstable discretizations (i.e., without addition of artificial diffusion) when using conventional lower order techniques. Higher order methods, on the other hand, are now known to give spectral accuracy when the order of the interpolant is enhanced [1,2]. The p-version of the finite element method is thus well suited for viscoelastic flow simulation, in addition to the fact that is extremely flexible in terms of space and variable discretization.

The problem of interest, here, is the two-dimensional transverse flow of viscoelastic fluids through different configurations of non-deformable, cylinders fixed in space. This choice of geometry is motivated by an idealization of fiber beds to understand the role of extensional stresses in flow through porous media.

2. PROBLEM FORMULATION

The pertinent equations (i.e., equations of motion and continuity) for inertia-less steady flow in the absence of body forces are:

$$\nabla \cdot v = 0 \tag{1}$$

$$-\nabla P + \nabla \cdot \tau = 0 \tag{2}$$

in which **v**, P and τ denote the velocity vector, isotropic pressure and the deviatoric stress tensor, respectively. The constitutive equation used is the Oldroyd-B fluid model:

$$\tau + \lambda_1 \tau_{(1)} = 2\mu(\dot{\gamma} + \frac{\lambda_2}{\lambda_1}\dot{\gamma}_{(1)}) \tag{3}$$

where $\dot{\gamma}$ is the strain rate tensor, subscript (1) denotes the upper convected material derivative, and λ_1 and λ_2 are the characteristic relaxation and retardation times for the fluid. The Oldroyd-B constitutive model simplifies to UCM and Newtonian fluid models when $\lambda_2 = 0$ and $\lambda_1 = 0$, respectively.

A weak formulation of the governing equations is obtained by applying standard Galerkin procedures to the system of equations, and can be found in [3]. The domain is divided into a number of elements, so that each element has the basis functions defined over its space [4]. The approximate solution is then locally expressed as a linear combination of these functions.

In brief, the computational procedure involves the following steps:

1. Check for convergence of the set of non-linear equations. For a given value of the relaxation time, if convergence is not achieved, the p-level is raised in every sub-domain, and the equations are rediscretized.

2. If convergence is achieved, the degree of momentum and mass conservation is assessed. If these criteria are not satisfied, p-levels are raised in those sub-domains which reveal the most significant deviations from the exact solution.

3. The value of a predefined norm is confirmed to have converged to the required degree. In the context of this research, this norm is defined as the flow rate of the fluid.

Variations of the residues of the mass and momentum conservation equations are also monitored concurrently. Experiments with Navier-Stokes equations (ref 5) show these expressions to be reliable indicators. To confirm the degree of momentum conservation, stress jumps at interelement borders are confirmed to be small compared to the applied pressure drop.

3. RESULTS AND DISCUSSION

The numerical technique relies on the fact that higher order Galerkin methods are stable with respect to the characteristic Weissenberg number (We), hence the addition of artificial diffusion is obviated (ref. 2). The application of this technique, however, raises numerous questions regarding the different combinations of discretization procedures. Among these are the consequences of a disparity in the relative order of the interpolant used to discretize the deviatoric stress and velocity variables (the relative interpolant disparity between velocity and pressure is bounded, but not fixed, by the Brezzi-Babuška condition). Furthermore, the polynomial degree can vary in the interelement sense, so that higher order interpolants may be used only in sub-domains in which elastic effects are important.

The domain of computation is shown in Figure 1. The porosity of the medium is changed by altering the radius of the cylinders. The hatched domains represent unit computational cells, with the assumption that the flow is fully developed. The appropriate boundary conditions of no slip (at the cylinder walls), symmetry and periodicity (for fully developed flow) are enforced, along with the applied pressure drop. In this geometry, the Weissenberg number is defined as We = $\lambda_1 Q/R^2$, where Q is the computed flow rate. Clearly, the domain is devoid of any geometric singularity, hence avoiding the issue of stress nonintegrability and the consequent corruption of the numerical solution.

The flow resistance is computed by the formula:

$$fRe_{VE} = \frac{Q_N}{Q_{VE}} fRe_N \qquad (4)$$

where Q_N and Q_{VE} are the flow rates of Newtonian and viscoelastic fluids, respectively, ascribed to a given pressure drop. f is the friction factor, fRe_N is the flow resistance for the corresponding Newtonian fluid and it is normalized to one.

Previous investigation with the square cylinder geometry has [2] demonstrated that the flow resistance of both UCM and OLD-B fluids decreases as We is increased, for the range of We studied. The staggered square geometry, on the other hand, exhibits different flow kinematics (i.e., for a given porosity the ratio of elongational/shearing kinematics is enhanced in comparison to the square array). Hence, in this study we have mainly focused our attention on this arrangement. The domain of computation is discretized as shown in Figure 2. The flow resistance as a function of We in this geometry increases monotonically after initially falling, thereby passing through a minimum at We ~ 0.2 (see Figure 3). Experiments with Boger fluids (ref. 5) in the same geometry indicate that fRe increases dramatically at We ~ 0.02. The discrepancy has recently been attributed to be due to a purely elastic instability (ref. 6). Steady

4. CONCLUSION

An accurate higher order finite element discretization for steady state simulation of viscoelastic fluids has been presented. It has been shown that this procedure gives rise to exponential convergence toward the exact solution for viscoelastic flows. In addition, the efficiency of the technique is dependent on mesh design, however, interelement and intervariable disparities in discretization allow for sufficient flexibility and efficiency of the algorithm.

This method was successfully used to consider transverse flow of UCM and OLD-B fluids past a staggered array of infinitely long, rigid cylinders. By virtue of the formulation, an upper Weissenberg number limit was not observed in this study. However, the memory requirements for the discretization grows quadratically with the order of interpolant. Consequently, the maximum attainable We is determined by availability of computational resources.

In order to reduce the computational intensity of our higher order technique, we have utilized the streamline upwind technique in conjunction with our hp-type finite element procedure. To achieve improved stability (i.e., in comparison to lower order Galerkin based techniques) restrictions on order of interpolants for the velocity approximation and element size must be enforced. Based on these restrictions an upper bound on the interpolant order for velocity discretization and size of the elements is obtained. These issues (i.e., bounds of the velocity interpolant and element size) in terms of two different formulations (i.e., with or without decoupling the stress into a Newtonian and elastic component) will be discussed in more detail in our presentation.

5. REFERENCES

1. S. Pilitsis and A.N. Beris, J. Non-Newt. Fluid Mech., 31, 231 (1989)
2. K. Talwar, B. Khomami, "Application of Higher Order Finite Element Methods to Viscoelastic flow in Porous Media", to appear in J. Rheology.
3. M.J. Crochet, A.R. Davies, K. Walters, "Numerical Simulation of Non-Newtonian Flow", Elsevier Science Publishers (1984)
4. B. Szabo, I. Babuška, "Finite Element Analysis", John Wiley & Sons, Inc. (1990)
5. L. Skartsis, B. Khomami and J.L. Kardos, "Polymeric Flow Through Fibrous Media", to appear in J. Rheol.
6. L. Skartsis, "The permeation of fiber beds by Newtonian and Non-Newtonian fluids with applications to the autoclave and resin transfer molding processes", D.Sc. Thesis, Washington University, St. Louis, MO (1992)

6. FIGURES

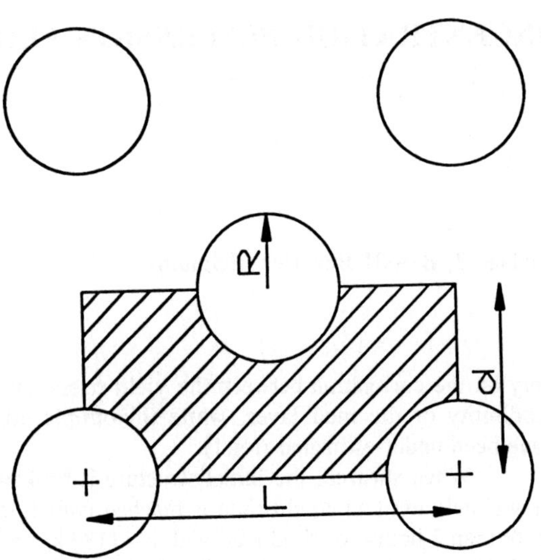

Figure 1 Geometry for Flow Past Staggered Cylinder Array

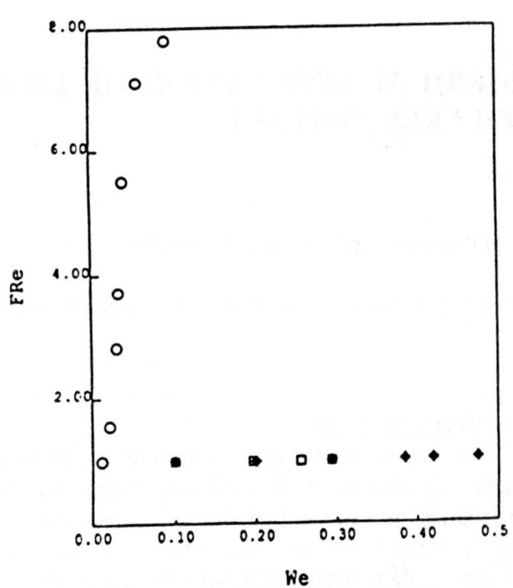

Figure 3(a) Flow Resistance as a function of Weissenberg number for Flow Past Staggered Cylinder Array
o experiments, □ UCM fluid, ◇ OLD-B fluid

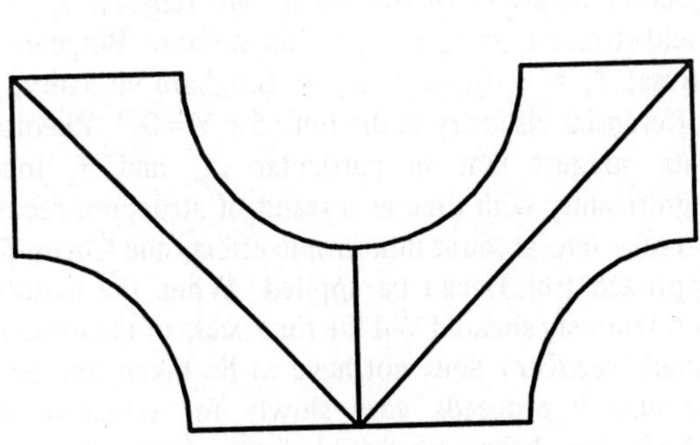

Figure 2 Mesh Design for Flow Past Staggered Cylinder Array

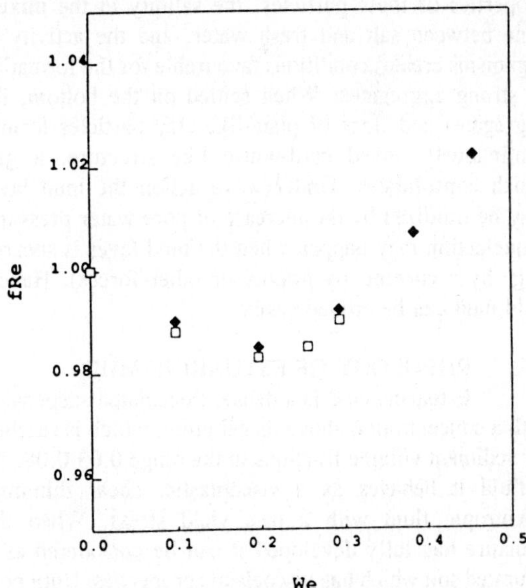

Figure 3(b) Fig. 3(a) with fRe scale magnified

Theoretical and Applied Rheology, edited by P. Moldenaers and R. Keunings
Proc. XIth Int. Congr. on Rheology, Brussels, Belgium, August 17-21, 1992

NUMERICAL SIMULATION OF HIGH CONCENTRATION SUSPENSION FLOW WITH FREE SURFACE.

E.A. TOORMAN and J.E. BERLAMONT

Hydraulics Laboratory, Katholieke Universiteit Leuven, de Croylaan 2, B-3001 Heverlee (Belgium)

1. INTRODUCTION

The study of the flow behaviour of cohesive sediment suspensions is of vital importance for the understanding of sediment transport in estuaries by gravity currents and for the design of dredging equipment. A numerical model has been developed for this purpose and has been applied to the simulation of mud pumping experiments (ref.1).

Estuarine fluid mud is a high concentration suspension of a mixture of fine, cohesive sediment (particle size $d<2\mu m$) with a small fraction of sand ($d>60\mu m$) and silt ($2<d<60\mu m$). Cohesive sediment contains a high fraction of clay-minerals, further some other minerals and organic matter. The colloidal properties of these particles, the salinity in the mixing zone between salt and fresh water, and the activity of organisms creates conditions favourable for the formation of strong aggregates. When settled on the bottom, the aggregates and flocs of plate-like clay particles form a continuously linked card-house like structure, a gel, which consolidates. Under wave action the mud layer may be fluidized by the increase of pore water pressure. Liquefaction may happen when the mud layer is sheared (e.g. by a current, by gravity or other forces). Hence, fluid mud can be eroded easily.

2. RHEOLOGY OF ESTUARINE MUD

Estuarine mud is a dense, flocculated suspension with a concentration above its gel point, which is reached for sediment volume fractions in the range 0.03-0.08. As a fluid it behaves as a viscoplastic, shear thinning, thixotropic fluid with a true yield stress. When the structure has fully developed it can be considered as a saturated soil which has viscoelastic properties. Until now the low shear rate flow behaviour of fluid mud has not been studied systematically, because of the present limitations of rheometry for cohesive sediment suspensions and because researchers in the field of cohesive sediment transport have focused their attention only on the correlation between the yield stress and the erodibility of the mud layer, while thixotropic effects have been under-estimated (ref.1).

When sheared, the initial structure is broken up very slowly until an equilibrium is reached (which takes more than 2 hours for fluid mud with $\rho=1189$ kg/m^3, or $\phi=0.11$, at a shear rate of $10s^{-1}$). Traditionally, it has always been assumed to be a Bingham fluid. However, for shear rates $>0.5s^{-1}$ the material could be described better by a Sisko power-law model with exponent 0.9. The equilibrium behaviour, as suggested by concentric cylinder viscometer data, can be described adequately by the Worrall-Tuliani model (ref.2):

$$\mu = \mu_\infty + \frac{\tau_y}{Y+Y_o} + \frac{\tau_o}{Y} \qquad (1)$$

where: μ = dynamic viscosity; Y = shear rate intensity (second invariant of the shear rate tensor); τ_o = true yield stress; $\tau_y = \tau_B - \tau_o$; τ_B = apparent or Bingham yield stress; $Y_o = \tau_y/(\mu_o-\mu_\infty)$; μ_∞ = Bingham viscosity; μ_o = differential viscosity in the limit for $Y=0s^{-1}$. Rheological data suggest that in particular μ_∞ and τ_o increase significantly with time as a result of structural recovery. To take into account thixotropic effects the Cheng-Evans approach (ref.3) can be applied. When the material is continuously sheared and the time scale of the experiment small, recovery does not have to be taken into account because it proceeds very slowly for estuarine muds, which has been concluded from consolidation tests (ref.1). The viscosity increases rapidly with concentration (power law exponent 3-5). With an increase of the sand fraction, the structural strength reduces and the gel point moves to higher concentrations.

3. FORCED MUD FLOW

3.1 Experiments

Since 1982 research has been going on for the design of a mud pumping installation for local maintenance dredging. Experiments have been carried out in a 12m long, 0.5m wide flume. Mud was being pumped out of a 0.3 m deep mud trap in the middle, forming only a crater in the 35 cm thick mud layer as a result of the rheological properties of mud.

3.2 Numerical simulation

To reduce experimental costs a numerical model has been developed. A mixed interpolation Galerkin finite element method has been used to solve the generalized Navier-Stokes equations in a 2-dimensional vertical plain. A mixed Euler-Lagrange approach, based on the Arbitrary Lagrangian-Eulerian method (ref.4), is used to calculate the displacements of the nodes of the free surface, which in the presented case study is the mud-water interface, by solving the kinematic condition equation. The highly non-linear viscous terms require an iterative procedure. The moving surface is updated at each iteration step. Remeshing of the total grid is carried out every time step in a controlled way by imposing nodal displacements which are defined as a certain fraction of the free surface nodal displacements. Hence, the distortion of the grid is minimized.

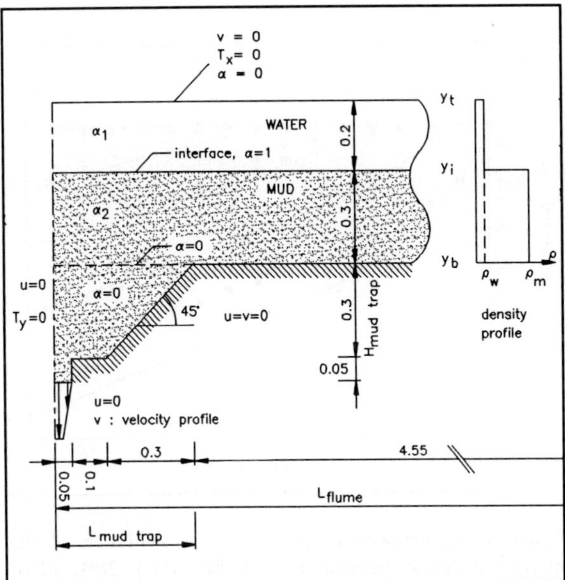

Figure 1: Problem description: domain and boundary conditions.

The model does not allow the exact determination of a yield surface. Instead it assumes that the material flows with very small velocities in the unyielded zone as a result of the high viscosity for Y approaching 0 (Eq.1). It has been shown in (ref.1) that there is little difference between the predicted flow fields obtained using this approach and one that does solve the flow only within the yielded region.

From experiments it was concluded that friction with the vertical walls and at the interface could be neglected. Consequently it sufficed to consider only the mud layer within the computational domain. The gravity constant then had to be corrected by a buoyancy factor $\Delta\rho/\rho$. Thixotropic effects and the consolidation, prior to the forced flow are also neglected in a first approximation. The density is assumed to be uniform, which is justified by measured density profiles in consolidating mud layers with initial concentrations above the gel point (ref.1).

4 DISCUSSION AND CONCLUSIONS

Results of the simulation of one of the experiments are shown in the Fig.2-4 (time step Δt=5s). For each time step 50 iterations were done (unless a tolerance of 0.5% was reached). The computational domain extends from the centre, a symmetry plane, up to 5 m distance. Qualitatively the flow pattern and the free surface profile are well in agreement with experimental observations. Fig.3 clearly shows the spreading of the liquefaction of the mud as a result of the forced flow.

By comparing the decrease rate of the mud volume with the known pump discharge, a measure for the error could be defined, which revealed that it stayed within the limit of 1%. Some oscillations occur in the free surface when the slope exceeds the equilibrium value. A smaller time step may improve the results, but has so far not been verified. Convergence is extremely slow due to the high viscosities. Note the triangular zone of hardly flowing material at the surface above the suction bell. The predicted initiation of the flow is more abrupt than was observed (Fig.4), which must be attributed to the neglect of thixotropy. Fig.3 also shows discrepancies between the observed and the predicted interface level in the centre as a result of a slight difference between the assumed and the true pumping discharge.

Because of the very low shear rate intensities ($<0.1s^{-1}$), the rheological behaviour is dominated by the Bingham term. Rheometrical data in this range, using a rotational viscometer, are unreliable due to slip and the narrow zone of flowing (i.e. yielded) material. Hence, calibration of the rheological model has to be performed by comparing computed and experimental data, in the presented case interface profiles. This revealed that a true yield stress had to be taken of the same magnitude as τ_B, which is significantly larger than the τ_o suggested by rheometrical data. The fully developed flow in the flume section corresponds well with a Bingham plug flow. Differences between simulations and experiments seem to be caused primarily by 3-dimensional effects.

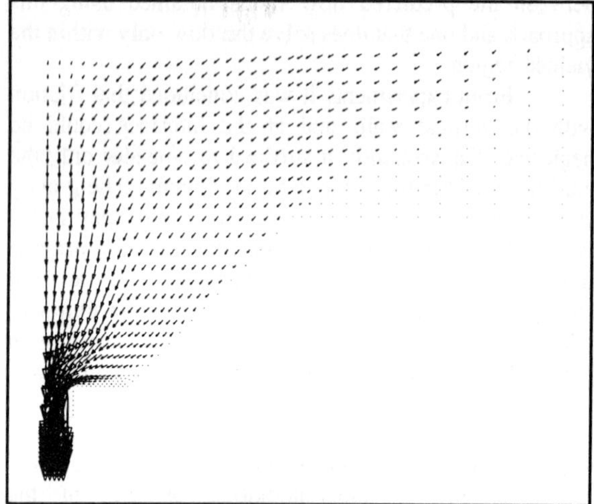

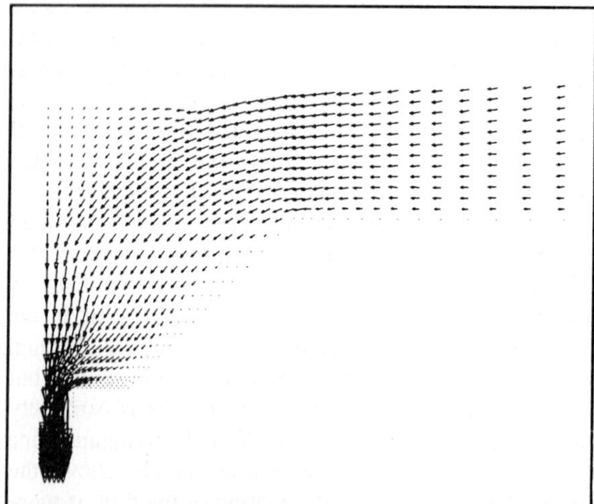

Figure 2: Computed velocity field after 1 and 10 min. pumping (ρ=1140 kg/m^3; τ_o = 9 Pa; μ_∞=0.035 Pa.s; pump discharge Q_p=0.26 l/s).

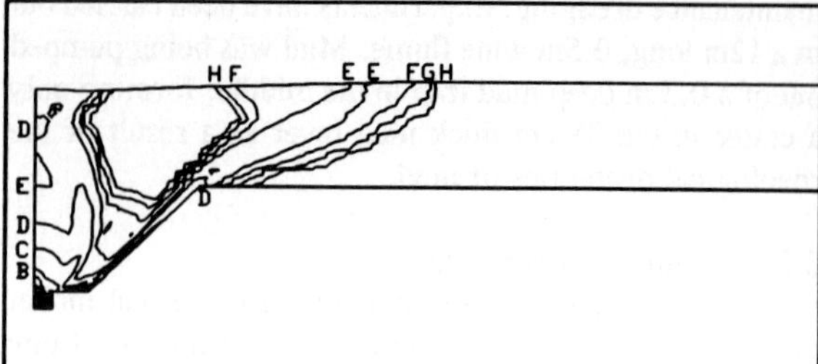

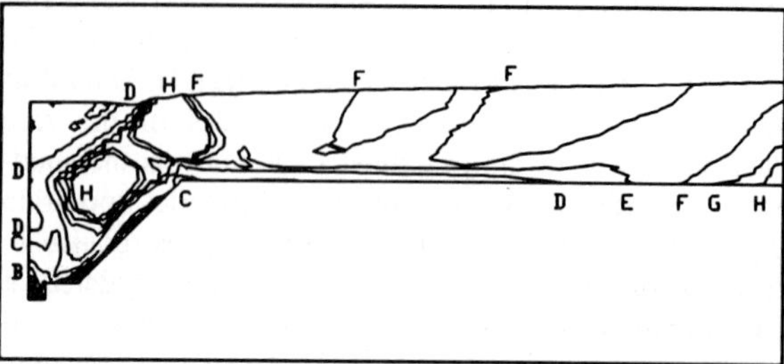

Figure 3: Computed shear rate intensity contours after 1 and 10 min. pumping, corresponding Fig.2. Legend: A=0.3, B=0.1, C=0.03, D=0.01, E=0.003, F=0.001, G=0.0003, H=0.0001s^{-1}.

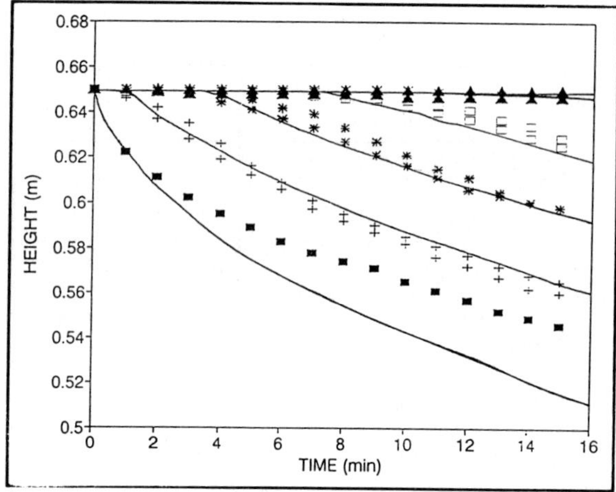

Figure 4: Experimentally recorded time evolution of the interface points at distances x = 0 (■), 1(+), 2(*), 3(□), 4(×) and 5(▲) m from the centre, compared with computed results (solid lines).

Future work will combine this flow model with a sediment transport model, presented in (ref.1), which can predict the settling and consolidation of a dense cohesive sediment suspension.

ACKNOWLEDGEMENTS

The first author received a grant as Research Assistant of the Belgian National Fund for Scientific Research.

REFERENCES

1. **E.A. Toorman**, Modelling of fluid mud flow and consolidation. PhD thesis, Civil Eng. Dept., K.U.Leuven, 1992.
2. **W.E. Worrall and S. Tuliani**, Trans. Brit. Ceramic Soc., 63, No.4 (1964) 167-185.
3. **D.C.-H. Cheng and F. Evans**, Brit. J. Appl. Phys., 16 (1965) 1599-1617.
4. **A. Huerta and W.K.Liu**, Comp. Meth. Appl. Mech. Eng., 69 (1988) 277-324.

Theoretical and Applied Rheology, edited by P. Moldenaers and R. Keunings
Proc. XIth Int. Congr. on Rheology, Brussels, Belgium, August 17-21, 1992

FINITE ELEMENT SIMULATION OF THE ROTATING FLOW OF A VISCOELASTIC FLUID

D. ANNE-ARCHARD[1], H.C. BOISSON[1] and H. HA MINH[1]

[1]Institut de Mécanique des Fluides de Toulouse, Avenue du Professeur Camille Soula, 31400 Toulouse, (France)

1. INTRODUCTION

Because of its difficulty, numerical simulation of viscoelastic fluid flow has been concentrated on a few test geometries (namely the 4:1 contraction). The aim of this paper is to discuss the effect of rotation on the flow inside a cylindrical vessel agitated by a paddle impeller. This flow is of practical interest, especially in the food industry. Viscoelasticity is, here, a large domain of investigation (ref. 1).

2. METHOD

The momentum, continuity and constitutive equations for the incompressible, isothermal flow of the viscoelastic fluid are expressed in a rotating frame. In this work, an Oldroyd "B" fluid is used, with its characteristic times λ_1 (relaxation) and λ_2 (retardation). We use the classical set $\lambda_2/\lambda_1 = 1/8$.

The two-dimensionnal flow is simulated by an unsteady finite element method converging to a final steady state. Complete biquadratic (nine nodes) approximations are used for velocities and stresses, and bilinear ones for pressure. Semi-implicit Euler time scheme is retained for temporal discretization and a decoupled approach is used for the hydrodynamic and constitutive equations.

3. RESULTS

Tests were performed on the 4:1 contraction with the Oldroyd model ($De_{lim} = \lambda_1 . \gamma_{wall} = 3.$) (ref. 2) and in the newtonian case for the vessel: the inertial effects are pointed out for Reynods (Re) number varying from 0.1 to 40 and compared with experimental data (ref. 3). For the viscoelastic case, the Deborah number $De=\lambda_1.\omega$ is increased gradually with a continuation technique. No stable solution was obtained beyond De=0.2 .

3.1 Velocities and power number

Our results show a weak effect of De on velocities (less than about 5%) and the flow pattern is not affected. The power number Np is proportional to the dissipated power P :

$$Np = P / \rho N^3 D^5$$

where $\omega=2\pi.N$ is the rotational speed of the impeller of diameter D. For newtonian fluids, the power consumption can be expressed as Np.Re=A where A is a geometrical constant. The product Np.Re for an Oldroyd "B" fluid is found to be slightly decreasing with De .

3.2 Extra Stress Tensor

The extra stress field is characterized by a sharp discontinuity at the impeller tips, which increases with De as shown on figure 1. Numerical oscillations dramatically grow with De, requiring stabilization processes in order to reach higher Deborah number.

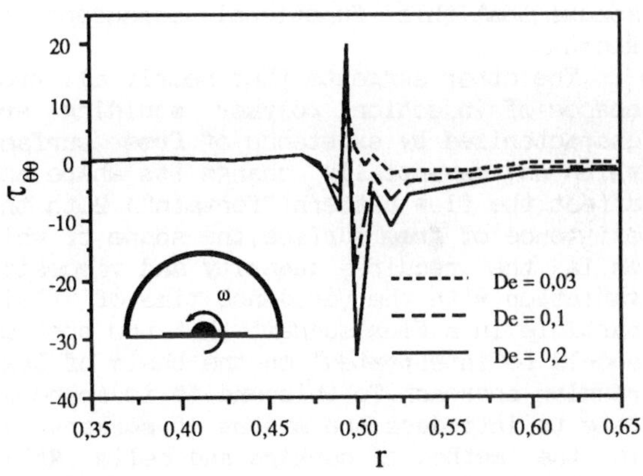

Figure 1 : tangential stress component in the leading edge plane (radius of the vessel: 1, radius of the impeller: 0.5, Reynolds number: 0.1).

ACKNOWLEDGMENTS: The authors wish to thank Pr R. Gaudu who has initiated this work.

REFERENCES

1. G. BOHME, M. STENGER, Chem. Eng. Technol., 11 (1988) 199-205.
2. D. ANNE-ARCHARD, H.C. BOISSON, R. GAUDU, Cahiers de Rhéologie, 8 (4) (1990) 219-236.
3. A. YOUCEFI, R. GAUDU, J. BERTRAND, D. ANNE-ARCHARD, M. SENGELIN, The 1992 IChemE Research Event, Manchester, U.K., January 9-10, 1992, Institution of Chemical Engineers, Rugby, U.K., 1992, pp. 377-379.

Theoretical and Applied Rheology, edited by P. Moldenaers and R. Keunings
Proc. XIth Int. Congr. on Rheology, Brussels, Belgium, August 17-21, 1992

NUMERICAL SIMULATION OF THE FLOW OF FLUIDS WITH TIME DEPENDENT VISCOSITY

I.K.BEREZIN

Institute of Continuous Media Mechanics,Korolev str.1,PERM,614061(Russia)

Industrial polymer processing, e .g . polymer injection moulding involves variety of hydrodynamic, thermal, physical-chemical phenomena,etc. In the present work emphasis is placed at the two significant aspects of this technological problem.One of them states that realistic polymer fluids are characterized by nonlinear dependence of viscosity on the shear rate and temperature, and that viscosity increases with time due to polymerization processes. In the following we shall abstract from the factors, responsible for viscosity increase with time and assume that this functional dependence is known.

The other suggests that nearly all processes of injection polymer moulding are characterized by existence of free surface which may essentially change its shape and affect the flow pattern("fontain").Both the existence of free surface,the shape of which is the required quantity and viscosity variation with the residence time of fluid particle in a flux suggest that the problem should be interpreted on the basis of Lagrangian approach.To this end it is appropriate to introduce the system of markers as in the method of markers and cells (MAC). Equations of motion and energy are solved on the basis of Galerkin finite-element method using the penalty function for meeting incompressibility condition. The pressure,velocity and temperature are defined on triangular finite-element mesh which may be refined during the flow process. The mesh includes a system of markers,which are referred to as mathematical points moving at a local fluid velocity. Calculation of non-Newtonian fluid is carried out in terms of iterative method of successive approximations.Viscosity variation with time is considered according to following procedure.Each marker is suppozed to carry information on its coordinates and time of residence in the flow. Every new particle entering the computational domain has its own reference time.Then, to calculate visco- of each element it is first necessary to define which particles have entered it at a given time. Further the number of element markers and the mean time of thier residence in a flow are to be estimated. Following the known functional viscosity-time relation one may determine the element viscosity and its dependence on the shear rate as difined in the previous iteration.

The developed algorithm has been employed for calculation of the plane channel filling process.Distribution of viscosity appears to be of complicated character,due to existence of two contradicting processes. On the one hand,the solid boundaries are seen to have the highest velocity gradient which reduces viscosity.On the other hand,the walls accumulates "the old" particles (non-slip condition), leading to increased viscosity in this region. This gives rise to irregular distribution of viscous properties over the flow volume which depends largely on time properties of the polymer material.

Theoretical and Applied Rheology, edited by P. Moldenaers and R. Keunings
Proc. XIth Int. Congr. on Rheology, Brussels, Belgium, August 17-21, 1992

Improved Bingham-body Computations in 3-D

C.R. Beverly and R.I. Tanner
Department of Mechanical Engineering, The University of Sydney, Australia

Summary

The present paper considers the problem of predicting motionless regions and true plug (constant velocity areas where the stress is below the yield stress) regimes for fully three-dimensional Bingham plastic flows. Numerical solutions are obtained using a finite element biviscosity formulation as adopted by Milthorpe and Tanner [1], O'Donovan and Tanner [2] Gartling and Phan-Thien [3] and Beverly and Tanner [4]. Comparisons are drawn to alternative numerical approximations and a review of common finite elements is made with a view to finding accurate, stable and economical schemes. A number of elements are compared and we conclude that some of the Fortin elements [5,6] are most useful on the grounds of computational overhead and solution accuracy. These are used to investigate three-dimensional axial flows of a Bingham body in an annulus. The numerical solutions were proved to produce features similar to the idealised Bingham response.

One problem of practical importance to the oil and gas industry is the characterisation of drilling mud as it moves axially along a borehole between the casing and the wellbore walls. The drilling mud is generally regarded as a non-Newtonian suspension exhibiting apparent yield stress and, to a first approximation, may be characterised by the Bingham model. Pearson [7] considered the idealisation of an eccentric annular gap between casing and formation, the inner metal cylinder neither reciprocating axially or rotating about its own axis. A more rigorous analytical solution was presented by Walton and Bittleston [8] based on the narrow-gap approximation. We show that the numerical solutions of the axial flow of a Bingham plastic in a narrow eccentric annulus show good agreement with observations and theoretical predictions derived by Walton and Bittleston [8] when the inner casing is fixed. The investigation has been extended to consider the new problem where the inner casing rotates about its own axis with a constant angular velocity.

In conclusion, results are presented for a series of Bingham materials extruded through a square die. It is shown that with increasing yield stress the extent of the plug region increasing migrates from the die walls towards the centre of the die, thereby suppressing the degree of extrudate swell.

References

[1] J.F. Milthorpe and R.I. Tanner, Third Int. Conf. on Num. Methods in Laminar and Turbulent Flow, University of Washington, Seattle, 1983.
[2] E.J. O'Donovan and R.I. Tanner, J. Non-Newtonian Fluid Mech., 15 (1984), 75.
[3] D.K. Gartling and N. Phan-T hien, J. Non-Newtonian Fluid Mech., 14 (1984), 347.
[4] C.R. Beverly and R.I. Tanner, *J*. Rheol., 33 (1989), 989.
[5] M. Fortin, Int. J. Num. Meth. Fluids, 1 (1981), 347.
[6] M. Fortin and A. Fortin, Int, J. Num. Meth. Fluids, **5** (1985), 911.
[7] J.R.A. Pearson, Xth International Congress on Rheology, Sydney University, Sydney, Australia, 1988, 1.73.
[8] I.C. Walton and S.H. Bittleston, J. Fluid Mech., 222 (1991), 39.

Theoretical and Applied Rheology, edited by P. Moldenaers and R. Keunings
Proc. XIth Int. Congr. on Rheology, Brussels, Belgium, August 17-21, 1992
1992 Elsevier Science Publishers B.V.

TIME-DEPENDENT FLOW OF FLUIDS OF INTEGRAL TYPE

Bruce Caswell, Division of Engineering, Brown University, Providence, RI 02912 USA.
Support for this work under NSFGrant CTS-8921668 is gratefully acknowledged.

In (ref 1.) an Eulerian-Lagrangian scheme was developed for the solution of time dependent flows of differential models. The essence of the method is the solution of the differential constitutive equation for the Piola stress by Lagrangian time differences; this step is followed by transformation of the result into Eulerian form. The effect of the Lagrangian time step is to uncouple the stress fields from the velocity-pressure fields. Following the solution for the velocity-pressure (V, P) fields the stress fields are then updated to the same time level. Thus at every time step the latter are obtained by solution of Eulerian spatial equations. For simplicity, the linear viscoelastic fluid will be used to illustrate the starting point for an integral model. The following is an exact difference identity for the Piola stress $\mathbf{S}_1(t)$ of a particle whose place at the current time t is $\mathbf{x}$ relative to the stress of the same particle at time $t-\Delta t$ when its place was $\mathbf{y}$,

$$\mathbf{S}_1(t) = \mathbf{S}_1(t-\Delta t)\cdot\frac{\partial \mathbf{x}^T}{\partial \mathbf{y}} + \mathbf{S}_2 + \int_{t-\Delta t}^{t} \Delta G(t,\Delta t,t')\dot{\mathbf{B}}(t',t)dt' \qquad (1)$$

Here $\Delta G = G(t-t') - G(t-\Delta t-t')$, $\dot{\mathbf{B}}$ is the strain rate, and the auxiliary stress $\mathbf{S}_2$ is given by the convolution integral

$$\mathbf{S}_2 = \int_{-\infty}^{t} \Delta G(t,\Delta t,t')\dot{\mathbf{B}}(t',t)dt' \qquad (2)$$

The last term of equation (1) is expanded in powers of Δt with coefficients which are expressible in terms of velocity fields $V(t-\Delta t)$(known) and $V(t)$(unknown), and these terms provide the main contribution to the matrix operator in the solution of the Eulerian field equations by the finite element method. Schemes for differential models have the virtue that the memory of the material is stored in the stress fields, and thus the need to compute memory integrals is obviated. The second term in equation (1) introduces a new memory integral (2). Expansion of ΔG in powers of Δt produces a time expansion of the integral (2) with memory integral coefficients which can be interpreted as auxiliary stresses. For linear viscoelastic fluids, in which the relaxation modulus G(t) is represented by a multimode Prony series, the auxiliary stresses are the modal stresses associated with each relaxation time λ_i. For such fluids the total stress can be calculated from an evolution formula such as (1) in terms of stored auxiliary stresses without the need to evaluate memory integrals. Likewise the auxiliary stresses can be updated without the evaluation of memory integrals. This property of the scheme is the definition of closure.

An evolution identity analogous to (1) can be written for non-linear BKZ fluids; schematically this amounts to the replacement of G(t) with a strain-dependent modulus in equation (1). However, the auxiliary stresses generated by use of the evolution identity do not satisfy closure. Approximate closure is achievable in some cases depending on the particular form of the strain dependent modulus. In all cases the evolution identity relegates to terms of higher order in Δt any irreducible memory integrals, and by expansion of the last integral in (1) provides a properly ordered operator for the solution of the field equations.

REFERENCES

1. Serdakowski, J. A. & Caswell B., J. Non-Newtonian Fluid Mech., 29(1988)217

Theoretical and Applied Rheology, edited by P. Moldenaers and R. Keunings
Proc. XIth Int. Congr. on Rheology, Brussels, Belgium, August 17-21, 1992

NUMERICAL CALCULATION OF FLOW OF A BINGHAM FLUID ON A ROTATING DISK

FAN CHUN[1] and CHEN YAOSONG[2]

[1]Institute of Mechanics, Chinese Academy of Scineces, Beijing 100080, China
[2]Department of Mechanice, Peking university, Beijing 100871, China

In many industries it is necessary to coat a thin film on a surface and spin-coating is a simple technique which is widely used. Chan Man Fong et.al.(ref.1) have described the flow of a Casson fluid on a rotating disk. In this paper, the coordinate system, the equations of motion and continuity, and boundary condition are as same as Chan's(ref.1) equation.

The constitutive equation of a Bingham fluid is

$$T_{ij}=\{ K(r,t)+\tau_y(r,t)/\left|A_{st}A_{ts}/2\right|^{1/2}\} A_{ij} \quad \text{for } T_{st}T_{ts}/2\geq\tau_y^2, \qquad (1)$$

$$A_{ij}=0 \quad \text{for } T_{st}T_{ts}/2<\tau_y^2$$

where $K(r,t)$ is the plastic viscosity. $\tau_y(r,t)$ is the yield stress. A_{ij} is the first Rivlin—Ericksen tensor.

We assume the plastic viscosity and the yield stress to be

$$K(r,t)=K_o\exp(\alpha rt),$$

$$\tau_y(r,t)=\tau_o\exp(\beta rt) \qquad (2)$$

where $K_o=K(r,t)\,|_{t=o,r=ro}$ is the initial plastic viscosity, and $\tau_o=\tau_y(r,t)\,|_{t=o,r=ro}$ is the initial yield stress.

In the case of spin-coating we can assume $h_o/a\ll 1$ and A_{ij} can be simplified, than constitutive equation (1) and equation of motion are also simplified. Final we obtain the distribution of velocity as follow:

$$u=R_e e^{-\alpha rt}r\{\xi^2-(\xi-z)^2\}/2 \qquad \xi\geq z\geq 0$$

$$u=R_e e^{-\alpha rt}r\xi^2/2 \qquad h\geq z>\xi \qquad (3)$$

where $R_e=\rho a h_o\Omega/K_o$ is the Reynolds number. ξ is the interface between fluid-like behaviour and solid-like behaviour:

$$\xi=h-\tau_y/R_e r \qquad (4)$$

Since ξ cannot be negative, so the flow region is restricted to $r\geq r_y$, where

$$r_y=\tau_y/R_e h \qquad (5)$$

To determine the thickness h of flim at any across the film at any fixed point r, that is, we integrate with respect to z, from z=0 to z=h . Using Eq.(3), after some calcuations, we obtaim the differential equation of film thickness as follow

$$(e^{\alpha rt}/R_e)\partial h/\partial t+rh(h-\tau_y/R_e r)\partial h/\partial r$$
$$+(h-\tau_y/R_e r)[4h^2+h\tau_y/R_e r+(\tau_y/R_e r)^2]/6$$
$$-[rt(h-\tau_y/R_e r)/3]\{\alpha h^2 \qquad (6)$$
$$+(3\beta/2-\alpha/2)[h\tau_y/R_e r+(\tau_y/R_e r)2]\}=0$$

The partial differential equation (6) is solved numerically using difference method, where the radius of disk a=20cm, the initial thickness h_o=20μ m, the initial plastic viscosity K_o=1.2 g/cm·s, the density ρ =1.2 g/cm3, the angular speed of disk Ω= 3000 rpm and the initial distribution of thickness of film $h(o,r)\equiv 1$ for the case of τ_o=0.01 ;0.1;0.2 and α,β being respectively 0.1;0.2;0.3;0.4;0.5.

(1) For the case of τ_o=0.01, the thickness h increases with increasing r. This result agrees with the experimental result of magnetic disk used in electronic computer. The inclination of thickness increases with increasing α.

(2) For the case of τ_o=0.1 and α=β=0, the thickness h decrease with increasing r. This result agrees with the experimental result of Jenekhe & Schuldt(ref. 2).

(3) For the case of τ_o=0.2, according to Eq.(5), β is restricted to β <0.2.

REFERENCES

1. C.F Chan Man Fong, Cai Fushi, Xu Yuanze, ACTA Mechanica Sinica, 3, (1987), 107
2. Jenekhe S. A., Schuldt S. B., Chen. Eng. Commun., 33, (1985), 135

Theoretical and Applied Rheology, edited by P. Moldenaers and R. Keunings
Proc. XIth Int. Congr. on Rheology, Brussels, Belgium, August 17-21, 1992

NUMERICAL SIMULATION OF TWO- AND THREE-DIMENSIONAL FLOWS BY THE STREAM-TUBE ANALYSIS

J.R. CLERMONT, M.E. DE LA LANDE, M. NORMANDIN

Laboratoire de Rhéologie, I.M.G, Domaine Universitaire, B.P. 53X, 38041 Grenoble Cedex (France)

1. INTRODUCTION

The stream-tube analysis has been recently introduced (ref.1-2) for the numerical computation of 2D and 3D flows . In contrast to classical flow simulations, the relevant unknowns are, together with the pressure p, a transformation, assumed to be one-to-one, between the physical flow domain D and a mapped computational domain where the streamlines are parallel and straight. The incompressibility condition is automatically verified and, in the isothermal case, only momentum conservation equations are to be written. The method may accomodate various constitutive equations and leads to reduction of CPU time and storage area, in relation to the following distinguishing features:

1° The mapped streamlines are parallel and straight in the computational domain D' and a rectangular mesh can be built on the rectilinear streamlines related to the main flow of the physical domain.

2°The classical particle-tracking problem is avoided: the mesh points in the computational domain correspond to successive positions of the particle on its pathline.

3° The computation may be carried out by considering successive sub-domains, the stream tubes, from the outer boundary to the central flow region.

The context of the stream-tube analysis implies that recirculating flow regions which may appear in complex geometries are not taken into account: only the main flow involving open streamlines is computed.

2. APPLICATIONS

The stream-tube analysis has been applied to the following cases, using various constitutive equations:

- a three-dimensional flow involving a three-fold rotational symmetry (ref.3) for a viscous fluid (two mapping functions),
- the circular swelling flow problem, using a modified integral Goddard-Miller fluid and the differential Oldroyd-B fluid (one mapping function).

In both cases, the flow is computed by considering successive stream tubes. .The determination of the swelling ratio is possible by only considering the "peripheral stream tube" .

3. RESULTS

The results are obtained with a limited number of unkowns (of order of 1000) in the three-dimensional case, using the Levenberg-Marquardt algorithm. Concerning the swelling problem, the program still converges for Weissenberg numbers of 20, with a reduced CPU time.

REFERENCES

1. J.R. Clermont, C.R.A.S. (Juillet 1983) 297,Série II.1, Paris
2. J.R. Clermont, Rheol. Acta 27 (1988),357
3. J.R. Clermont, M.E. de la Lande, submitted to Theor. and Comp. Fluid Dyn.

Theoretical and Applied Rheology, edited by P. Moldenaers and R. Keunings
Proc. XIth Int. Congr. on Rheology, Brussels, Belgium, August 17-21, 1992

SIMULATION OF FIBRE SUSPENSION FLOW

X.J. FAN
Department of Mechanics, Zhejiang University, Hangzhou 310027 (China)

1. INTRODUCTION

The constitutive equation for semiconcentrated fibre suspensions in the Oldroyd-B fluid has been used in the numerical simulation of such a suspension flow past a sphere in a tube in order to exam the performance of the equation. The numerical technique and typical results of the simulation are presented here.

2. GOVERNING EQUATIONS

If the inertial force is negligible and the body force is potential, the conservation equations for isothermal flow reduce to the following dimensionless form

$$-\nabla p+\eta_m \nabla^2 V+\nabla\cdot R=0 \tag{1}$$

$$\nabla\cdot V=0 \tag{2}$$

where **R** is related to the extra stress tensor and expressed for simeconcentrated fibre suspension in the Oldroyd-B fluid as follows

$$\tau=\eta_m\dot{\gamma}+R \tag{3}$$

$$R=\frac{\phi a_r^2\eta_r}{3\ln(\pi/2a_r\phi)}\{\frac{1}{35}(1-f)\,[5\dot{\gamma}:N\delta+10\,(N\cdot\dot{\gamma}+\dot{\gamma}\cdot N-2\dot{\gamma}]+f\dot{\gamma}:NN-2c_1\dot{\gamma}\,(\delta-3N)\}+(1-\eta_m)\,\dot{\gamma}+\tau_p \tag{4}$$

where **N** is the structure tenor of fibres, which is determined using the equation of change:

$$N_{(1)}=\frac{2}{35}(1-f)\,[\dot{\gamma}-5\,(0.5\dot{\gamma}:N\delta+N\cdot\dot{\gamma}+\dot{\gamma}\cdot N)\,]-f\dot{\gamma}:NN+2c_1\dot{\gamma}\,(\delta-3N) \tag{5}$$

and τ_p is the polymer contribution of the extra stress tensor, which can be calculated from the constitutive equation

$$W_i\tau_{p(1)}=-\alpha\sigma\tau_p-\frac{3}{2}\alpha\,(1-\sigma)\,(N\cdot\tau_p+\tau_p\cdot N)+(1-\eta_r)\,\dot{\gamma} \tag{6}$$

where W_i is the Weissenbuger number, α and σ are two parameter characterising the effect of fibres on the hydrodynamic force of polymer molecules.

3. NUMERICAL TECHNIQUE

A decoupled Galerkin finite element method is used to solve the conservation equations. The equation of change for the structure tensor of fibres and the constitutive equation for the polymer contribution to the extra stress tensor are both solved at each node of elements using the time-dependent method. The flow field is divided into the quadrilateral elements with nine nodes for each.

4. NUMERICAL RESULTS

Flow past a sphere in a tube has been simulated using the present method. The radii of the sphere and tube are 1.0 and 2.0, respectively, and the length of the tube is 12.0. The mesh consists of 252 elements and 1083 nodes. Some of results are shown in the following figures.

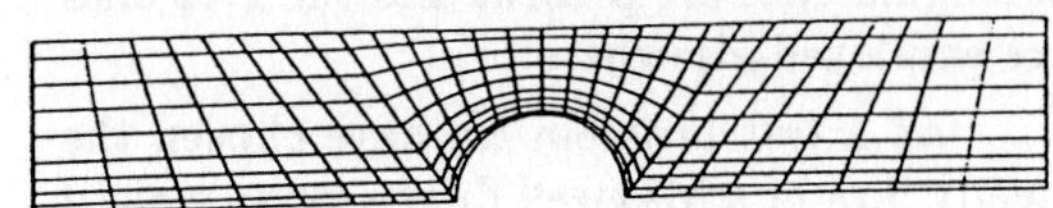

Fig. 1. the finite element mesh.

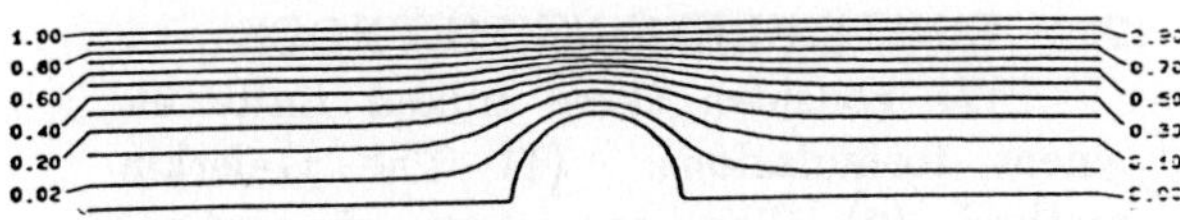

Fig. 2. the pattern of streamlines.

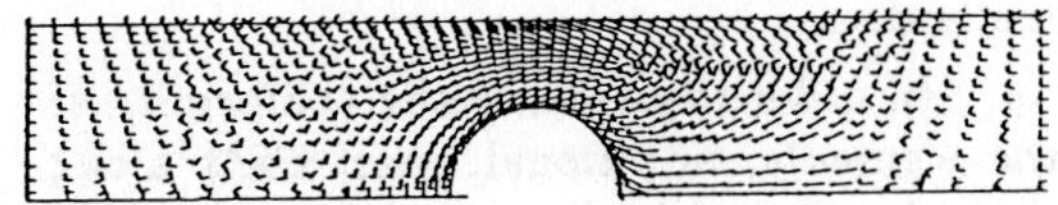

Fig. 3. the orientation of fibres.

ACKNOWLEDGMENT

Financial supports from the University of Melbourne, N.S.F. of China and of Zhejiang Province are greatly acknowledged.

Theoretical and Applied Rheology, edited by P. Moldenaers and R. Keunings
Proc. XIth Int. Congr. on Rheology, Brussels, Belgium, August 17-21, 1992

THE SIMULTANEOUS USE OF 4×4 AND 2×2 BILINEAR STRESS ELEMENTS FOR VISCOELASTIC FLOWS

G.C. GEORGIOU

Unité de Mécanique Appliquée, Université Catholique de Louvain, 2 Place du Levant, B-1348 Louvain-la-Neuve (Belgium)

1. INTRODUCTION

The introduction of the 4×4 sublinear stress elements by Marchal and Crochet (ref. 1) has been an important development for the numerical solution of viscoelastic flow problems. The objective of the present work is to examine the simultaneous use of 4×4 and 2×2 bilinear stress elements for the solution of viscoelastic flow problems in an attempt to reduce the total number of unknowns, and consequently the computational cost, without sacrificing the accuracy. The 4×4 bilinear elements are used only in those regions where stress boundary layers or singularities are present and the 2×2 ones are employed elsewhere.

As a test problem we have chosen the steady flow of a Maxwell fluid around a rigid sphere moving with a constant translational velocity along the axis of a cylindrical tube.

2. FINITE ELEMENT METHODS

We consider three mixed finite element formulations: (1) The Galerkin method (2) The consistent streamline-upwind/Petrov-Galerkin method, and (3) The non-consistent streamline-upwind method.

In order to preserve the conformity of the elements, additional constraints along the sides shared by 4×4 and 2×2 elements are required. Care is also taken to assure the C^o-continuity of both the finite element expansion and the discretized equations for the stresses.

3. RESULTS AND DISCUSSION

To make comparisons we employ the three meshes selected by Crochet and Legat (ref. 3). Representative results for the drag correction factor K obtained with different numbers of the 4×4-element layers (N_l) are given in Table 1. Our results show that a thin layer of 4×4 elements around the surface of the sphere (of thickness 0.2 times the radius of the sphere) is adequate to assure the accuracy and the stability of the solution.

Table 1: Calculated values of K with the Galerkin method and mesh II.

We	Full 4×4	N_l=4	N_l=0
0.0	5.9476	5.9473	5.9473
0.1	5.8622	5.8620	5.8619
0.2	5.6598	5.6596	5.6595
0.3	5.4198	5.4195	5.4193
0.4	5.1873	5.1870	5.1865
0.5	4.9803	4.9801	4.9808
0.6	4.8034	4.8032	Diverges
0.7	4.6553	4.6551	
0.8	4.5332	4.5318	
0.9	4.4346	Diverges	
1.0	Diverges		

REFERENCES

1. J.M. Marchal and M.J. Crochet, J. Non-Newtonian Fluid Mech., 26 (1987), 77-114.

2. M.J. Crochet and V. Legat, J. Non-Newtonian Fluid Mech., in press.

Theoretical and Applied Rheology, edited by P. Moldenaers and R. Keunings
Proc. XIth Int. Congr. on Rheology, Brussels, Belgium, August 17-21, 1992

A THEORETICAL AND NUMERICAL STUDY OF NON-FOURIER EFFECTS IN VISCOMETRIC AND EXTENSIONAL FLOWS OF AN INCOMPRESSIBLE SIMPLE FLUID

R. R. HUILGOL[1], N. PHAN-THIEN[2] and R. ZHENG[2]

[1] School of Information Science and Technology, Flinders University of South Australia, G.P.O. Box 2100, Adelaide 5001 (Australia)

[2] Department of Mechanical Engineering, University of Sydney, Sydney 2006 (Australia)

SUMMARY OF RESULTS

A continuum theory of viscometric and extensional flows of an incompressible simple fluid has been used (ref. 1) to discuss the effects a non-Fourier law of heat conduction may have on these flows.

In the viscometric flows, it is shown that the heat flux vector is not parallel to the temperature gradient which is assumed to be in the same direction as the velocity gradient. Hence there is the possibility of heat conduction along the flow as well as the usual heat transfer in line with the temperature gradient. It also turns out that there are two non-Fourier heat conduction coefficients both depending on the shear rate, the temperature and its gradient; that which influences the heat transfer across the flow can be indirectly determined by measuring the temperature profile in a Couette flow.

Assuming the viscosity to have a power law index of 0.5 and an exponential temperature dependence, the energy and momentum equations are solved numerically for the case of the plane Couette flow to determine the effect produced by the non-Fourier component parallel to the temperature gradient. While the bottom plate is assumed to be at rest and the top moves at a constant unit speed, two different sets of temperature boundary conditions are imposed. In the former, the two walls are at the same temperature so that the effect of viscous heating may be studied; in the latter, there is a global temperature gradient of unity. The effects caused by the temperature dependence of the viscosity and the non-Fourier coefficient on the temperature profile are examined thoroughly.

Next it is shown that uniaxial elongational flows are possible in all incompressible simple fluids if the temperature gradient exists along the flow only. The relevant energy equation is solved numerically by assuming specific forms for the internal energy, the tensile stress and the heat flux functionals.

Overall, it is found that non-Fourier effects are quite modest.

REFERENCES

1. R. R. Huilgol, N. Phan-Thien and R. Zheng, J. Non-Newtonian Fluid Mech., in press.

Theoretical and Applied Rheology, edited by P. Moldenaers and R. Keunings
Proc. XIth Int. Congr. on Rheology, Brussels, Belgium, August 17-21, 1992

SEDIMENTING SPHERES IN SQUARE AND CIRCULAR CONDUITS: EXPERIMENTS AND NUMERICAL PREDICTIONS

V. ILIC[1], D. TULLOCK[1], N. PHAN-THIEN[1] and A. L. GRAHAM[2]

[1]Department of Mechanical Engineering, The University of Sydney, NSW 2006, Australia
[2]Los Alamos National Laboratory, Los Alamos, New Mexico, 87545, USA. WI 53706, USA

SUMMARY

The terminal settling velocities and rotation rates of spherical particles settling in circular (60 mm dia) and square (50×50 and 200×200 mm) conduits were investigated experimentally and numerically with the aim to benchmark the numerical predictions using the boundary element method (BEM), (ref.1).

Spheres were allowed to settle in a Newtonian fluid (silicone oil, 14.6 Pas and 0.97 g/cm^3 @ 20°C) under conditions such that only hydrodynamic forces exerted an appreciable effect, with the sphere Reynolds number range 0.001-0.107. The terminal settling velocities and the rotation rates were measured as a function of the size (2-50 mm diameter) and density of the falling sphere, the drop position b (from tank axis) of the sphere in the conduit and the dimensions and geometry of the containing vessel or conduit. The experimental measurements were subjected to an exacting error analysis and compared with fully three-dimensional boundary element calculations.

We found that the experimental results and numerical (BEM) simulations showed remarkable agreement within the bounds of experimental error (less than $\approx \pm$ 2% for settling velocity). Typical data are shown in Figs. 1,2, while the variation of angular velocity across the section was qualitatively similar to Fig. 2. The theoretical result for the small sphere in Fig. 2 was also in excellent agreement.

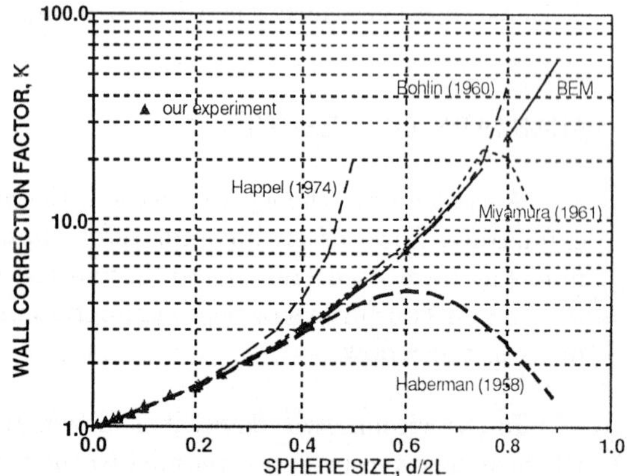

Figure 1. Comparison of the estimate of the Stokes' law wall correction factor K ($\equiv U_\infty/U$) using BEM, Bohlin, Haberman, Miyamura and Happel & Bart formulations with experimental data in a square tube for a range of sphere sizes.

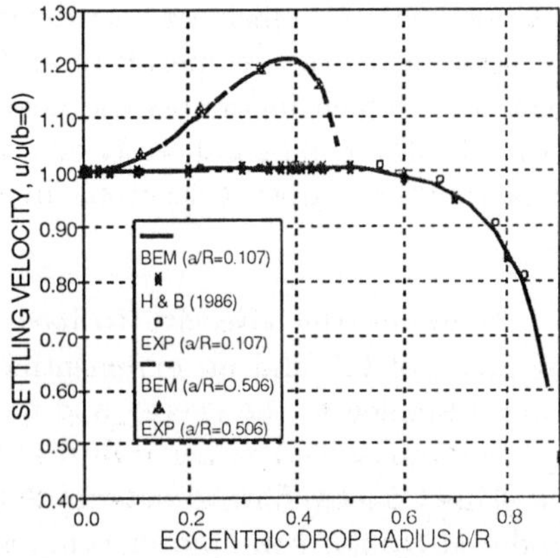

Figure 2. Variation of normalised settling velocity $u/u(b=0)$ with the dimensionless drop distance b/R from the axis of a circular cylinder, 59.15 mm diameter (=2R), for 6.35 and 29.95 mm diameter (ruby and Delrin, respectively) spheres.

Acknowledgements

This research was supported by the University of Sydney Special Projects Grant (NPT) and the US DOE at LANL under contract W-2405-ENG-36 with the University of California (ALG).

REFERENCE

1. D. Tullock, N. Phan-Thien and A. L. Graham. Rheo. Acta (to be published in 1992).

Theoretical and Applied Rheology, edited by P. Moldenaers and R. Keunings
Proc. XIth Int. Congr. on Rheology, Brussels, Belgium, August 17-21, 1992

A COMPARATIVE STUDY BETWEEN THE DISCRETIZED FINITE-ELEMENT AND CONTINUUM APPROACHES FOR INFLATION PROBLEMS

R.E. KHAYAT, A. DERDOURI, A. GARCIA-REJON

National Research Council of Canada
75 De Mortagne Blvd, Boucherville, Québec J4B 6Y4 CANADA

The free and confined inflation of hyperelastic membranes have been extensively examined in the past. Various numerical techniques have been devised, particularly the finite-element method, in an attempt to handle complex configurations. Unlike the continuum formulation, the finite-element method assumes at the outset that the membrane is discretized into a number of finite number of connected elements. From the wide body of related works in the literature, however, it is not difficult to observe that: (1) the confining geometries usually considered are relatively simple and (2) there is little, if any, work dealing with unstable confined inflation.

Many workers followed the continuum-based approach of Green and Adkins [1]. Feng & Huang [2] worked on the free and confined inflation of elastic flat membranes. Petrie & Ito [3] formulated the problem of confined inflation of a cylinder where the confining solid is an infinite circular cylinder. This problem was recently extended by Khayat et al. [4] to account for multiple contacts. The pioneering work of Oden and associates [5,6] have set the foundation of a Galerkin based finite-element procedure to handle problems in elastic deformation. In the work of Oden and Sato [5] the displacement fields corresponding to a given element were taken to be linear functions of the local nodes coordinates of that particular element. This is equivalent to assuming that straight lines connecting node points in the initial element remain straight after deformation. In a later paper [7] this assumption was removed by taking piecewise polynomial approximations for the displacements. Clearly, unless a sufficiently large number of elements is considered the assumption of linear dependence of displacements on the nodal coordinates restricts the problem to small deformation. Nevertheless, this assumption was later adopted by other workers who based their formulation on Oden's.

For instance, Charrier *et al.* considered the free and confined inflations of axisymmetric [8] and general flat membranes [9] using the neo-Hookean constitutive model. Moreover, the method used by Charrier *et al.* cannot account for any compressive stresses and only approximate boundary conditions were used whereby the condition of tangency at the point of contact between cylinder and solid boundary was not respected. DeLorenzi *et al.* [10] examined more complex confining configurations but adopted similar approximations, in particular that of taking linear interpolation functions. They later pointed out that higher order interpolation functions often lead to numerical instabilities. This is apparently due to thesevere deformation of the element during contact with the solid surface. On the other hand linear interpolation functions cannot handle the unstable regime or complex confined geometries (such as moderately sharp corners). It is under these conditions, however, that various nonlinear effects are bound to manifest themselves [11].

Regardless of the approach adopted, discretized finite-element or continuum based formulation, some kind of discretization will have to be employed if a numerical procedure is implemented. However, most of the parametric analyses so far conducted on stability and bifurcation of thin shell solutions were based on the continuum formulation. This is not surprising since existing stability and bifurcation theories have specifically been developed to handle a small number of (partial or ordinary) differential equations subject to appropriate boundary or initial conditions, rather than a set of a large number of nonlinear algebraic equations (in this case the stiffness relations). Therefore, a continuum formulation makes it easier to locate the physical singularities (limit pressure and bifurcation points) which are bound to arise for a large deformation. Moreover, existing methods to solve differential equations are easily implemented in the case of inflation problems once the class of governing equations is determined.

1. A.E. Green and J.E. Adkins, Oxford University Press, Oxford, 1970.
2. W.W. Feng and P. Huang, Int. J. Solids Struct., 11, 437 (1975).
3. C.J.S. Petrie and K. Ito, Plastics and Rubber Proc., June (1980) 68.
4. R.E. Khayat, A. Derdouri and A. Garcia-Rejon, (submitted to J. Mech. Phys. of Solids).
5. J.T. Oden and T. Sato, Int. J. Solids Struct., 3 (1967) 471.
6. J.T. Oden, McGraw Hill, N.Y., 1972.
7. T. Endo, J.T. Oden, E.B. Becker and T. Miller, Comp. Struc. 18 (1984) 899.
8. J.M. Charrier, S. Shrivastava and R. Wu, J. Strain Analysis 22 (1987) 115.
9. J.M. Charrier, S. Shrivastava and R. Wu, J. Strain Analysis 24 (1989) 55.
10. H.G. DeLorenzi, H.F. Nied and C. Taylor, Proc. Society of Plastics Engineers 46th Annual Conference, Atlanta, Georgia, 1988.
11. R.E. Khayat, A. Derdouri and A.Garcia-Rejon, Int. J. Solids Struct., 29, (1992) 69 .

Theoretical and Applied Rheology, edited by P. Moldenaers and R. Keunings
Proc. XIth Int. Congr. on Rheology, Brussels, Belgium, August 17-21, 1992

NUMERICAL SIMULATION OF A VISCOELASTIC FLUID FLOW PAST AN ELLIPTIC CYLINDER

Y. MOCHIMARU

Dept. of Mechanical Engineering, Tokyo Institute of Technology, Tokyo 152 (Japan)

1. INTRODUCTION

Two-dimensional, steady-state laminar flow fields past an elliptic cylinder placed in a uniform flow are analyzed.

2. NUMERICAL METHODS

The fluid is assumed to obey the constitutive equation $\tau^{ij} = \mu d^{ij} - \lambda(\delta/\delta t)d^{ij}$ (d^{ij}:rate of deformation tensor, $\delta/\delta t$:convective time derivative).Introduced is an elliptic coordinate system(α,β) such that

$$x + iy = e^{i\phi} \sinh(\alpha + i\beta)/\cosh\alpha_0 \quad , \quad (1)$$

where ϕ is the amplitude of the minor axis (ref.1). Then vorticity ζ (length is based on the semi-major axis a ; b: semi-minor axis) is given by

$$\frac{\partial}{\partial t}\left\{ \frac{1}{2\cosh^2\alpha_0}(\cosh 2\alpha + \cos 2\beta)\zeta + \varepsilon\gamma\left(\frac{\partial^2}{\partial\alpha^2} + \frac{\partial^2}{\partial\beta^2}\right)\zeta \right\} + \frac{\partial(\bar{\zeta},\psi)}{\partial(\alpha,\beta)} = \frac{1}{Re}\left(\frac{\partial^2}{\partial\alpha^2} + \frac{\partial^2}{\partial\beta^2}\right)\zeta \quad , \quad (2)$$

$$\bar{\zeta} = \zeta + \varepsilon \frac{2\cosh^2\alpha_0}{\cosh 2\alpha + \cos 2\beta}\left(\frac{\partial^2}{\partial\alpha^2} + \frac{\partial^2}{\partial\beta^2}\right)\zeta \quad (3)$$

where ε is an elasticity number, γ: parameter $\leqq 0$. Drag, lift, and moment coefficients C_D, C_L, C_M are given by

$$C_D + i\,C_L = \frac{i\,e^{i\phi}}{Re\cosh\alpha_0}\int_0^{2\pi}\left(\zeta\cosh(\alpha_0 + i\beta) - \frac{\partial\zeta}{\partial\alpha}\sinh(\alpha_0 + i\beta)\right)d\beta - \frac{\varepsilon\,e^{i\phi}}{2\cosh\alpha_0}\int_0^{2\pi}\zeta^2\cosh(\alpha_0 + i\beta)\,d\beta \quad (4)$$

$$C_M = \frac{1}{4\,Re\cosh^2\alpha_0}\int_0^{2\pi}\frac{\partial\zeta}{\partial\alpha}\cos 2\beta\,d\beta + \tanh\alpha_0 \times \frac{1}{Re}\int_0^{2\pi}\zeta\,d\beta + \frac{\varepsilon}{4\cosh^2\alpha_0}\int_0^{2\pi}\zeta^2\sin 2\beta\,d\beta \quad (5)$$

A Fourier spectral method is used for a solution procedure as in Ref.1.

3. RESULTS AND CONCLUSION

Streamlines and relative drag coefficients are shown in Figs.1 and 2 respectively. Large increases of drag due to viscoelasticity are found.

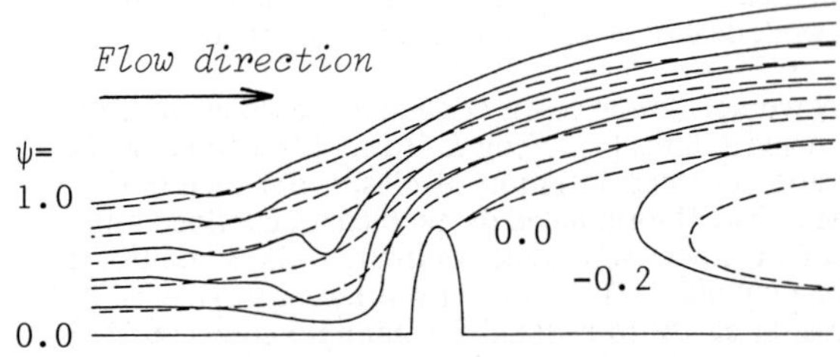

Fig.1 Streamlines at Re=50, b/a=1/4, ϕ=0. ——: ε=0.1, ----: ε=0.

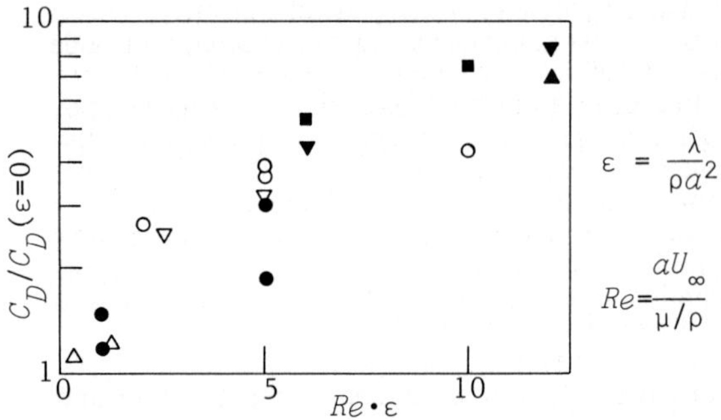

Fig.2 Drag coefficients (Re: 10～1200). $\phi=\pi/2$ (■:b/a=2/3, ○:b/a=1/2, ●:b/a=1/4, ▲:b/a=1/6, ▼:b/a=1/12); $\phi=0.475\pi$ (△:b/a=1/6); ϕ=0 (▽:b/a=1/4).

REFERENCE

1. Y.Mochimaru, in: K.R.Cho(Ed.), The 2nd KSME-JSME Fluids Eng.Conference,Seoul, Korea,October 10-13,1990, Korean/Japan Soc. of Mech.Engrs., Vol.1,pp.424-429.

Theoretical and Applied Rheology, edited by P. Moldenaers and R. Keunings
Proc. XIth Int. Congr. on Rheology, Brussels, Belgium, August 17-21, 1992

Boundary Element Modeling of Three-Dimensional Multiparticle Composites

T.D. Papathanasiou [1], M.S. Ingber [2], L.A Mondy [3] and A.L. Graham [1]

This paper presents a numerical analysis of a three-dimensional (3D) composite system consisting of up to 40 rigid rods embedded in an elastic cylinder. The whole system is subjected to uniform displacement at the two ends, simulating a bar tensile test. The method of boundary elements [1] used in this work reduces the dimensionality of the problem and allows for the modeling of systems (such as the ones presented in this and previous studies [2],[3]) for which finite element meshing can be extremely tedious.

The effective modulus of the composite cylinder is studied as a function of particle aspect ratio, particle volume fraction, and degree of ordering. Figure (1) shows a sample of particles distributed randomly in the matrix. Because of this random nature of the system the statistical averages obtained for a number of 10-20 simulations per volume fraction and the corresponding 95 % confidence intervals are reported. The results show that the effective modulus of the composite increases with the fiber aspect ratio (up to an aspect ratio of 10 studied in this work) for a given fiber volume fraction. The effect of fiber volume fraction is shown in Figure (2), for aligned and random systems. A comparison between the 3D multiparticle computations and an axisymmetric (2D) one-particle model (*fiber in cell model*) [4] is also given in Figure (2). At low fiber loadings the results of the 2D model and those of up to 40 three-dimensional fibers aligned with the tensile axis are statistically indistinguishable. At higher concentrations, the 2D model predicts higher values for the effective modulus. Over the range of our data, aligned systems are characterized by larger moduli than systems reinforced with spatially and orientationally random fibers.

[1] Los Alamos National Laboratory, MEE-9, MS G789, Los Alamos, NM 87545, USA

[2] Department of Mechanical Engineering, University of New Mexico, Albuquerque, NM 87131, USA

[3] Sandia National Laboratories, Division 1511, Albuquerque, NM 87185, USA

References

1. Youngren, G.K. and Acrivos, A, *J. Fluid Mech.* **69**, 377-403 (1975).
2. Ingber, M.S., *International Journal for Numerical Methods in Fluids*, **10**, 791-809, (1990).
3. Mondy, L.A., Ingber, M.S. and Dingman S.E., *Journal of Rheology*, **35**, 5, (1991).
4. Povirk, G.L., Needleman A. and Nutt, S.R., *Material Science and Engineering*, A132, 31, (1991).

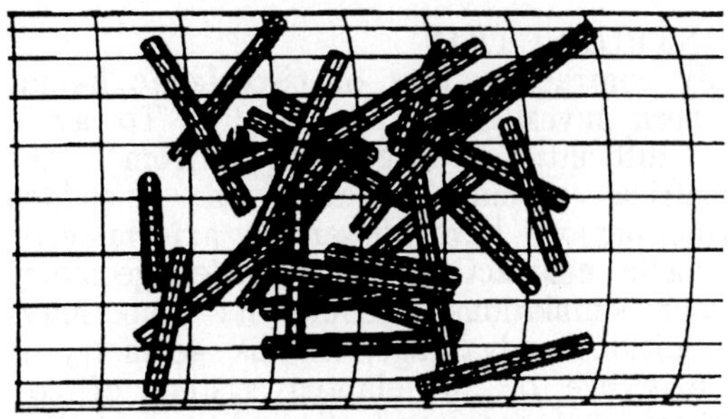

Figure 1: Sample random particle distribution in a finite cylinder.

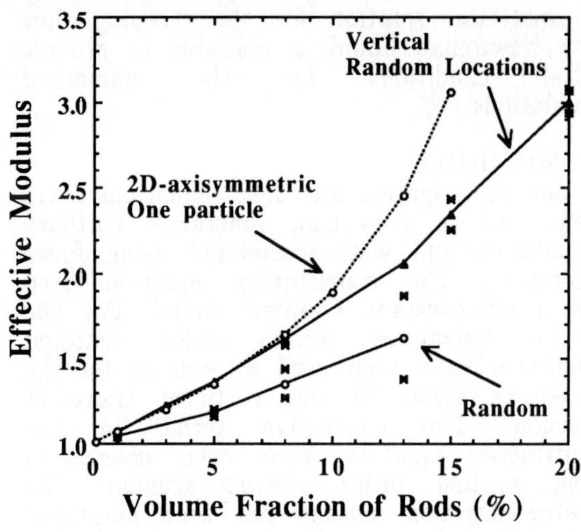

Figure 2: Predicted moduli for aligned and random systems of rods in an elastic cylinder as a function of volume fraction. Rod aspect ratio =3.

Theoretical and Applied Rheology, edited by P. Moldenaers and R. Keunings
Proc. XIth Int. Congr. on Rheology, Brussels, Belgium, August 17-21, 1992

NUMERICAL INVESTIGATION OF STEADY CONTRACTION FLOW OF VISCOELASTIC LIQUIDS

P. SCHÜMMER and C. BECKER

Institut für Verfahrenstechnik, RWTH Aachen, Turmstr. 46, D–W–5100 Aachen (Germany)

1. INTRODUCTION

Steady contraction flow of viscoelastic liquids has been investigated numerically. To avoid the difficulties originating from the geometrical singularities in sudden contractions we have chosen an axisymmetric hyperbolic contraction as a model geometry for our simulations. Besides its smoothness an additional advantage of this geometry is the existence of an adapted system of co–ordinates and its great flexibility including circular aperture and pipe geometries as limiting cases. Furthermore the wall of the hyperboloid is always a line of constant co–ordinate so that boundary conditions can be specified in a relatively simple way. Finally an analytical solution for the creeping flow of a Newtonian liquid is available to provide initial conditions for the numerical calculations.

2. METHODS

In our investigation the field equations have been used in a stream function– vorticity formulation and were discretized using finite differences. The constitutive equation used was a six–constant Oldroyd model. For the second derivatives second order centered differences have been used as well as for the convective terms in the vorticity transport equation. The convective terms in the constitutive equation have been discretized using a first order upwind scheme. The Newton–Raphson method has been employed to linearize the resulting system of nonlinear algebraic equations. The linearized equations were solved using a preconditioned conjugate gradient method. In contrast with this we have used a Picard–type iteration in our previous work. Another approach we have pursued is the direct introduction of the constitutive equation of a second–order–fluid (SOF) in the vorticity transport equation. This has the advantage of reducing the system of equations from six to two with the drawback that the highest order of the derivatives of the vorticity function is shifted from two to three.

3. RESULTS

Starting from the given analytical solution for the creeping flow some calculations have been made for Newtonian fluids. Convergence was assumed when the maximum residual was smaller than 10^{-6}. With the Newton–Raphson method converged results could be obtained for Reynolds numbers up to 100 in at most five iterations. As a first test problem for viscoelastic fluid flow the upper–convected Maxwell model has been investigated. The Weissenberg number (We) was increased in steps of 0.25 . With increasing Weissenberg number the convergence properties of the Newton algorithm get worse and worse. At We=1.0 about 100 Newton iterates were required. For Weissenberg numbers greater than one converged results could only be obtained with some modifications of the Newton algorithm. These modifications essentially consist in adding a scalar matrix to the Jacobian used in the usual Newton algorithm. We were then able to obtain results up to We=7. Even for higher Weissenberg numbers the algorithm doesn't diverge but the residual oscillates between values which practically don't change.

Theoretical and Applied Rheology, edited by P. Moldenaers and R. Keunings
Proc. XIth Int. Congr. on Rheology, Brussels, Belgium, August 17-21, 1992

Numerical Solutions for Viscoelastic Fluid Flow without Decomposition of Extra-Stresses

K. Strauß, S. Koch

Lehrstuhl Energieprozeßtechnik, Universität Dortmund, Postfach 500500, D-4600 Dortmund, Germany

1 Introduction

Numerical instabilities often occur in flow simulation of viscoelastic fluids when the Weissenberg-number exceeds a certain value. The assumption of this contribution is that the numerical behaviour can be improved by formulating the equations without decomposition of extra-stresses. The resulting equations must be transformed in order to solve them with Picard iteration in their discretized form.

2 Method

The set of differential equations contains the governing equations for a two-dimensional stationary flow of a Maxwell-B fluid in primitive variables neglecting inertia force. These equations are discretized with finite differences. A decoupled relaxation of the resulting equations is not possible. The problem can be solved by transforming the variables to get another set of equations which can be treated with Picard iteration. At a given point the differential equations can be formulated with the linear operator $\mathcal{L}$ as $\mathcal{L} \cdot U = 0$. With the transformation operator $\mathcal{T}$ one might get a 'better' equation in the dependent variable W: $\mathcal{L} \cdot \mathcal{T} \cdot W = \mathcal{M} \cdot U = 0$. For the given problem a transformation $\mathcal{T}$ was found which yields to an operator matrix $\mathcal{M}$ with the highest derivatives on the diagonal elements. Picard iteration can now be used in the sense of distributive relaxation [1]. The new variable W may never be calculated. The obtained system of equations is only used to construct an iteration scheme.

3 Numerical Results

The resulting Picard scheme was successfully applied to the stick-slip flow up to $De = 2.4$. De is calculated with the shear-velocity at the wall of the inlet-flow. The used boundary conditions describe the transition from stick to slip at the lower bound of channel flow. Figure 1 shows the increase in normal stress with the Deborah-number. The behaviour of the numerical scheme can be evaluated by figure 2. The shear stress along the lower wall does not show oscillations behind the singularity. Beyond $De = 2.4$ oscillations appear at the wall in the inlet region.

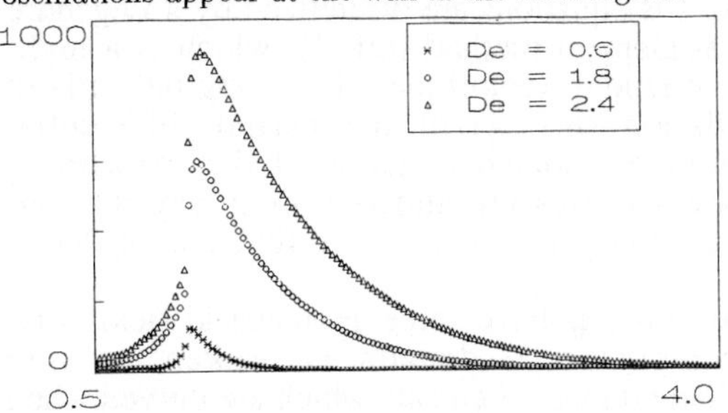

Fig. 1: Normal stress along the wall

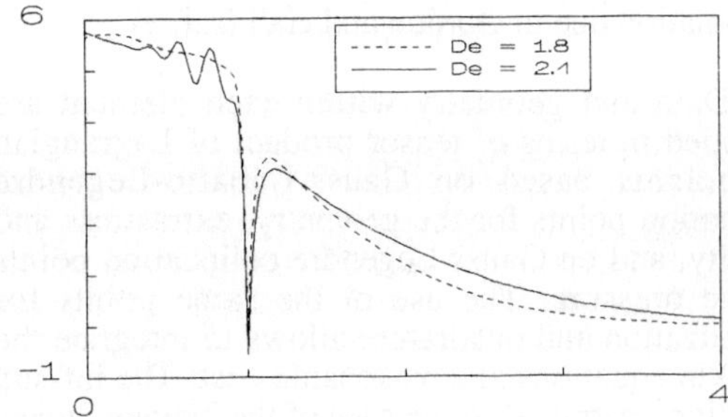

Fig. 2: Shear stress near the wall

4 Conclusion

In this contribution a new algorithm for flow simulation of a Maxwell-B fluid is presented which is based on transformation of the original equations and the method of distributive relaxation. The resulting equation is treated with finite differences and solved with a Picard iteration for the stick-slip flow. Stable results can be obtained for De up to 2.4.

[1] A. Brandt; Multigrid Techniques: 1984 Guide with Applications to Fluid Dynamics

Theoretical and Applied Rheology, edited by P. Moldenaers and R. Keunings
Proc. XIth Int. Congr. on Rheology, Brussels, Belgium, August 17-21, 1992

LEGENDRE SPECTRAL ELEMENTS FOR NON-NEWTONIAN FLUID FLOWS

Vincent Van Kemenade[1], and Michel Deville[1]

[1]Applied Mechanics, Université Catholique de Louvain, Place du Levant 2, 1348 Louvain-la-Neuve (Belgium)

The two-dimensional steady incompressible Maxwell-B equations are discretized by a Legendre spectral element method (ref. 1), which is a high-order element technique. The original mixed formulation (ref. 2) of the partial differential equations are solved in terms of the extrastress, velocity and pressure allowing us to say that our method is general for all kind of differential models.

In order to have some geometrical flexibility, the computational domain is broken up into quadrilateral macroelements which are curved when needed. The mapping technique used to handle curved geometries is based on a transfinite interpolation due to Gordon and Hall (ref. 3).

Data and geometry within each element are expanded in terms of tensor product of Lagrangian interpolants based on Gauss-Lobatto-Legendre collocation points for the geometry, extrastress and velocity, and on Gauss-Legendre collocation points for the pressure. The use of the same points for discretization and quadrature allows to integrate the Galerkin equations at a reasonable cost. The inf-sup condition, restricted to the case of the Navier-Stokes equations, imposes to choose the velocity in P^N and the pressure in P^{N-2}. Extrastress is chosen on the same grid as the velocity, i.e. P^N. The set of equations, linearized by Newton's technique, is solved by means of the frontal method with static condensation inside each element.

We consider the steady creeping flow (Re=0) of an upper convected Maxwell fluid through a plane channel with smoothly and periodically perturbed boundaries. We show that stable results are obtained for high Weissenberg numbers (We ≈ 10, fig. 1) with a limited number of degrees of freedom (4 elements with degree 15 interpolants, i.e. 4567 variables). Moreover, mesh refinement for the particular Weissenberg number of 4 shows that our method is able to produce accurate results without resorting to an upwind scheme to stabilize the solutions. Indeed, if we consider the pressure drop resulting of the computation on a mesh composed of 4 elements with degree 20 interpolants, we see that results are accurate until the 7th and 9th digit for the discretizations based on degree 10 and 15 interpolants respectively.

REFERENCES

1. Y. Maday, A.T. Patera, State-of-the-Art Surveys on Computational Mechanics, Eds. A.K. Noor, J.T. Oden, ASME, New York, 1989, pp. 71-143.
2. M.J. Crochet, A.R. Davies, K. Walters, Numerical Simulation of Non-Newtonian Flow, Elsevier, 1984.
3. W.J. Gordon, C.A. Hall, Int. J. Num. Meth. Eng., 7 (1973), pp 461-477.

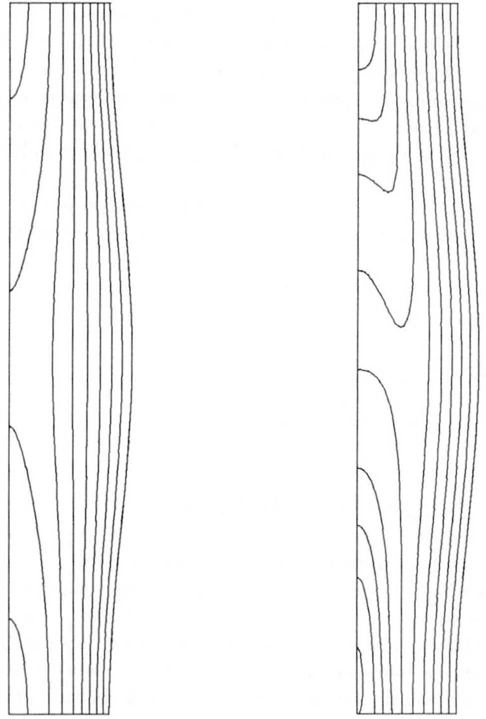

Figure 1 : Axial velocity for the inertialess periodic plane flow of Newtonian (left) and Maxwell-B (right, We = 10) fluids.

Theoretical and Applied Rheology, edited by P. Moldenaers and R. Keunings
Proc. XIth Int. Congr. on Rheology, Brussels, Belgium, August 17-21, 1992

A BOUNDARY ELEMENT/PARTICULAR SOLUTION APPROACH FOR NON-NEWTONIAN FLOW PROBLEMS

R. ZHENG[1], N. PHAN-THIEN[1], R.I. TANNER[1] and C.J. COLEMAN[2]

[1]Department of Mechanical Engineering, The University of Sydney, NSW 2006, Australia
[2]Surveillance Research Laboratory, DSTO, Salisbury, SA 5108, Australia

1. INTRODUCTION

A significant advantage of the boundary element method (BEM) over the domain type methods is the possibility of discreting only the boundary of the problem. However, nonlinear problems usually require an integration over the whole domain (ref. 1). Recently, Zheng *et al* (refs. 2,3) developed a new technique for solving the generalized Poisson equation and the Navier-Stokes equations. The approach avoids domain integrals by replacing the inhomogeneity within the domain in terms of a linear combination of radial basis functions. The present paper extends this method to solve non-Newtonian fluid flow problems.

2. FORMULATION

The solution schemes for non-Newtonian flow problems are iterative in nature and produce inhomogeneous linear partial differential equations at each stage of iteration. We write the governing equations as

$$-\nabla p + \eta_N \nabla^2 \mathbf{u} = \mathbf{f}, \quad \nabla \cdot \mathbf{u} = 0, \tag{1}$$

where η_N is a Newtonian-like viscosity, and

$$\mathbf{f} = \nabla \cdot (2\eta_N \mathbf{D} - \boldsymbol{\tau}) \tag{2}$$

is known as a pseudo-body force evaluated from previous step of iteration. In eqn. (2) $\mathbf{D} = \nabla \mathbf{u} + (\nabla \mathbf{u})^T$ is the strain rate tensor, and $\boldsymbol{\tau}$ is the model-dependent extra stress tensor. We now replace $\mathbf{f}$ by a linear combination of a radial basis function, i.e.,

$$\mathbf{f}(\mathbf{r}) \approx \sum_{i=1}^{N} \boldsymbol{\alpha}_i \exp\{-(|\mathbf{r} - \mathbf{r}_i|/\beta_i)^2\}, \tag{3}$$

where the $\mathbf{r}_i$'s are chosen to be points distributed within the interior domain, and the β_i's are chosen to be the average distance of the adjacent nodes to $\mathbf{r}_i$. These choices for the parameters lead to a good interior approximation for $\mathbf{f}$ which can be determined from an effectively sparse system of equations for the $\boldsymbol{\alpha}$'s. Thus an approximate particular solution of the governing equations can be derived (ref. 4). The remaining part of the solution of the given problem will then satisfy a set of linear homogeneous equations which lead to a set of boundary integral equations without the domain integral term.

3. EXAMPLE

As an example, we consider the driven cavity flow of a Carreau fluid with a shear-rate-dependent viscosity given by

$$\eta(\dot{\gamma}) = \eta_\infty + \frac{\eta_0 - \eta_\infty}{(1 + \lambda^2\dot{\gamma}^2)^{(1-n)/2}}, \tag{4}$$

where $\dot{\gamma} = \sqrt{2\mathrm{tr}\mathbf{D}^2}$ is the general shear rate. The parameters chosen for the simulation are: $\eta_0 = 1$, $\eta_\infty = 0$, $\lambda = 1$ and $n = 1, 0.5, 0.2$. The fluid is contained in a square cavity ($x \in [0,1]$, $y \in [0,1]$), and the upper wall at $y = 1$ moves with unit velocity.

Results show that, as the degree of shear thinning increases, the centre of the circulation moves closer to the sliding lid. The position of the centre of the circulation measured from the bottom is found to be 0.76, 0.79 and 0.81 for $n = 1$, 0.5 and 0.2, respectively. A mesh refinement study has indicated that the solution converges as $O(h^{2.6})$, where h is the element size. The CPU time required by the new method is found to be about 40% less than using the the old BEM involving domain integrals.

This research was supported by an Australian Research Council Grant.

REFERENCES

1. M.B. Bush and R.I. Tanner, in P.K. Bannerjee and L. Morino (Ed.), Boundary Element Methods in Nonlinear Fluid Dynamics, Elsevier Applied Science, New York, 1990, pp. 285-317.
2. R. Zheng, C.J. Coleman and N. Phan-Thien, Computational Mechanics, 7 (1991) 729-288.
3. R. Zheng, N. Phan-Thien and C.J. Coleman, Computational Mechanics, 8 (1991) 71-86.
4. R. Zheng, Ph.D. thesis, The University of Sydney, 1991.

CONTRIBUTED PAPERS

MELTS AND POLYMER PROCESSING

Theoretical and Applied Rheology, edited by P. Moldenaers and R. Keunings
Proc. XIth Int. Congr. on Rheology, Brussels, Belgium, August 17-21, 1992

Linear Rheology of Copolymer Modified Blends: Experiments and Predictions of Emulsion Models.

A. Ait-Kadi, A. Ajji and B. Brahimi.

University Laval, CERSIM-Chemical Engineering Department, Faculté des sciences et de génie, STE-FOY, QUEBEC, CANADA, G1K 7P4.

1. INTRODUCTION

Blending and alloying of immiscible polymers have been extensively studied in the last decades. The objectives of these operations are first to obtain materials with tailored properties, and second to find two or more polymers whose mixture will have improved properties not attainable by the components of the blend separately. The principal limitation in these operations is the fact that the majority of polymers are immiscible. To obtain a material with desired properties, one has then to use modifiers in order to control the morphology of the blend.

In a previous study of the particular blend of high density polyethylene, HDPE, and high impact polystyrene, HIPS, (refs. 1-2), we have found that the addition of block copolymers to this particular system led to an enhanced phase dispersion and interphase interactions. Our conclusions were mainly based on the dynamic mechanical and thermal behavior (ref.1) and the linear viscoelastic rheology (ref.2) of the blends. Similar conclusions have been drawn on the basis of scanning electron microscopy observations by other authors (ref. 3). The copolymers are also known to first locate at the interface and to modify the interfacial tension between the blend components (ref.4). Since our experimental results with well designed copolymers of pure diblock hydrogenated polybudiene-b-polystyrene (HPB-b-PS, H77) and tapered HPB-b-PS (H35) showed that only small quatities are required to decrease the size of the dispersed phase and the interfacial tension, the addition of copolymer was seen as a suitable method to study the effect of the minor phase dimensions and the interfacial tension on the rheological properties of these immiscible blends (ref. 2). The model used was that developped by Oldroyd (ref.5) and used by Graebling et al. (ref.6) for an emulsion of two incompressible fluids. Model predictions were found to fit correctly the experimental results of unmodified blends but failed in predicting the effect of copolymers on the dynamic viscosity and the storage modulus of copolymer modified blends. This was mainly attributed to the fact that the model used incorporates the interfacial tension and the size of the dispersed phase but did not take into account the effect of phase interactions due to the presence of the copolymer in the blends. Moreover, and since the predictions were obtained using a wide range of the dispersed size and interfacial tension without any significant effect on linear rheological properties of the copolymer modified blends, it was concluded that this parameter (interactions) should be included in any rheological model for copolymer modified immiscible blends.

In this paper we present the predictions of a rheological model of viscoelastic emulsions with an interfacial tension that depends on shear deformation and variation of area recently developped by Palierne (ref.7). The theoretical results are compared to the experimental linear viscoelastic properties of HDPE/HIPS blends and copolymer modified HDPE/HIPS blends (ref. 2).

2. RESULTS AND DISCUSSION

The rheological model presented in (ref.7) was first developped for a single spherical inclusion of a viscoelastic material in a viscoelastic matrix. The effect of concentration of the minor phase (non dilute solutions) was introduced through the mechanical interactions between inclusions using a self consistent treatment similar to the Lorentz sphere in electricity. The model is developped in the linear viscoelastic regime with no effect of the imposed deformation neither on the morphology of the system nor on the modification of the size and the distribution of the inclusions. In this regime and at first order in deformation, the interface tension was written as:

$$\alpha_{ab} = \alpha^0\delta_{ab} + \beta_{ab} \tag{1}$$

The first term in the right hand side is the equilibrium tension, independent of the frequency, while the second term is a contribution to the interfacial tension which oscillates at the same frequency, ω, as the interface strain, γ_{ab}, of the interface. This contribution was further decomposed to write:

$$\beta_{ab} = \frac{1}{2}\beta'\gamma_{cc}\delta_{ab} + \beta''(\gamma_{ab} - \frac{1}{2}\delta_{ab}\gamma_{cc}) \tag{2}$$

The first term in the right hand side of (eq.2) is the isotropic part of β_{ab}, proportional to γ_{cc}. It takes into account the relative variation of the interface area. The second term, proportional to the interface strain deviator, reflects the effect of shear without area variation. Note that the summation convention on the repeated indices is used in the above equations. β' and β'' are both complex and frequency dependent. $\beta'(\omega)$ is interpreted as the surface dilatation modulus and $\beta''(\omega)$ is interpreted as the surface shear modulus. The subsequent developments led to a set of equations (eqs. 2.5, 2.6 and 2.7 of ref.7) giving the macroscopic modulus of the emulsion, at linear order in concentration, as a function which takes into account the radius, the viscoelastic modulus and the interfacial tension of each kind of inclusion. In the case of non dilute systems, these expressions were modified by considering the fact that the actual deformation experienced by an individual inclusion is no longer equal to the macroscopic deformation but rather influenced by the neighboring inclusions. This effect was taken into account by using the Lorentz sphere concept (see ref. 7 for details). For a monodisperse inclusion type system the appropriate equations are:

$$G^* = G_M^* \left[\frac{1 + \frac{3}{2}\frac{\Phi E}{D}}{1 - \frac{\Phi E}{D}} \right] \tag{3}$$

where G^* and G_M^* are the complex moduli of the system and the matrix respectively and ϕ is the volume fraction of inclusions. The expressions of E and D appearing in the abve equation are given by:

$$\begin{aligned} E = {} & 2(G_I^* - G_M^*)(19G_I^* + 16G_M^*) + 48\beta'\alpha^0/R^2 \\ & + 32\beta''(\alpha^0 + \beta')/R^2 + 8\alpha^0/R(5G_I^* + 2G_M^*) \\ & + 2\beta'/R(23G_I^* - 16G_M^*) \\ & + 4\beta''/R(13G_I^* + 8G_M^*) \end{aligned} \tag{4}$$

$$\begin{aligned} D = {} & (2G_i^* + 3G_M^*)(19G_I^* + 16G_M^*) + 48\beta'\alpha^0/R^2 \\ & + 32\beta''(\alpha^0 + \beta')/R^2 + 8\alpha^0/R(5G_i^* + 2G_M^*) \\ & + 2\beta'/R(23G_I^* - 16G_M^*) \\ & + 4\beta''/R(13G_I^* + 12G_M^*) \end{aligned} \tag{5}$$

where G_I^* and R are the complex modulus and the radius of the inclusion respectively and α^0 is the equilibrium interfacial tension.

Several particular cases can be recovered from eq. (3) including the Oldroyd equation for the complex modulus of the blend ($\beta' = \beta'' = 0$).

In the most general case, β' and β'' are complex functions of the frequency ω which can be written as:

$$\beta' = f'(\omega) + ig'(\omega) \tag{6}$$

$$\beta'' = f''(\omega) + ig''(\omega) \tag{7}$$

As a first approximation of the functions β' and β'', we have used real constants say β' = a' and β'' = a". This choice is motivated by the fact that the addition of copolymer decreases the interfacial tension (refs.4,8). The copolymer first locates at the interface and then disperses selectiveley in one phase or in both phases (ref.3). Mechanical deformations which may contribute to surface dilatation will then lead to surface tension variations due to the surface variation (the surface concentration of the copolymer will change due to the surface dilatation). The equilibrium surface tension $\alpha^0(c)$ will then decrease to ($\alpha^0(c) - c(\gamma - 1)/\gamma\, \partial\alpha^0(c)/\partial c$) at first order in concentration. γ represents the dilatation ratio

Figure 1 shows model predictions for several combinations of model parameters. Predictions when β' and β'' are both set equal to zero are those presented earlier in (ref.2) (Oldroyd model). The experimental results presented in this figure are those obtained for blends composed of 20 wt% of a high density polyethylene and 80 wt% of a high impact polystyrene. Two different diblock copolymers where used as compatibilzers for these blends (the results in fig. 1 are for H77 only). Details on material preparation and characterization are given elsewhere (refs. 1-2). The results obtained show that a non-zero value of the dilatation modulus, β', results in an increase of the storage modulus, G', over the whole range of the frequency as compared G' values obtained with Oldroyd model ($\beta' = 0$). However, no effect of the real and non-zero values of β' was observed on the predicted values of G' in the low frequency region. Only a slight effect can be detected in the high frequency region. The experimental data of the unmodified blend and the 3% copolymer modified blend are well predicted when a non zero value of β' is used. For the 1% modified blend, model predictions in both cases are still far from the experimental results. Other more sophisticated expressions for the β' and β'' must be used. However, we should mention here that this two model parameters are related to the properties of the interface which are not obvious to obtain experimentally.

3. CONCLUSIONS

The approximation of the surface dilation and the surface shear moduli functions by real constants did not lead to any significant enhancement of the predictions of the linear viscoelastic properties of a copolymer modified immiscible blend. More work is still needed in order to predict correctly the effect of copolymers on the rheological properties of immiscible blends. The effect of phase interactions promoted by the presence of the copolymer should be explicitly introduced.

REFERENCES

1. B. Brahimi, A. Ait-Kadi, A. Ajji and R. Fayt, J. Polym. Sci.: Part B, Polymer Physics, **Vol 29**, (1991), 945-961.
2. B. Brahimi, A. Ait-Kadi, A. Ajji, R. Jérôme and R. Fayt, J. Rheol **35(6)**, (1991), 1069-1090.
3. R. Fayt, R. Jérôme and P. Teyssié, Makromol. Chem., **187**, (1986), 837.
4. S.H. Anastasiadis, I. Gancarz and J.T. Koberstein, Macromolecules, **22** (1989), 1449-1453.
5. J.G. Oldroyd, Proc. R. Soc. London Ser. A **218**, (1953), 122-132
6. D. Graebling and R. Muller, J. Rheol., **34** (1990), 193-205.
7. J.F. Palierne, Rheol. Acta, **29** (1990), 204-214
8. J. Noolandi and K.M. Hong, Macromolecules **15** (1982), 482-492.
9. J.F. Palierne, private communication.

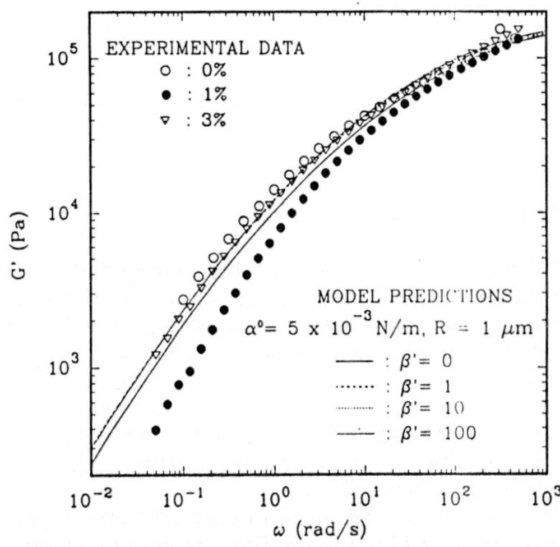

Fig I: Comparaison of model predictions with the experimental data for unmodified and copolymer (H77) modified 20/80 (HDPE/HIPS) blends.

Theoretical and Applied Rheology, edited by P. Moldenaers and R. Keunings
Proc. XIth Int. Congr. on Rheology, Brussels, Belgium, August 17-21, 1992

NUMERICAL SIMULATION OF MIXING

Th. Avalosse[1], M. Crochet[1], A. Fobelets[2] and C. Dehennau[2]

[1]Unité de Mécanique Appliquée, Université Catholique de Louvain, Place du Levant 2, 1348 Louvain-la-Neuve (Belgium)

[2]Solvay & Cie, Laboratoire Central, rue de Ransbeek 310, 1120 Bruxelles (Belgium)

1. INTRODUCTION

Mixing flows are known for their complexity. Even in the case of laminar mixing, the presence of moving (generally rotating) boundaries and complex flow domains prevents the use of analytical methods for predicting the flow and the kinematic parameters describing the quality of mixing. A detailed description of the kinematics of mixing may be found in Ottino [1] while the significance of mixing parameters for polymer processing is described by Tadmor and Gogos [2].

We wish to use the performance of finite element methods for describing mixing flows in detail and for an accurate calculation of mixing parameters. The present paper is limited to a two-dimensional analysis of flows around rotating cams.

2. STRETCHING AND EFFICIENCY OF MIXING

We consider in Fig. 1 the motion of a viscous fluid occupying in space a volume Ω_0 at the reference time t_0 and Ω at time t. Let X denote a material point which occupies the position $\mathbf{X}$ in Ω_0 and $\mathbf{x}$ in Ω, and let $d\mathbf{X}$ denote a material vector in Ω_0 which transforms into $d\mathbf{x}$ at time t. The orientation of $d\mathbf{X}$ is $\mathbf{M} = d\mathbf{X}/|d\mathbf{X}|$, while its orientation in Ω is $\mathbf{m} = d\mathbf{x}/|d\mathbf{x}|$.

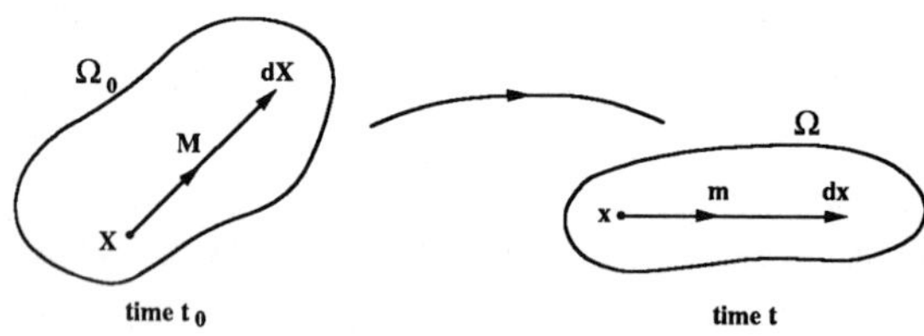

Fig. 1.- Motion of a continuum and deformation of a material vector.

The lineal stretch is defined as the ratio

$$\lambda = \lim_{|d\mathbf{x}| \to 0} \frac{|d\mathbf{x}|}{|d\mathbf{X}|} = (\mathbf{M} \cdot \mathbf{C}\,\mathbf{M})^{1/2}, \tag{1}$$

where $\mathbf{C}$ is the right Cauchy-Green strain tensor at time t with respect to the reference configuration Ω_0. The new orientation of $\mathbf{M}$ at time t is given by

$$\mathbf{m} = \mathbf{F}\,\mathbf{M}/\lambda\,, \tag{2}$$

where $\mathbf{F}$ is the deformation gradient tensor with respect to Ω_0. The instantaneous efficiency of stretching for a given material point is defined as follows,

$$e_\lambda = \frac{\dot{\lambda}/\lambda}{(\mathbf{d}:\mathbf{d})^{1/2}} = \frac{\mathbf{m}\cdot\mathbf{d}\,\mathbf{m}}{(\mathbf{d}:\mathbf{d})^{1/2}}\,, \tag{3}$$

where $\mathbf{d}$ is the rate of deformation tensor and a dot denotes the material derivative.

It is important to determine whether the efficiency of mixing is maintained throughout the mixing process. We are thus led to define, for a given material point, the time averaged stretching efficiency which is defined as follows,

$$\langle e_\lambda \rangle = \frac{1}{t}\int_0^t e_\lambda \, dt. \tag{4}$$

where we have explicited the fact that e_λ is calculated for the same material point for which a material fiber orientation $\mathbf{M}$ has been selected in the reference configuration.

The reader is referred to Ottino [1] for additional parameters characterizing mixing. In the present paper, we limit our analysis to the evolution of λ, e_λ and $\langle e_\lambda \rangle$ through a complex periodic motion.

3. NUMERICAL METHOD

We wish to evaluate the lineal stretch and the mixing efficiency in a flow domain bounded by a moving boundary. Typical examples are the single screw and the corotating twin screw extruders. Here, we limit our investigation to a two-dimensional analysis : the flow domain is bounded by an external barrel and by one or two rotating cams. In order to evaluate λ and e_λ, it is thus necessary to calculate pathlines of material particles in deformable domains.

Consider in Fig. 2 a typical configuration of the flow domain between the outer barrel and the rotating cams. The domain inside the barrel is fully covered by a finite

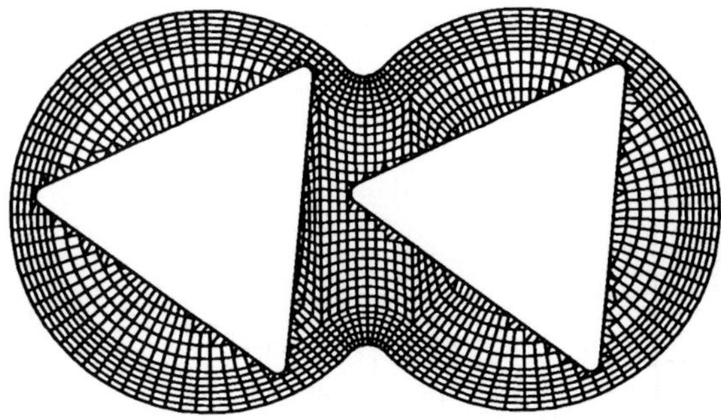

Fig. 2.- Typical finite element mesh for twin cams.

element mesh. For a number of time steps selected at the outset (120 for a full rotation of the cams), the cams are applied on the finite element mesh. An automatic mesh generation procedure [3] creates a finite element mesh compatible with the shape of the boundary ; the new mesh contains as many elements as possible from the original mesh. For every time step, it is thus easy to impose a rigid rate of rotation of the cams and to calculate a velocity field by means of a finite element formulation. In the present problem, we neglect inertia terms in Navier-Stokes equations in view of the very low Reynolds number.

In a preliminary approach, we had developed the capability of calculating finite element representations of λ and e_λ and to consider them as field variables. More precisely, by calculating **F** and **C** as time dependent fields over deforming meshes, our goal was to obtain nodal values of λ and e_λ. However, such a field representation would require a smooth variation between neighbouring nodes of the mesh. With the present moving geometrical configuration, such a requirement is generally not satisfied. The resulting finite element representations of the fields are then fraught with spurious wiggles.

The only valid approach is to calculate λ and e_λ along the pathline of material particles. We select at the outset a grid of material point X_i in Ω_0 (of the order of 9000 for a configuration such as that of Fig. 2). For every material point, we calculate the pathline by integrating the velocity field (from one mesh to the next) while the deformation gradient is calculated on the basis of the relationship

$$\dot{\mathbf{F}} = \mathbf{L}\,\mathbf{F}. \qquad (5)$$

where **L** is the velocity gradient tensor. In order to preserve accuracy after several rotations of the cams, it was found necessary to use an integration scheme of the Crank-Nicholson type. For obtaining the new location of a point we solve for example the equation

$$\frac{\mathbf{x}^{t+\delta t} - \mathbf{x}^{t}}{\delta t} = \frac{1}{2}\left[\, \mathbf{v}^{t+\delta t}(\mathbf{x}^{t+\delta t}) + \mathbf{v}^{t}(\mathbf{x}^{t})\,\right]. \qquad (6)$$

while a similar procedure applies for **F** . On the basis of (1-4) we can thus calculate for each material point the values of λ, e_λ and $\langle e_\lambda \rangle$.

The fields of λ, e_λ and $\langle e_\lambda \rangle$ are then represented through graphic maps. Using a Lagrangian description, we surround the pixel at point X_i with a colour corresponding to the current value of the field.

4. COMPARISON BETWEEN A SINGLE CAM AND A COROTATING CAM MIXER

Let us now analyze the mixing parameters in two configurations : a single rotating cam surrounded by a circular barrel, and the system shown in Fig. 2. For the single and the twin cam systems, we follow respectively 1400 and 9000 material points through ten full rotations (20π) of the cams.

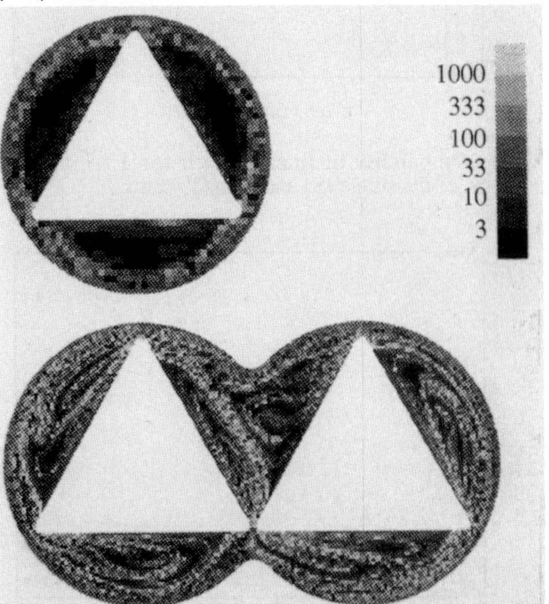

Fig. 3.- Lagrangian distribution of lineal stretch after two rotations of the cams.

Fig. 3 shows a Lagrangian description of the lineal stretch after two rotations of the cams (4π) for an initial orientation **M**, such that the instantaneous efficiency is maximum initially for each point . For the single cam, it is found that λ remains relatively small in recirculating zones between the sides of the cam and the barrel. Large values of λ are found in the trail of the edges of the cam. The distribution of λ is much more uniform in the twin cam system where matter is exchanged from one side of the barrel to the other. It is found that, after a couple of rotations of the cams, the initial orientation **M** of the material fibers has essentially no impact upon the distribution of λ.

The use of an average value of λ is useless for a quantitative evaluation of the information contained in Fig. 3. It is found indeed that the very large values of λ associated with some material points renders the average meaningless.

However, it is quite useful to produce a distribution function $\phi(t,\lambda)$, which represents the probability of finding at time t a lineal stretch lower than λ for a given

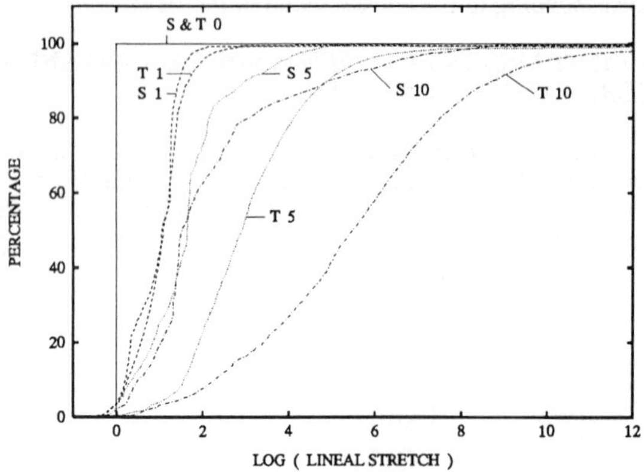

Fig. 4.- Probability of lineal stretch for 1, 5 and 10 rotations of the single (S) and twin (T) cams.

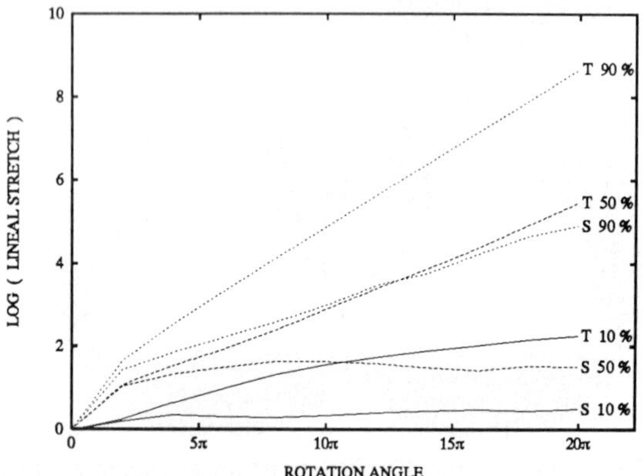

Fig. 5.- Lineal stretch as a function of rotation angle as in Fig. 4.

material point. Fig. 4 shows curves of $\phi(t,\lambda)$ after 1, 5 and 10 rotations of the cams for single (S) and twin (T) corotating cams. As expected, the probability of finding λ below some pre-selected value decreases with the number of rotations while the stretching is more intense with corotating cams than with a single one. A similar observation can be made on Fig. 5 which shows, for a given probability (10, 50 and 90 %) the maximum stretch as a function of the rotation angle. As expected, the lineal stretch increases as a function of the rotation angle while it is definitely larger with twin than single cams. Calculations show that, after moderate rotations of the cams, the curves of Fig. 4 and 5 depend little upon the initial orientation **M** of the material vector.

We have also calculated for every point of our sample the probability of finding a time averaged efficiency below some value selected at the outset. Fig. 6 shows a typical evolution of the efficiency as a function of the rotation angle. The efficiency decreases with the rotation angle. However, it decreases faster for the single cam than for the twin cam system.

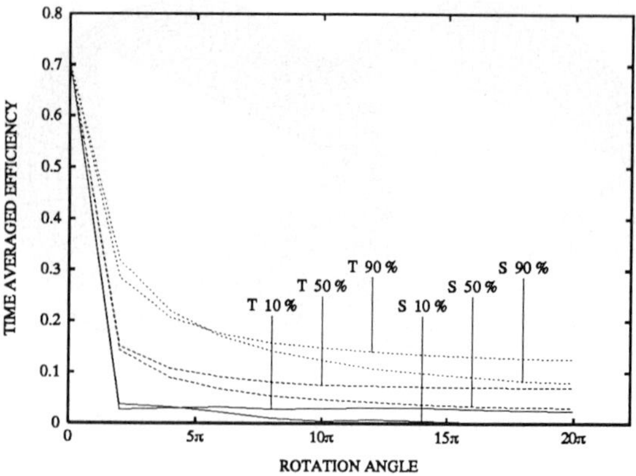

Fig. 6.- Time averaged efficiency as a function of rotation angle for single (S) and twin (T) cams.

5. CONCLUSION

The example of section 4 shows the capability of numerical solutions for quantifying the efficiency of a model mixer. One can easily evaluate the effect of geometrical parameters (such as the effect of the cams) and of material parameters (such as the shear-thinning behavior).

We have found that a major ingredient of the evaluation of mixing is the accuracy of the calculation. While it is easy to obtain an accurate velocity field over a moderately refined mesh, the calculation of pathlines requires the use of Crank-Nicholson technique. First-order Euler methods generate major inaccuracies. We have also found that a finite element representation of the kinematic fields is meaningless because such fields exhibit strong variations over lengths much smaller than the element size. The only accurate method consists of solving ordinary differential equations along pathlines and of representing the field by means of pixels and statistical distributions.

The method is presently extended to three-dimensional realistic configurations in single and twin screw extruders.

ACKNOWLEDGEMENTS

This work is supported by the "Région de Bruxelles-Capitale" through the "Institut pour l'Encouragement de la Recherche Scientifique dans l'Industrie et l'Agriculture (I.R.S.I.A.)".

REFERENCES

1. J. M. Ottino, "The Kinematics of Mixing : Stretching, Chaos and Transport", Cambridge University Press, 1989.
2. Z. Tadmor, C. G. Gogos, "Principles of Polymer Processing", Wiley, 1979.
3. A. Couniot, PhD Thesis, Louvain-La-Neuve, 1991.

Theoretical and Applied Rheology, edited by P. Moldenaers and R. Keunings
Proc. XIth Int. Congr. on Rheology, Brussels, Belgium, August 17-21, 1992

LINEAR VISCOELASTIC PROPERTIES OF A MISCIBLE POLYMER BLEND SYSTEM

H.C. BOOIJ and J.H.M. PALMEN

DSM Research, P.O. Box 18, 6160 MD Geleen (The Netherlands)

1. INTRODUCTION

Several investigations (refs. 1-5) have established that monomeric friction coefficients of individual macromolecules in melts of binary mixtures of miscible but chemically dissimilar polymers can differ strongly, and that the values as well as their ratio depend on composition and temperature. Also, the molar mass between entanglements is not simply the average of those of the constituents. This results in a breakdown of superposition of curves of dynamic viscoelastic moduli vs logarithm of the frequency measured at various temperatures. Data on melts of blends of poly(ethylene oxide) and atactic poly(methyl methacrylate) described by Colby (ref. 4) have been supplemented by us to illustrate some points not yet clear.

2. EXPERIMENTAL

Polymers from Polymer Laboratories Ltd., with reported peak molar masses of 227,000 for PMMA and 280,000 for PEO and polydispersities <1.1, were used to make blends of 10, 20, 35 and 50% wt.(± 0.1%) PEO by dissolution in chloroform at 25°C. Thoroughly dried foils were investigated with Rheometrics' Mechanical Spectrometer.

Blends of PMMA and PEO are known to form single-phase melts. The miscibility of these polymers has been documented in detail (refs. 4,6).

Linear viscoelastic quantities not containing units of time do not shift along their axis when the temperature is changed, so that a plot of e.g. the phase shift δ vs the absolute value of the dynamic shear modulus Gd, measured at various frequencies ω, is temperature-invariant (refs. 7,8). This has been proved to be true for polydisperse homopolymers, but is also claimed (ref. 9) for compatible blends.

Indeed, perfect superposition of data of δ vs Gd measured between 130 and 230°C for PMMA and 70 and 290°C for PEO has been obtained. However, dramatic deviations were found for the blends, as is shown in Figures 1 and 2 for the 80/20 and 50/50 blends. All data relating to the softening dispersion at low T and high ω have been omitted for clearness' sake. At low ω and

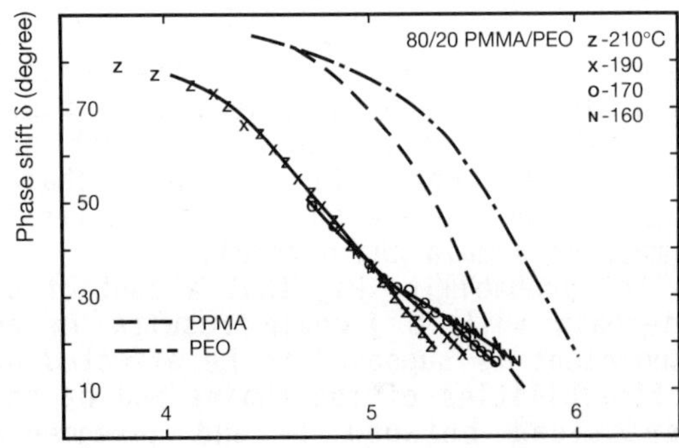

Figure 1 Phase shift vs dynamic modulus of a 80/20 PMMA/PEO blend at the indicated temperatures

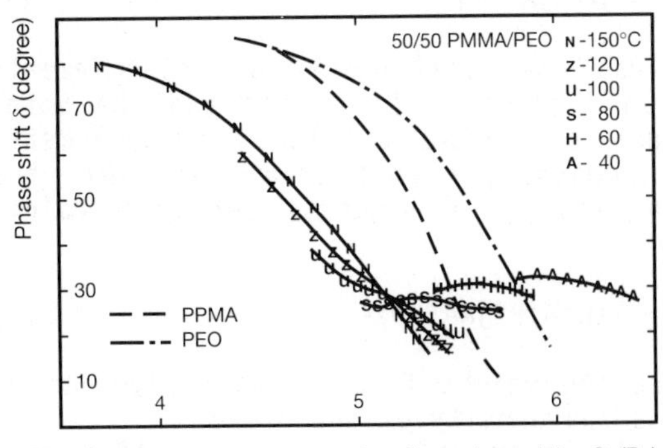

Figure 2 Phase shift vs dynamic modulus of a 50/50 PMMA/PEO blend at the indicated temperatures

high T a very reasonable superposition has been found for the 80/20 blend but not for the other blends, and at lower T the superposition was totally lost for all blends. In these figures the master curves of PMMA and PEO are also sketched, which by extrapolation to $\delta = 0$ yield a value of the rubber plateau modulus GNo of PMMA which is a factor 3 smaller than that of PEO, but the blends appear to have values which can be smaller than that of PMMA, between those of PMMA and PEO, or larger than that of PEO, depending on composition and T. This is consistent with data of Colby (ref. 10), but contrary to predictions of Wu (ref. 11) that only values between those of the components are to be expected.

3. THEORY

In a blend of two miscible heteropolymers with molar mass Mi, density ρi and volume fraction φi (i = 1,2), the number of i-molecules per unit volume is $\nu i = \rho i \varphi i NA/Mi$. The number of contacts of an i-chain with j-chains (j = 1,2) is assumed to be proportional to the contour length of the i-chain (MiLoi/Moi) times the total uncrossable contour length per unit volume (ref. 12) of the j-chains ($\rho j \varphi j NALoj/Moj$, where Lo and Mo are length and mass of a main chain bond).

The probability Pij that a contact of an i-chain with a j-chain results in an entanglement is supposed to be affected by the flexibilities of the chains and by the interactions between i- and j-segments (ref. 13), so that Pji is unequal to Pij. Including the factor NALoiLoj/(MoiMoj) into Pij, we derive for the number of times an i-chain is entangled by a j-chain

$$Nij = \rho j \varphi j MiPij \qquad (1)$$

Because the total number of entanglements on an i-chain is $Ni = \Sigma Nij$ and the total number of segments between entanglements on i-chains per unit volume is νiNi, the total number of segments per unit volume is

$$Ne = \Sigma\Sigma wiwjPijNA(\rho b)^2 \qquad (2)$$

since the weight fraction $wi = \rho i \varphi i/\rho b$, if ρb is the density of the blend.

In binary blends the relaxation rate of a chain is determined by the total friction it encounters. This makes the relaxation times τi dependent on molar mass, composition, interaction and temperature, so that two different relaxation functions Fi(t) have to be anticipated. Assuming that every segment between entanglements gives a contribution kT to GNo, we find for the shear modulus of the blend in the terminal flow region

$$G(t) = \Sigma\Sigma wiwjGijFi(t) \qquad (3)$$

where $Gii = RTPii(\rho b)^2 = GNoi$ indicate the plateau moduli of the components and $Gij = RTPij(\rho b)^2$. Tsenoglou (ref. 14) has proved that if entanglements arise perfectly randomly (theta state), $Pij = \sqrt{(PiiPjj)}$, i.e. $Gij = \sqrt{(GiiGjj)}$. Generally this will not be the case and therefore we have to accept Gij as adjustable parameters.

In this investigation anionic polymers of high molar mass and narrow molar mass distribution were used. Therefore it has been supposed that the relaxation in the terminal region is described reasonably well by the Doi-Edwards reptation theory, yielding in first approximation

$$Fi(t) = \exp(-t/\tau i) \qquad (4)$$

The complex shear modulus is then given by

$$G^*(\omega) = \Sigma\Sigma wiwjGij\omega\tau i/(1 + i\omega\tau i) \qquad (5)$$

Four variables, $\omega\tau 1$, $\tau 2/\tau 1$, G12/G11, and G21/G11, depend on the composition of the blend and the temperature, and two constants, G11 and G22, are specific for the constituents.

For the quotient G22/G11 a value of 3 will be used in subsequent calculations. It is now possible to calculate with eq. (5) at many values of $\omega\tau 1$ or $\omega\tau 2$ the reduced dynamic modulus Gd1/G11 and the phase shift δ, or Gd2/G11 and δ, resp. The corresponding plots for the 1- and 2-melts are represented in Figure 3.

The lowest value of GNob, with G12 = G21 = 0, for a 80/20 blend is 0.76*G11. With $\tau 2/\tau 1 = 4$, a δ vs Gd curve is calculated similar in shape to the experimental curve at 210°C. The curves at lower T require larger values of G12, G21 and $\tau 2/\tau 1$. With trial and error the set of parameters given in Table 1 has been obtained, yielding the family of curves of Figure 3, which qualitatively shows the same effects as Figure 1. Slight variations of the parameters do not change the shape of the curves essentially. The effects have been established rather firmly: G12 is about three times G21 and increases at lower T, as does $\tau 2/\tau 1$, i.e. PEO molecules do entangle more with PMMA molecules than vice versa, and below 160°C even more than with themselves, and the relaxation of the more flexible and more entangled PEO molecules becomes more and more slower than that of the PMMA chains with decreasing T. This last finding is not consistent with experimental results of Colby (ref. 4).

For a 50/50 blend the curves in Figure 4 have been calculated, reproducing the main features of Figure 2. Now G12 is a factor 4 to 5 smaller than G21 and $\tau 1$ is much longer than $\tau 2$, especially at low T, and GNo of the blend becomes at low T twice that of PEO. This type of analysis appears to be much less conclusive for a 65/35 blend due to the fact that $w1^2$ nearly equals $3w2^2$.

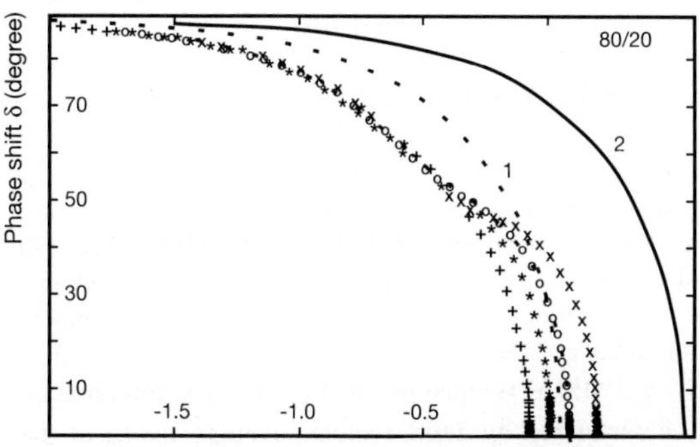

Figure 3 Calculated curves of phase shift vs reduced dynamic modulus for the polymers 1 and 2 and for a 80/20 blend with the parameters given in Table 1

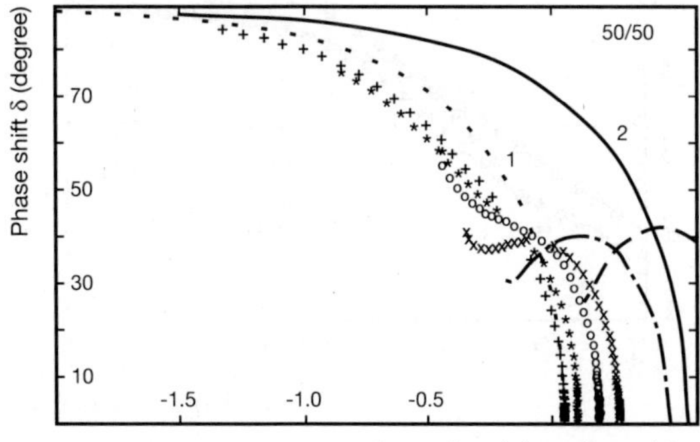

Figure 4 Calculated curves of phase shift vs reduced dynamic modulus for the polymers 1 and 2 and for a 50/50 blend with the parameters given in Table 1

Table 1 Set of parameters used to calculate the curves in Figures 3 and 4

w1	G12/G11	G21/G11	$\tau 2/\tau 1$	symbol
0.8	0	0	4	+
0.8	0.75	0.25	6	*
0.8	1.5	0.5	7	o
0.8	3	1	10	x
0.5	0	0	1/4	+
0.5	0.1	0.4	1/6	*
0.5	0.3	1.2	1/10	o
0.5	0.5	2	1/20	x
0.5	1.1	5	1/40	-·
0.5	3	15	1/100	--

Summarizing we conclude that the temperature dependencies of the relaxation times of the individual molecules in a miscible blend can be quite different and dependent on the composition, as are the degrees of mutual entanglement and hence the rubber plateau modulus.

REFERENCES

1. L.H. Wang and R.S. Porter, J. Polym. Sci., Polym. Phys., 21 (1983) 1815-1823.
2. D. Lefebvre, B. Jasse and L. Monnerie, Polymer, 25 (1984) 318-322.
3. R.J. Composto, E.J. Kramer and D.M. White, Polymer, 31 (1990) 2320-2328.
4. R.H. Colby, Polymer, 30 (1989) 1275-1278.
5. G.G. Fuller, 14th Discussion Conference, Prague, July 15-18, 1991.
6. S. Cimmino, E. Martuscelli, M. Saviano and C. Silvestre, Makromol. Chem., Macromol. Symp., 38 (1990) 61-80.
7. J.M. Dealy and K.F. Wissbrun, Melt Rheology and Its Role in Plastics Processing, Van Nostrand Reinhold, New York, 1990, p. 90.
8. H.C. Booij and J.H.M. Palmen, Rheol. Acta, 21 (1982) 376-387.
9. C.D. Han and H.K. Chuang, J. Appl. Polym. Sci., 30 (1985) 4431-4454.
10. R.H. Colby, in P.H.T. Uhlherr (Ed.), Xth International Congress on Rheology, Sydney, Australia, August 14-19, 1988, Australian Society of Rheology, 1988, Vol, 1, pp. 278-281.
11. S. Wu, J. Polym. Sci., Polym. Phys., 25 (1987) 2511-2529.
12. W.W. Graessley and S.F. Edwards, Polymer, 22 (1981) 1329-1334.
13. P. Cifra, F.E. Karasz and W.J. MacKnight, J. Polym. Sci., Polym. Phys., 29 (1991) 1389-1395.
14. C. Tsenoglou, J. Polym. Sci., Polym. Phys., 26 (1988) 2329-2339.

Theoretical and Applied Rheology, edited by P. Moldenaers and R. Keunings
Proc. XIth Int. Congr. on Rheology, Brussels, Belgium, August 17-21, 1992

RHEOLOGICAL PROPERTIES IN THE MELT AND MORPHOLOGY OF IMPACT PMMA

M. BOUSMINA and R. MULLER

Institut Charles Sadron (CRM-EAHP), 4 rue Boussingault, 67000 Strasbourg (France)

1. INTRODUCTION

If the melt rheology of incompatible polymer blends is compared to that of individual phases, it is quite generally observed that multiphase polymer melts show pronounced elastic properties and very long relaxation time processes. For blends forming a dispersion, recent theoretical developments (ref.1) on emulsion models and experimental results on selected blends (ref.2-3), have shown that the deformability and relaxation of shape of the inclusions of the minor phase give rise to mechanical relaxation times involving the ratio of interfacial tension to droplet radius and having the order of magnitude of those observed experimentally. It is the purpose of the present paper to verify if the emulsion-type analysis is able to account for the rheological behaviour in the melt of impact PMMA, which is a rubber-toughened polymer. We first recall the expression of the complex shear modulus as given by an emulsion model developped for viscoelastic phases. We then compare this model to experimental data for a blend of two thermoplastic polymers, and for the impact PMMA.

2. EMULSION MODEL

Palierne (ref.1) worked out the linear constitutive equation of a dispersion of viscoelastic incompressible spherical inclusions with complex shear modulus G^*_i in a viscoelastic incompressible matrix with complex modulus G^*_m. The model takes into account the polydispersity of particle sizes. If ϕ_i is the volume fraction of inclusions of radius R_i and α the interfacial tension, the complex modulus of the blend has the following expression:

$$G^*(\omega) = G^*_m(\omega)\frac{1+3\sum_i \phi_i H_i(\omega)}{1-2\sum_i \phi_i H_i(\omega)}$$

where:

$$H_i = \frac{4\left(\alpha/R_i\right)(2G^*_m+5G^*_i)+(G^*_i-G^*_m)(16G^*_m+19G^*_i)}{40\left(\alpha/R_i\right)(G^*_m+G^*_i)+(2G^*_i+3G^*_m)(16G^*_m+19G^*_i)}$$

all moduli being taken at the same frequency. If the viscoelastic behaviour of the phases, the interfacial tension and the morphology are known, the dynamic moduli G' and G" of a two-phase polymer blend in the melt may be easily calculated from the above expression.

3. COMPARISON WITH DATA ON PS/PMMA BLEND

All dynamic measurements have been carried out on a Rheometrics RMS-605 mechanical spectrometer. Figure 1 compares experimental data on G' and G" with the values calculated from the emulsion model, for a PMMA/PS 70/30 blend at 200°C (ref.4). The dynamic moduli of the phases were determined experimentally by independent measurements; the ratio of zero-shear viscosities η_0(PS)/η_0(PMMA) is equal to 0.035 (<<1) . The morphology was characterized by TEM (volume-average radius of the PS inclusions: 0.25μm) and the interfacial tension has been taken as 1.5 mN/m.

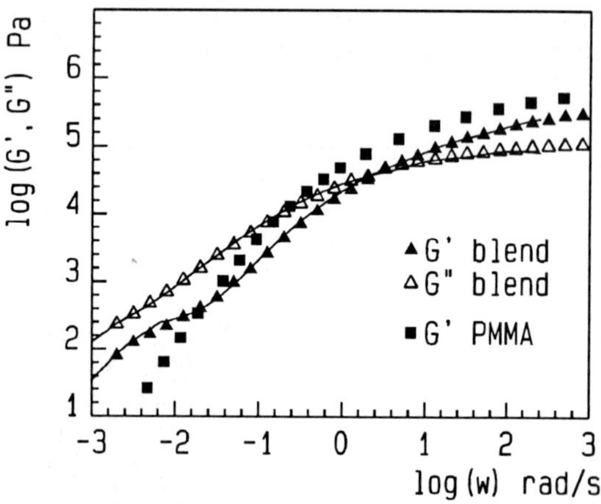

Figure 1: G' and G" data at 200°C vs. frequency for a 70/30 PMMA/PS blend, compared to G' of the PMMA matrix and to the emulsion model prediction (full lines).

The low-frequency secondary plateau in G' is characteristic of a long-time relaxation process at about

100s, whereas the terminal relaxation times of the phases are between 0.1s and 1s. The emulsion model describes quantitatively this phenomenon and ascribes it to the deformability and shape relaxation of the suspended PS droplets. Actually, the secondary plateau in G' does not appear that clearly for blends were the inclusions are much more viscous than the matrix and deform only slightly during a macroscopic strain (ref. 5).

4. COMPARISON WITH DATA ON IMPACT PMMA

The same PMMA than in the preceeding section has been blended with rubber particles in a batch mixer. The latex particles have a core-shell poly(butylacrylate-co-styrene)-PMMA structure and the volume fraction ϕ of the rubber core in the blends range from 5% to 50%. The micrography in Figure 3a shows the type of morphology obtained by melt-blending for volume fractions ϕ above 0.15: although the latex particles are well-dispersed in the PMMA matrix, they seem to be connected to each other and to form a network-type structure.

The G'(ω) curves at 200°C have been plotted for these blends in Figure 2. A plateau appears at low frequencies for volume fractions of rubber phase higher than 15%, which is also the concentration at which the latex particles start to form a three-dimensional network structure.

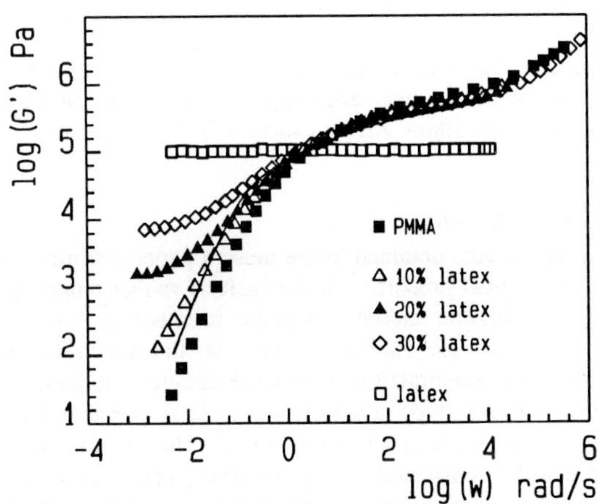

Figure 2: Storage modulus G' vs. frequency at 200°C for rubber-toughened PMMA at various rubber concentrations. Full lines are the emulsion model prediction for ϕ=30%.

The predictions of the emulsion model have been compared to these data. The complex moduli of the phases were determined by separate measurements (for the dispersed phase, a rubber-type behaviour with $G^i \sim 10^5$ Pa has been found). The particle radius is close to 80 nm. The interfacial tension is unknown, due to the presence of the PMMA grafts at the surface of the particles. However, it appears that the value of interfacial tension does not affect significantly the model predictions since in the frequency range where the low-frequency plateau is observed, $G^*_i >> G^*_m$ which means that the inclusions remain almost undeformed in a macroscopic strain.

The emulsion model describes the experimental G' data for ϕ lower than 0.15, but does not account for the low-frequency plateau appearing at higher concentrations. This means that the corresponding long time relaxation mechanisms are not only due to the deformability of the dispersed rubber particles (like for the PS inclusions in section 3). In fact, one of the basic assumptions in the emulsion model is that there exist only hydrodynamic interactions between the particles of the dispersed phase (i.e. through the deformation of the matrix). According to the morphology in Figure 3a, this is probably not the case for the concentrated blends (ϕ>=0.15).

5. INFLUENCE OF PARTICLE AGGREGATION

In order to verify if the low-frequency plateau in the rubber-toughened PMMA samples is related to the connectivity between the latex particles, blends with different morphologies but identical content of dispersed phase (ϕ=0.15) have been prepared. The well-dispersed melt-blended sample (Fig.3a) has been compared to a blend obtained by freeze-drying a suspension of the latex in a solution of the PMMA in benzene. This way of preparation of the blend leads to particle aggregation (Fig. 3b).

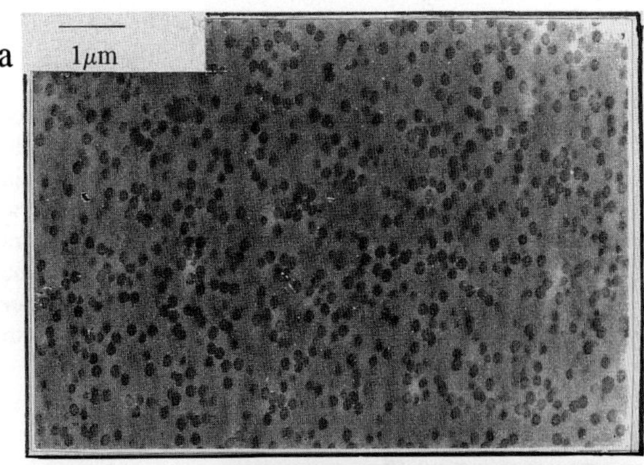

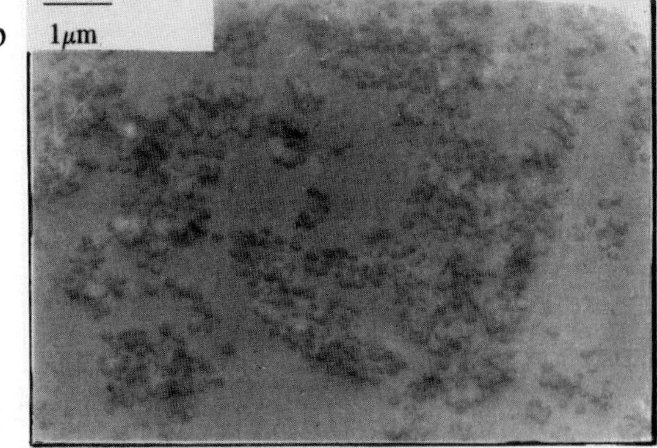

Figure 3: TEM micrography of rubber-toughened PMMA containing 15% rubber. 3a: melt-blended sample. 3b: sample obtained by freeze-drying of a solution.

The viscoelastic data of these two blends are shown in Figure 4: while the loss modulus G" is almost identical for both samples, G' values at low frequencies are lower for the aggregated morphology. This result supports the assumption that the network structure that is formed in concentrated melt-blended systems, contributes significantly to their viscoelastic behaviour in the melt.

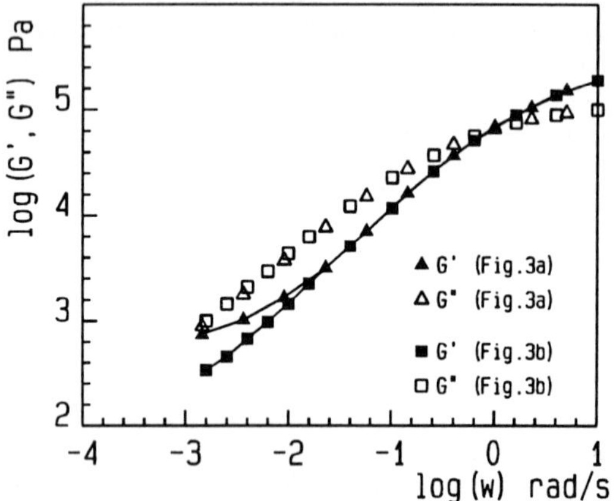

Figure 4: Dynamic mmoduli at 200°C for samples of Figures 3a (melt-blended) and 3b (solution-blended).

6. CAPILLARY FLOW MEASUREMENTS

The response of the melt-blended rubber-toughened PMMA samples to a steady shear flow has been determined on a ROSAND RH7-2 capillary rheometer. The main results are the following: i) the steady shear viscosity of the blends is lower than their dynamic viscosity (unlike for pure PMMA, the Cox-Merz is not verified for the blends); ii) the steady shear viscosity decreases as a function of rubber content (see Figure 5); iii) the die-swell also decreases with increasing rubber content.

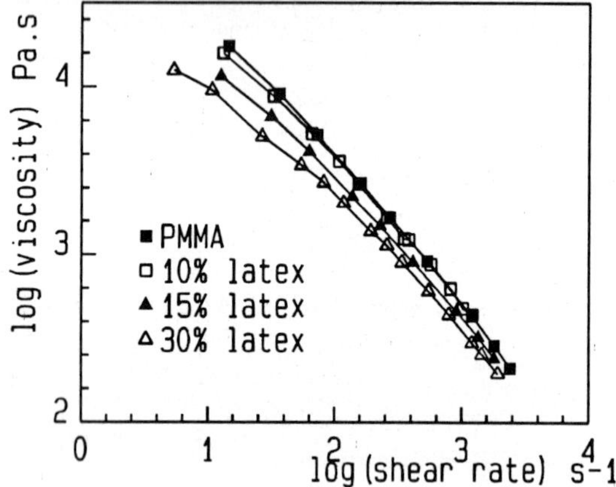

Figure 5: Steady shear viscosity at 200°C of blends at various rubber content compared to data on pure PMMA.

This last result seems to be in contradiction to the high G' values of concentrated blends, since die-swell is usually associated with elastic properties of the melt. To clarify the observed behaviour, the morphology in the extrudate (for ϕ=0.15) has been characterized. Depending on the distance from the wall, two types of morphologies were found: near the centre of the extrudate, the morphology is close to that in Figure 3a (network-type structure) whereas near the wall, the particles lign up in the direction of extrusion as shown in Figure 6. At the particular distance from the wall where the oriented morphology appears, the shear stress is close to the value of G' as determined from dynamic measurements at the low-frequency plateau.

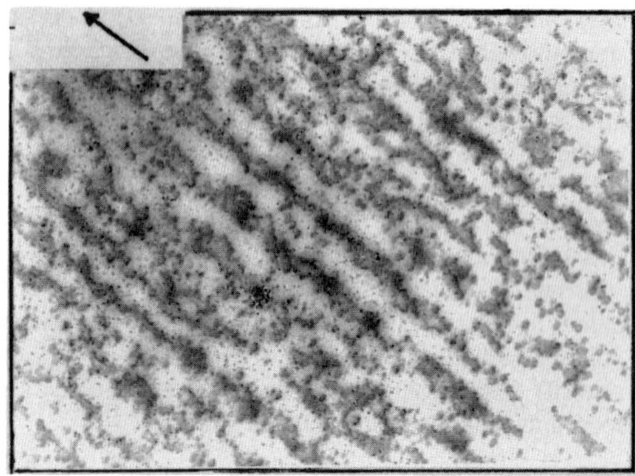

Figure 6: TEM micrograph of rubber-toughened PMMA containing 15% rubber after capillary extrusion. Distance from the wall: 50μm. Arrow indicates the flow direction.

7. CONCLUSION

The results obtained in the present paper confirm that the rheological properties in the melt of rubber-toughened polymers depend closely upon the morphology, i.e. the dispersion of the rubber particles in the thermoplastic matrix. In concentrated melt-blended samples, the particles form a network-type structure, which is responsible for a low-frequency plateau modulus in G'. This plateau value can also be considered as a yield stress, above which the network structure is replaced by alignements of the particles in the direction of flow. This oriented morphology seems to be responsible for the low viscosity and die-swell values obtained for concentrated blends.

ACKNOWLEDGEMENT: Financial support from G.I.S. "Alliages de Polymères" (MRT, France) is gratefully acknowledged. We also thank NORSOLOR for having made the latex and PMMA samples available.

REFERENCES

1. J.F. Palierne, Rheol. Acta, 29 (1990) 204
2. D. Graebling and R. Muller, J. Rheol., 34 (1990) 193
3. D. Graebling and R. Muller, Colloids and Surf. 55 (1991) 89
4. M. Bousmina, Thesis, Université Strasbourg, 1992
5. D. Graebling, R. Muller and J.F. Palierne, in preparation

Theoretical and Applied Rheology, edited by P. Moldenaers and R. Keunings
Proc. XIth Int. Congr. on Rheology, Brussels, Belgium, August 17-21, 1992

CHEMORHEOLOGY OF THERMOSETS USED FOR THE ENCAPSULATION OF MICROELECTRONIC DEVICES

L.BOUTIN[‡], A.AJJI[*] and L.CHOPLIN[†]

CERSIM, Département de génie chimique, Université Laval, Ste-Foy, Québec, G1K 7P4

[‡]IBM Canada Ltée, 23 Boul. de l'aéroport, Bromont, Québec, J0E 1L0.

[†]ENSIC, 1 Grandville, Nancy, France.

[*]To whom all correspondance should be addressed.

1. INTRODUCTION

Encapsulation of microelectronic devices involves transfer molding of a thermoset into a cavity that contains a lead frame, some wires and a device (ref.1). To minimise flow problems during this process, it is important to understand the rheology of thermosets used which is known to be closely related to their chemical reaction (ref.2).

In this study, the chemorheology of two highly filled epoxy resins used for the encapsulation of microelectronic devices was investigated and compared. Using DSC technique, a kinetic equation was determined for the two resins (ref.3). Rheological measurements were carried out in the dynamic and steady modes. Gel times were determined by the crossover of the loss tangent at different frequencies (ref.4). Finally, the measured viscosity was compared with the predictions of two different models: double-Arrhenius model (ref.5) and a model developed for the reaction injection molding (RIM) process (ref.6).

2. MATERIALS AND METHODS

Two pre-mixed highly filled epoxy compounds differing only by the amount of catalyst were used. The compounds were solid and pre-formed in pellets. To analyse the compounds, the pellets were crushed to have a homogeneous powder.

Thermal analysis was used to find a kinetic equation for the cross-linking reaction. Isothermal curves were carried out at different temperatures on a Mettler-20. For nth order system, the equation used was:

$$\frac{d\alpha}{dt} = k(T)\,(1-\alpha)^n \tag{1}$$

where k(T) is the rate constant for a given temperature; n, the reaction order and α, the conversion. In equation 2, we can see that if $\ln(d\alpha/dt)$ as a function of $\ln(1-\alpha)$

$$\ln\left(\frac{d\alpha}{dt}\right) = \ln k(T) + n \ln(1-\alpha) \tag{2}$$

is linear, then we have an nth order system.

Rheological measurements were carried out in the dynamic and steady modes on a Rheometrics System IV. Parallel plates configuration was used. Powder of the compounds was compressed in a mold of the appropriate dimensions to form a disc for rheological measurements. Grooved plates were used to avoid slippage with samples of approximately 2,5 mm of thickness. Frequencies from 0,3 rad/s to 10 rad/s were used. A temperature profile was used: the sample was introduced into the rheometer oven at 60°C and after stabilization, the temperature was raised as quickly as possible to the desired one.

Comparison of the rheological results were performed with the predictions of two different models: double-Arrhenius model and a model developed for the RIM process. The equation used for the double-Arrhenius model was (ref.5):

$$\ln \eta(t) = \ln \eta_\infty + \frac{\Delta E_\eta}{RT} + t k_\infty e^{\frac{\Delta E_k}{RT}} \tag{3}$$

and the one developed for the RIM process was (ref.6):

$$\eta = \eta_o(T)\left[\frac{\alpha_{gel}}{(\alpha_{gel}-\alpha)}\right]^{C1+C2\,\alpha} \tag{4}$$

$$where \quad \eta_o(T) = A_\eta e^{\frac{\Delta E}{RT}} \tag{5}$$

3. RESULTS

Plot of $\ln(d\alpha/dt)$ as a function of $\ln(1-\alpha)$ for the two systems studied were linear, thus indicating that the reaction kinetics of the compounds was of nth order. Table 1 presents the parameters found for the two resins with isothermal DSC curves at different temperatures. Figure 1 illustrates some results of the conversion as a function of time at different temperatures calculated from the kinetic equations found for the two resins. We can see that resin B reacts faster than resin A and that both resins react very quickly at high temperatures. The results of the calculation for isothermal temperatures and temperature profiles obtained with the rheometer oven were compared. An induction time for the reaction in the case of temperature profiles was observed. This indicates that, even if it takes a certain time at 60°C before we start the measurement in the rheometer, no reaction has occured.

Table 1: Parameters found for nth order systems

$\ln \frac{d\alpha}{dt} = \ln Z + \frac{E}{RT} + n \ln (1 - \alpha)$			
COMPOUNDS	n (∓6%)	Ln Z (∓1%)	E/R (°K) (∓1%)
A	0.62	16.9	- 9286.6
B	0.60	17.7	- 9416.5

Figure 2 presents the results obtained for the viscosity as a function of time at different frequencies for resin A. We can observe the melting, the minimum viscosity and the beginning of the reaction of the compound. These steps are all important to define the parameters of the transfer molding process such as the rate of transfer and the time before transfer. We can also observe the shearthinning character of the resin. Representation of the viscosity as a function of frequency instead of time, yielded a power law behavior in the frequency range studied. Figure 3 shows the effect of the temperature on the viscosity. It can be seen that, as the temperature is increased, the reaction occurs faster and the duration of the viscosity minimum become smaller.

We can also represent the different curves obtained at different frequencies in the form of loss tangent as a function of time. We can then find the gelification time (ref.4). Table 2 shows the comparison of the gelification times found with the crossover of the loss tangent and the ones found with the crossover of the moduli G' and G". If we compare both results for the same temperature profile, we can see that for high temperatures, the results of the gel times obtained at G'=G" are close to those of tan δ crossover, but for low temperatures, the results are different. For these resins, we can conclude that the crossover of the modulus cannot be an approximation of the gelification times. As expected, we can also observe that the gelification time is higher with the decrease of the temperature.

Table 2: Gelification times for different temperature profiles.

	COMPOUND A		COMPOUND B	
	Gelification time (min.) (±4%)			
Temperature (°C)	Crossover tan δ	G' = G"	Crossover tan δ	G' = G"
60° to 120°C	11.5	4.0	11.0	6.5
60° to 170°C	3.5	2.5	2.5	2.7
60° to 180°C	2.2	1.9	2.0	2.0

Some experiments were also done in the steady mode. The results showed that, in the steady mode, the deformation of the sample is so important that we never know if the fragile structure of the gelification was intact. The curves obtained had the overall shape of the one obtained in the dynamic mode except that the results were going up and down indicating the breaking of the structure.

Finally, figure 4 shows the results of the models' predictions for the isothermal part of the viscosity curve. We can clearly see that the predictions of the RIM model are following the experimental results much better than the double-Arrhenius model. The major disadvantage of the double-Arrhenius model is that it does not predict the final abrupt rising of the experimental curves.

4. CONCLUSION

As a conclusion, we can say that it is possible to obtain satisfactory kinetic predictions without knowing the overall reaction of the system studied. It is shown that, for these resins, the crossover of the moduli is not an adequate method for the determination of the gelification time, the crossover of the loss tangent is more apropriate. We have observed that the rheological measurements can be done in the dynamic mode only, the steady mode breaks the gelification structure. Finally, the predictions of the viscosity with an empirical model was satisfactory.

REFERENCES

1. Manzione, L.T., Plastic packaging of microelectronic devices, Van Nostrand & Reinhold, NY, 1990, 377p.
2. Gillham, J.K., Polym. Eng. and Sci., **26**, 1429, (1986).
3. Shim, J.S., Lee, W. and Jang, J., Polym. J., **23**, 911, (1991).
4. Venkataraman, S.K. and Winter, H.H., Rheologica Acta, **29**, 423, (1990).
5. Roller, M.B., Polym. Eng. and Sci., **26**, 432, (1986).
6. Castro, J.M. and Romagnoli, J.A., Dev. Plast. Technol., **2**, 43, (1985).

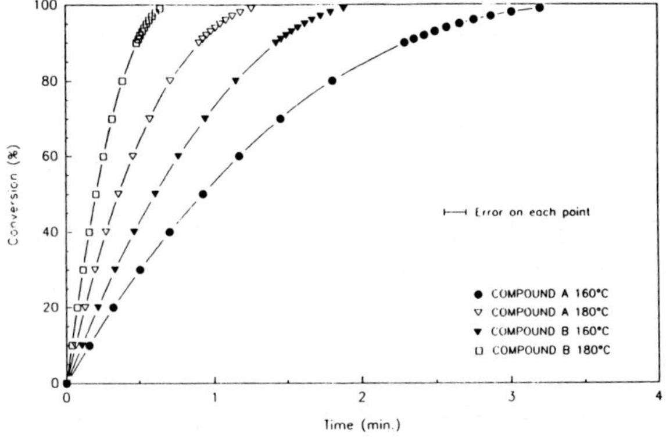

Figure 1: Results obtained with the kinetic equations.

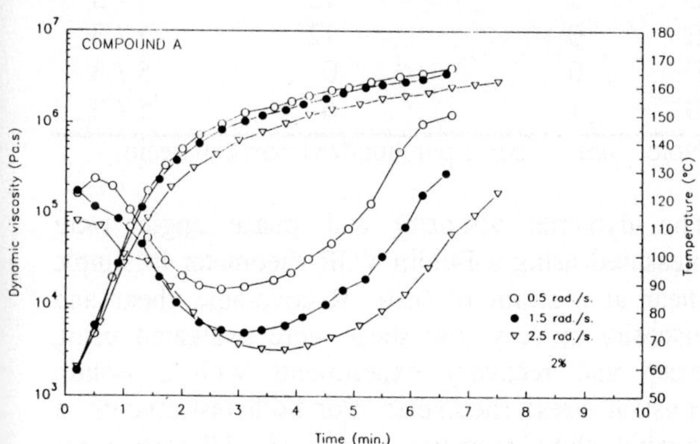

Figure 2: Rheological experimental results obtained at different frequencies (COMPOUND A).

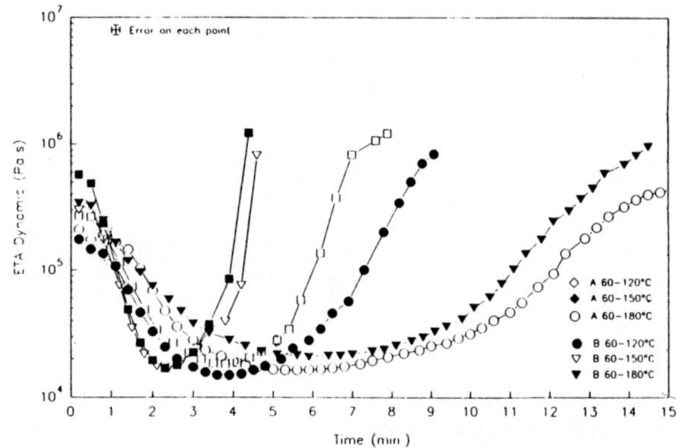

Figure 3: Rheological experimental results obtained at different temperatures (COMPOUNDS A and B).

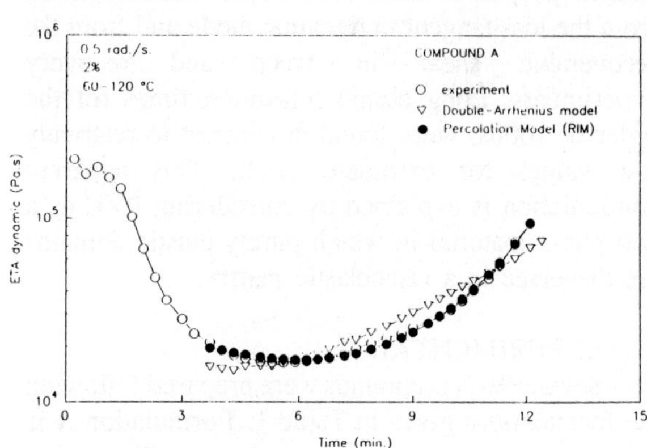

Figure 4: Isothermal predictions of different models of the rheological results (COMPOUND A).

Theoretical and Applied Rheology, edited by P. Moldenaers and R. Keunings
Proc. XIth Int. Congr. on Rheology, Brussels, Belgium, August 17-21, 1992

EXTRUDATE SWELL OF RIGID PVC COMPOUNDS

P. J. Carreau[1], P. G. Lafleur[1], M. A. Huneault[1] and V. P. Gupta[2]

[1]CRASP, Ecole Polytechnique de Montréal, P.O.Box 6079, Station "A", Montréal,Québec, H3C 3A7, Canada
[2]Synergistics Industries ltd., 177 St-André, St-Rémi , Québec, J0L 2L0, Canada

1. INTRODUCTION

The extrudate swell of rigid polyvinyl chloride (PVC) compounds is a complex viscoelastic phenomenon. It is of particular interest in rheology as well as in polymer processing (e.g. for profile extrusion die design). Rigid PVCs are non-homogenous fluids and typical formulations include processing aids, impact modifiers, lubricants, and fillers. Contrarily to other thermoplastics, the PVC resin is not fully molten at processing temperature. Extrudate swell is strongly dependent on formulation and degree of fusion. In this paper, extrudate swell data for several PVC compounds are presented. Elastic properties (time constants) were obtained from the loss tangent in dynamic mode and from the recoverable shear in creep and recovery experiments. Long elastic relaxation times (of the order of 1000s) were found in contrast to relatively low values for extrudate swell. This apparent contradiction is explained by considering PVC as a two-phase material in which purely elastic domains are dispersed in a viscoelastic matrix.

2. EXPERIMENTAL

Seven PVC compounds were prepared following the formulations given in Table 1. Formulation A is the base compound with only stabilizer and lubricants. The resin is a medium molecular weight PVC (k=66) except for formulation G in which a high molecular weight PVC resin (k=77) has been used. In formulations B to G, fillers were added to the base compound and contents of processing aid and impact modifier were varied. The processing aid, Paraloïd K125™, is an all-acrylic polymer used extensively for the extrusion of profiles. The impact modifier, Paraloïd KM 334 ™, is an all-acrylic elastomer.

TABLE 1 : PVC Formulations

A	(Base formulation)		
	PVC resin		100 parts
	Stabilizer		2.5 phr
	Calcium stearate		1.2 phr
	Stearamide		1.2 phr

	Process. aid (phr)	Impact Modifier (phr)	Filler (phr) CaCO3/TiO2
B	1	6	5 / 3
C	2	0	5 / 3
D	2	12	5 / 3
E	0	12	5 / 3
F	0	0	5 / 3
G	1	6	5 / 3

Note: phr = parts per hundred parts of resin.

The dynamic viscosity and phase angle were measured using a Bohlin VOR rheometer in simple shear at a strain of 0.01. Recoverable shear and viscosity at very low shear were evaluated using creep and recovery experiments with a Bohlin constant stress rheometer. For both instruments, a parallel plate geometry was used. All tests were made in a nitrogen atmosphere to delay degradation. Samples with diameters of 25 mm were compression moulded under a pressure of 70 MPa at 180°C for 4 min. A 45mm single screw extruder was used with different capillary dies for the determination of the swelling ratio. The swelling ratio (d/D) was obtained directly by measuring the diameter of extrudates after cooling at room temperature.

3. RESULTS AND DISCUSSION

In order to understand the rheology of PVC compounds, it is important to understand its unique morphology. PVC has a particulate structure: grains consist of subgrains formed themselves by primary particles. The primary particles are made up of crystalline domains which bound the PVC chains into a tridimensional network. Through processing, PVC is broken down to the primary particles level. In order to obtain good mechanical properties, the crystalline network must be partly molten. This occurs in the usual processing temperature range (170-190°C). Complete fusion of PVC occurs only at temperatures higher than 220°C. Consequently, the rheology of PVC at usual operating temperatures can be compared to that of a filled polymer in which the "concentration" of the filler is temperature dependent (ref. 1). The effect of temperature on the swelling ratio for PVC compound B is presented in Figure 1. Swelling increases significantly with temperature up to 200°C because of the increasing degree of fusion. Above that temperature, the compound behaves as a homogeneous polymer melt and swelling becomes more or less independent of temperature.

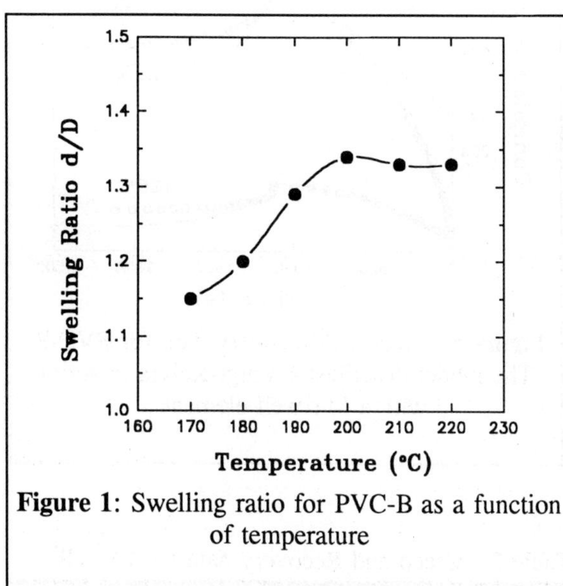

Figure 1: Swelling ratio for PVC-B as a function of temperature

The extrudate swell ratio for a given temperature has been shown to depend mainly on the residence time in the die (ref. 2). In Figure 2, the swelling ratios for PVC compounds B and F at 190°C are compared for residence times between 0.2 and 20 s. The only difference between compounds B and F is the resin molecular weight. Clearly, compound F containing the high molecular weight resin exhibits much less swelling. This is explained by an indirect effect of molecular weight which is responsible for increased crystallinity and higher fusion temperature of PVC resins. Hence, for given extrusion conditions, the degree of fusion of PVC-G is lower and thus its swelling is smaller.

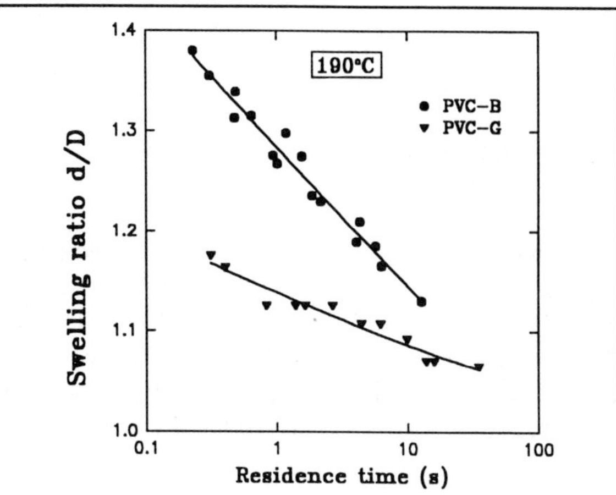

Figure 2: Swelling ratio for PVC-B and PVC-G as a function of residence time.

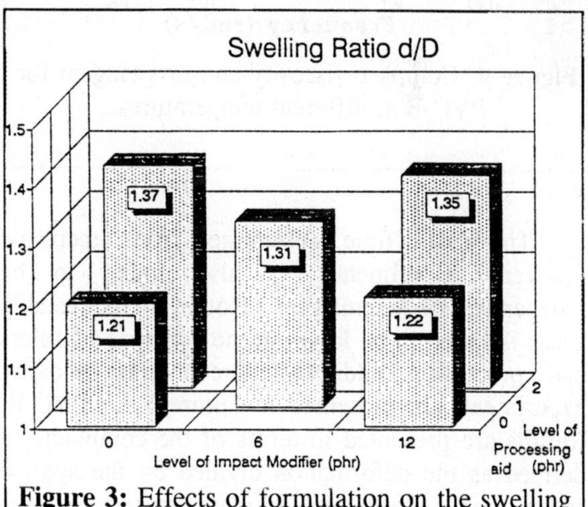

Figure 3: Effects of formulation on the swelling ratio for a residence time of 1 s.

The effect of the processing aid and impact modifier at 190°C and for a residence time of 1 s is presented in Figure 3. By promoting the fusion of PVC, the processing aid has a similar effect as temperature and thus has a positive effect on swelling. The impact modifier has no effect on the swelling ratio even at levels of 12 parts per hundred of resin (phr).

The dynamic properties are barely affected by formulation. Figure 4 presents the complex viscosity η^* and tangent δ as a function of frequency for PVC-B. Tangent δ is not affected by the frequency even at low frequency. The effect of temperature is to increase tangent δ, hence reducing elasticity.

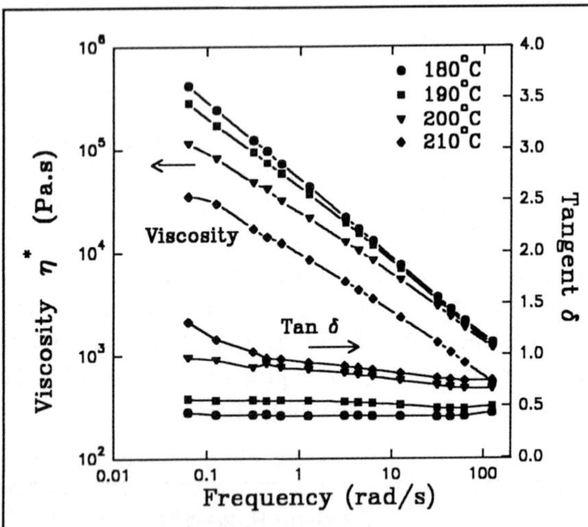

Figure 4: Complex viscosity and loss tangent for PVC-B at different temperatures.

The elastic time, determined from creep and recovery experiments, is also found to be independent of formulation. However, recoverable shear is found to be lower in the presence of fillers and processing additives. Creep and recovery experiments are presented in Figure 5 for PVC-B. Results are presented in terms of the compliance J defined as the deformation divided by the applied shear stress. Creep and recovery parameters for PVC-B are presented in Table 2. Recoverable compliance increases by a factor of 8 when temperature is increased from 190 to 220°C. However, the elastic time defined as the product of viscosity and recoverable creep decreases with temperature. Recoverable compliance is thus a better indicator of the swelling behaviour of PVC. The elastic time decrease and swelling increase with temperature is at first glance contradictory. However, at low temperature, the PVC compounds behave almost as elastic solids. For these materials, the recovered deformation is equal to the creep deformation. Elastic time of a viscoelastic solid is only a measure of the speed of the recovery while for a viscoelastic fluid it is a measure of the recoverable shear. Therefore, elastic times for PVC cannot be used as a measure of elasticity in the way elasticity is generally understood for polymer melts.

REFERENCES
1) Utracki L.A., J.Vinyl Technol., 8, 156 (1986).
2) Huneault M.A., Lafleur P.G., Carreau P.J., Polym. Eng. Sci., 30, 1544 (1990).

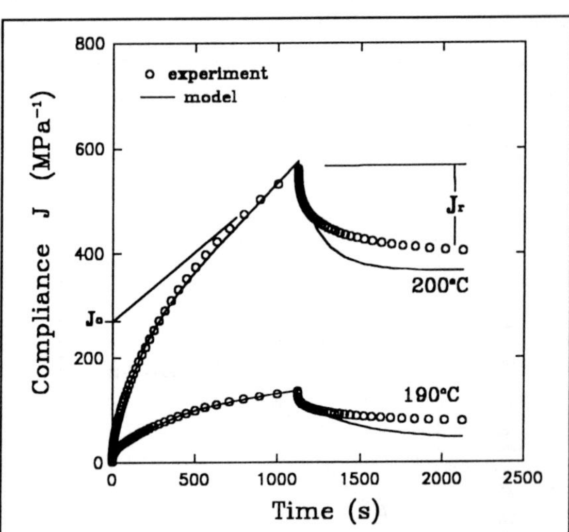

Figure 5: Creep and recovery data for PVC-B. The model describes a Voigt-Kelvin in series with a Maxwell element .

Table 2. Creep and Recovery data for PVC-B

Temp. (°C)	Jr (MPa^{-1})	Viscosity (MPa.s)	Elastic time (s)
190	54	23	1200
200	156	4.0	620
210	380	0.17	65
220	410	0.028	11

Theoretical and Applied Rheology, edited by P. Moldenaers and R. Keunings
Proc. XIth Int. Congr. on Rheology, Brussels, Belgium, August 17-21, 1992

Two-Dimensional Deuteron Exchange NMR Studies of the Dynamics of Individual Species in Miscible Blends

Geunchang Chung and Julia A. Kornfield
California Institute of Technology, Pasadena, CA 91125

1 Broad Glass Transition of Miscible Blends

In miscible blends, the glass transition in both mechanical and dielectric spectra tend to be significantly broadened. In certain pairs such as polystyrene/poly(vinyl methyl ether) (PS/PVME) [1,2] and polyisoprene/poly(vinyl ethylene) (PI/PVE) [3,4], the glass transition occurs over a range of temperature as broad as 50K. A broad glass transition is a manifestation of dynamic heterogeneity. A broad distribution of rates of conformational rearrangement can arise from both local composition fluctuations and intrinsic molecular differences between the components.

There are a few experimental studies that give direct evidence of such dynamic heterogeneity. Dielectric spectroscopy is used to monitor the effect of blending on the dynamics of one species in a blend (e.g. the PVME in PS/PVME); the dielectric loss spectrum is not only shifted, but broadened [2]. Carbon-13 (^{13}C) NMR can be used to observe the effect of blending on the dynamics of each component. The change in ^{13}C NMR line-widths in magic angle spinning (MAS) experiments on a PI/PVE blend [5] shows that there is a dramatic difference in the temperature dependencies of dynamics between the two components in a blend. Such differences in the dynamics of each species may also underlie the breakdown of time-temperature superposition that has been observed for poly(ethylene oxide)/poly(methyl methacrylate) (PEO/PMMA) blends [6].

In order to examine the connection between molecular motions and the unusual glass transition and thermorheological behavior of miscible blends, it is desirable to quantitatively characterize the dynamics of each species in the blend as functions of composition and temperature. The ideal method for this purpose is two-dimensional broadline deuteron exchange NMR (2D ^{2}H NMR), which quantitatively determines the type and rate of molecular motion in polymer melts near the glass transition temperature (T_g) [7,8,9] and is intrinsically amenable to isotopic labeling.

2 Two-Dimensional Deuteron Exchange NMR

The basis of 2D ^{2}H exchange NMR is the sensitivity of the resonance frequency of a deuteron to the orientation of its carbon-deuterium (C-D) bond in the labeled polymer. Reorientation of C-D bonds acts as a probe of back-bone motion. The direct relationship between the resonance frequency of a deuteron and the orientation of its C-D bond stems from the quadrupolar interaction between local electric field gradient, mainly along the C-D bond, and the quadrupole moment of the deuterium nucleus (Figure 1).

Reorientational motion occurring in a controlled time period called the "mixing time" produces exchange from one resonance frequency to another. The 2D exchange spectrum displays the joint probability distribution of finding a resonance frequency $\omega(t)$ at time t, starting from a resonance frequency of $\omega(0)$ at time zero [10] (Figure 2 a). This, in turn, is related to the dynamic evolution of the orientation distribution, or the joint probability of finding a C-D bond that had orientation $\theta(0)$ at time zero oriented at $\theta(t)$ at time t. By iteratively solving the spectra for the evolution of orientation, we can determine the mean reorientational time and the width of the reorientation time distribution [11].

3 Experimental Work to date

We have chosen blends of PI/PVE and PS/PVME as our model systems. The PI/PVE pair is known to form a nearly ideal blend in which miscibility occurs without any specific interaction due to their chemical similarity [4]; and the PS/PVME pair is one of the most extensively studied miscible blends to date. There are many recent studies on these miscible blends, including dielectric, DSC, FTIR and ^{13}C linewidth measurements. For our NMR work, the pairs of deuterated and normal polymers of each of the species are under preparation.

The 2D ^{2}H exchange spectra are obtained using a pulse sequence developed by Spiess and coworkers [12] as shown in Figure 2 b. By recording exchange spectra at various mixing times, the exchange of resonance frequency, which is in turn directly related to the reorientational motion, is monitored in real time. To solve the set of experimental spectra for the dynamic evolution of the orientation distribution, it is assumed that the reorientation is described by rotational diffusion. Previous studies have established that this is an appropriate approximation for amorphous polymers [8,9,11]. The characteristic time of rotational diffusion and its distribution are determined such that they fit the observed exchange spectra at all mixing times.

To confirm the correct implementation of the technique, we have been reproducing the poly(ethylene propylene) (PEP) spectra originally taken by Schaefer and Spiess [13]. Our spectra capture the general shape of their results (Figure 3). The reorientation angle distributions and corresponding calculated spectra that match the experiments are shown in Figure 4. These are generated using a scheme developed by

Spiess and coworkers [11]. To capture the detailed features of the spectra, we are improving our experimental protocol.

At the Congress we will present our initial results on the dynamics of PI and PVE homopolymers and each component in a 50/50 PI/PVE blend at a series of temperatures near the broad glass transition region.

Acknowledgements : This work is supported by the NSF Presidential Young Investigator (PYI) program.

4 References

1. M. Bank, J. Leffingwell, and C. Thies, Macromolecules, 4, (1971), p. 43.
2. A. Zetsche, F. Kremer, W. Jung, and H. Schulze, Polymer, 31, (1991), p. 1883.
3. C. A. Trask and C. M. Roland, Macromolecules, 22, (1989), p. 256.
4. C. M. Roland, J. Poly. Sci.: Poly. Phys. Ed., 26, (1988), p. 839.
5. J. B. Miller, K. J. McGrath, C. M. Roland, C. A. Trask, and A. N. Garroway, Macromolecules, 23, (1990), p. 4543.
6. R. H. Colby, Polymer, 30, (1989), p. 1275.
7. C. Schmidt, S. Wefing, B. Blumich, and H. W. Spiess, Chem. Phys. Lett., 84, (1971), p.130.
8. D. Schaefer, H. W. Spiess, U. W. Suter, and W. W. Fleming, Macromolecules, 23, (1990), p. 3431.
9. U. Pschorn, E. Rossler, H. Sillescu, S. Kaufmann, D. Schaefer, and H. W. Spiess, Macromolecules, 24, (1991), p. 398.
10. S. Wefing and H. W. Spiess, J. Chem. Phys., 89, (1988), p. 1219.
11. S. Wefing, S. Kaufmann, and H. W. Spiess, J. Chem. Phys., 89, (1988), p. 1234.
12. C. Schmidt, B. Blumich, and H. W. Spiess, J. Magn. Reson., 79, (1988), p. 269.
13. D. Schaefer, Dissertation, U. Mainz, (1992).

5 Figures

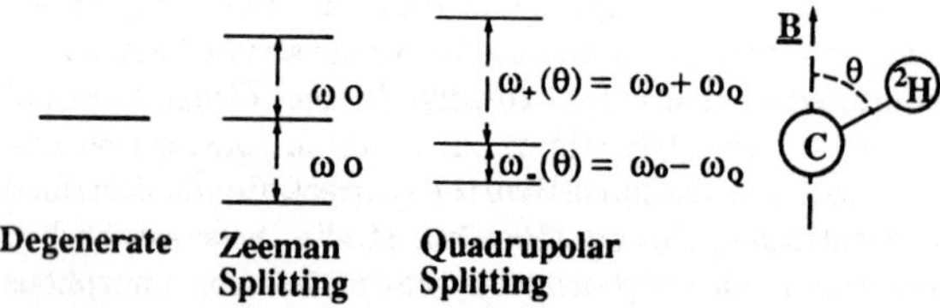

Figure 1: Energy states at various degrees of external field interaction. Zeeman splitting caused by the strong external magnetic field (**B**) is altered due to quadrupole coupling of the deuteron with the local electric field gradient. This gives rise to two resonances separated from the unperturbed frequency ω_o by $\omega_Q = \delta/2 \cdot (3\cos^2\theta - 1)$, where δ is the quadrupole coupling constant.

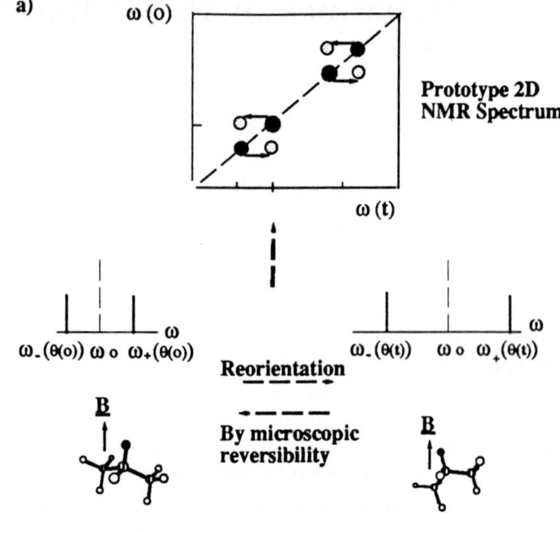

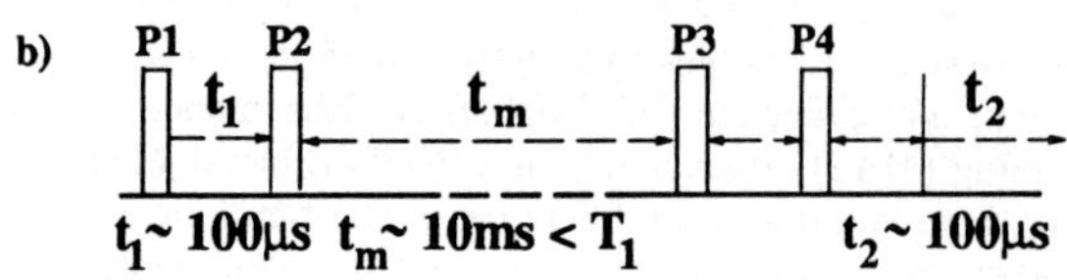

Figure 2: Schematics of the 2D deuteron exchange NMR experiment. a) A prototype 2D exchange spectrum for the case where there are only two accessible orientation angles. Reorientation during the mixing time results in the corresponding change in resonance frequency. The joint probability of resonance frequencies $\omega(\theta(0))$ *and* $\omega(\theta(t))$ is proportional to the intensity of the exchange signal at the corresponding point in the 2D spectrum. b) The four pulse deuteron exchange NMR experiment [12]. The pulses P1 and P2 create coherent spin order that persists in the time-scale of spin-lattice (T_1) relaxation time. The resonance frequency before the mixing time, ($\omega(\theta(0))$, is stored as a sinusoidal modulation in $\omega(\theta(0)) * t_1$. After the mixing time, the P3 and P4 detection sequence turns the spin order into an observable magnetization, which is modulated with frequency $\omega(\theta(t_m))$ during the detection period t_2.

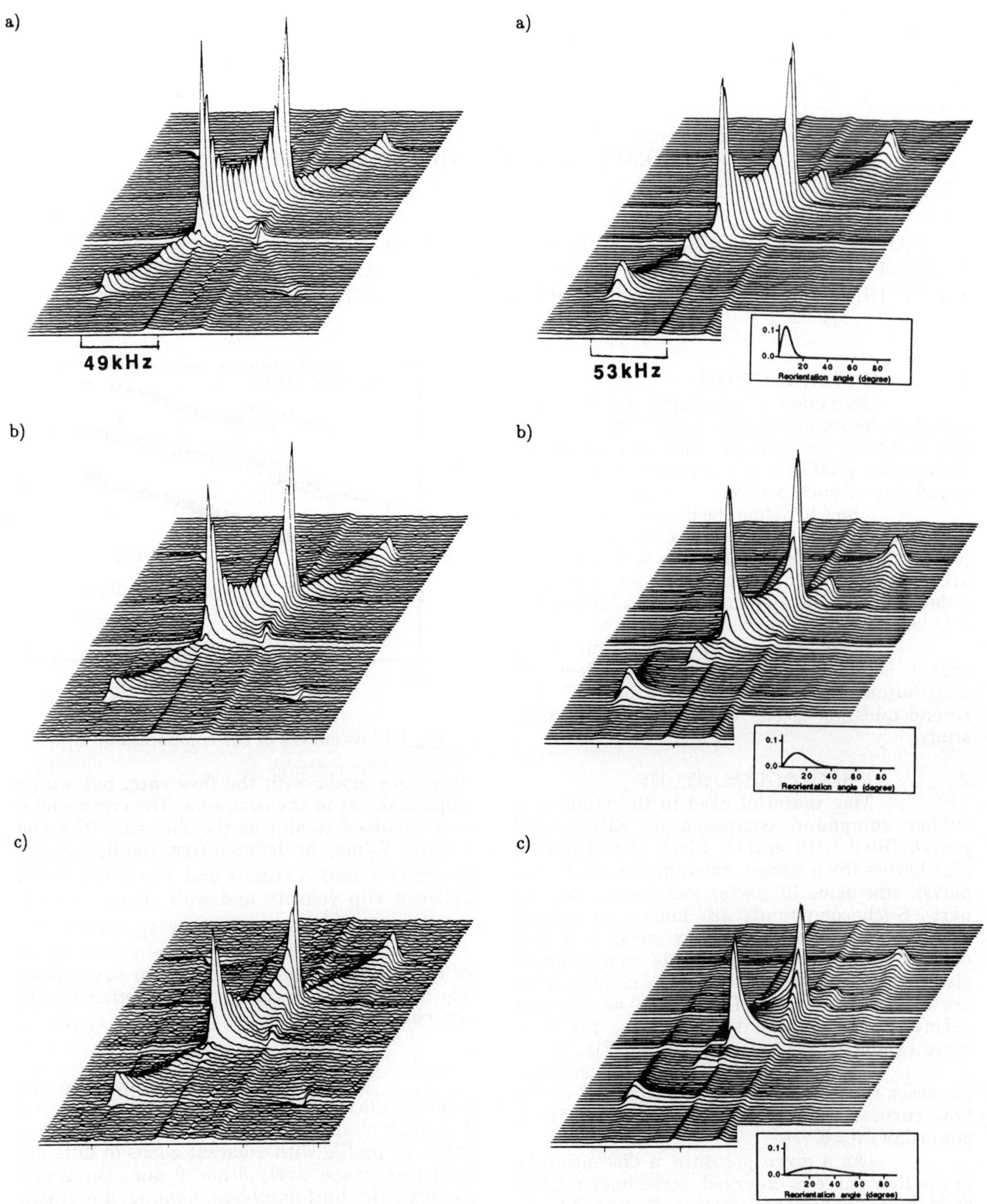

Figure 3: Experimental spectra for PEP at 225K at three different mixing times, t_m: a) 1 ms, b) 5 ms and c) 20 ms. Off-diagonal features grow with increasing mixing time as a result of progressive reorientation of the C-D bonds. These correspond to the spectra in Figure 3 if $t_c \simeq 250$ms.

Figure 4: Calculated spectra and corresponding reorientation angle distribution based on a rotational diffusion model of C-D bond reorientation for three different diffusion times $\tau = t_m/t_c$, where t_c is the characteristic time constant of reorientation. Here τ is 0.025, 0.125, 0.5 respectively.

Theoretical and Applied Rheology, edited by P. Moldenaers and R. Keunings
Proc. XIth Int. Congr. on Rheology, Brussels, Belgium, August 17-21, 1992

EXTRUSION OF ELASTOMERS IN PROFILE DIES: 3-D COMPUTATIONS AND EXPERIMENTS

S. d'HALEWYN[1], M.F. BOUBE[2]., B. VERGNES[1], and J.F. AGASSANT [1]
[1]. CEMEF. Ecole des Mines de Paris. BP 207 - 06904 SOPHIA ANTIPOLIS Cedex (FRANCE)
[2]. HUTCHINSON, Centre de Recherches, BP 31 - 45120 CHALETTE SUR LOING (FRANCE)

1. INTRODUCTION

Extruded elastomeric profiles with complex shapes are largely used in the automotive industry, for example for car door joints. Two main problems are encountered for the processing of such profiles:

- how to define the die section in order to obtain at the die exit the required shape of the profile? This problem, which needs to take into account elastic properties of the compound in order to predict die swelling, is not addressed in this work.

- how to control the flow conditions in order to obtain at the die exit a uniform flow rate distribution and then a balanced profile? This second question is the purpose of the present study.

2. RHEOLOGICAL STUDY

The material used in the study is a rubber compound composed of EPDM (130 parts), filled with carbon black (126.5 parts), plasticizers (52.5 parts), calcium carbonate (47 parts), zinc oxide (8 parts) and stearic acid (1 part). Such compounds are known to exhibit complex rheological behaviour (refs. 1-2). The detailed characterization of this compound is given elsewhere (ref. 3). Only the main results are summarized here. Depending on the die geometry and on the temperature, the flow curve exhibits three successive areas (Fig. 1).

- at low shear rates, the pressure increases with the flow rate and this part of the flow curve may be classically described by a power-law: $\eta = K \dot{\gamma}^{m-1}$.

- for a given pressure, a discontinuity in the flow curve is observed, sometimes related at high L/D values to unstable flow conditions and regular pressure oscillations. This second area, in which the flow conditions are modified, could be interpreted as a transition from sticky to wall slip boundary conditions.

- at higher shear rates, the pressure increases again with the flow rate, but with a slope lower as in the first area. The compound is now assumed to slip at the die wall. The slip velocity V_s may be deduced from the flow curves in areas 1 and 3 (ref.4) and the relationship between slip velocity and wall shear stress is expressed by another power-law: $\tau_p = \alpha K V_s{}^p$.

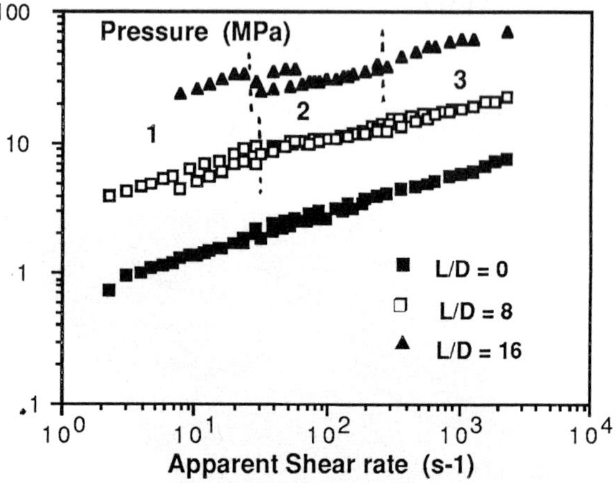

Fig.1 Flow curves of the compound at 100°C

To summarize, the behaviour of the EPDM compound is defined by two power-laws, one for the bulk viscosity and the other for the slip boundary condition.

3. EXPERIMENTS

Two profile dies were used in the experiments. Die 1 is a rectangular symmetric dumbbell-shaped die, which will provide a balanced profile with adjacent zones of different thickness, respectively 3 and 8 mm. Die 2 is a asymmetric half-dumbbell shaped die (thick zone: 8 mm, thin zone: 4 mm), leading to unbalanced flow conditions.

Flush-mounted pressure transducers located at different positions along the thin and thick zones of both dies permitted to define precisely the local flow conditions. Numerous experi- ments were carried out at different flow rates. In all circumstances, a correct profile was obtained with the dumbbell die, whereas a twisted one was produced with the half-dumbbell die. Concerning the pressure measurements, following tendencies were observed:

- the pressure was always higher in a thick zone than in a thin one.
- in the upstream part of the die land, the pressure gradient was higher in a thin zone than in a thick one.
- the pressure varied linearly in a thick zone and exponentially in a thin one.

All these results were more pronounced in the case of the dumbbell die, for which the material in the thin zone seemed to be "drag" by the two adjacent thick zones.

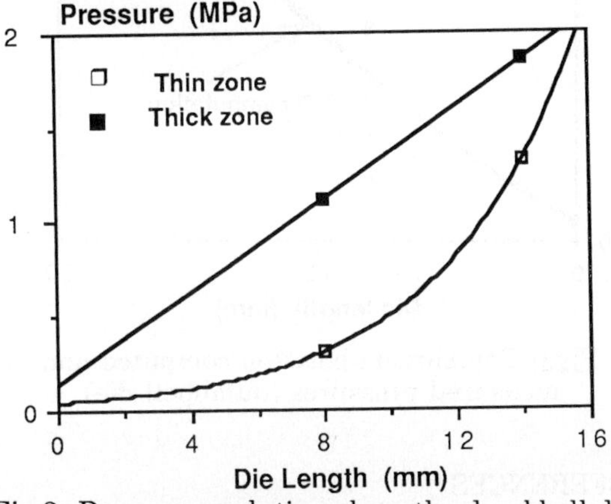

Fig.2 Pressure evolution along the dumbbell die

4. 3-D COMPUTATION

Starting from the software Forge3 (ref. 5), a finite element computation was developed in order to predict the flow of the compound through the reservoir and the profile die. The velocity field is solved by minimizing the following functional, which includes the slip behaviour and insures the incompressibility through a penalty method:

$$\Phi_\chi(\underline{u}) = \int_\Omega \frac{K}{m+1} \dot{\gamma}^{m+1} d\Omega + \int_{\partial\Omega_c} \frac{\alpha K}{p+1} \|\underline{u}\|^{p+1} dS$$

$$+ \frac{\chi}{2} \int_\Omega K (\operatorname{div} \underline{u})^2 d\Omega$$

An example of mesh is given in Fig.3 for the half-dumbbell die. Due to the symmetry of the problem only a half-die is computed.

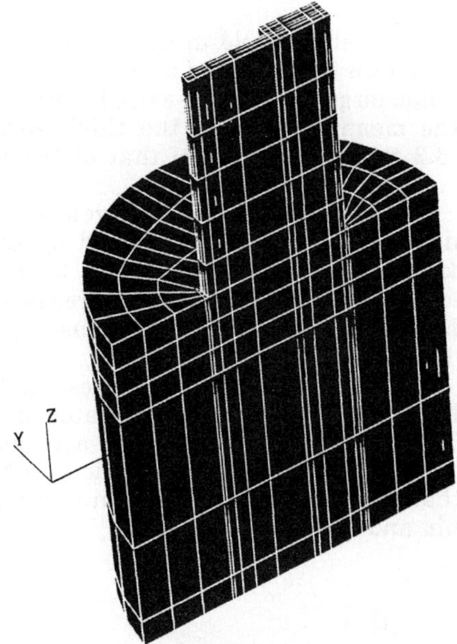

Fig.3 Mesh of the half-dumbbell die

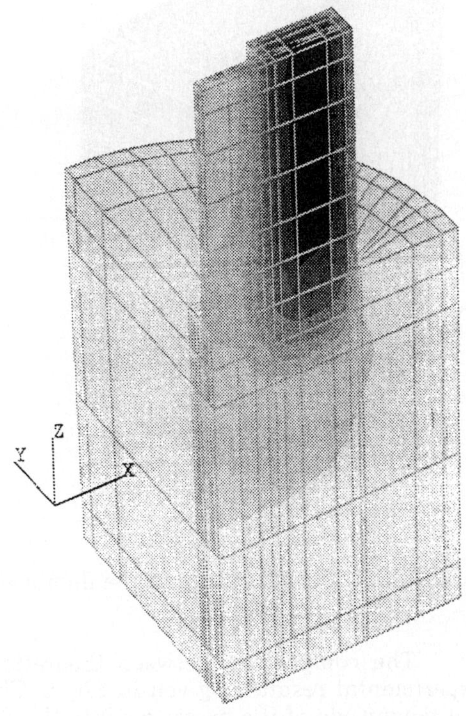

Fig.4 Isovalues of the velocity in z-direction

Depending on the mesh refinement, the CPU times are in the range 5-12 h, on a Vax Station 3520.

The velocity field in the z-direction for the dumbbell die is presented on Fig. 4. We may observe that large differences exist between the zones: the mean velocity in the thick zone is around 3.3 times higher than that of the thin zone.

The derivatives of the velocity are obtained using the method developed by Liska and Orkisz (ref. 6). The extra-stress tensor is then deduced and the hydrostatic pressure is finally derived by a mixed Galerkin-least square method (ref. 7).

The pressure field is is presented on Fig.5: the pressure drop is concentrated in the die land and the isobar lines are quite perpendicular to the flow direction, which means that the pressure gradients are identical in the thin and thick zones.

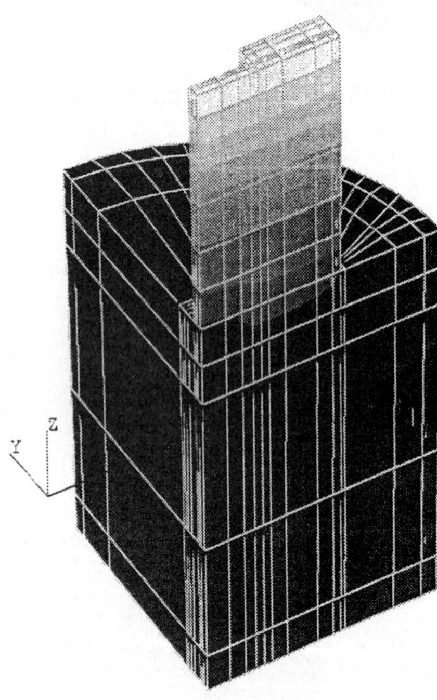

Fig.5 Isovalues of the pressure in the dumbbell die

The comparison between theoretical and experimental results is given in Fig. 6. The order of magnitude of the pressures into the die is correctly given by the computation. However, the value of the pressure into the reservoir is largely underestimated: 3.9 MPa are obtained by the 3-D computation, instead of 8.3 MPa measured. Moreover, the differences between the thin and thick zones are not described by the theoretical model.

Different modifications were tested in order to improve the theoretical computation. Introducing a slip velocity depending on the local gap thickness or taking into account the existence of a yield stress allowed to reduce the gap between theory and experiments. In fact, the main improvement should be in a much better description of the rheological behaviour of the rubber compound, including viscoelastic effects in the abrupt transition between reservoir and die.

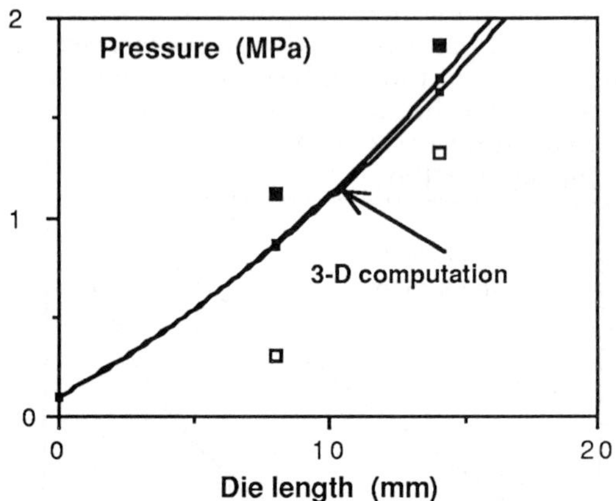

Fig.6 Comparison between computed and measured pressures (dumbbell die)

REFERENCES

1- C. Jepsen and N. Räbiger, Kautschuk Gummi Kunst., 41 (1988) 342-352

2- K. Geiger, Kautschuk Gummi Kunst., 42 (1989) 273-283

3- B. Vergnes, S.d'Halewyn and M.F. Boube, XIth International Congress on Rheology, Brussels (1992)

4- P. Mourniac, J.F. Agassant and B. Vergnes, Rheol. Acta, submitted for publication

5- J.P. Cescutti, Doctoral Thesis, Ecole des Mines de Paris (1989)

6- T. Liska and J. Orkisz, Comput. and Struct., 11 (1980) 83-95

7- B. Neyret, Doctoral Thesis, Ecole des Mines de Paris (1985)

Theoretical and Applied Rheology, edited by P. Moldenaers and R. Keunings
Proc. XIth Int. Congr. on Rheology, Brussels, Belgium, August 17-21, 1992

DYNAMIC PROPERTIES OF MODEL FILLED POLYMER MELTS IN THE LINEAR VISCOELASTIC REGION

J. V. DeGroot, Jr[1] and C. W. Macosko[1]

[1]Department of Chemical Engineering and Material Science, University of Minnesota, Minneapolis, MN 55455

1. INTRODUCTION

Two of the most industrially important reinforcing fillers are carbon black and fumed silica. Much work has been done studying the effects of these fillers on the rheological properties of suspension in polymer melts (refs. 1-4). However, one of the significant difficulties encountered in interpreting the data from these systems is determining the effective volume fraction of the filler. Since these fillers have very complicated structures, one must be concerned with the effects of occluded and bound rubber which increases the effective volume. Some work has been done on systems with model filler particles (refs. 5,6), but the problem with these is the particles tend to be fairly large (> 1 μm). At these sizes the effects of the bound rubber layer tend to be obscured. Very little study has been done on model particles of colloidal size in polymer melts.

The system considered here consists of linear polydimethylsiloxane (PDMS) which has been filled with colloidal silica particles produced by precipitation. Studies of the activity of these particles show that the strength of adsorption of the polymer on to the surface is essentially the same as that for fumed silica. The advantage of using these particles is they are spherical and have a very narrow size distribution which makes it fairly simple to estimate the effective volume fraction and amount of bound rubber.

The preliminary data from this model system indicates that strongly interacting fillers form a gel network above a critical volume fraction of filler and the networks formed appear to follow a scaling relationships above the gel point similar to those found in unfilled, chemically crosslinked systems.

2. EXPERIMENTAL PROCEDURE

2.1 Materials

The polymer used for this study was a methyl terminated linear polydimethylsiloxane viscosity fluid (M_n=67,700 , M_w/M_n~ 2) obtained from Petrarch (Huls of America) and was used as received.

The model silica particles were were formed by the method of Stober et. al (ref. 7). The size was controlled by using the empirical correlations reported by Bogush et. al (ref. 8). The particle size and size distribution were determined from SEM pictures taken on the Jeol 840 II Scanning Electron Microscope. The pictures were analyzed both by measuring by hand and using image analysis software by GrafTex. The results from both methods are in good agreement and the average particle radius is 39 nm. The ratio of weight average diameter to the number average diameter was found to be 1.02. The weight percent solids of the silica in ethanol was determined to be 3.56% .

2.2 Compounding

The polymer and filler were mixed in solution to prevent possible aggregation caused by removing the silica from solution. The polymer was weighed into a flask and dissolved in a 1/1 by volume mixture of heptane and toluene (technical grades). A mixture of solvents was used because reduced problems with flocculation and more uniform dispersions resulted. The appropriate amount of silica solution was added to achieve the desired filling level and the mixture was stirred for approximately 6 hours. The solvents were partially removed using rotary evaporation. The mixture was then transferred to a crystallization dish and placed in an oven at 60°C under vacuum (>30 inches Hg). The samples were left in the oven until a constant weight was achieved (approximately 10 days).

2.3 Rheological Characterization

After the samples reached constant weight G' and G' were measured on 50 mm diameter parallel plate fixtures with a gap of 1-2 mm (Rheometrics System IV and RMS-800: 2-2000 g-cm transducers). Samples were loaded on to the plates with care taken to avoid air entrapment. The edges were trimmed and the sample was allowed to rest for 5 hours before any testing was done to allow any residual stesses from loading to relax.

Strain sweeps were run to determine the length of the linear viscoelastic region, then frequency sweeps of the samples were done in the LVE region. The agreement of the results from both instruments confirmed the reproducibility of the sample loading procedure as well as the calibration of each instrument. Due to the sensitivity in the low frequency region of the RMS-800, only that data will be shown here.

3. RESULTS AND DISCUSSION

The results from the frequency sweeps in the LVE region are shown in Figures 1 and 2. When no filler is present we can see the terminal response of the polymer. As the filler level increases, G' and G" increase as expected. Then at some critical loading level a definite plateau appears in the data. This data has the characteristic trends observed for chemically gelling networks.

Percolation theory has been used to explain the scaling behavior of gelling systems as well as flocculated suspension in Newtonian fluids and aggregation of colloidal particles (refs. 9,10). According to the theory above the gel point the modulus should scale as

$$G \sim (P - P_c)^n$$

where G is the modulus
P is the conversion of functional groups
P_c is the conversion at the gel point
and n is the critical exponent.

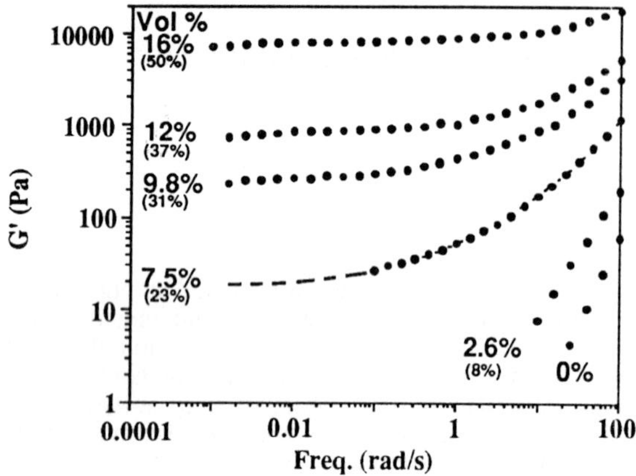

Figure 1: G' vs frequency for various vol% 39nm radius silica spheres. Strains <2% and in LVE region for all curves.

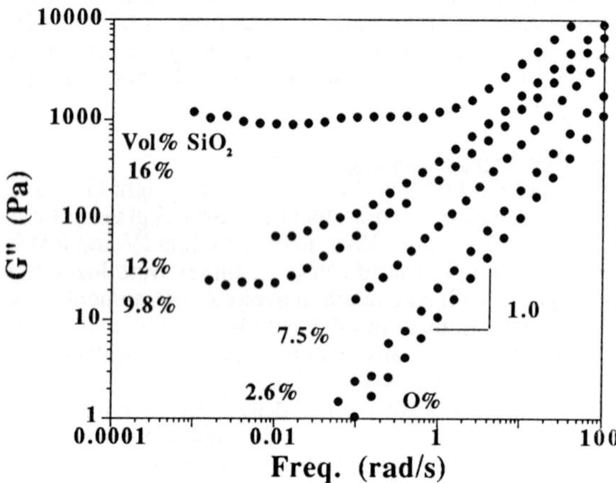

Figure 2: G" vs frequency for various vol% 39nm radius silica spheres. Strains <2% and in LVE region for all curves.

In the filled system the volume fraction of filler added to the matrix can be thought of as serving the same function as crosslinking. At some volume fraction of loading, a network of filler-polymer-filler contacts will form which spans the sample and can transmit stress (ref. 11). Since the colloidally sized silica particles used here are spherical, it is a simple task to calculate the volume occupied by the silica. However, since the polymer and filler are strongly interacting, one must consider also the contribution of the bound polymer layer to the effective volume.

The thickness of the bound layer is reported to scale as the molecular weight to the 1/2 power (ref. 12). The prefactor for the scaling has not been reported. So as a first approximation we used the correlation reported by Flory (ref. 13) for the root mean square end-to-end distance of an unperturbed coil to estimate the thickness of the bound layer. The volume percentages reported on figures 1 and 2 are for just solid silica and the values in parenthesis include the bound layer estimate. For small particles the bound layer contributes significantly to the effective volume. The critical volume fraction for percolation was estimated to be approximately 15% by a method similar to that descibed by Pouchelon (ref. 11).

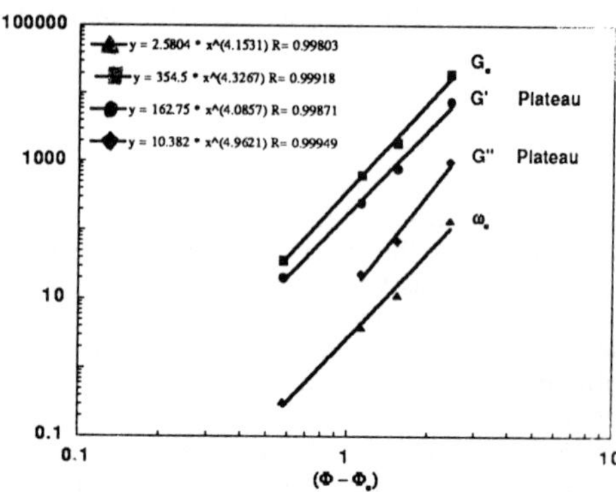

Figure 3: Plateau and crossover power law fits for suspensions of 39nm radius silica spheres in 68K M_n PDMS.

Figure 3 shows the power law fits of the low frequency G' and G" plateaus as well as the level and frequency at which G" crosses over G' as a function of ($\phi - \phi_c$). Excellent fits to the data are seen and the while the exponent for each of these curves varies slightly, the value of the exponent appears to be approximately 4.2. This exponent is in agreement with those observed by others (ref. 9, 10). The uncertainty for G" is considerable because only 3 points were obtained.

Adolf and Martin (ref. 14) report a time-cure superposition for chemically crosslinking systems and the overlap obtained by the curves is quite remarkable. In order to investigate whether we had a similar phenomenon occuring in the filled systems, the 9.8 volume % suspension was chosen as the reference and the crossover points were normalized so that they would fall on top of the 9.8 volume % crossover. Figure 4 shows the superposition of the data with the horizontal and vertical shift factors, a and b respectively. These shift factors were computed to get the best superposition. However, they can be related to the distance from the critical volume fraction using Figure 3.

The superposition of the curves is fairly good and it appears a time-filling level superposition may be valid. Some deviation in the G" plateau is noted, however this data is at the very low end of the torque transducer sensitivity and therefore there may be some problems with the resolution of tan δ in this region. Also, the high frequency region of the 7.5 vol % silica sample shows some deviation but is still in fairly good agreement with the other samples. The slope of the curves in the high frequency region is approximately 0.7-.8 which is in good agreement with the value Adolf and Martin report, 0.72, for the exponent on the frequency dependence of the crosslinked gels. However, from Figure 3, the dependance of G' is found to be 4.1. For chemically crosslinked gels the value was found to be 2.8 so the agreement of the exponent in the high frequency region may be a coincidence. More work on the calculation of the exponents is currently being done to determine whether the filled systems follow the same scaling as crosslinked gels, but the initial results indicate that the scaling is very similar.

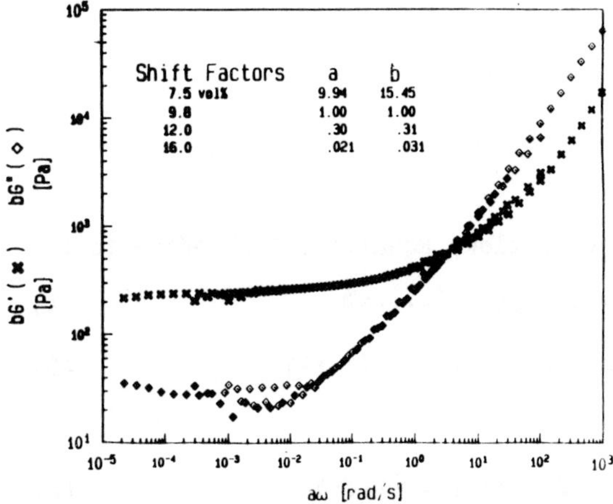

Figure 4: Superposition of Curves above ϕ_c

The question as to whether suspensions made from more complex particles such as fumed silica display similar behavior now arises. The superposition of curves in Figure 5 of data taken by Aranguren (ref. 15) on fumed silica suspensions shows that this principle appears to be valid for more complex systems. Only two curves are shown because they were the only ones to display the G" G' crossover. There is not enough data to determine the exponents for these systems.

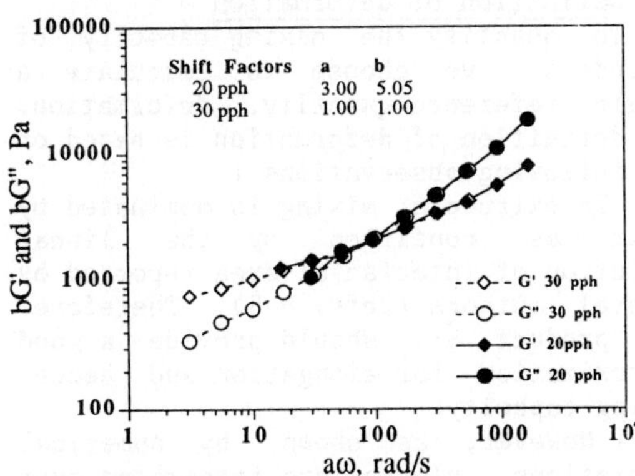

Figure 5: Superposition of fumed silica data.

It should be noted here that all of the samples were tested within one week of them attaining constant weight in the vacuum oven and all were aged at the same temperature. Some recent data indicates that there is a significant effect of aging on the modulus of the suspensions prepared by the solution method. This is believed to be due to the slow kinetics of adsorption of PDMS onto silica. The time to reach equilibrium in these systems can be quite large (ref. 16). The consequences of this aging effect on the validity of the time-filler level superposition are not well understood and this time, but they are being studied. The results from these studies will be reported at a later date.

4. CONCLUSIONS

Initial work using suspensions of model filler particles in elastomeric melts indicate that above a critical loading level a physical network of polymer-filler-particle bonds occurs and the scaling of the dynamic moduli in the LVE region of these networks appears to follow trends similar to those seen for covalently bonded (unfilled) polymer networks above the gel conversion.

Work is currently under way to evaluate the other exponents and the regions near and below the gel point to see if they also follow the findings of Adolf and Martin (ref. 14). The effects of particle size, surface treatment and polymer molecular weight are also being studied.

ACKNOWLEDGEMENTS
We would like to thank Dow Corning Corporation (Midland, MI) and the Center for Interfacial Engineering at the University of Minnesota for their financial support.

REFERENCES

1. L.E.Kosinski and J.M.Caruthers, Rheol. Acta, 25 (1986) 153-160.
2. R. S.Zieglebaur and J.M.Caruthers, J. Non-Newtonian Fluid Mech., 17 (1985) 45-68.
3. Y.G.Yanovskii, G.V.Vinogradov, and V.V.Barancheyeva, Vysokomol. soyed. A28:No. 5 (1986) 983-990.
4. M.I.Aranguren,E.Mora,J.V.DeGroot, Jr, and C.W.Macosko, submitted to J. Rheol. (1992).
5. R.E.S.Bretas and R.L.Powell, Rheol. Acta, 24 (1985) 69-74.
6. A.B.Metzner, J. Rheol., 29 (1985) 739-775.
7. W.Stober, and A.Fink, J.Colloid Interface Sci., 26 (1968) 62-69.
8. G.H.Bogush, M.A.Tracy, and C.F.Zukoski IV, J.Non-Cryst. Solids, 104 (1988) 95-106.
9. R.Navarette, Ph.D. Thesis, University of Minnesota (1991).
10. R.Buscall, P.D.A.Mills, J.W.Goodwin, and D.W.Lawson, J. Chem. Soc., Faraday Trans. 1, 84 (1988) 4249-4260.
11. A.Pouchelon, and P.Vondrecek, Rubber Chem. Technol. 62 (1989) 788-799.
12. J.P.Cohen-Addad, Polymer, 30 (1989) 1820-1823.
13. P.J.Flory, Principles of Polymer Chemistry, Cornell University Press, Ithaca, New York, 1953.
14. D.Adolf and J.E.Martin, Macromolecules, 23 (1990) 3700-3704.
15. M.I.Aranguren, Ph.D. Thesis, University of Minnesota, 1990.
16. O.Girard, and J.P.Cohen-Addad, Polymer, 32 (1991) 860-863.

Theoretical and Applied Rheology, edited by P. Moldenaers and R. Keunings
Proc. XIth Int. Congr. on Rheology, Brussels, Belgium, August 17-21, 1992

APPLICATION OF THE KINEMATIC THEORY OF MIXING TO THE DEVELOPMENT OF POLYMER BASED BLENDS.

A. FOBELETS

Solvay et Cie, Laboratoire Central, Bruxelles.

1. INTRODUCTION

Ottino has redefined various aspects of mixing in laminar flows within the frame of continuum mechanics theory (ref. 1). Ottino's analysis is completely compatible with earlier but less general descriptions of mixing in polymer processing equipment (refs. 2-5). We have applied the same fundamental principles to quantify the mixing capacity of extruders, single screw, as well as co- and counter- rotating twin-screw. The results of our evaluation are used to compare extrusion/ coumpounding results obtained on extruders that differ in size and/or in type.

2. DEFORMATION : A MEASURE OF MIXING CAPACITY

2.1 Fundamental principles

The concept of elongation constitutes the basis of our analysis. Following Ottino's formalism, the elongation of a line element in the vicinity of a point in the flow is :

$$\lambda = (\mathbf{C}:\mathbf{MM})^{1/2} \tag{1}$$

where **C** is the left Cauchy-Green tensor and **M**, the orientation for the line element subjected to stretching in the flow. The elongation rate is given by :

$$(\dot{\lambda}/\lambda) = \mathbf{D}:\mathbf{MM} \tag{2}$$

where **D** is the rate of deformation tensor (or stretching tensor) and the efficiency of mixing is defined as :

$$e = (\dot{\lambda}/\lambda)/(\mathbf{D}:\mathbf{D})^{1/2} \tag{3}$$

For the particular case of a simple Couette flow, Equations 1-3 reduce to :

$$0 \leq \lambda \leq (1 + \dot{\gamma}^2 \cdot t^2)^{1/2} \tag{4}$$

$$(\dot{\lambda}/\lambda) = \dot{\gamma}^2 \cdot t/(1 + \dot{\gamma}^2 \cdot t^2) \tag{5}$$

and

$$e = \sqrt{2} \cdot \dot{\gamma} \cdot t/(1 + \dot{\gamma}^2 \cdot t^2) \tag{6}$$

where $\dot{\gamma}$ and t are the velocity gradient and time, respectively. Equations 4-6 express the known results that, in shear dominated flows, the distance between 2 points (and similarly the amount of interfacial area) increases linearly with the product $\dot{\gamma} \cdot t$, and that both the rate of elongation and the mixing efficiency decrease as 1/t for sufficiently long times.

2.2 Definition of deformation

To quantify the mixing capacity of extruders, we choose to calculate a single reference quantity, deformation. Our definition of deformation is based on the following observations :

(1) In extruders, mixing is dominated by shear as confirmed by the linear evolution of interfacial area reported by several authors (refs. 6-8). Therefore, the product $\dot{\gamma} \cdot t$ should provide a good approximation for elongation and hence, mixing capacity.

(2) However, as shown by numerical simulations, elongations integrated over different streamlines differ by several orders of magnitude, for example, between 10^2 and 10^6 (ref. 9). Very high elongations are due to the contribution of regions in the vicinity of screw flights that are characterized by both high shear and long residence time.

Thus, defining a reference quantity for elongation, a variable ranging over several orders of magnitude does not have much physical meaning. Therefore, we calculate a reference quantity for the elongation $\dot{\gamma}\cdot t$ that is based on average values for the velocity gradient $\dot{\gamma}$:

$$Def = \Sigma_i \langle\dot{\gamma}\rangle_i \cdot t_i \quad (7)$$

where $\langle\dot{\gamma}\rangle_i$ and t_i are the average shear rate and residence time in the segment i of the screw profile. The average $\langle\dot{\gamma}\rangle_i$ is computed over the flow cross section, perpendicular to the screw flights. This method of averaging is often referred to as "mixing-cup average" in the literature (ref. 10). Though the procedure is similar, our averaging method is different from that proposed by Tadmor for the Weighted Average Total Strain (ref. 4).

2.3 Implementation

Using the fluid mechanics software POLYFLOW©, average shear rate can be computed from 2-D or 2½-D flow calculations for any screw cross section (single and twin- screws, co- and counter-rotating). For the calculation of deformation imparted by an extruder, we choose to include the contribution of the feed zone, to neglect the contribution of clearances between barrel and the top of screw flights and to run the calculations for a newtonian, isothermal fluid (the objective is to quantify the amount of kinematic mixing that an extruder can provide, NOT to predict the final state of the mixture).

Residence time calculation is based on measured or imposed flow rates when available, on the best engineering evaluation otherwise (for example, on criteria for the throttle ratio in the pumping section of the extruder (ref. 11)).

3. RESULTS

3.1 Computation of residence times

The first figure shows the correlation between calculated and measured residence times. Data include 25 experimental conditions recorded on a 25 mm, 40 L/D co-rotating twin-extruder. The 25 operating conditions are representive of an experimental design that included screw profile, screw speed, flow rate and barrel temperature as

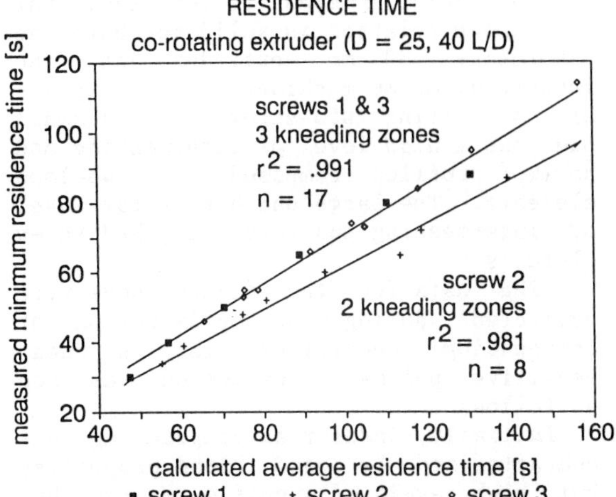

independent variables. On the abscissa, the residence time is an average value calculated from the known flow rate (an independent variable in this case). The experimental values for residence time are measured as the time necessary for colored granules to reach the outlet of the machine. It is thus a measure of the minimum residence time.

From the first figure, one can also derive the incremental residence time due to back filling of the screw upstream of a kneading zone.

3.2 Computation of deformation

On the second figure, we show the relation between deformation, our calculated estimator of mixing capacity for extruders, and measured specific energy.

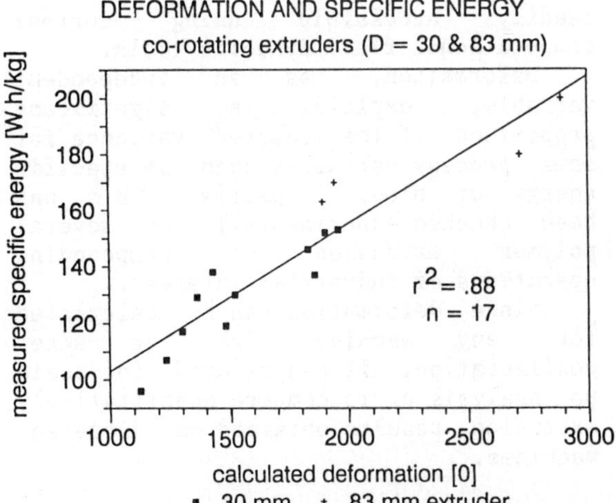

Of particular interest is the fact that the above relation is valid for both the laboratory 30 mm extruder and the industrial 83 mm machine. Both extruders are co-rotating twin-screw, but the small one has a high level of intermeshing and screw profiles composed of two-lobe elements. The large one has a low level of intermeshing and three-fligthed screw elements.

The data for the figure above were collected during the optimization of compounding conditions for a heat sensitive polymer. The method was used as follows :

- Initially, industrial compounding was characterized by good blend properties but high levels of specific energy that compromised heat stability of the compound during injection (2 points to the far right of the figure).
- Next, using experimental design procedures, the laboratory extruder was used to test if new operating conditions could decrease the specific energy while preserving optimum blend properties.
- Finally, the industrial process was modified to produce a blend with good properties and higher heat stability.
- Calculated deformation was used to extrapolate operating conditions between the two machines and to choose the most favorable levels in the experimental design on the laboratory extruder.

4. CONCLUSION

We define a parameter - deformation - which is a measure of the total shear imparted by an extruder. As defined, deformation also measures the mixing capacity of the extruder. Following our simplifying hypotheses, deformation is readily accessible using current computational fluid dynamics tools.

Deformation, as an independent variable, explains a significant proportion of the measured variance for some process variables such as specific energy or blend properties. This has been checked experimentally for several polymer extrusion or compounding operations of industrial interest.

Since deformation can be calculated for any machine size or screw configuration, it can be used for scale up analysis or to compare quantitatively extrusion results obtained on different machines.

5. REFERENCES

1. J.M. Ottino, The Kinematics of Mixing: Stretching, Chaos, and Transport, Cambridge University Press, Cambridge 1989.
2. W.D. Mohr, R.L. Saxton and C.H. Jepson, Mixing in Laminar Flow Systems, Industrial and Engineering Chemistry, 49 (1957), 1855-1856.
3. W.J. Schrenk, K.J. Cleereman and T. Alfrey, Continuous Mixing of Very Viscous Fluids in an Annular Channel, Transactions of the Society of Plastics Engineers, 3 (1963) 192-100.
4. Z. Tadmor and I. Klein, Engineering Principles of Plasticating Extrusion, Van Nostrand Reinhold, New York, 1970.
5. L. Erwin, Theory of Laminar Mixing, Polym. Engr. and Sci., 18, (1978) 1044-1048.
6. D. Bigio and S. Zerafati, Parametric Study of a 2-D Model of the Nip Region in a Counter-Rotating, Non-Intermeshing Twin-Screw Extruder, Polymer Engineering and Science, 31 (1991), 1400-1410.
7. B. David and Z. Tadmor, Laminar Mixing in Co-rotating Disk Processors, Intern. Polymer Processing 3 (1988), 38-47.
8. K.Y. Ng and L. Erwin, Experiments in Extensive Mixing in Laminar Flow. I. Simple Illustrations, Polymer Engineering and Science, 21 (1981), 212-217.
9. T. Avalosse and M. Crochet, Numerical Simulation of Mixing, published elsewhere in these proceedings of the XIth International Congress on Rheology.
10. D.E. Rosner, Transport processes in Chemically Reacting Flow Systems, Butterworths, Boston, 1986.
11. C. Rauwendael, Polymer Extrusion, Hanser Publishers, Munich, 1986.

This work is supported by the "Région de Bruxelles-Capitale" through the "Institut pour l'Encouragement de la Recherche Scientifique dans l'Industrie et l'Agriculture (I.R.S.I.A.)".

Theoretical and Applied Rheology, edited by P. Moldenaers and R. Keunings
Proc. XIth Int. Congr. on Rheology, Brussels, Belgium, August 17-21, 1992

SWELL AND SAG DURING THE PARISON FORMATION STAGE OF THE BLOW MOULDING PROCESS AND ITS RELATIONSHIP WITH WALL THICKNESS DISTRIBUTION IN THE FINAL OBJECT

A. GARCIA-REJON AND R.W. DiRADDO

National Research Council of Canada
75 De Mortagne Blvd, Boucherville, Québec J4B 6Y4 CANADA

INTRODUCTION

In the last thirty years many attempts have been made to study parison behaviour and its relationship to basic rheological properties. These investigations can be classified in three groups: 1) swell behaviour of extruded polymer melts from simple annular dies under controlled experimental conditions [1]; 2) numerical simulation of swell behaviour for a number of constitutive equations and simple cylindrical and annular dies [2]; and 3) parison behaviour of polymers extruded on actual blow moulding machines [3]. The latter studies have been centered on developing techniques to quantify the effects of processing parameters and different resin characteristics on swell and sag. Of particular interest are the recent non-contact, on-line techniques developed by Swan *et al* [4,5] and DiRaddo *et al* [6]. These techniques together with the relevant rheological properties of the material, the inflation dynamics and the properties of the final part can be used to optimize the programming of the parison.

The final characteristics such as thickness profiles and mechanical properties of a blow moulded part depend strongly on the material behaviour during the extrusion of the parison and its subsequent inflation and cooling in the mould. Parison flow is governed by two distinct phenomena, swell and sag. These phenomena are coupled and by their nature oppose each other. Their relative importance determines the dimensions of the parison and consequently the properties of the final part. Conservation of mass dictates that parison thickness profiles directly affect final part thickness distributions after inflation. Diameter profiles also have an effect on the blow-up ratio and therefore on the strain given to the inflating part.

This work presents some preliminary results of a study geared towards the understanding of parison behaviour during extrusion as related to the final characteristics of the blown part.

EXPERIMENTAL SET-UP

The experimental set-up consisted of an industrial blow moulding machine (Battenfeld-Fisher FBZ 1000 equipped with a MACO 8000 parison programmer). The parison was extruded (at constant screw RPM) through a converging die having an outer diameter of 1.7 cm and a flush gap of 1.0 mm. A CCD (Charged Coupled Device) video camera was employed to film the parison behaviour during extrusion. Total parison length as a function of time and diameter profiles at the end of extrusion were measured from a calibrated monitor. The parison melt temperature near the die and at the bottom of the extrudate were measured with a Kane-May infrared pyrometer. These data were obtained in order to estimate the cooling effects during extrusion into ambient air.

A 1.5 l bottle mould was used in this study. The bottle thickness distribution was measured at selected points along the arc length of the bottle starting at the neck and ending at the bottom of the part.

MATERIALS

Three high-density polyethylene (HDPE) blends were used in this study. Resin A is a typical blow moulding resin. Resins B and C have 40 and 20% by weight of Resin A respectively. The balance is made up of a low molecular weight HDPE (injection moulding grade). For details regarding the blending process the reader is referred to Reference [4]. These materials were chosen for this study because they exhibit a broad range of rheological properties.

The stress growth function in uniaxial extension, $\eta_T^+(t)$ (Figure 1) was measured in a Rheometrics Extensional Rheometer (RER). The diameter and thickness swells (equilibrium values in the absence of sag and isothermal conditions) were measured on a converging die following the technique described by Swan *et al* [4]. Table 1 gives a summary of some relevant properties of these blends.

TABLE 1: HDPE Characteristics

	RESIN A	RESIN B	RESIN C
Polydispersity (M_w/M_n)	10.3	6.0	4.6
Solid density (g/cm^3)	0.945	0.954	0.957
Shear viscosity (Pa - s) @ 0.1 s^{-1}	40 000	6 500	2 600
@ 100 s^{-1} @ 190°C	1 500	700	520
Diameter swell*	1.7	2.15	1.98
Thickness swell*	1.85	2.62	2.30

* ultimate swell and in the absence of sag (190°C)

RESULTS AND DISCUSSION

From the rheological characteristics given in Table 1, Resin B exhibits the highest equilibrium swell in both diameter and thickness followed by Resins C and A respectively. The occurrence of the maximum swell values for Resin B has been explained elsewhere [6]. On the other hand the stress growth function $\eta_T^+(t)$ (Figure 1) indicates that Resin C has a higher tendency to sag followed by Resins B and A.

Figure 2 represents the length vs time behaviour of the extruded parison. The length was monitored up to the point in which the bottom of the parison disappeared from the camera's field of view (approx. 35 cm). Due to the fact that screw RPM was maintained constant, the mass flowrate of the three resins was different ($W_C > W_B >$. W_A). In order to obtain a constant length parison (Lp = 30 cm) at the moment of mould closing, the extrusion time was adjusted accordingly (7 to 10 sec). As could be inferred from the $\eta_T^+(t)$ values Resin C exhibits the highest sag. It should be mentioned that the elongational strain rates encountered during the extrusion of the parison fall within the range of strain rates measured in the RER.

Within experimental error Resins A, B and C have the same diameter profile curve. The diameter increases sharply in the region closest to the die and reaches a plateau half way between the die exit and the tip of the parison. Resin A sags the least and its extrusion time is the longest. Therefore, the material has more time to swell unaffected by the sagging phenomenon. On the other hand, the combination of greater sag potential, shorter extrusion times and higher equilibrium swell values reduces the parison diameter profile of Resins B and C to that of Resin A.

The thickness profiles of the parison are given in Figure 3. The local thickness values are computed from L_p vs t and D_p vs L curves using the following expression [6]

$$h_p(i)\,[D_p(i) - h_p(i)] = \frac{w t_x(i)}{\pi \rho(T)\, L\rho(i)} \qquad (1)$$

where $h_p(i)$ is the segment thickness ; Dp(i) is the segment diameter; tx(i) is the segment hang-time; $L_p(i)$ is the segment length; W is the segment mass flowrate; and $\rho(T)$ is a temperature dependent melt density.

Resins A and B show a similar thickness profile with Resin B having a higher thickness overall. It is interesting to note that the lower end of the parison (region under no extensional stresses) exhibits a thickness swell close to its ultimate value. Resin C having a greater sag potential shows an important thickness decrease in the middle section of the parison. The bottom portion of the parison will not sag while the top portion being the last extruded does not have the time to sag significantly. The same parisons described in Figures 2 and 3 were blown to the same final bottle. Figure 4 shows the thickness distribution along the bottles made out of Resins A and B. It was not possible to form a bottle from Resin C due to rupture of the parison. Inflation visualisation experiments using a transparent mould have confirmed the fact that the rupture occurs at the parison mid-section where it is the thinnest.

At mould closing the parison is squeezed both at the top and at the bottom. Therefore the part of the parison actually being blown extends between 6 and 30 cm from the die exit. Once inflated into the final part, Resin B, as expected, has a larger thickness than Resin A in the top and main body of the bottle. The minimum thickness corresponds to the bottom corner of the mould (arc length ~ 28 cm). However, this behaviour is reversed in the bottom of the bottle. The pinching action at the bottom can affect considerably the local dimensions of the parison. Resin B being less viscous might be squeezed more than Resin A therefore reducing both diameter and thickness at the pinch-off. From visualisation experiments the inflation patterns for both resins are similar in the main body of the bottle. However, the inflation patterns to fill the bottom corners are quite different. It is in this region where the material is stretched the most and the deformation is typically biaxial.

CONCLUSIONS

The flow behaviour of three HDPE blends during extrusion and subsequent inflation has been studied. The results show that parison thickness distribution is the parameter that relates best with the thickness distribution in the final object. Furthermore the stress growth function in uniaxial extension appears to be the material parameter best suited to predict final bottle characteristics.

Acknowledgements

The authors wish to thank Mr. Daniel Poirier for his contributions in developing the visualisation techniques and for carrying out the experimental program.

REFERENCES

1. J.M. Dealy, K.F. Wissbrun, Melt Rheology and its Role in Plastics Processing, Van Nostrand, NY (1990).
2. A. Polinski, J. Tsamopoulos, AICHE Journal, 36 (1990), 1837-1850.
3. R.W. DiRaddo, W.I. Patterson, M.R. Kamal, Intem. Polymer Processing, 6, (1991) 271-224.
4. P.L. Swan, J.M. Dealy, A. Garcia-Rejon, A. Derdouri, Polym. Eng. Sci., 31 (1991) 705-710.
5. P.L. Swan, P. Cielo, A. Garcia-Rejon, M.R. Kamal, Industrial Automation Conference Proceedings, (1992) Montreal, June 1-3.
6. R.W. DiRaddo, A. Garcia-Rejon, Polym. Eng. Sci. (in press), (1992).

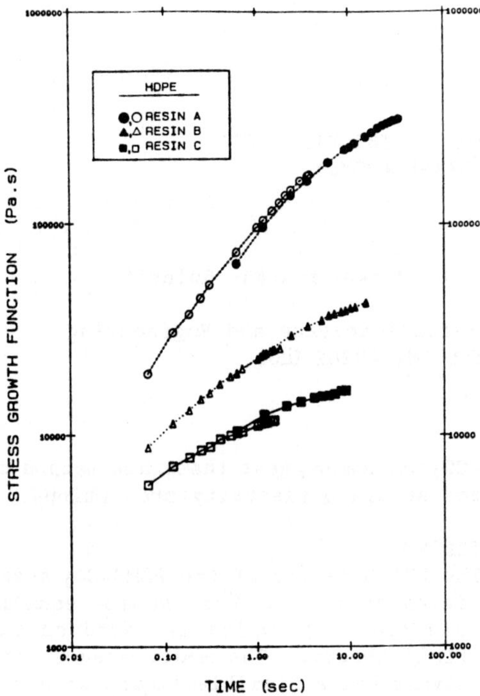

Figure 1: Stress Growth Function
(Strain rate = 0,05 - 1 $^{1}/s$

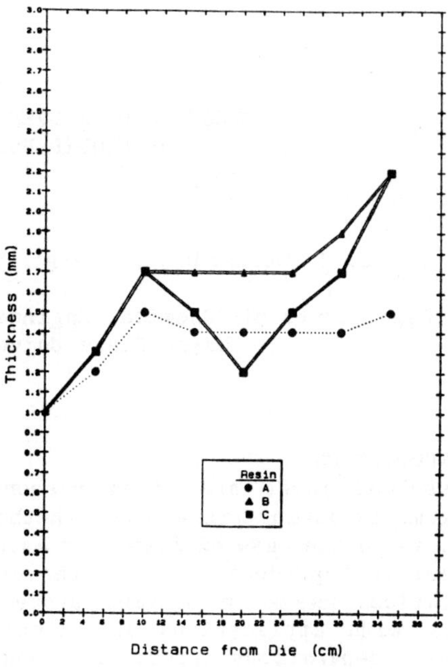

Figure 3: Parison Thickness Distribution

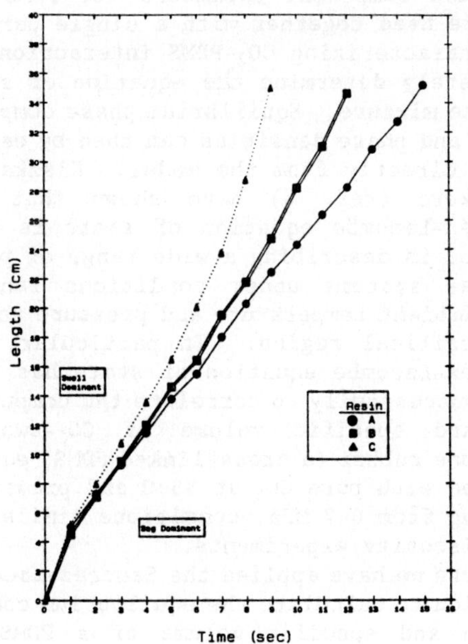

Figure 2: Parison length as a function of time

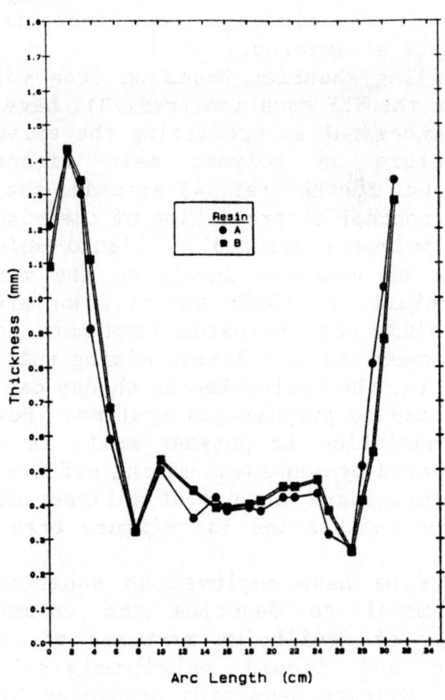

Figure 4: Wall Thickness Distribution in the Moulded Bottle

Theoretical and Applied Rheology, edited by P. Moldenaers and R. Keunings
Proc. XIth Int. Congr. on Rheology, Brussels, Belgium, August 17-21, 1992

EFFECT OF DISSOLVED CARBON DIOXIDE ON THE VISCOSITY OF LIQUID POLYDIMETHYLSILOXANE (PDMS)

L. J. Gerhardt[1], A. Garg[1], Y. C. Bae[1], C. W. Manke[2] and Es. Gulari[1]

Departments of [1]Chemical Engineering and [2]Materials Science and Engineering
Wayne State University, Detroit, MI 48202 USA

1. INTRODUCTION

Dissolved gases are often present in polymer melts during processing, whether by design, as in the case of foams, or as evolved volatile byproducts. Also, the use of supercritical gases in polymer systems is emerging with applications in paints and coatings, separations (ref. 1), and in composite fiber impregnation (ref. 2). In all of these applications, the presence of dissolved gas can be expected to modify the viscosity of the polymer melt. Suitable theoretical methods for predicting the effect of dissolved gas on viscosity are therefore of interest.

Scaling theories based on free volume, such as the WLF equation (ref. 3), have been very successful in predicting the effect of temperature on polymer melt viscosity. Kelley and Bueche (ref. 4) extended the free volume concept to prediction of the viscosity of polymers diluted by liquid solvents through an equation involving the mixture free volume. In their theory, the mixture free volume was evaluated from pure component properties by a linear mixing rule. In principle, the Kelley-Bueche theory can also be applied to polymer-gas systems. However gas dissolution in polymer melts is often accompanied by non-ideal mixing effects that would invalidate the use of a linear mixing rule for calculating the mixture free volume.

Here we have employed an equation-of-state model to describe the volumetric behavior of equilbrium mixtures of carbon dioxide and liquid polydimethylsiloxane (PDMS). Mixture densities predicted by the equation of state are then used to calculate the mixture free volume in the Kelley-Bueche equation directly, thereby eliminating the need for a mixing rule. Comparison of these predictions with viscosity data for the PDMS-CO_2 system suggest that this method can produce accurate viscosity predictions.

2. THEORY

The PVT behavior of the PDMS-CO_2 system is modeled by the Sanchez-Lacombe equation of state (ref. 5), which is based on lattice-fluid theory. The model assumes that the polymer has a flexible liquid structure and occupies r lattice sites. The equation of state relates reduced pressure $\tilde{P}$, temperature $\tilde{T}$, and density $\tilde{\rho}$:

$$\tilde{\rho}^2+\tilde{P}+\tilde{T}[\ln(1-\tilde{\rho})+(1-\frac{1}{r})\tilde{\rho}]=0 \qquad (1)$$

Pure component paramters for PDMS and CO_2 are used together with a single parameter characterizing CO_2-PDMS interactions to completely determine the equation of state for the mixture. Equilibrium phase compositions and phase densities can then be calculated directly from the model. Kiszka and coworkers (ref. 6) have shown that the Sanchez-Lacombe equation of state is successful in describing a wide range of polymer-gas systems under conditions ranging from ambient temperature and pressure to the supercritical region. In particular, the Sanchez-Lacombe equation of state has been used successfully to correlate the composition and specific volume of CO_2-swollen silicone rubber (a cross-linked PDMS) equilibrated with pure CO_2 at 35 C and pressures ranging from 0-7 MPa, conditions similar to our viscosity experiments.

Here we have applied the Sanchez-Lacombe equation to calculate the equilibrium composition and specific volume of a PDMS-CO_2 lower phase in equilibrium with pure CO_2 at 25 C and at pressures ranging from 0-3 MPa, corresponding to our viscosity data. Predicted values for the specific volume of the

$PDMS-CO_2$ lower phase, taken from the Sanchez-Lacombe equation of state, are then used in the Kelley-Bueche equation to predict the effect of dissolved CO_2 on the mixture viscosity.

In terms of free volume, Kelley and Bueche (ref. 4) express the viscosity of a polymer diluted by solvent as:

$$\eta = BC_p^4 \exp\left(\frac{1}{f}\right) \tag{2}$$

where B is a constant, C_p is the polymer concentration. The fractional free volume f is given by:

$$f = \frac{(v - v_o)}{v} \tag{3}$$

where v is the specific volume and v_o is the occupied specific volume. Kelley and Bueche continued the development of their theory by evaluating f for the mixture from the free volume of the pure components by employing a mixing rule that was linear in the volume fraction of the constituents. Their mixing rule gave successful viscosity predictions for liquid solvents, where mixing is nearly ideal. In polymer-gas systems, however, mixing often departs strongly from ideality, and negative values for the excess volume of mixing are typical at low to moderate pressures. For this reason, we extend the Kelly-Bueche theory to polymer-gas mixtures by employing eqs. 2-3, but evaluate the mixture specific volume v directly with the Sanchez-Lacombe equation of state, thereby avoiding the mixing rule.

3. RESULTS

We first applied eqs. 2-3, together with the equation of state model, to pure PDMS and determined the constants B and v_o by an optimized fit to viscosity of pure PDMS measured as a function of temperature. Viscosity data are shown in fig. 1 for two PDMS grades (M=308K and M=204K) at temperatures ranging from 10 C to 70 C. Using optimized values for B and v_o, the combined Kelley-Bueche/equation of state (KBEOS) model is in excellent agreement with the experimental data. The value v_o=0.79 cm^3/g obtained from the data fit is very close to values of 0.82-0.85 cm^3/g reported for polystyrene (ref. 7).

The success of the KBEOS model in describing the temperature dependence of pure PDMS is encouraging because it indicates that the Sanchez-Lacombe equation of state accurately represents the temperature dependence of specific volume for pure PDMS.

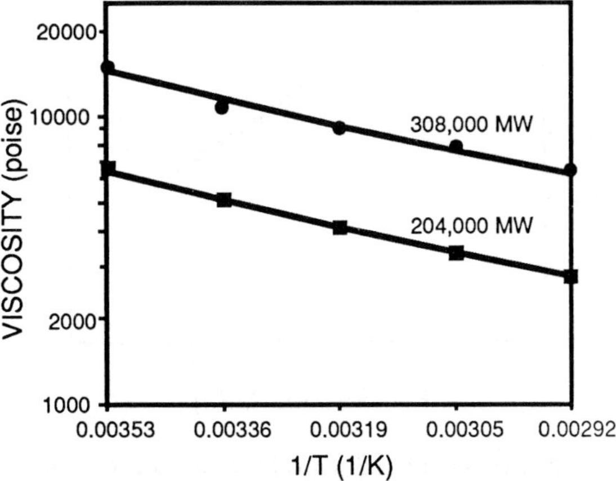

Fig. 1 Variation of pure PDMS viscosity with temperature. Solid curves represent the KBEOS predictions. Data (symbols) were taken with a Carri-Med controlled stress rheometer.

The KBEOS model was then applied to $PDMS-CO_2$ mixtures equilibrated with pure CO_2 at 25 C and pressures ranging from 0-3 MPa. The Sanchez-Lacombe equation was used to predict both the composition of the CO_2-swollen lower phase and its specific volume v. The occupied specific volume of the mixture, needed in eq. 3, was then calculated from pure component occupied specific volumes by assuming a linear mixing rule:

$$V_o = W_{PDMS} V_o^{PDMS} + W_{CO2} V_o^{CO2} \tag{4}$$

where the weight fractions of the components w_{PDMS} and w_{CO2} were obtained from the predicted equilibrium lower phase composition. Since the occupied volume represents an incompressible component of specific volume, the linear mixing rule assumed in eq. 4 should be a very good approximation. The viscosity predictions of the KBEOS model are compared to viscosity measurements performed by Bae (ref. 8) in fig. 2. Again, the agreement of the KBEOS predictions with the experimental data is remarkably good.

The predicted excess volume of mixing, $v^E = v - w_{CO2}v_{CO2} - w_{PDMS}v_{PDMS}$, shown in fig. 3, illustrates the extent of nonideal behavior in equilibrium $PDMS-CO_2$ mixtures, especially at low pressures. While representation of this behavior emerges naturally from the equation of state model, such large deviations from ideal mixing would be very difficult to accommodate with free volume mixing rules.

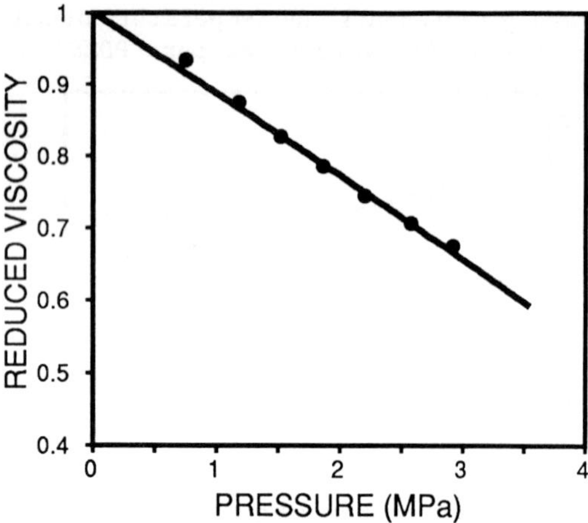

Fig. 2 Variation of equilibrium lower phase PDMS(M=204K)-CO_2 viscosity with pressure. Solid curve is the KBEOS prediction. Data are taken from Bae (ref. 8). Viscosity values are reduced by the viscosity of pure PDMS.

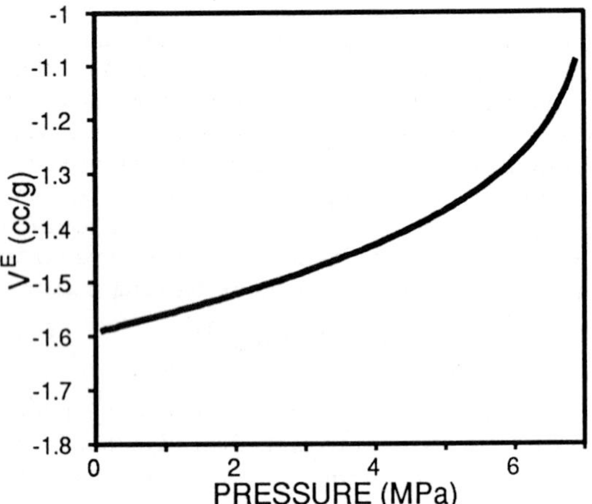

Fig. 3 Variation of excess volume of mixing with pressure for equilibrium PDMS(M=204K)-CO_2 mixtures, as predicted by KBEOS.

The large negative values of v^E exhibited at low pressures limit the extent to which dissolved CO_2 can reduce the mixture viscosity by added free volume. However as pressure is increased to about 6 MPa the magnitude of v^E begins to decrease sharply. The corresponding effect on mixture free volume is shown in fig. 4. Free volume begins to increase at about 6 MPa, reflecting both high CO_2 solubility and more moderate values for v^E. Unfortunately, this region is beyond the pressure range of our present viscosity data.

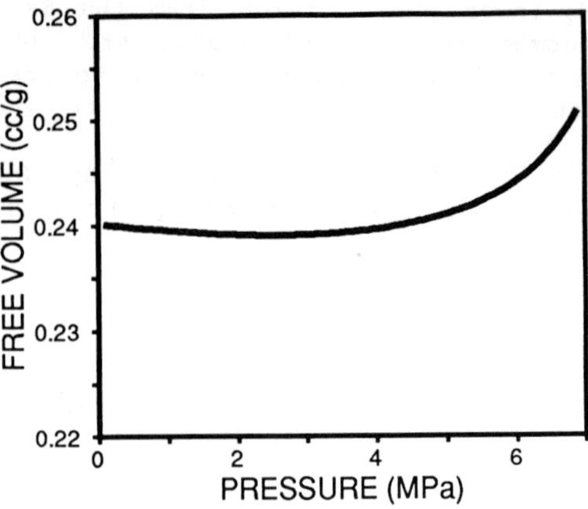

Fig. 4 Variation of free volume with pressure for equilibrium PDMS(M=204K)-CO_2 mixtures, as predicted by KBEOS.

However, we have performed CO_2 solubility measurements at pressures up to 26 MPa (ref. 2), which are well within the supercritical region. Under supercritical conditions, CO_2 is highly soluble in PDMS (25 wt% at 26 MPa) and the large magnitude of v^E observed at lower pressures is greatly reduced. Both effects should lead to large reductions in viscosity according to eq. 2. New experiments are in progress employing a specially adapted capillary rheometer that will extend our viscosity data for equilibrium PDMS-CO_2 mixtures into the supercritical region. These data will probe the effects of free volume introduced by dissolved gas, and they will provide a more stringent test of the method for viscosity prediction presented here.

REFERENCES

1. M. Daneshvar and E. Gulari, J. Supercritical Fluids, (1992) in press.
2. A. Garg, L. Gerhardt, E. Gulari, and C. Manke, in M. McHugh (Ed.), 2nd International Symposium on Supercritical Fluids, Boston, MA, May 20-22, Johns Hopkins Univ., Baltimore, 1991, pp 233-236.
3. M. Williams, R. Landel, and J. Ferry, J. Am. Chem. Soc., 77 (1955) 3701.
4. F. Kelley and F. Bueche, J. Polym. Sci., 50 (1961) 549.
5. I. Sanchez and R. Lacombe, Macromolecules, 11 (1978) 1145-1156.
6. M. Kiszka, M. Meilchen, and M. McHugh, J. Appl. Polym. Sci., 36 (1988) 583-597.
7. R. Haward, J. Macromol. Sci.-Revs. Macromol. Chem. C4(2) (1970) 191-242.
8. Y. C. Bae, Ph.D Dissertation, Wayne State University, 1989.

Theoretical and Applied Rheology, edited by P. Moldenaers and R. Keunings
Proc. XIth Int. Congr. on Rheology, Brussels, Belgium, August 17-21, 1992
1992 Elsevier Science Publishers B.V.

Simulating the Mixing of Polymer Blends Using the Boundary Element Method

Paul J. Gramann, J. Christoph Mätzig and Tim A. Osswald
Polymer Processing Research Group, Department of Mechanical Engineering
University of Wisconsin-Madison, Madison, WI 53706, U.S.A.

INTRODUCTION

The mixing of filled and unfilled polymer blends is an important issue in the plastics industry. The quality of the finished product in almost all polymer processes stems back in part to how well the material was mixed. Processing difficulties with polymer blends have been encountered in the mixing quality as well as in the thermal degradation due to viscous heating. A better understanding of the mixing process and control of the viscous heating during mixing will help us achieve optimum processing conditions and increase the quality of the final part. Quantifying the mixing inside an extruder or an internal mixer and predicting the thermal degradation due to viscous heating is an extremely difficult task. Various researchers have simulated mixing processes using the finite difference technique and the finite element method [1-4]; none included viscous dissipation. From the literature one can see that the moving boundary nature of mixing processes makes the finite element method cumbersome to use. This is due to the difficulty encountered when re-organizing the finite difference grid and finite element mesh to rapidly fit different mixer geometries or after every consecutive time step when the computational domain changes shape. This paper presents the boundary integrals needed to solve mixing and heat transfer effects when processing Newtonian fluids. The simulation is compared to realistic mixing processes which serves to demonstrate some of the basic phenomena that occur during the mixing of filled and unfilled polymer compounds.

BOUNDARY INTEGRAL EQUATIONS

The flow equations were derived assuming an incompressible, Newtonian fluid where the viscous forces are much larger than the inertia effects. The boundary conditions are known tractions, t_α, or velocities, U_α. The following boundary integral equation for Newtonian, isothermal creeping flows results from a weighted residual statement that satisfies the momentum and mass balance equations [5]

$$c_i U_k^i + \int_\Gamma [\, t_\kappa^\alpha U_\alpha - t_\alpha U_\kappa^\alpha \,]\, d\Gamma = 0 \qquad (1)$$

In Eq. (1) c_i is unity inside the domain and 1/2 on the boundary, t_κ^α and U_κ^α are fundamental solutions which represent velocities and tractions in an infinite domain, caused by concentrated forces acting in the two perpendicular directions on a singular point "i" [6]. Since the integrals in Eq. (1) cannot be evaluated analytically, the boundary is discretized into a finite number of elements. Figure 1 shows a typical boundary element discretization of a Banbury type mixer. In discretized form, Eq. (1) becomes a system of independent linear equations which can be solved for the unknown velocities and tractions on each element. Internal values of velocity, strain rates and stresses can be determined by substituting the U_α's and t_α's back into Eq.(1) and moving the source point "i"

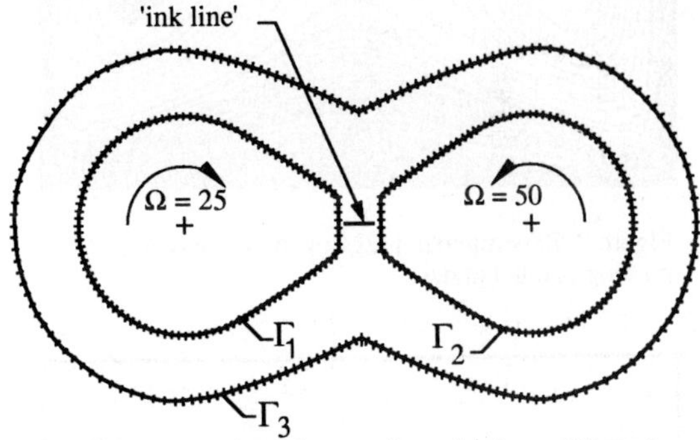

Figure 1. Boundary element discretization and boundary conditions of the Banbury type mixer.

to the location where the internal value is sought. A more in depth derivation of the equations can be found in [6]. To compute the heat transfer during typical mixing processes an energy balance for isotropic cases was used with boundary conditions of known temperature, T, or the heat flow, q. A boundary integral equation that satisfies the energy balance and meets the required boundary conditions was derived by applying the dual reciprocity method discussed in [7] and [8] .

$$c_i T_i - \int_\Gamma T_i^* q d\Gamma + \int_\Gamma q_i^* T d\Gamma =$$

$$\sum_{j=1}^{N+L} \alpha_i \left(\int_\Gamma T_i^* \hat{q}_j d\Gamma - \int_\Gamma q_i^* \hat{T}_j d\Gamma - c_i \hat{T}_j \right) \quad (2)$$

where T* and q* are fundamental solutions which represent temperature and heat flow caused by a concentrated heat source on point "i" and the terms in the right hand side result from the dual reciprocity method and represent internal heat generation, transient effects and non-linearities [8].

RESULTS AND DISCUSSION

Flow Between moving parallel plates
The first process simulated shows the complex mixing phenomena that occur in an experimental set-up found in the literature [12]. Figure 2 shows the complex deformation a diagonal tracer line undergoes when opposite sides of a rectangular cavity move at equal speeds in opposite direction. Figure 3 shows that the simulated pattern agrees very well with the experiment. In the simulation, the tracer line was represented with 950 equidistant tracer points.

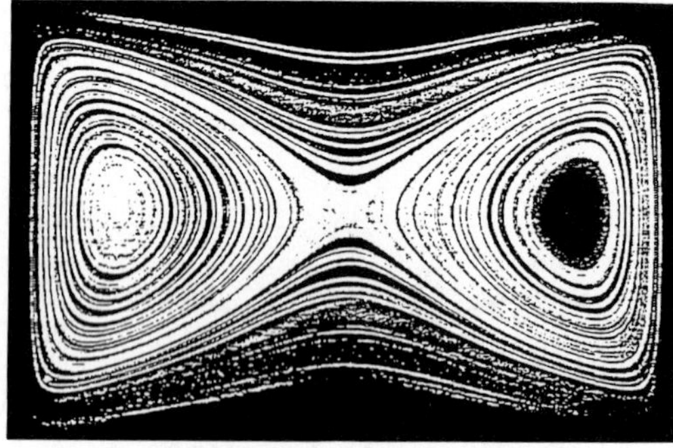

Figure 2. Experimental [12] flow between two moving parallel plates.

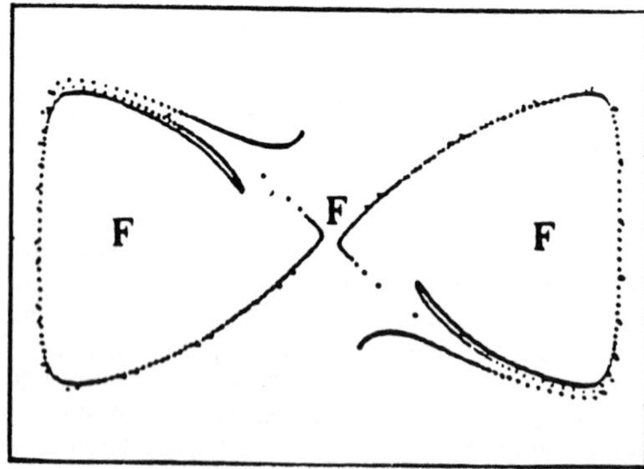

Figure 3. Simulated flow between two moving parallel plates.

Flow Inside a Banbury Type Mixer
The second process simulated with our boundary element program was the mixing inside a Banbury type mixer. The boundary element discretization and the boundary conditions used in our computation are shown in Fig. 1. The simulation assumes that the chamber of the mixer is completely filled and also neglects three dimensional flow characteristics. The right rotor rotates counter-clockwise at twice the speed as the left rotor which rotates in a clockwise manner. To help visualize the mixing that occurred, an "ink" line made up of 198 points was placed between the two rotors, as shown in Fig. 1. As the process advanced, the points moved with the fluid. A computer simulated flow of the mixer is shown in Fig. 4. The final location of the tracer points shows the complex recirculation pattern that occurred inside the mixing chamber, caused by combinations of drag and pressure flows. For each tracer point a strain, vorticity and stress history can be computed to quantify the mixing inside the chamber [6].

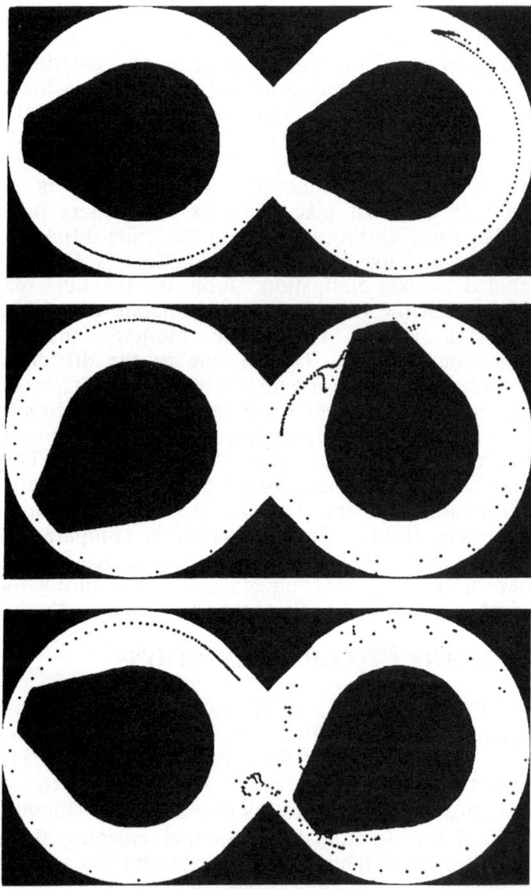

Figure 4. Simulated flow of the Banbury type mixer.

Flow and Heat Transfer Inside an Eccentric Cylinder

The next step within the project was to combine the flow and heat transfer algorithms and simulate more complicated flow problems, where the velocities and the velocity gradients cannot be computed analytically. The flow and heat transfer inside eccentric cylinders was analyzed. The diameter of the inner and outer cylinders were 1 and 3 cm, respectively. The inner cylinder rotated clockwise at 5 rad/s. Figure 5 displays the geometry and resulting streamlines of this process. These agree well with values from the literature [13]. The boundaries were kept at a constant temperature of 0°C on both cylinder walls. Figure 6 displays the temperature profile in the narrow gap between the spinning cylinder and the fixed cylinder. The maximum temperature is slightly closer to the moving boundary which agrees with the higher velocity gradient in this region. The temperature history of a particle moving on the streamline marked with "S" in Fig. 5 is shown in Fig. 7. The particle heats up while it is traveling along the streamline due to viscous dissipation and the minimum temperature occurs in the narrow gap due to the streamline's proximity to the cooled wall.

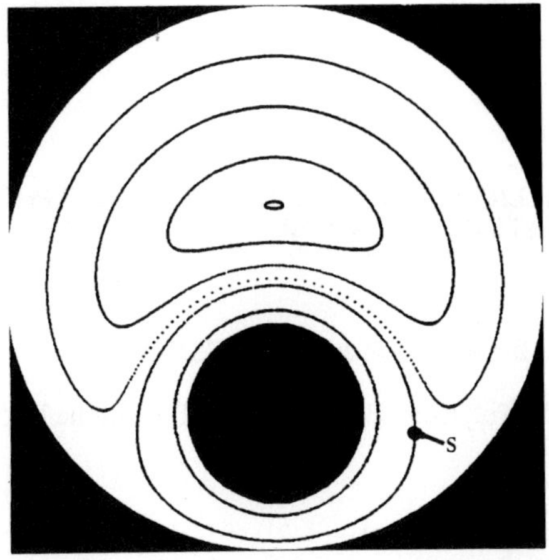

Figure 5. Simulated streamlines of eccentric rotating cylinders.

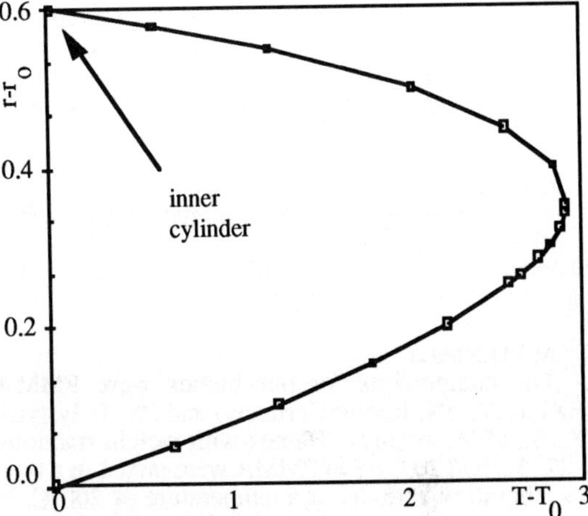

Figure 6 Temperature profile between the eccentric cylinder's narrow gap.

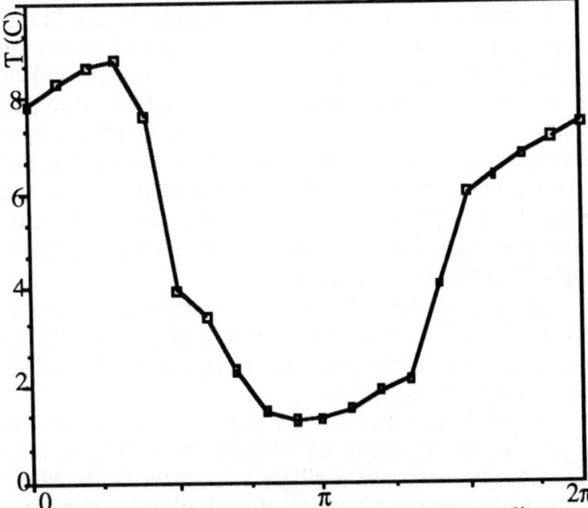

Figure 7. Temperature history of a particle traveling clockwise along the streamline marked "S" .

ACKNOWLEDGEMENTS

The authors would like to thank the National Science Foundation under grant number DDM-9158145 and 3M Engineering Systems and Technology for their financial support. Acknowledgement is also made to the University-Industry Research program of the University of Wisconsin-Madison for supporting this research project in its early stages.

REFERENCES

1. David,B.,T.Sapir,A.Nir, Z.Tadmor, *Int. Pol. Process.* ,5,3, 155-163,(1990).
2. Yagii, K. and K. Kawanish, *Int.Pol.Process.*, 5, 3, 164-172, (1990).
3. Cheng, J. and I. Manas-Zloczower, *Int.Pol.Process.*, 5, 3, 178-183, (1990).
4. Yagii, K. and K. Kawanish, *Int.Pol.Process.*, 5, 3, 173-177, (1990).
5. Brebbia, C.A., *The Boundary Element Method for Engineers,* Wiley, New York, (1978).
6. Gramann, P.J. , M.S. Thesis, University of Wisconsin-Madison, (1991).
7. Partridge, P. W., Brebbia, C. A., *8. International Conference on Computational Methods in Water Resources,* Venice, June 1990.
8. Mätzig, J. C., M.S. Thesis, University of Wisconsin-Madison, (1991).
9. Shearer, C. J., *Chem Eng. Sci.*, 28, 1091-1098 (1973).
10. Ybarra, R. M., Eckert, R. E., *AiChE Journal*, 26, 5, 1980.
11. Prins, J. A., Mulder, J., Schenk, J., *Appl. Sci. ^Res.*, A2, (1950).
12. Leong, C.W. and J.M. Ottino, *J.Fluid Mech.*, 209, 463-499, (1989).
13. Ottino, J.M., *The Kinematics of Mixing: Stretching, Chaos and Transport,* Cambridge University Press, Cambridge, (1989).

Theoretical and Applied Rheology, edited by P. Moldenaers and R. Keunings
Proc. XIth Int. Congr. on Rheology, Brussels, Belgium, August 17-21, 1992

THE INTERDEPENDENCE BETWEEN MORPHOLOGY AND MELT RHEOLOGY OF POLYMER BLENDS IN ELONGATION AND SHEAR

H. GRAMESPACHER and J. MEISSNER

Institut für Polymere, Eidgenössische Technische Hochschule, ETH-Zentrum, CH-8092 Zürich (Switzerland)

1. INTRODUCTION

The rheological properties of melts of immiscible polymer blends are influenced by their morphology. Conversely, the morphology of the blends depends on the rheological properties of their components and the deformation history. Therefore, when rheological measurements are performed with multiphase polymer melts, one has to consider the morphology and the change in morphology during the measuring process. Consequently, for rheological measurements of such systems two pre-conditions are essential:

- simple, well defined deformation modes (e.g.simple elongational flow)
- possibility of determining the morphology at each state of deformation.

A rheometer for which these conditions are satisfied is the elongational rheometer for polymer melts (ref.1).

2. MATERIALS

The components for our blends were PMMA (PLEXIGLAS 5N, Röhm, Germany) and PS (Polystyrol 158K, BASF, Germany). Blends with weight fractions of 8, 12, 16, and 20% PS in PMMA were mixed in a twin screw laboratory extruder at a temperature of 200°C. In order to establish the same thermal and rheological history for all the samples to be tested the pure components were extruded under the same conditions as the blends.

3. ELONGATIONAL FLOW

To study the rheological properties of the blends and the components we use the elongational rheometer with the rotary clamp technique. The rheometer and the method have been described in refs.(1,2). One advantage of this rheometer is the possibility to measure the recovery of the melts at the end of the elongation. Another advantage is that the morphology at various instants of deformation can be studied by quenching the samples in water. The morphology of the quenched samples was determined by light-microscopy and SEM.

The elongational tests were performed with the strain rate $\dot{\varepsilon}_0 = 0.1\ s^{-1}$ at a temperature of 170°C (ref. 2). The time-dependent elongational viscosities of PS, PMMA and the blends are shown in fig.1.

For small elongations the curves of the blends are very similar to the curve of PMMA; it is only at large strains that the viscosities slightly increase with the PS content. Although the viscosity of each blend is similar to that of PMMA, the behavior in recovery is quite different.

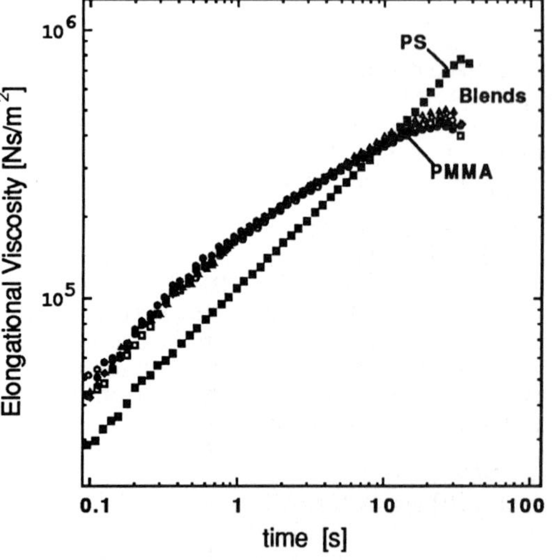

Fig.1: Elongational viscosities of PS, PMMA and their blends, $\dot{\varepsilon}_0 = 0.1\ s^{-1}$, T_0=170°C

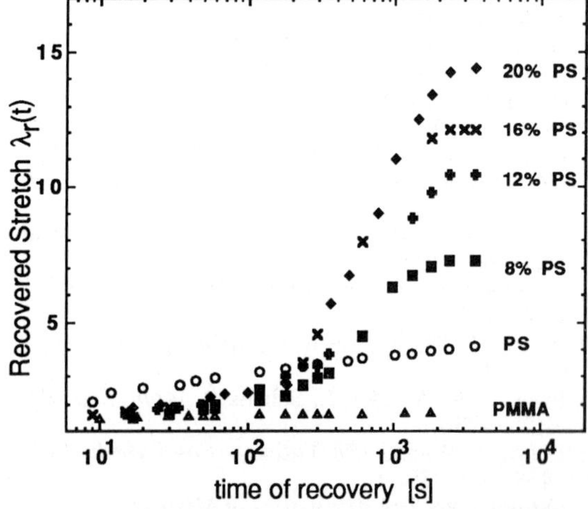

Fig.2: Time dependence of the recovered stretch $\lambda_r(t)$ for PS, PMMA and the blends; after elongation with strain rate $\dot{\varepsilon}_0 = 0.1\ s^{-1}$ and a total strain of $\varepsilon = 3.5$; $T_0 = 170$°C.

In fig. 2 the recovered stretch $\lambda_r(t)$ is plotted as a function of the recovery time t. The recovered stretch $\lambda_r(t)$ is the ratio of the length of the cut-off at the instant of cutting and the length after recovery of duration t.

We notice that up to a recovery time of 200 s the values of $\lambda_r(t)$ of the blends are between the values of the pure components. At longer recovery times, $\lambda_r(t)$ of the blends increases remarkably and gets much larger than that of PMMA and PS. Morphological studies at various instants of deformation show that the dispersed PS droplets (which are spheres of approximately 1 micron in diameter) are stretched to long ellipsoids during the elongational period (refs. 2) (fig.3). During the recovery process the PS-ellipsoids change back into spheres.

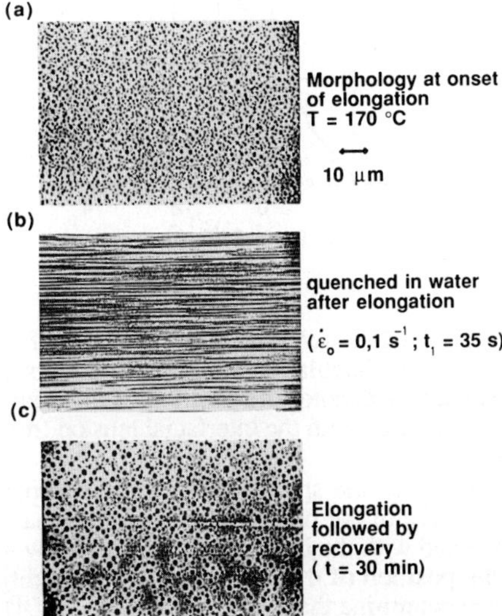

Fig. 3: Morphology of PS/PMMA blend with 12% PS at different instants of the deformation history

For long recovery times the interfacial tension α between the melts of the components is the essential driving force for this change in the shape of the PS droplets. A model could be developed that describes the measured recovery for all the blends. Comparison of the theoretical predictions with experimental results allows one to determine the interfacial tension α. For the interfacial tension between the melts of PS and PMMA at the temperature of 170°C we found $\alpha = (2.2 \pm 0.2) \times 10^{-3}$ N/m.

4. SHEAR OSCILLATIONS

The shear oscillation experiments were performed in a cone-and-plate oscillatory system (Rheometrics Mechanical System RMS 800) at 170°C and 190°C (refs. 3,4). The shear oscillations had an amplitude of 0.1 shear units. All results shown were shifted to 170°C.(Within experimental accuracy both components, PS and PMMA, have the same temperature shift factor.) Figures 4 and 5 show the storage moduli G' and loss moduli G" for the components and the blends.

At low frequencies the storage moduli G' of the blends increase with the PS content even above the curve for the pure PS. At high frequencies the storage moduli of the blends are approximately the same and are between the curves of the components.

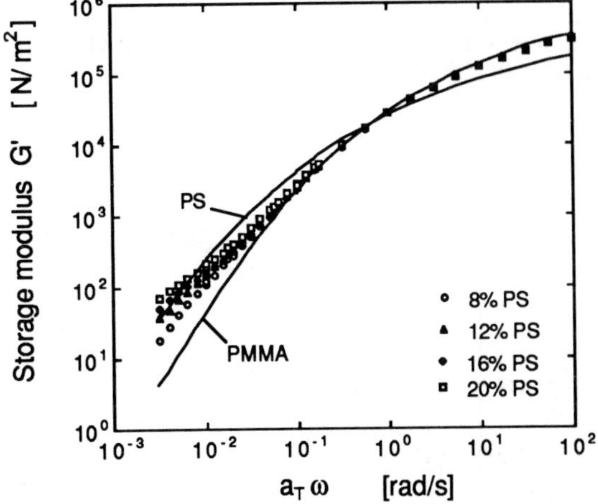

Fig. 4: Measured storage moduli G'(ω) of PS, PMMA, and the blends with 8, 12, 16, and 20% (weight) of PS. Within the filled squares shown at high frequencies are the individual results for all the blends. T_O = 170°C.

The reason for this behavior at low frequencies is the change of the shape of the PS droplets in the PMMA matrix during oscillatory shear deformation. Obviously, the total interface area is changing and so is the interfacial energy. The interfacial energy yields an additional contribution to the stored elastic energy. At high frequencies this additional energy portion is negligible compared with the elastic energy portion required for the viscoelastic deformation .

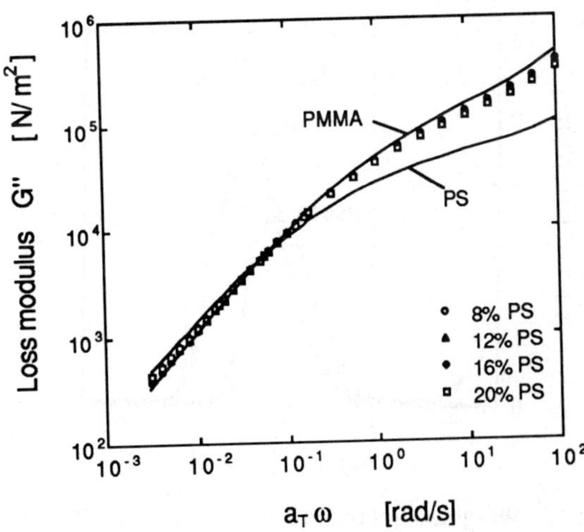

Fig. 5: Measured loss moduli G"(ω). Same materials and conditions as in figure 4. The filled circles cover the individual results of the blends with 8, 12, and 16% PS content.

A theoretical model that takes into account the contribution of the interfacial tension was used to describe the frequency dependence of the storage and loss moduli for all the blends (see ref. 4 and refs. cited therein). The agreement between calculated and measured data for all the blends is very good. For instance, figure 6 shows the experimental and calculated values of G' and G" for the blend with 20% PS content.

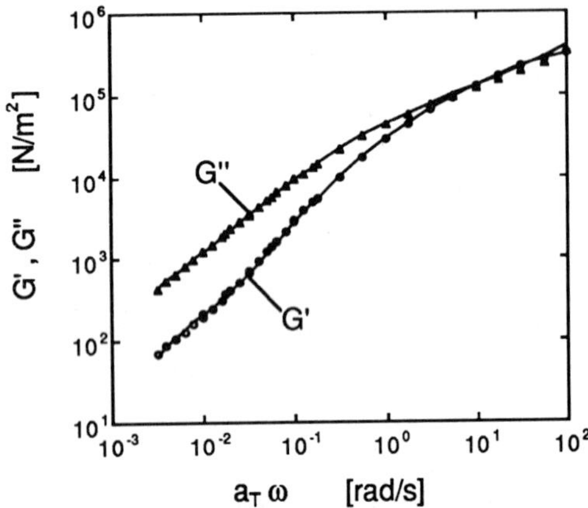

Fig.6: Dynamic moduli G' and G" obtained by measurement (symbols) and predicted by theory (curves), PS/PMMA blend with 20% PS, T_0=170°C; (ref.4).

We want to point out that there is no fit parameter in the theoretical description. The value for the interfacial tension α was obtained from measuring the recovery after elongational flow. Conversely, if the interfacial tension α is unknown it can be determined by a theoretical analysis of the measured dynamic moduli.

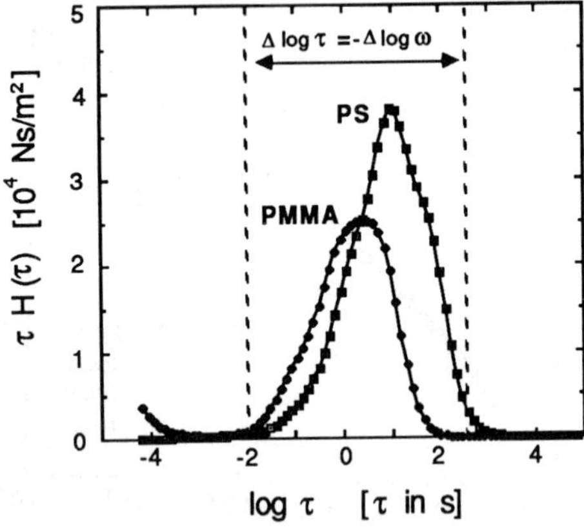

Fig. 7: Weighted relaxation time $\tau H(\tau)$ for PS and PMMA; calculated by J. Weese and J. Honerkamp from measured dynamic moduli; The range between the hatched lines refers to the frequency range of the measurements with $\omega = 1/\tau$. T_0 = 170°C; (ref. 4).

One method of analysis is the calculation of the relaxation time spectra of the blends. In figure 7 and 8 the calculated relaxation time spectra (weighted with the relaxation time τ) of the pure components and the PS/PMMA blend with 8% PS are shown (refs.4,5).

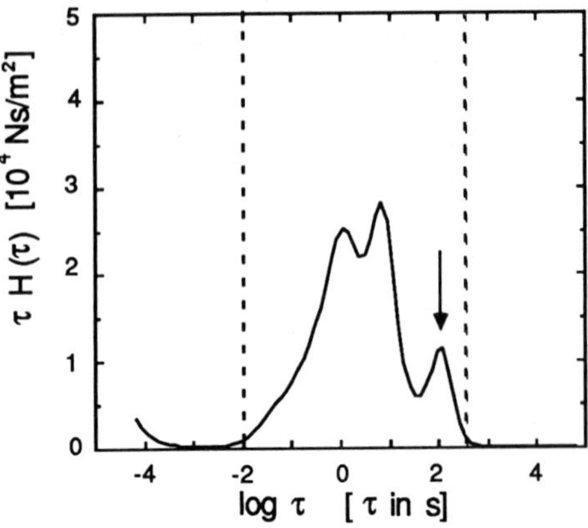

Fig. 8: Weighted relaxation time $\tau H(\tau)$ for the PS/PMMA blend with 8% PS. Calculation and test conditions as for figure 7. The arrow denotes the position of the relaxation time that is connected with the interfacial tension (ref.4).

In contrast to the spectra of the pure components (fig.7) the spectra of the blends show an additional peak that is connected with the interfacial tension (arrow in fig. 8). From the position of the maximum of this additional peak one can determine the interfacial tension (ref.4). For the melts of PS and PMMA we found $\alpha = (1.9 \pm 0.3) \times 10^{-3}$ N/m, which is in good agreement with the value we obtained from recovery measurements after melt elongation.

5. CONCLUSIONS

It is shown that the rheological behavior of immiscible polymer blends depends on the morphology that is changing during the measuring process. This change in morphology is connected with a change in the interfacial energy between the phases. Therefore, if the morphology is known, the interfacial tension between the melts of the components can be determined from rheological measurements of their blends.

REFERENCES

1. J. Meissner, Trans. Soc. Rheol,16 (1972) 405-420
2. H. Gramespacher, J. Meissner, in preparation
3. J. Meissner, R.W. Garbella and J. Hostettler, J. Rheol., 33 (1989) 843-864
4. H. Gramespacher, J. Meissner, submitted for publication to J. Rheol.
5. J. Weese, J. Honerkamp, personal communication (1991)

Theoretical and Applied Rheology, edited by P. Moldenaers and R. Keunings
Proc. XIth Int. Congr. on Rheology, Brussels, Belgium, August 17-21, 1992

VISCOELASTIC BEHAVIOR OF FILLED AND UNFILLED SILICON OILS

D. Hadjistamov, Ciba-Geigy AG, 4133 Schweizerhalle, Switzerland

1. Introduction

Silicon oils were investigated with a Weissenberg Rheogoniometer, model R18, Sangamo Ltd., with cone-plate (θ=5 cm, 6°) and plate-plate (θ = 5 cm) at 25 ± 0.2°C. The flow or viscosity curves and normal stress curves represent the steady state values from the stressing experiments. We investigated the following silicon oils (poly-dimethyl-siloxane) - M20000, M50000, M80000, M100000 and M500000. The weight-average molecular weight of the silicon oils was received by Bayer AG - 68'000, 85'000, 97'000, 104'000 and 164'000, i.e. above the critical molecular weight value M_c of appr. 30000.

2. Experimental

2.1 Silicon oils

The viscosity curves of these silicon oils, measured with cone-plate and plate-plate arrangements show a very good agreement (fig.1). The examined silicon oils

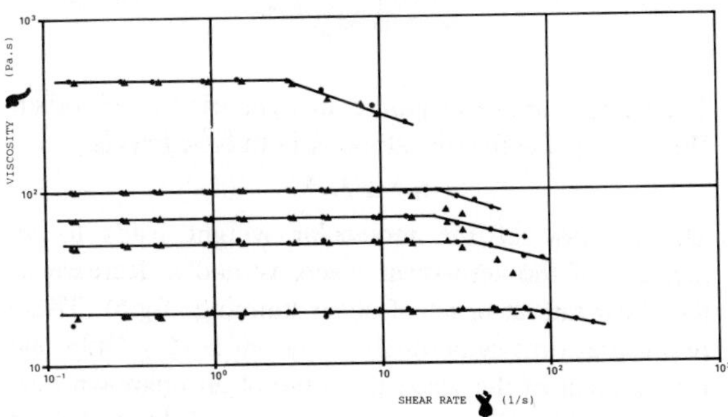

Fig.1 Viscosity curves of silicon oils (• plate-plate, ▲ cone-plate)

exhibit in accordance with the shape of the viscosity curves structural-viscous flow behavior (1,2,3,4). The first Newtonian region reaches almost the same shear stress:

M20000	$1.4.10^3$ Pa	at	70 1/s
M50000	$1.5.10^3$ Pa	at	30 1/s
M80000	$1.45.10^3$ Pa	at	21 1/s
M100000	$1.6.10^3$ Pa	at	16 1/s
M500000	$1.45.10^3$ Pa	at	3.3 1/s

We assume that this shear stress or viscosity at onset of shear thinning is an important value.

The viscosity master curve is usualy defined as $\lg \eta/\eta_o = f(\lg \dot\gamma\eta_o)$. This method leads to a very good master curve (fig.2). The viscosity master curve has a value of 1 up to

$$\dot\gamma\eta_o = 1500 \text{ Pa} \pm 150 = \tau_c$$

(τ_c - shear stress at onset of shear thinning). This is an important consideration - the onset of shear thinning

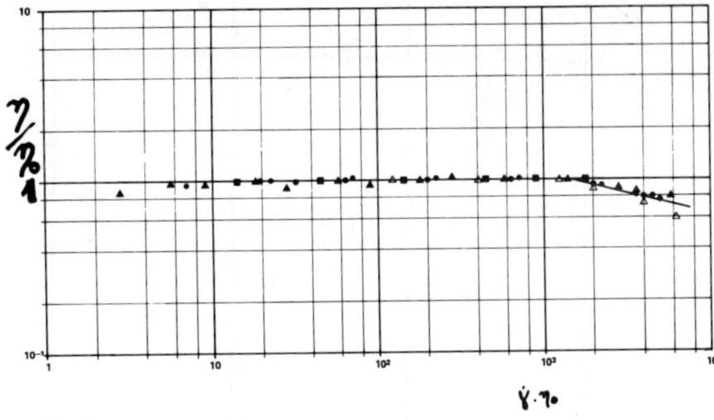

Fig.2 Viscosity master curve of silicon oils (▲M20000, + M50000, ■M80000, * M100000, Δ M500000)

begins at a certain very special shear stress, independent of the molecular weight.

The first normal stress difference (cone-plate arrangement) shows nearly a quadratic dependence on the shear rate (fig.3). The values of -N2/N1 are rather small - below 1. They seem to decrease with increasing shear rate.

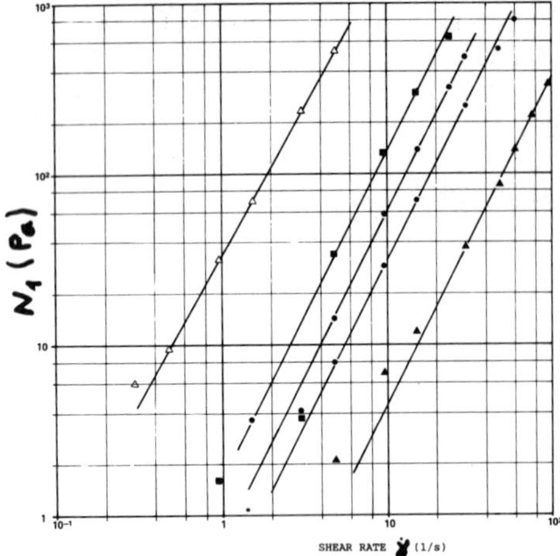

Fig.3 Normal stress curves (▲ M20000, · M50000, * M80000, ■ M100000, Δ M500000)

The modified first normal stress coefficient

$$\Theta' = \Theta_1 \cdot \dot{\gamma} = N_1/\dot{\gamma}$$

can be represented as a function of shear rate

$$\lg N_1/\dot{\gamma} = f(\lg \dot{\gamma}) \text{ (fig.4).}$$

This is a similar function as the viscosity dependence on the shear rate.

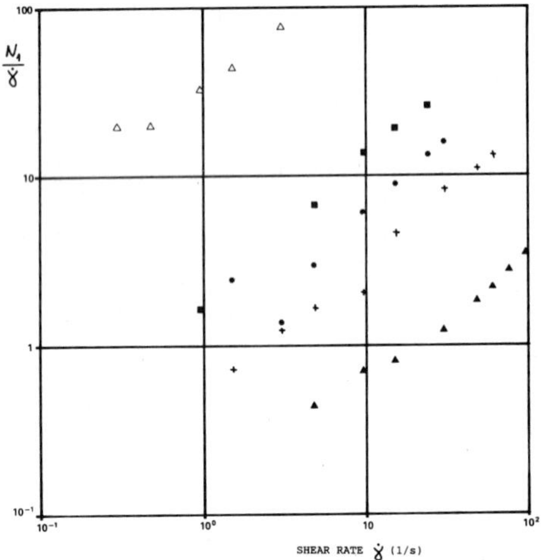

Fig.4 Dependence of the modified first normal stress coefficient on shear rate (▲ M20000, + M50000, ·M80000, ■ M100000, Δ M500000)

We can use the modified first normal stress coefficients to represent a normal stress master curve using the method for the viscosity master curve (fig.5). In this way, we obtain evidently a well formed normal stress master curve (with exeption of the points at lower values) with n = 1 and A = 1.2.10^{-4}:

$$N_1/\dot{\gamma}\eta_o = A * (\dot{\gamma}\eta_o)^n \qquad \text{or}$$

$$N_1 = 1.2.\ 10^{-4} \cdot \tau_o^2$$

with τ_o- shear stress for the zero-shear viscosity. It is possible to calculate with this equation the values of the first normal stress difference at known shear stress.

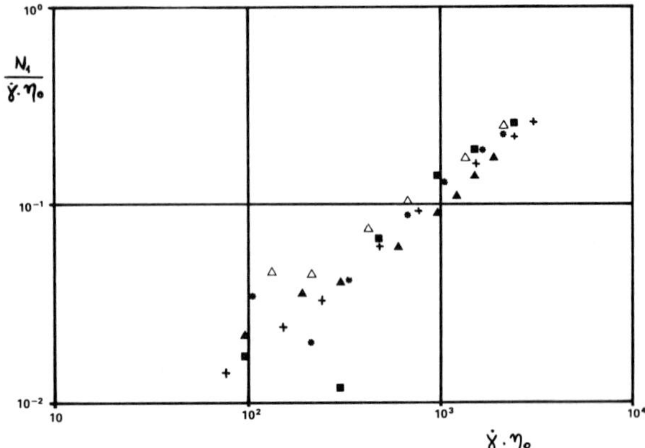

Fig.5 Normal stress master curve (▲ M20000, + M50000, ·M80000, ■ M100000, Δ M500000)

There is a power law relation between the zero-shear viscosity and the weight-average molecular weight (fig.6): $\eta = A.M_w^{3.48}$.

The slope of this function is n = 3.48 and A= 3.4*10^{-16} for our measurments. The slope value corresponds with the theory of Bueche (5).

The shear rate at onset of shear thinning depends also on the weight-average molecular weight with a slope of n = -3.48 and B = 4.57*10^{18} (fig.6):

$$\dot{\gamma}_c = B.M_w^{-3.48}$$

The straight lines of figure 6 are symetric to each other. We can assume that the slope of both functions is

$$n = \pm 3.48.$$

The increase of the molecular weight leads to an increase of the zero-shear viscosity and a decrease of the shear rate at onset of shear thinning (fig.6). These two values drift symetricaly away (90°± 16°). Only one measurment of the viscosity curve of an unknown new product lead to the zero-shear viscosity and to the shear stress at onset of shear thinning. If we know the molecular weight of the product, only two straight lines with 74° and 104° need to be drawn through these two points. With only one measurement we can also obtain the zero-shear viscositiy and shear rate at onset of shear thinning of the product with different molecular weights.

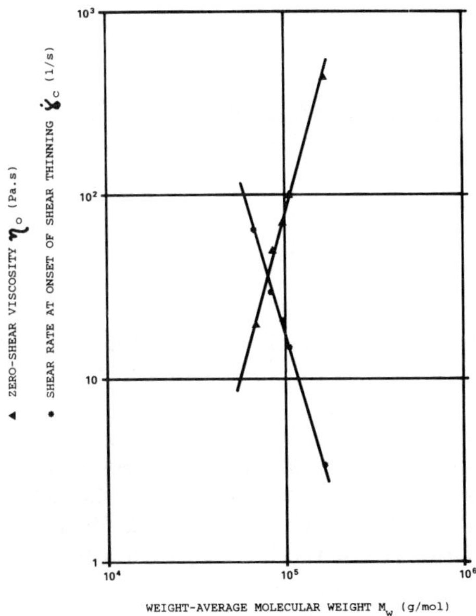

Fig.6 Dependence of the zero-shear viscosity (▲) and shear rate at onset of shear thinning (●) on weight-average molecular weight

The shear stress at onset of shear thinning is equal to:

$$\tau_c = \eta_o \dot{\gamma}_c = A.M_w^{3.48} * B.M_w^{-3.48} =$$

$$= A.B.1 = 1554$$

We can now postulate that the shear stress at onset of shear thinning does not depend on the weight-average molecular weight. The first Newtonian region will end or the shear thinning region will begin for a product with different molecular weights at the same shear stress - the shear stress at onset of shear thinning. This proves the observation of Vinogradov (6), that different polymer melts with different molecular weights have the same master curve.

2.2 Filled silicon oils

The silicon oil M 20000 was filled with the thixotropic agent Cab-o-sil TS-720. The increase of the Cab-o-sil concentration leads to a rise in viscosity. The first normal stress difference rises also with increasing concentration of Cab-o-sil (fig.7).

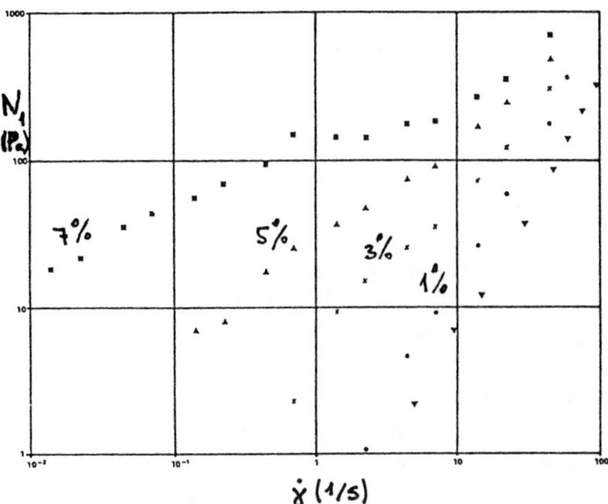

Fig.7 Normal stress curves of M20000 (▼) and M20000 with different concentration of Cab-o-sil

3. Literature:

(1) Hadjistamov, D., K. Degen, Rheol. Acta 18, 168 (1979)
(2) Hadjistamov, D., Rheol. Acta 19, 345 (1980)
(3) Hadjistamov, D., Proceedings of the 8th Int. Congress on Rheology, Naples, 1980, P. 609
(4) Hadjistamov, D., Proceedings of the 9th Int. Congress on Rheology, Acapulco, Mexico, 1984, Vol. 2, P. 277
(5) Bueche, F., J.Chem.Phys. 20, 1959 (1952), 25, 599 (1956)
(6) Semjonov, V., Fortschr.Hochpolymer-Forschung 5, 387 (1968)

Theoretical and Applied Rheology, edited by P. Moldenaers and R. Keunings
Proc. XIth Int. Congr. on Rheology, Brussels, Belgium, August 17-21, 1992

RHEOLOGICAL CHARACTERIZATION OF POLYETHYLENE FRACTIONS

E-L. Heino[1], A. Lehtinen[1], J. Tanner[2*] and J. Seppälä[2]

[1]Neste Oy Chemicals, Polyolefins R&D, P. O. Box 320, SF-06101 Porvoo, Finland
[2]Technical University of Helsinki, Department of Chemical Engineering, Kemistitie 1, SF-02150 Espoo, Finland

1. INTRODUCTION

Melt viscosity and elasticity of polyethylene are dependent on molecular structure, such as molecular weight (MW), molecular weight distribution (MWD) and degree of branching. Especially long chain brancing (LCB) has a great influence on rheological properties. However, when branched polyethylenes are studied, it is often difficult to distinguish different structural features and their influencies from each other. This is especially true in the case of low density polyethylene (1, 2), because with increasing degree of LCB usually also MWD gets broader. The structure of linear low density polyethylene is not so complicated but although different LLDPE grades have been studied and for example, the influence of comonomer length has been evaluated, the effect of short chain branching is not very clear (3,4).

In this paper rheological data on some linear and branched polyethylenes and their fractions are reported. To obtain more detailed information about the influence of different structural parameters, the polymers were fractionated acc. to molecular weight and acc. to structure.

2. MATERIALS

The polymers studied consisted of one high density (HDPE), one linear low density (LLDPE) and two low density polyethylene (LDPE) samples. The branched polyethylenes were all commercial film grade resins, but the HDPE studied was an injection moulding grade. Their molecular weight averages and distributions were near each other as shown in Table 1.

All samples were fractionated acc. to molecular weight using direct extraction technique (5) and the branched polymers also acc. to structure using successive extraction with different hydrocarbon solvents (n-hexane, n-heptane, n-octane, xylene). The fractionated amount was 40 - 50 g of polymer and it was divided in 4 - 6 fractions. The molecular weight distributions of the whole polymers and the fractions were determined with a Waters 150C ALC/GPC - Viscotek differential viscometer combination in TCB at 135°C. The degrees of branching were calculated from ^{13}C NMR spectra obtained on a Jeol GSX400 spectrometer.

Table 1. Properties of the polymers investigated

Polymer	M_w	M_w/M_n	η_0[1]	MFR_5
LDPE 1	137000	4.4	19100	6.4
LDPE 2	108000	3.8	15600	9.1
LLDPE	124000	4.4	14400	3.3
HDPE	95600	4.3	4050	2.9

[1] at 170°C.

3. EXPERIMENTAL

The dynamic mode of Rheometrics System Four was used for rheological characterization. Measurements were made under nitrogen atmosphere at 170°C. In this way, storage modulus (G'), loss modulus (G") and complex viscosity (η^*) were obtained as a function of frequency (ω). The LLDPE and LDPE grades were also tested at two other temperatures (150°C and 190°C) and flow activation energies were determined. Because of too small amount of sample, measurements could not be made on all fractions (usually the first and/or the last fraction not measured).

4. RESULTS

Zero viscosities were calculated using complex

* The present adress: Neste Oy, Innopol, Tekniikantie 12 P. O. Box 356, SF-02151 Espoo, Finland

fluidity defined as the reciprocal of complex viscosity. Its real and imaginary parts are thus expressed by:

$$f'(\omega) = \eta'(\omega)/[\eta'(\omega)^2 + \eta''(\omega)^2] \qquad (1)$$

$$f''(\omega) = \eta''(\omega)/[\eta'(\omega)^2 + \eta''(\omega)^2]. \qquad (2)$$

where η' and η'' are real and imaginary parts of the complex viscosity ($\eta' = G''/\omega$, $\eta'' = G'/\omega$). According to Verney and Michel (6) there is the following simple relation between $f'(\omega)$ and $f''(\omega)$:

$$f'(\omega) = 1/\eta_0 + \tan((h\pi/2)f''(\omega) \qquad (3)$$

where h is the parameter of the relaxation time distribution. Obviously expression (3) describes a linear relation between $f'(\omega)$ and $f''(\omega)$ in the complex plane with zero ordinate value of $1/\eta_0$.

The zero viscosities obtained for different linear low density and low density fractions are seen in Tables 2 and 3 together with molecular weights (M_w), molecular weight distributions (M_w/M_n) and short (C_2) or long chain branching (LCB). Since the polymers were divided in so few fractions, the molecular weight fractions were still reasonably broad (MWD 3 - 4). The molecular weight distributions of the first structural fractions of LDPE 1 and LDPE 2 were bimodal and therefore, broader than those of the whole polymers and the other fractions.

To study the influence of molecular weight distribution or branching on the viscosity behavior, the complex viscosity function is presented in a reduced form, $\eta^*(\omega)/\eta_0$ vs. $\omega\eta_0$ (Figures 1 and 3).

According to Han (7), elasticity can be evaluated by plotting G' vs. G". This function is independent on molecular weight (with high M_w) and very slightly

Table 2. Properties of LLDPE fractions

Polymer	M_w	M_w/M_n	C_2 /1000C	η_0 (1
LLDPE,whole	124000	4.4	14	14400
According to MW				
fraction 2	31100	3.2	-	17
fraction 3	50100	3.0	20	124
fraction 4	116000	2.9	12	1820
fraction 5	173000	2.7	10	42600
According to sructure				
fraction 1	84300	4.1	27	5680
fraction 2	111000	4.2	10	16300
fraction 3	131000	4.2	8	23000

(1 at 170°C.

Table 3. Properties of LDPE fractions (fractionation acc. to structure)

Polymer	M_w	M_w/M_n	LCB /1000C	η_0 (1
LDPE 1, whole	137000	4.4	2.5	19100
fraction 1	103000	6.6	2.1	2120
fraction 2	138000	4.4	2.2	14200
fraction 3	159000	3.2	2.6	70900
LDPE 2, whole	108000	3.8	2.1	15600
fraction 1	75300	5.7	4.6	990
fraction 2	104000	3.6	2.9	17600
fraction 3	150000	3.3	1.6	66100

(1 at 170°C.

dependent on temperature. It is, however, srongly influenced by molecular weight distribution. The elasticities of the whole polymers are compared in Figure 2 and three LLDPE fractions in Figure 4.

5. DISCUSSION

The dependence of zero viscosity on molecular weight is usually expressed by the well-known equation:

$$\eta_0 = KM^a, \text{ when } M > M_c \ (\approx 3000 \text{ for PE}). \qquad (4)$$

where $a \approx 3.4$ for monodisperse linear polymers and K depends both on material and temperature. Figure 5 shows that η_0 vs. M_w is linear on a logaritmic scale. The point for HDPE lies very near to this correlation which means that short chain branching has little influence on zero viscosity.

Because all three polyethylene grades have similar molecular weight distributions, the differences in reduced viscosity functions are caused by differences in branching. Figure 1 indicates that short chain branching in LLDPE has little influence on the viscose behavior. On the other hand, effect of long chain branching is pronounced and complex viscosity of LDPE decreases with frequency much more sharply than that of the other grades. LLDPE resembles HDPE also in elasticity which is much higher for LDPE (Figure 2).

The apparent flow activation energies obtained for LLDPE and LDPE (29 kJ/mol and 56 kJ/mol) are well within the range found in literature (3, 8).

There are only slight differencies in the reduced viscosity functions of LLDPE fractions with similar MWD. The curve of the fraction with highest short chain brancing has somewhat different shape (Figure 3). Also the differencies in elasticity are small, but elasticity seams slightly to decrease with increasing SCB.

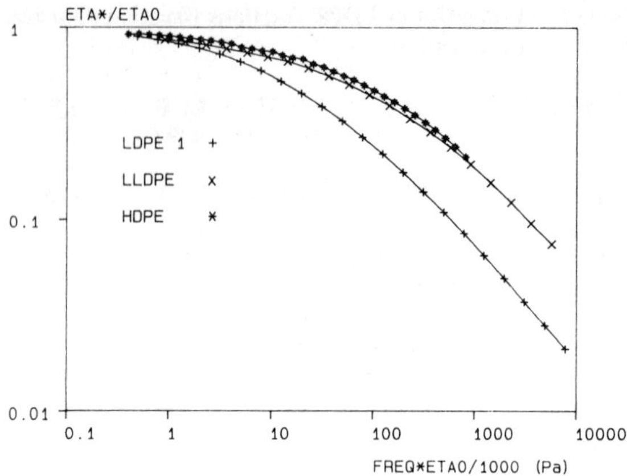

Figure 1. Reduced complex viscosity functions of PE grades.

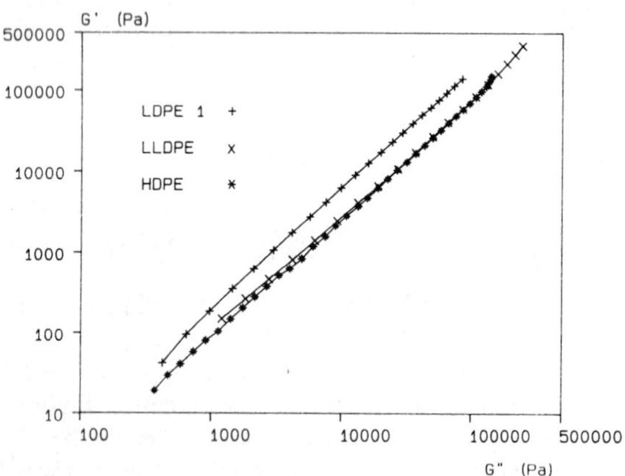

Figure 2. Elasticity of PE grades.

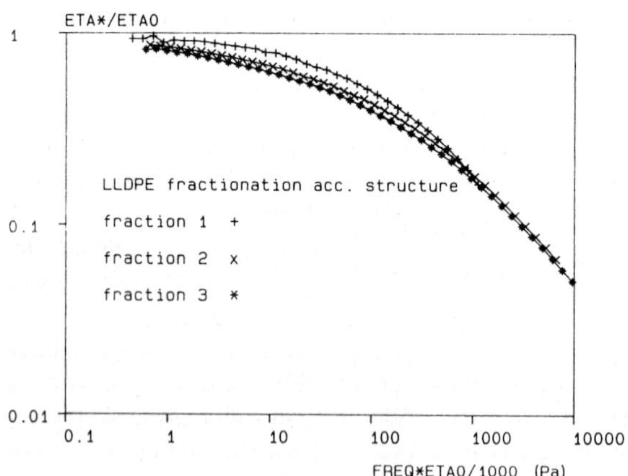

Figure 3. Reduced viscosity functions of LLDPE fractions.

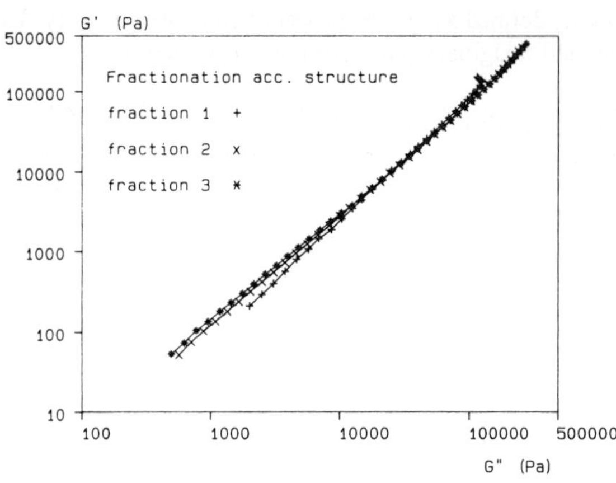

Figure 4. Elasticity of LLDPE fractions.

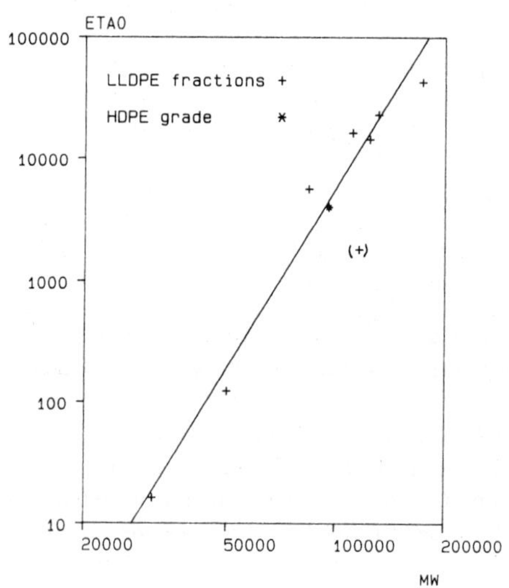

Figure 5. Zero viscosity as a function of molecular weight.

REFERENCES

1. A. Santamaria, Materials Chemistry and Physics, 12 (1985) 1-28.
2. W. Minoshima and J. L. White, J. Non-Newtonian Fluid Mech., 19 (1986) 251-274.
3. D. M. Kalyon, D. W. Yu and F. H. Moy, Polym. Eng. Sci., 28 (1988) 1542-50.
4. A. Valenza, F. P.La Mantia and D. Acierno, Eur. Polym. J., 24 (1988) 81-85.
5. W. Holtrup, Makromol. Chem. 178 (1977) 2335-2349.
6. V. Verney and A. Michel, Rheol. Acta, 28 (1989) 54-60.
7. C. D. Han, J. Appl. Polym. Sci., 35 (1988) 167-213.
8. B. H. Bernsted, J. Appl. Polym. Sci. 30 (1985) 3751-3765.

Theoretical and Applied Rheology, edited by P. Moldenaers and R. Keunings
Proc. XIth Int. Congr. on Rheology, Brussels, Belgium, August 17-21, 1992

ELONGATIONAL FLOW AND POLYMER COMPOUND MIXING

H.P. Hürlimann
Polymer Institute, ETH Zentrum, CH 8092 Zürich

1. INTRODUCTION

Besides the shear rheology of polymer melts, elongational flow rheology has long been neglected, despite its early application in rubber mixing mills, fibre spinning and film blowing. Two-roll mills remain popular for their simplicity and excellent mixing performance. They rely mainly on instable types of elongational flow that are hardly understood.

Zülle et al (ref.1) deducted from kinematically matched elongational and shear tests, that the classification of elongational flows as "strong" and of shear flows as "weak" is misleading. Wagner (ref.2) showed, that elongation at constant rate and at constant tensile force may both be described by the Lodge-Wagner equation. Hürlimann (ref.3,4) found from die inlet flow tests, that "melt-fracture" is an instable type of flow at constant tensile force.

The following analysis bases on experiences with an advanced elongational mixer prototype (ref.5) that offers good distribution and dispersion at a low energy level.

2. DISTRIBUTION EFFECT IN TWO DISCONTINUOUS SHEAR- AND ELONGATIONAL- MIXERS

To compare the competitiveness of shear- and elongational- mixers, two mixers of the same capacity were analysed. The continuous mixers are not discussed here, because elongational flow is inherently discontinuous. All components have the same rheological behaviour. The striation thickness is taken as the measure for distribution.

Figure 1 shows an idealized discontinuous shear mixer

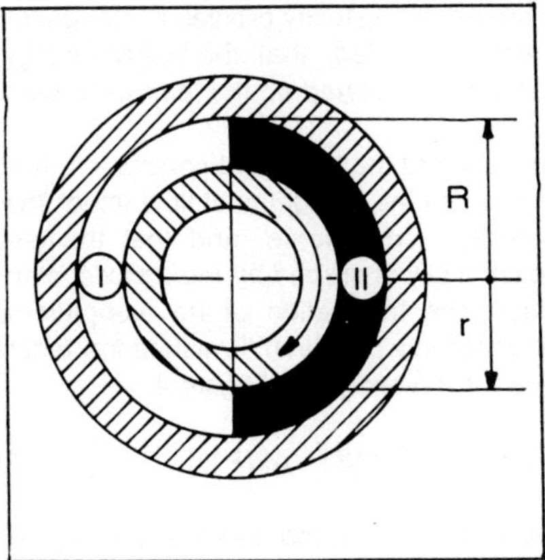

Figure 1

Idealized shear mixer.

R radius inside of outer cylinder
r radius outside of inner cylinder
Sj Striation thickness of spiral after j revolutions.

consisting of an inner rotating cylinder and a concentric stationary outer cylinder. At the start, equal quantities of components I and II fill the gap between the cylinders to both sides of the middle line. The components intermix in spirals. After j revolutions, the striation thickness is

$$Sj' = (R - r)/(j+1) \qquad (1).$$

This mixing process becomes less and less effective after some revolutions. To reactivate it, a superposed radial transport process must be generated by some device to reorient the striation pattern into a radial direction. It is assumed, that one quarter rotation is superposed every m revolutions. The striation thickness for this modification is

$$Sj'' = (R - r)/2 \quad [(R - r)/\pi(R + r)]^{(j-1)/m} \qquad (2).$$

m = 5 seems to be technologically feasible.

Figure 2 shows an idealized elongational mixer consisting of two cylinders with a radius R1 and cross-sectional area A1 that are pressed to the faces of a short die of the

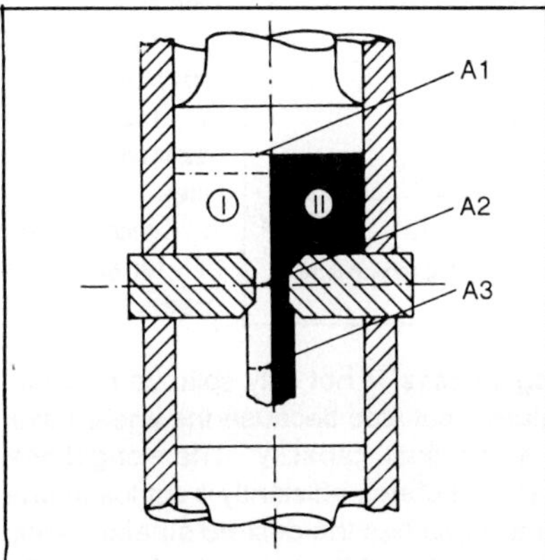

Figure 2

Idealized elongational mixer

Crosssectional areas of:
A1 cylinder,
A2 aperture,
A3 extrudate.
Sk Striation thickness after k extrusions.

aperture area A2. The elongational flow develops at the inlets of the die, as the two pistons push the compound back and forth. The striation thickness after k cycles is

$$Sk = R1\ (A3/A1)^{k} \qquad (3).$$

In figure 3 the mixing time is given versus the striation thickness S. The scales are logarithmic. An elongational mixer with a ratio A1/A3 = 16 needs 5 extrusion cycles to mix the two components down to the requested striation thickness. A modified shear mixer (m = 5) needs about 100 revolutions to arrive at the same distribution.

It is assumed from experience, that the dissipated energy for three revolutions equals that for one extrusion cycle. The temperature rises about half a degree for one rotation and about one and a half degrees for one extrusion cycle. The shear mixer therefore consumes three times more time and energy for the same distribution result.

For a perfectly cooled mixer, the compound temperature is constant. For poor cooling, the process is adiabatic. The temperature rise effects a decrease of the viscosity.

For the same duration, the adiabatic process saves mixing energy at the price of a reduced dispersion. Actual process tends more to adiabatic than to isothermal conditions. The process is stopped at the desired striation thickness or at a critical temperature. The temperature rise limit is assumed to be 100 K. The shear mixing process often is completed on a two roll elongational flow mill.

The inhomogeneities of concentration in the submicrometer range that remain after the distributive and the dispersive mixing processes are equalized by diffusion. This homogenisation should need only minutes. If the distribution is poor, it may take days. The diffusion may be stimulated by compatibilizers and the substitution of incompatible homopolymers by compatible copolymers.

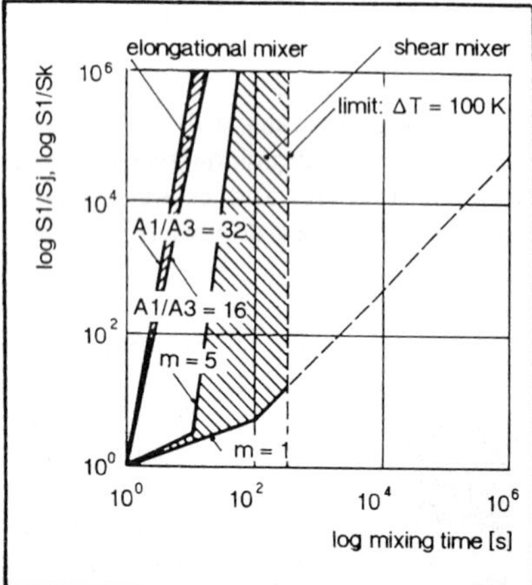

Figure 3

Compared mixing effectiveness

S striation thickness
j number of revolutions
k number of extrusion cycles
t mixing time
T temperature rise

The shear mixing process is not only split up to avoid excessive temperatures, but also because the shear mixer can operate only at nominal capacity. The elongational mixer on the other side operates efficiently from full to one third load. As it usually reaches the desired striation thickness before the temperature hits the critical point, the diffusive homogenisation follows immediately after the mixer is stopped and finishes in minutes. This opens new processing strategies.

3. ELONGATIONAL FLOW AT THE INLET OF DIES

For the development of an elongational mixer the knowledge of stable and instable elongational flow rheology is indispensable. The prominent pioneer of research on instable die inlet flow was Tordella (ref.6). He noted the transition from laminar to irregular flow at a critical flow rate in a glass capillary. He heard cracking noises in the melt that he interpreted as "melt fracture". He also observed the recoil of stretched melt on the middle line of the die and a periodic change of flow direction at the die inlet.

The mixing effectiveness in an elongational mixer can be considerably influenced by a competent control of the flow mechanisms. A proper knowledge of the parameters that control the transition from laminar flow to melt fracture is important for the effectiveness of elongational mixing. The transition to melt fracture can be monitored by a pressure transducer. The frequency and amplitude of the signal can be interpreted for control purposes.

Figure 4 shows a schematic capillary rheometer and the boundary flow lines in the fluid that separate the central elongational flow from the surrounding isolated cicular vortex. The flow is assumed laminar and stationary.

For the experimental investigation of the velocities, a stroboscopic light beam was directed on reflecting particles distributed in a LDPE-melt. A transparent capillary rheometer was designed to take undistorted pictures. The constant flow rate was just below the transition to melt fracture. The velocities in the elongational flow were constant throughout each cross-sectional area. The transition to the velocity profile of the circular vortex was narrow.

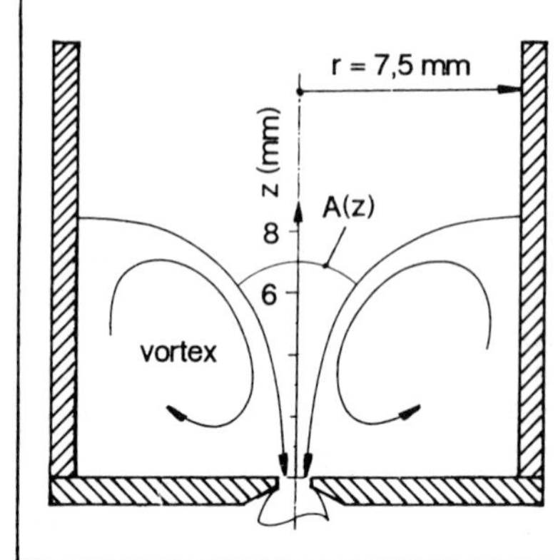

Figure 4

Boundary flow lines

LUPOLEN, 1810 H, 150°C critical flow rate 4,3 $mm^3 s^{-1}$.

z elongational flow coordinate
A(z) cross-sectional area

This result is in agreement with earlier measurements of inlet flows of LDPE-melt (LUPOLEN 1810H, 150°C). These were made with a more intricate procedure (ref.4). The tests are analysed below on the basis of an uniaxial spherical sink flow for which the rate of elongation was calculated as a function of time.

For a full rheologic analysis, this kinematic information had to be complemented by the corresponding kinetic one. The problem was solved in good approximation by the separation of the die inlet pressure loss into energy fractions that refer to the different distinguishable elements of the inlet flow. For example the energy dissipated in the circucular vortex, estimated by a finite element method, indicated, that this fraction is one order of magnitude smaller than the one for the elongational flow. This coincides with the mentioned observed flat velocity profiles in elongational flows. It was also calculated, that the kinetic energy and the surface energy generated at the inlet were both negligible (ref. 4).

These evaluations led to the important conclusion, that it is reasonable to analyse the elongational flow under the assumption of constant tensile force, and that the inlet pressure loss is essentially absorbed by the uniaxial elongational flow process. The integration of the elongational deformation work under this condition furnishes the tensile stresses σ required for the analysis of figure 4.

$$\sigma 2 = \sigma 1(A1/A2) = pe/(1 - A2/A1) \qquad (4)$$

pe is the inlet pressure loss. For the indices see figure 1

Figure 5 gives the full analysis of this elongational

flow at constant tensile force for LDPE (LUPOLEN 1810 H at 150 °C) (ref.4). The flow rate is just transient. The stress is given versus the rate of elongation in logarithmic scales. At the lower left end of the curve, the stress

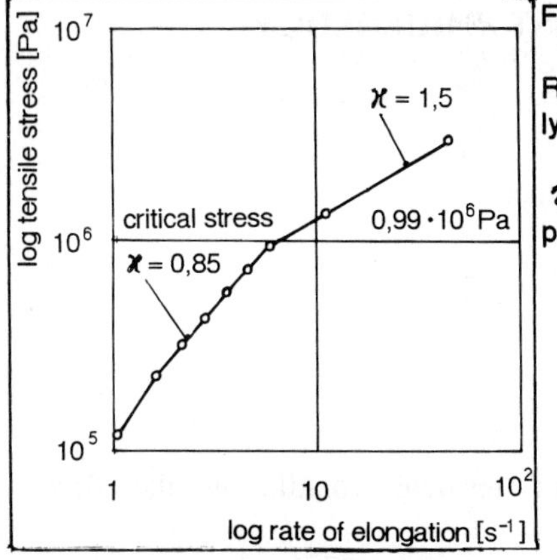

Figure 5

Rheologic analysis of figure 4

χ exponent of power - law

growth versus the rate of elongation is more than linear or dilatant. Near the critical stress, where the transition to melt fracture takes place, the stress growth is linear or Newtonian. This is in agreement with other tests discussed by Wagner (ref.2). At the upper right end of the curve, where the flow is instable, the stress growth is less than linear or pseudoplastic. The corresponding flow lines were straight and conically focussed to a point on the middle line at the die inlet.

To understand this phenomenology, the elongational behaviour of a power-law model fluid was inverstigated for constant tensile force. For this purpose, the cross-sectional area was calculated versus the flow coordinate. It resulted, that if the exponent is given a value higher than 1, the cross-sectional area becomes zero after a finite distance of flow. If it has the value of 1,5, the flow lines become straight and focus to a point at short distance.

The application of this theoretic result to the observations of the mentined sections in figure 5 indicates, that the value of the exponent shifts from 0,85 to 1 and 1,5. The elastic energy stored in the elongatial flow finally causes the retarded recoil of the melt from the point of constriction along the middle line. Afterwards the molecular reorientation by elongational flow starts for a new cycle. This mechanism is important for polymer mixing.

The critical stress decreases with increasing temperature and molecular weight.

4. ADAPTATION TO DECREASING VISCOSITY

To mix a compond without interruption of the process requires the adaptation of the elongational mixer to the decreasing viscosity. As this one is designed for a nominal extrusion pressure and rate, and as the number of apertures is limited by the size of the die, it remains only the geometry of the die apertures to be adapted. According to equation 5, a small convergence ratio A1/A2 with the largest possible radius R1 is suited for high viscosities and smaller radii for lower viscosities. This design permits to keep the extrusion pressure as high as the energy dissipation permits and as a good particle dispersion requires.

5. DISPERSION IN CRITICAL DIE INLET FLOWS

The melt fracture mechanism is more complicated for a polymer melt that contains viscous or solid components. The principles of flow behaviour however are the same as for the pure melt. The different components establish an interfacial stress equilibrium and the polymer components will stretch and orient the molecules in the flow.

Figure 6 shows a schematic flow configuration of a dispersed soft phase approaching the die inlet. This particle may have a lower critical tensile stress than the matrix or its cross-sectional area may be weakened by hard filler

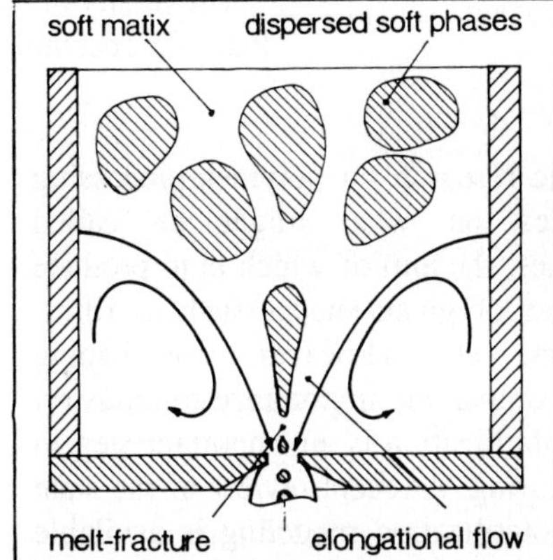

Figure 6

Schematic view of a soft phase dispersed at the die inlet by melt - fracture

particles. The soft phase may therefore fracture within the surrounding matrix. Elongational mixing therefore needs a sophisticated process control.

6. CONCLUSIONS

Compared to the shear mixer, the elongatial mixer offers high distributive and dispersive effectiveness at low energy dissipation. Because the critic temperature is usually not reached before the whole mixing process is finished, and because it operates efficiently at one third of the compound volume, it gives access to new operative strategies.

7. ACKNOWLEDGEMENTS

The author thanks Prof. Dr. W. Knappe for the opportunity to investigate the subject of melt fracture at the DKI in Darmstadt. He also thanks Prof. Dr. J. Meissner for the assignment to investigate laminar melt flow near die inlets at the ETH in Zürich and A. Speckert for the execution of some delicate measurements.

REFERENCES

1 B. Züllle, J.J. Linster, J. Meissner and H.P. Hürlimann, J. of Rheology, 31 (1991) 583-598.
2 M.H. Wagner, Rheol. Acta 18 (1979) 681-692.
3 H. Hürlimann and W. Knappe, Rheol. Acta 8 (1969) 384-392.
4 H. Hürlimann, Dr.mont.-thesis, Montanistic University Leoben, A (1981)
5 Swiss patent 542'699 dating from 15.10.1973
6 J. P. Tordella, Trans. Soc. Rheol., 1 (1957) 202-212.

Theoretical and Applied Rheology, edited by P. Moldenaers and R. Keunings
Proc. XIth Int. Congr. on Rheology, Brussels, Belgium, August 17-21, 1992

POLYMER DEVOLATILIZATION: HOW IMPORTANT IS RHEOLOGY?

Giovanni Ianniruberto, Pier Luca Maffettone
and Gianni Astarita
Department of Chemical Engineering
University of Naples Federico II
Piazzale Tecchio, 80125 Napoli, Italy

Academic research on polymer processing often concentrates on what might be called "shaping" processes, the aim of which is to produce polymer manufacts of given shapes such as films, extrudates, fibers, etc. However, non-shaping processes are of comparable importance to industry: for instance, devolatilization is an important step in polymer manufacturing. A recent review of the state of the art in devolatilization modeling is available (Astarita and Denson 1991). It is of interest to observe that most models of devolatilization processes are based on the approximation that the polymer phase is a Newtonian fluid, i.e., there is very little rheological content in them. In spite of this, the models are quite successful in correlating experimental data (Collins et al. 1985, Maffettone et al. 1991). Here we concentrate on one particular process, slit devolatilization, to illustrate the point - and to point out that there are aspects of the modeling procedure where rheological considerations may be quite important.

The real process of slit devolatilization is easy to describe. The polymer solution is fed through a large number of parallel slits, which are heated by way of hot oil circulating in the regions between slits. The slits discharge into a chamber where vacuum is exerted; the material is extracted from the bottom of this chamber with an extruder which discharges to atmospheric pressure.

When analyzing this process, it is natural to first focus attention to the (academic) case of a single slit, and this was in fact done. Experiments were carried out at the Montedipe Research Center on an instrumented single slit where temperature and pressure profiles could be measured as well as the concentration of residual volatile in the final product.

Maffettone et al (1991) have recently presented a (non-rheological) model of the single slit configuration which contains only one adjustable parameter; in order to fit the data, the value of the parameter only needs to exceed some minimum value established by comparison with experiments. A brief description of the Maffettone et al. model is given in the following.

The fluid mechanics are based on the assumption of a symmetric flow configuration, with a gas layer in the middle of the channel (an alternate possible flow configuration with the gas flowing near one of the walls gives a poorer fit to the data). The thickness of the gas layer is determined by the condition that the local pressure gradient has to be the same in the two phases. The gas layer thickness turns out to always be negligible as compared to that of the polymer layers, and the model is developed by neglecting the gas layer thickness when appropriate. The polymer is regarded as being Newtonian. Heat transfer from the solid wall to the polymer is assumed to be by pure conduction, and in the region where experiments were made the Nusselt number is constant. This is due to the fact that, since devolatilization occurs to a significant extent in the slit, only a small fraction of it is used for heating up the polymer, and hence the Graetz number based on slit length is small. The rate of mass transfer is assumed to be proportional to the difference between the vapor pressure of volatile corresponding to the local temperature and concentration in the polymer, and the absolute pressure at that section. The proportionality constant is the adjustable parameter.

The model may appear a very crude one, but two points should be kept in mind. First, a reliable correlation is needed for the vapor pressure as a function of temperature and concentration, and this is a (thermodynamic, but not rheological) matter of major relevance. Second, due to temperature and, more importantly, concentration variations along the slit, the polymer phase zero shear viscosity changes by four or five orders of magnitude from the entrance to the exit. It is this second point which makes consideration of non Newtonian effects for the single slit problem rather futile. Actual shear rates are rather low, probably in the Newtonian limit region, and any shear thinning effect is likely to be by far less important than variations of the zero shear viscosity due to temperature and concentrations changes. More importantly, available correlations for the zero-shear viscosity as a function of concentration and temperature are given in terms of its logarithm, and uncertainties about the actual values is much larger than any effect due to shear thinning would have. Unfortunately, viscosities are correlated logarithmically, but calculated pressure drops are proportional to the values of the viscosity one uses. The model correlates pressure drops as well as one could hope, given the intrinsic limitations discussed above, and inclusion of shear thinning effects would be unlikely to yield any significant improvement.

The model also predicts a somewhat unexpected feature of the experimental data, namely, that as the flowrate through the slit is increased, the pressure drop first increases, then reaches a maximum, and then starts to decrease again. The physical reason for this is that, at the largest flowrates considered, devolatilization is not very efficient, the volatile concentration is higher, and therefore the average viscosity is lower; the effect of a lower average viscosity more than overcomes that of an increased flowrate. Experiments were performed up to flowrates at which the leakage of volatile became significant, and the model is appropriate only in that range.

We have subsequently extended the model so that it is appropriate for arbitrary values of the flowrate (Ianniruberto 1992). This requires setting up a fully twodimensional model, because the Graetz number is not necessarily small, and the transient heat conduction equation needs to be integrated numerically. In this case the pressure drop, after going through a maximum, decreases until it reaches a minimum, and then starts increasing again, eventually reaching an asymptotic behavior which, if the fluid is regarded as being Newtonian, is one where the pressure drop is proportional to flowrate. This asymptotic condition is reached when the flowrate is so high that no devolatilization takes place, the polymer heats up negligibly, and the problem reduces to laminar flow of the inlet (low viscosity) solution. In this limit the shear rate in the slit may well be high enough that the fluid cannot be regarded any more as being Newtonian, (the volatile concentration is "high", but one is still considering a solution which is 50% by weight polymer and hence may be significantly shear thinning). We have not included this effect in our model in a quantitative way, but it is guaranteed to be of importance.

The discussion above shows that the plot of flowrate vs. pressure drop is three-valued in the region between the minimum and maximum pressure drops. This has an important consequence in the actual multislit configuration. The total flowrate is imposed on the system, and of course the pressure drop is the same through all slits. However, if the pressure drop is in the three-valued region, the system is unstable: part of the feed will move very slowly through some slits, and part of it will shoot through the other ones where almost no devolatilization takes place. Hence the largest pressure drop compatible with stable operation is the one corresponding to the minimum (but with the flowrate being the one obtained from the low flowrate branch of the curve), and this corresponds to a flowrate well below the one at which leakage becomes significant: the largest possible flowrate per slit in actual operation is much less than the one which optimizes single slit operation, as has been also observed experimentally. Our extended model predicts this result. Notice that shear thinning behavior, by lowering the high flowrate pressure drop asymptote, results in a lower value of the minimum of the pressure drop curve, and in this sense it makes multislit operation even more unstable: a quantitative estimate of the maximum possible flowrate per slit resulting in stable multislit operation would necessarily need to take into account the shear thinning behavior of the polymer solutions considered.

The agreement between model predictions and experiments for the single slit case can be shown to imply that the process is essentially governed by the rate of heat transfer. Since the latter is influenced only marginally by the detailed flow structure, it is not surprising that the rheological behavior of the polymer is of no relevance to this aspect of the problem at hand; it is, however, relevant to the stability analysis, as discussed above. By analogy, in the case of bubble-free devolatilization in a twin screw extruder, most of the devolatilization takes place in stagnant regions of polymer (Collins et al. 1985), and hence again the rheology of the polymer does not play any significant role: for the latter problem, no major concrete aspect has been identified for which the rheological behavior of the polymer phase might be of any great importance.

The conclusion from this work may be worded as follows. In real life situations, the rheological behavior of the materials considered may, or may not, be significant in determining the overall performance of a process. Modeling efforts should include rheological complications only when there are good grounds for believing that they are indeed of importance.

REFERENCES

Astarita, G., and C. D. Denson, "Polymer Devolatilization", Proc. Europ. Symp. Pol. Proc. Soc., Palermo, September 1991

Collins, G. P., C. D. Denson and G. Astarita, "Determination of Mass Transfer Coefficients for Bubble-free Devolatilization of Polymeric Solutions in Twin Screw Extrudeers", AIChEJ, 31 , 1288, (1985)

Ianniruberto, G., "Stabilita' della Devolatilizzazione di Polimeri in sistemi a multifessura", Senior Chem. Eng. Thesis, University of Naples, 1992

Maffettone, P. L., G. Astarita, L. Cori, L. Carnelli and F. Balestri, "Slit Devolatilization of Polymers", AIChEJ, 37, 724, (1991)

Theoretical and Applied Rheology, edited by P. Moldenaers and R. Keunings
Proc. XIth Int. Congr. on Rheology, Brussels, Belgium, August 17-21, 1992

MIXING OF IMMISCIBLE LIQUIDS

J.M.H. JANSSEN[1], G.W.M. PETERS[1], H.E.H. MEIJER[1], F.P.T. BAAIJENS[1,2]

[1]Centre for Polymers and Composites, Faculty of Mechanical Engineering, Eindhoven University of Technology, P.O. Box 513, 5600 MB Eindhoven (The Netherlands)
[2]Philips Research Laboratories, P.O. Box 80000, 5600 JA Eindhoven (The Netherlands)

1. INTRODUCTION

The 'dispersive mixing' of immiscible liquids is an essential subject in many industrial processes involving multiphase flow. For example, in the melt blending of incompatible polymers the morphology of the blend is of major influence on its mechanical (and other) properties. Therefore, it is important to model the development of morphologies during the mixing process; starting with large (mm) domains and ending with a distribution of small (μm) domains of the dispersed phase (ref. 1).

The deformation of a dispersed liquid domain in a flow field of the continuous phase is governed by the Capillary number, defined as the ratio of the (deforming) shear stress τ and the (conservative) interfacial tension σ scaled on the characteristic radius of curvature of the interface:

$$Ca = \frac{\tau}{\sigma/R} \qquad (1)$$

During the initial stages of mixing the radius R is large (typically order of mm) so that τ is dominant over σ/R. In this case, only distributive mixing occurs and the parameters of interest are the total deformation applied by the compounding equipment and the number of folds or reorientations. The dispersed drops deform affinely with the flow field of the matrix.

Once thin liquid threads have been formed by affine deformation of the dispersed phase, local radii have decreased such (μm) that the interfacial stress σ/R becomes important. Distortions at the interface grow and cause breakup of the threads into lines of small droplets (dispersive mixing). These droplets may deform and break up again in the flow field or they may resist further deformation if they are small enough (σ/R dominant over τ). In other words: Drops do not break up if Ca is below a certain critical value Ca_{crit}. This value is a function of the type of flow and the viscosity ratio p between dispersed and continuous phase:

$$p = \eta_d \, / \, \eta_c \qquad (2)$$

Apart from breakup of drops via threads into smaller drops also coalescence of colliding droplets may occur, coarsening the morphology.

A classical way to model the mixing process of immiscible liquids is to study the deformation and breakup of a single dispersed drop using model liquids (at roomtemperature) in well defined flows. This study presents some results on distributive ($Ca >> Ca_{crit}$) and dispersive ($Ca \approx Ca_{crit}$) mixing in plane elongational flow, generated in an opposed jets device. Special attention is paid to viscoelasticity of the dispersed phase.

2. THE OPPOSED JETS DEVICE

In studies on drop deformation and breakup, a variety of experimental devices is used to generate specific types of flow (e.g., a Couette device for simple shear flow or a four roll mill for plane hyperbolic flow). For the present study a so called opposed jets device was developed to generate plane hyperbolic flow (ref. 2). The principle is shown in Fig. 1 and is based on the stagnation flow of two opposed jets that are forced inbetween four solid blocks, fixed between two transparant parallel plates. Advantages over the four roll mill are the small size of the device and its simplicity. Numerical simulations have shown that in the central region the flow field is plane hyperbolic (2D elongational), obeying the velocity field:

$$u = Gx, \qquad v = -Gy, \qquad w = 0 \tag{3}$$

where G (s^{-1}) is the velocity gradient or rate of elongation. From the simulations it turned out that G is more uniform (as it should be) in an opposed jets device than in a four roll mill. Experimental verification of the velocity field, using laser Doppler anemometry, confirmed the central area of constant G.

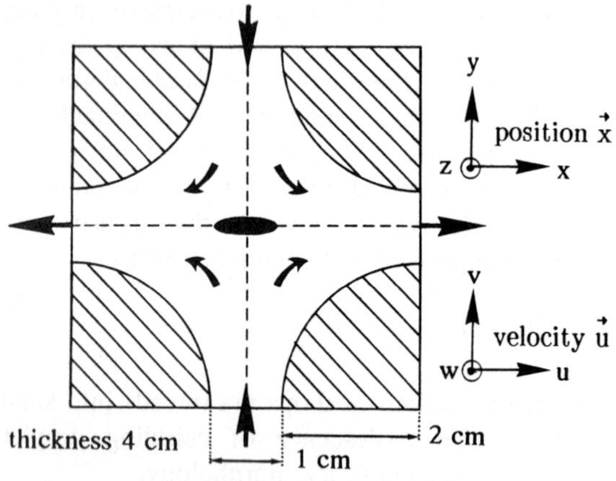

Fig. 1 Principle of the opposed jets device for drop deformation in plane hyperbolic flow.

An automatic control system is required to keep the deforming drop at its principally unstable position in the stagnation point, which it tends to leave. A control cycle consists of determination of the drop position (using a camera and an image processor), calculation of the required flow adjustment, and realization of the control action. Flow adjustment is carried out by decreasing the exit flow rate in the direction of the drop movement. In this way, the stagnation point is transferred beyond the drop forcing it to return to the centre. A sample rate of 25 control cycles per second results in a stable position of the deforming drop at the centre of the device at flow rates up to $G \approx 10\ s^{-1}$. Deformation of a dispersed (viscous or viscoelastic) drop or thread in elongational flow can now be studied using videorecordings through a microscope.

3. RESULTS ON DISTRIBUTIVE MIXING

If the applied Capillarynumber amply exceeds the critical value necessary for drop breakup, i.e. if $Ca \gg Ca_{crit}$, the drop is expected to deform affinely with the global flow field without any significant resistance of the interfacial tension. In practice, this is the case for large drops ($Ca \sim R$), thus in the initial stages of mixing. Using the opposed jets device, this can be simulated as follows.

A 1 mm drop of castor oil (Newtonian, viscosity 0.74 Pa·s) is subjected instantaneously to a steady plane hyperbolic flow of silicon oil (Newtonian, viscosity 0.94 Pa·s, interfacial tension to castor oil $4.1 \cdot 10^{-3}$ N/m). The velocity gradient G is made such that Ca exceeds Ca_{crit} with different ratios. In plane hyperbolic flow (ref. 1):

$$Ca = \eta_c 2GR / \sigma \tag{4}$$

A measure for the deformation of the drop, with long and short axes L and B, is usually chosen as:

$$D = (L-B) / (L+B) \tag{5}$$

and ranges from 0 for a sphere to 1 for an infinitely long and thin body. Upon integrating Eq. (3), it can be seen that if the drop deforms affinely with the matrix, it elongates exponentially with time (t); D is a function only of the applied deformation Gt:

$$D = (\exp(2Gt)-1) / (\exp(2Gt)+1) \tag{6}$$

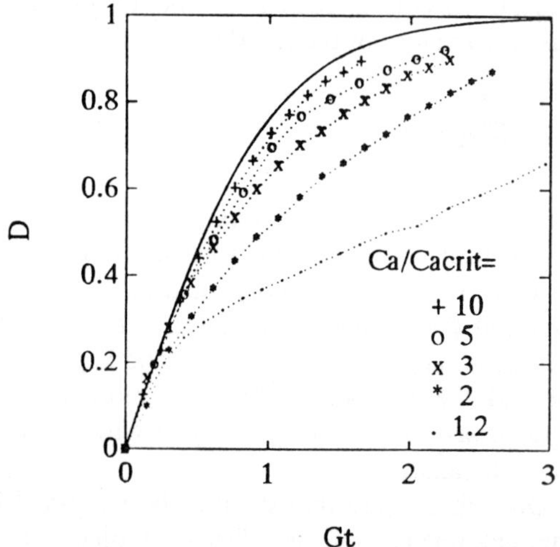

Fig. 2 Deformation of an initially spherical drop in a uniform plane hyperbolic flow at different ratios of exceeding Ca_{crit}; the full curve corresponds to affine deformation, Eq. (6).

In Fig. 2 the measured progress of D is shown. It can be seen that on exceeding Ca_{crit} the deformation tends to the theoretical curve for affine deformation. Using conservation of volume the width of the drop in z-direction can be estimated from L and B (measured directly). It seems that as the deformation progresses affinely this third drop axis remains constant (as it should, since w = 0). So the 3D shape of the drop in plane hyperbolic flow is like an ellipsoid with non-circular cross section.

4. RESULTS ON DISPERSIVE MIXING

As the dispersed drops have been elongated into long thin liquid threads, these threads become unstable. Very small distortions that are always present at the interface between thread and continuous phase are favoured to grow due to the interfacial tension. Depending on the viscosity ratio, distortions having one specific wavelength grow fastest and cause breakup of the thread into a line of small droplets. For the case of Newtonian liquids and without any external flow field the exponential growth of these sinusoidal 'Rayleigh distortions' can be calculated (ref. 3). An external flow field complicates the problem since the thread radius continuously decreases and the actual fastest growing wavelength is stretched so that it is no longer dominant: a new wavelength has to develop. As a consequence, breakup times of threads are increased by an external flow field (ref. 4).

The effect of viscoelasticity of the thread phase on the breakup of threads is even more dramatically. Initially, distortions grow even faster than in Newtonian systems (ref. 5, 6). But at a certain amount of deformation the amplitude almost stops growing and dumbbell shaped threads remain for a relatively long time (Fig. 3). This is due to built up elongational stresses in the connecting fibers, drastically increasing the elongational viscosity and thus inhibiting further drainage from the fibers into the bulbs. Breakup times of viscoelastic threads can be orders of magnitude longer than comparable (i.e. viscosity) Newtonian threads. Once a fiber has broken, the remaining parts elastically retract to neighbouring drops.

The influence of viscoelasticity of the dispersed phase on the value of Ca_{crit} (necessary for drop breakup) is much less. From experiments in the opposed jets device it appears that for viscosity ratios around unity Ca_{crit} does not significantly deviate from Newtonian systems (also ref. 6, 7).

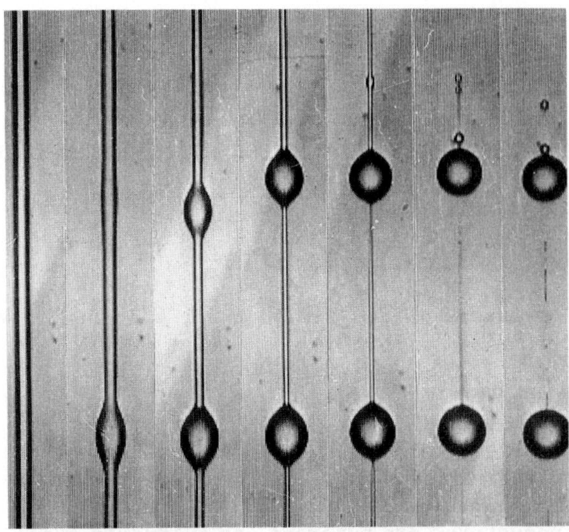

Fig. 3 Disintegration of a 0.07 mm viscoelastic thread (80% corn syrup/ 20% water/ 0.01% polyacrylamide) in a quiescent Newtonian matrix (silicon oil). Characteristic is the dumbbell shape. Photographs were taken after every 3 seconds.

5. NUMERICAL APPROACH

Progress is made in the numerical simulation of the drop / thread deformation problem. A finite element approach was chosen with an 'arbitrary Lagrange Euler' (ALE) mesh formulation. First the Newtonian problem has been solved whereafter a viscoelastic element is incorporated.

6. CONCLUSIONS

The opposed jets device is well suitable for drop deformation experiments. Upon sufficient excess of Ca_{crit}, drops deform nearly affinely. Viscoelasticity of the dispersed phase strongly decelerates the final stage of thread breakup.

REFERENCES

1. H.E.H. Meijer, J.M.H. Janssen; in: I. Manas-Zloczower (Ed.), Mixing and Compounding- Theory and Practice, Progress in Pol. Proc. Series, Carl Hanser Verlag, in press, Section 'Mixing of Immiscible Liquids'.
2. J.M.H. Janssen, G.W.M. Peters and H.E.H. Meijer; in: J. Laven, H.N. Stein (Ed.), IACIS conf. / event 439 of the EFChE 'The Preparation of Dispersions', Veldhoven, The Netherlands, October 14-16, 1991, 121-130. also: in press for Chem. Eng. Sci.
3. S. Tomotika; Proc. R. Soc., London, A 150 (1935), 322-337.
4. M. Tjahjadi, J.M. Ottino; JFM, 232 (1991), 191-219.
5. D.W. Bousfield, R. Keunigs, G. Marrucci, M.M. Denn; JNNFM, 21 (1986), 79-97.
6. H.B. Chin, C.D. Han, J. Rheol.; 24(1) (1980), 1-37.
7. W.J. Milliken, L.G. Leal; JNNFM, 40 (1991), 355-379.

Theoretical and Applied Rheology, edited by P. Moldenaers and R. Keunings
Proc. XIth Int. Congr. on Rheology, Brussels, Belgium, August 17-21, 1992

PREDICTING THE DYNAMICS OF EXTRUDATE SWELL FROM SHEAR VISCOSITY EXPERIMENTS.

R.J. KOOPMANS, D. PORTER

Dow Benelux N.V., P.O. Box 48, 4530 AA Terneuzen, The Netherlands.

1. INTRODUCTION

Predicting the swelling of a molten polymer resin emerging from a die is a challenge (ref.1). The difficulties encountered are numerous, as extrudate swell is a dynamic property depending on the molecular structure and the processing behaviour of the polymer resin in a rheometer or extrusion machine.

Although general agreement exists on the elastic nature of extrudate swell, none of the present day theories is capable of predicting the transient nature of the phenomenon accurately. Recently, the K-BKZ constitutive equation was solved for simulating industrial flows using numerical techniques (ref.2). However, certain "model" deficiencies will have to be sorted out still.

From an industrial point of view there is a need for simplified methods to capture, quantitatively, the physics of the transient nature of extrudate swell. Such methods are scarce (refs.3-5) and are not always applicable without difficulty. The present paper wants to bring a phenomenological model capturing the physics of polymer melt flow and allowing easy, quantitative prediction of the dynamics of extrudate swell of linear polyethylene.

2. THEORETICAL ASPECTS

The use of reduced variables is a well established procedure in science (refs.6-7). A shear viscosity - shear rate master curve can be obtained for linear polyethylene resins of different molecular weight as well as for data measured at different temperatures. The use of a topological model relating the deformation of a spherical fluid domain of radius R_0 into an ellipsoid with long axis R due to a shear field, allows in combination with a stress balance between viscous drag and elastic reaction on each domain, the definition of a master curve expression for normalised viscosity η/η_0 as a function of the normalized shear rate ($\dot{\gamma}\ \tau$) (ref.8).

$$\dot{\gamma}\tau=\left(\frac{\eta_0}{\eta}\right)\left(\left(\frac{\eta_0}{\eta}\right)^{\frac{1}{3}}-1\right) \qquad (1)$$

The polydispersity of polymer resins enters the mastercurve equation via τ (ref.9).
The maximum extrudate swell, Dm, of a polymer melt can be related to the recoverable strain of the fluid domain in the shear field and accordingly to the viscosity values via equation 2.

$$D_m=\left(\frac{R}{R_0}\right)^{\frac{1}{2}}=\left(\frac{\eta_0}{\eta}\right)^{\frac{1}{6}} \qquad (2)$$

The isothermal time dependent strain recovery of the fluid domain allows for a derivation of the extrudate swell as a function of the distance from the capillary die, z, in terms of maximum extrudate swell, Dm, and radius, r, of the capillary die.

$$z=rD_m^4(D_m^2-1)\left(\ln\left(\frac{(D_m+D)\ (D_m-1)}{(D_m-D)\ (D_m+1)}\right)-2\frac{(D-1)}{D_m}\right) \qquad (3)$$

Given the appropriate shear viscosity - shear rate mastercurve of a linear polyethylene resin the isothermal extrudate profile at any shear rate and capillary die geometry can be calculated.

3. RESULTS AND DISCUSSION

For two high density polyethylene (HDPE) resins - 801 and 802 - the isothermal extrudate swell, measured at 190 °C for a capillary die with L:D=30:2 mm, was compared to the model calculations. The influence of gravity was experimentally accounted for by measuring in oil of a density nearly equal to that of the polymer melt (ref.10). The zero shear viscosity of the resins were obtained via the HDPE mastercurve. The structural difference between the two resins was related to somewhat more high molecular weight material in 802. This was reflected in a higher low shear viscosity for 802. Analytical techniques such as Gel Permeation Chromatography and Low Angle Laser Light Scattering could not distinguish these samples unambiguously. The extrudate swell as function of time showed a lower value for 802 than for 801 at short times (seconds), while at longer times (minutes) the Dm of 802 was higher. The time at which the extrudate swell was equal for the two resins shifted to longer values when the shear rate or throughput increased. Similar experimental effects were observed for extrudate swell as function of distance from the die. The present model showed, within acceptable error, the same isothermal extrudate swell as function of distance features, for various shear rate or throughput.

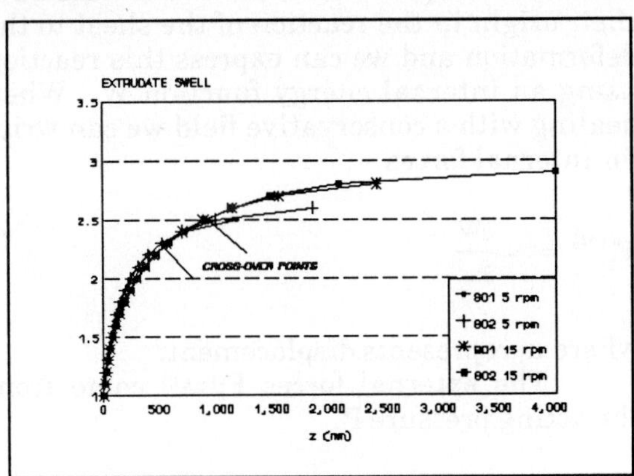

Figure 1: Calculation results of equation 3 for isothermal extrudate swell as function of distance from the die for two HDPE samples 801 and 802 at two laboratory extruder screw rpm.

Accordingly, a simple model has been developed, capable of quantitatively predicting isothermal extrudate swell, via directly accessible and measurable viscosity parameters. This is of special importance for understanding the short time dynamics of extrudate swell in fast operating machines e.g. extrusion blow moulding. Furthermore, the current use of normalised parameters emphasises the importance of time scaling and forms a basis for evaluating the performance of resins in industrial blow moulding operations.

REFERENCES

1. J.M. Dealy, K. Wissbrun, Melt Rheology and its Role in Plastics Processing, Van Nostrand, New York, 1990.
2. A. Goublomme, B. Draily, M.J. Crochet, in preparation.
3. F.N. Cogswell, Polymer Melt Rheology, John Wiley & Sons, New York, 1981.
4. R.I. Tanner, Appl. Polym. Symp., 20 (1973) 201.
5. J. Guillet, M. Seriai, Rheol.Act., 30 (1991) 540-548.
6. J.D. Ferry, Viscoelastic Properties of Polymers - 3rd ed., John Wiley & Sons, New York, 1980.
7. G.V. Vinogradov, A.Y. Malkin, J.Polym.Sci.Chem.Ed., 4 (1966) 135-154.
8. D. Porter, Polym. Eng. Sci., in press.
9. Y.H. Zang, R. Muller, D. Froelich, Polymer, 28 (1987) 1577-1582.
10. R.J. Koopmans, Polym. Eng. Sci., in press.

Theoretical and Applied Rheology, edited by P. Moldenaers and R. Keunings
Proc. XIth Int. Congr. on Rheology, Brussels, Belgium, August 17-21, 1992

MODELING OF THERMOFORMING AND BLOW MOLDING

K. Kouba and J. Vlachopoulos

CAPPA-D, Chemical Engineering, McMaster University, Hamilton, Ontario, Canada L8S 4L7

1. INTRODUCTION

Thermoforming and blow molding involve the biaxial deformation of molten or semimolten polymers. In thermoforming a thermoplastic heated sheet is shaped by applying pressure/vacuum or mechanical drawing for the fabrication of a variety of products. In blow molding usually a cylindrically shaped preform is heated and then inflated with or without plug assistance for the production of plastic bottles. The thickness of the sheet or parison is usually small when compared to other dimensions so that the thin membrane approximation can be applied. For modeling purposes a finite element grid can be created on the membrane surface and the deformation process can be studied in a Lagrangian frame of reference. Considerable difficulties are encountered in finite element modeling due to large strains, material nonlinearities, contact between polymer membrane and walls, and quite frequently physical or numerical instabilities.

A review of finite element modeling of thermoforming and blow molding has been published by deLorenzi and Nied (ref. 1). More recently, Song et al (refs. 2,3) Kouba et al (ref. 4) examined various aspects of thermoforming. Interesting analyses of blow molding were presented by Erwin et al (ref. 5) and Poslinski and Tsamopoulos (ref. 6).

Most of the published modeling studies use the Mooney or Ogden models of rubber elasticity. It is generally believed that viscoelastic effects can be neglected for certain problems, especially for thermoforming. However, a recent paper by Hylton (ref. 7) provides convincing evidence, based on practical experience, that viscoelastic effects are very important in thermoforming. Hylton gives several examples of how the thermoformability of a given resin may be assessed on the basis of several measurable viscoelastic properties. The same arguments can also be extended to blow molding. In this paper, we present both the rubber elasticity and viscoelasticity approach to modeling of thermoforming and blow molding.

2. RUBBER ELASTICITY MODELING

To model the inflation process we may assume that the polymer membrane goes through a set of quasistationary positions and that in each of these positions the acting forces are in equilibrium. Internal forces have their origin in the reaction of the sheet to the deformation and we can express this reaction using an internal energy function W. When dealing with a conservative field we can write for internal forces

$$F_i^{(int)} = -\frac{\partial W}{\partial u_i} \tag{1}$$

where u_i represents displacement.

The external forces $F_i^{(ext)}$ come from the acting pressure P

$$F_i^{(ext)} = pn_i \tag{2}$$

This force acts perpendicularly to the surface in the direction of the outward normal vector n_i. For the entire membrane, we have

$$\sum_i (F^{(ext)} + F^{(int)} = 0 \tag{3}$$

The summation is over all elements e.

Equation (3) represents a set of equations which are nonlinear due to both geometrical and physical considerations. The standard Newton-Raphson iterative method with a frontal equation solver were used.

For large deformations for rubber sheets the Mooney model is frequently used, with

$$W = C_{10}(I_1 - 3) + C_{01}(I_2 - 3) \tag{4}$$

where I_1, I_2 are the scalar invariants of the strain tensor. When $C_{01} = 0$ and $C_{10} = G$, this model is known as neo-Hookean.

$$W = G(\lambda_1^2 + \lambda_2^2 + \lambda_3^2 - 3) \tag{5}$$

where G is the elastic modulus. It appears that Ogden's model is more flexible. The strain energy density function is assumed to be in the form

$$W = \sum_{n=1}^{m} \frac{\mu_n}{\alpha_n} (\lambda_1^{\alpha_n} + \lambda_2^{\alpha_n} + \lambda_3^{\alpha_n} - 3) \tag{6}$$

Here, α_n, μ_n are material constants which fit the results of large stress-strain experiments and m is usually 1, 2, or 3.

The Ogden model was very suitable for the present calculations and was used for the majority of problems solved while the neo-Hookean model gave considerable numerical instabilities. The results shown in Figure 1 were obtained by fitting stress-strain data to the Ogden model and then using the finite element method described above.

3. VISCOELASTIC MODELING

The K-BKZ model (ref. 8) is used for the description of time dependent deformation of a polymeric sheet.

$$\sigma_{ij}(t) = \int_{-\infty}^{t} \mu(t-t')\, h(I_1, I_2)\, B_{ij}(t, t')\, dt' \tag{7}$$

$\mu(t-t')$ is a material memory function

$$\mu(t-t') = \sum_i a_i\, e^{-(t-t')/\tau_i} \tag{8}$$

$h(I_1, I_2)$ is an appropriate damping function of the two strain invariants, e.g. Wagner's type (ref. 9)

$$h(I_1, I_2) = \frac{1}{1 + a\sqrt{(I_1 - 3)(I_2 - 3)}} \tag{9}$$

a is a material parameter.
$B_{ij}(t, t')$ is the Finger strain tensor.

Assuming that a pressure is applied and the deformation of the planar sheet starts at time t = 0, we may write the stress in two principal directions as (for i = 1, 2)

$$\sigma_{ii}(t) = \int_0^t \mu(t-t')\, h(I_1, I_2)\, [L_i^2(t, t') - L_3^2(t, t')]\, dt' + h(t)\, [L_i^2(t) - L_3^2(t)]\, G(t)$$

$$G(t) = \int_{-\infty}^{0} \mu(t-t')\, dt' \tag{10}$$

$L_i(t, t')$ is the stretch ratio in time t, related to time t'.

With the above expression, the past history of deformation of each element must be taken into account when performing the balance of internal and external forces

$$\sum_i F_e^{(ext)} + F_e^{(int)} = 0 \tag{11}$$

These expressions were solved using the Newton-Raphson method with a frontal solver. Instability problems, including aneurysms (ref. 5), are overcome using a modification of Riks' method (ref. 10).

By carrying out the finite element calculations, we may examine the effects of process rate and strain hardening. For example, the thickness distributions for two different applied pressures are shown in Fig. 2. There are noticeable differences in deformed sheet thickness depending on whether the process is slow (5 kPa) or fast (20 kPa). A blow molding simulation is shown in Figs. 3. These simulations indicate that the wall thickness distributions are determined both by the viscoelastic material properties and the the operating conditions, i.e. whether the process is slow or fast and whether prestretching is applied.

4. REFERENCES

1. H.G. deLorenzi and H.F. Nied, "Finite Element Simulation of Thermoforming and Blow Molding", in Progress in Polymer Processing, A.I. Isayev (ed.), Hanser Verlag, Munich, 1991.
2. W.N. Song, F.A. Mirza and J. Vlachopoulos, J. Rheol., 35 (1991), p. 93.
3. W.N. Song, K, Kouba, F.A. Mirza and J. Vlachopoulos, 49th ANTEC Tech. Papers, 37 (1991), p. 1025.
4. K, Kouba, O. Bartos and J. Vlachopoulos, Polym. Eng. Sci., to be published (1992).
5. L. Erwin, M.A. Pollock and H. Gonzalez, Polym. Eng. Sci., 23 (1983), p. 826.
6. A.J. Poslinski and J.A. Tsamopoulos, AIChE J, 36 (1990), p. 1837.
7. D. Hylton, 49th ANTEC Tech. Papers, 37 (1991), p. 580.
8. R.I. Tanner, Engineering Rheology, Clarendon Press, Oxford (1985).
9. M.H. Wagner, J. Rheol., 34 (1990), p. 943.
10. B. Ramm in: W. Wunderlich, E. Stein and K.J. Bathe (eds.), Non-Linear Finite Element Analysis in Structural Mechanics, Springer, New York, 1981, pp. 63-89.

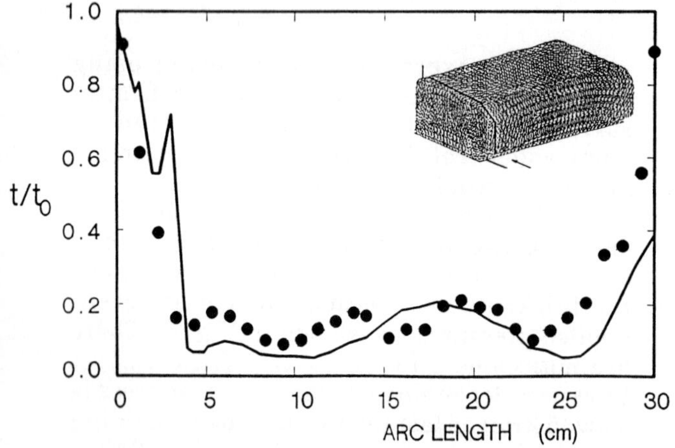

Fig.1. Comparison of predicted thickness distribution and the experimental data of deLorenzi et al. (1).

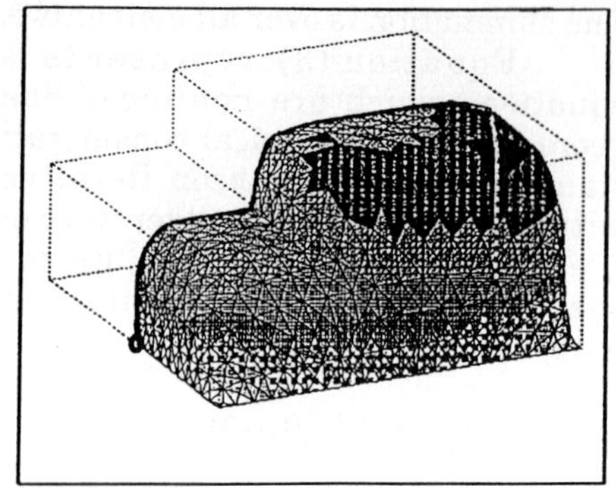

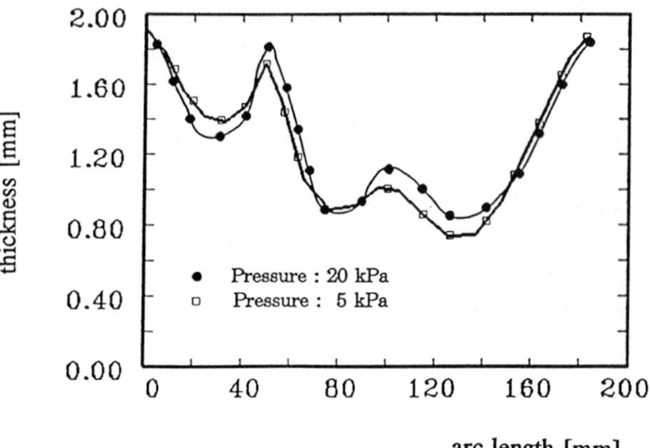

Fig.2. Modeling of thermoforming for K-BKZ material and corresponding thickness profile along the cut line obtained at pressure 5 [kPa] and 20 [kPa].

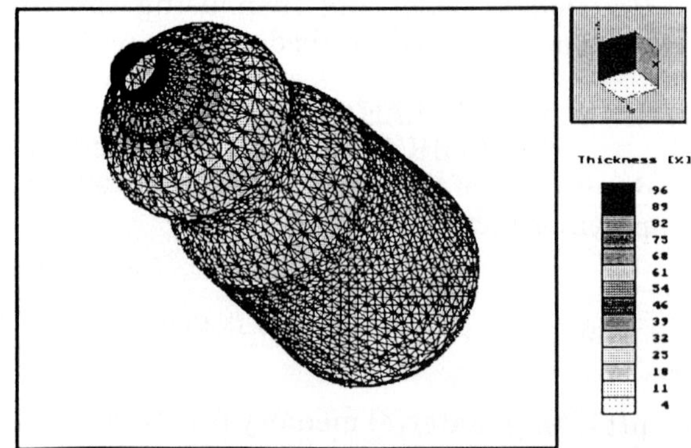

Fig.3. Blow molding simulation.

Theoretical and Applied Rheology, edited by P. Moldenaers and R. Keunings
Proc. XIth Int. Congr. on Rheology, Brussels, Belgium, August 17-21, 1992

THE DYNAMICS OF SHARKSKIN MELT FRACTURE: EFFECT OF DIE GEOMETRY

STUART J. KURTZ
Union Carbide Chemicals and Plastics Company Inc., Polyolefins Division Research and Development
P.O. Box 670, Bound Brook, New Jersey 08805 (USA)

SUMMARY

We have proposed a general mechanism for the process of sharkskin melt fracture which not only is consistent with observed frequency and depth but also suggests that some forms of gross melt fracture can be viewed as an extension of the sharkskin process. This mechanism is supported by studies on LLDPE as a model for narrow molecular weight linear polymers showing the effect of shear stress, molecular weight, temperature and, in the current study, die diameter, on the severity of sharkskin melt fracture. Experimentally, the frequency of SSMF ranged from 10 Hz for high molecular weight, low shear stress conditions to over 400 Hz for low molecular weight, high shear stress conditions. These frequencies appear to be unaffected by die diameter but strongly dependent on shear stress in the die. The fracture depth however is a strong function of die diameter, being directly related to the stretching field outside the die. The results of our study imply that explanations based solely on either land slip or exit stretching are inadequate.

1. INTRODUCTION

A full description of sharkskin melt fracture (SSMF) should include the ability to predict and explain not only its onset but the frequency at which ridges form and their depth as well. Over the last 8 years or so some key studies have tested concepts about mechanisms concerning the onset of SSMF. Some of the mechanisms include the effects of die geometry, process aids, adhesion at the wall, extensional flow at the exit and cavitation. The work of Kurtz (ref. 1) noted the coincidence of a change in slope of the log shear stress-log shear rate curves and the visual onset of sharkskin. In the same paper, under the section "A General Theory of Sharkskin", a critical wall shear stress and a critical acceleration and stretching were found to be necessary and sufficient to explain the onset of SSMF data. It was also shown that offset dies in a tubular blown film geometry can significantly reduce SSMF.

Recent studies (ref. 2) point to the effect of pressure on the sharkskin process and note that as the pressure increases in the upstream direction in the die, the degree of slip must decrease. The slip process is not uniform in the die.

Most studies have focused on the conditions necessary for the onset of SSMF. To understand the mechanism of sharkskin, we recently began looking at the dynamics of SSMF. In the first part of this study, we showed how the temperature, molecular weight, and shear stress at the die wall influence both frequency and depth (ref. 3). Figure 1 shows the important result that frequency does not go to zero as the depth goes to zero. Material emerging from the die land in a periodic manner due to an internal slip-stick process combined with a stretching distribution at the exit will determine crack growth and thus its depth. Stretching will be a function of the average velocity and die diameter.

Figure 1. Frequency versus Depth

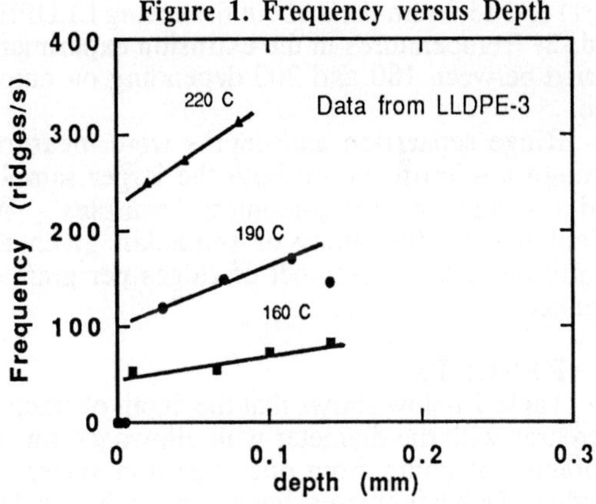

Based on studies by ourselves and from works published by others we propose an integrated model with the following elements: 1. Sharkskin is a fracture process that occurs during stretching. The fracture itself occurs in the region of separation of the polymer from the die land. 2. Fracture is dependent on the material state. Stretching histories, entanglement density, local compositional distribution of molecules and the type of molecules determine the propensity to fracture. 3. Exceeding a local critical shear stress, a discontinuous slip stick process occurs. This process of alternating slip-stick in the die provides an additional stretching at the die exit and governs the frequency of the ridges making up sharkskin. These frequencies are generally very high, extending at least to several hundred ridges/s. 4. Slip-stick in the die is a function of position in the die with the amplitude of the slip phase of the process increasing towards the die exit. (It is not clear whether the slip process is everywhere simultaneous in the die or whether it is a traveling wave type process). In capillary dies, the critical shear stress for slip increases as the pressure increases upstream in the die and will be dependent to some extent on the dynamic chemical and physical makeup of the die land surface. The overall phenomenological result can be seen as a change in slope of the log shear stress - log shear rate plot.

2. PROCEDURE

To test the proposed mechanism we sought to vary the exit velocity profile while keeping the shear stress in the die constant. We used various butene comonomer LLDPE resins having melt indices of 0.5 (LLDPE 1), 1.0 (LLDPE 2), and 2 (LLDPE 3). One set of tests was done in a capillary rheometer having 20:1 L:D dies of diameters 0.76, 1, 1.5, 2, and 2.5 mm using a LLDPE-2 at 220 C. Scale up tests on extrusion dies were done on a commercial 2.5 inch extruder with dies of 6.4 mm (L:D about 12:1) and 15.9 mm (L:D about 6:1) using LLDPE-1 and -2. Temperatures in the extrusion experiments varied between 180 and 200 depending on output rate.

Ridge separation and depths were measured through a microscope on both the larger samples and on some of the rheometer samples. We calculated the frequency of sharkskin given the output rate and the number of ridges per gram of sample.

3. RESULTS

Table 1 below shows that the depth of fracture increases with die diameter with allowance for the problems of going from one extrusion system to another. Depth increases, but at a rate which is less than proportional to the diameter so that it does not scale geometrically. There is significant scatter in the measure of frequency perhaps due to differences in the die surfaces. It appears that the frequency is not a strong function of the diameter but seems more closely related to the shear stress. The shear stresses for the extrusion dies are only approximate.

Studies done with the dies ranging from 0.76 mm to 2.5 mm in diameter showed essentially the same critical shear stress for sharkskin (0.12 MPa) and about the same critical shear stress for gross melt fracture (0.3 MPa).

TABLE 1 Frequency and Depth vs. Die Diameter

diam. (mm)	shearstress (MPa)	#ridges/s (@0.2MPa)	depth (mm)
0.76	0.2	120	0.04
1.0	0.2	47	0.04
1.5	0.2	50	0.08
2.0	0.2	150	0.15
6.4	0.2?	80	0.14
15.9	0.2?	80	>0.2

4. DISCUSSION

We feel we can now integrate the understanding of sharkskin to include flow processes before and after the critical shear stress. To do this we first focus on the reason for the change in slope at the onset of sharkskin. See Figure 2 below.

Following Hatzikirakos and Dealy (ref. 2), we take the slip process as beginning at a minimum shear stress and the lowest normal stress at the wall. In a capillary die this would occur first at the die exit and involve more of the upstream flow with increasing output rate. Slip will add to the nominal shear rate with a lesser effect on shear stress. Since this is a continuously cumulative process over larger areas, we would expect, and do see, a change in slope of the log shear stress - log shear rate curve after the onset of some slip at the critical shear stress. Locally, the amplitude of a slip wave should increase as more of the die land is involved in slip with increasing stress. We also note that the frequency increases with stress.

Figure 2. Generalized SSMF Behavior

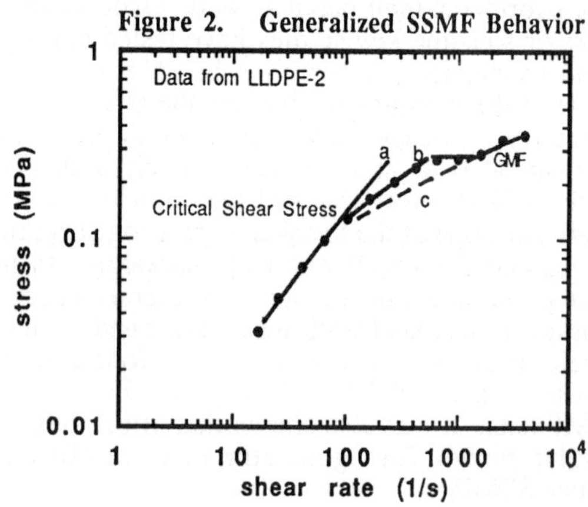

The extrapolation of the actual curve before the break point corresponds to the no slip condition. This is line (a) in Figure 2. The data in the figure shows typical LLDPE behavior. As shear stress is increased, an unstable region corresponding to gross melt fracture (GMF) occurs at about 0.3 MPa. In this region, the lower shear rate end is simply an extension of the sharkskin melt fracture slip stick process. At the upper end is the continuous frictional slip process. By connecting the slip line (c) with the critical point we have an extrapolated "slip" line. The actual curve (b) is thus an integrated response for flow between the slip and stick times at all the positions along the die land.

Gross melt fracture seen in this figure is known as slip-stick melt fracture. The period for "slip" and "stick" are often several seconds. Our interpretation is that the "slip phase" is actually continuous slip and the "stick phase" is an extension of the sharkskin high frequency alternating slip-stick.

Certain predictions of this model may be noted. At constant shear stress, an increase in capillary pressure should decrease sharkskin and the process of "slip-stick" gross melt fracture, but should not change the critical shear stress. Extrusion into a high pressure reservoir should delay and lessen the sharkskin process. Continuous slip should occur earlier in shorter dies at a given shear stress. The sharkskin slip-stick process in the die should have the same frequency as the ridges which appear on the sample.

5. CONCLUSIONS

The frequency of sharkskin measured in the extrudate samples and the slip-stick in the die appears to be controlled by the shear stress in the die. Combined with other studies we are led to speculate that the frequency of sharkskin is determined by the frequency of slip-stick in the die. In addition, the depth of sharkskin increases at constant die shear stress with increasing diameter dies. Our model suggests that the stretching process plays a key role in determining the severity of sharkskin.

We believe that the cumulative evidence from studies on sharkskin melt fracture using various materials, surfaces, geometries, stresses, temperatures, process aids, and deformation histories is consistent with our proposed model. This model is testable and can lead to a better understanding of the rheology of many polymers melts being processed under commercial conditions.

ACKNOWLEDGEMENTS

I wish to thank Union Carbide for permission to present and publish results of our studies on SSMF. I would also like to thank my colleagues both within Union Carbide and outside who have provided stimulating discussions; thus helping clarify in my mind some of the key issues. Both R. Vatalaro and W. Kendzulak did a superb job in assisting in running these rheological studies and in data reduction.

REFERENCES

1. S. J. Kurtz, in B. Mena, A Garcia-Rejon and C. Rangel-Nafaile (Eds.), Advances in Rheology, IX ICR, October 8-13, 1984, Vol 3, pp 399-407.

2. S. G. Hatzikiriakos and J. M. Dealy, "Wall Slip of Molten HDPE. II. Capillary Rheometer Studies" Submitted to Journal of Rheology.

3. S. J. Kurtz, Program and Abstracts: Polymer Processing Society 7th An. Meet. Hamilton, Ontario, April 21-24, 1991, pp 54-55

Theoretical and Applied Rheology, edited by P. Moldenaers and R. Keunings
Proc. XIth Int. Congr. on Rheology, Brussels, Belgium, August 17-21, 1992

MELT FLOW INSTABILITY OF LINEAR POLYETHYLENES WITH VERY BROAD MOLECULAR WEIGHT DISTRIBUTION

S.-Y. LAI and G.W. KNIGHT, Polyolefins & Elastomers R&D Dept., Dow Chemical U.S.A., B-1221 Bldg., Freeport, Texas 77541, U.S.A.

1. INTRODUCTION

The occurrence of non-smooth extrudate of melt flow instability has long been observed during extrusion of linear polyethylene melts at a high throughput rate. This phenomenon has also long been associated with some of the molecular structures such as very high molecular weight or narrow molecular weight distribution. It is also reportedly shown that the broadening of the molecular weight distribution of a resin would generally alleviate its susceptibility to the instability during extrusion. In many practical applications, very high and very low molecular weight polyethylenes are often blended to become very broad molecular weight distribution commercial polyethylene products. In this study, various GPC shapes of very broad molecular weight distribution linear low density polyethylene samples are obtained via solution blending. These samples were then used to perform melt instability experiments. Several unique melt flow characteristics of these samples were observed and are reported in this study.

2. MATERIAL AND EXPERIMENTAL

Three linear low density polyethylene (L-LDPE) polymers were used in this study. The melt flow index (MFI) and molecular weight distributions of these samples are listed in Table I. Molecular weight distribution curves obtained from Gel Permeation Chromatography (GPC) are shown in Figure 1. Samples B and C are solution blended broad molecular weight distribution samples. Sample A is a non-blended LLDPE resin for comparison purposes.

Table I.

Sample	MFI	Mw	Mw/Mn
A	0.58	134,000	2.89
B	0.51	163,000	5.17
C	0.66	137,000	15.0

Rheological characterizations were performed using a Gas Extrusion Rheometer (GER), and a Rheometrics Mechanical Spectrometer (RMS-800). The GER procedure was performed by extruding the material at 190 ^{o}C by pressurized nitrogen gas using a 0.0296" diameter, 20 L/D die. The RMS-800 experiments were carried out in the dynamic rate sweep mode ramping the shear rate from 0.01 to 100 radians per second.

3. RESULTS AND DISCUSSIONS

The RMS-800 dynamic viscosity vs. frequency plots for samples A-C are shown in Figure 2. It is expected that the sample with a broader molecular weight distribution will exhibit a higher shear sensitivity (with the order of A<B<C). The GER apparent shear stress vs. shear rate plots are shown in Figure 3.

It is clearly shown that the flow curve of sample C is very different from the flow curves of sample A and B. Surprisingly this broadest molecular weight distribution sample exhibits a fast increase in flow rate at an apparent shear stress of 0.03 * 10^6 Pa., which was observed as the onset of melt instability. The flow discontinuities for samples A and B were observed at the shear stress of about 0.35 * 10^6 Pa., substantially higher than the onset stress of melt flow instability for sample C. Moreover, the flow rate of sample C tends to increase at a slower rate when extruded at a higher shear stress of about 0.2 * 10^6 Pa. It is also unexpected since it has been commonly postulated that an adhesive failure between the polymer melt and the die wall results in a "stick-slip" phenomenon observed at this shear stress level [1].

The surface characteristics of the extrudates obtained from the GER for sample C were also analyzed using an optical microscope. It was very interesting that when the sample C extruded at a shear stress of 0.04 * 10^6 Pa. it exhibited a spiral-shape extrudate. The surface of this extrudate is rather smooth and the spiral flow pattern is very repetitive. When the extrusion stress increases, the frequency of the spiral shaped extrudate also increases. The surface of the extrudate finally shows non-repetitive distortion at a shear stress of 0.2 * 10^6 Pa. or higher.

Slip analysis was performed for sample C, extruded by the GER using three different diameter dies. The results are given in Figure 4. The slope of this plot is proportional to the "slip" velocity [2]. At shear stress of 0.03 * 10^6 Pa. or lower, the slope is zero, therefore there is no "slip". Conventional "slip" flow was reported only at shear stress of 0.1 * 10^6 Pa. or higher. However, some non-conventional "slip" flows were observed at shear stresses from 0.04 * 10^6 Pa. to 0.2 * 10^6 Pa. Those flows tend to increase as the extrusion stress increases (Figure 4). Further study of Sample A extruded at the same stresses confirms that the mechanism of "slip" does not occur at such a low shear stress of extrusion. A "cohesive failure" mechanism is believed to cause this type of melt instability. Note that Sample C consists of two distinct types of molecules with a substantial amount of low molecular weight fraction. This low molecular weight fraction tends to reduce the number of entanglements and highly reduces the strength of the melt. As a result of the weakness of melt entanglements, Sample C could suffer cohesive failure, which leads to a melt instability in the extrusion process. The melt instability due to cohesive failure is postulated at a low shear stress level as compared with the normally accepted adhesive failure (~ 0.1- 0.2 * 10^6 Pa.) for polyethylenes.

4. CONCLUSIONS

The melt flow properties of some broad molecular weight distribution polyethylenes were characterized. The low shear rheological properties of these samples show the consistent trend of higher shear sensitivity for broader molecular weight distribution material. However, a very unique melt flow instability phenomenon was observed for the sample with the broadest molecular weight distribution (Mw/Mn = 15). The onset of melt instability for this sample was observed at shear stress of 0.04 * 10^6 Pa., substantially lower than the commonly accepted critical shear stress of 0.1- 0.2 * 10^6 Pa. A non-conventional slip flow was also observed and calculated for this sample extruded at shear stresses between 0.04 * 10^6 Pa. to 0.2 * 10^6 Pa. A "cohesive failure" mechanism was postulated to account for this type of melt instability, based on the GPC curve, where the sample consists only two distinct amounts of high and low molecular weight fractions.

REFERENCES

1. A.V. Ramamurthy, "Wall Slip in viscous Fluids and Influence of Material Construction", J. of Rheology, 30(2), 337-357 (1986)
2. D.S. Kalika and M.M. Denn, "Wall Slip and Distortion in Linear Low Density Polyethylene", J. of Rheology, 31(8), 815-834 (1987)

4. D.M. Binding, J. Non-Newton. Fluid Mech. , 27(1988)173.
5. J.Z. Liang, Acta Mechanica Sinica, 1(1990)79.

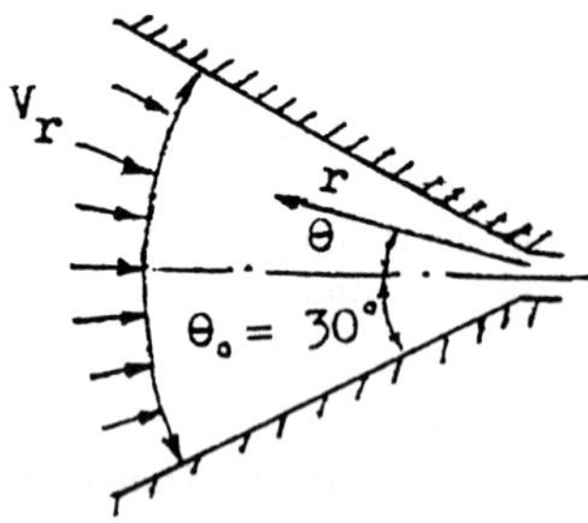

Fig.1 Diagram for a conical die flow.

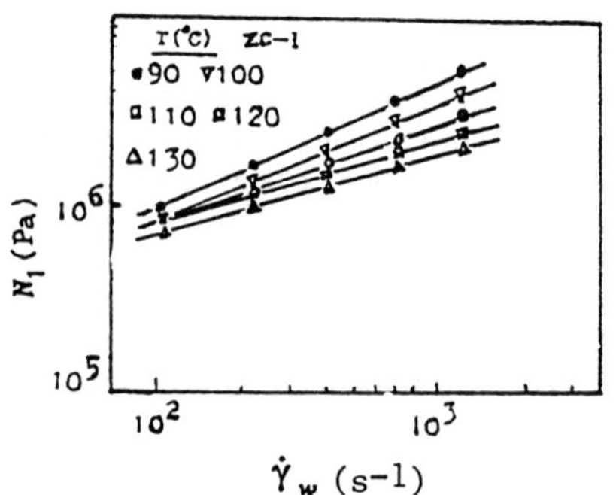

Fig.3 The first normal stress difference versus shear rate at different temperature for the sample.

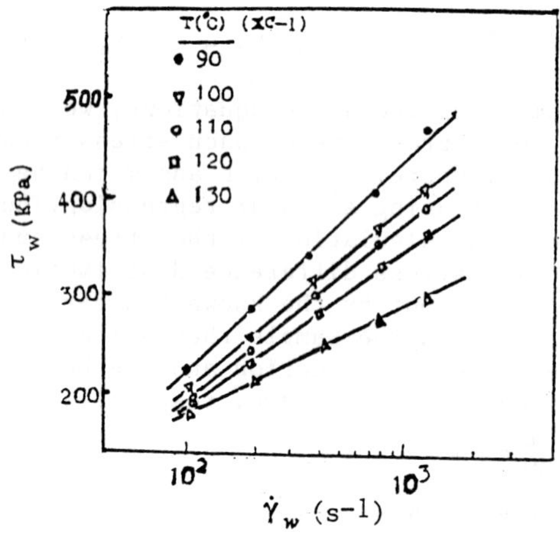

Fig.2 Flow curves of the sample.

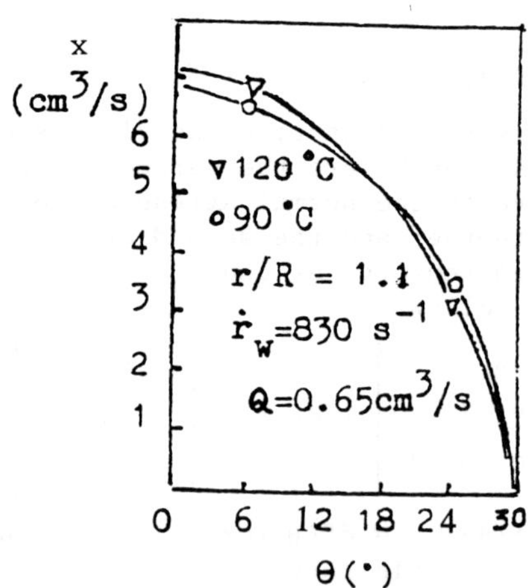

Fig.4 Dependence of flow rate on θ.

Theoretical and Applied Rheology, edited by P. Moldenaers and R. Keunings
Proc. XIth Int. Congr. on Rheology, Brussels, Belgium, August 17-21, 1992

EXTRUDATE SWELL IN POLYMER MELT FLOWS THROUGH SHORT AND LONG DIES

E. MITSOULIS and H.J. PARK

Department of Chemical Engineering, University of Ottawa, Ottawa, Ontario, K1N 9B4 (Canada)

1. INTRODUCTION

In extrusion of polymer melts through capillary and slit dies it is well-known that the ratio die-length/diameter (or slit gap) affects the extrudate swelling of the materials (1). This behavior is a manifestation of memory phenomena associated with the viscoelasticity of polymeric materials. Thus for short dies the travel time is shorter than the fluid memory, so that an increased swelling is observed as the fluid tends to recover the shape it had in the reservoir. With sufficiently long dies the fluid memory fades almost completely, and an asymptotic swelling ratio is reached.

A successful simulation of this phenomenon was recently achieved by Luo and Tanner (2), who used a K-BKZ integral constitutive equation for flow of an IUPAC-LDPE melt flowing from a 45°-entry tubular reservoir into capillary dies of different length. The extrudate swell results closely matched experimental findings (3).

This work presents some new results for an LLDPE melt flowing from a 90°-entry planar reservoir into a short ($L/2H_o = 2$) and a long ($L/2H_o = 8$) slit die. The emphasis is on determining extrudate swell as a function of flow rate (or equivalently, apparent shear rate), as well as entry flow patterns.

2. MATHEMATICAL MODELING

The flow is governed by the usual conservation equations of mass and momentum for an incompressible fluid under isothermal, creeping flow conditions ($Re \approx 0$). The constitutive equation that relates the extra stresses $\bar{\bar{\tau}}$ to the deformation history is a K-BKZ equation proposed by Papanastasiou et al. (4) and is written as:

$$\bar{\bar{\tau}}(t) = \int_{-\infty}^{t} \left[\Sigma \frac{a_k}{\lambda_k} \exp\left(-\frac{t-t'}{\lambda_k}\right)\right] \times \frac{\alpha}{(\alpha-3)+\beta I_{C^{-1}}+(1-\beta) I_C} \bar{\bar{C}}_t^{-1}(t')\, dt' \quad (1)$$

where λ_k and a_k are the relaxation times and relaxation modulus coefficients at a reference temperature T_o, respectively, α and β are material constants, and I_c, I_c^{-1} are the first invariants of the Cauchy-Green tensor $\bar{\bar{C}}_t$ and its inverse $\bar{\bar{C}}_t^{-1}$, the Finger strain tensor.

Special numerical schemes have been developed for the solution of the above integral constitutive equation (5). The finite element method is the method of choice and special integration procedures are necessary to deal with flows having both open and closed streamlines, e.g. regions with recirculation. Galerkin discretization is applied and the primary variables are the velocities and pressure (u-v-p formulation). The solution procedure advances slowly from low flow rates (Newtonian behavior) to higher ones by using either a flow rate increment scheme or an elasticity increment scheme (5).

3. RHEOLOGICAL CHARACTERIZATION AND MODELING OF LLDPE MELT

The melt under consideration is a linear low-density polyethylene (LLDPE), called Lotrex FC1010 and used in a previous experimental study by Beaufils et al. (6). Figure 1 gives the shear viscosity η_s and first normal stress difference N_1 as a function of shear rate $\dot{\gamma}$ at the temperature of the experiment (145°C). In the same figure we plot the model predictions obtained by best-fitting the data using Eq. (1) with a spectrum of relaxation times. The material parameters

Table 1: Material parameter values used in Eq. (1) for fitting data of LLDPE melt at 145°C ($\alpha = 5.0$, $\beta = 0.5$).

k	λ_k (s)	a_k (Pa)
1	1×10^{-4}	1×10^{5}
2	1×10^{-3}	7×10^{5}
3	1×10^{-2}	2.2×10^{5}
4	1×10^{-1}	7.7×10^{4}
5	1×10^{0}	1×10^{4}
6	1×10^{1}	2×10^{2}
7	1×10^{2}	9×10^{-3}
8	1×10^{3}	6×10^{-3}

obtained from the best fit are given in Table 1. For the elongational viscosity η_E and due to lack of measurements for this particular polymer melt, we have adopted a value of $\beta = 0.5$ in Eq. (1), so that a reasonable strain-thinning curve can be obtained for high elongational rates, in agreement with experiments for LLDPE melts (7).

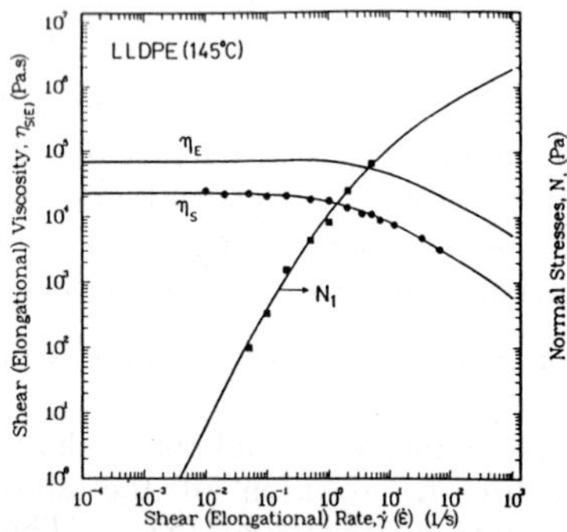

Fig. 1 Model predictions of shear viscosity η_s, first normal stress difference N_1, and elongational viscosity η_E for LLDPE at 145°C using Eq. (1). Symbols correspond to experimental data (6).

4. RESULTS AND DISCUSSION

The numerical simulation of the flow and birefringence experiments carried out previously with a particular LLDPE melt (6) involves planar entry and exit flow from a slit die, as shown schematically in Fig. 2. Dies of different lengths and different entry angles were used in the experimental setup. In this work we shall focus our attention on dies 1 and 2 (Fig. 3) for which extensive experimental data on birefringence have appeared in the literature (6).

In our computations we started as usual from the Newtonian creeping flow solution and increased the apparent shear rate value $\dot\gamma_a$ to reach higher shear rates (or equivalently higher flow rates). The streamline patterns for different apparent shear rates are shown in Fig. 4. We have also calculated

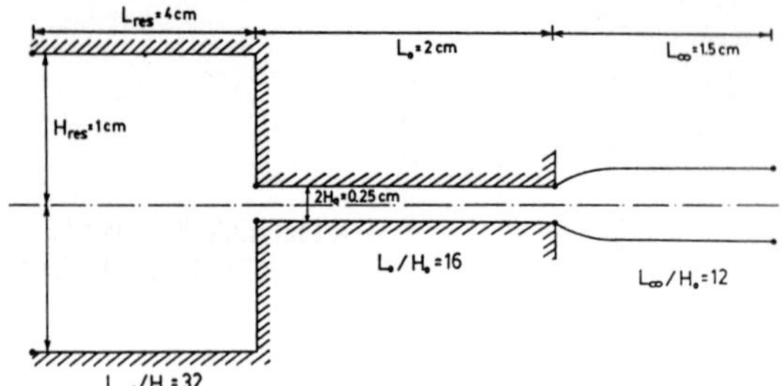

Fig. 2 Schematic diagram of the planar contraction geometry, definitions and dimensions of relevant lengths (die 1).

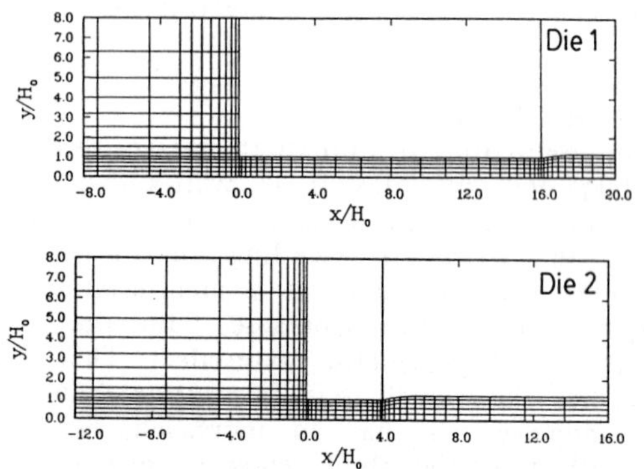

Fig. 3 Partial view of the finite element meshes used in the computations for the 8:1 planar abrupt contraction (die 1, upper half; die 2, lower half). The full meshes extend upstream from the contraction to $-32H_o$ and downstream after the die exit to $+12H_o$.

the corresponding stress ratio (or recoverable shear) S_R, defined by

$$S_R = \frac{N_1}{2\tau_w} \tag{2}$$

where N_1 is the first normal stress difference and τ_w is the shear stress at the die wall. Note that $S_R < 1$ for the range of experimental $\dot\gamma_a$ values.

A very small and weak vortex exists, which is reduced in intensity with higher flow rates. This is consistent with general trends found from numerical simulations in planar contractions (8) and experiments carried out for linear polyethylenes such as LLDPE (9).

The numerical simulations provide the extrudate swell for each shear rate, since they include the die exit region. The results are shown in Fig. 5. It is interesting to see that initially there is a contraction below the Newtonian value of 19% and then a slight increase, where the swell ratio finally reaches an asymptotic value and remains

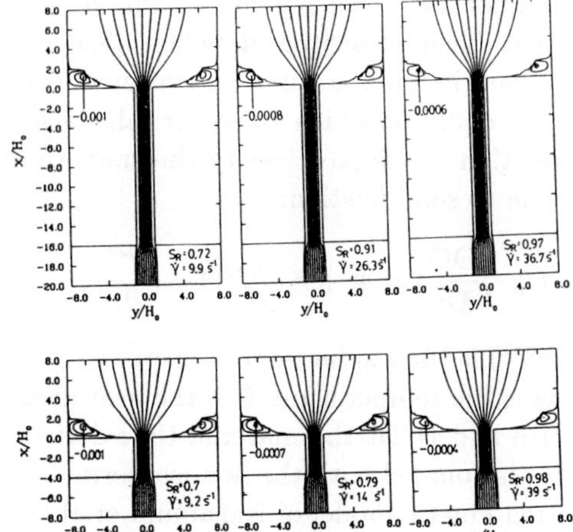

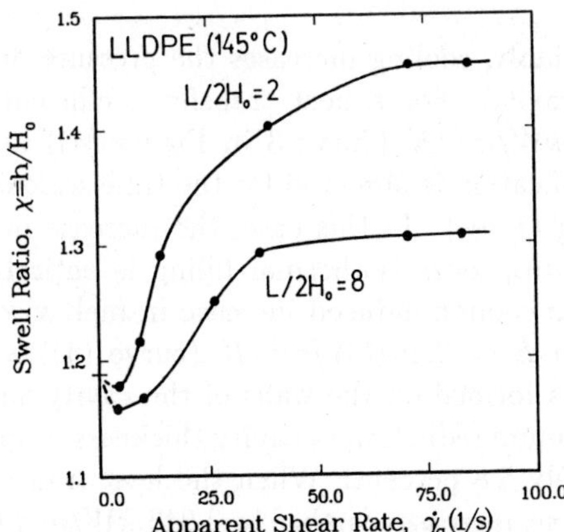

Fig. 4 Streamline patterns at different apparent shear rates for the flow of the LLDPE melt (145°C) in planar contractions (die 1, upper half; die 2, lower half).

unaltered. These results are consistent with previous simulations of viscoelastic constitutive equations that predict a slight decrease for low S_R values (10) and with experimental observations by Beaufils (11), who did not find any appreciable swelling of the LLDPE polymer. However, the well-known phenomenon of enhanced elastic recovery is very well captured when the melt is extruded from a short die ($L/2H_o = 2$) where the swelling reaches about 47% and is always higher than the swelling obtained from the long die ($L/2H_o = 8$) where it reaches about 35%. These results are also in agreement with similar previous computations by Luo and Tanner (3) for an LDPE melt through tapered contractions and circular dies.

Fig. 5 Extrudate swell as a function of apparent shear rate for the LLDPE melt extruded through a long ($L/2H_o = 8$) and a short ($L/2H_o = 2$) die at 145°C.

5. CONCLUSIONS

Numerical simulations for the flow of an LLDPE melt inside slit dies have been carried out to study the effect of die length on the viscoelastic behavior of the melt. The constitutive equation used is an integral equation of the K-BKZ type with a spectrum of relaxation times, capable of fitting viscosity and normal stress data measured in shear and also assuming a reasonable strain-thinning elongational viscosity for high elongational rates. The numerical solutions show that for the range of experimental apparent shear rates ($10 < \dot{\gamma}_a < 100\ s^{-1}$), there is no vortex growth in the reservoir and a moderate swelling which depends on the die length. For a short die it is always higher reaching asymptotic values of 47%, whereas for a long die it reaches about 35%. The viscoelastic constitutive equation is also capable of capturing such strong viscoelastic phenomena as the delayed relaxation of stresses inside the die and outside in the extrudate. This is in sharp contrast with previous purely viscous, inelastic simulations which predict a sudden relaxation of stresses in entry and exit flows.

ACKNOWLEDGEMENTS

Financial assistance from the Natural Sciences and Engineering Research Council of Canada (NSERC) is gratefully acknowledged.

REFERENCES

1. J. Vlachopoulos, Rev. Def. Beh. Mat., 3 (1981) 219.
2. X.-L. Luo and R.I. Tanner, Int. J. Num. Meth. Eng., 25 (1988) 9.
3. J. Meissner, Pure Appl. Chem., 42 (1975) 551.
4. A.C. Papanastasiou, L.E. Scriven and C.W. Macosko, J. Rheol., 27 (1983) 387.
5. X.-L. Luo and E. Mitsoulis, Int. J. Num. Meth. Fluids, 11 (1990) 1015.
6. P. Beaufils, B. Vergnes and J.-F. Agassant, Intern. Polym. Proc., 4 (1989) 78.
7. J.M. Lacaze, G. Marin and Ph. Monge, Rheol. Acta, 27 (1988) 540.
8. B. Debbaut, J.M. Marchal and M.J. Crochet, J. Non-Newt. Fluid Mech., 29 (1988) 119.
9. S.A. White, A.D. Gotsis and D.G. Baird, J. Non-Newt. Fluid Mech., 24 (1987) 121.
10. R.I. Tanner, J. Non-Newt. Fluid Mech., 7 (1980) 265.
11. P. Beaufils, Ph.D. Thesis, École des Mines de Paris, Sophia-Antipolis, France (1988).

Theoretical and Applied Rheology, edited by P. Moldenaers and R. Keunings
Proc. XIth Int. Congr. on Rheology, Brussels, Belgium, August 17-21, 1992

A Model for the Filling of Cold Cavities with Solidifying Semi-Crystalline Polymers

Thanasis D. Papathanasiou [1]

Background

Computer modeling of the filling stage of injection molding of plastics has been the subject of intense study and numerous publications in the past fifteen years[1-7]. Previous modeling work has identified and analyzed a number of important features of the process, such as the fountain flow, the effect of the rheological properties of the polymer and the presence of complex geometries. However, even though the intrinsically three dimensional nature of the heat transfer problem has been recognized in the past and solutions in three dimensions have been presented [6,7], wall solidification and its effect on the parameters of the filling process such as pressure drop and pressure build-up at the gate has not been addressed satisfactorily at the modeling level.

This article presents a model for the filling stage of injection molding that includes wall solidification and the development of crystallinity in the solidified polymer skin. The model considers a two dimensional flow analysis, a three dimensional stress analysis based on the White-Metzner constitutive model and a three dimensional thermal analysis. Solution of the model equations is sought numerically, using body fitted curvilinear coordinates (BFCCs) for the discretization of the flow field. The potential of geometry-adaptive BFCCs for the modeling of the filling of complex cavities with or without inserts has been demonstrated in the past [3,4,7]. In this work the filling of a simple rectangular cavity is presented, since this geometry facilitates the comparison with analytical solutions as well as the interpretation of the results. The model further considers temperature and pressure dependent material properties.

[1] Los Alamos National Laboratory
MEE-9, MS G736, Los Alamos, NM 87545, USA.

A two-dimensional version of the mathematical model has been presented elsewhere [3,4]. The present work includes a three-dimensional form of the energy equation which allows for calculation of the reduction in effective cavity thickness during filling due to solidification:

$$\rho \cdot C_p \cdot [\frac{\mathbf{D}\Theta}{\mathbf{D}t}] = \mathbf{\Phi} + \kappa \cdot [\nabla^2\Theta + (\frac{\partial^2\Theta}{\partial z^2})] \qquad (1)$$

where (∇^2) is the Laplace operator in two dimensions, Θ is the temperature, Φ is the heat dissipation term and D/Dt the material time derivative [4]. This is coupled with the non-isothermal crystallization kinetic model of Nakamura et.al [8]:

$$x_R(t) = 1.0 - exp[-(\int_0^t K(\Theta) \cdot dt)^m] \qquad (2)$$

where x_R is the relative fractional crystallinity, m is the Avrami exponent and K(Θ) is a non-isothermal crystallization rate constant. A sample mesh generated during the filling of a rectangular mold is shown in Figure (1).

Results and Discussion

Figure (2) shows the pressure contours in the region around the gate (including a part of the runner), while Figure (3) shows a comparison between numerical (lines) and analytical (+) pressure drops in the case of isothermal filling. Figure (4) analyzes the effect of cooling rate on the pressure drop in the cavity. The pressure drops for the isothermal and adiabatic cases are also shown for comparison.

Invariably, cooling increases the pressure drop in the cavity. For a heat transfer coefficient $h = 0.986 kW/m^2/K$ (curve 3 in Figure (4)) no wall solidification is observed for the time scale of this filling (1 sec). In this case, the increase in pressure drop over isothermal filling is entirely due to the cooling-induced increase in melt viscosity. When $h = 2.464 kW/m^2/K$ (curve (4)) a solid film is formed on the walls of the cavity and the maximum reduction in cavity thickness is approximately 5.8 percent. When the heat transfer coefficient increases further to 3.943 $kW/m^2/K$ the maximum reduction in cavity thickness increases to 11 percent and the corresponding pressure drop is shown in curve 7. Curves 3 and 4 in Figure (4) show approximately identical pressure drops, even though the values of the corresponding heat transfer coefficients are quite different. The formation of the solid skin in the case of curve 4 insulates the interior of the cavity and results in a higher gap-averaged temperature (and lower viscosity) in the flowing melt as compared to the case of curve 3 where, due to the smaller cooling rate, no wall solidification occurs. The net result is approximately identical pressure drops even though the cooling conditions are quite different. This example clearly demonstrates the complex character of the interaction between wall solidification and viscosity reduction, as factors affecting the pressure drop in the cavity.

The spatial distribution of the effective cavity thickness (after solidification) for uniform cooling conditions with $h = 3.943kW/m^2/K$ is shown in Figure (5). These results show that the thickness of the solidified layer is mainly affected by cooling to the mold, and, to a lesser extend, by convective heat transfer from the gate to the interior of the cavity.

The distribution of crystallinity along the centerline (y=0.0) at the surface of the rectangular plate for $h = 2.464kW/m^2/K$ is shown in Figure (6). The lower crystallinity near the melt front is the result of the smaller residence time of the melt in the forward part of the cavity and is clearly a kinetic effect. The crystallinity level attained at the surface of the article during filling is roughly equal to the ultimate crystallinity corresponding to the surface temperatures. Since the filling stage is followed by packing and cooling, if no post-molding annealing occurs the surface crystallinity of the final article will be the crystallinity which has developed during filling and *frozen* during subsequent molding stages. This clearly demonstrates that crystallinity predictions at the end of the filling stage are not only essential as an initial condition for post-filling modeling, but can, under certain conditions, have direct relevance to the properties of the final product.

Concluding Remarks

A model for the filling stage of injection molding which includes wall solidification and crystallization has been used to investigate the interaction between solidification and pressure drop during the filling of a cold cavity with high-density polyethylene. Predictions regarding the crystallinity distribution on the skin of the molded article were further presented.

Acknowledgement

This work was supported by the Office of the Director and by the MEE Division at Los Alamos National Laboratory.

References

1. Kuo Y. and Kamal M.R., *AIChE J.* , **22** 4, (1976).
2. Kamal M.R and Papathanasiou T.D. , *Polymer Engineering and Science*, in press.
3. Papathanasiou T.D. and Kamal M.R., *ANTEC'91* , pp.829-835.
4. Papathanasiou T.D. and Kamal M.R., *Polymer Engineering and Science*, in press.
5. Gogos C.G., C.F. Huang and L.R. Schmidt, *Polymer Engineering and Science* , **26,** 1457, (1986).
6. Shen S.F., *International Journal for Numerical Methods in Fluids,* **4**, 171, (1984).
7. Subbiah S., Trafford D.L. and Guceri S.I., *International Journal of Heat and Mass Transfer* , **32,** 3, 415, (1989).
8. Nakamura K., Katayama K. and Amano T., *Journal of Applied Polymer Science*, **17**, 1031, (1973).

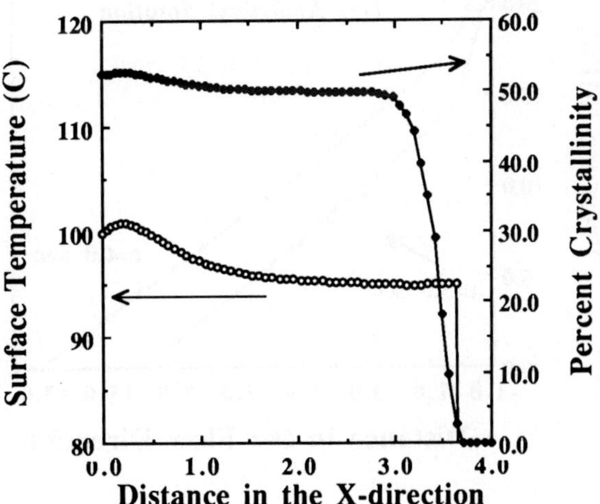

6. Variation of surface crystallinity and temperature along the line y=0.0 for $h = 2.464kW/m^2/K$.

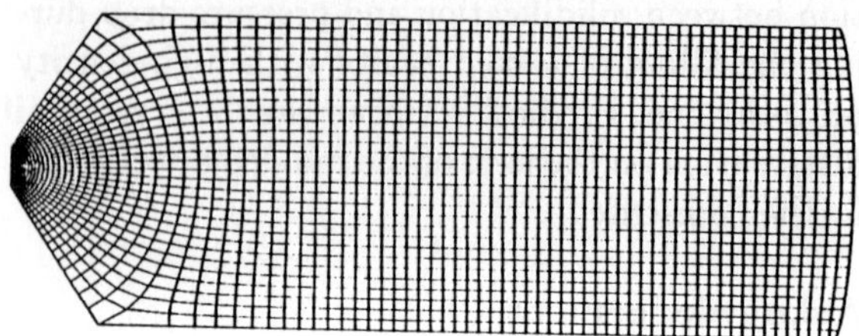

1. Curvilinear mesh created by the computer code during a filling simulation.

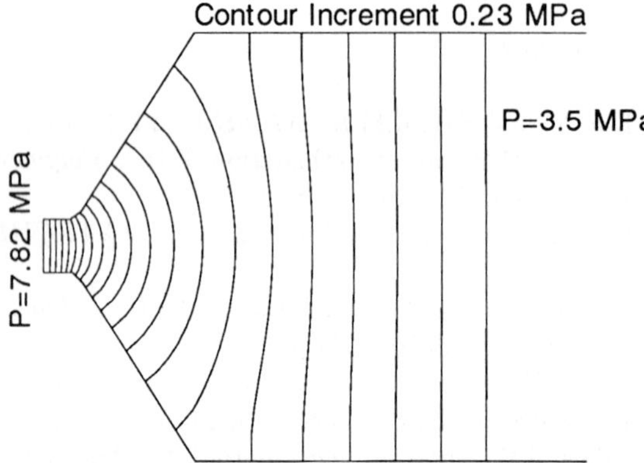

2. Pressure contours at the entrance section of the cavity showing the transition from a radial to a fully-developed flow.

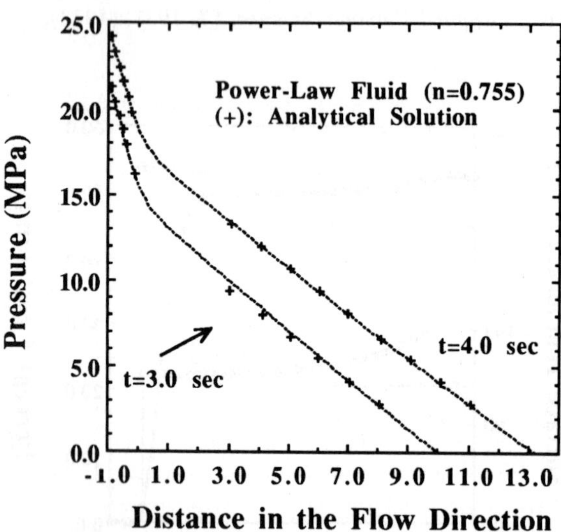

3. Comparison between numerical and analytical solution at two instances during filling.

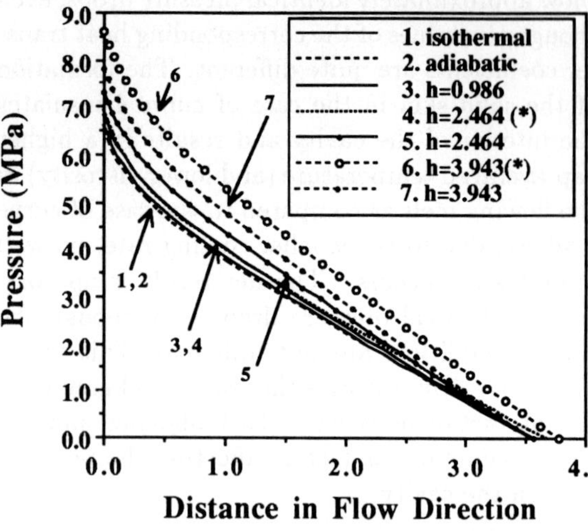

4. Pressure drop along the center of the cavity (line y=0.0) for various values of the heat transfer coefficient. (*) indicates consideration of cavity thickness reduction in the calculation of pressure.

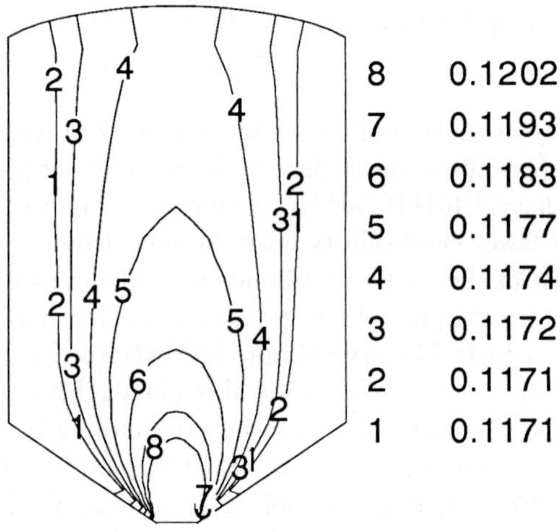

5. Spatial variation of the effective cavity thickness for $h = 3.943kW/m^2/K$.

Theoretical and Applied Rheology, edited by P. Moldenaers and R. Keunings
Proc. XIth Int. Congr. on Rheology, Brussels, Belgium, August 17-21, 1992

The Mechanics of Air-Gap Wet-Spinning of Fibers

Volker Simon

Technische Strömungslehre, Technische Hochschule Darmstadt, Petersenstr. 30
W 6100 Darmstadt (Germany)

1. INTRODUCTION

A simple model is given to describe the mechanics of steady air-gap wet-spinning of fibers. In this process the spinning solution is extruded through a spinneret hole as a hot melt. The fiber is first drawn through an air gap, where it is stretched and cooled and most of the radius reduction takes place, then it traverses the coagulation bath. Finally, the fiber is collected on a wind-up roll at a rate that exceeds the extrusion velocity. The process is shown schematically in Fig. 1. The flow in and near the spinneret is not considered in this analysis and the origin of the coordinate system is placed at a position beyond the point of maximum die swell.

The spinning solution is a ternary system composed of cellulose, solvent and nonsolvent. In the coagulation bath, solvent diffuses out of the fiber into the bath and nonsolvent diffuses from the bath into the fiber. Since the material properties of the spinning solution change with changing solvent concentration, the melt coagulates and a solidified outer skin is formed, resisting further deformation.

It would be desirable to determine the material properties characterizing the behaviour of the material as functions of solvent concentration. In the particular process investigated, however, the only reliable information available comes from simple shear flow measurements of the hot melt, indicating a liquid-like viscoelastic behaviour of the uncoagulated spinning solution. Likewise, determination of the properties of the solidified fiber shows solid-like viscoelastic behaviour of the freshly spun wet fiber. In previous analyses of wet-spinning (refs. 1-3), the fiber was always assumed to be a Newtonian material with a concentration dependent viscosity. This assumption appears to be inadequate in the present case. Since the ratio of shear modulus G_s of the solidified fiber to mean shear modulus $\bar{G}$ of the melt is rather large, it is assumed

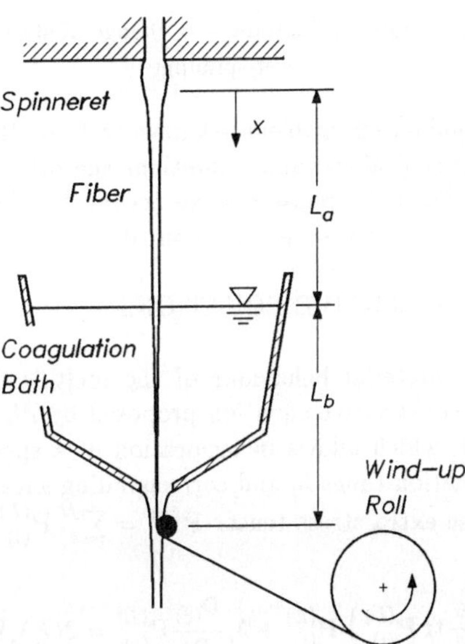

Fig. 1. Schematic of air-gap wet-spinning process.

that solidification occurs instantaneously at a critical concentration c_s, and further changes of material properties in the liquid and solid states are ignored. The critical nonsolvent concentration at which fully coagulated material is observed can be determined from a solubility diagram as shown in Fig. 2.

Instantaneous solidification at a given value of concentration implies a distinct boundary at radius $r = a_s(x)$ dividing the solid and liquid phases in the spinning process. This radius, and hence the thickness of the solid skin, may then be obtained from calculation of the concentration profile in a fiber with initial concentration c_0 moving through a fluid with concentration c_∞. It can be shown that the concentration profile in a slightly stretched fiber does not differ much from the corresponding profile in a fiber moving at constant velocity (ref. 4). For constant fiber velocity and large Schmidt-number, the solution of the analogous

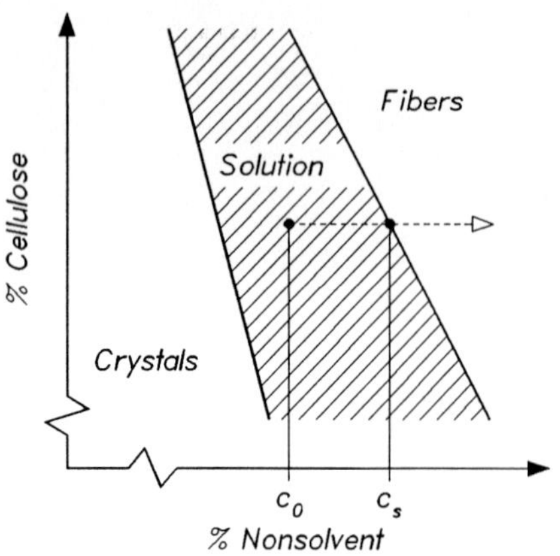

Fig. 2. Solubility diagram; - - - change of state during wet-spinning.

heat conduction problem is known (ref. 5). It is found that for typical spinning conditions the ratio of radius a_s, at which the phase change occurs, to fiber radius a at the end of the bath is about 0.8.

2. CONSTITUTIVE EQUATIONS

The material behaviour of the melt is described by the constitutive equation proposed by Phan-Thien (ref. 6), which allows incorporation of a spectrum of N relaxation times λ_i and corresponding shear moduli G_i. The extra stress tensor $\mathbf{P}^{(L)} = \sum_i^N \mathbf{P}_{(i)}^{(L)}$ is given by

$$\exp\left(\frac{\varepsilon}{G_i}\operatorname{tr}\mathbf{P}_{(i)}^{(L)}\right)\mathbf{P}_{(i)}^{(L)} + \lambda_i\frac{\mathcal{D}_{(\xi)}}{\mathcal{D}t}\mathbf{P}_{(i)}^{(L)} = 2G_i\lambda_i\mathbf{E} \tag{1}$$

with

$$\frac{\mathcal{D}_{(\xi)}}{\mathcal{D}t}\mathbf{P} \equiv \frac{\mathrm{D}}{\mathrm{D}t}\mathbf{P} + \boldsymbol{\Omega}\cdot\mathbf{P} - \mathbf{P}\cdot\boldsymbol{\Omega} - \xi(\mathbf{P}\cdot\mathbf{E} + \mathbf{E}\cdot\mathbf{P}), \tag{2}$$

where $\mathbf{E}$ and $\boldsymbol{\Omega}$ are the symmetric and antisymmetric parts of the velocity gradient, respectively, and ξ and ε are material parameters which can be determined from simple shear and elongational flow measurements. The variation of viscosity $\eta_i = \lambda_i G_i$ with temperature is approximated by an Arrhenius form, while the shear moduli G_i are relatively insensitive to temperature changes and are taken to be constant.

The solidified skin is assumed to behave like a nonlinear elastic solid,

$$\frac{\mathcal{D}_{(\xi=1)}}{\mathcal{D}t}\mathbf{P}^{(S)} = 2G_s\mathbf{E}\,. \tag{3}$$

Superscripts (L) and (S) indicate liquid and solid parts of the fiber, respectively.

3. SPINNING EQUATIONS

Since a typical fiber radius a_0 is small compared to the length $L = L_a + L_b$ of the air gap and coagulation bath, a one-dimensional approach is justified. Consequently, the axial velocity $u(x)$ is assumed to be constant over a cross section and to be the same for core and skin layers. Likewise, the components P_{xx} and P_{rr} of the stress tensor for the liquid and solid phases are assumed to be functions of x only. Then the volume flow rate $\dot{V} = ua^2\pi = const$ and the radial velocity $v = -r/2\,\mathrm{d}u/\mathrm{d}x$. Neglecting inertia, gravity, and surface tension, an integral momentum balance that incorporates the continuity of the stresses at the fiber surface and phase boundary gives

$$\frac{\mathrm{d}}{\mathrm{d}x}\left(\left(P_{xx}^{(L)} - P_{rr}^{(L)}\right)a_s^2\right) + \frac{\mathrm{d}}{\mathrm{d}x}\left(\left(P_{xx}^{(S)} - P_{rr}^{(S)}\right)\left(a^2 - a_s^2\right)\right) + 2a\tau = 0\,, \tag{4}$$

where τ is the shear stress acting on the fiber due to its motion through the bath. In the air gap, $a_s = a$ and $\tau = 0$, and the equations reduce to those employed for melt spinning calculations (refs. 7,8).

The constitutive equations (1)–(3) together with the momentum equation (4) can be manipulated to give a set of $2N+1$ and $2N+3$ coupled first order differential equations for the velocity and stresses in the air gap and coagulation bath, respectively. These equations are supplemented by boundary conditions. It is assumed that at $x = 0$ the velocity is known:

$$u = u_0\,, \tag{5}$$

and at $x = L$, we may either prescribe the take-up velocity:

$$u = u_e \tag{6}$$

or the force applied to draw the fiber:

$$\left(P_{xx}^{(L)} - P_{rr}^{(L)}\right)a_s^2 + \left(P_{xx}^{(S)} - P_{rr}^{(S)}\right)\left(a^2 - a_s^2\right) = F\,. \tag{7}$$

The elasticity of the melt requires specification of initial stresses, which are taken to be (ref. 7)

$$P_{xx(i)}^{(L)} = \frac{\lambda_i^2 G_i}{\bar{\lambda}^2\bar{G}}\frac{Fu_0}{\dot{V}} \quad\text{and}\quad P_{rr(i)}^{(L)} = 0\,, \quad i = 1, N\,. \tag{8}$$

Equations (8b) are in fact redundant, because they follow from equations (4), (7) and (8a).

In addition, the stresses $P_{xx}^{(S)}$ and $P_{rr}^{(S)}$ have to be prescribed along the phase boundary $r = a_s(x)$. We set $\mathbf{P}^{(S)} = \mathbf{P}^{(L)}$ at the beginning of the coagulation bath. The assumption of constant cross sectional stresses $P_{xx}^{(S)}$ and $P_{rr}^{(S)}$ in the solid phase implies

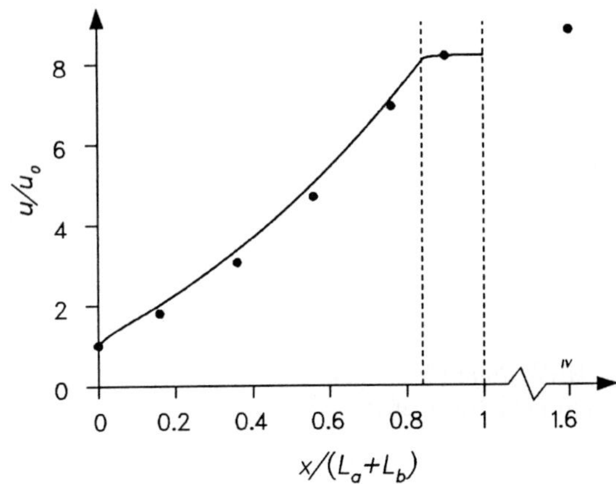

Fig. 3. Computed and experimental (•) velocity profile; - - - beginning and end of coag. bath; case *I*.

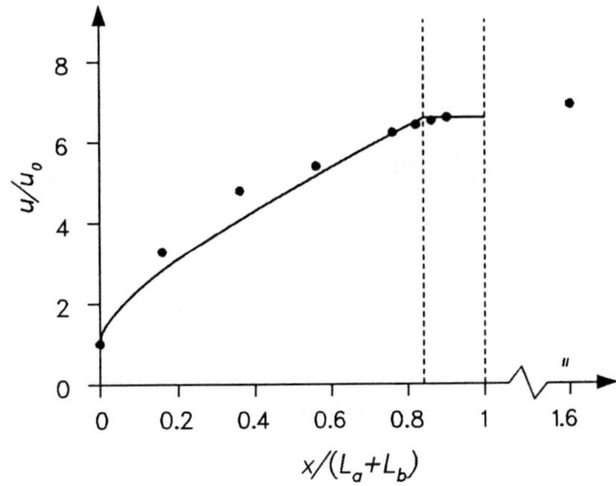

Fig. 4. Computed and experimental (•) velocity profile; - - - beginning and end of coag. bath; case *II*.

that these stresses are imposed along the phase boundary by the already solidified part of the fiber.

4. RESULTS AND DISCUSSION

The system of equations was solved numerically with a relaxation time spectrum containing $N = 12$ modes as an initial value problem. The applied winding force was then adjusted until a given velocity at some point downstream was attained. The shear stress τ was obtained from numerical boundary layer computation and the temperature profile was interpolated from measured data in the air gap and approximated by an analytic solution (ref. 5) in the bath. The relaxation spectrum was computed from measured dynamic data by the method outlined in (ref. 9) and ξ was determined as proposed in (ref. 6). Since no data on pure elongational flows are available, ε was given the value suggested in (ref. 6).

Figs. 3 and 4 compare the computed and experimental velocity profiles for two cases, denoted by *I* and *II*. The filament velocity has not been measured at the end of the coagulation bath but only at some distance between the bath and the wind-up roll. The experimental data suggest further stretching of the filaments in the air gap following the bath. Because the flow in this region has not been modelled, the boundary condition (6) was applied at the location of the last measured point in the bath. The two cases differ in volume flow rate and initial and take-up velocities; all other parameters are the same. This leads to $\dot{V}_I/\dot{V}_{II} = 13.8$, $De_I = \bar{\lambda}u_0/L = 3.33$ and $De_{II} = 0.86$. The lower volume flow rate in case *II* leads to a more rapid cooling of the fiber in the air gap and the bath and, hence, to a higher viscosity, which appears to be the main agent for the observed change of curvature of the velocity profile. Also the mass flux of solvent is higher and the area of the solid skin is larger in case *II*. The quantitative agreement between the model prediction and the experiment is encouraging, even though the change of filament velocity in the coagulation bath is slightly underestimated in both cases.

ACKNOWLEDGEMENTS

All experimental data were provided by a spinning factory. I am grateful to Prof. J.H. Spurk, who initiated this work, for his advice and helpful discussions. I have also benefitted from valuable discussions with Dr. G. Frischmann.

REFERENCES

1. C.D. Han and L. Segal, J. Appl. Polym. Sci., 14 (1970) 2999-3019.
2. J.L. White and T.A. Hancock, J. Appl. Polym. Sci., 26 (1981) 3157-3170.
3. A. Ziabicki, Fundamentals of Fibre Formation, John Wiley & Sons, London, 1976, p. 276
4. D. Wilhelm, Studienarbeit (unpublished), Technische Hochschule Darmstadt, 1990.
5. H.S. Carslaw and J.C. Jaeger, Conduction of Heat in Solids, Oxford University Press, London, 1959, p. 346
6. N. Phan-Thien, J. Rheol., 22 (1978) 259-283.
7. M.M. Denn and G. Marrucci, J. Non-Newtonian Fluid Mech., 2 (1977) 159-167.
8. J. Mewis and C.J.S. Petrie, in: N.P. Cheremisinoff (Ed.), Encyclopedia of Fluid Mechanics, Gulf Publishing Company, Houston, 1987, Vol. 6, pp. 111-139.
9. S.F. Gull and J. Skilling, I.E.E. Proc., 131F (1984) 646-659.

Theoretical and Applied Rheology, edited by P. Moldenaers and R. Keunings
Proc. XIth Int. Congr. on Rheology, Brussels, Belgium, August 17-21, 1992

THE EXPANSION OF THICK TUBES

J. SUN[1], X.-L. LUO[2] and R.I. TANNER[1]

[1]Department of Mechanical Engineering, University of Sydney, NSW 2006 (Australia)
[2]CSIRO, Division of Mathematics and Statics, P.O. Box 218, Lindfield NSW 2070 (Australia)

1. INTRODUCTION

The tube expansion process,as shown in Fig.1,is an industrially important process. Polymer is extruded from an annular die. Then it is expanded under the inner pressure and extended by the traction force on its end. There is a large tube which guides the wall of the expanded and elongated tube. This problem is similar to a biaxial stretching viscoelastic flow. But the tube is not thin,and so the usual shell theory can not be applied. We have not found any numerical simulation of this type of problem in literature. This paper develops an effective numerical simulation technique for thick tube expansion by using a Finite Element Method and presents simulation results for LDPE at 160^oC using Newtonian and K-BKZ viscoelastic models. These results show the relationship between the final shape of the tube and the pressure needed for its inflation.

2. GOVERNING EQUATIONS AND K-BKZ MODEL

For steady, isothermal and creeping flow of an incompressible fluid the governing equations are:

$$\nabla \cdot \mathbf{V} = 0 \tag{1}$$

$$-\nabla P + \nabla \cdot \tau = 0 \tag{2}$$

where $\mathbf{V}$ is the velocity vector,τ is the extra-stress tensor, and P is the scalar pressure. The constitutive equation that relates τ to the deformation history is the K-BKZ equation proposed by Papanastasiou et al (ref.1). It is:

$$\tau(t) = \int_{-\infty}^{t} [\sum \frac{a_k}{\lambda_k} exp(-\frac{t-t'}{\lambda_k})] \times \frac{\alpha}{(\alpha-3)+\beta \mathbf{I}_{\mathbf{c}-1}+(1-\beta)\mathbf{I}_{\mathbf{c}}} \mathbf{C}_{\mathbf{t}}^{-1}(\mathbf{t}')\mathbf{d}\mathbf{t}' \tag{3}$$

where λ_k and a_k are the relaxation times and relaxation modulus coefficients at the reference temperature T_o, respectively,α and β are material constants,and $\mathbf{I}_{\mathbf{c}}$, $\mathbf{I}_{\mathbf{c}-1}$ are the first invariants of the Cauchy-Green tensor $\mathbf{C}_{\mathbf{t}}$ and its inverse $\mathbf{C}_{\mathbf{t}}^{-1}$, the Finger strain tensor. The material parameter values of LDPE at 160^oC for the K-BKZ model can be found in (ref.2).

3. NUMERICAL SIMULATION TECHNIQUE

The numerical method used in this paper is based on the methods developed by Luo et al (refs.3-4) which are used for solving coextrusion and annular die swell. We do not repeat the details here. There are several differences between tube expansion and coextrusion and annular die swell. The first is that the boundary conditions are different,as shown in Fig.1. There is a pressure which is exerted on the inner surface. The end of the tube is elongated by the extension force. The outer surface is divided into two parts by the contact point B. The part AB is a free surface. The part BC is held by a cylinder whose radius is several times larger than the radius of the die. The inner surface on which a pressure is exerted can be treated as an interface between polymer and gas (ref.5) and in the absence of surface tension (ref.6) the boundary conditions are:

$$\boldsymbol{\sigma} \cdot \mathbf{n} = -P_a \mathbf{n} \tag{4}$$

$$\mathbf{V} \cdot \mathbf{n} = 0 \tag{5}$$

where $\boldsymbol{\sigma}$ is the total stress tensor, P_a is the ambient pressure, $\mathbf{n}$ is the outward pointing unit normal to the interface. The second is that the contact point is changeable. That means we do not know the contact point position while beginning to simulate the tube expansion. At first we assume a contact point position, then calculate velocity fields and the free surface shape (as the broken line in Fig.1). If $\Delta R = R_f - R_o > 0$,section B-E is moved backward. If $\Delta R < 0$, section B-E is moved forward. Using this method we can change the contact point continuously, which is the reason why we use this method. The third is the contact boundary condition, $\sigma_{rr} \leq 0$ on BC. This is important because if $\sigma_{rr} > 0$ that means some tensile forces are exerted on boundary BC and the pressure value is not sufficient for the tube expansion. The fourth is the solution procedure. Because tube expansion is a biaxial stretching flow and has complex boundary conditions, it is very difficult to get the convergent results even for Newtonian fluid flows. So we have had to develop a simulation technique for tube expansions. We begin the simulation from annular die swell. After a convergent result has been obtained, the pressure inside the tube is increased from zero and at the same time the conditions on the outer boundary (BC) and on the end of the tube (CD) are put in . The radius of the tube is increased little by little. Before the radius is increased from the value R_{oi} to R_{oi+1}, convergent results under R_{oi} should be achieved. Finally,the tube is expanded to the required size. For viscoelastic fluid flow,it is better to introduce all of the pseudo-body force before increasing the radius of the tube. Even with this procedure,it is necessary to be very careful while increasing pressure and radius of the tube if one is to get convergent results.

4. RESULTS AND DISCUSSION

In order to inspect the relationship between the shape of the tube and the inner pressure, the entrance boundary condition is fixed on fully developed Poiseuille flow and the flow rate is 0.0140 cm^3/s. The exit boundary condition is fixed on C=0.00883 cm/s. We use three hundred isoparametric quadrilateral elements for the domain shown in Fig.1. Every element has eight nodes and twenty degrees of freedom. The simulation of the tube expansion is begined at R_{o1} = 1.10 cm and the $R_{oi+1} - R_{oi}$ = 0.05 cm. Simulation results are shown in Fig.2 ~ Fig.11.The shapes of the expanded tube are shown in Fig.2 ~ Fig.9. After polymer is extruded from the die it behaves like annular extrudate swell. The thickness of the tube increases and then decreases as the tube is expanded by the inner pressure and elongated by the axial extension. From these results we can find that under the same exit velocity boundary condition and the same R_o/R_w, the viscoelastic flow exhibits larger extrudate swell and suffer a larger extension force along the axial direction than Newtonian flow. So the contact point is further away from the die in viscoelastic flow than in Nowtonian flow. The relationships between the inner pressure and the radius of the expanded tube are shown in Fig.10 and Fig.11. Both in Newtonian flow and in viscoelastic flow there is a point at which the pressure attains the maximum value. Before this point the pressure increases as the R_o/R_w increases. After this point, the pressure decreases as the R_o/R_w increases. For Newtonian flow this point is within R_o/R_w = 1.20 ~ 1.30 and for viscoelastic flow within R_o/R_w = 1.26 ~ 1.34.

REFERENCES

1. A.C. Papanastasiou, L.E. Scriven, and C.W. Macosko, J.Rheol.27 (1983), 387-410.
2. X.-L. Luo and E. Mitsoulis, J.Rheol. 34 (1990), 309-342.
3. X.-L. Luo and E. Mitsoulis, submitted for publication.
4. X.-L. Luo and E. Mitsoulis, J.Rheol. 33 (1987), 1307-1327.
5. S.F. Kistler and L.E. Scriven,in:J.R.A. Pearson and S.M. Richardson (Ed.), Computational Analysis of Polymer Processing, Applied Science Publishers, London and New York, 1984, pp.247-248.
6. R.I.Tanner, Engineering Rheology, Oxford University Press, New York, 1988, p.305-307.

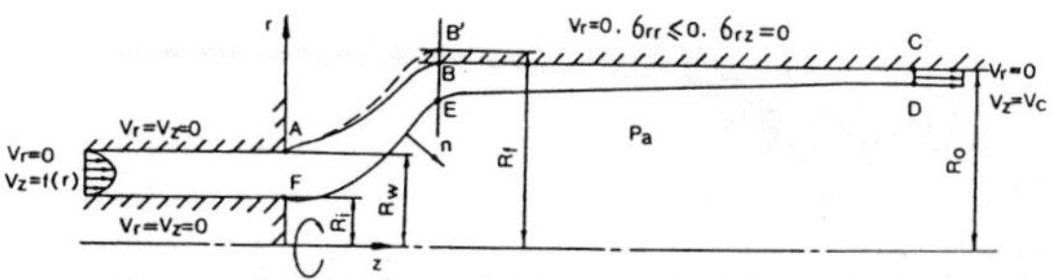

Fig.1 Tube expansion process

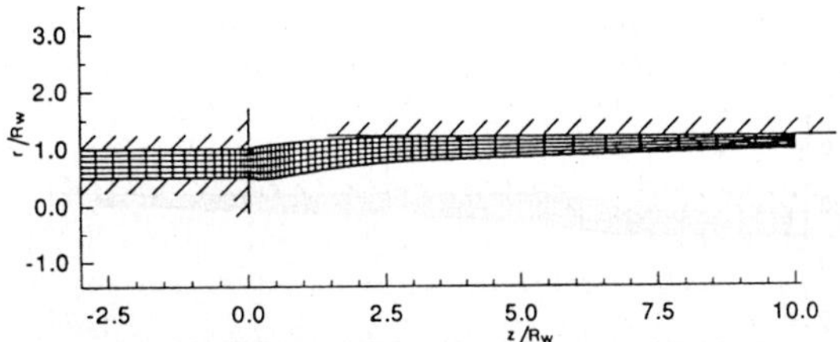

Fig.2 Final shape and finite element grids,Newtonian flow,R_o/R_w = 1.2

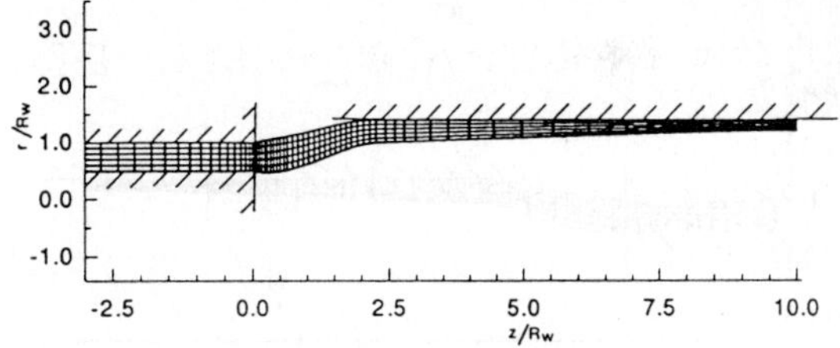

Fig.3 Final shape and finite element grids,Newtonian flow,R_o/R_w = 1.4

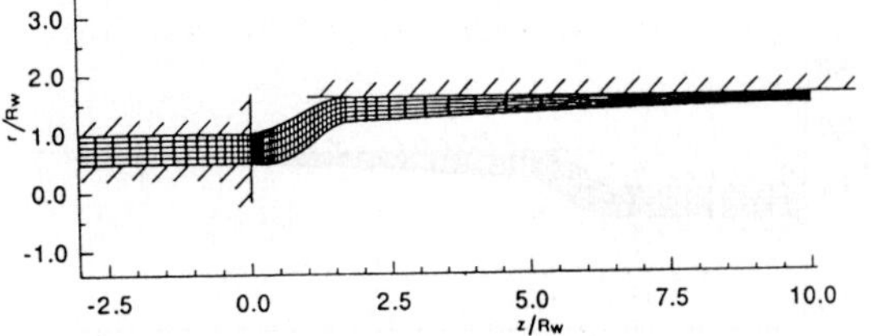

Fig.4 Final shape and finite element grids,Newtonian flow,R_o/R_w = 1.6

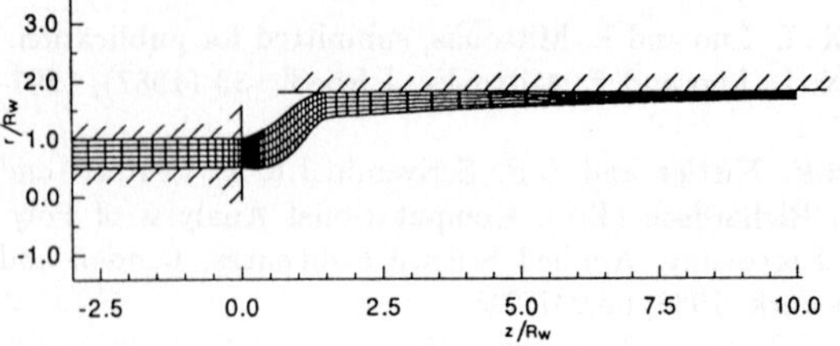

Fig.5 Final shape and finite element grids,Newtonian flow,$R_o/R_w = 1.8$

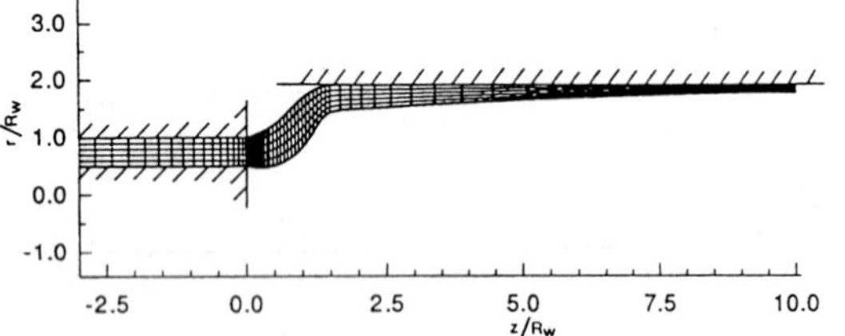

Fig.6 Final shape and finite element grids,Newtonian flow,$R_o/R_w = 1.9$

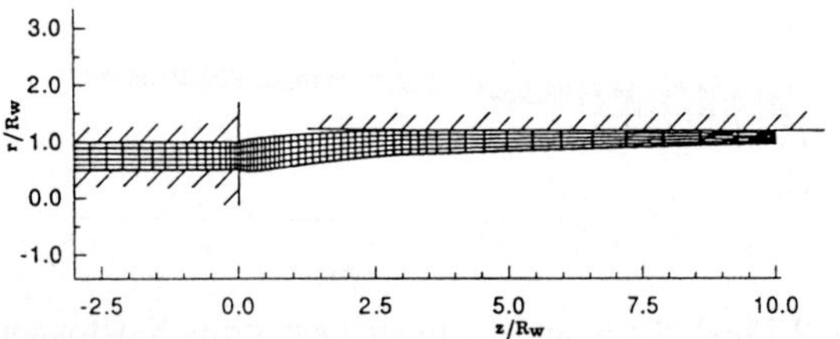

Fig.7 Final shape and finite element grids,viscoelastic flow,$R_o/R_w = 1.2$

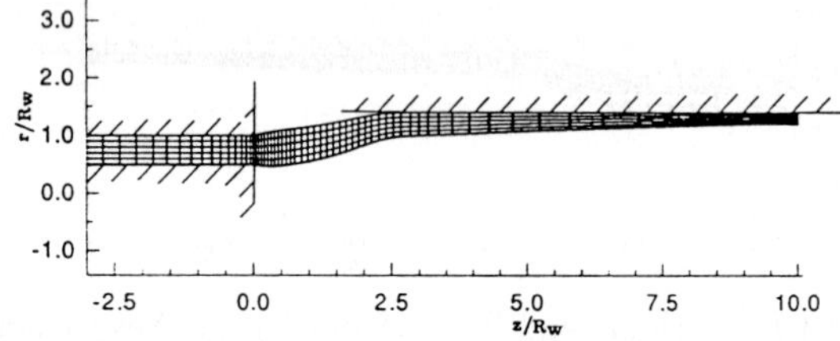

Fig.8 Final shape and finite element grids,viscoelastic flow,$R_o/R_w = 1.4$

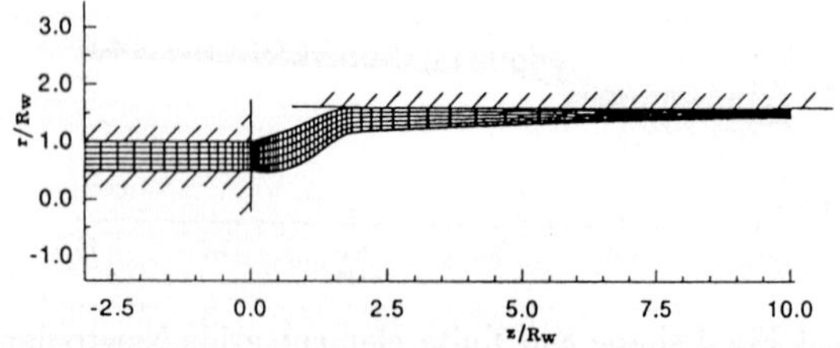

Fig.9 Final shape and finite element grids,viscoelastic flow,$R_o/R_w = 1.6$

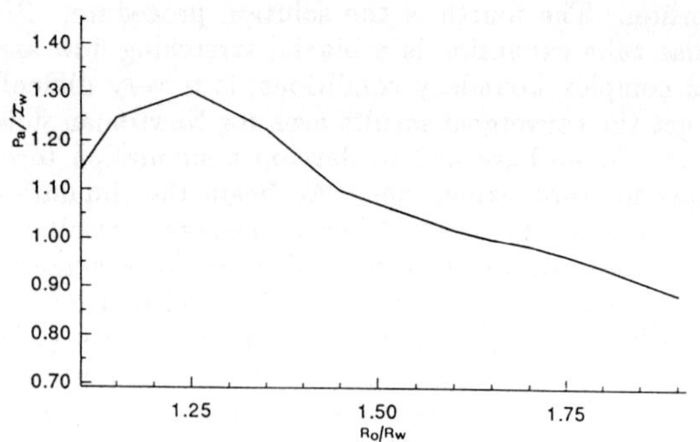

Fig.10 Pressure distribution as the tube is expanded from $R_o/R_w = 1.1$ to 1.9,Newtonian flow

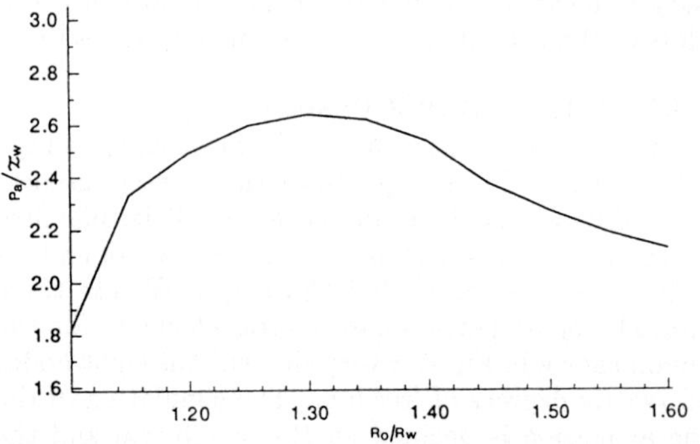

Fig.11 Pressure distribution as the tube is expanded from $R_o/R_w = 1.1$ to 1.6,viscoelastic flow

Theoretical and Applied Rheology, edited by P. Moldenaers and R. Keunings
Proc. XIth Int. Congr. on Rheology, Brussels, Belgium, August 17-21, 1992

RHEOLOGICAL MOLECULAR WEIGHT DISTRIBUTION DETERMINATIONS OF ETHYLENE/TETRAFLUORO-ETHYLENE COPOLYMERS: IMPLICATIONS FOR LONG-CHAIN BRANCHING

William H. Tuminello[1], Warren H. Buck, and Dewey L. Kerbow

[1]Du Pont Company, The Experimental Station, P. O. Box 80356, Wilmington, DE 19880-0356 USA
[2]P. O. Box 80353, Wilmington, DE 19880-0353
[3]Washington Works, R&D, P. O. Box 1217, Parkersburg, WV 26102

1. INTRODUCTION

Ethylene/tetrafluoroethylenes (PETFE) are nearly perfectly alternating copolymers. Like most fluoropolymers, they are difficult to dissolve, making classical molecular characterization techniques impractical. Even so, light scattering measurements have been taken of PETFE dilute solutions at high temperatures and molecular weight determinations made. (refs.1-3) Due to our success at making molecular weight distribution determinations from dynamic modulus measurements of polymer melts (refs.4-7), we attempted the same on commercial PETFE. Our primary objective was to compare molecular weight distributions determined from rheology with those determined by light scattering.

2. METHODS

Dynamic mechanical measurements were made on three commercial PETFE samples from the Du Pont Company (tradename Tefzel®). An additional sample, PETFE-4, was prepared without chain transfer agent to get the highest possible molecular weight. These were identical samples used in references 1-3. The apparatus was a Rheometrics System 4 rheometer. Details are given elsewhere. (ref.8)

The rheological determination of molecular weight distribution relies on the observations that the storage modulus versus frequency in the plateau and terminal zones is a mirror of the distribution. It has been shown that the frequency (ω) axis can be transformed to a molecular weight (MW), using Equation (1) and the storage modulus, G'(ω), can be transformed to a cumulative weight fraction (CUM W_i) using Equation (2). Thus, a relative molecular weight distribution is obtained. G_N^0 refers to the plateau modulus.

$$\frac{1}{\omega} \propto (MW)^{3.4} \qquad (1)$$

$$\left[\frac{G'(\omega)}{G_N^0}\right]^{0.5} \propto \text{CUM } W_i \qquad (2)$$

Same trends are obtained with the different geometries and at different temperatures. The pressure oscillations in zone 2 are similar to those observed with linear polymers (refs. 4 and 5), but, in our case, the aspect of the extrudate was totally smooth in the three zones and neither surface nor volume defects were observed.

4. FLOW CURVES

The first problem encountered for interpreting these experimental results was to apply the Bagley corrections. Generally speaking, the plots were more correct at high shear rate (zone 3) than at low shear rate (zone 1). However, in all circumstances, a linear fitting was more appropriate than an exponential one (Fig.2). For the different diameters, the Bagley corrections were finally made on the three shortest lengths, including the zero length capillary.

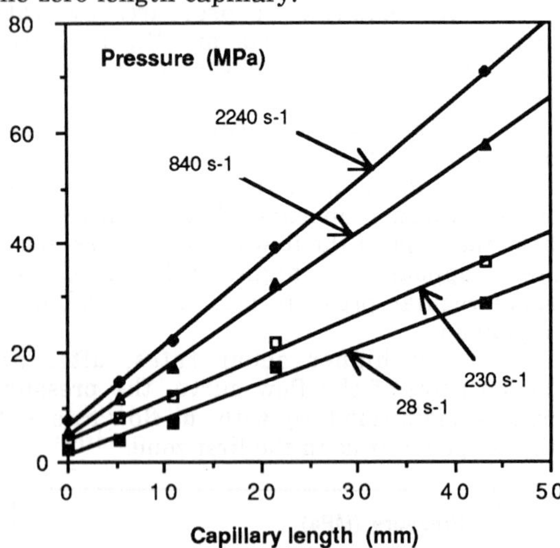

Fig.2 Bagley plot (100°C; D = 1.3 mm)

The corrected shear stress is plotted Fig. 3 as a function of the apparent shear rate. The three preceding zones are still present:

- under a critical shear stress of 0.223 MPa (zone 1), the shear stress increases with a dependence on the die diameter.

- in zone 2, the shear stress remains constant around 0.22 MPa, whatever the diameter.

- in zone 3, the stress increases again and is also a function of the geometry.

The influence of the temperature is shown on Fig. 4. In zone 1, the curves are distinct and in a classical ranking. For the same value of the shear stress and quite independently of the temperature, the zone 2 is observed. Finally, zone 3 is similar to zone 1.

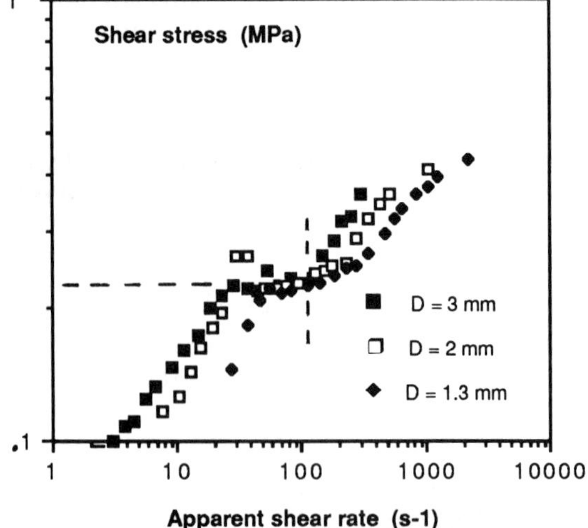

Fig. 3 Flow curve at 100 °C

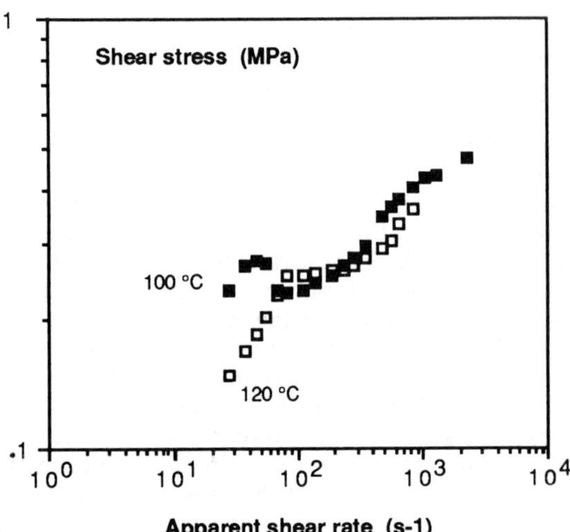

Fig.4 Influence of the temperature on the flow curve

5. DISCUSSION

The flow curves presented in Fig. 3 exhibit three distinct zones, in which the flow conditions are different. The zone 2, which begins by regular pressure oscillations, may be interpreted as a transition between slip and no-slip conditions or, more generally, between two regimes of wall slip. Effectively, obtaining flow curves depending on the die diameter is often considered as an evidence of slip conditions. The characterization of slip velocity is frequently made by applying the Mooney method (ref. 6). In our case, this led to incorrect results. Wiegreffe (ref. 3) and Geiger (ref. 2) have proposed modified Mooney method in order to include the dependence of the slip velocity on the die gap. Using our data, Wiegreffe method was also inappropriate, but Geiger method allowed to obtain physically correct results.

In zone 1, it was then possible to define the true viscous behaviour of the compound, expressed by a power-law, and the slip velocities for the different diameters (Fig.5). A second approach (ref. 7), based on the comparison between real and apparent viscous behaviour, was applied to the same data and gave similar results. In this first zone, the compound slip at the wall, may be on a thin lubricant film, with a slip velocity in the range 0.2 - 5 mm/s and strongly dependent on the die diameter.

The flow rate may be increased regularly, until a certain level of compressibility is attained. Then, the compound can no more be compressed and reacts by modifying drastically the wall boundary conditions. This change is all the more pronounced as the diameter is small and the capillary long. It leads to the zone 2, along which the slip velocity is assumed to increase (until 20 to 30 mm/s).

In zone 3, the same treatment as in zone 1 (ref. 7) may be applied. Higher values of slip velocities are obtained, in the range 15 - 165 mm/s. As indicated on Fig.5, the dependence on the die diameter is less important than in zone 1. The relationship between wall shear stress and slip velocity may be classically expressed by a power-law.

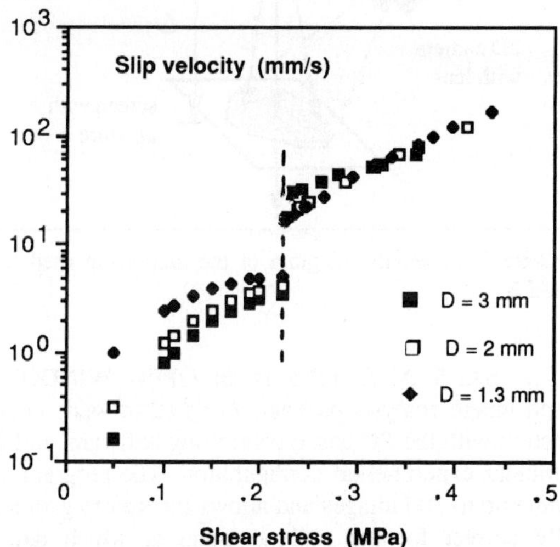

Fig.5 Slip velocity as a function of wall shear stress (T = 100 °C)

6. CONCLUSION

The EPDM compounds exhibit complex rheological behaviour, including wall slip phenomena and flow instabilities. It was shown that three main zones may be defined on the flow curve. Under a critical shear stress, the viscous behaviour of the compound is described by a power-law, with slip velocities depending on the die diameter. Above the critical shear stress, the flow becomes unstable, mainly for large L/D values, with drastic modifications of the slip conditions. At high shear rate, the slip velocities are higher, less dependent on the die geometry and can be described by a second power-law.

REFERENCES

1. C. Jepsen and N. Räbiger, Kautschuk Gummi Kunst., 41 (1988) 342-352
2. K. Geiger, Kautschuk Gummi Kunst., 42 (1989) 273-283
3. S. Wiegreffe, Kautschuk Gummi Kunst., 44 (1991) 216-221
4. E. Uhland, Rheol. Acta, 18 (1979) 1-24
5. J.A. Lupton and R.W. Regester, Polym. Eng. Sci., 5 (1969) 235
6. M. Mooney, J. Rheol., 2 (1931) 210
7. P. Mourniac, J.F. Agassant and B. Vergnes, Rheol. Acta, submitted for publication

Theoretical and Applied Rheology, edited by P. Moldenaers and R. Keunings
Proc. XIth Int. Congr. on Rheology, Brussels, Belgium, August 17-21, 1992

Flow Induced Phase Transitions for Incompatible Blends

Douglas E. Werner, Gerald G. Fuller and Curtis W. Frank

Department of Chemical Engineering, Stanford University, Stanford, CA 94305-5025

1.0 INTRODUCTION

Some of the first experiments in flow compatibilization of polystryrene/poly (vinyl methyl ether) were done by Mazich and Carr[1] where a sudden change in viscosity with constant shear indicated blend homogenization. Other work has been done using light transmission as a technique to detect both shear induced miscibility as well as shear induced phase separation.[2 3] Fluorescence quenching measurements with tagged polystyrene have been performed statically as well as dynamically to determine phase transitions.[4 5] Excimer fluorescence has also been used to examine phase transitions in the quiescent state.[6] The goal of our research is to dynamically probe the changes in morphology and phase induced by shear on a 50/50 blend of polystyrene/poly(vinyl methyl ether) using a new small angle light scattering technique to probe length scales on the order of the wave length of light as well as excimer fluorescence to probe more microscopic scales on the order of angstroms.

2.0 EXPERIMENTAL

2.1 Equipment

2.1.1 SALS Apparatus

In order to dynamically measure the changes in local morphology, a unique small angle light scattering (SALS) system was developed as shown in Figure 1. The 2D CCD array camera records images via an IBM PC equipped with a Data Translations DT2856 frame grabber. Using controlling software developed in our laboratory, it is possible to record up to 80 images with frame rates approaching 3 frames per second.

A 10 mW HeNe Laser with a wavelength of 632.8 nm was used. The incident light was passed through a parallel plate flow cell, described below. Light scattered by the sample was captured on a non-reflecting white screen. The transmitted beam passed through a 1 cm aperture. The camera then recorded the scattering pattern. The data was further analyzed using a software package, SALS ANALYSIS, running on a SUN SPARCstation.

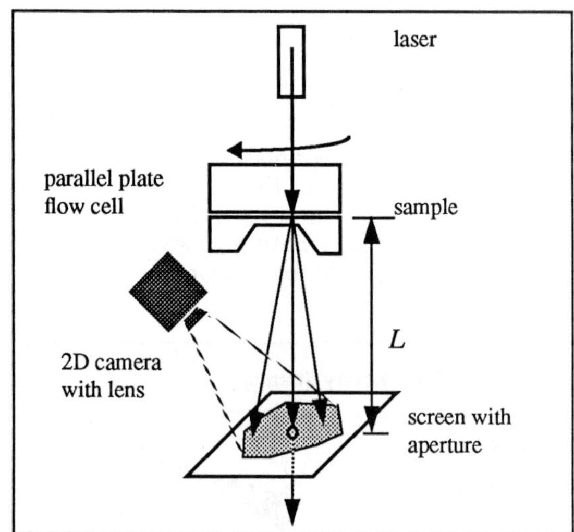

Figure 1. Schematic diagram of the instrument used for SALS.

SALS ANALYSIS is an OPEN WINDOW™ based image analysis package designed to work in conjunction with the PC based controlling software, and this particular experimental configuration. The program will handle up to 100 images and allows the user to geometrically correct for the oblique angle at which data is recorded. A variety of quantitative analyses can be performed on each time resolved image.

2.1.2 Flow Cell

The flow cell is a parallel plate cell designed to be used in both SALS experiments as well as in excimer fluorescence experiments. The aluminum cell is heated by four resistive heating cartridges and is capable of reaching temperatures approaching 200°C. Shear takes place between two optically smooth quartz plates. The quartz is free of fluorescence and will not interfere with the fluorescence spectra.

2.1.3 Fluorescence Apparatus

Fluorescence measurements were taken with the apparatus set up as shown in Figure 2. By monitoring changes in the relative intensities of the excimer and monomer fluorescence peaks, it is possible to monitor changes in miscibility.

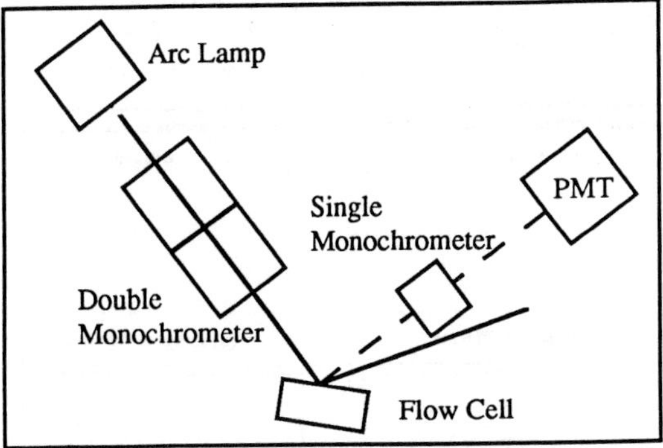

Figure 2. Fluorescence apparatus.

2.1.4 Sample

The sample consisted of a 50/50 wt% blend of polystyrene and poly(vinyl methyl ether). The polystyrene was obtained from Scientific Polymer Products, Inc.and had a molecular weight of M_W=280,000. The poly (vinyl methyl ether) was also obtained from Scientific Polymer Products, Inc. and had a M_W=100,000. Both polymers were dissolved in toluene then solution cast onto glass and dried in a vacuum oven for three days at 60°C to drive off the toluene. The resulting film was optically clear. The quiescent cloud point temperature was 120°C. After performing several experiments, the sample tended to thermally degrade due to the oxidation of PVME. The cloud point decreased to 105° C but the qualitative behaviour of the sample remained in tact. This was later verified using a fresh sample. Film thicknesses of 0.51mm were used.

2.2 Experiments

The first experiment consisted of a SALS experiment where the sample was sheared for 60 sec at a shear rate of $\dot{\gamma} = 8\text{sec}^{-1}$ at a temperature 9C° below the cloud point temperature. For the second experiment, a shear rate of $\dot{\gamma} = 16\text{sec}^{-1}$ was imposed upon the sample for 60sec at a temperature of 5°C above the cloud point temperature. Excimer fluorescence measurements were then conducted under these same conditions. At the lower temperature, the sample was homogenious in its quiescent state. At the higher temperature, the sample was phase separated.

3.0 RESULTS

There is an excellent correlation between the monomer fluorescence peak at 280nm and the phase of the system.[6] As the relative intensity increases, the system becomes homogenized. As this peak decreases, the system becomes phase separated. The results for shear rate of $\dot{\gamma} = 16\text{sec}^{-1}$ and a $\Delta T = 5°\text{C}$ is shown below in Figure 3. The intensity increases with shear, reaching a steady state value after approximately 20sec indicating homogenization. After the flow is switched off at $time = 60\text{sec}$, one can see an exponential decay indicating phase separation as it relaxes back to its original state.

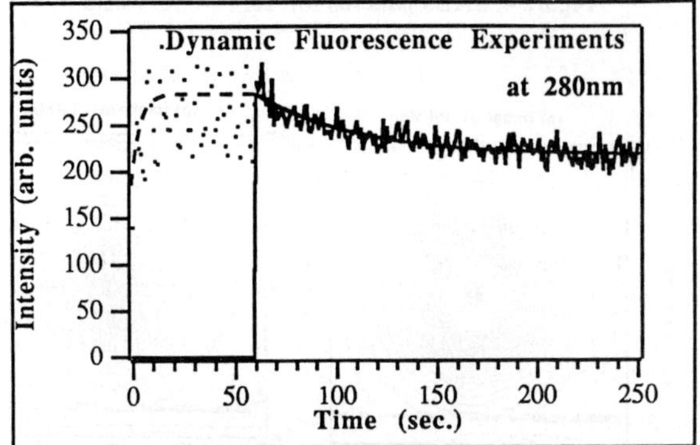

Figure 3. Monomer intensity as a function of time for $\dot{\gamma} = 16\text{sec}^{-1}$, $\Delta T = 5°\text{C}$.

The SALS patterns under the same conditions al indicated similar behaviour, (Figure 4). First, the images elongated and separated into a "butterfly"pattern with two lobes (Figure 4b). The elongation in fourier space is along the vorticity axis which indicates the polymer domains are elongating along the flow direction in real space. This pattern then deformed and disappeared (Figure 4c).Finally, a faint butterfly returned (Figure 4d). This faint butterfly is the result of scattering caused by oriented concentration fluctuations which arrise from the fact that the sample itself is a blend. At this point it is believed that the sample is completely homogenized on the order of the wavelength of light. The fluorescence data seems to confirm this.

For the lower temperature case, $\Delta T = -9°\text{C}$, where the sample was originally homogenious, an apparent phase separation was observed (Figure 5). The patterns first shrank indicating a growth in domain size (Figure 5b). The pattern then started to elongate in such an oreintation as to indicate that the polymer domains elongate in the direction of flow (Figure 5c). These patterns then fully developed again into the "butterfly"pattern (Figure 5d) at which time the aparrent phase separation was complete.

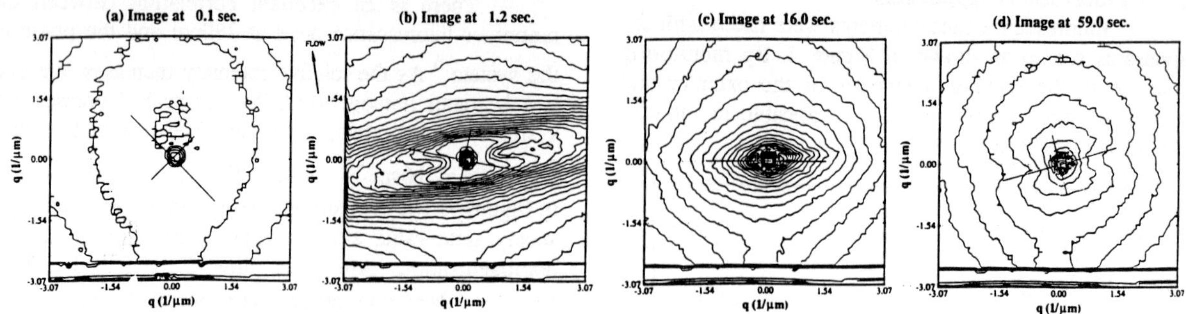

Figure 4. SALS patterns for $\Delta T = 5C°$ case.

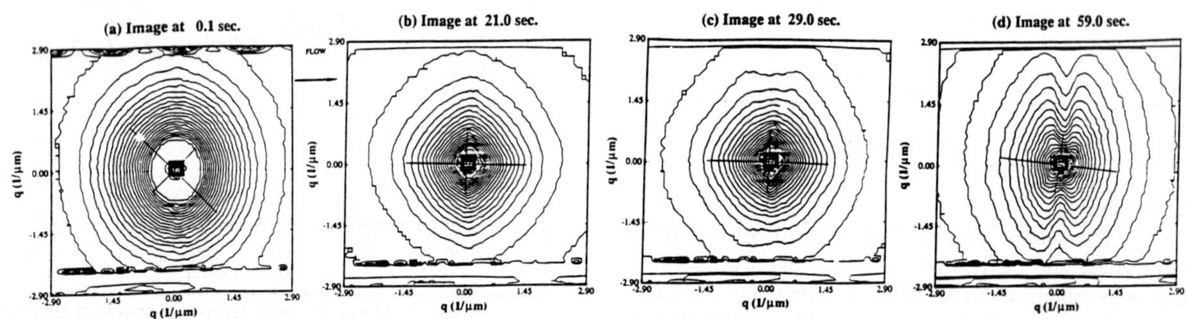

Figure 5. SALS patterns for $\Delta T = -9C°$ case.

References

1 Mazich, K .A. and Carr, S. H., J. Appl. Phys., 54, 5511 (1983).

2 Katsaros, J. D., Malone, M. F., and Winter, H. H., Polymer Bulletin, 16, 83 (1986).

3 Katsaros, J. D., Malone, M. F., and Winter, H. H., Poly. Eng. and Sci., 29, 1434(1989).

4 Halary, J. L., Ubrich , J. M., Nunzi, J. M., Monnerie, L., Stein, R. S. Polymer, 25, 956 (1984).

5 Mani, S., Malone, M. F., Winter, H. H., Halary, J. L. and L. Monnerie. Macromolecules, 24, 5451 (1991).

6 Gelles, R., "Photophysical Analysis of Miscible and Immiscible Polystyrene/Poly (Vinyl Methyl Ether) Blends" (Phd Thesis) 1983.

Theoretical and Applied Rheology, edited by P. Moldenaers and R. Keunings
Proc. XIth Int. Congr. on Rheology, Brussels, Belgium, August 17-21, 1992
1992 Elsevier Science Publishers B.V.

FLOW-INDUCED ORDERING AND ANISOTROPY OF A TRIBLOCK COPOLYMER STYRENE-BUTADIENE-STYRENE WITH CYLINDRICAL DOMAIN MORPHOLOGY

H.H. WINTER[1,2], D. B. SCOTT[1], A. J. WADDON[2], Y.G. LIN[1] and F. E. KARASZ[2]

[1]Department of Chemical Engineering and [2]Department of Polymer Science and Engineering, University of Massachusetts, Amherst MA 01002

1. SUMMARY

The anisotropic relaxation modulus (ref. 1) and shear induced orientation transitions (ref. 2) have been studied in a triblock copolymer styrene-butadiene-styrene with cylindrical domain morphology. The order in these polymers is defined by the direction of the cylinders and by the direction of the hexagon planes of the hexagonally packed cylindrical domains. Highly ordered samples with uniform orientations of domains and hexagon planes could be produced either with planar extension or with unidirectional shear at intermediate strains. The flow induced morphologies were confirmed with transmission electron microscopy (TEM) and small angle x-ray scattering (SAXS). A novel sample preparation technique made it possible to probe the anisotropy mechanical properties in the concentric-disk rheometer. In the melt state, shear moduli were found to be significantly lower when probing in direction of the cylinder morphology than perpendicular to it. The as-cast sample with random domain morphology had intermediate properties.

Flow-induced order transitions were studied by starting out with a highly ordered morphology and by imposing a well-defined shear flow with shear direction perpendicular to the director of the original orientation. Shear flow rotated the cylinder direction by 90^0 into the shear direction. Intermediate textures of reduced order were found at small strains, maximum alignment was achieved at intermediate strains, and the order deteriorated again at large strains. SAXS and TEM on sheared and quenched samples gave snapshots of this evolving morphology.

2. INTRODUCTION

Triblock copolymer molecules consist of three homopolymer blocks covalently bonded together. In the case of styrene-butadiene-styrene (SBS), the molecular structure is linear and consists of two blocks of polystyrene (PS) connected by a block of polybutadiene (PB). The thermodynamic incompatibility between the blocks results in a periodic domain structure with pure PS domains in a pure PB matrix. The size of the phase separated domains is prescribed by the radius of gyration of the blocks and the interaction parameter (ref. 3). With volume fractions between 21-34% (ref. 4), the minor component forms cylindrical domains which are periodically arranged (hexagonal packing) in a continuous three dimensional matrix of the major component. The cylinders and the hexagonal lattice planes locally align over the length scale of about 1 micron to form units denoted as grains. The director and the lattice planes vary randomly from grain to grain, giving a macroscopically isotropic structure.

The effect of domain orientation can best be seen in the most extreme/ideal structure where all the domains are aligned uniformly throughout the sample ("single crystal") (refs. 5-8). Single crystal structure ideally has a uniform director and uniform direction of the lattice planes. Deviations from this uniformity are fairly large in practical cases, expressing themselves in order parameters, S<1.

Anisotropic morphology greatly affects the relaxation behavior of the polymer in the solid state (ref. 7). This study is concerned with the molten triblock copolymer. The three major topics are:
(1) producing single crystal structure with uniform alignment of cylindrical domains and lattice planes
(2) measuring its anisotropic mechanical properties
and (3) inducing an order transition by shear flow.

3. EXPERIMENTAL

3.1. Materials and Sample Preparation.

The polymer used was a triblock copolymer SBS (Kraton 1102, Shell Development Company) containing 26 wt% polystyrene with molecular weight proportions of PS(10,600)-PB(60,100)-PS(10,600). The PS is well below and the PB is well above their respective entanglement molecular weight. At all accessible temperatures, the polymer is strongly phase separated. For cleaning, the polymer was twice precipitated into isopropanol from a 10% toluene solution. Films were cast from a 10%wt toluene solution with .5 wt% Irganox as an antioxidant and vacuum dried for 24 hours at 150 ^{0}C (ref. 9).

3.2. Domain Ordering in Planar Extension.

Domain orientation was produced with planar extension in a specially designed rectangular flow cell consisting of two lubricated parallel plates (refs. 10-12). This is a non-axisymmetric shear free flow in which the sample is compressed in one direction, stretched in the second direction (extension direction) and held at a constant width in the third direction. Lubricating the surfaces produces a slip condition with the normal stress on the plate surface being much larger than the shear stress. Rectangular films cut from the as cast material were stacked between the two metal plates of the flow cell, thinly coated with Viscasil 300,000, thermally equilibrated at 120^{0}C, slowly compressed to the desired thickness, and annealed for two hours at 120^{0}C. This annealing step is essential for healing the domain structure. The number of films stacked was varied to control the compression ratio, λ, which is defined as the initial height, H_0, divided by the final height, H_1.

3.3. Domain Ordering in Shear Flow.

Samples were subjected to a constant shear rate ($\dot{\gamma}$= 0.01/s) in cone and plate geometry using a Rheometrics Mechanical Spectrometer (RMS-800) at 140^{0}C. Single crystal structure was obtained by shearing to 15-30 strain units and annealing followed by cooling to room temperature in a few minutes.

3.4. Orientation Determination.

SAXS patterns were taken using a pinhole collimated Rigaku-Denki camera operating with a Ni filtered Cu Kα radiation source. Patterns were taken with the x-ray beam parallel to all 3 axes of the sample to obtain a full textural characterization: direction of cyliders, direction of lattice planes, and degree of order (from arcing)

For TEM, thin samples were cut with a AO/Reichert FC 4 cryo-microtome using a diamond knife both at -110 ^{0}C. The PB phase was selectively stained with osmium tetraoxide to provide mass thickness contrast between phases for viewing in the JEOL 100 kV electron microscope. On all micrographs the PB phase appeared black from the osmium stain.

3.5. Rheometry.

The rectangular sheet from planar extension does not suit the axisymmetric geometry required for rotational rheometry. Therefore, a special sample preparation was required to axisymmetrically align the cylinders. For this purpose, oriented films were cut in an alternating zig-zag pattern with respect to the cylinder orientation. A total number of 18 triangularly cut pieces were re-constructed to form a disk shaped sample (like pieces of a pie). The cut segments were mended together with toluene, a common solvent for both phases. The zig-zag pattern was cut with an apex angle of 20^0, giving a root-mean-squared error of 5.8^0 in the cylinder mis-alignment between cut segments. The structural discontinuity at the boundaries between cut segments was assumed to be negligible. Dynamic mechanical properties were measured in two directions with respect to the cylinder orientation using the RDS-7700 in small amplitude oscillatory shear (SAOS) with parallel plate fixtures both above and below the glass transition temperature of poly(styrene) in the frequency range of 0.01 rad/s $< \omega <$ 100 rad/s.

3.6. Shear-induced orientation transition

We start out with an initially ordered structure. The rectangular sheet from planar extension was used to construct a circular sample in which the cylinders were essentially aligned radially. To induce the order transition, the disk-shaped samples were subjected to an axisymmetric shear flow at a constant rate ($\dot{\gamma}$= 0.01/s) using parallel plate geometry at 140^{0}C. The direction of shear was circumferential so that the shear was perpendicular to the initial cylinder orientation. The amount of shear strain, γ, at a radius, r, is given by $\gamma = \theta$ r/H, where θ is the angular displacement in radians and H is the sample height. Thus, with one sample we were able to study a range of applied strains by looking at different radial positions. The displacement was held constant during rapid cooling to room temperature.

4. RESULTS

4.1. Anisotropic properties (ref. 1)

Both, planar extension and unidirectional shear were found to successfully produce single crystal structure, but only under the appropriate conditions. Shear moduli of highly ordered samples were measured in two directions with respect to the domain orientation. In the melt, high frequency data were unaffected by domain orientation. The most dramatic effects of domain orientation in the spectra were seen in the low frequency regime. Shearing parallel to the direction of domain orientation was characteristic of liquid behavior at low frequency. There was less resistance to flow since the molecules can move without having to overcome the network structure. We believe that this is the only orientation where the material can still flow and remain phase separated. However, the time constants for the flow are very long.

When the cylinder orientation is perpendicular to the shear direction, the shear deformation is restricted by the phase separated network structure. During large shear strains, shearing perpendicular to domain orientation must overcome the network structure, pulling PS segments from domains. In a strongly phase separating system as the one studied here, only a fraction of the PS end blocks is pulled out of domains and it quickly rejoins the persisting domain pattern which most likely is oriented in flow direction.

4.2. Shear-induced order transition (ref. 2)

Shear flow introduces irreversible textural changes. Affected are the direction of the cylindrical PS domains, the direction of the hexagon planes, and the degree of order. With increasing strain, the order increases at first and then deteriorates. We identified the structuring processes which occur in the early stages of shear flow. The shear-induced morphology is retained by rapidly cooling the samples to room temperature. SAXS and TEM give sufficient information to propose models for the evolving domain orientation by analyzing intermediate states of orientation.

The effect of a sufficiently large shear is to align the cylinders parallel to the direction of shear. Hexagon planes align with the 1-3 planes (shear planes). The proposed structuring process requires the cylindrical domains to break temporarily and heal again. However, the length scale of the broken structure could not be identified with the available techniques.

This study suggests that the processes during shear are highly influenced by initial morphology, and it would be informative to conduct analogous studies on samples of different starting textures.

ACKNOWLEDGMENTS

H. H. Winter, D. Scott, and Y.G. Lin gratefully acknowledge the support of the Materials Research Laboratory (MRL) of the University of Massachusetts. F. Karasz and A. Waddon gratefully acknowledge the support of grant AFOSR 91-001.

REFERENCES

1. Y.G. Lin, A.J. Waddon, D.B. Scott, F.E. Karasz, and H.H. Winter, *submitted to* Macromolecules, (1992).
2. D.B. Scott, A.J. Waddon, Y.G. Lin, F.E. Karasz and H.H. Winter, *submitted to* Macromolecules, (1991).
3. Helfand, E. and Z. R. Wasserman, Macromolecules, 9 (1976) 879-888.
4. G. Molau, in: S.L. Aggarwal, ed., Plenum Press: New York, (1970).
5. A. Keller, E. Pedemonte, F.M. Willmouth, Nature, 225 (1970) 538.
6. J. Dlugosz, A. Keller and E. Pedemonte, Kolloid Z. Z. Polym., 242 (1970) 1125-1130.
7. J.A. Odell and A. Keller, Polym. Eng. Sci., 17 (1971).544-559.
8. G. Hadziioannou, A. Mathis, A. Skoulios, Colloid Polym. Sci., 257 (1979).136-139.
9. F.A. Morrison and H.H. Winter, Macromolecules, 22 (1989) 3533-3539.
10. Sh. Chatraei, C.W. Macosko and H.H. Winter, J. Rheol., 25(4) (1981) 433-443,.
11. P. Soskey and H.H. Winter, J. Rheol. 29(5) (1985) 493-517.
12. S.A. Khan and R.G. Larson, Rheol. Acta, 30(1) (1991) 1-7.

Theoretical and Applied Rheology, edited by P. Moldenaers and R. Keunings
Proc. XIth Int. Congr. on Rheology, Brussels, Belgium, August 17-21, 1992

3D FLOW FIELD ANALYSIS OF A BANBURY MIXER

Haur-Horng Yang and Ica Manas-Zloczower

Department of Macromolecular Science, Case Western Reserve University, Cleveland, OH 44106

Introduction

One of the most common high intensity batch mixers used in the plastics and rubber industries is the Banbury mixer. The Banbury mixer consists of a mixing chamber shaped like a figure eight with a spiral lobed rotor in each chamber. The two rotors can be operated either at even speed or at different speeds. In the small clearances between the rotors and between the rotors and the chamber wall dispersive mixing takes place. Materials are fed through a vertical chute in which an air or hydraulic driven ram is located. The very complex geometry and the transient character of the flow make the flow field analysis of the Banbury mixer very difficult.

Description of Method

The approach is similar to that of the previous studies (1-4), namely a number of sequential geometries have been chosen to represent a complete mixing cycle. For polymer processing operations with laminar flow of highly viscous materials, the overall effect caused by a changing geometry can be analyzed from the results obtained separately in selected sequential geometries. A fluid dynamics analysis package - FIDAP (5), using the finite element method was employed to simulate the flow patterns in the Banbury mixer.

In this paper, we choose to simulate the flow patterns in a BB-2 type Banbury mixer with two-wing rotors, maintaining the exact dimensions of the machine. In a BB-2 type Banbury mixer, the two wings of the rotor are not equal in length. One wing is about 9.5 cm long and the other one has a length of about 7.0 cm. The total length of the rotor is 15.0 cm. The two wings of the rotor are overlapping with each other in the middle region of the rotor.

The rotors were operated at even speed. To illustrate one complete revolution of the rotors we selected 6 different geometries in a sequential manner. The geometries analyzed representing a complete cycle are shown in Figure 1 and are labeled by the angles α_1 and α_2, where α_1 is the angle between the left rotor tip and the X-axis and α_2 is the angle between the right rotor tip and the X-axis. In Figure 1, the long rotor wing of the left rotor is in the front, and so is the short rotor wing of the right rotor. The elements used were brick elements with 8 nodal points in each element. The total number of elements and nodal points were 16,416 and 18,984 respectively.

The field equations (Eqs. (1) and (2)) for the steady-state, isothermal flow of an incompressible fluid were solved for each geometry.

$$\underline{\nabla} \cdot \underline{v} = 0 \qquad (1)$$

$$\underline{\nabla} \cdot \rho \underline{v}\,\underline{v} = -\underline{\nabla} P + [\underline{\nabla} \cdot \underline{\underline{\tau}}] \qquad (2)$$

The material used in our analysis was a masterbatch rubber compound with 65 parts SBR 1502 and 35 parts carbon black N550. The rheological behavior of this compound was studied by Freakley and Patel (6), and was described in terms of a power law model fluid with temperature and time dependent parameters. For our analysis we choose a fixed temperature of 80°C (isothermal flow simulations) and a mixing time of 8 minutes, since the isothermal assumption is less dramatic after

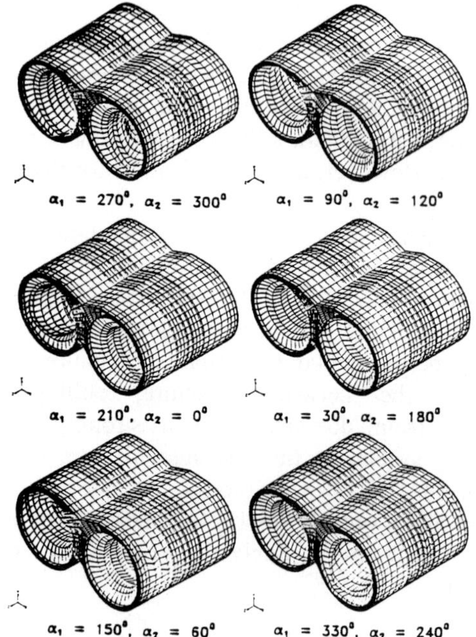

Figure 1. Mesh design for a complete cycle.

long enough mixing times. Under these circumstances, the consistency index, m in the Eq. (3) below, is $9.87\mathrm{x}10^4$ $\mathrm{N{\cdot}s^{0.22}/m^2}$ and the power law index, n, is 0.22.

$$\underline{\underline{\tau}} = m\,|\dot{\gamma}|^{n-1}\underline{\underline{\dot{\gamma}}} \qquad (3)$$

No-slip boundary conditions for the chamber wall and rotor surfaces were employed. There are two small clearances (about 1 mm) between the edges of the rotors and the chamber wall. These clearances were incorporated into our mesh design with the respective no-slip boundary conditions. In the simulations, the rotational speed was set at 60 rpm and a filling factor of 1 was employed. The simulations were carried out on a CRAY Y-MP supercomputer. The cpu time required to perform the flow simulations for one geometry was around 24000 s.

The characteristics of the flow field were analyzed in terms of a parameter λ defined by;

$$\lambda = E \; / \; (E + \omega) \qquad (4)$$

where E is the magnitude of the rate of deformation tensor and ω is the magnitude of the vorticity tensor. The parameter λ quantifies the elongational and rotational flow components and it can assume values between 0 for pure rotation and 1 for pure elongation. A value of λ of 0.5 is the familiar case of simple shear.

The parameter λ is important in assessing the efficiency of a flow field in dispersive mixing. Based on experimental observations, Elmendorp (7) reported that elongational flows are more effective than simple shear flows for the dispersion of liquids with high viscosity ratio and low interfacial tension. Also Manas-Zloczower and Feke (8,9) posited the same conclusion for the dispersion of solid agglomerates into liquids. However, in interpreting the efficiency of different flow fields for mixing, one should consider not only the values for the parameter λ but also the magnitude of shear stresses generated in the flow field.

Results and Discussion

The velocity profiles were visualized from two kinds of 2D velocity plots, the X-Y plane projection of the velocity vectors and the Y-Z plane projection of the velocity vectors. The former shows the fluid motion in the X-Y direction (rotational direction). The latter portraits the axial motion of the fluid. The flow profiles for the geometry with α_1 = 90^0 and α_2 = 120^0 are shown in Figures 2.

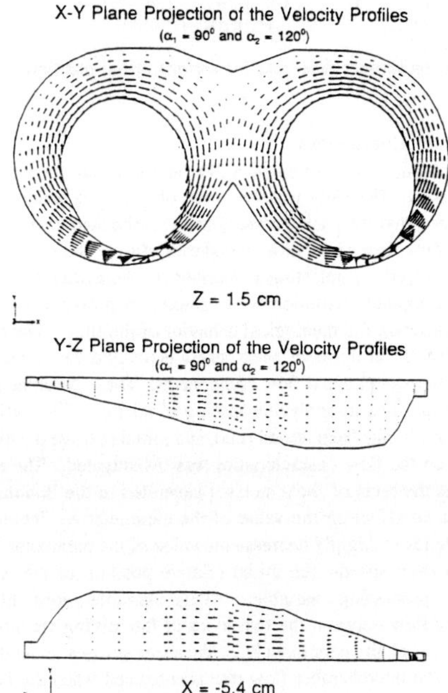

Figure 2. X-Y and Y-Z plane projection of the velocity vectors.

Table 1 shows the average values of the parameter λ, shear rate, and shear stress for all the 6 geometries in a complete cycle of the Banbury mixer. The average values were obtained by weighing the corresponding parameters for each element by the volume of the element itself for the entire flow domain. The various geometries for a complete cycle are labeled by angles α_1 and α_2 (see Figure 1). Also shown in Table 1 are the values for the interchamber flow rate, defined as the flow rate across the boundary between the left chamber and the right chamber. This quantity is related to the homogeneity of the material between the two chamber lobes (distributive mixing). Note that all the average characteristic values, except for the interchamber flow rate, remain almost unchanged when going from one geometry to another. This result implies that it is possible to conduct a process optimization study by analyzing the flow simulation results obtained for only one single geometry. The average values of the parameter λ are close to 0.5 (simple shear flow). However, the volumetric distribution of the parameter λ (Figure 3), is quite broad, with a relatively high volume percentage of material experiencing elongational flow. Figure 4 shows the volumetric distribution of the shear stress for the same geometry. The shear stress distribution is considerably more uniform than the parameter λ distribution.

Table 1. Average values for various flow characteristics

Geometry		$\dot{\gamma}$	τ	λ	Q (cm^3/s)
α_1	α_2				(interchamber flow)
270^0	300^0	19.671	1.774x10^5	0.5855	35.599
210^0	0^0	19.509	1.768x10^5	0.5775	78.176
150^0	60^0	19.740	1.772x10^5	0.5710	119.991
90^0	120^0	19.739	1.776x10^5	0.5883	36.683
30^0	180^0	19.568	1.764x10^5	0.5794	37.922
330^0	240^0	19.784	1.775x10^5	0.5770	64.234

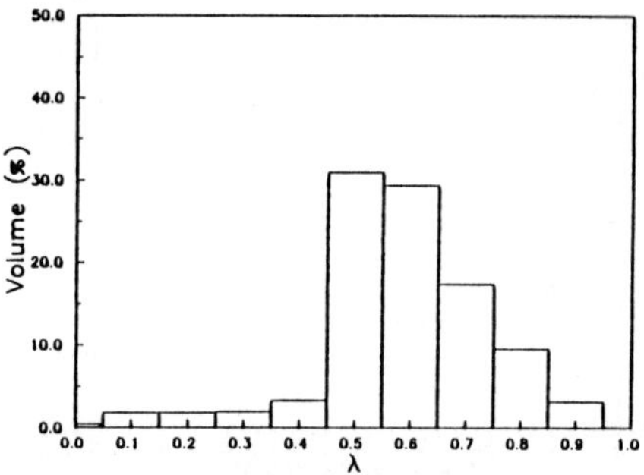

Figure 3. Volumetric distribution of the parameter λ for the geometry with α_1 = 270^0 and α_2 = 300^0.

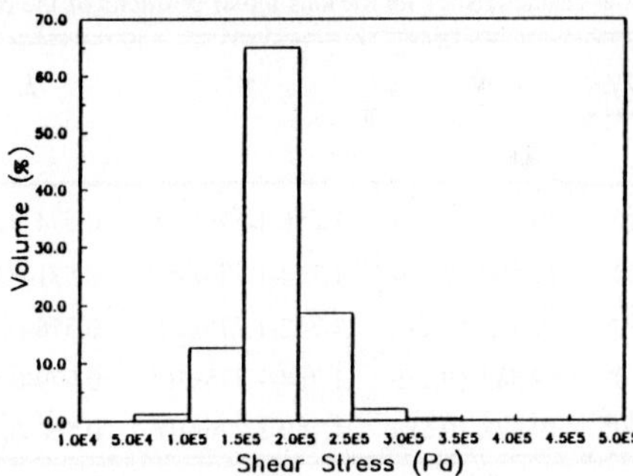

Figure 4. Volumetric distribution of the shear stress for the geometry with α_1 = 270^0 and α_2 = 300^0.

The simulation results were used to study the influence of processing variables on the flow field characteristics. In a Banbury mixer, we can either change the rotational speed of the rotors or the speed ratio between the two rotors. Figure 5 shows the influence of rotational speed on the flow field characteristics (parameter λ and shear stress), while operating at a speed ratio of 1 for the two rotors. The shear stress increases as the rotational speed increases, whereas the parameter λ shows no change. To study the influence of the speed ratio, we kept the left rotor at 60 rpm, and varied the rotational speed of the right rotor from 10 to 60 rpm. The results are shown in Figure 6. The average values of the parameter λ slightly decreases as the speed ratio increases from 1 to 6. The average shear stress also decreases since the rotational speed of the right rotor is reduced.

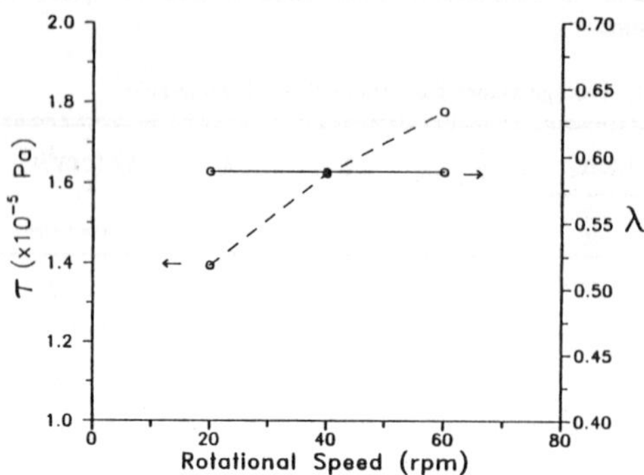

Figure 5. The influence of rotational speed on the average values for the parameter λ and shear stress (speed ratio = 1).

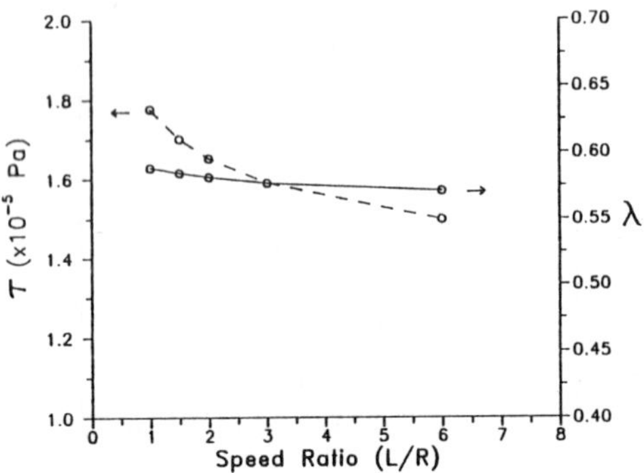

Figure 6. The influence of speed ratio on the average values for the parameter λ and shear stress.

Table 2. Flow characteristics for various initial positions of the rotors

Geometry α_1	α_2	$\dot{\gamma}$	τ	λ
90^0	90^0	19.517-19.761	1.766-1.779x10^5	0.574-0.588
90^0	120^0	19.509-19.784	1.764-1.776x10^5	0.571-0.588
90^0	150^0	19.395-19.711	1.762-1.775x10^5	0.576-0.584
90^0	180^0	19.482-19.673	1.764-1.775x10^5	0.579-0.582
90^0	270^0	19.378-19.684	1.759-1.776x10^5	0.566-0.594

When running the Banbury mixer at even speeds for the two rotors, the initial relative position of the rotors might become an important parameter. With a speed ratio other than 1, one rotor will always sweep the other, enhancing the interchamber flow between the two lobes of the mixing chamber. This phenomenon is no longer granted when the two rotors operate at the same speed. Therefore, the influence of the initial relative position of the two rotors on the flow characteristics should be analyzed. Since the interchamber flow rate is not a constant for the various geometries of a complete cycle (see Table 1), we must look at the complete cycle for a correct interpretation of the results. Figure 7 shows the interchamber flow rate for the 6 different geometries of the cycle (position 1 and 7 are the same), starting with different relative positions of the rotors. Average values for the interchamber flow rate for a complete cycle are also shown in the figure. The highest values for the interchamber flow rates are obtained with an initial position of the rotors represented by $\alpha_1 = 90^0$ and $\alpha_2 = 270^0$. Table 2 lists the range of average values for the parameter λ, shear rate and shear stress for a complete cycle (6 different geometries) when starting at various relative positions of the two rotors. The effect of the initial relative position of the rotors on these flow characteristics is insignificant.

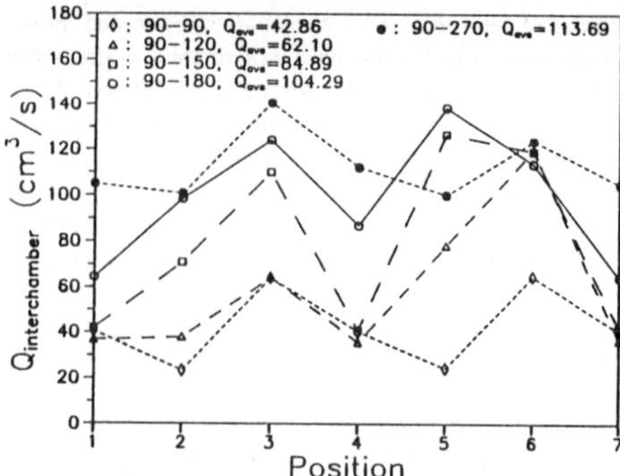

Figure 7. Interchamber flow rate for various initial positions of the two rotors.

Summary and Conclusions

In this paper we presented a 3D flow analysis for a BB-2 type Banbury mixer. The flow patterns were obtained by using FIDAP, a fluid dynamics analysis package based on the finite element method. The problem of time dependent flow boundaries, due to the rotations of the rotors, was solved by selecting a number of sequential geometries to represent a complete revolution of the rotors. A power law model was used to characterize the rheological behavior of the fluid. The flow field was characterized in terms of shear stress generated and a parameter λ quantifying the elongational flow components. These parameters are the most important ones in analyzing mixing efficiency. The influence of processing conditions (rpm, speed ratio, and initial relative position of the two rotors) on the flow characteristics was investigated. The rotational speed affects the level of shear stresses generated in the Banbury mixer, but it shows no effect on the value of the parameter λ. Increasing the speed ratio tends to slightly decrease the value of the parameter λ. When operating at even speeds, the initial relative position of the two rotors becomes a processing variable. This variable does affect the interchamber flow between the two lobes of the mixing chamber, but it shows no effect on the parameter λ, or the shear stresses generated in the flow field. The interchamber flow rate is enhanced when the two rotors are initially opposite to each other ($\alpha_1 = 90^0$; $\alpha_2 = 270^0$).

Acknowledgement

The authors would like to acknowledge the use of computing services from the Ohio Supercomputer Center.

References

1. J. J. Cheng and I. Manas-Zloczower, *Polym. Eng. Sci.*, **29**, 701 (1989).
2. J. J. Cheng and I. Manas-Zloczower, *Polym. Eng. Sci.*, **29**, 1059 (1989).
3. J. J. Cheng and I. Manas-Zloczower, *J. Appl. Polym. Eng. Sci.: Appl. Polym. Symp.*, **44**, 35 (1989).
4. J. J. Cheng and I. Manas-Zloczower, *Int. Polymer Processing*, **V**, 178 (1990).
5. FIDAP Package, Fluid Dynamics International, Inc., Evanston, Ill.
6. P. K. Freakley and S. R. Patel, *Rubber Chem. Technol.*, **58**, 751 (1985).
7. J. J. Elmendorp, *Polym. Eng. Sci.*, **26**(6), 418 (1986).
8. I. Manas-Zloczower and D. L. Feke, *Int. Polym. Proc.*, **II**, 185 (1988).
9. I. Manas-Zloczower and D. L. Feke, *Int. Polym. Proc.*, **IV**, 3 (1989).

Theoretical and Applied Rheology, edited by P. Moldenaers and R. Keunings
Proc. XIth Int. Congr. on Rheology, Brussels, Belgium, August 17-21, 1992

SOME APPROACHES OF CONSTRUCTION OF GENERALIZATION OF RHEOLOGICAL CHARACTERISTIC OF POLYMER BLEND MELT VISCOSITY

E.K. BORISENKOVA, YU.G. YANOVSKY

Institute of Applied Mechanics of Russian Academy of Sciences. Leninsky Prospect 32A, 117334 Moscow (Russia)

A great deal of attention has been given to the problem of investigating and practically employing immiscible polymer blends in view of using liquid crystalline (LC) thermoplastic materials as the disperse phase, which considerably decrease the blend viscosity and improve simultaneously the mechanical characteristics of final articles.

Taking into consideration the generality of rheological behavior and structure of different immiscible polymer blends, we formulated the problem of finding the means for the invariant presentation of the rheological data and for establishing their relation to the stream structure.

A very simple reduction procedure is suggested for the blend viscosities of different polymer pairs [1]. This procedure is based on the comparison of the blend viscosity, normalized either to the matrix or to the disperse phase viscosity, with the viscosities ratio of the initial polymers. We have obtained the universal linear dependencies, mutual analysis of which allows connection of their special points with the stream morphology. The fibrillous morphology takes place in the range of the ratio of the initial components viscosities equal 0,1-5. Simultaneous, the thin skin consisting of the disperse phase polymers is formed. The above-indicated dependencies are plotted for different polymer pairs. Both the ordinary thermoplastic polymers and mesophase and the liquid crystalline polymers are used as the disperse phase.

The previously mentioned method was used for description of rheological behavior of miscible polymer, namely the narrow faction of polybutadiene and polyisoprene. The results showed that this factions are relaxation immiscible. The Coule-Coule diagrams confirm such conclusion.

REFERENCES

1. E.K. Borisenkova, V.G. Kulichikhin and N.A. Platé, Rheol Acta, 30:581-584, (1991).

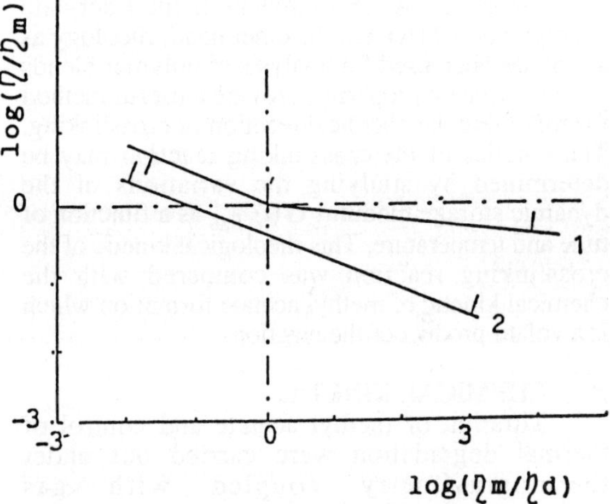

Fig. The dependences of the polymer blend viscosity, normalized to the matrix viscosity on the viscosities ratio of the initial polymers for miscible (1) and immiscible (2) polymers. Arrows show the direction of mentioned dependencies in case of decrease of polymer disperse phase concentrations.

Theoretical and Applied Rheology, edited by P. Moldenaers and R. Keunings
Proc. XIth Int. Congr. on Rheology, Brussels, Belgium, August 17-21, 1992
1992 Elsevier Science Publishers B.V.

CO-CROSSLINKING OF EVA/EMA COPOLYMERS: CHEMICAL AND RHEOLOGICAL STUDIES OF KINETICS

P.CASSAGNAU, M. BERT,
V. VERNEY and A. MICHEL

Laboratoire des Matériaux Organiques - CNRS
BP 24 - 69390 VERNAISON (FRANCE)

1. INTRODUCTION

The chemical reaction of transesterification catalyst by organometallic compounds (dibutyltin oxide) can be used to co-crosslink blends of copolymers such as ethylene vinyl acetate copolymers (EVA) and ethylene methyl acrylate copolymers (EMA). On the other hand, rheology as a tool has been used for analysis of polymer blends and viscoelastic properties provide a useful method for following the chemical reaction of crosslinking. The kinetics of the crosslinking reaction may be determined by studying the variations of the dynamic storage modulus $G'(t)_{T,\omega}$ as a function of time and temperature. This rheological kinetic of the crosslinking reaction was compared with the chemical kinetic of methyl acetate formation which is a volatil product of the reaction.

2. CHEMICAL KINETIC

Titration of methyl acetate and control of thermal degradation were carried out under thermogravimetry coupled with gas chromatography. The determination of the extent p of the chemical reaction in isothermal experiments has been made in the range of temperatures 170<T°C<230. The extent of the reaction at the end $(t\rightarrow\infty)$ of the crosslinking reaction is low $(p\infty\rightarrow0.22)$. Indeed, the motions of the polymer chains are slowed down by the chemical network and the realaxation times of the chains tend to infinite value with the degree of crosslinking.

Consequently, we assume that the kinetic equation may be expressed as a kinetic equation of second order coupled whith a "damping function" which takes into account the effects of the crosslinking bonds over the mobility of copolymer chains:

$$p=\frac{ak1t}{1+ak1t}p\infty(1-\exp(-t/\tau)) \tag{1}$$

where a is the initial concentration of reactive groups, k1 is the kinetic constant and τ is a damping parameter of the network. These parameters have been adjusted through a non linear reaction.

3. RHEOLOGY KINETIC

When the reaction of crosslinking occurs, the storage and loss moduli increase with the time as the network resulted. At a particulary point, the storage and loss moduli cross and define the gel point. As defined by the range of frequencies $0.2<\omega rad/s<1$, the storage modulus G'(t) may be assumed to be the equilibrium shear elastic modulus Ge(t) defined by the theory of rubber elasticity. However, this assumption is not valid at the vicinity of the gel point (t<tgel).

A review of the theoretical relation for the equilibrium shear modulus was made by Gottlieb et al and Pearson et al. The expression generally admitted is the following:

$$G'(t)=Ge(t)=[\nu(t)-\mu(t)h]RT+G_e^{max}\ Te(t) \tag{2}$$

t>tgel

The first term is the chemical contribution due to crosslinking where ν and μ are the respective concentrations of elastically active strands and junctions, h is an empirical parameter 0<h<1, here h=0. The second term is the contribution of the portion of topological interactions or entanglements between chains which are permanently trapped by the chemical network.

The storage modulus G'(t) calculated through the relation (2) and from experimental chemical kinetic and from chemical kinetic law (relation 2) is in good agreement with the values of the storage modulus measured from rheology experiments.

4. CONCLUSION

Co-crosslinking of EVA/EMA blends through transesterification of ester groups has been followed by classical route through assessment of methyl acetate as a volatil product of the reaction and by dynamic rheological measurements. Dynamic rheology through storage modulus is also a useful method for following the crosslinking reaction. The knowledge of the extent of the crosslinking reaction permits to calculate the variations of the storage modulus versus time from theories for elastic modulus. Then, there is a good agreement between the kinetic data supplied by dynamic rheology and kinetic data supplied by chemical assesment.

Theoretical and Applied Rheology, edited by P. Moldenaers and R. Keunings
Proc. XIth Int. Congr. on Rheology, Brussels, Belgium, August 17-21, 1992

EFFECTS OF REPETITIVE EXTRUSION UPON THE MOLTEN AND SOLID-STATE PROPERTIES OF POLYETHYLENES

F. Chambon, J.P. Autran

Exxon Chemical International, Machelen Chemical Technology Center, Hermeslaan 2, B-1831 Machelen

1. INTRODUCTION

New experiments have been performed on polyethylenes to better understand the nature and the origin of the molecular changes brought about by repetitive extrusion. We have been trying to uncover the underlying phenomenon of this process, also often referred to as "shear refining" (1).

2. EXPERIMENTAL

The polyethylene grades used in this study have the following characteristics:

Polymer	Grade	ρ (g/cm3)	MI	Mw (kg/mole)	Hexyl+ /1000 C
LDPE	LD451	0.917	1.2	182	4.75
LDPE	LD165	0.9226	0.3	90.2	2.35
LLDPE	LL1001	0.918	1	96	1

Repetitive extrusion was done on a Leistritz corrotative twin screw extruder. Extrusion temperature was 250°C. 0.5wt% Irganox 1044 was added before the first pass in the extruder to limit oxydative degradation as much as possible. Up to 5 consecutive passes were made.

Elongational flow behavior was characterized via Cogswell analysis of zero-length die capillary data as well as with non-isothermal spinning experiments taken with the Rheotens device attached to a Gottfert 2002 rheograph.

Shear flow response was analyzed with a Rheometrics Mechanical Spectrometer, RMS800, utilizing a cone & plate geometry and small amplitude oscillatory deformation mode.

Solid-state mechanical testing of the repetitively extruded samples was performed on compression molded sheets and films using a Zwick machine model 1445-03 for tensile tests and a Zwick REL1852 for dart impact evaluations. Additional creep data were obtained by hanging free weights to film strips.

Memory effects in "shear refined" samples were studied by dissolving the polymer in p-xylene at 120°C, at concentrations below and above the overlap. The polymer was then recovered either by slow cooling to room temperature or by quenching in cool methanol with subsequent filtration/drying.

3. RESULTS

It was found that with Low Density Polyethylene (LDPE), strain hardening almost completely disappeared after 4 passes in the extruder. However, the shear viscosity, at low deformation rates, decreased by less than 40%. The decay in elongational properties, or otherwise melt elasticity, was the most sensitive for the LDPE with the highest degree of long chain branching, ie. LD451, whereas it was almost none for the Linear Low Density Polyethylene (LLDPE). However, the changes observed are reversible, either partially as after annealing at high temperature for very long times or fully as after dissolution/recrystallization treatments below or above the chain overlap concentration.

As far as the mechanical properties of the "shear refined" polymers are concerned, it was found that they are also affected by repetitive extrusion, but to a much lesser extent than melt rheology. Furthermore, in an attempt to evaluate the efficiency of "shear refining" versus solid-state deformation, the melt rheology of repetitively extruded LD451 and that of highly drawn LD451 at temperatures $T_g<T<T_m$ were compared. No indication of reduction in melt elasticity was found with these latter samples.

4. CONCLUSIONS

Our results strongly support the hypothesis of long-range configurational changes rather than of a gross loss of entanglements upon "shear refining". The proposal by Munsted (2) of side-chains alignment along the backbone appears as the best qualitative representation of the phenomenon. This new configuration describes a different state of intermolecular/intramolecular inter-actions which, in our view, are at the origin of the rheological changes observed with LDPE but not with LLDPE. It implies no significant changes in the thermodynamic state of the melt; otherwise annealing would be fast. Also, the longest relaxation time derived from linear viscoelastic tests is only slightly decreased. Therefore there is no contradiction (3) between measurements and reptation models (4).

REFERENCES

1. A.Rudin, H.P. Schreiber, Polym. Eng. Sci. , **23**, 8,(1983)
2. H. Munstedt, Rheol. Acta, **14**, 1077, (1975)
3. P.J.R. Leblans, C. Bastiaansen, Macromolecules, **22**, 3312, (1989)
4. H.H. Wagner, J. of Rheol., **36**(1), 1992

Theoretical and Applied Rheology, edited by P. Moldenaers and R. Keunings
Proc. XIth Int. Congr. on Rheology, Brussels, Belgium, August 17-21, 1992

MELT FLOW-INDUCED ANISOTROPY IN AMORPHOUS POLYMERS

C. COURMIER[2], J. LADEVEZE[1], R.MULLER[1] and J.J.PESCE[1]

[1]Ecole d'Application des Hauts Polymères, 4 rue Boussingault, 67000 Strasbourg (France)
[2]PSA, 18 rue des Fauvelles, 92250 La Garenne-Colombes (France)

1. INTRODUCTION

Many properties of polymers in the glassy state (modulus, thermal expansion,...) become anisotropic when the chains are oriented. This is the case in samples subjected to melt flow and quenched before complete relaxation of the chain orientation. For some amorphous polymers (e.g. PS), anisotropic properties in the glassy state could be directly related to the birefringence Δn. It is the aim of the present study to investigate in more detail the thermal expansion below T_G for PS samples elongated in the melt.

2. STRESS-OPTICAL LAW ABOVE T_G

Specimens have been stretched at constant strain rate in the temperature range [102-125°C] (T_G is close to 95°C) with an extensional rheometer designed at our laboratory, allowing to follow Δn during the whole test and to quench the stretched specimens within a few seconds. For temperatures above T_G+20, a one-to-one relationship between true stress σ and Δn has been found (Fig. 1). For σ lower than 2MPa, the ratio $\Delta n/\sigma$ tends to the stress-optical coefficient of molten PS (4.8 10^{-9} Pa^{-1}). Below T_G+20, Δn is not directly related to σ. As shown in Fig.1, σ at 103°C reaches about 3 MPa before Δn starts to increase. When the elongation is stopped, this contribution to σ relaxes very quickly and the equilibrium relation between σ and Δn is recovered.

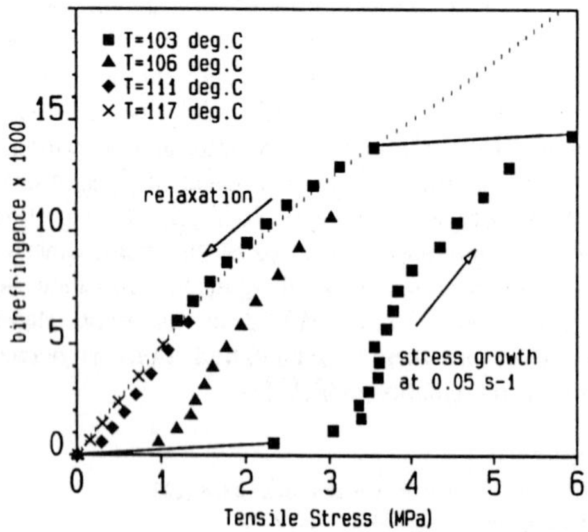

fig. 1: Birefringence vs. true stress at various temperatures

3. ANISOTROPY IN THE SOLID STATE

Figure 2 shows the thermal expansion data below T_G in the direction of stretching for specimens elongated at 102.2°C with a strain rate of 0.05 s-1 and an extension ratio of 2.5. The difference between the three samples in Figure 2 is the way of cooling the stretched specimens. Sample (a) has been quenched at constant elongation just after the end of the stretching. Since the quenching takes a few seconds, a significant part of the stress has probably relaxed (see Figure 1). For sample (b), the elongation up to λ=2.5 was followed by a creep test at constant true stress during 20s, and the quenching takes then place at constant elongation. Finally, sample (c) was cooled down below T_G <u>under constant true stress</u> (non-isothermal creep). The final birefringence is the same in samples (b) and (c) (Δn=0.0176) and lower (Δn=0.0122) in sample (a).

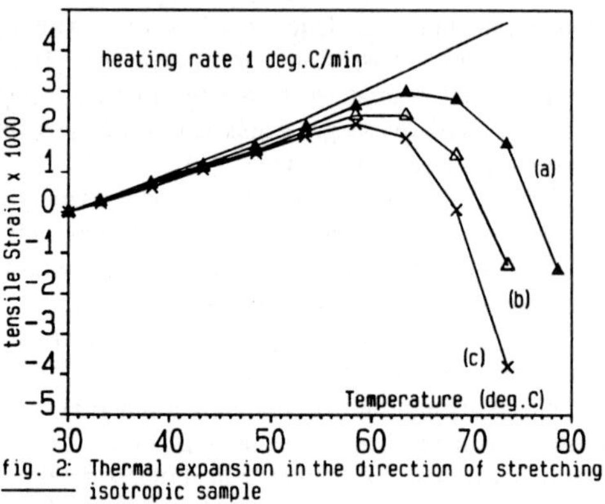

fig. 2: Thermal expansion in the direction of stretching
——— isotropic sample

4. DISCUSSION

Below 60°C, a thermal expansion coefficient, lower than that of an isotropic sample, can be determined for the oriented samples. But the major effect is a strong recovery of the elongated samples above 60°C (T_G-35°). This effect is most pronounced for the specimen cooled down under constant stress, and seems to depend not only on the birefringence, but on the whole thermomechanical history.

ACKNOWLEDGEMENT: This work has been funded by C.L.I.P. (Club Logiciels pour l'Industrie Plastique).

Theoretical and Applied Rheology, edited by P. Moldenaers and R. Keunings
Proc. XIth Int. Congr. on Rheology, Brussels, Belgium, August 17-21, 1992
1992 Elsevier Science Publishers B.V.

RHEOKINETICS OF NETWORK FORMATION IN ELASTOMER COMPOSITIONS

DONSKOY A.A[1]., KULICHIKHIN S.G[2]., SHERSHNEV V.A.[1], YULOVSKAJA V.D.[1], MALKIN A. YA.[2]

[1]Moscow Institute of Fine Chemical Technology, 117571, Moscow, pr. Vernadscogo, 86, Russian Federation.
[2]Research Instiute for Plastics, NPO "Plastmassy", 111112, Moscow, Perovsky proezd, 35, Russian Federation.

1. INTRODUCTION

The possibility to control the process of network formation in elastomers have a great importance for elastomer processing. Some papers described the overall process of network formation from reactive oligomer compounds. At the same time analogous data for high molecular elastomers are not sufficient. The purpose of the present work is the study of rheological behavior of model compositions on the base of cis-I,4-polyisoprene (PI) during crosslinking by the rheological methods.

2. METHODS

The viscometer "Rheomat-30" was used for viscosity of elastomer compositions determination. The gel-fraction contents were determined by the sol-gel analysis method in toluene. The molecular weight (MW) of PI were determined through intrinsic viscosity of polymer's solution. The MW of PI by light-scattering method were determined on "Sofica" nefelometer. The dynamical characteristics of elastomer compositions (storage modulus, loss modulus, tan δ) were obtained by TBA-methods at $\approx$ 1 Gz.

3. RESULTS AND DISCUSSION

When comparing compositions characterised by prolonged induction time at crosslinking, good agreement in the change of viscosity, storage modulus and tan δ was noted. For all sudied compositions extremal shape of tan δ -vs-time curve was found in the region adjoining to increasing part of storage modulus curve.

Comparing viscometric data with tan δ changes it was noted that the maximum of the latter correlates with the moment of flow cessation. It allows us to identify it as a gel-point. Using crosslinking system of high activity, rate of crosslinking turned up to be so high, that experimental determination of induction period becames impossible. The possibiliy to observe the change of rheological characterisitics during the whole cure process in this case appears only when a special crosslinking inhibitor was added and temperature of a sample was decreased to 120°C. It was found that three main rheological effects : increasing viscosity during induction period connected with the growth of polymer MW, existance of the induction period on the G'(t) function and appearance of the tan δ maximum are typical for all systems under investigation.

On the base of experimental results the "rheological" measure of convertion (β) were determined. The shape of the $\beta(t)$ curves proves that the polyisoprene crosslinking process can be satisfactory described by the equation of self acceleration type.

4. CONCLUSIONS

It is necessary to stress that the presented results justify the idea about identity of physical-chemical processes which determine the process of material with network stucture formation. Peculiarities of this structure formation connected with the stages of accumulation of reaction active products leading to autocatalysis of crosslinking reactions, although these systems can be different in nature.

Theoretical and Applied Rheology, edited by P. Moldenaers and R. Keunings
Proc. XIth Int. Congr. on Rheology, Brussels, Belgium, August 17-21, 1992

INFLUENCE OF THE MOLECULAR WEIGHT ON THE OSCILLATING FLOW OF HDPE MELTS

V. DURAND[1], B. VERGNES[1], J.F. AGASSANT[1], and R.J. KOOPMANS[2].
[1]. CEMEF. Ecole des Mines de Paris. BP 207 - 06904 SOPHIA ANTIPOLIS cedex (FRANCE)
[2]. DOW BENELUX N.V. Box 48. 4530AA TERNEUZEN (THE NETHERLANDS)

The purpose of this work is to study a type of melt flow instability which is usually called "stick-slip" flow and is characterised by an oscillating flow. In order to study the molecular weight influence on this melt flow defect and on its appearance conditions, the rheological behaviour of three high density polyethylenes was studied in capillary rheometry.

1.Experimental: Three HDPE with an Mw/Mn of 5 and different Mw 170000 (A) - 129000 (B) and 107000 (C) were characterised at imposed flow rate at 160°C with different dies of diameter:

D = 0.93 - 1.3 - 2 - 3 mm
and L/D = 0 - 4 - 8 - 16 - 32

2.Flow curves: Excepted for zero length dies, the pressure-flow rate relationships exhibit the form of a double branched curve. The two branches are respectively limited by (Q1,P1) and (Q2,P2) with Q2 > Q1 but P2 < P1. When the imposed flow rate is between Q1 and Q2, no stable flow can be obtained: the pressure oscillates between P1 and P2 and the die exit flow rate varies in a discontinuous way. Video investigation shows that oscillation maxima correspond to a sudden jump of the die exit flow rate, and minima to a sudden decrease. The pressure increase phases correspond to low die exit flow rates, the pressure decrease phases to high die exit flow rates. These observations allow us to describe this instability as an hysteresis loop between the two branches of the flow curve.

Both die length and diameter affect the flow curve discontinuity. When the length is increased, the flow curve point (Q1,P1) shifts to a lower flow rate. An increase of die diameter dramatically decreases the oscillations amplitude and the oscillations tend to disappear.

With the same die, increasing the HDPE's molecular weigth shifts the instability to lower flow rate (fig.1) but the critical pressure P1 is about the same.

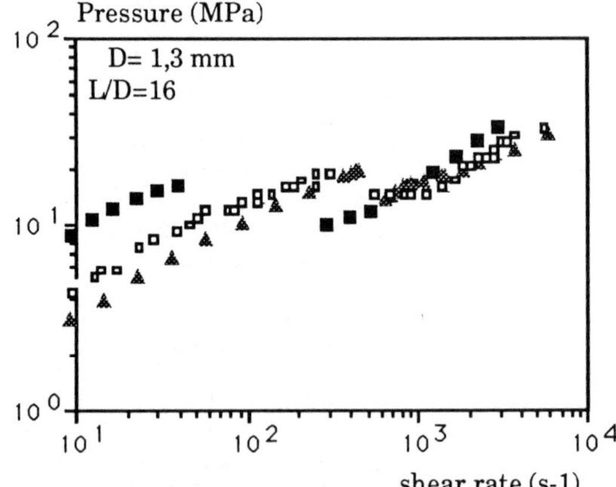

Fig.1 : Pressure versus apparent shear rate for the three HDPE: A: ■ B: ▫ C: ▲

3.Pressure oscillations: Because of polymer compressibility, the period of the pressure oscillations is linearly decreasing when the rheometer barrel is emptied. Nevertheless, if the oscillations are normalized in order to give them a "normalized period" of 1 and a "normalized amplitude" of 1, we find an oscillation shape which characterises the imposed flow rate (Fig. 2). For each HDPE, the same change in the shape of the normalized oscillation is observed when the mean flow rate is increased in the oscillating zone.

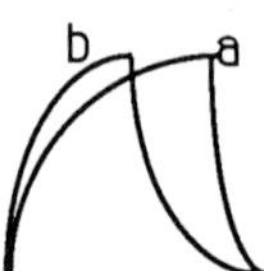

Fig.2: Schema of two typical normalized pressure oscillations. The imposed flow rate of "a" is lower than the imposed flow rate of "b".

Theoretical and Applied Rheology, edited by P. Moldenaers and R. Keunings
Proc. XIth Int. Congr. on Rheology, Brussels, Belgium, August 17-21, 1992

SOME ASPECTS OF THE VISCOELASTICITY OF POLYMER BLENDS

L. FAITELSON[1] and E. JAKOBSON[1]

[1]Latvian Academy of Science, Inst. of Polymer Mechanics, 23 Aizkraukles st., 226006 Riga (Latvia)

The frequency relaxation spectra N(s) for melts of HDPE and LDPE and their binary blends pass through a maximum at a frequency s_0. The value $N(s)|_{s_0}$ increases and s_0 decreases with an increase of the fraction of HDPE in the blend. It was found that $s_0 \sim \eta_0^{-1.4}$ and it is possible to conclude that the blends melts form a single-phase system. A similar conclusion follows also from examination of the viscoelastic relaxation spectra lgH(lg λ) curves: the values of H monotonicaly decrease in the terminal zone, while the more or less pronounced secondary plateau, characteristic for blends of incompatible components, is absent. The characteristic relaxation time $\lg \lambda_0 = w_1 \lg(\eta_{0(HD)} J_{0(HD)}) + w_2 \lg(\eta_{0(LD)} J_{0(LD)})$, and the dynamic viscosity vs. blend composition dependences (at a fixed value of ω) obeys the logarithmic law of summation. The linear viscoelasticity of HDPE and LDPE melts and their binary blends can not be described by the Havriliak–Negami equation. The dependences of lgG'(lgG") present graphically the viscoelasticity of melts better than the Coule–Coule plot (ref.1,2).

Analysis of the results of nonlinear viscoelasticity gives a different conclusion (ref.3). In distinction from homopolymers the flow curves of blends, particularly for the composition of LDPE:HDPE=8:2, reveal two regions of non–Newtonian flow with distinguish index of flow. For given σ_{12} the melt viscosities of the blends deviate from the values calculated according to the logarithmic law of summation, and $\gamma_e = N_1/\sigma_{12}$ vs. the composition of blend (at a fixed value of σ_{12}) pass through the maximum. This allows us to suppose that in the last case the shear flow promotes the phase separation and increases the interfacial tension (ref.4).

REFERENCES

1. E.E.Jakobson, L.A.Faitelson, Mechanics of Composite Materials, 26 (1990) 130–140.
2. S.Havriliak, S.A.Negami, Polymer, 8 (1964) 161–210.
3. E.E.Jakobson, L.A.Faitelson, Mechanics of Composite Materials, in press.
4. H.Vanoene, J. Coll. and Interface Sci., 40 (1972) 448–467.

Theoretical and Applied Rheology, edited by P. Moldenaers and R. Keunings
Proc. XIth Int. Congr. on Rheology, Brussels, Belgium, August 17-21, 1992
1992 Elsevier Science Publishers B.V.

A LOG-NORMAL MODEL OF THE MELT VISCOELASTIC RELAXATION TIMES SPECTRUM FOR POLYDISPERSE POLYMERS. INFLUENCE OF MOLECULAR PARAMETERS.

R. FULCHIRON, V. VERNEY,
P. CASSAGNAU, A. MICHEL

Laboratoire des Matériaux Organiques - CNRS
BP 24 - 69390 VERNAISON (FRANCE)

1. INTRODUCTION

The relaxation spectrum ($H(\tau)$) is the fundamental kernel of the melt viscoelastic behavior of polymer materials. From the terminal zone (long relaxation times) till to the transition zone (mid relaxation times, rubbery behavior), the molecular dynamics of the whole chain and of chain segments with respect to the molecular environment is concerned. It contains through its various integrals all the fundamental viscoelastic parameters such as the newtonian viscosity η_0, the steady-state compliance $J_e°$ or the plateau modulus $G_N°$. Moreover it can be considered in close relation with the molecular weight distribution. Several methods based upon approximation of nth derivatives of loss or storage modulus have been proposed to determine $H(\tau)$ (refs. 1,2). Another method is to have an analytical function of one of the components of either the complex modulus or the complex viscosity and to use the complex algebra inversion formulæ (ref. 3).

At last another way is to have an analytical function of the spectrum itself and it is the purpose of this paper to propose a new model based upon a log-normal law.

2. ANALYTICAL MODEL

One major consideration to take into account is to be sure that any model will respect theoretical considerations about the limit slopes of loss and storage moduli versus frequency. Mathematically it can be shown that if the higher and higher order integrals of the spectrum converge this theoretical consideration is respected. For this purpose we used a log-normal law for the spectrum according to:

$$H(\tau)=\frac{\eta_0}{\sigma\sqrt{\pi}}\cdot\frac{\exp[-(1/\sigma.\mathrm{Ln}(\tau/\tau_m))^2]}{\tau} \quad (1)$$

with: $\tau_m=\sqrt{\tau_w.\tau_n}$ and $\sigma=\sqrt{2.\mathrm{Ln}(\tau_w/\tau_n)}$

Thus the spectrum is defined with three parameters: the newtonian viscosity η_0, the weight average relaxation time τ_w and the number average relaxation time τ_n. The major interest of the log normal law is that each function $\tau^n.H(\tau)$ will be a bell-shaped curve with a finite integral. The calculation of the spectrum is based upon a non linear least square adjustement of the model onto experimental measurements according to:

$$\eta^*(\omega)=\int_0^{+\infty}\frac{H(\tau).d\tau}{1+j\omega\tau} \quad (2)$$

3. CONCLUSION

The validity of the model is checked a posteriori by calculation of loss and storage modulus frequency variations and comparison with experimental points. The molecular structure relationships are established with polystyrene and polypropylene samples with various average molecular weights and molecular weight distributions. The method provides the plateau modulus value, even in the case of polydisperse polymers (polypropylene) while its experimental vizualization is very doubtful because of coupling of Tg molecular chain dynamics. At last a unique relation is proposed to correlate the molecular weight distribution (M_w/M_n) to the breadth of the relaxation times distribution (τ_w/τ_n).

REFERENCES

1. J. D. FERRY
Viscoelastic properties of polymers, 2nd Ed., J. Wiley & Sons, New-York, 1971
2. N. W. TSCHOEGL
The phenomenological theory of linear viscoelastic behavior, Springer Verlag, BERLIN, 1989
3. B. GROSS
Mathematical structure of the theory of the viscoelasticity, Hermann, PARIS, 1953

Theoretical and Applied Rheology, edited by P. Moldenaers and R. Keunings
Proc. XIth Int. Congr. on Rheology, Brussels, Belgium, August 17-21, 1992

The Investigation on Melt Pumping Mechanisim of Counter–rotating Intermeshing Twin Screw Extruders

Xiaozheng Geng Ruiju Sang

Beijing Institute of Chemical Technology, Beijing, 100029, China

Counter–rotating intermeshing twin schew extruders is a great class of screw extruder. But the complexity of its geometry makes it rather difficult to carry out the precise analysis of melt pumping characteristics. The present models describing the melt pumping are far from perfection, because they do not including intermeshing zone and four clearances (i. e. flight clearance, side clearance, calender clearance and tetrahedron clearance).

This thesis gives a new physical model describing melt pumping based on socalled C– chamber, which links the channel with four clearances of intermeshing zone.

The method of solving the physical model is as follows:along the channel (X direction), the C –chamber is cut into a series of thin slices . Although the variables of flow field in different slice (three velocities and one pressure) are a function of coodinate X, in a fixed slice, they are almost independent of coodinate X. In this way, the Navier– Stokes equation describing the flow field is divided into two groups. To solve the equation in every slice, Finite Element Method is adopted. According to the condition of constant flow rate along channel,the relation between neighbour slices can be obtained by means of iteration. If every slice is thin enough,the solutions can approach to the three dimesional flow field described by physical model. By iterating the viscosity at the same time, this model is suitable for other non –Newtonian fluid,say power–law fluid.

With the computer program we given, the distribution of velocity,pressure,shear rate and shear stress can be easily determimed at the difinite screw geometry, processed material and operating conditions,so can the output– pressure charateristics, the torque and power consumption of a single C–chamber. More importantly and practically,this thesis puts forward a way of determining the number of fully filled C– chamber and operating point of the pumping zone from the output–pressure characteristics of a C–chamber and the die, which makes it possible to calculate the total pressure drop, the total torque and the total power consumption of the melt pumping zone. Comparision between the experimental results and the theoretical calculating results proves that this theoretical model is a great improvement on the present model in determining the pressure drop,the torque, the consumption and number of fully filled C–chamber.

Theoretical and Applied Rheology, edited by P. Moldenaers and R. Keunings
Proc. XIth Int. Congr. on Rheology, Brussels, Belgium, August 17-21, 1992

RHEOLOGICAL PROPERTIES OF POLYMER BLEND CONTAINING A COPOLYMER

WON HO JO and SUK HOON CHAE

Department of Textile Engineering, Seoul National University, Seoul 151-742 (Korea)

Together with the effort on the miscibility and phase behavior of polymer blends, rheological properties of miscible blends have been reproted by many investigator[1-3]. Although several mixing rules for rheological properties of polymer blends have been reported[4-6], these don't take account of interaction parameter which is very important for miscible polymer blends. In this study, rheological properties of polymer blends containing a copolymer was examined as a function of copolymer composition in order to interpret the effect of the interchain interaction on the rheological properties of polymer blends. The comonomer content provides a useful way of systematically varing the strength of interaction of a polymer-polymer pair. The blend of poly(ε-caprolactone) (PCL) and poly(styrene-*co*-acrylonitrile)(SAN) was chosen for that purpose, since it is well known that the SAN copolymers are miscible with PCL over a certain range of the AN content. The SAN copolymers have similar molecular weight which excludes the effect of molecular weight on zero shear viscosity.

Chiu and Smith[7] reported the miscibility of PCL/SAN blends over the entire range of blend compositions. Very recently, Jo and Kim[8] estimated the interaction parameters for PCL/SAN blends. Figure 1 shows the values of interaction parameters for PCL/SAN blends as a function of copolymer composition. Figure 2 shows the dependence of $1/G_{Nb}^{0}$ which corresponds to the entanglement molecular weight for the blends. As compared with Figure 1, $1/G_{Nb}^{0}$ is inversely proportional to the interaction parameter, which indicates that the entanglement probability between dissimilar chains becomes smaller as the interchain interaction becomes stronger (i. e. more negative χ). This phenomenon is well explained by Wu's interpretation[9]. In other words, the specific interactions, responsible for the miscibility of polymer blends, tends to locally to align the chain entanglements for association, and thus stiffen the chain and reduce thier convolution, resulting in reduced entanglement between dissimilar chains. It is also observed that the zero shear viscosity show strongly negative deviation and minimum point, which qualitatively corresponds to the deviation of interaction parameter, so that zero shear viscosity becomes smaller as the interchain interaction becomes stronger.

REFERENCES

[1] W.W. Prest and R.S. Porter, J. Polym. Sci., PartA-2, 10 (1972) 1639.
[2] H.K. Chuang and C.D. Han, J. Appl. Polym. Sci., 29 (1984) 2205.
[3] Y. Aoki, Polymer J., 16 (1984) 431.
[4] C.D. Han and H.H. Yang, J. Appl. Polym. Sci., 33 (1987) 1199.
[5] E.K. Harris, Jr., J. Appl. Polym. Sci., 17 (1973) 1676.
[6] S. Wu, J. Polym. Sci.: Part B: Polym. Phys., 25 (1987) 2511.
[7] S.C. Chiu and T.G. Smith, J. Appl. Polym. Sci., 29 (1984) 1781.
[8] W.H. Jo and H.G. Kim, J. Polym. Sci.: Part B: Polym. Phys., 29 (1991) 1579.

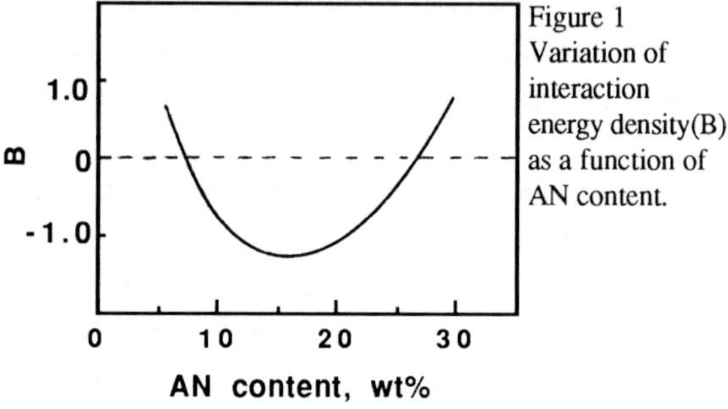

Figure 1 Variation of interaction energy density(B) as a function of AN content.

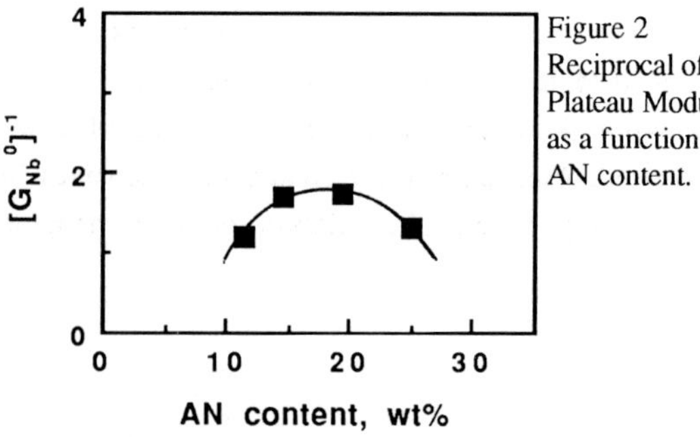

Figure 2 Reciprocal of Plateau Modulus as a function of AN content.

Theoretical and Applied Rheology, edited by P. Moldenaers and R. Keunings
Proc. XIth Int. Congr. on Rheology, Brussels, Belgium, August 17-21, 1992

THE EFFECT OF POLYMER MELT RHEOLOGY AND MELT SPINNING DYNAMICS ON THE ULTIMATE FINENESS OF PET FIBERS IN HIGH-SPEED SPINNING PROCESSES.

C.T. KIANG[1], J.A. CUCULO[2]
[1]Rhodia S.A., Polymer Tech. Dept., Santo Andre, SP (Brazil)
[2]North Carolina State Un., Fiber and Polymer Science, Raleigh, NC (USA)

1. INTRODUCTION

The extensive work on the high-speed spinning process has motivated further studies on the polymer melt spinnability, due to the growing interest on microdenier fibers, obtained in the direct extrusion methods. Therefore, this study was aimed at determining the effect of the polymer melt rheology and melt spinning conditions on the minimum attainable as-spun fiber denier.

2. EXPERIMENTAL

Linear PET polymers (IV around 0.6), from different suppliers, were characterized in an Instron capillary rheometer. Apparent elongational viscosities were obtained in contraction flow experiments, according to the model proposed by Binding(ref.1). These polymers were melt spun in a Fourne extruder and taken-up at speeds ranging from 2000 to 6000 m/min with as-spun fiber denier below 1 denier per filament (dpf) down to 0.2 dpf. The spinline dynamics were calculated from the measured threadline diameter profile and from the take-up tension.

3. RESULTS AND DISCUSSIONS

The polymers presented similar behavior in shear but differed in elongational flow. The polymer with the lower apparent elongational viscosity level, at a given deformation rate and temperature, allowed the spinning of the finer denier fibers. This is consistent with the lower spinline tension level when spinning such a polymer.

In order to reduce the spinline tension by modifying the threadline dynamics(ref.2), the spinning conditions were modified by the use of heating device, insulation plate and convergence guide. The largest decrease of the threadline cooling rate was obtained with the use of the heating device and the least with the convergence guide. However, the largest decrease of the spinline tension level, and therefore the decrease of the minimum attainable denier, was achieved when the insulation plate was used simultaneously with the convergence guide. These effects are more remarkable at higher take-up speeds, when the spinline tension level is higher. At low take-up speeds, it was determined that the decrease of the as-spun fiber denier is mainly achieved by modifying the spin block, in order to homogenize the melt temperature as well as to improve the polymer melt flow distribution inside the spin pack.

4. CONCLUSIONS

At speeds up to 4000 m/min,the ultimate fineness are more dependent on the conditions of the uniformity of polymer melt flow. At higher speeds, the important parameter is the spinline tension level, which must be kept low in order to obtain finer fibers. Higher apparent elongational viscosity level or faster spinline cooling generates higher spinline tension level, resulting in higher values of the ultimate fiber fineness.

REFERENCES

1. D.M. Binding, J. Non-Newtonian Fluid. Mech.,27 (1988) 173-189.

2. A. Ziabicki, in A. Ziabicki and H. Kawai (Ed.), High Speed Fiber Spinning, J. Wiley & Sons, 1985, pp. 21-62.

Theoretical and Applied Rheology, edited by P. Moldenaers and R. Keunings
Proc. XIth Int. Congr. on Rheology, Brussels, Belgium, August 17-21, 1992

VISCOUS BEHAVIOR OF POLYMERIC MELTS. INVESTGATED BY NEMD-SIMULATIONS AND A FOKKER-PLANCK EQUATION.

M. KRÖGER and S. HESS

TU Berlin, Inst. f. Theoretische Physik, Sekr. PN 7-1, D - 1000 Berlin 12, Germany

1. INTRODUCTION

The viscoelastic behavior of polymeric fluids as measured by oszillatory shear flow experiments is described by the complex viscosity η which contains a contribution η_a associated with the symmetric traceless segment-alignment tensor $\overline{a} \equiv \langle \overline{\underline{u}^\sigma \underline{u}^\sigma} \rangle$, cf. [1]: $\eta_a \sim \int_0^1 \overline{a}\, d\sigma$. A unit segment vector at relative position σ within a polymer chain is denoted by $\underline{u}^\sigma$. For a plane Couette geometry with γ as shear rate, only 3 of the 5 independent components of the alignment tensor are nonzero, viz.
$a_+ = \langle u_x u_y \rangle$, $a_- = \frac{1}{2}\langle u_x u_x - u_y u_y \rangle$, $a_0 = \langle u_z^2 - \frac{1}{3} \rangle$.

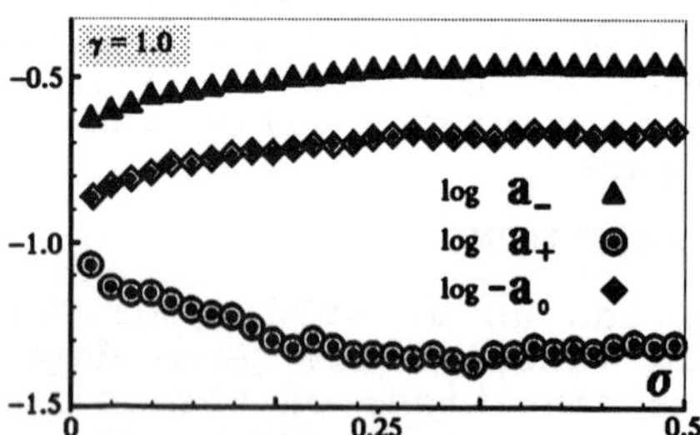

Fig. 1. The segment-alignment tensor has been extracted from nonequilibrium molecular dynamics (NEMD) computer simulations and substantiates the nonvanishing anisotropy of chain ends ($\sigma = 0$) under shear flow, which has usual been neglected, e.g. [1], but leads to a qualitative change of the viscoelastic behavior as calculated from a Fokker-Planck (FP) equation.

With the solution $\overline{a}$ from the FP-eqn. [2] follows the shear modulus $G := G' + iG'' := i\omega\eta_a$;

$$G \sim (1 + ig\omega\lambda)\left(\frac{2\tanh(\frac{\sqrt{-i\omega\lambda}}{2})}{\sqrt{-i\omega\lambda}} - 1\right), \qquad (1)$$

with a relaxation time λ and the parameter $g \equiv \lambda_{end}/\lambda$, where λ_{end} is connected with the flow-induced alignment of chain ends [2]. Here, we assume, that $\lambda_{end} \sim M \sim \lambda_{Rouse}$ and $\lambda \sim M^{3.4}$ [3] in order to derive a dependency of the complex shear modulus on both the frequency ω and the molecular weight M. For $g = 0$, eq. (1) reduces to the expression given by Doi and Edwards [1].

2. INFLUENCE OF THE MOL. WEIGHT

Our result (1) yields the width of a plateau region (WPR) for $g \neq 0$, which is defined by the ratio of two limiting frequencies ω_{low} and ω_{high}: $[\lim_{\omega\to 0} G']_{\omega_{low}} = [\lim_{\omega\to 0} G'']_{\omega_{low}} =: G_s$ and $[\lim_{\omega\to\infty} G'(resp.\, G'')]_{\omega_{high}} = G_s$.

The analytic expression for the dependency of the WPR on the molecular weight is given in Fig. 2 together with experimental results [3]. Notice, that every dot represents the full molecular weight <u>and</u> frequency dependency of both the real and imaginary part of the complex viscosity, because a given value for the WPR is strongly connected with the quantity g and thus (eq. 1 or [2]) with a characteric viscous behavior.

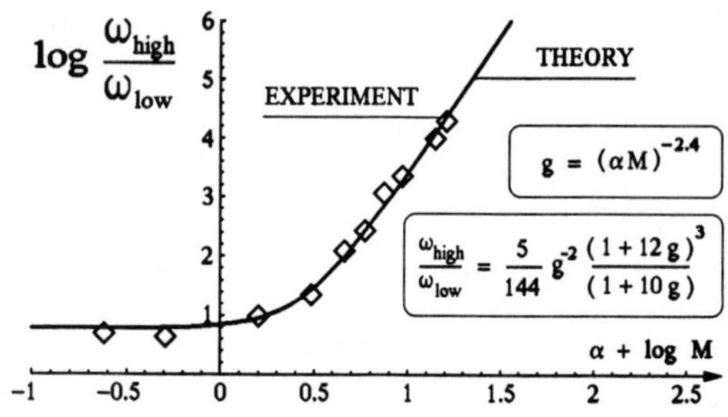

Fig. 2. The plateau region - especially the WPR - of the complex shear modulus depends on the molecular weight, as measured by Onogi et al. ([3] with $g \equiv (\alpha M)^{-2.4}$, $\alpha = 4.96$) and calculated (drawn line) from eq. (1).

REFERENCES

[1] M. Doi and S.F. Edwards, J. Chem. Soc. Faraday Trans. 74 (1978) 1789, 1818; 75 (1979) 38.
[2] M. Kröger and S. Hess, in preparation.
[3] T. Masuda, K. Kitagawa, I. Inoue and S. Onogi, Macromolecules 3 (1970) 109.

Theoretical and Applied Rheology, edited by P. Moldenaers and R. Keunings
Proc. XIth Int. Congr. on Rheology, Brussels, Belgium, August 17-21, 1992
1992 Elsevier Science Publishers B.V.

RHEOLOGY OF SILICONORGANIC REACTIVE OLIGOMERS.

S.G. KULICHIKHIN, A.S. REUTOV,
I.I. MIROSHNIKOVA, A.Ya. MALKIN.

Research Institute of Plastics, 35, Perovskii proezd, Moscow, 111112, (Russia).

1. INTRODUCTION

Rheological properties are very sensitive to change of molecular weight and physical structure of the reactive systems in the curing process (refs.1,2). General picture of the rheological change in the curing materials reflects different stages of the process-gelation as the first stage and curing after gel-point as the second stage. Gel-point is the critical moment which divides the whole process in two stages. The present paper discusses specific features of the rheological properties change in the curing process of siliconorganic reactive oligomers with different chemical structure.

2. MATERIALS AND METHODS

Siliconorganic oligomers of different chemical structure - methyl, methylphenyl and phenyl - have been used. Concentration of hydroxyl groups in the studied samples was 2-4 %. Different rheological methods (viscometry and dynamic mechanical analysis) have been applied depending on the relaxation state of the curing siliconorganic oligomers. To estimate the molecular characteristics and phase structure of the reactive system optic, sol-gel and GPC analysis are used.

3. RESULTS AND DISCUSSIONS

Curing process of siliconorganic reactive oligomers is characterazed by a number of interesting rheological effects. Experimental tests show that viscosity change of the gelation stage is determined, firstly, by change in molecular weight of the oligomer and, secondly, by microphase separation of the reactive system before gel-point (ref.3). As a result decrease of viscosity after the phase separation instead of its continuous growth up the gel-point becomes possible. Molecular weight of the sol-fraction of the oligomer chages in extremal manner and attains its maximum value at the phase separation point.

Inhomogeneity of the silicononorganic reactive system influences rheokinetics of the process after gel-point when dynamic modulus components are used as basic rheological characteristics. In general cose change of storage modulus at the two-phase curing process is expressed by the equation with self-acceleration member. Self-acceleration is observed regardless of the form of the kinetic function used. General rheokinetic equation taking into account superimposed influence of self-acceleration and self-retardation effects at siliconorganic oligomers curing is proposed. Self-retardation is conditioned by the influence of physical aspects - vitrification of the reactive system and topological problems arising in the process.

4. REFERENCES

1. C.W. Macosko, Brit.Polymer J., 17 (1985) 239-245.
2. A.Ya. Malkin, S.G. Kulichikhin, Adv.Polymer Sci., 101 (1991) 217-258.
3. S.G. Kulichikhin, in: D.R.Oliver (Ed.), Third European Rheology Conference, Edinburgh, U.K., September 3-7, 1990, Elsevier Applied Science, London, 1990, pp.289-291.

Theoretical and Applied Rheology, edited by P. Moldenaers and R. Keunings
Proc. XIth Int. Congr. on Rheology, Brussels, Belgium, August 17-21, 1992

DYNAMIC MECHANICAL ANALYSIS OF POLY(ETHER ETHER KETONE) AND POLY ETHERSULFONE

Tariq M. Malik

Research and Development Center, Tremco Inc., A BFGoodrich Co., 3777 Green Road, Beachwood, Ohio, 44139, USA.

INTRODUCTION

Polymer Blending is a proven tool to obtain new type of materials with a wide diversity of properties intermediate between those of the pure components along with an economic advantage. As these materials can be developed by simply melt mixing the homopolymers and because of the high productivity and design flexibility offered, the polymer blends are increasingly replacing commodity polymers in engineering applications. Most of the polymer blends reported in the literature are two phase systems, morphology of which depends on the arrangement of the phases and their interphase. The case of semicrystalline-amorphous polymers blend, the crystallization which in turn controls the mechanical properties, takes place during the molding cycle. The blending of semi-crystalline engineering thermoplastic such as PEEK with an amorphous and relatively less expensive polymer such as poly(ether sulfone)(PES) could not only reduce the cost of finished products but also would facilitate development of a new material with possibly combined characteristics of both the PEEK and PES. In the present paper, we have analyzed blends of PEEK/PES. Dynamic Mechanical and thermal methods are used to characterize the blends.

EXPERIMENTAL

The Blends were prepared by mixing the two resins in Brabender Plasticorder at 335°C. The samples were then compression molded in a press at 50 MPa and 355°C. The mold was cooled at room temperature keeping the pressure constant. A DuPont Dynamic Mechanical Analyzer was used to measure the viscoelastic response of PEEK, PES and other blends between -100°C to 350°C. A limited number of measurements were also made at Rheometries Mechanical Spectrometer between same temperatures. For thermal analysis, a DuPont differential scanning calorimeter was used.

RESULTS

Dynamic mechanical and thermal properties show that PEEK and PES form semi-miscible blends at all proportion. Dynamic modulus increase significantly with increasing concentration of PEEK in the blend, reaching a maximum at 40% PEEK concentration and then drop to the values predicted by theory. This partial miscibility, we believe, arises from the similarity in the structure of two polymers. Thermal analysis showed a single Tg for the blends, higher than theoretical predictions.

Theoretical and Applied Rheology, edited by P. Moldenaers and R. Keunings
Proc. XIth Int. Congr. on Rheology, Brussels, Belgium, August 17-21, 1992

FLOW - INDUCED CRYSTALLIZATION IN POLYMER MELTS

A.J. McHUGH[1] AND R.K. GUY[2]

[1]Department of Chemical Engineering, University of Illinois, 1209 West California, Urbana, Illinois, 61801 (USA)
[2]Exxon Product Research, Houston, Texas (USA)

1. INTRODUCTION

Flow-induced crystallization, long known as a technique for producing oriented crystal growth in polymers, has only recently been studied in-situ under controlled flow conditions. As a result our understanding of the mechanisms associated with this process has been undergoing dramatic changes. Crystal growth has generally been ascribed to the same entropic, melting point elevation effects which are believed to drive strain-induced growth of crosslinked systems. However, kinetic data show that this view is incomplete. We now realize that stress and flow can enter through structuring, as well as thermodynamic interactions in ways that are not as yet completely understood quantitatively.

2. METHODS AND RESULTS

In order to optimize the use of flow-induced crystallization (FIC) and structure formation in melt processing, a quantitative understanding of how the nucleation, growth, and crystallization kinetics are influenced by the orienting effects of the flow is needed. To accomplish this we have developed a technique for the study of FIC and structure formation in the droplet phase of a two-phase flow. The primary components consist of a 4-roll mill along with flow visualization, digital imaging and optical birefringence hardware to monitor the flow fields in the two phases and to quantify crystallization as a function of time in the suspended drop. The carrier phase is a LLDPE with a melting point of 122° C and the suspended (crystallizing) phase is either a pure HDPE or various molecular weight blends of HMWPE's. The near zero surface tension reduces form effects and allows large deformations from modest extension rates in the carrier phase (typically ~0.1 s^{-1}). In-situ measurement of the birefringence and dichroism are combined with dynamic measurements of the droplet shape to construct crystallization isotherms as a function of the flow field stress and droplet strain at various temperatures above the normal quiescent crystallization conditions.

One finds a window of flow and temperature histories (i.e., extension rate, extensional strain, and interrupted or continuous flow) within which the induced crystallization can be monitored as a function of droplet stress and strain. In the case of interrupted flow, after an initial decay due to amorphous relaxation, the birefringence increases due to crystallization resulting from the "frozen-in" orientation. Delay times are on the order of seconds to minutes depending on temperature, stress, and strain. Initial crystallization rate curves show stress and strain patterns in a temperature window of 130°C to 135°C, suggesting a type of viscoelastic shifting behavior.

Our results show that the initial crystallization rate depends on both stress and strain in a fashion suggestive more of a viscoelastic process rather than a simple nucleation and growth mechanism. Likewise, concepts related to the melting point elevation associated with chain deformation are unable to provide an explanation for these results, either quantitatively or qualitatively. A phenomenological explanation based on molecular rheology gives insights into a possible mechanism as well as pointing out the areas where further theoretical work is needed.

ACKNOWLEDGEMENTS

The authors wish to acknowledge support for this work from the National Science Foundation through the University of Illinois Materials Research Laboratory (DMR 89-20538).

Theoretical and Applied Rheology, edited by P. Moldenaers and R. Keunings
Proc. XIth Int. Congr. on Rheology, Brussels, Belgium, August 17-21, 1992

EFFECTS OF THERMOMECHANICAL HISTORY IN THE MELT ON THE CRYSTALLIZATION KINETICS OF SOME ISOTACTIC POLYPROPYLENE SAMPLES

J.C.MICHEL MONTEFINA Laboratoire de recherche polypropylène, Zone Industrielle C 7181 FELUY, BELGIUM

1. INTRODUCTION

Besides well-known extrinsic variable effects of supercooling and pressure on the quiescent crystallization kinetics of semi-crystalline polymers, the state of the melt, even above the ill-defined themodynamic melt temperature (reported between 184 and 220°C for isotactic polypropylene, iPP), sometimes appears to strongly influence the crystallization process from the supercooled melt. Recent reports by Rybnikar (ref.1) have confirmed the importance of the thermomechanical history in the melt for iPP and iPP composites. Despite this, conditions of melting (when reported) vary widely from a few minutes at 185°C up to 10 min at 250°C, and no experimental evidence is given with regard to the independence of crystallization kinetics to a further stay at, or above that melt temperature.

2. EXPERIMENTALS

Overall crystallization kinetics was followed by DSC (standard conditions : 2.5 min at 228°C and Tc of 128°C). Other isothermal crystallization temperatures (Tc), melt temperature (T_F) and times at T_F were also tried. The atmosphere was dry nitrogen. Hot stage polarized microscope studies complemented the DSC work. Dynamic mechanical spectroscopy studies of the crystallization process were conducted with a Rheometrics RDA at 135°C under different small shear strains at fixed frequencies after spending 3 minutes at 210°C. Controlled thermomechanical histories were obtained from extrusion of reactor grade VALTEC$^{(R)}$ (HIMONT) samples of MFI 11 on a Brabender laboratory extruder equipped with a single screw (T_F 210 or 250°C, 50 or 125 RPM) or a twin screw (230°C, 20 RPM). The MFI of the pellets was unaffected by the extrusion under nitrogen blanketting.

3. RESULTS AND DISCUSSIONS

Compared to other reactor grades, iPP samples (pellets) of MFI 12 displayed a strong melt memory effect on the overall crystallization rate. Thus, the time at the maximum of the crystalization exotherm, t (max), could vary between 2.5 and 15 min (very reproducible runs) for a same iPP grade (same stabilizer package, no nucleants). No difference in molecular weight or atactic content could explain this phenomenon. As the time elapsed in the melt or as T_F increased the fast crystallizing samples crystallized more slowly pointing on towards the existence of polymeric metastable nuclei. However, the fastest crystallizing samples degraded before erasure of these metastable nuclei.

A four-to-five fold difference in $t_{(max)}$ (at different Tc from 125 to 141°C) was observed for the extruded VALTEC. The samples which experienced lower residence time and T_F crystallized the fastest. Under mild oscillating shear as encoutered between the parallel plates of the rheometrics, the same ranking was found and even enhanced. A qualitative correlation with the total strain experienced by the melt and the melt temperature is proposed.

4. REFERENCES

1. F.RYBNIKAR, J. Appl. Polym. Sci., 27 (1982) 1479 ; Eur.Polym. J. 27 (1991) 549

Theoretical and Applied Rheology, edited by P. Moldenaers and R. Keunings
Proc. XIth Int. Congr. on Rheology, Brussels, Belgium, August 17-21, 1992

THE METHODS OF MODELLING AND CALCULATION OF THE PROBLEMS OF RHEOLOGICALLY COMPLEX MEDIA IN SCREW CONVEYER CHANNELS

Yu.G.NAZMEEV[1], M.A.RAIMOV[1]

[1]Kazan Branch of Moscow Power Engineering Institute, Krasnoselskaya street, 51, 420066 Kazan, CIS (ex.USSR)

1. INTRODUCTION

The representation in the form of inverse problem is the traditional setting the problem of fluid motion in a screw channel of extrusion machine. The fixed feed screw and revolving housing are considered in such setting the problem, and as a result the problem is reduced to motion in a rectangular channel with the upper wall moving at an angle to the longitudinal axis.

Another direct setting is possible.Then the revolving feed screw and fixed case are considered, and the solution of problems is carried out by means of screw coordinates introduced differently. The self-similarity of velocity vector relative to the third screw coordinate has been proved. The impossibility to introduce orthogonal coordinates, in which a screw shift would become a coordinate shift, also has been proved (ref.1).

2. METHODS

2.1. Equations

Let us denote the region taken by fluid at the moment t by D_t, a part of boundary D_t belonging to the housing - by K_t, a part of boundary D_t belonging to the feed screw by L_t ;the vector of current velocity at the point $\chi(x,y,z)$ at the moment t-by $V(\chi,t)$

Then boundary-value problem of viscous-elastic fluid motion (the fluid of Reiner-Rivlin) being taken as an example) in the screw channel of an extrusion machine may be written after transformations in the following way:

$$\rho[-2W_o(y)A+W_o(y)(\nabla W_o)(y)]-(\nabla^T P_2)(y)+\nabla^T c(W_o)(y)=0 \qquad (1)$$

$$S_p(\nabla W_o)=0, \qquad y\in D_o, \qquad (2)$$

$$-\lambda(I_2(W_o))(W_o+V_L)=pr_{Nk}\ N_k c(W_o)\big|_{K_o}, \qquad (3)$$

$$-\lambda(I_2(W_o))W_o=pr_{NL}\ N_L c(W\ )\big/_{Lo} \qquad (4)$$

Here in $B(V)= V+(\nabla V)^T$; $I_2(V)=SpB^2(V)$; $N_k(\chi)$-identity vector of inner normal to K_t at the point $\chi\in K_t$; $N_L(\chi)$-the identity vector of the external normal to L_t at the point $\chi\in L_t$; $V_L=w*\chi$ - rotation velocity of the point $\chi\in L_t$; T-the sign of transposition; pr_N- the orthogonal projection on the plane, which is perpendicular to N;
$c(V)=\varphi_1(I_2(V))B(V)+\varphi_2(I_2(V))B^2(V)$; $W=V-V_L$;
$W_o(\chi R_{-t})R=W(\chi,t)$; $y=\chi R_{-t}$;
R_t-the matrix of rotation during the time t with the angular velocity arround the screw axis; φ_1,φ_2-material functions of rheological model;B(V)- the first kinematic tensor of the White-Metzner; $A=R_t\frac{d}{dt}R_{-t}$; P,ρ-pressure. density.

2.2. Results

The discontinuity approximation of the velocity vector has been carried out by means of cos function (ref.2).The epures of the velocity vector components in different cross-sections of extrusion feed screw channels have been obtained. The equations for currents in the gap on the extruder walls has been obtained.The member $-\rho W_o(y)A=K$ is a Coriolis force. For some working media,the value of the criterion K=0.25-0.4 which denotes the necessity of taking into account the Coriolis forces.

Thus, the presence of the member $-2\rho W_o(y)A$ in equation (1) shows that the situations "the rotating feed screw" (the direct setting) and "the rotating housing" (the reverse setting) are not equivalent.

REFERENCES

1. Nazmeev Yu.G., All-Union Scientific-Technological J. of Engineering Physics, 60(1991), 277-284.
2. Nazmeev Yu.G.,Bobrov V.F., Dits V.G., Vachagina E.K., All-Union Scientific-Technological J. of Engineering Physics, 61(1991), 392-399.

Theoretical and Applied Rheology, edited by P. Moldenaers and R. Keunings
Proc. XIth Int. Congr. on Rheology, Brussels, Belgium, August 17-21, 1992

Effect of Mixing on Particle Dispersion in Ceramic Injection Molding Mixtures

Kenji Okada, Shin Akasaka, Yasuharu Akagi and Naoya Yoshioka
Department of Applied Chemistry Okayama University of Science, 1–1 Ridaicho, Okayama 700, Japan

1. INTRODUCTION

Injection molding technique has been applied to form complex shaped ceramic articles. In the process the mixing of ceramic powders with binder is an important processing because poor dispersion of the powders allow the survival of agglomeration which may cause irreparable defect on the microstructure of the final ceramic products. However, very few studies have been published on the particles dispersion in the mixtures "(refs.1–2)".

In the present study, the effect of mixing on the particles dispersion in aluminum oxide powders and polyethylene melts mixtures has been investigated.

2. EXPERIMENTAL

The binder was low–density polyethylene with a melt index of 200. Al_2O_3 powder was mixed with the binder using a twin screw extruder at 40 vol% powder loading. The dispersion of the samples was evaluated using a parallel plate viscometer with oscillatory mode.

3. RESULTS AND DISCUSSIONS

Dispersion of Al_2O_3 particles in the mixtures was strongly affected by the mixing temperature. Figure 1 shows the viscosity, η_{ri}, of the samples mixed at temperatures of 130, 150 and 180 °C. In the figure total energy, W, consumed for mixing is also plotted. In this study we defined η_{ri} as the value representing the degree of dispersion in the mixtures. It is seen that η_{ri} increases as the temperature increases. Figure 2 shows the effect of mixing speed, ω_M, on dispersion. For the samples mixed at temperature of 150 °C, η_{ri} decreases with increasing ω_M. On the other hand, at higher mixing temperature(180 °C), η_{ri} increases with increasing ω_M. This indicates that the mixing with high mixing speed at higher temperature causes shear thickening.

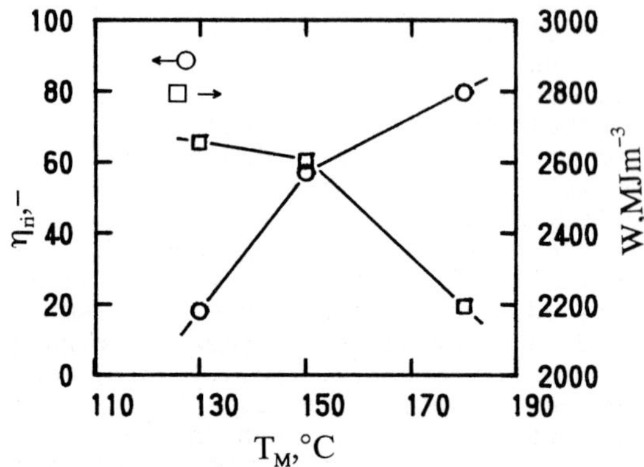

Fig. 1. Effect of mixing temperature T_M on η_{ri}(○) and W(□) for 0.2 μm Al_2O_3 mixtures mixed at ω_M=60 r.p.m..

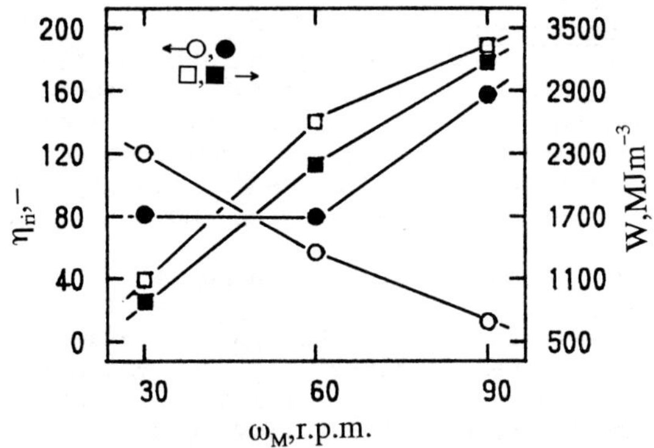

Fig. 2. Effect of mixing speed ω_M on η_{ri}(○,●) and W(□,■) for 0.2 μm Al_2O_3 mixtures mixed at T_M=150 °C(○,□) and 180 °C(●,■).

REFERENCES

1. K. Okada, Petrotech, 15(1992) 66–70.
2. J.H. Dow, M.D. Sacks and A.V. Shenoy, Ceramic Powder Science, 1 Part A(1988) 380–388.

Theoretical and Applied Rheology, edited by P. Moldenaers and R. Keunings
Proc. XIth Int. Congr. on Rheology, Brussels, Belgium, August 17-21, 1992

MELT FLOW SYNERGISM IN POLYMER MIXTURES: THE CONCOMITANT MORPHOLOGY

A.P. Plochocki, Dept. of Chemistry and Chem. Eng., STEVENS INSTITUTE OF TECHNOLOGY, Castle Point, Hoboken, NJ 07030, US.

The synergy of properties is the driving force of rapid, sustained growth in the development of "polyblends". A general theory of the synergism [1] which does not have to account for the stress dependence of the viscosity ratio (**VR**) nor the significance of elasticity in shaping the phases is therefore attractive. Its principal premise of the emulsion <-> slip mechanism, however leads to an untenable notion of the synergism caused by the 'interfacial slip'.While it is a factor in the rotational rheometry of polyblends [2] a recent comprehensive study confirmed that the synergism in melt flow results from a quantitative transition in their morphology [3]. At the synergistic composition of a polyblend flow under the shear stress,$\mathbf{p}_{12}$ **gradient** (in a capillary) the distribution of size of the minor phase domains changes from normal into the bimodal. The morphology remains disperse yet there is a change in density and the specific composition depends on $\mathbf{p}_{12}$ [4,5].Here the larger domains form a core (**c**) surrounded by a sheath (**s**) of the fine domains[4]. The subphases **c,s** meet at the location **L** defined by the ratio of the core radius,$\mathbf{r}_c$ to that of the capillary,**r**. The viscosity functions of the subphases, approximated with the power law are:

$$\mathbf{p}_{12} = \mathbf{V}^{o}_{c,s} * \dot{\mathbf{V}}^{\mathbf{n}_{c,s}} \quad (1)$$

where $\mathbf{V}^{o}$ -standard viscosity, $\dot{\mathbf{V}}$ - apparent shear rate and **n** - the viscosity exponent. Boundary conditions for integrating equatn of motion of the **c/s** flow [6] with respect of the velocity $\mathbf{v}_z$ are: $\mathbf{v}_{zs} = 0$ at $\mathbf{r}_c = \mathbf{r}$ and

$\mathbf{v}_{zc} = \mathbf{v}_{zs}$ at $\mathbf{r}_c$ i.e. radius of the core.The pressure gradient $\mathbf{G}_p$ imposes flow of the melt with the vol. rate,$\dot{\mathbf{Q}}$. Introducing terms **D, E** and **q** = 1/**n** we have:

$$\mathrm{D} = \mathrm{Pi}*\mathbf{r}\exp(\mathbf{q}_s+3)*[2\mathbf{V}^{o}_{s}\exp(\mathbf{q}_s)*(\mathbf{q}_s+3)]^{-1}$$

$$\mathbf{E} = \mathrm{Pi}*\mathbf{r}\exp(\mathbf{q}_c+3)*[2\mathbf{V}^{o}_{c}\exp(\mathbf{q}_c)*(\mathbf{q}_c+3)]^{-1}$$

$$\dot{\mathbf{Q}} = \mathbf{D} * \mathbf{G}_p^{q_s} *[1-\mathbf{L}^{(q_s+3)}]+\mathbf{E} * \mathbf{G}_p^{q_c} * \mathbf{L}^{(q_c+3)} \quad (4)$$

From eqn.(4) it follows that **L** depends on **n** which is a measure of the rate dependence of the shear viscosity [7]. Hence the synergism seems to result from Denson-type,radial migration of the minor phase domains which is confined within **L** specified boundary. In the steady flow it corresponds to the truncation of the overall velocity profile and is determined by the viscosity functions of **c** and **s**. It explains the stress **and** composition dependence of the synergism [3,5]. Hence neither the simplistic notion of a mixing-rule-controlled emulsion nor the embellishing slip concept are adequate here. The latter implies a stratification of **phases** in the polyblend - a morphology inherent to the **rare** systems in which elasticity of the matrix is higher than that of the minor phase (at a given composition and size of the domains).Even then, as Doppert&Overdiep have had shown some 20 years ago [7] simple mixing rules are ineffective.

REFERENCES

1. L.A. Utracki, J.Rheology,**35**('91),1615
2. T.A. Huang et al.,SPE **TP36**('90),921
3. MELT RHEOLOGY AND CONCOMITANT MORPHOLOGY OF A POLYBLEND, a summary of the interlaboratory (IUPAC) program, in submission to the Pure Appl.Chem.J.
4. A.E. Woodward,Atlas of Polymer Morphology, Oxford (1989),p.312
5. S.K. Dey et al.,Intl.Polymer Processing J.,**4**(1989),119
6. C.D. Han,Multiphase Flow in Polymer Processing, Acad.Press (1981)
7. Kohoudic & Finlayson,Eds.,Adv.Polymer Blends/Alloys Technology,Technomic, Lancaster,**2**(1989),162ff.

Theoretical and Applied Rheology, edited by P. Moldenaers and R. Keunings
Proc. XIth Int. Congr. on Rheology, Brussels, Belgium, August 17-21, 1992

"UNIVERSAL" RELATIONS FOR VISCOSITY/SHEAR-RATE AND EXTRUDATE SWELL PROFILE OF A POLYMER MELT

D. PORTER, R. J. KOOPMANS

Dow Benelux N.V., P.O. Box 48, 4530 AA Terneuzen, The Netherlands.

A simple topological model for the recoverable strain deformation in a spherical domain of fluid in a shear field is presented to derive a consistent set of predictive equations for some important rheological properties of a polymer melt.

A spherical domain of fluid has a radius R_o at rest, and deforms to the equilibrium shape of a prolate ellipse of long axis radius R in a linear uniaxial shear gradient. The power dissipated per domain passing event is used to derive a relation between the viscosity, η at a shear-rate $\dot{\gamma}$, and the zero shear viscosity η_o.

$$\frac{\eta}{\eta_o} = \left(\frac{R_o}{R}\right)^3 \qquad (1)$$

A stress balance of viscous drag and elastic reaction on each domain in the fluid gives is used in combination with the standard relation of η_o=Gτ (where τ is the base relaxation time and G the plateau modulus) to give a "universal" expression for the normalised viscosity, η/η_o, as a function of the normalised shear-rate, $\dot{\gamma}\tau$

$$\dot{\gamma}\tau = \left(\frac{\eta_o}{\eta}\right)\left(\left(\frac{\eta_o}{\eta}\right)^{\frac{1}{3}} - 1\right) \qquad (2)$$

The rheological parameters η_o, G, and τ can be related to the structure of the polymer or processing conditions (molecular weight, temperature) either empirically or by structure/property techniques to allow the correct set of parameters to be assigned for any required flow profile.

The maximum extrudate swell, D_m, of a polymer melt as it leaves a die exit can be related to the recoverable elastic strain of the fluid domains in the shear field, and to the normalised viscosity as a function of shear-rate via equations (1) and (2).

$$D_m = \left(\frac{R}{R_o}\right)^{\frac{1}{2}} = \left(\frac{\eta_o}{\eta}\right)^{\frac{1}{6}} \qquad (3)$$

A "master-curve" of viscosity and maximum extrudate swell as a function of normalised shear-rate is shown in Figure 1, and fits experimental flow profiles very well.

A more general model treatment of the time-dependent deformation of the fluid domains allows the profile of isothermal extrudate swell, D, as a function of distance from the die exit, z, to be described, where r is the radius of the die

$$z = rD_m^4\,(D_m^2-1)\left(\ln\left(\frac{(D_m+D)}{(D_m-D)}\,\frac{(D_m-1)}{(D_m+1)}\right) - 2\,\frac{(D-1)}{D_m}\right)$$

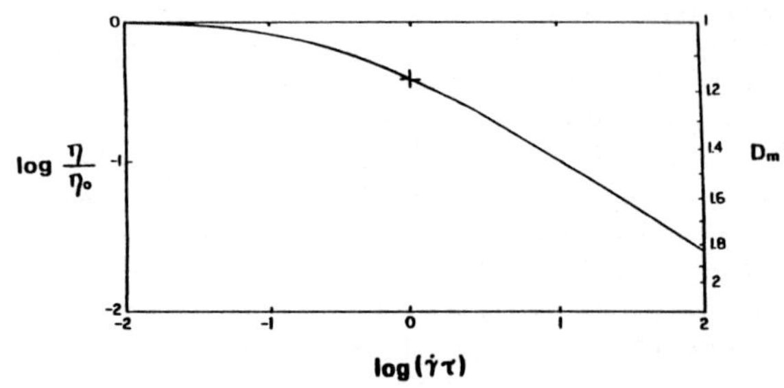

Figure 1: A "master curve" showing the normalised viscosity and maximum extrudate swell as a function of shear-rate from equations (3) and (4).

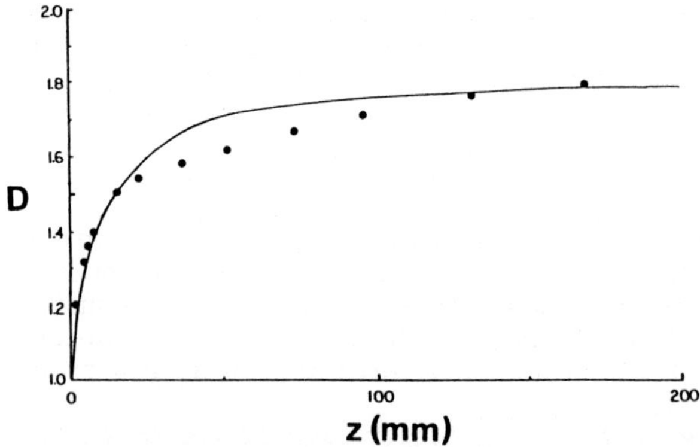

Figure 2: Comparison of the predictions of equation (4) (line) for the isothermal extrudate swell profile of polyethylene (D_m=1.8 and r=1mm) with experimental observation under non-isothermal conditions at 190°C (points)

Theoretical and Applied Rheology, edited by P. Moldenaers and R. Keunings
Proc. XIth Int. Congr. on Rheology, Brussels, Belgium, August 17-21, 1992

MOLECULAR ASPECTS OF PRESSURE AND SHEAR FLOW INDUCED MASS DENSITY CHANGES IN POLYMER MELTS

P. ŘÍHA[1] and J. KUBÁT

Department of Polymeric Materials, Chalmers University of Technology, 41296 Gothenburg (Sweden)
[1] Institute of Hydrodynamics, 166 12 Prague 6 (Czechoslovakia)

1. INTRODUCTION

Investigations of density changes in polymer melts are mostly focused on pressure and temperature dependent changes for equilibrium system. The non-equilibrium conditions are considered in (ref.1,2) for dilatational motion.

The constitutive equation involving possible hydrostatic pressure and shear flow induced melt density changes may be conveniently expressed as

$$t_{kl}=(P+p+ \lambda d_{mm}) \delta_{kl}+2 \mu d_{kl} , \qquad (1)$$

where t_{kl} denotes stress tensor, $P(\rho, \theta)$ is the hydrostatic equilibrium pressure, p, λ, μ are, in general, functions of mass density ρ, temperature θ and of invariants of shear rate tensor d_{kl} and denote the deviation from pressure P under non-equilibrium conditions, dilatational and shear viscosity, respectively.

Since molecular theories are probably the most suitable for quantifying various aspects of the organization of macromolecules influenced by hydrostatic pressure and motion, the Curtiss-Bird theory (ref.3) is used to express the contribution of terms in (1) to the stress tensor t_{kl}.

2. MOLECULAR THEORY

Following the steps explained in (ref.3), but for a compressible material, the single-link orientational distribution function f for a homogeneous strain and homogeneous retarded-motion is

$$f(u, \sigma ,t)=P_1 +(3/2) \lambda P_2 (\gamma^{(1)}:uu)+..., \qquad (2)$$

$$P_n(\sigma)= \sum_{\alpha} (\alpha\sin \pi\alpha\sigma)/(\pi^2\alpha^2 + \chi\lambda)^n ,$$

where the notation of (ref.3) is used and χ denotes the time rate of change of mass density.

When the solution (2) is used to determine the expansion for the stress tensor (ref.3), the terms in (1) can be specified. For instance, the isotropic term p

$$p=(NnkT/3) [1-S_1+(ma^2/kT)T_1 \gamma^{(1)} \gamma^{(1)}]+...,$$

$$\text{where } S_n=(2/\pi) \sum_{\alpha,odd} 1/(\pi^2\alpha^2+\chi\lambda)^n , \qquad (3)$$

$$T_n=(4/\pi^3) \sum_{\alpha,odd} 1/[\alpha^2 (\pi^2\alpha^2+\chi\lambda)]^n ,$$

and the notation of (ref.3) is used.

The shear- and mass density change rate-dependent contribution p to the shear stress t_{kl} is probably the consequence of the variation in molecular packing since the motions prevent molecules from packing as efficiently as they might at equilibrium. The prediction (3) shows that at constant temperature and pressure the mass density of a polymer melt decreases with increasing mass density change rate and shear rate.

REFERENCES

1. G. Schenkel, Kunstofftechnik, 12 (1973) 1-7.
2. J.V. Aleman, in: D.R.Oliver (Ed.), Third European Rheology Conference, Edinburgh, U.K., September 3-7, 1990, Elsevier Applied Science, London 1990, pp. 16-17.
3. C.F. Curtiss and R.B. Bird, J.Chem.Phys., 74 (1981) 2016-2033.

Theoretical and Applied Rheology, edited by P. Moldenaers and R. Keunings
Proc. XIth Int. Congr. on Rheology, Brussels, Belgium, August 17-21, 1992

MELT RHEOLOGY OF HIGH DENSITY POLYETHYLENE/ETHYLENE-VINYL ACETATE COPOLYMER BLENDS

MA RONGTANG[1], XU SHIAI[1], KOU XICHUN[1], LI JIGEN[2] AND YIN MINGDONG[2]

[1]Department of Chemistry, Jilin university, P. O. Box 130023, Changchun (P. R. China)
[2]Qiqihaer Institute of Light Industry, P. O. Box 161006, Qiqihaer (P. R. China)

1. EXPERIMENTAL

A commercial high density polyethylene (HDPE) was blended with three types of ethylene-vinyl acetate copolymer (EVA) in a Brahender-like apparatus. EVA_1 and EVA_2 were ordinary copolymer containing 16 and 33 percent of vinyl acetate respectively. EVA_3 was acid-modified, i. e. Elvalay 741. The rheological measurements were carried out by a capillary rheometer using a die with a diameter D=1mm and L/D=40 at shear rate of $10s^{-1}$ to $500s^{-1}$ and four temperatures: 160, 170, 180 and 190℃. The data were corrected according to the Bagley procedure (ref. 1).

2. RESULTS AND DISCUSSION

Figure 1 shows a set of flow curves of HDPE/EVA_1 blends at 190℃. The apparent shear viscosity η_a decreases with increasing shear rate $\dot{\gamma}$ and EVA content. The viscosity results of HDPE/EVA_1(75/25) blend are presented as a function of temperature in Fig. 2. The apparent viscosity decreases with increasing temperature at all shear rates investigated.

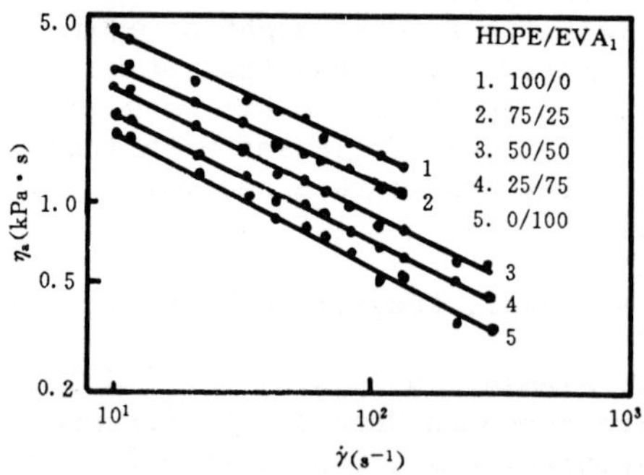

Fig. 1 Flow curves for HDPE/EVA_1 blends.

In order to compare the influences of structure and vinyl acetate of EVA on the rheological behavior, the plots of activition energy of flow E_a versus shear stress τ and η_a versus $\dot{\gamma}$ of HDPE/EVA(60/40) blends are depicted in Fig. 3 and Fig. 4 respectively. It is shown that in all the cases E_a decreases with increasing τ and η_a decreases with increasing $\dot{\gamma}$. HDPE/EVA_3 shows much lower values of E_a, as we as η_a. The flow behavior shows good agreement between HDPE/EVA_1 and HDPE/EVA_2.

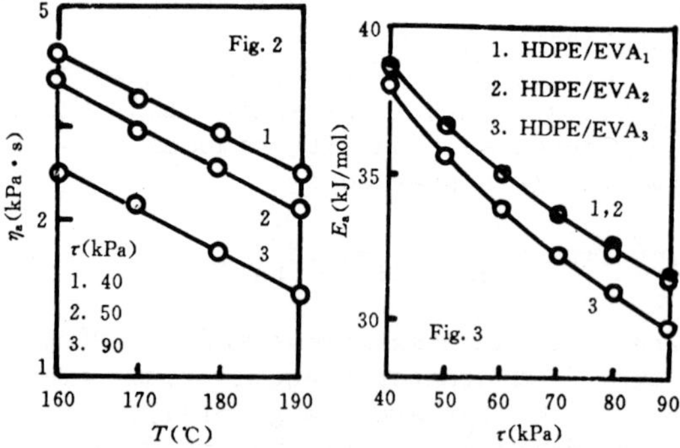

Fig. 2 η_a vs T for HDPE/EVA_1(75/25) blends.
Fig. 3 E_a vs τ for HDPE/EVA(60/40) blends.

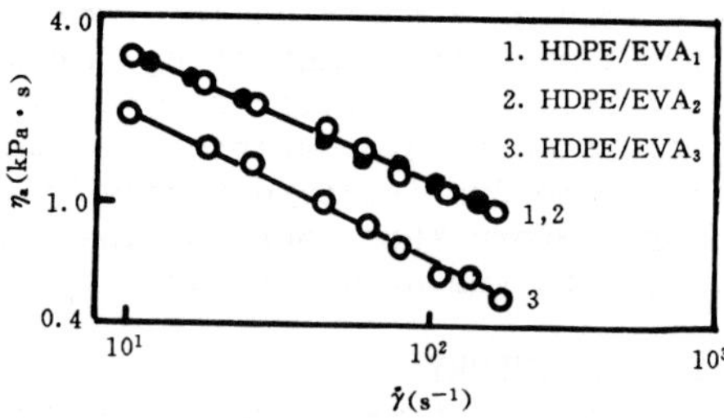

Fig. 4 Flow curves for HDPE/EVA(60/40) blends.

It may conclude that the rheological properties of HDPE/EVA blends are markedly affected by the structure rather than vinyl acetate content of EVA.

REFERENCES

1. E. B. Bagley, J. Appl. Phys., 28(1957)624.

Theoretical and Applied Rheology, edited by P. Moldenaers and R. Keunings
Proc. XIth Int. Congr. on Rheology, Brussels, Belgium, August 17-21, 1992

ELONGATIONAL FLOW OF POLYMER MATERIALS RHEOTENS: EXPERIMENT AND THEORY

V. SCHULZE[1], J. BERGMANN[2], R. SCHNABEL[3], E.-O. REHER[3]

[1] GÖTTFERT Werkstoff-Prüfmaschinen GmbH, Siemensstr. 2, Postfach 1261, 6967 Buchen/FRG
[2] Pädagogische Hochschule Halle, Kröllwitzer Str. 44, 4090 Halle/FRG
[3] Technische Hochschule Merseburg, Geusaer Str., 4200 Merseburg/FRG

In the processing of polymer materials, elongational flow is a partial process occurring in the production sequence prior to or during the forming and hardening processes (which for the most part proceed simultaneously) and hence has an essential influence on product quality. To achieve optimal quality of final products, material properties have to match the processing parameters. The paper discusses the curves of technical elongation diagrams obtained by the Rheotens elongational tester as related to the characteristics of thread diameter and thermographic records. Mathematical models for describing uniaxial and strip biaxial elongation of viscoelastic materials (Jeffrey) under non-isothermal process conditions are presented. Under the simplified conditions generally used for stationary, axially symmetric die flows, the cylindrical coordinates are obtained from the balance equations of mass, momentum and energy.

$$\frac{1}{r}\frac{\partial(r\rho v_r)}{\partial r} + (\rho v_z)' = 0$$

$$\rho\left(v_z v_z' + v_r \frac{\partial v_z}{\partial r}\right) = \rho g + P_{zz}' + \frac{1}{r}\frac{\partial}{\partial r}(r\, P_{rz})$$

$$\rho\left(v_z v_r' + v_r \frac{\partial v_r}{\partial r}\right) = P_{rz}' + \frac{1}{r}\frac{\partial}{\partial r}(r\, P_{rr}) - \frac{P_{\varphi\varphi}}{r}$$

$$\rho c_p\left(v_z T' + v_r \frac{\partial T}{\partial r}\right) = \frac{1}{r}\frac{\partial}{\partial r}\left(r\lambda \frac{\partial T}{\partial r}\right)$$

$$\mathcal{T} + \lambda_1(\dot{\mathcal{T}} - a(\mathcal{T}D + D\mathcal{T})) = 2\eta\,(D + \lambda_2(\dot{D} - 2a\,DD))$$

The material parameters ρ , c_p, λ and η are asumed to be temperature dependent, heat conduction in axial direction as compared to that in radial direction is neglected, ' denotes the derivation with respect to z. The constitutive equations for viscoelastic fluids of the differential and integral types are estimated with respect to their suitability for describing elongational flows. The efficiency of the program developed for simulating the field distributions of the process variables in the Rheotens experiment is demonstrated with the help of experimentally determined field distributions. The modelling of multilayer elongational flows is the subject of further theoretical and experimental discussions. New applications of the Rheotens will be seen in combination with conventional capillary and on-line rheometry for the simultaneous determining of elongational and shear viscosities with the same material.

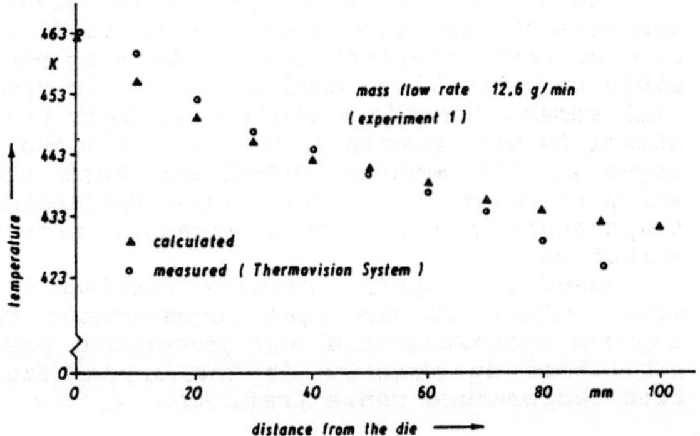

Theoretical and Applied Rheology, edited by P. Moldenaers and R. Keunings
Proc. XIth Int. Congr. on Rheology, Brussels, Belgium, August 17-21, 1992

INFLUENCE OF SHEAR RATE AND TEMPERATURE ON THE CRYSTALLIZATION OF MULTICOMPONENT MELTS

Dr. Th. Seidel, Dr. L. Heymann, E. Noack

Group "Flow behaviour of non-Newtonian fluids", KAI e.V., Reichenhainer Str. 88, O-9022 Chemnitz (Germany)

1. INTRODUCTION

Crystallization of multicomponent melts plays an important role in processing of various materials i.e. glass-ceramics, metals, polymers, and food. In this paper the influence of shear rate and temperature on both the rheological behaviour in the melt state during crystallization and on the microstructure and the mechanical behaviour in the solid state of metal alloys and a special glass-ceramic is presented.

2. RESULTS CONCERNING METAL ALLOYS

First investigations of the rheology of metal alloys in the range between solidus and liqidus temperatures were made by Flemings et. al. (ref. 1) in 1972. The Sn-15% Pb alloy has been investigated using a Couette type viscometer. A main result is that the viscosity, which depends on fraction solid, decreases with increasing shear rate. The initial dendritic crystals were fragmented and the microstructure altered into a globulitic one by shearing.

In the last few years many further results have been obtained concerning various alloys. Generally speaking the sheared alloy exhibits better and homogeneous mechanical properties. Strength, ductility, melleability, hardness, impact toughness, and elongation can be influenced (i.e. ref. 2 and 3). New metal forming processes like Stircasting, Rheocasting, Thixocasting, Thixoforging, Compocasting and so on are examples for the world-wide development.

The results concern alloys with low melting temperatures. Therefore the alloys with melting temperatures above 1000°C are also to be investigated in the future.

3. RESULTS CONCERNING GLASS-CERAMIC

Usually the crystallization of glass-ceramics occurs after glasslike solidification by heat treatment, but it is also possible to crystallize during cooling in special cases. Therefore additional heat treatment is not necessary. However, disadvantages of the second method are both the sharp increase of viscosity with decreasing temperature and the coarse-grained microstructure.

Based on hopeful results reached for metal alloys it has been investigated to improve microstructure and processing possibilities by shearing in the crysallization temperature range (ref. 4).

The microstructure of the used phlogopit glass-ceramic at room temperature consists of glimmer crystals and surrounding quenched matrix. The experiments were made using a Searle-viscometer specially developed for this purpose.

Results similar to metals are obtained. The viscosity decreases with increasing shear rate. The microstructure as well as the rheological behaviour is markedly changed by shearing. The initial platelet - structure is entirely replaced by a structure with broken short crystals. This structure was not previously known for this material. First investigations of the compressive stength of cubes (5 mm) show that this property can be improved by shearing.

The mathematical model to calculate the crystal distribution function n(L,f,t) of the suspension based on the following dimensionless population balance for crystals (ref. 5):

$$\frac{\partial n}{\partial t}+\nabla(r_L n)+D-B=0 \quad (1)$$

L is the diameter, and f describes the form of crystals. r_L is is a function of crystal growth velocity. B and D represent birth and deth density functions at a point in the phase space. The apparent viscosity η is calculated as a function of the glass viscosity η_g and n at each time point.

$$\eta=\eta(\eta_g, n) \quad (2)$$

The experimental and theoretical results open new ways for processing spontaneous crystallizing glass-ceramics.

REFERENCES

1. D.B. Spencer, R. Mehrabian and M.C. Flemings, Metall. Trans. 3(1972) 1925-1932
2. N.A. El-Mahallawy, N. Fat-Halla and M.A. Taha, 4th Internat. Conf. on Mechanical Behaviour of Materials, Stockholm, Schweden, 1983
3. K. Ichikawa and S. Ishizuka, Trasact. of the Japan Inst. of Metals 28(1987)2 145 - 153
4. T. Seidel, C. Friedrich, J. of Mater. Sci. 27(1992) 263-269
5. A.D. Randolph and M.A. Larson, Theory of Particulate Processes, Academic Press, Inc., New York, 1988

Theoretical and Applied Rheology, edited by P. Moldenaers and R. Keunings
Proc. XIth Int. Congr. on Rheology, Brussels, Belgium, August 17-21, 1992

ELONGATION DEFORMATION TEST FOR POLYMER PROCESSING

V.D. SEVRUK

Scientific-Industrial Association "Plastik", Berezkovskaia nab.20, 121059 Moscow (Russia)

1. INTRODUCTION

Shear as well as elongation deformation of polymer melts take place in most of polymer processing methods. But practically polymer melts are usually tested only under shear deformation for determination of melt index (MI) and ratio MI with different loads. In some polymer processing methods it is possible to manufacture articles using one type of polymer raw materials, while it is impossible to use other raw materials, through they have the same PMI and ratio PMI.

The purpose of this article is to determine the rheological characteristics which correlate the processing features.

2. METHODS AND MATERIALS

We used Sangamo cone-plate viscosimeter for producing low density polyethylene (LDPE) and polyisobutylene (PIB) full flow curves.

We also used Gettfert capillary viscosimeter Reograph-2000 for thermoplast trade marks, namely high density polyethylene (HDPE) 288, 203 and 205, LDPE 16803-070, 11503-070, Alcaten WNC-71, Bailon 19N430, Peten B-8015 and B-4524, polystyrene PS-168N and PS-151, polyvinilchloride (PVC).

Elongation deformation was made on a model laboratory equipment at a constant force regime. All above mentioned raw materials were also investigated under elongation.

3. RESULTS

We simulated free extrudate swelling and extrudate extension from the capillary by uniform elongation. We obtained a good correspondence between the swelling extrudate profile photographs and profile estimated by the uniform extension data was obtained.

It gave us a possibility to carry out a model test. We used two quite different polymers, LDPE and PIB, and estimated the temperature, when the polymers had the same flow curves. We found that these polymers had not only the same flow curves and also the analogical largest Newtonian viscosities and relaxation times. But the behavior of these polymers was quite different under elongation. PE elongated till the same deformation six times slower than PIB IB elastic and viscous parts of deformation increased constantly during the elongation. LDPE had a part of curve, when viscous deformation didn't change and this time only elastic part increased (For the first this effect was seen in (1)). These polymers strain rate changing at elongation was also quite different.

Thus for polymer processing it is necessity to take into account not only shear but also elongation behavior of polymer melts.

From this point of view we researched 1. manufacturing of HDPE oriented planar thread; 2. LDPE laminated cover of paper and boards for milk packaging; 3. polysterene planar films and threads for capacitors; 4. HDPE oriented tubes; 5. PVC pipes and fittings for heat supply.

The analysis of all these processing methods under shear as well as elongation deformation showed the necessity to take into consideration both kinds of deformation.

REFERENCES

1. A.N. Prokunin Nonlinear elastic effect at the elongation of elastic polymer liquids. Preprint Problem Mechanic Institute of Academy of Science USSR N104, 1978.

Theoretical and Applied Rheology, edited by P. Moldenaers and R. Keunings
Proc. XIth Int. Congr. on Rheology, Brussels, Belgium, August 17-21, 1992

RHEOLOGICAL PROPERTIES OF POLYBUTYLENE TEREPHTHALATE/POLYPROPYLENE BLENDS

Dacheng Wu and Yuzhong Wang*

Textile College, Chengdu University of Science and Technology, Chengdu 610065, P. R. China

1. EXPERIMENTAL

The polybutylene terephthalate (PBT) /polypropylene (PP) blends at several compositions were prepared by melt-mixing PBT and PP with a laboratory scale single screw extruder to which a static mixer was attached. An Instron 3211 Capillary Rheometer was used to determine the rheological properties of the polymer blends at shear rates ranging from 10^1 to 10^3 sec^{-1}. The temperatures used in the experiment were 240, 255 and 270℃.

2. RESULTS AND DISCUSSION

The blends of various compositions exhibit the flow behavior which is pseudoplastic and simillar to that of pure samples at all temperatures. Their melt viscosities lie between those of pure homopolymers, and increase as the amount of PBT increases. On the other hand, the viscosity difference between the blends of different compositions gradually reduces with a rise in temperatures.

PBT/PP blends are incompatible (ref. 1), and therefore exhibit microphase seperation. When a finely dispersed two-phase incompatible polymer blend is extruded from a large reservior into a capillary, the blend forms a morphology depending upon the relative viscosities and the size of the droplets. When the dispersed phase is more viscous than the continuous phase and is present in minor or equal amount, the droplets maybe elongated at the entrance region. These elongated droplets snap back in the stress relaxation region and form randomly distributed" fibrils" of thin continuous layers parallel to the flow direction in developed flow region. In this case, Alle et al. (ref. 2) theoretically derived the following blending rule:

$$1/\eta = \varphi_1/\eta_1 + \varphi_2/\eta_2 \qquad (1)$$

where η is the viscosity of the blend, η_1 and η_2 are the viscosities of the components 1 and 2 respetively, and φ_1 and φ_2 are the volume fractions of the two dividual components. When high viscous PBT is minor component which acts as dispersed phase, the theoretical values calulated by Eq. 1 coincide closely with the experimental values at all temperatures. At a lower temperature (say 240℃, see Fig. 1), however, the calculated values deviate gradually from the experimental values with the increase of the amount of PBT which becomes major component. This indicates that the varations in the volume ratio of high visous component PBT and less viscous component PP can cause the changes in the phase morphology of blends. On the other hand, when the less viscous component acts as minor one which is disperete phase, the dispersed droplets that are elongated at the entrance region of capillary either recoil or breakup into smaller droplets in the stress relaxation region. In this case, the relationship between the viscosity and the composition of blends should be different from the Eq. 1, consequently the experimental values deviate from the theoretical values calculated by Eq. 1.

REFERENCES

1. Y. Wang, M. S. Theis, Chengdu Univ. of Sci. Technol., 1987.
2. N. Alle, et al., *Rheol. Acta*, 19 (1980), 104.

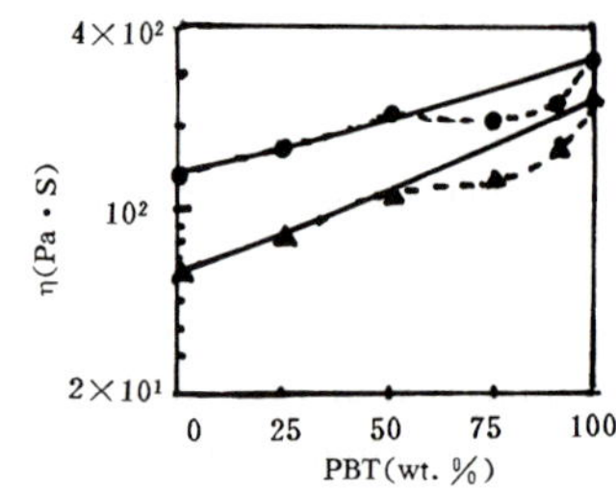

Fig. 1 Viscosity of PBT/PP blends as a function of PBT content at different shear stresses (Pa): (●) 5×10³; (▲) 3×10⁴; (---) experimental values; (—) values calculated from Eq. 1

* to whom correspondence should be addressed

Theoretical and Applied Rheology, edited by P. Moldenaers and R. Keunings
Proc. XIth Int. Congr. on Rheology, Brussels, Belgium, August 17-21, 1992

RHEOLOGICAL PROPERTIES OF POLYOLEFINES — THE RELATIONSHIP BETWEEN η_a AND $\dot{\gamma}_N$

WANG ZIJIE and LI SHUQUAN

Department of Chemical Engineering, Dalian Institute of Light Industry, Dalian 116001 (The China)

Rheological properties of linear low density polyethylene (LLDPE), low density polyethylene (LDPE), their blends and polypropylene (PP) was studied. It was found that the relationship between η_a and $\dot{\gamma}_N$ appears anti—S curve. The relationship was simulated by the use of modified Weibull's model haveing three parameters. (ref. 1)

$$\frac{1}{\eta_a}=\alpha-\beta\exp(-\theta\dot{\gamma}_N) \tag{1}$$

$$\alpha=\frac{1}{\eta_\infty} \tag{2}$$

$$\beta=\frac{1}{\eta_\infty}-\frac{1}{\eta_0} \tag{3}$$

Where α, β and θ are the model parameters, η_a is the apparent viscosity, η_0 the zero—shear viscosity, η_∞ the infinite shear viscosity and $\dot{\gamma}_N$ the apparent shear rate.

The volume flow rate, θ, of above mentioned various materials was determined by means of a CFT—500 capillary rheometer of SHIMADZU (Japan) under the different pressures, and then the $\dot{\gamma}_N$, η_a and the shear stress were calculated. According to these data experimentally obtained, three parameters α, β and θ of this model, were calculated by the aid of the curve fitting method of unary nonlinear model. (ref. 2) Upon the substituting them and $\dot{\gamma}_N$ into Eq. 1, the relationship could be obtained. The values calculated from Eq. 1 are in agreement with the experimental results. (Fig. 1 and 2)

The equation expressed Non—Newtonian Index (n) was also derived from Eq. 1.

$$n=1-\frac{\beta\theta\dot{\gamma}_N\exp(-\theta\dot{\gamma}_N)}{\alpha-\beta\exp(-\theta\dot{\gamma}_N)} \tag{4}$$

Therefore, η_∞ may be obtained from Eq. 2, η_0 from Eq. 3 and n from Eq. 4.

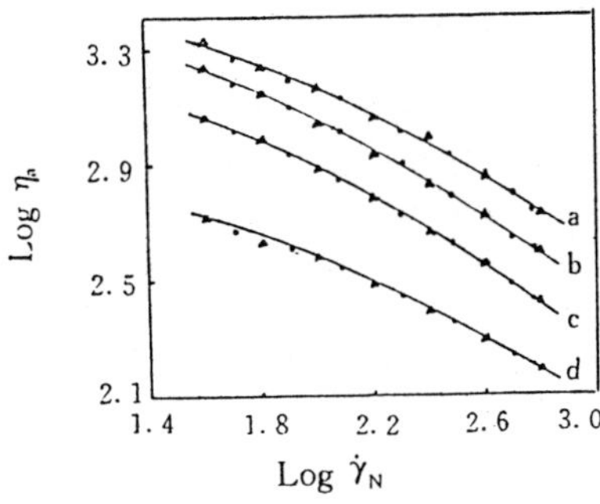

Fig. 1 Log(η_a) *Versus* Log($\dot{\gamma}_N$) for samples of LLDPE, LDPE and their blends (experimental (▲) and calculative (•)) at various blended ratioes of LLDPE/LDPE: (a) 100/0, (b)70/30, (c)30/70, (d)0/100.

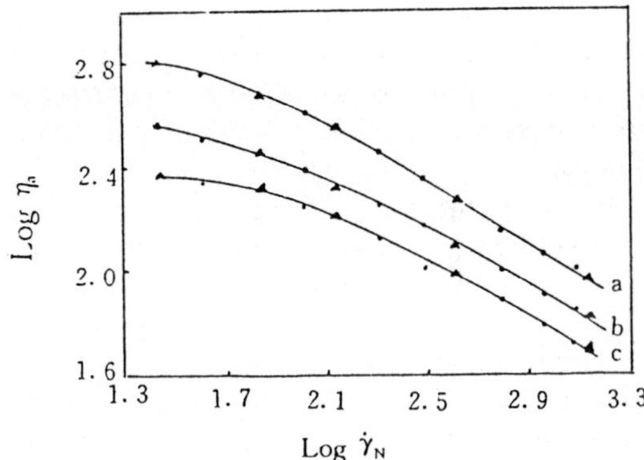

Fig. 2 Log(η_a) *versus* Log($\dot{\gamma}_N$) for PP samples (experimental (▲) and calculative (•)) at various temperatures: (a)180, (b)210, (c)240 C.

REFERENGES

1. M. Chang, X. Q. Zhang, Y. H. Cao, Z. J. Wang and S. J. Wang, Sci. Bll., 32(1987)1230—1233.
2. K. T. Fang, H. Quan and Q. Y. Chen, Applied Regression analysis, 1989, P168—183.

Contributed papers

POLYMER SOLUTIONS

Theoretical and Applied Rheology, edited by P. Moldenaers and R. Keunings
Proc. XIth Int. Congr. on Rheology, Brussels, Belgium, August 17-21, 1992

COATING FLOW INSTABILITY AND NON-NEWTONIAN FLOW OF CELLULOSE SOLUTIONS

S. K. AHUJA[1], M. STEVANOVIC[2] and L. E. SCRIVEN[3]

[1]Xerox Corporation, Bldg. 139-64A, 800 Phillips Rd., Webster, NY 14580, U.S.A.
[2]Xerox Corporation, Speakman Drive, Mississauga, Ontario, Canada
[3]Chemical Engineering Dept., University of Minnesota, Minneapolis, MN, U.S.A.

1. INTRODUCTION

Barring, ribbing, and light spots as coating defects are not acceptable in electrophotographic or photographic products. Coating defects in dried coatings can come about from instability in coating flows which are aggravated by the shear and the extension rates in excess of 10^3 s^{-1}. Schunk and Scriven (ref. 1) have reported that the local deformation rates in a slide coater can be as high as 10,000 s^{-1} immediately at the dynamic contact line, while shear rates downstream of the Skiadis boundary layer can be nearly 10,000 s^{-1}. For Newtonian fluids, coating flows can have imperfections and failures arising out of too little vacuum which causes ribbing. This coating instability increases with increase in web speed or increase in viscosity of the liquid. There is also the problem of air entrainment as the web speed is increased resulting in nonuniformity of the coating. In the case of non-Newtonian fluids, which shear thin without thickening extensionally, the coating bead droops, whereas extensional thickening can cause liquid to be pulled onto the coating web.

2. EXPERIMENTAL

The viscoelastic behavior of the solutions was measured using Couette flow geometry in Rheometrics RFS8400 and extensional flow was obtained from the Rheometrics using RFX. The flow instability was visually observed by video taping the ripples observed upstream from the gap during multislide coating and then photographing the ripples by freeze frame technique (ref. 2). The flow stability analysis results were obtained by solving the Navier-Stokes system by the Galerkin/finite element method, discretization of free surface flows, and computation of eigen values by Arnoldi's method (ref. 3).

3. RESULTS AND DISCUSSION

On comparing zero shear rate viscosity and shear rate at the onset of thinning in the polymers and their mixtures used in the study, the carboxy methyl cellulose is more concentration dependent and is more shear thinning than either hydroxy propyl methyl cellulose or polyethylene oxide. The linear viscoelastic response of the carboxy methyl cellulose and the mixture show deviation from Rouse-like behavior at 1% concentration, whereas the hydroxy propyl methyl cellulose and polyethylene oxide follow Rouse-like behavior up to 2%, showing no appearance of plateau indicative of polymer chain entanglement (Fig. 1).

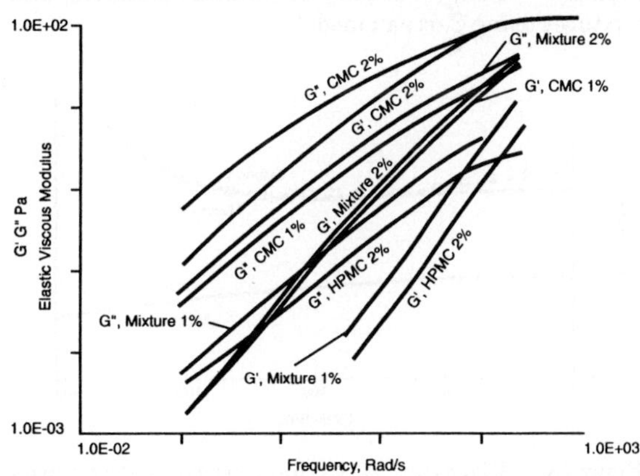

Figure 1. Dynamic sinusoidal linear viscoelastic deformation.

The five parameter model of Carreau (ref. 3) was used to fit cellulose solutions and their mixture in both shear and extension. The equation has been reported to be valid for polyacylamide solution (ref. 1) and is given by

$$\eta-\eta_{\infty}=\eta_{0}-\eta_{\infty}\left[1+\lambda\dot{\gamma})^{a}\right], \qquad (1)$$

where

$\eta, \eta_{\infty}, \eta_{0}$

are the viscosities in shear or extension at a particular strain rate, at the infinite strain rate, and at the zero strain rate. The curve fittings of the experimental data are given in Figs. 2 and 3. Using the Segalman memory function in a potential function for a factorized K-BKZ constitutive equation leads to expressions for viscosity and normal stress coefficient (ref. 3). Substitution of η_0 of 0.5 Pa.S, λ of 1.0 S, n = 0.49 at 1% CMC and η_0 of 0.8 Pa.S, λ of 10.0 S, n = 0.48 at 2% CMC gives BKZ functional behavior with varying parameters of p and v.

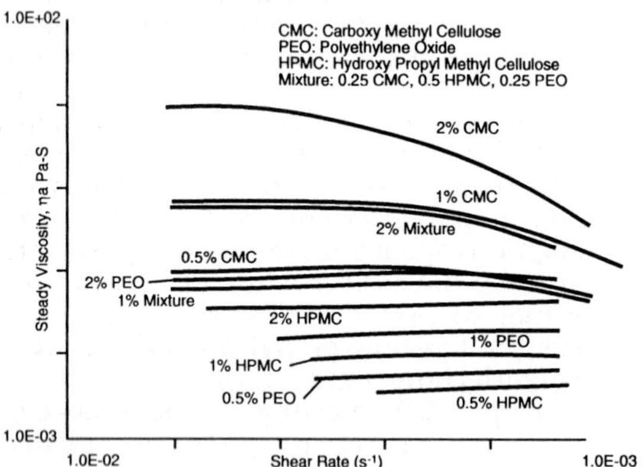

Figure 2. Steady viscosity of cellulose solutions and mixtures using Carreau model.

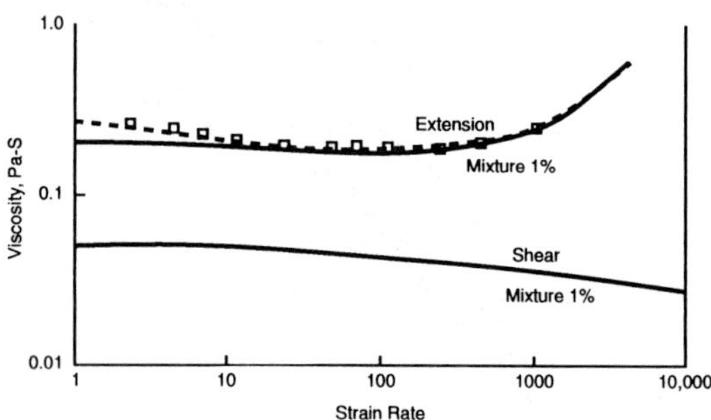

Figure 3. Viscosity at various strain rates using Carreau model.

Flow visualization study of the mixtures of cellulose solutions showed flow instability upstream from the gap in a multislide coater. This flow instability had the topography of ripples which became more intense as the concentration of the solutions increased. The ripples visibly distort the upper free surface in the area upstream from the gap where the layered structure was present. If only one layer was coated, no ripple was observed and when two layers were coated simultaneously, resulting in an interface, a barely visible ripple was observed. Intensity of the ripples increased as the viscosity gradient between the two superposed layers increased. It appears that increase in concentration of our polymer solutions increases non-Newtonian behavior resulting in more pronounced rippling. The analysis of ripples is complicated as it involves the result of forces at the interfaces; the liquid/liquid interface, the interface mode, and the liquid/air interface, or the free surface mode. The randomness of the ripples may depend on the competition between the interface and the free surface modes. Reflection of a straight white strip on the upper free surface of a three layer structure is shown in Fig. 4.

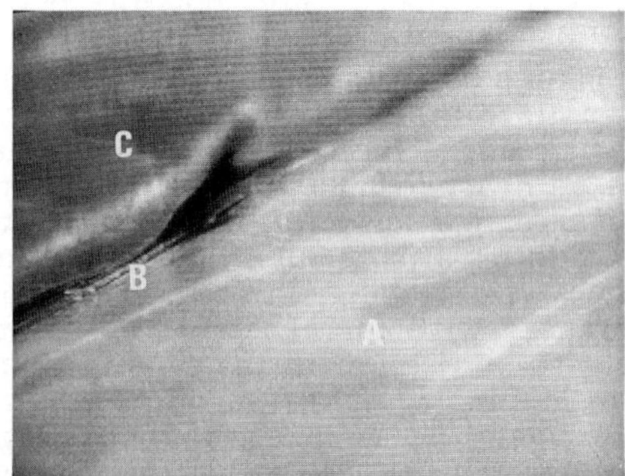

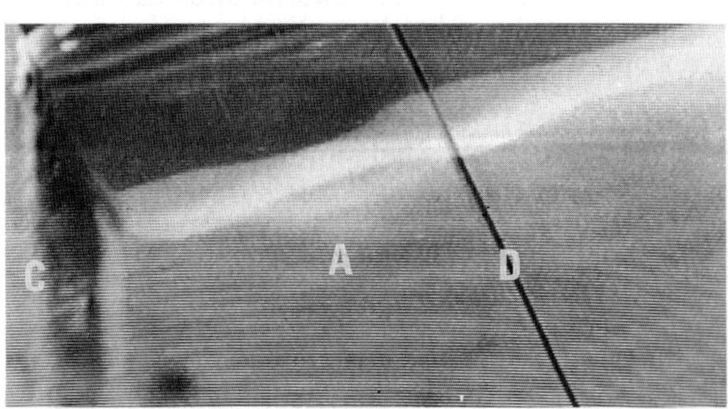

Figure 4a, 4b. Flow visualization of ripples. In the figures, A is backing roll, B is the surface of front block, C is the gap, and D is the downstream area from the first slot.

Flow stability of a two-dimensional flow in a slide coater subjected to two or three-dimensional disturbance can be investigated using the Galerkin/finite element method and Arnoldi's method to determine growth rates of disturbances through eigen values and disturbance shapes through eigen modes (ref. 4). Using this method of stability analysis, flow profiles are found to be quite modified from a circular to a flat profile when the top layer viscosity is reduced from 0.8 Pa.S to 0.004 Pa.S, while keeping the bottom layer viscosity constant, 0.004 Pa.S (Fig. 5). The flow profile acquires two folds when the bottom layer viscosity is reduced from 0.087 Pa.S to 0.004 Pa.S (ref. 5). For a more concentrated, shear thinning fluid, lowering of the web speed would increase the viscosity of the fluid creating more viscosity stratification with lower concentrated fluid, thus increasing flow instability. The three-dimensional stability analysis for calculating air film breakdown close to the apparent dynamic contact line using a very robust elliptical mesh generation scheme would help quantify ripples and quantify operating windows for non-Newtonian fluids (Fig. 6).

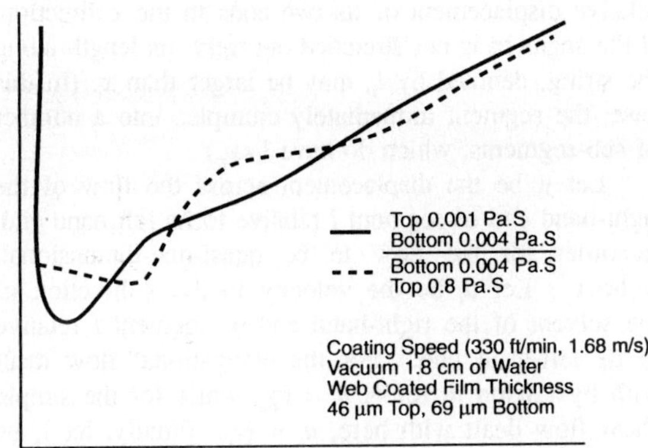

Figure 5. Effect of top layer viscosity on flow profile.

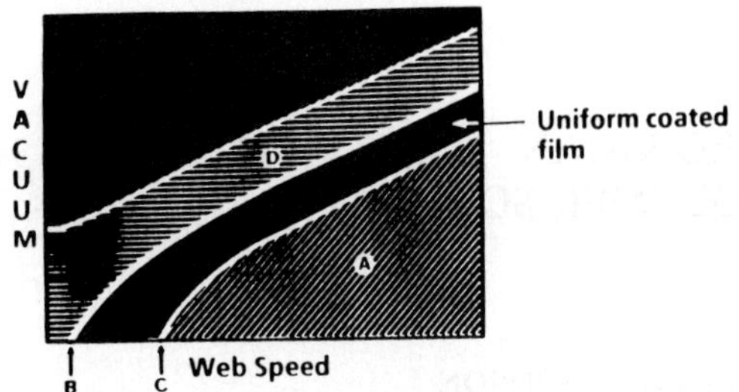

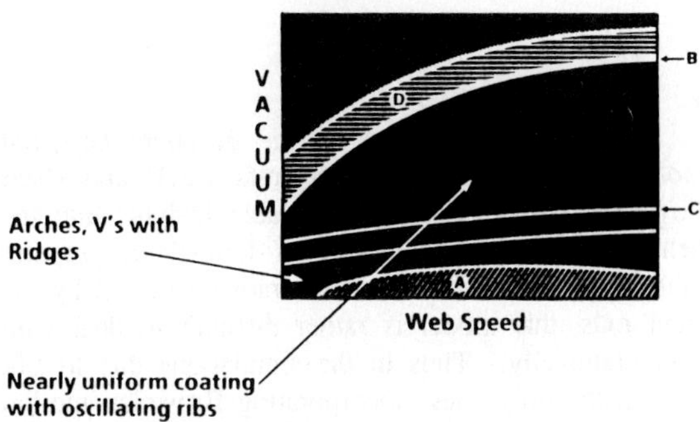

Figure 6. Typical operability windows.

REFERENCES

1. P. R. Schunk and L. E. Scriven, J. Rheol, 34, 1085 (1990).
2. M. Stevanovic, A. Pundsack and L. E. Scriven, Annual Meeting of AICHE, Orlando, Florida, 1988.
3. R. Bird, R. Armstrong and O. Hassager, Dynamics of Polymeric Liquids (John Wiley and Sons, N. Y., 1987).
4. K. N. Christodoulou, Ph.D thesis, University of Minnesota, 1990.
5. K. S. A. Chen and L. E. Scriven, Annual Conference of the Society for Imaging Science and Technology, 1991.

Theoretical and Applied Rheology, edited by P. Moldenaers and R. Keunings
Proc. XIth Int. Congr. on Rheology, Brussels, Belgium, August 17-21, 1992

BROWNIAN KINK DYNAMICS APPLIED TO SHEAR FLOWS OF DILUTE POLYMER SOLUTIONS

JOHN D. ATKINSON

Department of Mechanical Engineering, The University of Sydney, N.S.W. 2006 (Australia)

1. INTRODUCTION

Computer simulations of the Kramers bead-rod molecule in elongational flow (refs 1,2,3) and shear flow (ref 3) have produced valuable insights into the behaviour of dilute solutions of long-chain polymer molecules. Because of the constraints imposed by the rigid rods, this model is rather difficult to deal with computationally. Thus in the simulations due to Liu (ref 3), the only ones incorporating Brownian motion effects over long time periods, which is essential for the treatment of shear flow, results are given only for cases with 20 or less beads.

The method of kink dynamics, recently developed by Larson (ref 4), models a long-chain polymer molecule in dilute solution in an extensional flow as a kinked string. The segments of string between the kinks are in tension, and the kinks themselves migrate and coalesce as the molecule is "pulled out" by the flow. The governing equations are considerably easier to simulate on the computer than those of bead-rod or non-linear bead-spring models, and the number of degrees of freedom involved is less. However the model is less general, since it applies only when the molecule is in a quasi-one-dimensional configuration, that is, when its dimensions in one direction, say the x-direction, are much larger than in the other two. In this paper, it is shown how the model can be extended to deal with cases when some of the segments are not in tension, and therefore crumple into several smaller segments. In addition, Brownian motion effects are added, so that the method can be applied to shear flows.

2. GOVERNING EQUATIONS

Consider a chain of total length L, consisting of a large number of freely hinged rods of length a. In the kink dynamics model, it is assumed this can be replaced by a flexible string, and that the flow is such that the string is in a quasi-one-dimensional kinked state, with N straight segments directed alternatively right and left. The kinks are numbered 1 to N-1, "kinks" 0 and N representing the string ends. If $\boldsymbol{R}_i$ is the position of kink i, then $x_i = |X_i - X_{i-1}|$ is the absolute value of the relative displacement of its two ends in the x direction. If the segment is not stretched out tight, its length along the string, denoted by l_i, may be larger than x_i. (In this case, the segment immediately crumples into a number of sub-segments, which do have $l_i = x_i$.)

Let y_i be the displacement across the flow of the right-hand end of segment i relative to its left-hand end. In order for the flow to be quasi-one-dimensional, $|y_i| << x_i$. Let u_i be the velocity in the x direction of the solvent of the right-hand end of segment i relative to its left-hand end. For the elongational flow dealt with by Larson in ref 4, $u_i = \dot{\varepsilon}\chi_i$, while for the simple shear flow dealt with here, $u_i = \dot{\gamma}y_i$. Finally, let v_i be the difference of the velocity of the solvent past the chain at the right-hand end of segment i relative to that at the left-hand end. For a segment in tension with $u_i > 0$, $v_i = u_i$, while for $u_i < 0$, $v_i = 0$, since the segment cannot support compression.

As shown by Larson, the kinks in the chain migrate according to the *kink evolution equations*, which in the present notation take the form

$$\frac{dx_i}{dt} = \frac{1}{4}(2v_i - v_{i-1} - v_{i+1}) \ . \qquad (1)$$

For simple shear, this becomes

$$\frac{dx_i}{dt} = \frac{1}{4}\dot{\gamma}(2y_i - y_{i-1} - y_{i+1}) \qquad (2)$$

with y_i replaced by zero when it is negative (and for $i = 0$ and $i = N+1$). When any segment reaches zero length, it *pulls out* and is absorbed by the adjoining segments, the total number N being reduced by two (or one if it is an end segment). In our simulation, we remove a segment as soon as its length drops below the minimum rod length a.

Brownian motion causes the chain to diffuse so that the relative displacement of the two ends of a rod changes by about $D^{1/2}$ per unit time, where

$$D = 2kT/\zeta' a \ , \qquad (3)$$

where ζ' is the solvent-polymer friction coefficient per unit length of chain. Thus two neighbouring kinks, with $n = l_i/a$ rods between them, will diffuse at a rate of $(nD)^{1/2}$ per unit time. For large $\dot{\gamma}$, this is much less than the change in x displacement due to convection; thus the x component of Brownian displacement can be neglected. The z component is uncoupled to the motion in the x-direction, and thus also need not be considered. However the y component moves the ends of the segment across the shear flow, and thus strongly affects the tension in the segment, and thus the evolution of the x_i, as seen from equation (1). The evolution equations for y_i are

$$dy_i = (Dl_i/a)^{1/2} dw_i(t) - (y_i/x_i)v_i dt \ , \qquad (4)$$

where $w_i(t)$ is a Gaussian random process (Weiner process) with $<dw_i(t)> = 0$ and $<dw_i(s)dw_i(t)> = dt\delta(s-t)$. The first term on the left-hand side of (4) represents displacement due to Brownian forces, and the second term the tendency for a segment in tension to realign itself with the flow due to drag forces.

The stationary solution to equation (4) (neglecting the buckling which occurs if $y_i<0$) has rms value of order $(Dx_i^2/\dot{\gamma})^{1/3}$. Thus the quasi-one-dimensional requirement for $(y_i)_{rms}$ to be much smaller than x_i for all $x_i \geq a$ means that the shear rate must satisfy

$$\dot{\gamma} \gg D/a^2 \ . \qquad (5)$$

The final effect to be incorporated into the model is the *folding up* of a segment which is not in tension. In this case, equation (1) still holds (with $v_i = 0$), but the length l_i along the chain between the ends of the segment evolves according to

$$\frac{dl_i}{dt} = \frac{1}{4}(2v_i - v_{i-1} - v_{i+1}) - u_i \qquad (6)$$

- that is, since u_i is negative, the segment becomes slack and crumples, a process which is almost instantaneous in terms of the kink dynamics time scale. The length of each of the subsegments into which the segment folds should be about the minimum segment length, namely a, and the subsegments must be directed alternately right and left in order for the kink dynamics model to be applicable to them. Thus if the segment to be folded has length x_i, there should be m subsegments, where m is an odd integer approximately equal to x_i/a. However, the total length of the subsegments must equal the length of the original segment, as must the distance between the beginning of the first subsegment and end of the last. This necessitates that subsegments number $1,3,...,m$ (those directed in the same direction as the original segment) must have length $(l_i+x_i)/(m+1)$, while subsegments number $2,4,...,m-1$ have length $(l_i-x_i)/(m-1)$. The way in which the y displacement of the original segment is distributed among the subsegments is somewhat arbitrary, but it was found that this did not significantly affect the subsequent evolution of the segment lengths as governed by equation (1).

3. RESULTS

Simulations were performed for L/a of 20, 50 and 100. These should correspond to roughly the same value for the number of rods in the Kramer model. Time averages were taken over time t equal to $5000(\zeta' a^2/kT)L$ or greater, which led to statistical errors of about 5% in viscosity. Time steps of $0.001(\zeta' a^2/kT)L$ were used for most cases; considerably smaller steps were necessary for the larger values of $\dot{\gamma}$.

The polymer contribution to the shear stress is given by

$$\tau_p = n_p \langle \sum_i \int_0^{y_i} \zeta' v \, dy \rangle = \frac{1}{12} n_p \zeta' \sum_i \langle v_i x_i y_i \rangle \qquad (7)$$

where n_p is the number density of the polymer molecules. The figure below shows the polymer contribution to the shear viscosity, $\eta_p = \tau_p/\dot{\gamma}$, obtained by kink dynamics. Results obtained by Liu (ref 3) for the Kramers model using Brownian dynamics are also shown. The two agree quite well for larger shear rates.

As predicted by equation (5), the kink dynamics model is not valid for low shear rates, giving viscosity values which are too high, and no constant viscosity (Newtonian) region. However for moderate and high shear rates there is a shear-thinning region with viscosity decreasing like the nth power of shear rate, where n

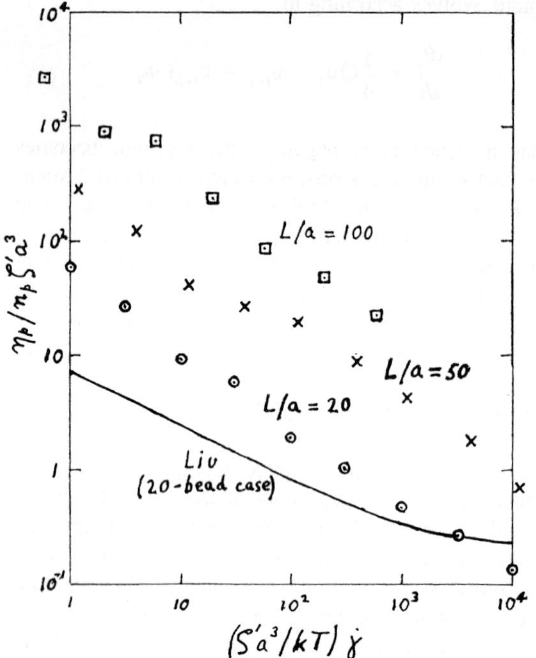

is about *0.6*. (Liu gets *0.4* to *0.5* for the Kramers model.) The constant shear rate regime at very high shear rates obtained by Liu and others for bead-rod molecules does not occur in the kink dynamics model. This is to be expected if this phenomenon is due to the necessity of a rod which is changing ends extending across the flow a distance of order a while it is tumbling, whereas a section of chain in the kink dynamics model will reverse direction, by migration of a kink, due to a very small y displacement if $\dot{\gamma}$ is large.

Computer animation showed that the chain tended to fluctuate between stretched-out and coiled-up states, as observed by Liu. The chain's contribution to shear viscosity occurred almost entirely when it was fully extended. More often than not, the chain did not flip ends while it was coiled up, but stretched out again in the same direction as before. At all shear rates the molecule was fully stretched out for about half the time, or somewhat more than this at the lower shear rates. At high shear rates, the chain became fully extended more often, but, contrary to expectation, the proportion of time it spent fully extended, as well as its mean end-to-end distance, actually decreased somewhat. This appeared to be due to the increased torque on the fully stretched molecule in this case (the last term in equation (4)), so that it more rapidly aligned itself with the flow and then folded up again. The figure opposite shows the dependence on shear rate of $<t_S>$, the mean time spent stretched out each time the chain was fully extended.

REFERENCES

1. J.M. Rallison and E.J. Hinch, J. Non-Newtonian Fluid Mech., 29 (1988) 37-55.
2. D. Acierno, G. Titomanlio and G. Marucci, J. Polym. Sci., Polym. Phys. Ed., 12 (1974) 2177-2187.
3. T.W. Liu, J. Chem. Phys., 90 (1989) 5826-5842.
4. R.G. Larson, Rheol. Acta, 29 (1990) 371-384.

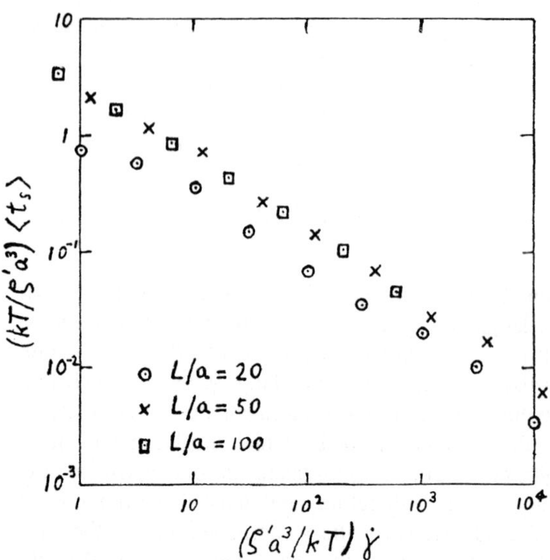

Theoretical and Applied Rheology, edited by P. Moldenaers and R. Keunings
Proc. XIth Int. Congr. on Rheology, Brussels, Belgium, August 17-21, 1992

INFLUENCE OF THERMAL HISTORY ON THE FLOW PROPERTIES OF CONCENTRATED POLYMER SOLUTIONS WITH RESPECT TO WALL SLIP AND SHEAR-THICKENING

H. BAUER[1], N. BÖSE[1], P. STERN[2]

[1] Physikalisch-Technische Bundesanstalt, Bundesallee 100, 3300 Braunschweig (Germany)
[2] Institute of Hydrodynamics of the Czechoslovak Academy of Science, Prague (CSFR)

1. INTRODUCTION

Under suitable conditions solutions of a styrene-isoprene di-block copolymer in hydrocarbons form a micelle system with slightly solved polystyrene cores. These solutions exhibit viscoplastic behaviour with a yield value (ref. 1) and in some cases a pronounced shear-thickening plateau in the flow curve (ref. 2). Since understanding of this phenomenon is not satisfactory at present, an experimental investigation was carried out, especially with respect to the influence of the thermal history of the samples and with respect to a possible influence of wall slip on the shear-thickening behaviour.

2. EXPERIMENTAL

The average molar mass of the linear di-block copolymer is about 100 000 g/mol; the mass distribution is characterized by a ratio of mass average to number average of 1.04. The polymer was dissolved at temperatures between 40°C and 130°C in a paraffinic hydrocarbon mixture of 1.3 mPa's viscosity at 20°C. The polymer mass fraction was 6% throughout the experiments. Flow curves were measured using controlled shear and controlled stress rotational viscometers with concentric cylinder and cone and plate systems. For the study of possible wall effects, cylinder systems with rough surfaces (sand-blasted) and with profiled surfaces were also used. In order to gain information about the microstructure of the micelle system, wide-angle laser light scattering measurements were carried out.

Density measurements with pycnometers and specific heat determinations using a Setaram C80-type differential scanning calorimeter (DSC) were performed to detect possible phase transitions.

3. RESULTS

The surface structure of the profiled systems has an influence on the flow field which can be taken into account by an increased effective gap width (correction factor 1.018, independent of viscosity above some 100 mPa's) as determined with Newtonian liquids. This correction has been applied to all measurements with profiled systems referred to here. At low viscosities (typically 5 mPa's), this correction factor does not compensate this influence completely; in addition, the onset of secondary flow is shifted to lower shear rates.

Flow curves measured at 20°C using smooth and profiled cylinder systems for samples that have been prepared at 40°C and heated to various temperatures are shown in fig. 1. There is no difference for the 40°C sample when it is measured in smooth and profiled systems. After heat treatment at 60°C and above, the flow curves start with a significantly lower viscosity followed by a sudden shear-induced transition to higher viscosity, which appears however, only when profiled systems are used. These measurements were performed in a controlled stress instrument. When a controlled shear viscometer is used, the transition step in the flow curve is always horizontal.

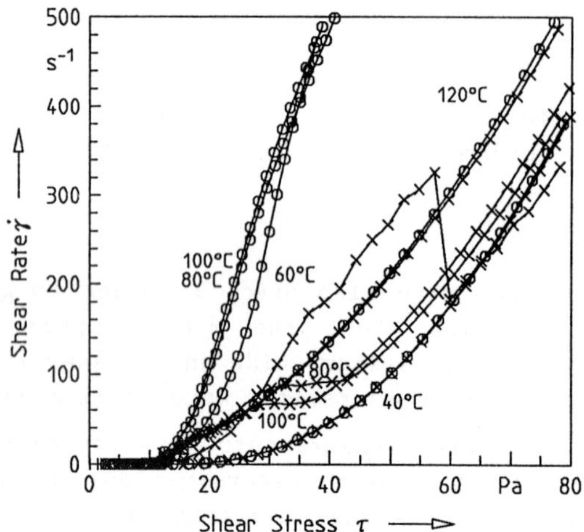

Fig. 1
Flow curves at 20°C for samples having undergone different heat treatment.
o cylinder systems with smooth surfaces
x cylinder systems with profiled surfaces

When smooth systems are used, the sample remains in the low-viscosity state up to the highest shear rates of 5 000 s^{-1} that were applied using cone and plate geometries. After the fluid has been heated to 120°C, the flow curves measured with smooth and profiled systems again coincide.

Fig. 2 shows the same flow curves in the low shear-rate range. In the region just above the yield point, only the 60°C sample shows an influence of the surface structure of the measuring system. Whereas the low-temperature samples (40°C and 60°C) start with a Casson-like flow curve, a sharp onset with Bingham-like behaviour is observed after heat treatment at 80°C and above. The 120°C sample shows less pronounced shear thickening at slightly different shear rates around 15 s^{-1} in smooth and profiled systems.

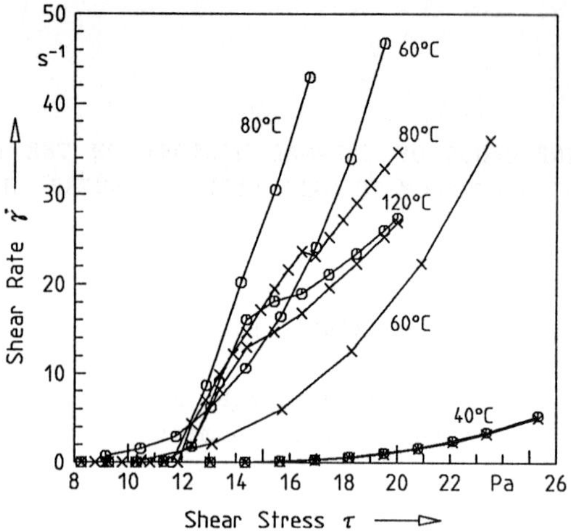

Fig. 2
Flow curves at low shear rates at 20°C for samples having undergone different heat treatment
o cylinder systems with smooth surfaces
x cylinder systems with profiled surfaces

The temperature dependence of a sample that has been heat-treated at 120°C is shown in fig. 3. The step in the flow curves becomes more pronounced with increasing temperature and is observed also with smooth systems. Below and above the transition range the flow curves are roughly the same in both types of measuring systems. At 120°C the solutions become Newtonian.

Flow curves measured with cone and plate systems and with cylinder systems with rough surfaces coincide with those of smooth cylinder systems (including the step in the flow curve).

These results indicate that the micelle system undergoes transitions in the microstructure between 40°C and 60°C and in the temperature range around 120°C. Density measurements as a function of temperature showed that the thermal expansion coefficient changed from $9.97 \cdot 10^{-4}$ K^{-1} to $10.12 \cdot 10^{-4}$ K^{-1} at 50°C. With the DSC method an endothermic peak in the $C_p(T)$-curve was observed at 110°C.

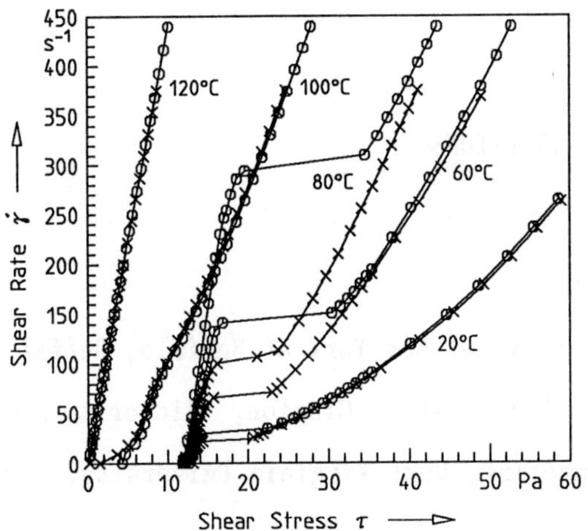

Fig. 3
Temperature dependence of the flow curve for a sample that has been heated to 120°C
o cylinder systems with smooth surfaces
x cylinder systems with profiled surfaces

On the basis of the light scattering measurements, an average mass of the micelles of about $25 \cdot 10^6$ g/mol for the 40°C and the 60°C samples was calculated, the radius of gyration being about 70 nm and 140 nm, respectively. For the sample heated to 120°C the corresponding values are $7 \cdot 10^6$ g/mol and 100 nm. These measurements were of course performed with highly diluted samples. Flow curve and viscosity measurement with samples prepared and diluted in different ways indicate, however, that the properties of the micelles are not affected by dilution at room temperature or only slightly so.

4. DISCUSSION

These results show that wall slip effects are not involved in the observed flow behaviour reported here and are not, in particular, responsible for shear thickening.

The endothermic transition observed at 110°C must be interpreted as destruction of the micellar structure. There is no longer an aggregation of polymer molecules above this temperature.

From the temperature dependence of the flow curves above the shear-thickening transition it must be concluded that the regular flow mechanism is effective only above the transition step. At lower shear rates below the transition the microstructure of the micellar network enables shear flow with less energy dissipation. In principle, the situation is similar to shear-thickening effects observed at concentrated dispersions (ref. 3) which were explained by the formation of a shear-induced superstructure at low shear rates and a breakdown of the long-range order at higher shear rates causing an increase in viscosity.

The changes observed in the properties of the fluid around 50°C cannot only be explained by an increase in the size of the micelles due to an improvement of the degree of solvatization. The abrupt change in the coefficient of thermal expansion and in the flow properties points to a structural change which is likely to be a transition from a spherical to a rod-like shape. One could then assume that at low shear rates the rotation of the micelles in the field of shear flow is blocked by a structural arrangement, thus causing a slipping of ordered layers of micelles. Only when the difference between the shear stresses acting upon opposite sides of a micelle reaches a certain threshold value is this dissipation mechanism activated. Distortions in the flow field caused by the ripples of the profiled system favour this transition.

REFERENCES

1. H. Bauer, N.Böse, in: D.R. Oliver (Ed.), Third European Rheology Conference, Edinburgh, U.K., September 3-7, 1990, Elsevier Applied Science, London, 1990, pp. 37-40
2. H. Bauer, G. Meerlender, Progr. Coll. & Pol. Sci., 72, 1986, pp. 106-111
3. H.M. Laun, Xth International Congress on Rheology, Sydney, Australia, August 14-19, 1988, Vol. 1, pp. 37-42

Theoretical and Applied Rheology, edited by P. Moldenaers and R. Keunings
Proc. XIth Int. Congr. on Rheology, Brussels, Belgium, August 17-21, 1992

DEFORMATION-INDUCED PHASE SEPARATION IN POLYMER SOLUTIONS

F.S. CAI[1], T. SRIDHAR[2] AND R.K. GUPTA[1,3]

[1]Department of Chemical Engineering, State University of New York at Buffalo, Buffalo, NY 14260 (USA)
[2]Department of Chemical Engineering, Monash University, Clayton, Victoria 3168 (Australia)
[3]Present address: Department of Chemical Engineering, West Virginia University, P.O. Box 6101, Morgantown, WV 26506 (USA)

1. INTRODUCTION

It has been demonstrated by a large number of researchers that optically-clear single-phase polymer solutions often turn turbid on being mechanically deformed (ref. 1 and other references therein). This has been interpreted (refs. 1,2) as a liquid-liquid phase separation, i.e., a shift in the phase transition to higher temperatures. This is, however, not universally accepted. Helfand and Fredrickson (ref. 3), in particular, have stated that this explanation has no basis in statistical mechanics and it neglects much of the essential physics. Instead, they consider the phenomenon to merely reflect scattering that is greatly enhanced by flow.

In this paper we present new data in steady shear as well as in constant stretch rate uniaxial extension on the previously-studied polystyrene-in-dioctyl phthalate system. The appearance of turbidity is found to follow very well-defined trends with polymer molecular weight, concentration and deformation rate in both flow fields. Additionally, simultaneous stress growth measurements during the stretching experiments are particularly revealing. These results should be explainable on a theoretical basis and should help in discriminating between the two theories referred to earlier.

2. MATERIALS USED

Four different polystyrenes were employed in this work. Two of these were high molecular weight ($9x10^5$, $1.86x10^6$) monodisperse polymers purchased from the Pressure Chemical Co. while the other two were low molecular weight polydisperse materials obtained from the DOW Chemical Co. and having weight average molecular weights of $2x10^5$ (Styron 675 APR) and $3x10^5$ (Styron 685D) respectively. The solvent used was dioctyl phthalate from Alpha Products having a molecular weight of 390.6 and a density of 0.981 g/cc.

2.1 Quiescent Cloud Points

These were determined visually by immersing a vial of polymer solution into a constant-temperature glass water bath whose temperature was lowered at the rate of 1 to 2°C per hour. The temperature at which cloudiness was observed was noted, and the average of four runs was reported as the cloud temperature. For the 300,000 molecular weight polymer, the cloud points corresponding to concentrations of 0.035, 0.05, 0.08, 0.10, 0.12 and 0.15 g/cc were 5.6, 5.4, 3.7, 3.3, 3.5 and 2.7°C. In general, an increase in the polymer molecular weight resulted in an enhancement of the cloud temperature.

3. SHEARING EXPERIMENTS

3.1 Apparatus Used

A transparent parallel plate instrument was built from poly methyl methacrylate for the steady shearing experiments. It was essentially a cylindrical jacketed cavity of 4 1/8" internal diameter. The base of the cavity formed the lower plate while the upper plate was a 3 1/2" diameter disk that could be inserted into the cavity to yield gap thicknesses of 0.2 to 0.5 cm. the disk itself was attached to a motor fitted with a speed controller; rotation of the disk allowed for the shearing of a polymer sample placed between the disk and the cavity base. Temperature was controlled by circulating water through the jacket surrounding the plates.

3.2 Procedure

When the sample had been loaded and the temperature stabilized, the upper plate was rotated and the speed increased. When the angular velocity reached a critical value, turbidity could be observed at the outer edge of the disk. This was best observed by carrying out the experiment in a darkened room and shining a flashlight at the polymer sample. The boundary between the clear and the milky solution was usually quite sharp and would move inwards as the speed of rotation was further increased; the movement of the front appeared to be instantaneous. The critical shear rate, $\dot{\gamma}_c$, for phase separation was computed from a knowledge of the angular velocity, Ω, the gap thickness and the distance r_c between the disk center and the boundary demarcating the clear from the opaque region.

3.3 Results

Typical results obtained with the 300,000 molecular weight polymer at a concentration of 0.1 g/cc and a gap width of 0.45 cm are listed in Table 1. Here t is the time taken for the turbidity to disappear on cessation of shearing.

It should be noted that these results were independent of the speed of rotation and the gap width, H.

Results at other polymer concentrations are shown in Figure 1 while the influence of polymer molecular weight, at a fixed concentration, is revealed in Figure 2. General conclusions that one can draw are that the cloud point temperature increases with increasing shear rate, polymer concentration and polymer molecular weight.

Table 1. Influence of shear on the cloud point.

No.	$\dot{\gamma}_c$ (s^{-1})	T (°C)	t (s)
1	31.0	5.0	30
2	37.5	7.0	25
3	57.8	8.8	5
4	71.7	10.5	3
5	96.8	12.8	2
6	103.3	14.3	1
7	153.1	18.0	0
8	190.8	20.5	0
9	251.3	23.5	0
10	298.0	30.5	0

4. STRETCHING EXPERIMENTS

4.1 Apparatus Used

The method of stretching initially stress-free filaments of polymer has been described recently in ref. 4. Constant stretch rate extension is achieved by holding a vertical liquid sample between two coaxial teflon disks, one of which is stationary while the other one is moved outwards at an exponentially increasing velocity. In the present instance, the bottom disk is moved downwards by a MTS material testing machine while the fixed upper disk is attached to a Cahn microbalance which measures the stretching force in the filament. In order to both eliminate the effect of gravity and to achieve various constant temperatures below room temperature, stretching was carried out in a transparent plastic tank filled with water.

4.2 Results

Turbidity was observed whenever the stretch rate was above a critical value, and it disappeared as soon as the experiment was stopped. Observations were again visual and done as in the case of the shearing experiments. For the 0.1 g/cc solution of the 300,000 molecular weight polymer, the critical stretch rate at 8.5°C was found to be 1.35 s^{-1}, and this stretch rate went up as the temperature was raised.

Since the apparatus could not achieve stretch rates above 2 s^{-1} for sustained periods of time and since the critical deformation rates were expected to go down with increasing polymer molecular weight, most of the extensional experiments were done using the two monodisperse high molecular weight polymers. For a 0.08 g/cc solution of the 900,000 molecular weight polymer the critical stretch rates at temperatures of 12.8, 14 and 17°C were observed to be 1.2, 1.35 and 1.55 s^{-1}. When stress growth was monitored during the stretching of solutions which showed turbidity and contrasted with those that did not, it was found that the stress went through a maximum for the turbid solutions; the clear solutions however, displayed monotonic stress growth. This is shown in Figure 3 for the stretching of a 0.08 g/cc solution of the 900,000 molecular weight polymer at 11.2°C. Results for the 1,860,000 molecular weight polymer were similar.

5. CONCLUDING REMARKS

New data have been presented on the apparent phase separation of polymer solutions under the influence of mechanical deformation. Although data in shear fields have been available for some time, these have been limited to very high polymer molecular weights. Present results are for polymers having molecular weights as low as 200,000. The extensional data obtained here are entirely new and have become available for the first time. An analysis of all these data should help in putting to rest the controversy surrounding the origin of this phenomenon.

6. ACKNOWLEDGEMENT

Acknowledgement is made to the donors of the Petroleum Research Fund, administered by the American Chemical Society, for partial support of this research.

REFERENCES

1. C. Rangel-Nafaile, A.B. Metzner and K.F. Wissbrun, Macromolecules, 17 (1984) 1187-1195.
2. D.F. Massouda, A. parametric study of stress-induced liquid-liquid phase separations in polymer solutions, PhD thesis, Chemical Engineering, University of Delaware, 1986.
3. E. Helfand and G.H. Fredrickson, Phys. Rev. Letters, 62 (1989) 2468-2471.
4. T. Sridhar, V. Tirtaatmadja, D.A. Nguyen and R.K. Gupta, J. Non-Newtonian Fluid Mech., 40 (1991) 271-280.

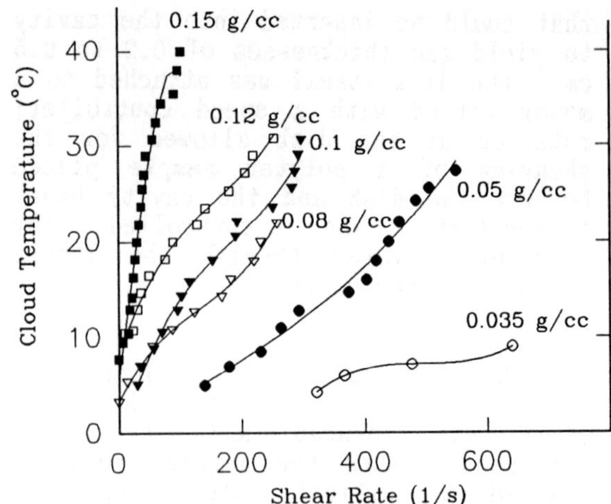

Figure 1. Influence of Shear rate and polymer concentration on the cloud point.

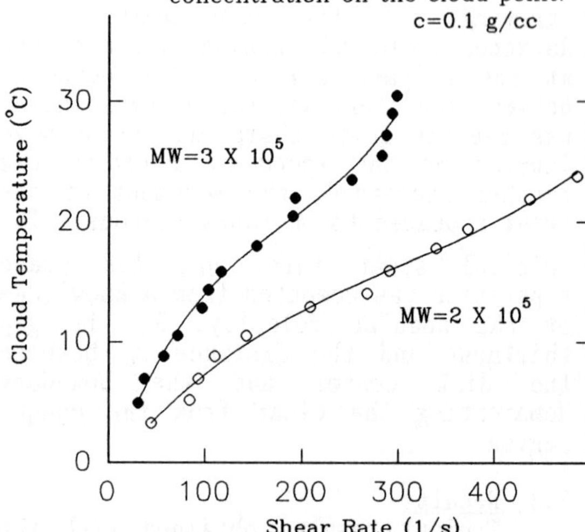

Figure 2. Influence of molecular weight on the cloud point

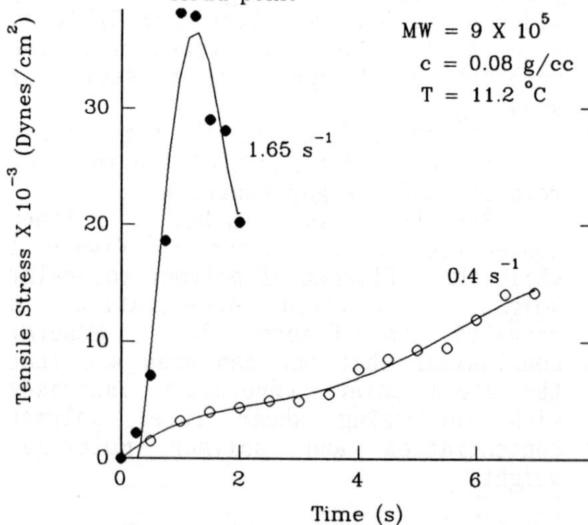

Figure 3. Stress growth during uniaxial extension above and below the critical stretch rate

Theoretical and Applied Rheology, edited by P. Moldenaers and R. Keunings
Proc. XIth Int. Congr. on Rheology, Brussels, Belgium, August 17-21, 1992

ELASTIC AND DISSIPATIVE STRESS-PRODUCING MECHANISMS IN CONTRACTION FLOWS OF POLYMER SOLUTIONS AS A FUNCTION OF FLOW STRENGTH AND CONCENTRATION

U. CARTALOS[1] and J.M. PIAU

Laboratoire de Rhéologie, I.M.G., B.P. 53X Domaine Universitaire, 38041 Grenoble Cedex (France)
[1]present address : Institut Français du Pétrole, B.P. 311, 92506 Rueil Malmaison Cedex (France)

1. INTRODUCTION

Due to their high elongational component, contraction flows are convenient geometries for generating and studying important elongational effects. As shown in (ref. 1), the particular advantage of orifice flows is that the pressure drop P_g across the orifice is related to energy losses in such a way that experimental flow curves can be analysed in terms of tensile stress growth.

Relating P_g to the fluid's elongational properties is not straightforward though, since shear effects are also present and the velocity gradient is not homogeneous. Interpretation may, however, be attempted by similarity-an analysis technique that is commonly applied to field equations and boundary conditions in fluid dynamics-provided that a realistic constitutive model is available.

Orifice flow experiments with dilute polymer solutions in a thick solvent (ref. 1), in which flow parameters were varied in the whole range, showed that the flow curves are characterised by 3 patterns : a low and a high flow pattern where P_g is linear in flow rate q_v and an intermediate flow pattern where P_g varies as q_v^2.

These scalings are concentration independent in the range examined (20-500ppm). New experimental data reported here show that an additional scaling occurs when polymer concentration is increased.

Our objective is to work out a similarity analysis in order to relate the observed scalings to the main trends of structural constitutive equations developed for polymer solutions. The particular structural model used here accounts for elastic effects due to chain unfolding as well as dissipative effects due to hydrodynamic interaction among elongated chains.

It is shown that in the case of a thick solvent, for which the description of molecular deformation is considerably simplified, flow patterns can be interpreted in terms of molecular mechanisms governing tensile stress growth.

2. FLOW CURVES

PEO (polyethylene oxide, M_w~4x10^6) solutions in a viscous solvent (a mixture of sugar-53% and water-47%) were pushed through an orifice, at imposed pressure, under inertialess flow conditions. A description of the experimental set-up can be found in (refs. 1,2).

Polymer concentrations of 100-500-1000-2000 and 5000ppm were tested. From these and previous experiments, two classes of flow curves can be identified, depending on polymer concentration. They are illustrated in figure 1 and can be discussed as follows

<u>Low (dilute) concentration range</u>

The flow curve of the 500ppm PEO solution is characterised by 2 regions-at low and high q_v-where P_g is linear in q_v and 1 region-at intermediate q_v-where P_g varies as q_v^2. The 100ppm PEO solution showed similar behaviour.

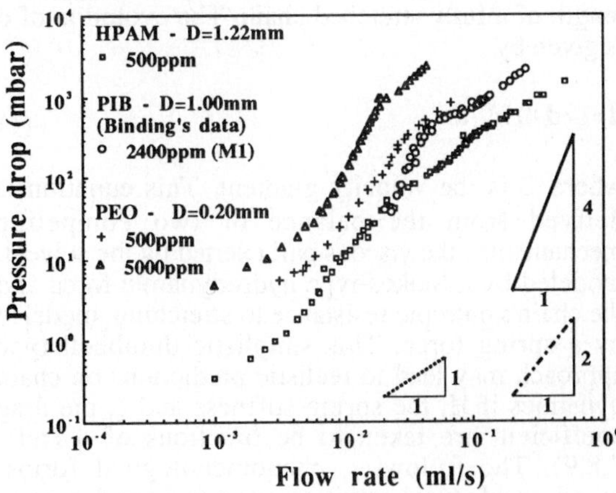

<u>Figure 1</u> Pressure drop versus flow rate in orifice flow for HPAM (20g/l of NaCl) and PEO solutions in thick solvents. Binding's data (ref. 4) for the M1 fluid are also plotted. D is the orifice diameter.

These patterns are the same as the ones reported in (refs. 1,2,3) for dilute HPAM (partially hydrolysed polyacrylamide, M_w~$7x10^6$) solutions in thick solvents. In (refs. 1,2) polymer concentration was varied from 20 to 500ppm. A representative curve of this series of experiments, the one for the 500ppm solution, is also plotted in figure 1. Note that HPAM solutions contained 20g/l of NaCl, a salinity at which electric charges along the chain backbone are neutralized and the chain behaves like a flexible coil at rest.

Intermediate concentration range

The flow curve of the 5000ppm solution, plotted in figure 1, shows all the qualitative aspects observed with the 1000 and 2000ppm solutions.

The low and high flow rate linear scalings remain unchanged, but a new pattern is observed at intermediate flow rates. After a quadratic region, P_g now varies as q_v^4.

The power 4 scaling region is detected in a small region of the flow curve of the 1000ppm PEO solution. As concentration is increased, the quadratic scaling region becomes shorter, whereas the power 4 scaling is observed in a wider region.

The power 4 scaling can also be found in Binding's data (ref. 4) for the M1 fluid (a 2400ppm solution of Polyisobutilene-M_w~$2x10^6$ in a thick solvent) which are plotted in figure 1.

The above scalings will be examined from the point of view of the molecular dynamics predicted by a structural model.

3. STRUCTURAL INTERPRETATION

3.1. The constitutive model. Case of a thick solvent

According to the structural model developed in (refs. 2,5,6), in a highly elongational flow chain conformation can be described by a vector **d**. The modulus d of **d** measures the mean length of a chain. d varies from d_0, the rest-state length to d_{max}, the length of a fully stretched chain. The evolution of **d** is given by

$$\dot{\mathbf{d}}=\mathbb{L}\cdot\mathbf{d}-(H/\zeta)\mathbf{d} \qquad (1)$$

where $\mathbb{L}$ is the velocity gradient. This equation is derived from the balance of two competing mechanisms : the viscous pull exerted by the solvent, modeled by a Stokes-type hydrodynamic force and the chain's entropic resistance to stretching, modeled by a spring force. This simplistic dumbbell type approach may lead to realistic predictions on chain dynamics if H, the spring stiffness and ζ, the drag coefficient are taken to be functions of d (refs. 7,8,9). The following phenomenological forms-based on physical intuition-are considered here in order to capture the essential aspects of chain deformation far away from equilibrium

$$H = H_0(1-d_0^2/d^2)^{1/2}(1-d^2/d_{max}^2)^{-q} \qquad (1a)$$

$$\zeta = \zeta_0(1-d^2/d_{max}^2)^h \qquad (1b)$$

Finite extensibility is incorporated in (1a). Introducing the exponent q results in a major improvement. If $0<q<1$, the energy stored in a chain remains finite when d approches d_{max}. On the contrary, for the usually considered Warner spring law (q=1) an infinite energy is needed to stretch a chain completely. The Brownian motion effect in producing a non-zero d_0 is also included in (1a).

The solvent action is exerted on packs of monomers on both ends of a chain. So, it varies with their size. The expression (1b) accounts for the fact that, as the chain unravels, the size of these packs decreases and so the solvent action diminishes. The ratio H_0/ζ_0 of H and ζ values at small departures from rest provides the model time constant τ_0.

The extra stress $\mathbb{T}$ is the sum of three terms

$$\mathbb{T} = 2\eta_s\mathbb{D} + nH(\mathbf{d}\otimes\mathbf{d})_{dev} + n\eta_s f_{int}((\mathbf{d}\cdot\mathbb{D}\cdot\mathbf{d})/d)(\mathbf{d}\otimes\mathbf{d})_{dev} \qquad (2)$$

The first term is the Newtonian contribution of the solvent, with η_s the solvent viscosity and $\mathbb{D}$ the rate of deformation tensor. The second term is the usual Kramers expression, accounting for elastic stresses transmitted through an elementary surface cutting a "dumbbell" (ref. 7). The third term arises from hydrodynamic interactions among highly stretched chains occuring when their length becomes much larger than their lateral spacing. Inspite it's significant importance at high rates of deformation, even for solutions that are dilute at rest, this term is frequently neglected in molecular theories. It's form is inspired by the one derived by Batchelor (ref.10) for closely spaced fibrous particles. f_{int} is a model parameter of order unity when hydrodynamic interactions occur.

The constitutive model given by (1), (2) obeys invariance and is compatible with Thermodynamics. Its predictions on material functions can be found in (ref. 2). The elongational viscosity, in particular, reaches a high elongational rate plateau value $\eta_{E\infty}$ that is due to the interaction term. For fluids considered in this study, $\eta_{E\infty}/(3\eta_s)$ is of the order of 10^5 to 10^7.

Equation (1) is considerably simplified for solutions in a thick solvent. In this case, the viscous pull, which is proportional to η_s, can dominate the resisting mechanism and so govern the conformational changes. This implies that the second term in the RHS of (1) is negligible (note that when $d_0^2<<d^2<d_{max}^2$, H/ζ is close to τ_0^{-1}. But τ_0, being proportional to η_s, is large and so the term in H/ζ can be neglected in (1)). The evolution of **d** is consequently given by

$$\dot{\mathbf{d}}=\mathbb{L}\cdot\mathbf{d} \qquad (3a)$$

However, when d approches d_{max}, H/ζ becomes

large whereas the derivative of **d** diminishes, since chains are rigidified. So, close to d_{max}, (1) becomes

$$\mathbb{L}\cdot\mathbf{d}=(H/\zeta)\mathbf{d} \tag{3b}$$

3.2. Similarity analysis for contraction flows

In the absence of inertia, the momentum balance reduces to

$$\text{div}\,\Sigma = 0 \tag{4}$$

where $\Sigma=-p\mathbb{1}+\mathbb{T}$ is the full stress tensor. For large values of the contraction ratio, the problem at hand has one length scale, the orifice diameter D. Equation (4) shows that all components of Σ scale in the same way with the characteristic parameters of the problem. So, p and the components of $\mathbb{T}$ scale similarly, and so does P_g, which is a linear combination of them. Consequently, the order of magnitude analysis of P_g can be given by the one of $\mathbb{T}$, i.e.

$$P_g \sim S+El+Int$$

S, El and Int stand respectively for the order of magnitude of the solvent, elastic and interaction terms in (2). For a given fluid, S varies as $\|\mathbb{D}\|$, the magnitude of $\mathbb{D}$, El as Hd^2 and Int as $\|\mathbb{D}\|d^3$. $\|\mathbb{D}\|$ varies as V/D with V the discharge velocity on the orifice plane. For a given D, $\|\mathbb{D}\|$ and V vary as q_v. The magnitude of d is given by the evolution equation. The following cases can be distinguished.

Low flow rates

Chain deformation is restricted to small departures from the rest state ($d\approx d_0$). Under these conditions, the elastic and ineraction terms make negigible contributions, of the order of the small polymer volume fraction. P_g grows then as S and so as q_v. Consequently, the linear scaling at low flow rates arises from dominant viscous stresses due to the straining motion of the Newtonian solvent.

Intermediate flow rates

Chains are subjected to flow conditions that lead to important departures from the rest state length. Conformational evolution is described by (3a), according to which **d** deforms like a vector joining two neighbouring material points. The separation in time of such two points is proportional to the local velocity. Consequently, $d\sim V$. This leads to $El\sim q_v^2$ and $Int\sim q_v^4$.

This implies that the quadratic scaling observed in the intermediate region can be explained by the fact that elastic stresses, due to important chain distortions predominate. For dilute solutions, the whole of the intermediate pattern is governed by elastic stresses. However, for more concentrated solutions, as q_v increases, chains are stretched to a certain length beyond which hydrodynamic interactions prevail.

High flow rates

Chains may be stretched close to their full length in which case (3b) is valid. It can be shown that (3b) leads to $El\sim q_v^{\alpha}$ ($\alpha=q/(q+h)<1$) and $Int\sim q_v$.

The linear scaling observed at high q_v for all polymer concentrations tested can be attributed to the fact that hydrodynamic interactions between fully stretched chains govern the flow.

4. CONCLUSION

This work shows that the orifice flow of polymer solutions in a thick solvent is characterised by 3 flow regimes, which occur successively as flow strength is increased. Conformational evolution and associated tensile stress growth are established in each regime.

The low and high flow rate regimes are governed by dissipative stresses, Newtonian ones in the former and stresses arising from hydrodynamic interactions among fully stretched chains in the latter. In these 2 regimes, the steady state elongational viscosity should be the important parameter, since chains effectively reach steady state-very close to d_0 at low flow rates and very close to d_{max} at high ones.

In the intermediate regime, elastic, but also dissipative (hydrodynamic interaction) stresses may prevail, the latter occuring beyond a certain polymer concentration as flow strength increases. But the relevant material function is the transient elongational viscosity, since chains do not reach a steady state length (see equation (3a) which holds in this regime).

Realistic predictions on elongational material functions can be readily obtained from the model used here (ref. 2). But an interesting point of the present analysis is that transient elongational effects are easier to study experimentally in the case of a thick solvent, due to the resulting high time scales.

REFERENCES

1. U. Cartalos and J.M. Piau, J. Non-Newtonian Fluid Mech., submitted for publication
2. U. Cartalos, Ph.D. Thesis, INP Grenoble, 1989
3. M. Moan, G. Chauveteau and S. Ghoniem, J. Non-Newtonian Fluid Mech., 5 (1979) 463-474
4. D.M. Binding, D.M. Jones and K. Walters, J. Non-Newtonian Fluid Mech., 35 (1990) 121-135
5. J.M. Piau and P. Doremus, Rheol. Acta, 23 (1984) 465-476
6. U. Cartalos and J.M. Piau, J. Th. and Applied Mech., special issue (1985) 101-140
7. R.B. Bird, C.F. Curtiss, R.C. Armstrong and O. Hassager, Dynamics of Polymeric Liquids, John Wiley and Sons, NewYork, 1987, vol 2, 2nd ed., pp 55-92
8. G.G. Fuller and L.G. Leal, J. of Non-Newtonian Fluid Mech., 8 (1981) 271-310
9. R.G. Larson and J.J. Magda, Macromolecules, 22 (1989) 3004-3010
10. G.K. Batchelor, J. Fluid Mech., 46 (1971) 813-829

Theoretical and Applied Rheology, edited by P. Moldenaers and R. Keunings
Proc. XIth Int. Congr. on Rheology, Brussels, Belgium, August 17-21, 1992

THE RHEOLOGY OF MIXTURES: RODS AND COILS

L.C. CERNY[1], E.L. CERNY[1], J.K. SUTTER[2], D.C. MALARIK[2], D. SCHEIMAN[2]

[1]Chemistry Department, Utica College of Syracuse University, Utica, NY 13502 USA

[2]NASA-Lewis Research Laboratory, MS49-1, 21000 Brook Park Road, Cleveland, OH 44135 USA

1. INTRODUCTION

There is an increasing interest in the aerospace industry toward the research and development of thermoplastics in both molding and as matrices for fiber reinforced composites. Since most thermoplastics are soluble, it is possible to characterize them with respect to molecular weight, solution viscosity and molecular size and shape. The processability, the structure integrity and the mechanical performance are closely related to the molecular parameters of the polymer. Therefore the information concerning the molecular weight and the molecular weight distribution will greatly enhance the research on new polymers with improved properties.

With all new polymeric materials, there is a fundamental need to bridge the discipline of polymer synthesis, characterization, processing and mechanical properties. The information obtained from dilute solution property measurements will play a critical role in this unification.

The proposed investigation will be a cross-section of analytical techniques which are currently available to the molecular characterization of advanced thermoplastics. A variety of methods is suggested in order to provide a real sense of the molecular weight distribution of new thermoplastic polymers.

2. METHODS

2.1 Polymers

All of the resins in this study were supplied by the Materials Division of the Lewis Research Center or prepared by the principal investigator according to Roberts and Lauber (ref.1) or Sutter (ref. 2).

2.2 Solvents

The solvents which are effective in dissolving some of these polymers are tetrahydrofuran (THF), N, N-Dimethylformamide (DMF) and N, N-Dimethylacetamide (DMA).

2.3 Intrinsic Viscosity

Intrinsic viscosity $[\eta]$ at 25C was determined for each of a series of polymers. Cannon-Ubbelohde dilution viscometers were employed. The measurements were made with a Wescan Automatic Timer.

2.4 Vapor Pressure Osmometry

The measurement of the number average molecular weight characterization of samples by the technique of vapor pressure osmometry were undertaken. The application is restricted to rather low molecular weight and finds particular use in examination of oligomer extracts. The choice of solvent is dictated by the solubility characteristics toward given polymers of interest and by a number of other practical considerations. Generally, solvents of moderate vapor pressure, low viscosity, and low heats of vaporization are desirable. A Wescan instrument thermostatted at an appropriate temperature was employed for our measurements.

3. CONCLUSIONS

The molecular weight, M_n, by vapor osmometry, and the intrinsic viscosity, $[\eta]$ have been determined for more than a dozen resins of various compositions. For one series of resins in dmf at 25C

$$[\eta] = 9.7 \times 10^{-6} M_n^{1.23}$$

It should be noted here that K value of 9.7×10^{-6} is two order of magnitude less than most high polymers. Also a normal value of a is in the range of 0.6 to 0.9. The results of some of this research are presented in Table I. All measurements were made in DMF. The temperature for the Vapor Pressure Osmometry was 60C, as suggested by the instruments manufacturer. The intrinsic viscosity was determined at 25C. It is worth noting that in all cases, the experimental molecular weights are less than the formulated values. This could mean that perhaps the reactions did not go to completion as predicted by the reaction mechanism or that indeed there is a distribution of polymeric materials. Vapor pressure osmometry serves as a guide but could not make this distinction or separation. For molecular weights in this range, the intrinsic viscosity - M_n data indicate a stiff, rod-like polymer. In order to validate the accuracy of the instrumentation, two low molecular weight, homogeneous polystyrene standards were examined, PS-577 And PS-578. Note that for M_n's in the same range as the resins, the intrinsic viscosities were much lower. These polystyrenes are short, but are flexible chains. For further comparison, two highly branched polymers are included, Dextran T-70 and a hydroxyethyl starch. In these cases, a match of intrinsic viscosities was attempted for much higher molecular weight polymers. Finally, samples of collagen are presented. This is a naturally occurring biopolymer with a triple helical rod-like structure. All of these data illustrate the novel nature and complex structure of the resins being studied. It is a significant and challenging problem.

REFERENCES

1. G.D. Roberts and R.W. Lauver, J. Appl. Sci., 33(1987) 2893-2931.
2. J.K. Sutter, "Long Term Isothermal Aging and Thermal Analysis of N-CYCAP Polyimides" NASA Technical Memorandum 104341.

Table I

Summary of Results

Sample	Formulated n value	Formulated M.W.	V.P.O. M.W.	[η] dl/gm	End cap	Dimethyl Ester	Diamine
KC-I-16 V-cap-12F	9	7198	3358	0.218	PAS	HFDE	HFDA
KC-I-18A PMR-12F-71	9	7164	1937	0.114	NE	HFDE	HFDA
KC-I17 V-cap-50	9	5290	4055	0.266	PAS	HFDE	DA
KC-I-19 PMR-12F-85	11	8520	2636	0.127	PAS	HFDE	HFDA
KC-I-62 ARF-II-50	9	5290	1840	0.120	NE	HFDE	DA
KC-I-59 V-cap-12F(X_n)	9	7198	3291	0,188	PAS	HFDE	HFDA
KC-I-61 X_n-II-50	9	5044	1989	0.115	NE	HFDE	DA
KC-I-60 PMR-II-50	9	5044	2474	0.150	NE	HFDE	DA
KC-I-58 V-cap-12-71	9	7198	3065	0.176	PAS	HFDE	HFDA
PMR-II-50	9	5044	1829	0.100	NE	HFDE	DA
Cy-cap	6	3950	1285	0.144			
Cy-cap	10	6014	1419	0.186			
PS-577	20	2000	2021	0.036			
PS-578	52	5200	5283	0.066			
P-84	n=1	2336	4122	0.390 0.440	(DMF) (DMAC)		

Sample	M.W.	[η] (dl/gm)
Dextran T-70	63,600 (M.O.)	0.260
Hydroxyethyl Starch	250,000 (L.S.)	0.190
Collagen (young)	300,000 (L.S.)	10.0
Collagen (middle)	125,000 (L.S.)	2.4
Collagen (old)	54,000 (L.S.)	0.40
Collagen (very old)	28,000 (L.S.)	0.12

<u>Resins</u>: $[\eta] = 9.7 \times 10^{-6} M_n^{1.23}$ DMF 25C

<u>Collagen</u>: $[\eta] = 8.8 \times 10^{-8} M_w^{1.46}$ water 25C

Theoretical and Applied Rheology, edited by P. Moldenaers and R. Keunings
Proc. XIth Int. Congr. on Rheology, Brussels, Belgium, August 17-21, 1992

ON THE DETERMINATION OF ELONGATIONAL MATERIAL PROPERTIES IN POLYMER SOLUTIONS FROM SPINNING FLOW

A.E. CHAVEZ[1], O. MANERO[1], H.Y. BAO[2], AND R.K. GUPTA[2]

[1]Instituto de Investigaciones en Materiales, UNAM, A.P. 70-360, México, D.F. 04510 (México)

[2]Department of Chemical Engineering, West Virginia University, Morgantown, WV, 26506, (U.S.A.)

1. INTRODUCTION

Elongational flows are important in many polymer processing operations. At present, determination of elongational material properties represents a difficult task. There are many problems involved in determining exactly the flow situation and state of stress along the spinline, such as the appearance of end effects induced by the flow transition at the extremes of the filament: preshearing history and external forces at the upper and lower ends. The objectives of this investigation is firstly, to determine to what extend is possible, by means of a molecular model, to evaluate all those external factors which make this flow difficult to be controlled experimentally. And secondly, to obtain material elongational properties in extensional flow from spinning flow data and their comparison with purely elongational flow.

2. THE MODEL

The fundamental idea of our model is to develop a set of fiber spinning equations which includes the flow parameters through a momentum balance and a molecular nonlinear model. Combination of these macro- and microscopic approaches allows one to describe the instantaneous dynamic situation at each point of the fluid filament.

The molecular model used is an anisotropic FENE dumbbell model, with conformation-dependent hydrodynamic friction. For uniaxial elongational flow in steady state (ref. 1) the velocity components are

$$v_x = -\dot{\varepsilon}x/2, \quad v_y = -\dot{\varepsilon}y/2, \quad v_z = \dot{\varepsilon}z \qquad (1)$$

where $\dot{\varepsilon}$ is the constant elongation rate. The corresponding governing equations are

$$X\langle x^2\rangle = (2/3N) - \dot{\varepsilon}\langle x^2\rangle N^{1/2} r$$

$$X\langle z^2\rangle = (2/3N) - \dot{\varepsilon}\langle z^2\rangle N^{1/2} r \qquad (2)$$

$$r^2 = 2\langle x^2\rangle + \langle z^2\rangle$$

where $X=X(N,r)$. N is the number of statistical subunits and it is proportional to molecular weight, r is the magnitude of the end-to-end vector $\underline{r}$. The tension in the spring force is assumed to vary according to the Warner expression. The elongational stress and the elongational viscosity are defined as

$$\tau_{zz}-\tau_{xx} = (1-r^2)^{-1}(\langle z^2\rangle-\langle x^2\rangle) \qquad (3)$$

$$\eta_E = (\tau_{zz}-\tau_{xx})/\dot{\varepsilon} \qquad (4)$$

In the momentum balance, effects such as gravity, surface tension and air drag are neglected (ref. 2). Using a difference scheme, this balance is given by (ref. 3)

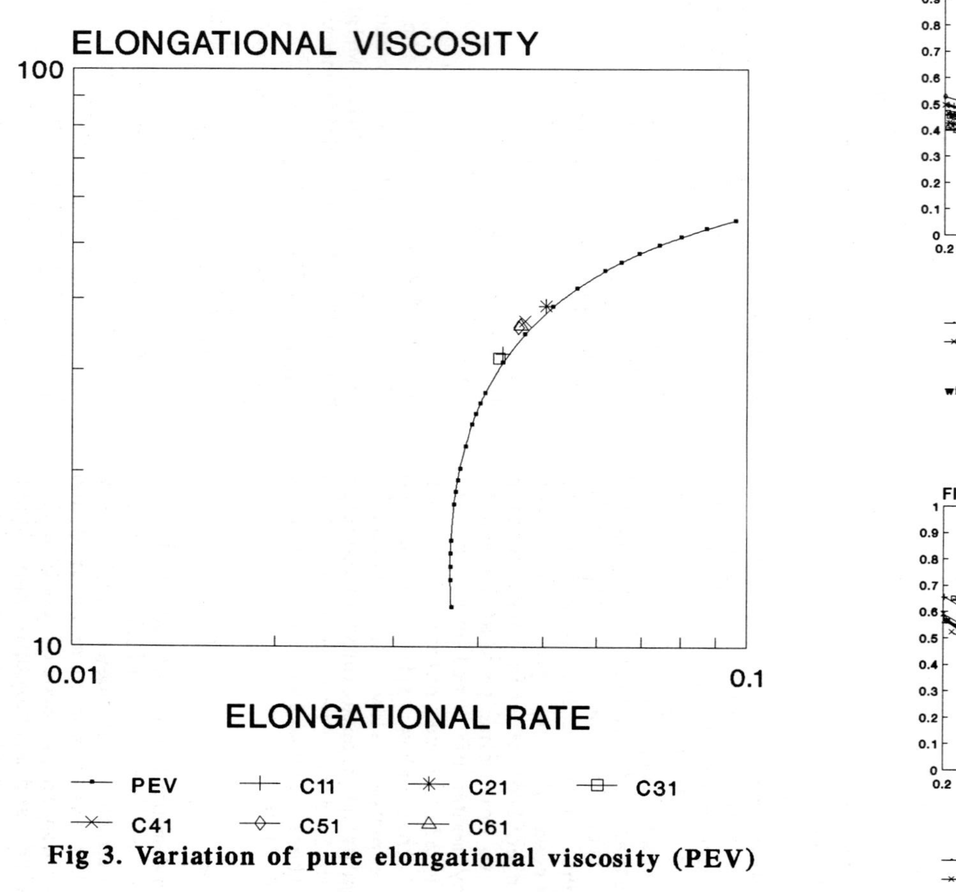

Fig 3. Variation of pure elongational viscosity (PEV) with elongational rate and experimental results

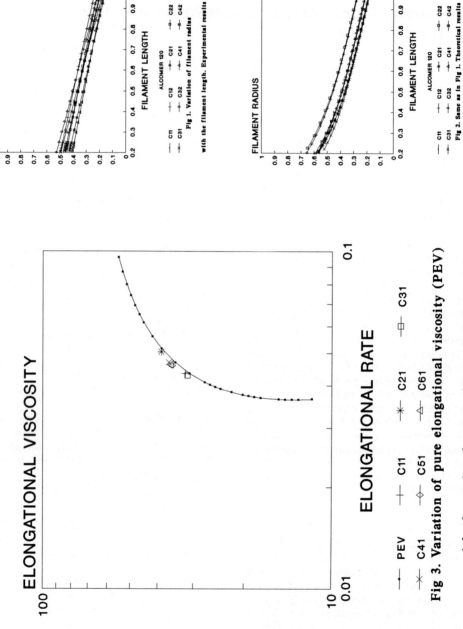

Fig 1. Variation of filament radius with the filament length. Experimental results

Fig 2. Same as in Fig 1. Theoretical results

$$(\tau_{zz}-\tau_{xx})_{n+1} = v_{n+1}\{[(\tau_{zz}-\tau_{xx})_n/v_n] + \rho(v_{n+1}-v_n)\} \quad (5)$$

$$z_{n+1} = z_n + (v_{n+1}-v_n)/(\dot{\varepsilon}/\lambda_H) \quad (6)$$

where $\dot{\varepsilon}$ is the non-dimensional strain-rate and λ_H is the relaxation time of the polymer.

From the initial values of velocity V_0 and stress τ_0 at z=0, a numerical solution of equations (3) and (4) together with (5) and (6), gives rise to velocity and stress profiles computed at each point along the filament. Point calculations continue until boundary conditions at the lower end are satisfied.

3. EXPERIMENTAL SECTION

The Gupta-Sridhar extensional viscometer was used (ref.2,4). The test fluids were aqueous solutions of Alcomer 120 (a copolymer of sodium acrylate and acrylamide) and Polyox (MW: $4x10^6$), with concentrations of 200 and 300 ppm respectively.

Experimental runs at room temperature were carried out to investigate effects arising from variation of the flow rate Q, the spinning force F, the initial stress and filament length L. Additional parameters such as N and λ_H were determined from simple shear flow data.

4. DISCUSSION AND RESULTS

In a previous work (ref. 5), the influence of Q, F and τ_0 was determined through two dimensionless numbers, the Weissenberg number $We=\lambda_H V_0/L$ and the reciprocal force parameter $\zeta=Q\tau_0/FV_0$. Moreover, it was shown that this model allows the separation of the filament length in three well-defined regions by comparing its predictions with the experimental results. The first shows the effects of preshearing as a consequence of the flow in the upper capillary and the latter is influenced by the acceleration produced by the suction force at the lower end. Nevertheless, in the intermediate region the variation of filament radius with the spinning coordinate z^* (expressed in non-dimensional form) is almost linear. Both the experimental and theoretical results match quite well for z^* greater than 0.5(figs.1,2). Further calculations show that there exists a range ($0.2<z^*<0.5$) where the elongational rate $\dot{\varepsilon}$ and elongational stress τ are almost constant. This behavior suggests that a real material property can be calculated provided that changes with z^* are so small that representative values for $\dot{\varepsilon}$ and τ may be obtained. Comparison between the pure elongational viscosity (PEV) calculated from the model (Eqs.2-4) and experimental values of extensional viscosity calculated as stated above confirms the validity of the procedure (fig.3).

5. CONCLUSIONS

Comparison between experimental and theoretical results of filament radius as a function of spinning coordinate shows a small region of fully developed elongational flow where end effects are not predominant. This selection is based on model predictions which take into account flow parameters and molecular information. In this way, the calculation of the spinning viscosity renders values that are similar to those of the steady-state viscosity corresponding to pure elongational flow.

REFERENCES

1. Phan-Thien N., Manero O., and Leal L.G., Rheol. Acta, 23(1984)151-162.
2. Sridhar T., and Gupta R.K.,Rheol. Acta, 24(1985)207-209.
3. Mackay, M.E., and Petrie Ch.J.S., Rheol. Acta, 28(1989)281-293.
4. Gupta R.K., and Sridhar T., Elongational Rheometers, in Rheological Measurement, A.A. Collyer and D.W. Clegg (Eds), Elsevier Applied Science Publishers, London, (1988)211-245.
5. Chavez A.E., Manero O., Bao H.Y., and Gupta R.K., Appl. Mech. Rev. 44(1991)S46-S50.

Theoretical and Applied Rheology, edited by P. Moldenaers and R. Keunings
Proc. XIth Int. Congr. on Rheology, Brussels, Belgium, August 17-21, 1992

THE CONCENTRATION EQUATION OF POLYMER SOLUTION IN NONHOMOGENEOUS FLOW FIELD

Hyoung Jin Choi[1] and Myung S. Jhon[2]

[1]Department of Polymer Science and Engineering, Inha University, Inchon, Korea

[2]Department of Chemical Engineering, Carnegie Mellon University, Pittsburgh, PA15213, U.S.A.

1. INTRODUCTION

We provide a systematic procedure for the calculation of observables characterized by polymer configuration and deal with the general case of nonhomogeneous flow with hydrodynamic interactions including bead-wall and bead-bead interactions. To avoid the complexity that arises from the extra bookkeeping that an analysis of the n-bead chain[1], we restrict ourselves to the dumbbell model. The direct computation of the probability function is all but impossible for this complicated system, however its first few moments are obtainable through the projection operator method.

By adapting the relevant space as the zeroth moment, one can obtain a formally exact equation for the concentration, using a projection operator method (initial state basis). A similar procedure could be adapted for higher moments. Instead of choosing the initial state (equilibrium) as the projection basis, we generalize the method by introducing the time dependent projection operator. This scheme could be particularly useful in studying polymer dynamics near steady-state. The concentration equation developed here has potential application to polymer migration problems because it holds all the dynamical information necessary to relate the flow field to the concentration profile.

2. DILUTE POLYMER SOLUTION THEORY INVOLVING NONHOMOGENEOUS FLOW

The dynamics of the polymer molecules immersed in a viscous media with an imposed flow field can be approximately described by the Kirkwood-Riseman equation [2-4].

The time rate of change of the position vector of the ith bead is written as:

$$\dot{r}_i = v(r_i) + \sum_k^n \zeta_{ik}^{-1} \cdot [- \frac{\partial}{\partial r_k} \ln\Psi - \frac{1}{k_B T} F_K] \quad (1)$$

We shall explain our procedure by considering a simple situation: neglecting external forces and only considering two bead dumbbell molecules. The equation of motion for the dumbbell can be rewritten in terms of the center of mass and relative coordinates:

$$\dot{r}_c = V - \frac{1}{2}(\zeta_{11}^{-1} + \zeta_{12}^{-1}) \cdot \frac{\partial}{\partial r_c} \ln \Psi \ , \quad (2.a)$$

$$\dot{R} = v - 2(\zeta_{11}^{-1} - \zeta_{12}^{-1}) \cdot (\frac{\partial}{\partial R} \ln \Psi + \frac{1}{k_B T} F^{(C)}) \ , \quad (2.b)$$

The probability function, Ψ, is obtained by substituting Eq. (2) into the continuity equation [5], yielding

$$\frac{\partial \Psi}{\partial t}(R, r_c, t) = - \frac{\partial}{\partial r_c} \cdot (V\Psi) + \frac{1}{2} \frac{\partial}{\partial r_c} \cdot (\zeta_{11}^{-1} + \zeta_{12}^{-1}) \cdot \frac{\partial}{\partial r_c} \Psi$$

$$- \frac{\partial}{\partial R} \cdot (v\Psi) + 2 \frac{\partial}{\partial R} \cdot (\zeta_{11}^{-1} - \zeta_{12}^{-1}) \cdot [\frac{\partial}{\partial R} + \frac{1}{k_B T} F^{(C)}] \Psi$$

$$\equiv D\Psi \quad (3)$$

The Eq. (3) contains abundant information on conformational changes in nonhomogeneous flow. Since an exact solution to the above equation does not exist, we must rely on approximate calculations. The experimentally observable quantities are typically the first few moments of $\Psi(R, t)$ [e.g., the zeroth moments (concentration ; $C(r_c)$) and second moments ($\langle RR \rangle (r_c, t)$: related with stress tensor [5] or intrinsic viscosity for the zero shear rate)]. It is desirable to obtain an explicit equation of motion for the physically observable quantities (first few moments). We shall adapt the projection operator method to achieve this goal. In our initial study we shall limit ourselves only to the zeroth moment to obtain an explicit equation for the concentration, generalization of this scheme to higher moments is straightforward.

3. FORMAL EQUATION OF MOTION FOR THE CONCENTRATION

We shall use conventional bra and ket notation to establish the equation of motion for the concentration. The scalar product is defined as

$$\langle A | M | B \rangle = \int d^3R \, A(R) \, \hat{\Psi} (R : t_0) \, B(R) \ , \quad (4)$$

and the time dependent projection operator is defined as $P(t_0) \equiv |0\rangle_{t_0} \langle 0| \equiv 1 - Q(t_0)$. Here, we shall use unsymmetric bra and ket notation [6]. For this procedure the bra is $\langle 0| \equiv \int d^3R$ and the ket is $|0\rangle_{t_0} \equiv \hat{\Psi}(R, t_0)$, where $\hat{\Psi}(R, t_0)$ is a normalized probability function.

By using $P(t_0) + Q(t_0) = 1$, we get two coupled equations for $P(t_0)\Psi$ and $Q(t_0)\Psi$ [7]. Recognizing that $P(t_0)=|0\rangle_{t_0}\langle 0|$ and $\langle 0|\Psi(t)= C(r_c, t)$, we can derive an explicit formal equation for the concentration:

$$\partial_t C(t) = \langle 0|D|0\rangle_{t_0} C(t) + \int_{t_0}^{t} \Sigma_{t_0}(t-t')\, C(t')dt' \qquad (5)$$

Here $\Sigma_{t_0}(t)$ is called a "memory function". It is clear that although Eq. (5) is exact, an explicit formulation of the memory function is all but impossible.

For initial state basis, if we assume the system is initially at equilibrium and choose $|0\rangle_0 = \Psi(t=0)$, and set $\Sigma_0 \equiv 0$, Eq. (5) becomes,

$$\partial_t C(t)=\langle 0|D|0\rangle_0 C(t) \quad . \qquad (6)$$

This initial state projection method is a natural way of studying the transient behavior of the concentration profile. By neglecting the memory function we have, however, excluded some important dynamic processes which occur at large t.

However, for the final state basis, the memory function is difficult to handle at large time using the initial basis formulation. An alternative method of dealing with long time behavior is to choose a different basis such that the memory effect becomes small. As t becomes large the memory function term becomes less significant. In other words, the final state basis equation can be given as:

$$\partial_t C(t) = \langle 0|D|0\rangle_\infty C(t) \quad . \qquad (7)$$

Eqs. (6) and (7) will be used to study polymer molecule migration phenomena in a nonhomogeneous flow field.

4. EQUATION FOR THE LOCAL POLYMER CONCENTRATION AND MIGRATION OF POLYMER IN NONHOMOGENEOUS FLOW FIELD

To illustrate our procedure, we will choose plane Poiseuille flow. Then the probability equation becomes:

$$\frac{\partial \Psi}{\partial t} = -\frac{\partial}{\partial X}(k + 2Lz_c)Z\Psi + \frac{1}{2}\frac{\partial}{\partial r_c}\cdot(\underset{\approx}{\zeta}_{11}^{-1} + \underset{\approx}{\zeta}_{12}^{-1})\cdot\frac{\partial}{\partial r_c}\Psi + 2\frac{\partial}{\partial R}\cdot(\underset{\approx}{\zeta}_{11}^{-1}+\underset{\approx}{\zeta}_{12}^{-1})\cdot\left[\frac{\partial}{\partial R} + \frac{F(C)}{k_BT}\right]\Psi \equiv D\Psi \quad (8)$$

Then the memoryless equation for the initial state basis becomes

$$\partial_t C(t) = \langle 0|D|0\rangle_0 C(t) \quad . \qquad (6)$$

Here $\langle 0|D|0\rangle_0$ is easily computed as

$$\langle 0|D|0\rangle_0 = \frac{k_BT}{2\zeta}\frac{\partial}{\partial r_c}\cdot\frac{\partial}{\partial r_c} + \frac{k_BT}{2}\frac{\partial}{\partial r_c}\cdot\langle 0|\underset{\approx}{T}_{12}(R, r_c)|0\rangle_0\cdot\frac{\partial}{\partial r_c} \qquad (9)$$

From Eqs. (6) and (8), the concentration equation can be expressed as:

$$\partial_t C(t) = \frac{\partial}{\partial r_c}\cdot j(r_c, t) \quad . \qquad (10)$$

Here, j is flux of beads and

$$j(r_c, t) = \left\{\frac{k_BT}{2\zeta}\underset{\approx}{I} + \langle 0|\underset{\approx}{T}_{12}(R, r_c)|\rangle_0\right\}\cdot\frac{\partial}{\partial r_c}C \quad . \qquad (11)$$

Therefore, by neglecting the memory function hydrodynamic interaction alters only the diffusive character and there are no outright "source terms". Put simply, this means that if the initial concentration is uniform the time derivative of concentration will be zero and remains zero forever after. As far as polymer migration is concerned the most salient information must be contained in the memory function Σ_0.

The memory function contains the coupling between the internal and external coordinates so this result is not surprising. Therefore it is important to incorporate the memory effect in the study of the initial state basis formulation to ensure the inclusion of all relevent dynamic information.

On the other hand, the final state basis concentration equation is the quickest way of studying the concentration profile near the steady state. However, it has a deficiency in that it is not easy to estimate transients far from steady states. For plane Poiseuille flow, it is easy to calculate the concentration equation

$$\partial_t C(t) = \langle 0|D|0\rangle\, C(t) \qquad (7)$$

Here, $\langle 0|D|0\rangle_\infty$ is easily computed as

$$\langle 0|D|0\rangle_\infty = \frac{k_BT}{2\zeta}\frac{\partial}{\partial r_c}\cdot\frac{\partial}{\partial r_c} + \frac{k_BT}{2}\frac{\partial}{\partial r_c}\cdot\langle 0|\underset{\approx}{T}_{12}(R, r_c)|0\rangle_\infty\cdot\frac{\partial}{\partial r_c} + \frac{k_BT}{2}\frac{\partial}{\partial r_c}\cdot\left\{\left[\frac{\partial}{\partial r_c}\cdot\langle 0|\underset{\approx}{T}_{12}(R, r_c)|0\rangle\right]_\infty - \langle 0|\frac{\partial}{\partial r_c}\cdot\underset{\approx}{T}_{12}(R, r_c)|0\rangle\right\} \qquad (12)$$

A more conventional formulation is same as Eq. (10) with the following flux of beads,

$$j(r_c, t) = \left\{\frac{k_BT}{2\zeta}\underset{\approx}{I} + \langle 0|\underset{\approx}{T}_{12}(R, r_c)|0\rangle_\infty\right\}\cdot\frac{\partial}{\partial r_c} + \delta H(r_c)C \qquad (13.a)$$

with

$$\delta H(r_c) = \frac{k_BT}{2}\left\{\left[\frac{\partial}{\partial r_c}\cdot\langle 0|\underset{\approx}{T}_{12}(R, r_c)|0\rangle_\infty\right] - \langle 0|\frac{\partial}{\partial r_c}\cdot\underset{\approx}{T}_{12}(R, r_c)|0\rangle_\infty\right\} \qquad (13.b)$$

Here, the hydrodynamic interaction not only alters the diffusive character but contributes a source term δH as well. In this case, even if the initial concentration is uniform there will be some propagation away from the initial state, as will be shown below.

So far we have confined ourselevs to Poiseuille flow: it is easy to derive the governing equations for the general case. The formal equation based on the final state basis may be written:

$$\partial_t C(t) = \langle 0|D|0\rangle_\infty C(t)$$

Here,

$$D = -\frac{\partial}{\partial r_c}\cdot V + \frac{1}{2}\frac{\partial}{\partial r_c}\cdot(\underset{\approx}{\zeta}_{11}^{-1} + \underset{\approx}{\zeta}_{12}^{-1})\cdot\frac{\partial}{\partial r_c}$$

$$- \frac{\partial}{\partial R_c} \cdot v + 2 \frac{\partial}{\partial R} \cdot (\underset{\approx}{\zeta}^{-1}_{11} - \underset{\approx}{\zeta}^{-1}_{12}) \cdot [\frac{\partial}{\partial R} + \frac{F^{(C)}}{k_B T}] \quad (14)$$

$$\langle 0|D|0 \rangle_\infty = - \frac{\partial}{\partial r_c} \cdot \langle 0|V|0\rangle_\infty + \frac{\partial}{2\zeta} + \frac{\partial}{\partial r_c} \cdot \frac{\partial}{\partial r_c}$$

$$+ \frac{k_B T}{2} \frac{\partial}{\partial r_c} \cdot \langle 0|\underset{\approx}{T}_{12}(R, r_c)|0\rangle_\infty \cdot \frac{\partial}{\partial r_c} + \frac{\partial}{\partial r_c} \cdot \delta H(r_c) \quad (15)$$

Therefore, the concentration equation for the general flow situation is written in conventional diffusion equation type form [8] and

$$j(r_c, t) = \underbrace{\{ \frac{k_B T}{2\zeta} \underset{\approx}{I} + \langle 0|\underset{\approx}{T}_{12}(R, r_c)|0\rangle_\infty \} \cdot \frac{\partial}{\partial r_c} C(r_c, t)}_{\text{Diffusive term}}$$

$$+ \underbrace{V(r_c) C(r_c, t)}_{\text{Convective term}}$$

$$\underbrace{\{ \delta H - 2 \langle 0|\exp(r_c \cdot \nabla') \sinh^2(\frac{R \cdot \nabla'}{4}) v(r_c')|_{r'=0}| 0\rangle_\infty \} C(r_c, t)}_{\text{Source term (source of migration)}} \quad (16)$$

The convective term does not change the migration velocity across streamlines, whereas the source term leads to migration across streamlines if it has components which are not parallel to the direction of flow. Even if the second source term, involving the fluid velocity, does not possess perpendicular component, the hydrodynamic interaction term, δH , could provide perpendicular contributions. The exact form of hydrodynamic interactions in a confined tube is extremely complicated. To simplify our discussions we use the well-known results of hydrodynamic interaction for unbounded fluid. However, the final state preaveraged form of the Oseen tensor is still dependent on r_c because of shear induced conformational changes. The calculation of the above quantity requires information on final state (steady state) probability function $\hat{\Psi}$. Calculation of $\hat{\Psi}$ is difficult in general, but it is rather easy to calculate the exact steady-state probability function for the free draining limit of quadratic velocity profile. It can be shown for the plane Poiseuille flow that δH has a net z-directional contribution. That is, there may be the migration of polymer molecule across the streamlines due to hydrodynamic interactions.

It has been found that even after neglecting hydrodynamic interaction the dumbbell molecules will cross streamlines for flows involving curved stremlines. The source term in Eq. (16) can be approximated as:

$$\{ \delta H - 2 \langle 0|\exp(r_c \cdot \nabla') \sinh^2(\frac{R \cdot \nabla'}{4}) v(r_c')|_{r'=0}| 0\rangle_\infty \}$$

$$= \delta H - \frac{1}{8} \langle RR \rangle_\infty : \nabla' \nabla' v(r')|_{r'=0} \quad (17)$$

In the absence of hydrodynamic interaction ($\delta H=0$) we must look at the second term for components perpendicular to the fluid velocity. We will consider a coordinate system that has one unit vector (δ_v) pointing along a particular streamline. The other two unit vectors are perpendicular to the streamline and to each other in the usual way. It is clear from vector calculus that in general $\nabla'\nabla' \delta_v$ will in general have components perpendicular to δ_v if δ_v is defined along a curved path. Therefore $\nabla'\nabla' v(r')\delta_v|_{r'=0}$ will have components in a direction other than the streamline (δ_v) if the streamline is curved. By the same token there will be no migration if the streamline is not curved, at least in the free draining limit.

5. SUMMARY

We have derived a formally exact equation for the concentration of polymer using a time-dependent projection operator formalism and demonstrated the utility of the memoryless equation based on both the initial and final states when applied to polymer migration in nonhomogeneous flow fields. We have given special attention to the role hydrodynamic interactions have in migration and the coupling of the dumbbell internal structure with the flow environment. However, one important aspect we have ignored until now is the effect of a wall.

In applying the general concentration equation to polymer migration, we chose two complementary sets of projection bases (initial and final state basis). The initial-state basis form is suitable for studying transient behavior, however our lack of understanding in regards to the memory function limits our application to short time. The example in section 3 shows that the memoryless equation fails to provide correct asymptotic long-time behavior. Therefore, some way of accounting for memory effects is needed. The calculation of the exact memory function is not trivial; however, there exists several likely approximate methods[9-11] do exist. In this paper we limited our interest to the concentration equation.

REFERENCES

1. M. S. Jhon and K. F. Freed, *J.Polym. Sci. Polym. Phys. Ed.*, 23, 955 (1985)
2. J. G. Kirkwood and J. Riseman, *J. Chem. Phys.*, 16, 565 (1948)
3. R.Zwanzig, *Adv. Chem.Phys.*, 15, 325 (1969)
4. K.F. Freed, in "Progress in Liquid Physics", Croxton, C.A. ed., Wiley, New York, p.343(1978)
5. R. B. Bird, C. F. Curtiss, R. C. Armstrong and O. Hassager, "Dynamics of Polymeric Liquids", Vol. 2, Wiley, N.Y. (1987)
6. M. S. Jhon and J. S. Dahler, *J. Chem. Phys.*, 69, 819 (1978)
7. R. Zwanzig, *J. Chem. Phys.*, 33, 1338 (1960)
8. J. O'M. Bockris and A. K. N. Reddy, "Modern Electrochemistry", Vol. 1, Plenum, New York (1970)
9. M.S. Jhon and D. Forster, *Phys. Rev.*, A12, 254 (1975)
10. M.S. Jhon, S. Fesciyan and J.S. Dahler, *J. Chem.Phys.*, 74, 4106 (1981)
11. J.J. Ou, J.S. Dahler and M.S. Jhon, *J. Chem. Phys.*, 74, 1495 (1981)

Theoretical and Applied Rheology, edited by P. Moldenaers and R. Keunings
Proc. XIth Int. Congr. on Rheology, Brussels, Belgium, August 17-21, 1992

RHEOLOGICAL PROPERTIES OF ROD-LIKE POLYMER SOLUTION IN ISOTROPIC AND NEMATIC PHASES

S. M. EOM and I. J. CHUNG

Chemical Engineering Department, KAIST, 373-1 Kusung-dong, Yusung-gu, Taejeon 305-701, KOREA

1. INTRODUCTION

Onsager presented an equilibrium theory in a nematic phase of a rod-like polymer solution (ref.1). The theory is modified to include the contribution of an elongational flow (ref.2) in the Helmholtz free energy.

Kazuu-Doi theory (ref.3) is extended to calculate rheological properties for the high flow rates with the Onsager's integral equation including the Helmholtz free energy under the flow system.

2. THEORIES

2.1 Modified Onsager's equation

Consider a solution of N rod-like molecules at a dialytic equilibrium with a solvent in a steady elongational flow field. Each molecule has the length of l and the diameter of d. The interaction of mole-cules is assumed to be caused not by the attraction but by the pure steric repulsion. Following the Onsager's theory, the Helmholtz free energy can be written as

$$\frac{F}{NkT} = \frac{\mu_o(T)}{kT} - 1 + \ln c + \sigma(f) + bc\,\rho(f) + \frac{1}{2}Gx^3\,\alpha(f) \tag{1}$$

where $\mu_o(T)$ is the chemical potential of one rod-like molecule in a solvent at the temperature T, $c = (N/V)$ is the number concentration and $b = \pi dl^2/4$. G is the dimensionless stretching rate (ref.2) and x is the axial ratio (l/d) of a rod. The functional s(f) of the orientational distribution function $f(\boldsymbol{u})$ is defined in the following form:

$$\sigma(f) = \int \ln[4\pi f(\boldsymbol{u})]\, f(\boldsymbol{u})\, d\boldsymbol{u} \tag{2}$$

describing the orientational entropy per one molecule.

$\rho(f)$ is defined as

$$\rho(f) = \frac{4}{\pi}\iint |\sin\gamma(\boldsymbol{u},\boldsymbol{u}')|\, f(\boldsymbol{u})\, f(\boldsymbol{u}')\, d\boldsymbol{u}\, d\boldsymbol{u}' \tag{3}$$

describing the excluded volume effect of two rods. The last function a(f) obtained from the flow field is defined as

$$\alpha(f) = \int \sin^2\theta\, f(\boldsymbol{u})\, d\boldsymbol{u} \tag{4}$$

The unit vector $\boldsymbol{u}$ specifies the direction of a given rod-like molecule and the angle θ describes the angle between the rod axis and the the flow axis. $\gamma(\boldsymbol{u},\boldsymbol{u}')$ in the equation (3) is the angle between $\boldsymbol{u}$ and $\boldsymbol{u}'$.

By minimizing the equation (1) with respect to $f(u)$, the integral equation is obtained:

$$\ln[4\pi f(\boldsymbol{u})] = \text{const} - \frac{8q}{\pi}\int |\sin\gamma|\, f(\boldsymbol{u}')\, d\boldsymbol{u}' - \frac{1}{2}Gx^3\sin^2\theta \tag{5}$$

Here $q = (bc)$ is a dimensionless concentration.

The equation (5) is solved numerically by the iterative method in order to obtain the accurate orientational distribution function (ref.4).

2.2 The Kazuu-Doi theory

The constitutive equation proposed by Leslie (ref.5) and Ericksen (ref.6) is written as

$$\sigma = \alpha_1'(\boldsymbol{nn}:\boldsymbol{A})\boldsymbol{nn} + \alpha_2'\boldsymbol{nN} + \alpha_3'\boldsymbol{Nn} + \alpha_4'\boldsymbol{A} + \alpha_5'\boldsymbol{nn}\cdot\boldsymbol{A} + \alpha_6'\boldsymbol{A}\cdot\boldsymbol{nn} \tag{6}$$

where σ is stress tensor, $\boldsymbol{A} = (\kappa + \kappa^+)/2$, $\boldsymbol{N} = d\boldsymbol{n}/dt + \Omega\cdot\boldsymbol{n}$, $\Omega = (\kappa - \kappa^+)/2$ and κ is the velocity gradient tensor. The six coefficients are called the Leslie coefficients. In general, a complete description of hydrodynamic behavior requires the specification of at least five independent coefficients for the nematic liquid crystal if Onsager's reprocity relation is assumed to be applicable.

Kazuu and Doi (ref.3) obtained the Leslie coeffi-cients from Doi's kinetic equation (ref.7) as follows:

$$\alpha_1 = -10C^3\rho^2 R \qquad \alpha_2 = -5C^3\rho\left(1+\frac{1}{\lambda}\right)S$$
$$\alpha_3 = -5C^3\rho^2\left(1-\frac{1}{\lambda}\right)S \qquad \alpha_4 = \frac{2}{7}C^3\rho^2(7-5S-2R)$$
$$\alpha_5 = \frac{10}{7}C^3\rho^3\,(5S+2R) \qquad \alpha_6 = -\frac{20}{7}C^3\rho^2(S-R) \tag{7}$$

where ρ is defined in the equation (3) and $C = q/q^*$, α_i's denote the dimensionless Leslie coefficients α_i'/η^* and η^* is the viscosity at the concentration of unstable isotropic phase ($q^* = 3.96$).

The order parameters are defined as

$$S = \int f(\boldsymbol{u})\, P_2(\boldsymbol{u}\bullet\boldsymbol{n})\, d\boldsymbol{u} \tag{8}$$

$$R = \int f(\boldsymbol{u})\, P_4(\boldsymbol{u}\bullet\boldsymbol{n})\, d\boldsymbol{u} \tag{9}$$

where $P_2(\boldsymbol{u}\cdot\boldsymbol{n})$ and $P_4(\boldsymbol{u}\cdot\boldsymbol{n})$ are the second and fourth order of Legendre's polynomials. The parameter λ is related to the stability of the flow system and obtained by solving the differential equation for λ (ref.3) with the accurate orientational distribution function. The dimensionless Leslie coefficients can be calculated when the distribution function $f(\boldsymbol{u})$ is obtained from the equation (5).

3. APPLICATION OF THEORY TO A STEADY FLOW

In an elongational flow, uniaxial symmetry is preserved and the director is given as $\boldsymbol{n} = (1, 0, 0)$ for the theory. The Leslie-Ericksen equation (6) gives the elongational viscosity in the form

$$\eta_e = \alpha_1 + \frac{3}{2}\alpha_4 + \alpha_5 + \alpha_6 \tag{10}$$

The equation (10) can be applied to the flow with high stretching rates as well as low stretching rates, because the flow effect is considered in the equation (1) by adding an appropriate potential term to the equilibrium free energy.

The concentration dependence of elongational viscosity is shown for four different stretching rates in the figure 1. In the case of a low stretching rate (Gx^3=0), the viscosity is double valued in the stable biphasic region. The peak of the viscosity shifts toward a smaller concentration and its magnitude becomes smaller as the stretching rate increases. It is worth noting that no biphasic region is observed for the flow with high stretching rates. It is also obvious that the nematic ordering is induced by the flow field.

Figure 2 shows the stretching rate dependence of elongational viscosity for the different concentrations. The observed regions are plateau and thinning regions. However, this theory does not predict the thinning region called the region I (ref.8), which is associated with a piled polydomain structure at low stretching rates because the system is considered to be spatially uniform. The plateau region becomes wider and its value becomes smaller as the concentration increases.

4. CONCLUSIONS

This paper presents the possibility to extend the Kazuu-Doi theory with the modified Onsager's equation and obtain the viscosity for the high stretching rates. The theory is also applicable for the steady shear flow.

Acknowledgement The authors are grateful to Korea Science & Engineering Foundation for the support of this work through Grant No. 911-1005-024-1.

REFERENCES

1. L. Onsager, *Ann. Acad. Sci. (N.Y.)*, 51 (1949) 627.
2. G. Marrucci, *Mol. Cryst. Liq. Cryst.*, 72 (1982) 153.
3. N. Kazuu and M. Doi, *J. Phys. Soc. Jpn.*, 53 (1984) 1031.
4. S.D. Lee, *J. Chem. Phys.*, 88 (1988) 5196.
5. F.M. Leslie, *Arch. Ration. Mech. Anal.*, 28 (1968) 265.
6. L.J. Ericksen, *Arch. Ration. Mech. Anal.*, 4 (1960) 231.
7. M. Doi, *J. Polym. Sci., Polym. Phys. Ed.*, 19 (1981) 229.
8. Y. Onogi and T. Asada, in: G.A. Astarita, G. Marrucci, and L. Nicholais (Eds.) Rheology (Vol.1), Plenum, New York, 1980, p.127.

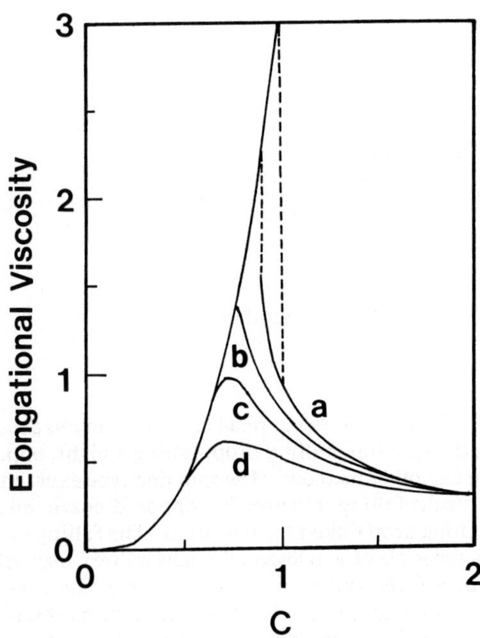

Figure 1. Concentration dependence of the elongational viscosity; the vertical dashed lines indicate the stable biphasic region: (a) $Gx^3 = 0.0$, (b) $Gx^3 = 0.5$, (c) $Gx^3 = 1.0$, and (d) $Gx^3 = 2.0$.

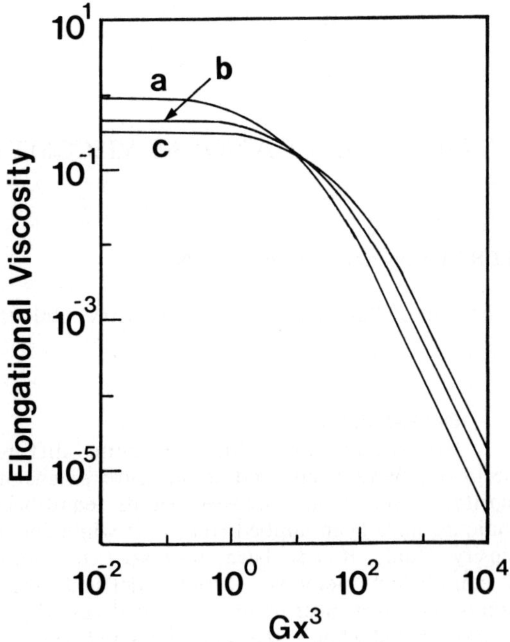

Figure 2. Elongational viscosity versus dimensionless stretching rate: (a) $C = 1.0$, (b) $C = 1.5$, and (c) $C = 2.0$.

Theoretical and Applied Rheology, edited by P. Moldenaers and R. Keunings
Proc. XIth Int. Congr. on Rheology, Brussels, Belgium, August 17-21, 1992

TRANSIENT ELONGATIONAL VISCOSITY MEASUREMENT

J. FERGUSON and N.E. HUDSON

Department of Pure and Applied Chemistry, University of Strathclyde, Glasgow, Scotland.

1. INTRODUCTION

Historically, two types of quite different experiments have been used to measure extensional viscosity. For high viscosity fluids equilibrium techniques have been applied (refs 1-3) while for low viscosity fluids it has been necessary to employ non-equilibrium methods. Strictly defined, theory requires that a true measure of extensional viscosity can only be obtained when extensional stress and strain rate are constant. For low viscosity fluids this condition can only be met in the Eulerian sense. That is, where conditions do not vary with time at, say, a single point on an extending threadline. This is not sufficient as would appear from the scatter of experimental data in the literature. However, much of the data quoted has been from different fluids examined at quite different sets of experimental conditions. To overcome this problem, a Boger fluid designated M1 was produced at Monash University and subjected to detailed shear and extensional flow study in over 25 laboratories worldwide. The results have been published and show no consistency except where the same technique (for example, the spin line method) was used in different laboratories (ref 4). The question being addressed in this paper is why this should be the case, especially since shear flow measurements were remarkably consistant from laboratory to laboratory. A similar pattern was found when the exercise was repeated with a second viscoelastic fluid, A1.

2. RESULTS

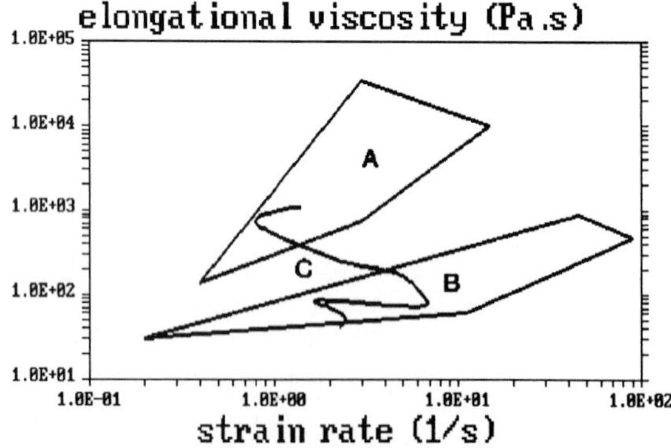

Fig. 1 Elongational viscosity of M1

Figure 1 shows an overall view of the results obtained using the spin line, falling drop, falling weight, opposed nozzle and other methods. The spin line type experiments gave results falling into area "a", opposed nozzle into "b" and falling drop linked the two areas. The falling weight with a timescale of a millisecond produced very high values. In contrast, the falling drop, with a timescale covering the relaxation time of the fluid produced the most startling results with a pronounced maximum in extensional viscosity as the filament was elongated. It is this discrepancy that provides the explanation of the different results.

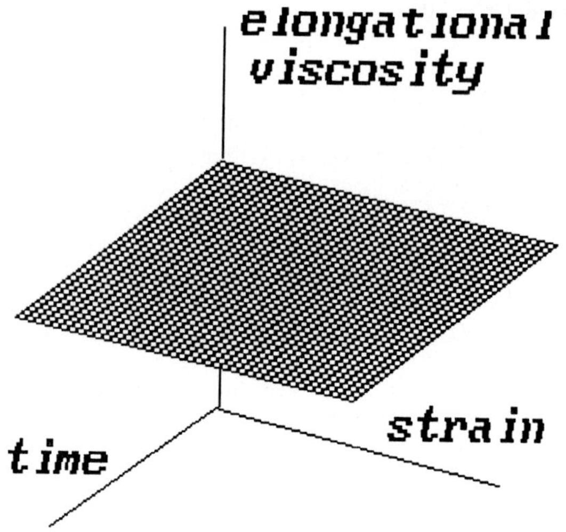

Fig. 2 Newtonian fluid

When "extensional viscosity" is plotted against strain and time over which an element of fluid is subjected to stretching flow it is found that a surface is produced characteristic of the particular fluid. Figure 2 shows this plot for a Newtonian fluid. The flat surface means that regardless of how the measurements are made, when projected on to a two dimensional extensional viscosity, total strain (or for that matter strain rate) plot, the same line of constant extensional viscosity corresponding to the Trouton Ratio of three will be obtained.

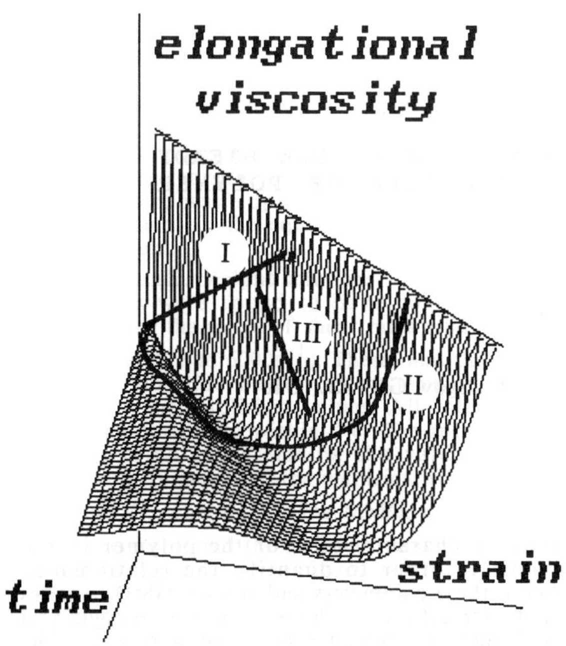

Fig. 3 Non-Newtonian polymer fluid

Consider now experiments on a viscoelastic material. Figure 3 shows a surface typical of many high molecular weight polymer fluids. From the Newtonian values at low strains, extensional viscosity rises to a maximum then decreases to a point at which elastic deformation becomes dominant, whereupon viscosity rises towards some "large" value at a limiting strain.

If we now examine the types of experiment used on M1 (and A1), it is found that they trace out different paths on the surface.

FALLING WEIGHT: High strain is produced over a short timescale (curve I).

SPIN LINE RHEOMETER: There are two modes of operation. In the first the filament diameter is measured down its length (consequently time is increasing) and strain will also increase although strain rate may go through a maximun in a strain hardening fluid (curve II). The second type of experiment involves measuring the stress at an average strain and strain rate. As strain rate is increasing, the time of extension is decreasing (curve III). Typical traces of the points on a surface show that a different area is being traversed in each case.

OPPOSING NOZZLE: Because of the complication of shear flow and the difficulty of precisely estimating extensional strain rate, this is the most difficult of the experiments to visualize. Nevertheless, as strain rate and strain are increased, time of extension will be decreasing (curve IV).

EQUILIBRIUM MEASUREMENTS: Time is extremely long in comparison with the other measurement techniques. The curves of extensional viscosity so produced represent the end boundary of the surface at long times (curve V).When these curves are projected back on to the two dimensional plot of strain vs extensional viscosity an apparently unrelated series of curves are obtained. Provided, however, the spin line experiment is carried out over a wide enough range of strain the same shape of curve as that found for the equilibrium measurement will be, and indeed has been, observed. At high strains a monotonically increasing extensional viscosity will be found.

3. CONCLUSION

It will be apparent from this paper that it is now possible to assess the extensional viscosity of a fluid by using non-equilibrium experiments. This requires the mapping of the extensional viscosity/strain/time surface, a task of some magnitude. Work is now in progress in doing this for a candidate polymer fluid and in determining the appropriate constitutive equation which will allow prediction of extensional viscosity under any given set of transient flow conditions.

Finally, it should be pointed out that this should enable measurement of the extensional viscosity contribution to many of the flows encountered in industry.

REFERENCES

1. H. Munstedt, Rheol. Acta, 14 (1975) 1077.
2. H. Munstedt and H.M. Laun, Rheol. Acta, 20 (1981) 211.
3. J. Meissner, Rheol. Acta, 8 (1969) 78.
4. Twenty seven papers in J. Non-Newtonian Fluid Mechanics, 35(2) (1990).

where X_r is the relative MMD. Hence, the intrinsic MMD can be correlated with the molecular weight, Figure 3. An example for PVP is

$[X] = 3.9 \times 10^{-8} M_W^{1.226}$ dl/g

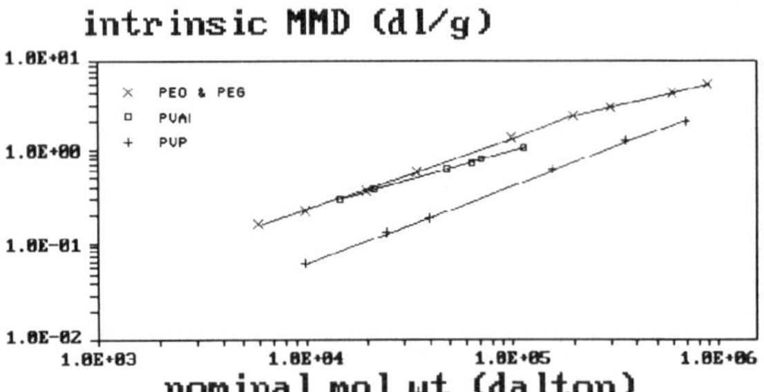

Figure 3 Intrinsic MMD as a function of molecular weight for polymers, PVAl, PVP, PEO and PEG.

Molecular weights were then determined from these formulae and compared to the results obtained from the intrinsic viscosity. It was seen that almost all values corresponded to within 5%. It was also possible to correlate the relative MMD directly with concentration and molecular weight. These correlations were of the form

$X_r - 1 = A c^n$ where $A = D M_W^b$

From these correlations it is possible to predict the MMD from a knowledge of molecular weight and the polymer concentration in solution.

Selected samples were tested to determine their properties in elongation. It is now generally agreed that extension is the dominant form of deformation for the fluid break up. In testing Newtonian fluids, this does not affect any formula since the ratio between elongational and shear viscosities is constant and equal to 3. However, this does not apply for most non-Newtonian fluids, where the ratio varies considerably with strain rate, and is usually very large. In all previous attempts to correlate the physical properties of a fluid with its MMD, the viscosity taken has been the shear viscosity. For the non-polymer solutions used above, good correlation was obtained using an exponent of 0.091 for the shear viscosity. It has been seen that PEG and the lower molecular weight PVAl solutions are Newtonian, and that apart from the initial rapid rise in MMD with concentration, the MMD scales as the 1/11th power of the viscosity. However, for the other polymer solutions it was quite common to produce aerosols whose MMD was, say, 4 times as large as that for the solvent. On the above scaling, this would suggest a viscosity of about 5000 Pa.s, whereas the measured shear viscosity was of the order of 10 mPa.s. The former value is of the same order as the elongational viscosity of the polymer solution at some given strain rate. Hence the elongational viscosity should be chosen as the correlating viscosity parameter. This would not affect the results for non-polymer solutions, since the pre-factor need only be changed by a factor of 0.9 for this scaling. Assuming that higher molecular weight polymers follow this same scaling, it was possible to determine an expected value of elongational viscosity as a function of concentration. However, these are only single values of elongational viscosity at an unknown strain rate. Moreover, prior to achieving their final sizes, the droplets will have been part of larger ones that had been subject to much higher strain rates. At the point of injection, the strain rate is likely to be very high, and we have observed with most materials tested that the elongational viscosity is a rapidly increasing function of strain rate is this region. Nevertheless, it was possible to determine the order of magnitude of final strain rates by moving the aerosol generating equipment nearer to the Malvern sizer. At distances greater than the final break up point, the particle size and distribution should remain fairly constant. The point at which particle size begins to increase can be taken as the break up point. Since the relative velocity is known, final strain rates can be determined from the ratio relative fluid velocity/break up length. Since the relative velocity was usually about 2 m/s, and break up lengths were in the range 30 to 200 mm, strain rates varied between 10 and 70 s^{-1}. Techniques to measure the elongational viscosity of highly mobile polymer fluids have been developed recently. Reproducible results were achieved at concentrations of 0.25 g/dl and above (cf Figure 4).

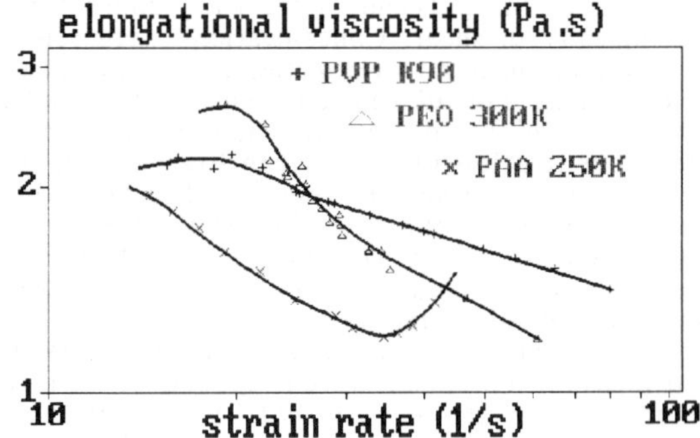

Figure 4 Elongational viscosity as a function of strain rate for polymers at 1.0 g/dl

At the low concentrations, strain rates were above 100 s^{-1}. The range of strain rates of interest was attained using concentrations of 1.0 g/dl and higher.

4. CONCLUSIONS.

From the results cited above, it is now possible to control, for certain polymers, the aerosol particle size distribution, and the mass median diameter as functions of molecular weight and concentration. The scalings involved are related to molecular weight distribution and polymer conformation. The tendency for certain polymers to complex (associate) in the high elongational flow field present during aerosol generation is likely to have important implications in a number of areas.

Theoretical and Applied Rheology, edited by P. Moldenaers and R. Keunings
Proc. XIth Int. Congr. on Rheology, Brussels, Belgium, August 17-21, 1992

RHEOLOGY OF FULLERENE SOLUTIONS

R. KAMATH and W. E. VANARSDALE

Department of Mechanical Engineering, University of Houston, Houston, TX 77204-4792 (USA)

1. INTRODUCTION

Fullerenes are carbon clusters shaped much like a soccer ball. Their existence was postulated and confirmed by Smalley and others in 1985 (ref.1). However, they could only isolate small quantities of this material (nanograms). Krätschmer, Huffman and others (ref.2) later developed a technique for producing larger quantities (grams). This technique involves arcing graphite rods in a helium atmosphere of about 100 torr. Nearly 8% of the resulting carbon soot is in the form of fullerenes (ref.3). Typically, this mixture is more than 75% C_{60} with 20-23% C_{70} and traces of C_{84}, C_{92} and larger clusters (ref.4). An aromatic solvent like toluene is used to extract these molecules from the soot. This solvent is removed through a rotary evaporation process leaving crystalline forms of solid C_{60} and C_{70}. Solvent which remains trapped in the lattice may be removed using heat. Smalley and co-workers (ref.5) have refined this technique to obtain yields in excess of 10%.

We are currently building a semi-automated device to produce fullerene-rich soot based on this technique. A schematic drawing of this apparatus is shown in Fig. 1. Mechanisms in two arms of the 5-way cross are used to feed a pair of graphite rods with length 15.25 cm and diameter 0.635 cm through high current connections. These rods are rotated to insure a consistent rate of material processing at the arc. These feed mechanisms allow up to 25 pairs of rods to be processed over a four hour period without opening the vacuum chamber. A fan is used to drive soot up through a cooled copper chimney into a dust separator for collection. Fullerenes are extracted from this soot using toluene and a rotary evaporator.

Properties of C_{60} suggest applications in several areas including lubrication and high strength materials. This nearly spherical molecule is extremely stable based on surface dissociation studies (ref.6). The crystalline solid has a face-centered cubic structure with a nearest neighbor packing distance of about 1 nm (ref.3). This solid has a density of 1.7 g/cm^3 and a bulk modulus of 18 GPa (ref.3). These values are high relative to other polymeric solids (ref.7). We are interested in determining to what extent fullerenes influence rheological properties of polymeric liquids. Here we report on some steady shear data for solutions in a commercial lubricant base stock.

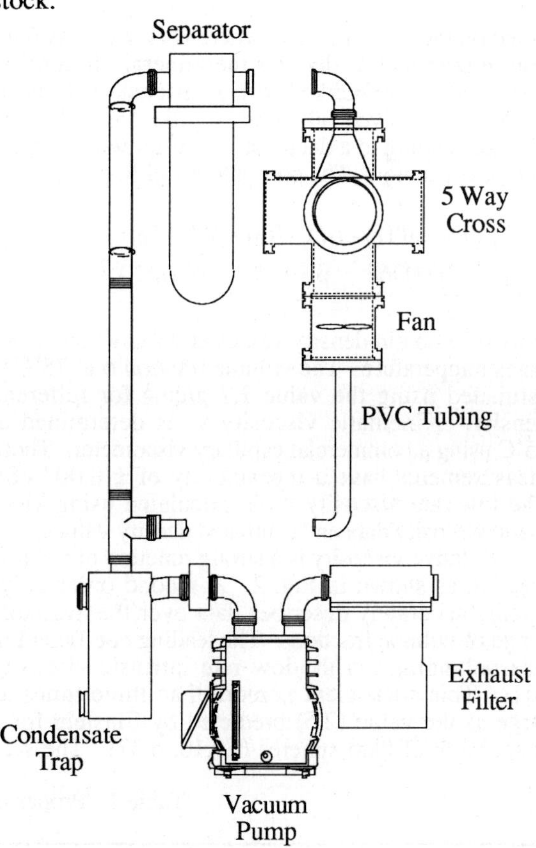

Figure 1: Schematic drawing of a device for producing fullerene-rich soot.

2. FULLERENE SOLUTIONS

This initial study involves fullerenes extracted from commercial soot. The resulting crystalline solid is assumed to contain 75% C_{60} and 25% C_{70} with a mean molecular weight of 750 g/mol. This material readily dissolves in a clear mineral oil at mass fractions up to 700 ppm. Such solutions are magenta in color just like solutions in toluene. Higher concentrations of fullerenes precipitate out of solution at room temperature. This low solubility may be due to the limited amount of aromatic hydrocarbons (less than one percent) typically present in lubricant base stocks (ref.8). Five solutions were prepared by diluting a 700 ppm solution with mineral oil. Properties of these solutions and the mineral oil are listed in Table 1.

Density was measured at 30°C for the base stock and two fullerene solutions using ASTM procedure D 4052. These measurements have a repeatability of ±0.0001 g/cm3. For the other solutions, density at 30°C is determined using the correlation

$$\frac{\rho}{\rho_s} = 1.0 - (2.4)10^{-7}\Phi - (1.4)10^{-9}\Phi^2, \qquad (1)$$

based on the available data where Φ is the mass fraction in ppm and ρ_s denotes the mineral oil density. These values are marked with an asterisk in Table 1. Note that dissolved fullerenes act like "holes" in the solution leading to a decrease in density with increasing mass fraction. The correlation (ref.9, p.145)

$$\rho(T) = \rho(T_0) - \alpha(T - T_0) - \beta(T - T_0)^2, \quad \alpha = 0.00066 \; ; \; \beta = (15.4 - 19T_0)10^{-7}, \qquad (2)$$

is used to obtain density values at 75°C where T denotes temperature. The volume fraction ϕ at 75°C is estimated using the value 1.7 g/cm3 for fullerene density. Kinematic viscosity ν_0 is determined at 75°C using a commercial capillary viscometer. These measurements have a repeatability of ± 0.003 cSt. The low rate viscosity η_0 is calculated using kinematic viscosity data and estimated density values.

Relative viscosity is a strong function of volume fraction as shown in Fig. 2. A second order polynomial accurately describes data over the available range of volume fractions. The leading coefficient in this polynomial is the low rate intrinsic viscosity $[\eta]_0$. This coefficient is more than three times as large as the value (2.5) predicted by Einstein for a suspension of hard spheres (ref.10, p.37). The second coefficient is orders of magnitude larger than typical values (5-8) determined in practice (ref.11, p.220). This strong dependence on volume fraction suggests that fullerenes do not behave like hard spheres in solution.

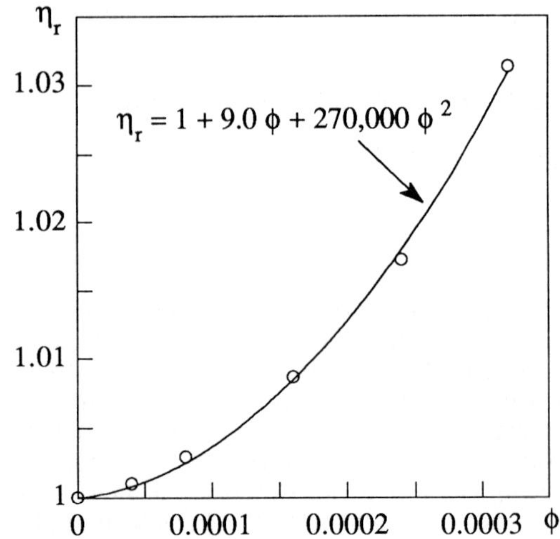

Figure 2: Relative viscosity $\eta_r \equiv \eta_0/\eta_s$ as a function of volume fraction ϕ for five fullerene solutions.

Empirical correlations (ref.12) suggest a different expression for relative viscosity

$$\eta_r = \left(1 - A\phi - B\phi^2\right)^{-2}, \qquad (3)$$

where the coefficient B is typically zero and A is $1/\phi_m$ in terms of a maximum volume fraction. However, B must be positive for (3) to accurately describe data in Fig. 2. This correlation gives a slightly better fit than the second order polynomial and predicts a maximum volume fraction of 0.27%. This estimate is significantly smaller than the value 74% associated with a face-centered cubic lattice. While this result may be unrealistic, it does suggest that the effective diameter of fullerenes may be larger than their molecular diameter.

Table 1: Properties of Fullerene Solutions

Φ (ppm)	ρ(30°C) (g/cm3)	ρ(75°C) (g/cm3)	ϕ(75°C) (ppm)	ν_0(75°C) (cSt)	η_0(75°C) (cP)
0	0.8076	0.7779	0	7.455	5.80
87.5	0.80758*	0.77787	40	7.467	5.81
175	0.80754*	0.77783	80	7.503	5.84
350	0.8074	0.7777	160	7.561	5.88
525	0.80719*	0.77748	240	7.658	5.95
700	0.8069	0.7772	320	7.755	6.03

A possible explanation for the unusually strong dependence on volume fraction is that fullerenes are stabilized by aromatic hydrocarbons in the mineral oil. These hydrocarbons absorb onto the fullerene surface causing disintegration of the initial crystalline structure. Isolated clusters are dispersed throughout the oil resulting in a color change from clear to magenta. Absorption of aromatic hydrocarbons also increases the effective diameter d_e of fullerenes in solution. One estimate of this change (ref. 10, p.477)

$$\frac{d_e}{d} \approx \left(\frac{[\eta]_0}{2.5}\right)^{1/3}, \tag{4}$$

gives an increase of 53%. While this estimate does not imply a very low value of ϕ_m, these results are consistent with trends for stabilized PMMA spheres given in (ref.10, p.478).

Fullerene solutions also exhibit shear thinning at high rates as shown in Fig. 3. A TBS viscometer was used to obtain this data using ASTM procedure D 4683. The solution viscosity decreases from its low rate value to the mineral oil viscosity for shear rates above 500,000 s^{-1}. This onset of shear thinning occurs at a Peclet number (ref.10, p.464) which is much smaller than one. The Cross viscosity model (ref.13) adequately describes data in Fig. 3, but the exponent is much larger than typical values ($1<m<2$) for polymeric liquids. This abrupt shear thinning may indicate a temporary shedding of aromatic hydrocarbons from the fullerene surface.

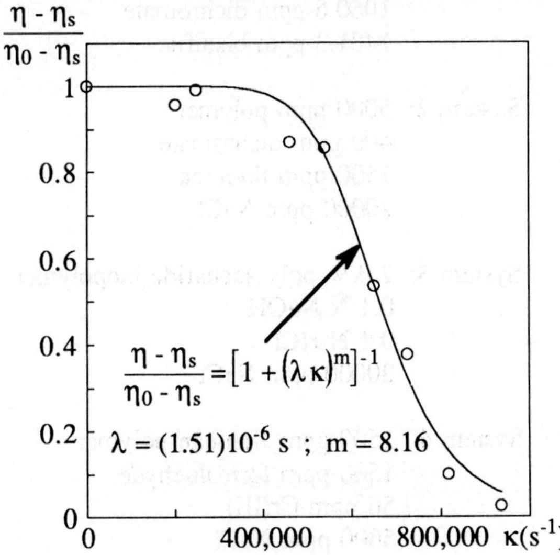

Figure 3: Reduced viscosity as a function of shear rate κ for the 700 ppm fullerene solution.

ACKNOWLEDGEMENTS

Support for this work was provided by the Texas Center for Superconductivity at the University of Houston. The authors would like to thank Lila Anderson (Texas Fullerenes Corporation) for her help in obtaining the C_{60} rich soot and Mark Bouldin (Shell Development Company) for providing the mineral oil. Jagadish Sorab (Ford Scientific Research Laboratory) provided the density values given in Table 1. The authors also benefited from discussions with Jim Cox (Chemistry) and Raj Rajagopalan (Chemical Engineering) at the University of Houston.

REFERENCES

1. H.W. Kroto, J.R. Heath, S.C. O'Brien, R.F. Curl and R.E. Smalley, "C_{60}: Buckminsterfullerene", Nature, 318 (1985) 162-163.
2. W. Krätschmer, L.D. Lamb, K. Fostiropoulos and D.R. Huffman, "Solid C_{60}: a new form of carbon", Nature, 347 (1990) 354-358.
3. D.R. Huffman, "Solid C_{60}", Physics Today, 44(11) (1991) 22-29.
4. R.F. Curl and R.E. Smalley, "Fullerenes", Sci. Amer., 265(4) (1991) 54-63.
5. R.E. Haufler, Y. Chai, L.P.F. Chibante, J.J. Coneicao, C. Jin, L.-S. Wang, S. Maruyama and R.E. Smalley, "Carbon Arc Generation of C_{60}" Mat. Res. Soc. Symp., Boston, 1990.
6. S.W. McElvany, M.M. Ross and J.H. Callahan, "First and Second Ionization Potentials, Reactions and Surface Collisions of C_{60} and Related Clusters", Mat. Res. Soc., 206, (1990).
7. M.F. Ashby and D.R.H. Jones, Engineering Materials, Pergamon Press, New York, 1980.
8. J.G. Wills, Lubrication Fundamentals, Marcel Dekker, New York, 1980.
9. E.W. Washburn (Ed.) International Critical Tables of Numerical Data, Physics, Chemistry and Technology, McGraw-Hill, New York, 1927.
10. W.B. Russel, D.A. Saville and W.R. Schowalter, Colloidal Dispersions, Cambridge University Press, Cambridge, 1989.
11. D.J. Shaw, Introduction to Colloid and Surface Chemistry, Butterworths, Boston, 1986.
12. A.B. Metzner, "Rheology of Suspensions in Polymeric Liquids", J. Rheology, 29(6) (1985) 739-775.
13. M.M. Cross, "Kinetic Interpretation of Non-Newtonian Flow", J. of Col. Inter Sci., 33(1) (1970) 30-35.

Theoretical and Applied Rheology, edited by P. Moldenaers and R. Keunings
Proc. XIth Int. Congr. on Rheology, Brussels, Belgium, August 17-21, 1992

RHEOLOGICAL BEHAVIOR OF THE GEL SYSTEMS USED IN ENHANCED OIL RECOVERY

R. KHARRAT[1] and S. VOSSOUGHI[2]

[1]University of Petroleum Industry, Department of Petroleum Engineering, Abdullah Coute, Ahwaz, (Iran)
[2]University of Kansas, Department of Chemical and Petroleum Engineering, 4006 Learned Hall, Lawrence, Kansas, 66045-2223 (U.S.A.)

1. INTRODUCTION

Gelled polymers are being used increasingly to redirect or modify reservoir fluid movement in the vicinity of the injection or production wells for the purpose of improving oil recovery. Gel treatments are also aimed at blocking fractures and/or high permeability zones. Briefly, the process consists of injecting a gelling fluid into the high permeability zones. This causes the injected fluid to divert into the lower permeability zones, thereby improving sweep efficiency. The gel can be formed on the surface and sheared so as to enable it to be pumped into the reservoir, or the gelling mixture with a pre-defined gelation time is injected into the reservoir to achieve *in situ* gelation. The rate of gelation is controlled by generating the crosslinking agent *in situ* through a redox reaction.

The rheological characterization of the gelling system constitutes an important part of the overall design of reservoir treatment. Most of the studies mainly involve the measurement of the shear viscosity. Other aspects of the rheological characterization have been mostly ignored. The purpose of this work is to provide a more complete rheological characterization of the commonly used gelling systems for the purpose of enhanced oil recovery.

2. EQUIPMENT

Weissenberg rheogoniometer model R19 was used in this work. The unit was equipped with supplemental equipments such as a constant temperature circulator, proportional temperature controller, and a personal computer based data acquisition and data treatment system. Experiments were performed under different shearing modes such as steady state, oscillatory, and oscillatory superimposed on steady state shear.

3. GELLING SYSTEMS

Four different gelling systems were studied. The composition of each gelling system is given below:

System 1: 2500 ppm polymer
1060.6 ppm dichromate
1481.3 ppm bisulfite

System 2: 5000 ppm polymer
400 ppm dichromate
1500 ppm thiourea
20000 ppm NaCl

System 3: 2.0 % polysaccharide biopolymer
0.1 N NaOH
0.1 N HCl
20000 ppm NaCl

System 4: 1500 ppm Xanthan polymer
1500 ppm formaldehyde
50 ppm Cr(III)
5000 ppm NaCl

Gelling systems 1 and 2 are based on a polyacrylamide-type polymer using bisulfite and thiourea as reducing agent respectively. Gelling systems 3 and 4 are based on two different biopolymers.

The gelling process of system 1 was performed under a constant shear. In this case, the components were mixed in a beaker then transferred to the cone-and-plate assembly of the rheogoniometer. A constant shear was applied while the variation of shear stress was monitored. A sudden rise in shear stress was an indication of gelation time. In the case of oscillatory shear environment (system 4), storage modulus with time was monitored. Formation of gel was detected by sudden rise in storage modulus. In the batch method (systems 2), the components were mixed in a beaker and left in a constant temperature bath. Samples were drawn for viscosity measurements. Sudden viscosity buildup was an indication of gel formation. System 3 gelled instantly by reducing its pH below a gelation pH of approximately 9.

4. RESULTS and DISCUSSION

Results presented here is an excerpt from an earlier work (Ref. 1). Detailed presentation of the experimental data is omitted for the sake of page limitation.

4.1 Gel System 1

The gelation process was carried out at a shear rate of 5.92 s^{-1}. The apparent viscosity remained constant during the initial period, followed by a sharp increase and then gradually levelled off. Post gelation study revealed a large overshoot phenomenon. The stress growth at the initiation of 5.92 s^{-1} shear rate was four times the steady state shear stress. A plot of steady state shear stress versus shear rate in logarithmic scale produced a straight line with a slope of 0.75 compared to the slope of 0.5 for the polymer solution of the same concentration. This indicates that the shear thinning behavior of the polymer solution is dampened significantly upon gelation. This phenomenon has also been reported in literature for other types of gel systems (Ref. 2). The post gelation shear stress/shear rate data also revealed a yield stress in the order of 4 dyne/cm^2. That is a Bingham plastic attribute is evolved as a result of gelation process.

The dynamic properties of the gel were also investigated using the oscillatory mode of the rheogoniometer. The storage modulus and the dynamic viscosity were determined as functions of frequency. The storage modulus was slightly frequency dependent, whereas the dynamic viscosity was found to be a strong function of frequency. The same observation was also made by Kote (Ref. 3) who studied polyacrylamide-chromium (III) gels under small amplitude oscillatory shear. The frequency range studied by Kote (Ref. 3) was smaller than that in this work.

4.2 Gel System 2

In this system the polymer/dichromate/thiourea solutions were mixed in a beaker and placed in a constant temperature bath at 25°C. The viscosity build up of the mixture was monitored with time. A gelation time of around 125 hours was determined. A post gelation study revealed dynamic properties similar to those of system 1. However no meaningful steady state data could be obtained because of the slippage between the sample and the plate surface.

4.3 Gel System 3

A newly discovered biopolymer was used in this system. The polymer is an exopolysaccharide produced by Cellulomonas flavigena strain KU. Production of the polymer was patented under U.S. patent number 4,908,310 (Ref. 4) and its application in subterranean permeability modification under U.S. patent number 4,941,533 (Ref. 5). The rheological characterization of the polymer was investigated by Vossoughi and Buller (Ref. 6) in their work on the permeability modification by *in situ* gelation. The polymer is soluble in alkaline solution and gels upon its titration by an acid. The gelation process is reversible and the gelation pH is around 9. The dynamic properties measurement of this gel revealed a frequency dependent storage modulus and zero dynamic viscosity. In fact, zero phase shift was observed for all the frequencies tested. This is attributed to the purely elastic nature of this gel system. The gel, tested under steady state shear environment, was found to be shear sensitive and its viscosity decreased continuously with time reaching an equilibrium value. The equilibrium shear stress/shear rate data produced a straight line on a logarithmic scale. This generated a power-law index of 0.47 which is an indication of strong shear thinning behavior of the gel.

4.4 Gel System 4

In this system a Xanthan polymer, formaldehyde, and dichromate solutions were mixed in a beaker. Small samples were removed from the mixture and tested under oscillatory mode at a frequency of 0.628 Hz. The storage modulus, G', rheogram versus time was then generated. The storage modulus provides information about gel structure forming. The rate increase in storage modulus is believed to be proportional to the rate of formation of elastically effective crosslinks (Ref. 7). In the first couple of hours little evidence of gel formation was observed. Subsequently, it increased sharply and continued increasing at a lower rate for a period of more than 50 hours of observation. The post gelation dynamic properties measurements revealed a storage modulus independent of frequency, a zero phase shift, and a zero dynamic viscosity. The frequency range studied was 0.2-5.0 Hz. The breakdown of the gel at a constant shearing environment prevented rheological measurement under steady state mode. In fact, the gel formation can not be completed if a continuous shear is applied to the sample (Ref. 8).

5. CONCLUSIONS

The rheological data presented here for the four gel systems studied tend to suggest that each gel system is unique in their rheological behavior and generalization of their rheological characterization does not seem to be warranted. The first two gel systems which were produced from the same polyacrylamide polymer behaved similarly. They both showed significantly lower shear thinning behavior compared to their polymer solutions of the same concentration. Their dynamic properties measurements reflected low frequency-dependent G' and strong frequency-dependent non-zero dynamic viscosity. The gel system 3 showed much stronger shear thinning behavior than its polymer solution of the same concentration. In fact, the freshly produced polymer solution behaves like Newtonian with no shear thinning behavior at all. Its frequency-independent G' and zero dynamic viscosity are also different than those observed for the first two systems. The gel system 4 is similar to the gel system 3 in terms of the frequency-independent G' and zero dynamic viscosity. However, its G' value being in the order of 300 was significantly lower than that of the gel system 3 being in the order of 4000.

6. REFERENCES

1. R. Kharrat, Ph. D. Dissertation, University of Kansas (1989).
2. S. Aslam, S. Vossoughi and G. P. Willhite, J. Chem. Eng. Comm., 48 (1986) 287-301.
3. K. Kote, M. S. Thesis, University of Kansas (1985).
4. C. S. Buller, U. S. Patent 4,908,310 (1990).
5. C. S. Buller and S. Vossoughi, U. S. Patent 4,941,533 (1990).
6. S. Vossoughi and C. S. Buller, SPE Reservoir Engineering J. (November 1991) 485-489.
7. W. W. Graessley, Advances in Polymer Sci. J., 16 (1974) 1.
8. M. D. Dolan, M. S. Thesis, University of Kansas (1989).

Theoretical and Applied Rheology, edited by P. Moldenaers and R. Keunings
Proc. XIth Int. Congr. on Rheology, Brussels, Belgium, August 17-21, 1992

FLOW PROPERTIES AND ELECTRICAL NOISE GENERATED DURING CAPILLARY FLOW OF NEWTONIAN LIQUIDS

C. KLASON[1], J. KUBÁT[2] AND O. QUADRAT[3]
[1]Polymer Centre, University of Lund, Box 118, 22100 Lund (Sweden)
[2]Department of Polymeric Materials, Chalmers University of Technology, 41296 Gothenburg (Sweden)
[3]Institute of Macromolecular Chemistry, 1888 Petriny, Prague 6 (Czechoslovakia)

1. INTRODUCTION

As reported earlier, easily measurable electrical noise signals are generated during capillary flow of polymer solutions (refs. 1-6) and dispersions (ref. 7). Despite intensive efforts, the origin of this effect remain unexplained. It was believed that the main role in this context was played by the elasticity of the polymer liquid, but to perform a more detailed characterization of such viscoelastic material from capillary measurements due to flow anomalies was impossible.

Our recent investigation (ref. 8) of aqueous high molecular weight poly(ethylene oxide), PEO, solutions demonstrated that at a critical pressure across the capillary, where the noise begins to attain measurable levels, an increase in the apparent viscosity occurs. We believed that this phenomenon is caused by the transition from laminar to turbulent flow. However, later experiments revealed that a strong signal is obtained also with non-elastic Newtonian liquids showing no increase in apparent viscosity. To elucidate this behaviour, in this study we have concentrated on a more detailed examination of the dependence of the noise on the length of capillary and viscosity of the liquid as two basic factors influencing the noise effect. For this purpose, a series of water-glycerol mixtures with different viscosity was used.

2.EXPERIMENTAL

The measuring equipment (ref. 6) consisted of two glass reservoirs connected to each other by a glass capillary. Three capillaries (0.3 mm in diameter; 2 mm, 8 mm and 22 mm long) with polished end faces were used. The flow of the liquid through the capillary was brought about by applying a defined constant pressure of nitrogen (maximum 6.5 x 10^4 Pa) in one of the reservoirs. The electrical signal induced by the flowing liquid was detected by platinum electrodes connected to a low noise preamplifier (Princeton Applied Research Co., model 113) followed by a frequency analyzer (PAR model 4513). The volumetric flow rate Q of the liquid through the capillary was measured using a special burette. All measurements were performed at 25°C.

Redistilled water (viscosity 0.89 mPas) and its mixtures with 15, 45 and 50 % glycerol (viscosity 1.31, 3.88 and 5.21 mPas, respectively) were used as model liquids. To reach the optimum resistance , 5 Mohm, between the electrodes, which secured a good matching with the input impedance of the preamlpfier, 60 ppm of KCl was added to each sample.

The frequency analysis of the noise signal by means of the FFT technique in the range 1 - 200 Hz provided spectra having the $1/f^{\alpha}$ character with a plateau at the low frequency end (Fig. 1), similar to those obtained previously with polymer solutions. The noise at rest corresponded to the thermal voltage fluctuation. The total noise power $U=U_f^2/U_r^2$ was expressed as the ratio of the mean square noise voltage during flow, U_f^2, and at rest, U_r^2, over the above frequency range.

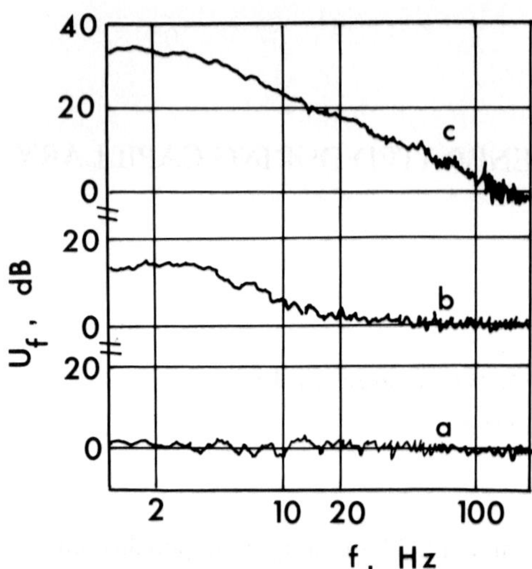

Fig. 1. Electrical flow noise spectra for water at different flow rates (applied pressure). Applied pressure [Px10^{-4} Pa]: a - at rest; b - 2.59; c - 6.51. Noise level at rest normalized to 0 dB.

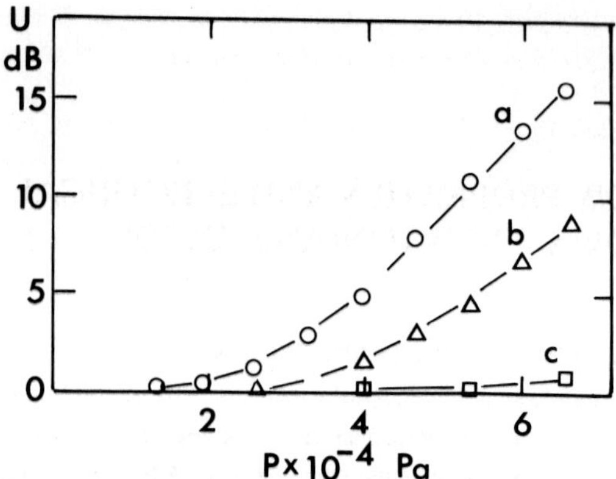

Fig. 2. Dependence of the total noise power ratio U (in dB) on the applied pressure P [x10^{-4} Pa] for water. Noise ratio level at rest: 0 dB.
Capillary length: a - 2mm; b - 8 mm; c - 22mm.

3. RESULTS AND DISCUSSION

The total noise power, integrated over the frequency range covered by the measurement, increases with the pressure difference across the capillary. In general, a certain minimum pressure is to be exceeded before a noise signal can be detected. With water we observed that the noise power decreased with increasing length of the capillary (Fig. 2). However, if the noise power U is plotted against the volumetric flow rate Q, one common U(Q) dependence is obtained, irrespective of the length of the capillary (Fig.3). This indicates that the flow rate is one of the main factors determining the noise intensity.

For water-glycerol mixtures, the appearance of the U(Q) plots depends on their viscosity. The lowest critical flow rate Q_c producing a measurable noise signal, and the steepest increase of U with Q is observed for pure water. At higher viscosities (glycerol contents) an increase in Q_c is recorded, while U rises with Q at a slower rate. At about 50 % glycerol the viscosity is so high that a noise signal exceeding the thermal background noise signal does not appear even at the highest pressure used.

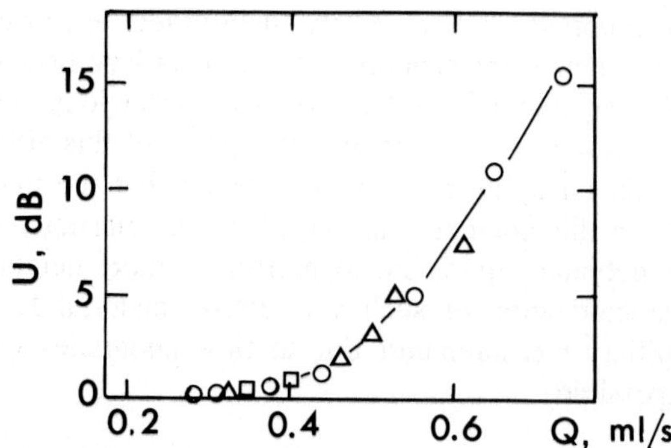

Fig. 3. Dependence of the total noise power ratio U on the volumetric flow rate Q [ml/s]. Notations as in Fig. 2.

In a recent study (ref. 8) we demonstrated that the noise observed with flowing PEO solutions is independent of the value of the streaming potential measured simultaneously. The noise signal thus does not arise in the electrical double layer at the capillary wall, a finding pointing towards a bulk effect. However, the results of the present study show that at constant flow rate the noise intensity observed with water does not depend on the capillary length. This appears to rule out possible association of the noise level with processes inside the capillary, at least for low molecular weight fluids, indicating instead that the effects observed are the result of the motion of the liquid in the entry/exit regions. Such a view is supported by the above results showing a noise reduction at higher viscosities, where also a reduction of possible irregularities of the entry/exit flow may be expected.

An example of a different behaviour with regard to the mechanism underlaying the flow noise generation is the behaviour of PEO solutions in mixtures of water and isopropyl alcohol. Such solutions, having a varying solvent power, influences the dimensions of the polymer coils in systems consisting of entangled macromolecules. Measurements on such samples showed clear positive correlation between the flow noise level and the increase in elasticity of the system caused by a higher density of entanglements due to overlapping in better solvents. In such entangled structures, it is likely to expect that the flow turns turbulent inside the capillary at relatively low flow rates. This then would represent another significant source of the noise phenomenon.

The experiments carried out hitherto thus show that flow generated excess noise, despite similarities in its 1/f like frequency distribution, may be associated with different mechanisms. Such an assumption is certainly not contradicted by the considerable amount of data on 1/f phenomena in other physical systems.

ACKNOWLEDGEMENT

The authors wish to thank the Swedish Research Council for Engineering Sciences for financial support.

REFERENCES

1. K. Hedman, C. Klason and J. Kubát, J. Appl. Phys. 52 (1979)8102.
2. K. Hedman, C. Klason and D. Popětová, Rheol. Acta, 22 (1983)440.
3. K. Hedman and J. Kubát, Rheol. Acta, 23 (1984) 417.
4. C. Klason and J. Kubát, J. Appl. Phys., 58 (1985) 2060.
5. K. Hedman and J. Kubát, Rheol. Acta, 25 (1986) 36.
6. C. Klason, J. Kubát and O. Quadrat, Polymer, in press.
7. C. Klason, J. Kubát and O. Quadrat, Rheol. Acta, 30 (1991) 180.
8. C. Klason, J. Kubát and O. Quadrat, submitted for publication.

Theoretical and Applied Rheology, edited by P. Moldenaers and R. Keunings
Proc. XIth Int. Congr. on Rheology, Brussels, Belgium, August 17-21, 1992

DIVERGENCE AND CROSSOVER IN CONCENTRATION DEPENDENT COMPLIANCE OF POLYMER NETWORK SYSTEMS

T. MASUDA[1], T. TAKIGAWA[1], M. TAKAHASHI[2] and K. URAYAMA[1]

[1]Research Center for Biomedical Engineering, Kyoto University, Shogoin, Sakyo-ku, Kyoto 606-01 Japan
[2]Department of Polymer Chemistry, Kyoto University, Yoshida, Sakyo-ku, Kyoto 606-01 Japan

1. INTRODUCTION

Structure and mechanical properties of physical gels have been investigated by many researchers (ref.1). The concentration (c) dependence of the modulus (E) is an interesting subject. Recently, the modulus near the gelation concentration was measured for various gels (ref.2), and the experimental results were analyzed on the basis of the percolation theory (refs.3,4). In order to understand the elastic behavior and the structure of physical gels, investigations on properties of the sol phase near gelation point might be necessary.

In this study, we will examine the critical behavior of compliance (J) in the region of c slightly below and above the gelation threshold (c_G) using a simple model of physical gelation, and will estimate the profile of J over a wide range of c. In addition, we will try to describe the c dependence of the steady-state compliance (J_e^o) in entanglement polymer systems based on the model.

2. A MODEL OF PHYSICAL GELATION

Physical gels are usually obtained by quenching a concentrated polymer solution at a certain temperature (T_Q) below the critical temperature (T_G) at which the gelation occurs. An ideal sol-gel transition curve for polymeric systems employed in the present paper is shown in Fig.1, which was drawn on the basis of an experimental study (ref.5). We also assume that the gelation of a polymer solution occurs due to the transfer of entanglements to physical crosslinks when the solution is quenched at $T_Q \leq T_G$. The gel might be composed of small crosslink domains and long flexible chains. When the quenching is deep ($T_Q << T_G$) enough, the most of entanglement points could be fixed as the crosslink points.

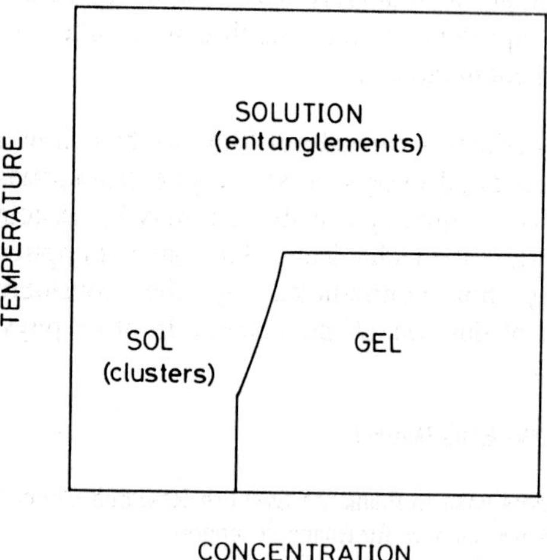

Fig.1. A sol-gel transition curve for physical gelation

3. SCALING

We adopt the compliance (J) to describe elastic properties of sols and gels, as defined by

$$J = \lim_{\omega \to 0} J'(\omega) \qquad (1)$$

Here, $J'(\omega)$ is dynamic storage compliance at an angular frequency ω. We also define the relative distance in concentration(c) from the gelation threshold (c_G).

$$\varepsilon = \frac{|c - c_G|}{c_G} \qquad (2)$$

At concentrations slightly higher than c_G, a cluster of infinite size is formed in the system. The infinite cluster in the gel is a fractal of dimension D. The elastic properties of the system near the gelation point are described by a certain fractal dimension of the multifractal (ref.6) defined for the infinite cluster. The back-bone cluster of dimension $D' \leq D$, a cluster excluding the dangling chains, may dominate the elastic properties of the system. However, we deal here with the simplest case of $D'=D$. The percolated fractal can be considered as an assembly of fractal spheres (ref.4), which cover the entire fractal. The dimension of the spheres is D, and the correlation length is ξ. J of the system can be expressed by

$$J \sim \frac{1}{N} \quad (3)$$

with the number (N) of fractal spheres covering the percolated cluster. N and J scale as (ref. 7)

$$N \sim \xi^{-D}, \qquad J \sim \xi^{D} \quad (4)$$

Near the gelation point, ξ can be written as $\xi \sim \varepsilon^{-\nu}$ with a critical exponent ν, and J scales as

$$J \sim \varepsilon^{-D\nu} \sim \varepsilon^{-t} \quad (5)$$

D and ν are respectively estimated to be 2.5 and 0.89 in 3-dimensional space by the percolation theory (ref.4), and then t=2.2.

As the concentration increases further, D approachs to the space dimension (d=3) at another critical concentration (c_c'). Above c_c', J can be expressed by

$$J \sim \xi^{d} \quad (6)$$

The network structure becomes uniform at the crossover concentration (c_c'), which is identical to the critical concentration for a uniform entanglement structure in polymer solutions. In this case, c-dependence of J for gels at $c > c_c'$ is identical to that of the steady-state compliance (J_e^o) for solutions before quenching. ξ scales as $\xi \sim c^{3/4}$ (ref.3) in good solvents, and

$$J \sim c^{-9/4} \quad (7)$$

On the other hand, the most experimental results for polymer solutions of $c > c_c'$ show that $J_e^o \sim c^{(2\sim3)}$ (ref.8).

At concentrations slightly below c_G, gelation does not occur but clusters are formed. In this region, J can be written as J_e^o of dilute solutions, with number (n) of clusters in unit volume and the molecular weight (M).

$$J \sim \frac{1}{n} \sim \left(\frac{M}{c}\right) \quad (8)$$

Compliance can also be expressed by a power law relation, $J \sim \varepsilon^{-t'}$, in this region of c. Although t' should be exactly distinguished from t in eq. (5), we here assume $t'=t=D\nu$. J in this c-region diverges as $J \sim \varepsilon^{-D\nu}$, and $M \sim \varepsilon^{-D\nu}$ which suggest that M corresponds to the z-average moleculan weight of clusters.

At much lower concentrations (outside the critical region), we can expect J nc^{-1}, because the effect of cluster formation on J is small.

The profile of J as a function of c is shown in Fig.2. J shows a λ-type transition at $c=c_G$. At $c>c_G$, J is governed by the network character, and J at $c<c_G$ by the dilute solution character. A crossover behavior in c-dependence of J is seen around $c_c' > c_G$.

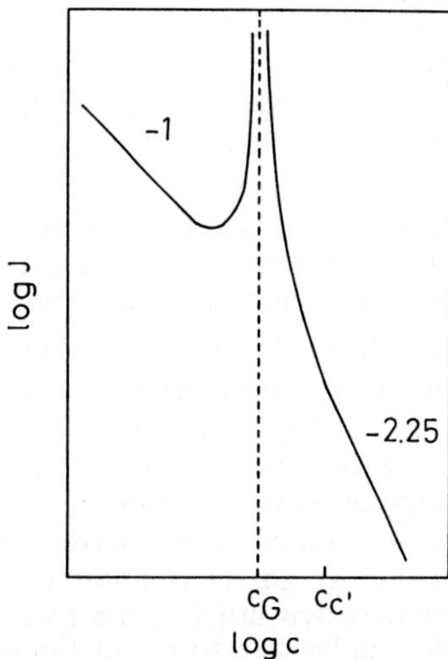

Fig.2. Predicted concentration dependence of the compliance for physical gels

4. SOLUTION SYSTEMS

The discussion on the divergence of J has been limited to c-dependence of J for systems at T_Q but J_e^o for solutions is defined by the same equation, eq.(1). It is not surprising if J_e^o for solutions at $T>T_G$ (in the solution regime in Fig.1) also shows a divergence behavior around c_G. At c slightly higher than c_G, a transient infinite cluster is formed in solutions. J_e^o for the solutions in this concentration range will be expressed by eq. (3) if N is regarded as the number of blobs in the transient cluster. In solutions at c slightly lower than c_G, transient clusters will be formed by the entanglement couplings. The clusters may be the same as those formed by crosslinks in structure, size, and size distribution. Then, J_e^o in this region of c can also be expressed by eq. (8) by regarding n as the number of clusters and M as that for the transient clusters. Based on the above consideration, it is suggested that solutions show a divergence of J_e^o at c_G, although c_G can not be defined for solutions. The threshold c_G for solutions corresponds to the concentration at which a infinite transient entanglement network is formed.

5. DISCUSSSION

Fig.3 shows plots of J against c for PVA gels. The data were re-calculated from those in ref.9. This figure also contains plots of J vs. ε. The plots of J vs. c clearly show the crossover of J at about $c=75kg/m^3$, which corresponds to $\varepsilon \simeq 5$. At $c>75kg/m^3$ J can be expressed by $J \sim c^{-2.4}$, which is also close to the prediction in eq. (7). Data at lower c deviate from the line, suggesting that they show an anomalous characteristic expected in the critical region. The concentration ($c=75kg/m^3$) seems to corresponds to c_c'. The crossover should be associated with the structural change of gels from a fractal to 3d network. As can be seen from the J vs. ε plots, J can be expressed as $J \sim \varepsilon^{-2.4}$ for all data points. The plots show that the critical behavior of J, which is estimated from the data in the critical region, is characterized by t=2.4. The value is close to the predicted value , $t=D\nu=2.2$.

In order to examine the divergence in J_e^o vs. c of entangled polymer systems, we checked again the previous data. Fig.4 shows plots of the reduced steady-state compliance (RTJ_e^o/M_w) as a function of c for solutions of polystyrenes having narrow molecular weight distribution. Data used here are those in review article (ref.8) written by Graessley. As can be seen from this figure, data reported by Einaga et al. (ref.10) from the creep recovery expriments (marked by ȯ) show a peak at $c \simeq 200kg/m^3$, although data reported by

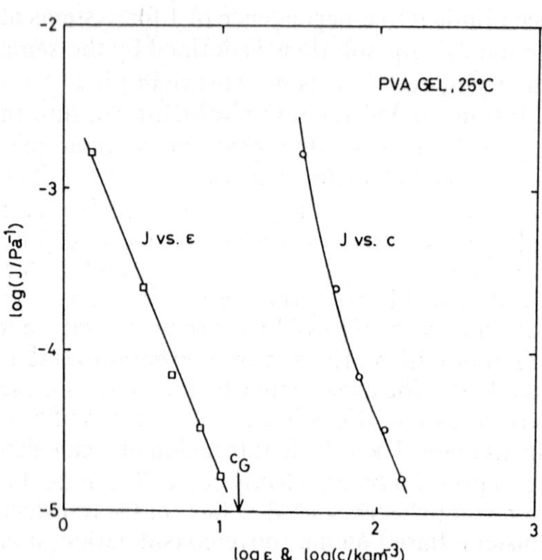

Fig. 3. Plots of J vs. c and J vs. ε for poly(vinyl alcohol) (PVA) gels

the other authers do not show a peak. The creep recovery measurement is the most direct method to obtain J_e^o and the data shown in Fig.4 would strongly suggest that the peak of the compliance actually exists. The peak appears reflecting the divergence character of the compliance. The reason why the divergence of J_e^o was not clearly observed may be due to the weak structure of the entanglement network.

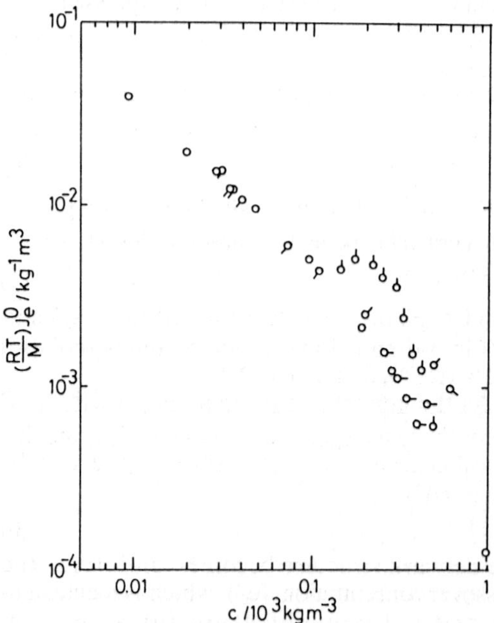

Fig.4. Concentration dependence of steady-state compliance for polystyrenes

REFERENCES

1. A. H. Clark and S. B. Ross-Murphy, Adv. Polym. Sci., 83 (1987) 57.
2. M. Tokita et al., J. Chem. Phys., 83 (1985) 2583; Phys. Rev., A35 (1987) 4329.
3. P.G. de Gennes, Scaling Concept in Polymer Physics, Cornell Univ. Press, Ithaca, 1979.
4. D. Stauffer, Introduction to Percolation Theory, Taylor and Francis, London, 1985.
5. K. Kaji, Annual Review of Res. Group on Polymer Gels. SPSJ, Tokyo, 1989, and personal communication.
6. T. Vicsek, Fractal Growth Phenomena, World Scientific, Singapore, 1989.
7. W. Hess, T. A. Vilgis and H. H. Winter, Macromolecules, 21 (1988) 2536.
8. W. W. Graessley, Adv. Polym. Sci., 16 (1974) 3.
9. T. Takigawa, H. Kashihara, K. Urayama and T. Masuda, J. Phys, Soc. Jpn., 59 (1990) 2598.
10. Y. Einaga, K. Osaki, M. Kurata and M. Tamura, Macromolecules, 4 (1971) 87.

Theoretical and Applied Rheology, edited by P. Moldenaers and R. Keunings
Proc. XIth Int. Congr. on Rheology, Brussels, Belgium, August 17-21, 1992

FLOW - INDUCED STRUCTURE FORMATION IN SOLUTIONS

A.J. McHUGH[1] AND A.J. KISHBAUGH[2]

[1]Department of Chemical Engineering, University of Illinois, 1209 West California, Urbana, Illinois, 61801 (USA)
[2]W.R. Grace & Co.-Conn., Washington Research Center, Columbia, Maryland 21044 (USA)

1. INTRODUCTION

Flow-induced structure formation in polymer solutions has become a topic of considerable interest in recent years. The work discussed in this paper has evolved form studies in our laboratories of flow-induced crystallization (FIC) from solution which document the liquid phase separation which occurs under flow in both crystallizable and amorphous systems (1-3). In addition our studies of shear thickening (4) clearly suggest the likelihood of a one-to-one relationship between the chain entanglement conditions leading to liquid phase separation and the existence of a clearly defined pattern in the rheological behavior of these same solutions. In this paper we wish to present the results of an in-in-situ rheo-optical analysis we have carried out of the shear-thickening behavior of polystyrene solutions in the semi-dilute region under very high shear rates.

2. METHODS

2.1 Materials

Solutions of four polystyrene fractions in either decalin or bromobenzene were used. Concnetration ranges were in the dilute and sem-dilute region. Data on their molecular weight and polydispersity along with the concentration ranges and solvents are given elsewhere (6). Bromobenzene was used as a near matching solvent.

2.2 Rheo-optical measurements

Dichroism, birefringence and the associated orientation angles were monitored using the polarization modulation technique developed by Frattini and Fuller (5). Rheological measurements were made using a couette flow cell with the outer cylinder rotating. This device is capable of laminar-flow, shear rates from 500 to 10,000 s^{-1}. Torque is measured at the stationary inner cylinder and the optical property of interest is simultaneously monitored by passing a laser beam axially down the gap. Complete details on the equipment design, procedures,and data analysis are given elsewhere (6).

3. RESULTS

Under all conditions in which shear-thickening occurs the corresponding dichroism exhibits a maximum at approximately the same shear rate as the viscosity minimum and undergoes a sign change at high shear rates. If the solution only shear thins, the dichroism saturates at high shear rates and shows no maximum or sign change. On the other hand, birefringence and the associated orientation angle of these same solutions do not show a sensitivity to the shear-thickening. Transient dichroism passes through all the steady state values during ramping up or down of the shear rate, achieving a psuedo-steady state at each point. This behavior demonstrates that the kinetics of the structure formation responsible for shear-thickening (and the corresponding dichroism) are essentially instantaneous and reversible.

4. ANALYSIS

4.1 Dichroism and shear-thickening

The behavior of the dichroism can be quantified in terms of a modification of the anomalous diffraction approximation (ADA) discussed by Meeten (7). We assume that the dichroism is due to the presence of oriented spheroidal particles composed of an isotropic, nonabsorbing material. Spheroid orientation is governed by the equations for dilute solutions of non-interacting rigid particles in simple shearing flow. Fits of the optical data to the modified theory indicate that dichroism and shear thickening can be attributed to the production and growth of micron-size, optically isotropic spheroidal particles during flow. On the other hand, the dichroism signal for the non shear-thickening solutions showed a saturation behavior with shear rate and a lower signal magnitude which could not be predicted by the ADA nor could the magnitude and experimentally observed non-colinearity of the birefringence signal be accommodated. In the case of the non shear-thickening solutions it is clear that the dichroism is due to structures larger than single chains but smaller than the wavelength of light. This behavior can be modeled by assuming the scattering structures are spheroidal in nature but in the Rayleigh size range.

The resulting expressions for the dichroism and birefringence are similar to those of the ADA. Both have the same shear rate dependence and the anisotropy in both is assumed to be due solely to shape effects. A fundamental difference is the phase shift and resultant sign change which occurs in the ADA as a result of the much larger particle size. In consequence of the smaller particle size, the Rayleigh theory allows the saturation behavior of the shear rate dependence in the dichroism signal to be modeled in terms of a much higher concentration of submicron-size particles which remain relatively constant in size. These results taken together indicate that the size of the structure which forms under shear determines both the viscosity and optical behavior. If the structures remain submicron in size, simple shear thinning and a saturation behavior in the dichroism occur. If the particles continue to grow to super-micron sizes then the solution shear-thickens and the dichroism shows a maximum followed by an eventual sign change. However, in neither case can the magnitude and sign of the birefringence be accommodated. This is consistent with our data which demonstrate that the birefringence is dominated by the majority of the molecules which remain dissolved in the solution.

To model the birefringence behavior, calculations were carried out using the flexible macromolecule model modified to account for macroform and microform effects. Chain extension under the high shear rates used was evaluated in terms of the Peterlin non-linear Langevin modification (8). The resulting zero-free-parameter calculations show that the model is capable of predicting the magnitude, sign, and shear rate dependence of the birefringence data. Likewise, the dichroism predicted from these calculations is both orders of magnitude lower than that observed experimentally and of the wrong optical sign. This offers convincing proof for the internal consistency of our phenomenological interpretation of the optical and rheological behavior of these solutions.

Further evidence of phase separation and potential model explanations for the phenomena discussed comes from our studies of flow-induced crystallization. These results taken together strongly suggest that molecular chain entanglements generated during coil deformation under flow can lead to the eventual development of network structures which are quasi-liquid-like in their properties. Such structures can either reversibly redissolve into solution or phase separate, depending on the particulars of the flow conditions. As such we believe a direct continuum from network formation to particulate structure formation and eventual liquid phase separation exists. In such a fashion all of the previous rheological and flow-induced phase separation can be put onto a common phenomenological basis. The development of first principles theories to describe the rheological, optical and thermodynamic behavior of these systems, including the strong molecular weight fractionation associated with the phase separation (9), remains as a major challenge.

ACKNOWLEDGEMENTS

The authors wish to acknowledge support for this work from the National Science Foundation through the University of Illinois Materials Research Laboratory (DMR 89-20538) and the Petroleum Research Foundation of the American Chemical Society.

REFERENCES

1. J. Rietveld and A.J. McHugh, J. Polym. Sci. Polym. Phys. Ed., 23 (1985) 2339-2385.
2. A.J. McHugh and R.H. Blunk, Macromolecules, 19 (1985) 1249-1255.
3. A.J. McHugh and J.A. Spevacek, J. Polym. Sci.: Part B: Polym. Phys., 29 (1991) 969-979.
4. E.P. Vrahopoulou and A.J. McHugh, J. Non-Newtonian Fluid Mech., 25 (1987) 157-175.
5. P.L. Frattini and G.G. Fuller, J. Colloid Interface Sci., 100 (1984) 506-518.
6. A.J. Kishbaugh, Ph.D. Thesis, University of Illinois (1992).
7. G.H. Meeten, J. Colloid Interface Sci., 87 (1982) 407-415.
8. A. Peterlin, Pure Appl. Chem., 12 (1966) 563-586.
9. A.J. McHugh, E.P. Vrahopoulou and B.J. Edwards, J. Polym. Sci.: Part B: Polym. Phys., 25 (1987) 953-956.

Theoretical and Applied Rheology, edited by P. Moldenaers and R. Keunings
Proc. XIth Int. Congr. on Rheology, Brussels, Belgium, August 17-21, 1992

NANORHEOLOGY OF CONFINED POLYMER SOLUTIONS.

E. PELLETIER[1], J.P. MONTFORT[1], J.L. LOUBET[2], S. MILLOT[2], A. TONCK[2], J.M. GEORGES[2]

[1] Université de Pau, L.P.M.I, 64000 Pau, France.
[2] Ecole Centrale de Lyon , L.T.S, B.P 163, 69131 Ecully Cedex, France.

1 . INTRODUCTION

The behavior of a polymeric liquid interacting with a surface differs from that of the bulk. Knowledge of this phenomenon is fundamental in all the fields dealing with interfaces such as adhesion processes, polymer blends and composite materials. The comprehension of chain conformations close to surfaces has made hudge progess both theorically and experimentally but in spite of this, studies about dynamics of adsorbed layers have been so far limited.

In this paper, the viscoelastic properties of a polybutadiene solution confined between a sphere and a plane are investigated. Oscillatory measurements are carried out with a new surface force apparatus adapted to operate as a rheometer. The results showing a liquid-solid like transition are presented and discussed.

2 . EXPERIMENTAL SECTION

Figure 1 shows a diagram of the experimental unit described in more details elsewere (ref.1). Experiments are conducted in air. A drop of the solution under study is confined between plane and sphere. The sphere of radius R is moved by a piezoelectric crystal.The plate is supported by a ajustable double-cantilever spring of stiffness K. A capacitive displacement transducer C_1 measures the relative displacement H of the sphere and the plane while C_2 measures the absolute motion X of the plane. The thickness D is determined within a resolution of 1 Å by using the capacitance C of the conductible surfaces constituting a capacitor.

When carring out dynamic experiments by imposing a sinusoïdal tension of frequency ω to the piezoelectric crystal, the motion of the plane is governed by the equation :

$$m\ddot{X} + C_2\dot{X} + KX + C_1\dot{H} + F_t(D,\dot{D}) = 0 \quad (1)$$

where m is the mass of the plate and its stand; C_1 and C_2 are the viscous coefficients associated with the capacity of same name; Ft represents the total force between the sphere and the plane including wetting force Fw, surface forces Fs and hydrodynamic forces F_h. The static forces are offset by the average spring force $K\overline{X}$, then we obtain the relation:

$$\left(\frac{F_h}{D}\right)^* = \left[\frac{X^*}{H^*}\left(m\omega^2 - i\omega C_2 - K\right) - i\omega C_1\right]\frac{H^*}{D^*} \quad (2)$$

where D*, X*, H* are the dynamic parts of the related distances.

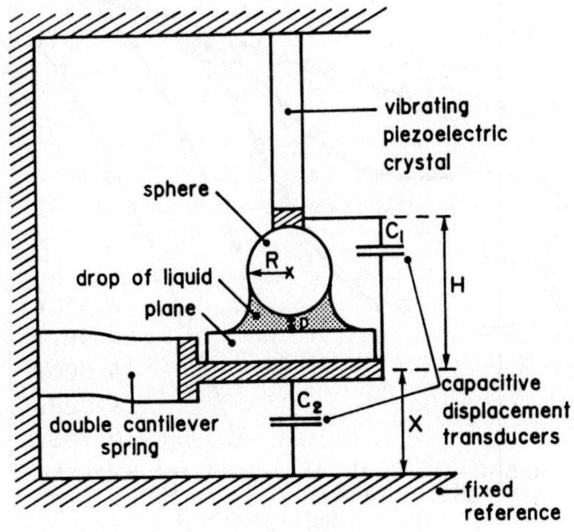

Fig. 1 : Sketch of the experimental equipment.

Experimentally X*/H* corresponds to the ratio of the force to the displacement and is provided by a gain-phase analyzer. The second transfer function H*/D* stands for the elastic deformation of the solids including the deflection of the stand and the elastohydrodynamic flattening of the solids.
On the other hand, the viscoelastic properties of the fluid are related to the dynamic transfer function $(F_h/D)^*$, within the framework of continuum mechanics by (ref.2):

$$\left[\frac{F_h}{D}\right]^* = 6\pi \frac{R^2}{\overline{D}} G^*(\omega) \qquad (3)$$

with $\overline{D}$ being the average thickness of the interface and G* the complex elastic modulus.

Then, from the relations 2 and 3, we derive the experimental values of $G^*(\omega)$.

The liquid sample is a solution of linear polybutadiene (weight-average molecular weight Mw=330 000, polydispersity index p= 1.04) diluted in a near-theta solvent (commercial hydrocarbon oil Flexon 391 from Exxon Chemical Co.) with a volume concentration of 7‰. The radius of gyration Rg of the chains is approximatively 215 Å.

A drop of solution was introduced between the sphere and the plane. The two surfaces were then kept away at a distance of about 50 Rg and let to incube half a day in order to reach an equilibrium adsorption (ref.3).

RESULTS AND DISCUSSION.

Experiments have been conducted at fixed distances $\overline{D}$ ranging from about one micrometer to the vicinity of Rg.

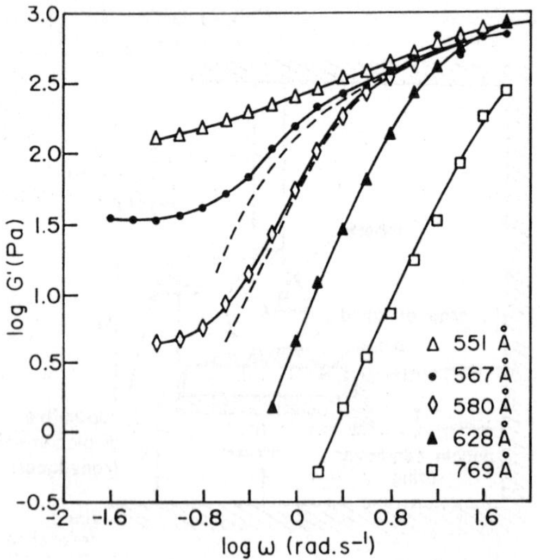

Fig. 2 : Storage modulus G' of the solution at various distances. The dashed lines are for G'-Gc.

At large distances, we expect the viscoelastic properties of the solution not to be perturbed by the adsorbed layers whereas at very short distances the interconection between those two layers should be emphasized. We have investigated the terminal and plateau regions of the relaxation domain of the solution in the range of frequencies 0.025 to 1000 S^{-1}. The amplitude of the oscillations has heen monitored in the range 1 - 15 Å so as to deal with linear viscoelasticity.

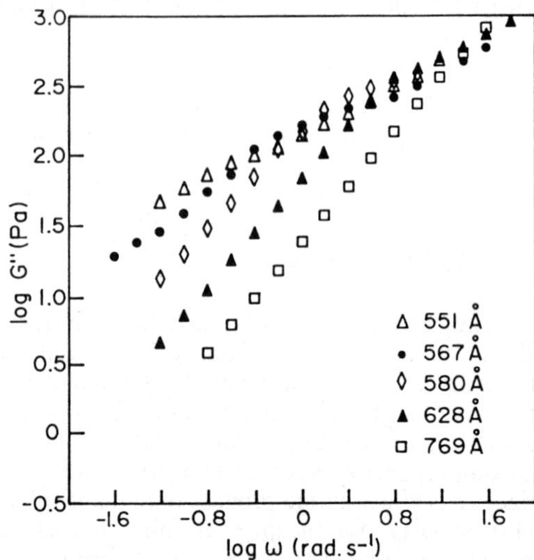

Fig. 3 : Loss modulus G" of the solution at various distances.

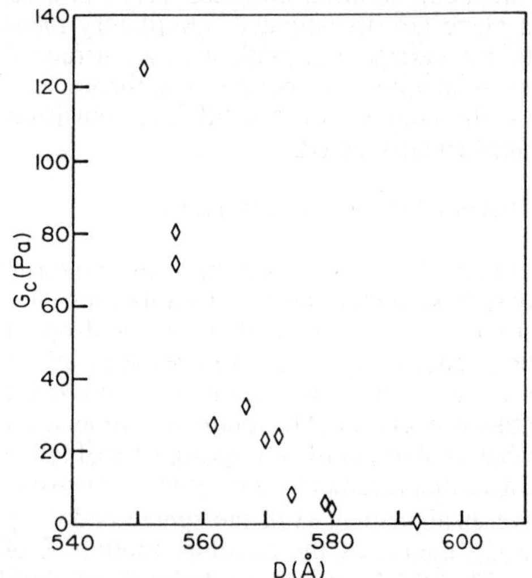

Fig. 4 : Variations of the compression modulus Gc as a function of distance $\overline{D}$.

From the loss modulus curves (Fig.3), we see that the viscosity increases as the gap decreases and at short distances no zero-shear viscosity can be determined. Furthermore, when bringing the two surfaces closer, a plateau emerges from the storage modulus plots (Fig.2). The modulus Gc of that low frequency plateau increases at narrower gaps (Fig.4). Then, a characteristic distance Dc is shown up and can be compared to the radius of gyration. In this experiment Dc ($\cong$ 3 Rg) is comparable to the hydrodynamic thickness attributed to adsorbed layers (ref.4). Then, we can picture the interface in that way : - for D > Dc, two adsorbed layers of average thickness Dc/2 separated by a confined solution of free chains ;

- for D < Dc, two compressed layers with a thickness varying between Dc/2 and D/2 according to their interpenetrating degree.

For D > Dc, the adsorbed layers are not interconnected. The overall behavior of the interface is qualitatively comparable to that of the solution with an increase of viscosity due to the hindered motions of the adsorbed molecules and of the chains trapped in the vicinity of the layers.

The most striking result is for D < Dc. In that case, the shear behavior of the compressed layers is becoming overwhelmed by the compression component in the terminal zone. Solvent molecules and no adsorbed chains are free to flow and have been driven away from the interface at larger distances. The adsorbed polybutadiene layers are compressed , interpenetrate each other and eventually some chains are extending themselves from one surface to another (bridging effects) leading to a restoring force responsible for the low frequency plateau modulus. That description of the dynamics of confined solutions of polymers is an experimental confirmation of very recent theoretical calculations by Fredrickson and Pincus (ref.5).

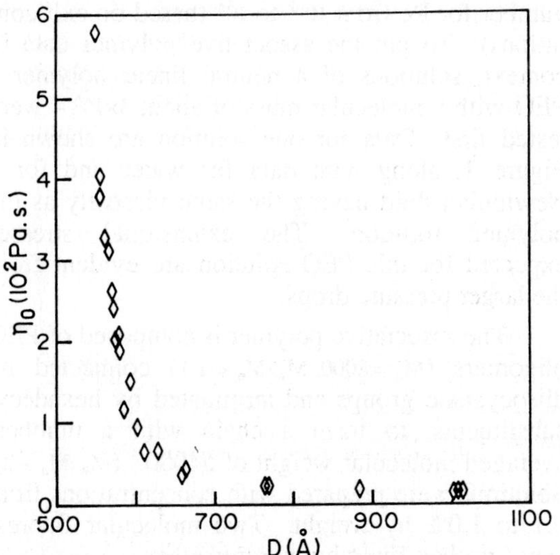

Fig. 5 : Variations of zero-shear viscosity.

Substracting the compression modulus from the overall complex modulus allows us to analyse the shear response of the confined solution. The terminal part of the storage shear modulus is restored as seen on the dashed lines of figure 2 allowing the determination of the steady-state compliance J_e^o and of an average relaxation time $\tau_o = \eta_o J_e^o$. The main feature is a sharp increase of η_o and J_e^o (Fig.5 and 6) at short distances from around Dc, connected to the hindered motion of the compressed layers.

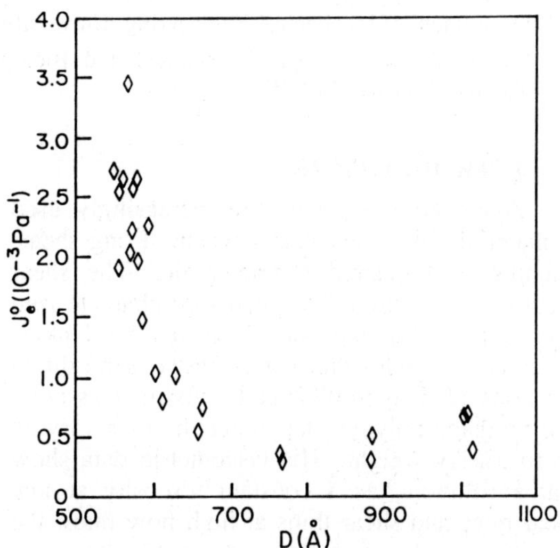

Fig. 6 : Variations of steady-state compliance.

CONCLUSIONS

The nanorheology of confined solutions and adsorbed layers of polymers reveals a complex behavior due to the specific change of conformation and dynamics of chains in the confined space. That kind of experimentation opens up the way for investigating the variations of the dynamics of thin films involved in adhesion, lubrification or colloïd stabilization problems.

REFERENCES :

1. J.P. Montfort, A. Tonck, J.L. Loubet, J.M. Georges, J. Polym. Sci, B, (1991) , 29 , 677-682
2. J.P. Montfort, G. Hadziioannou, J.Chem.Phys, (1988) , 88 , 7187-7196.
3. H.Terashima, J. Colloid Interface Sci, (1988) , 125 , 2 , 444-455.
4. M. Cohen-Stuart, F. Waajen , T. Cosgrove, B. Vincent , T. Crowley , Macromolecules , (1984) , 17 , 1825-1830.
5. G.H. Fredrickson , P. Pincus , Langmuir , (1991) , 7 , 786-795.

Theoretical and Applied Rheology, edited by P. Moldenaers and R. Keunings
Proc. XIth Int. Congr. on Rheology, Brussels, Belgium, August 17-21, 1992
1992 Elsevier Science Publishers B.V.

Hydrodynamic and Spectroscopic Measurements of Associative Polymer Solutions in Extensional Flow

T.D. STANTON[1†], D.F. JAMES[2] AND M.A. WINNIK[3]

[1,3] Department of Chemistry, University of Toronto, Toronto, Canada M5S 1A1

[2] Department of Mechanical Engineering, University of Toronto, Toronto, Canada M5S 1A4

[†] Present Address: Applied Chemical and Biological Sciences Department, Ryerson Polytechnical Institute, Toronto, Canada M5B 2K3

1. INTRODUCTION

Associative polymers are increasingly used in materials like latex paints where strong shear-thinning is a desired characteristic. The shear rheology of solutions has been well characterized from studies like the one conducted by Jenkins (reference 1), who thoroughly tested semi-dilute solutions of a hydrophobically-modified associative urethane polymer, the concentrations varying up to 5% by weight. His viscometric data show that a solution has a constant viscosity at low shear rates and shear thins at high flow rates; the transition is abrupt, and the shear thinning is so severe that the shear stress is virtually constant or, equivalently, the viscosity varies inversely with shear rate. This behaviour is believed to be related to the breakup or reorganization of networks of associative clusters formed by the hydrophobic end groups.

For these fluids, little is known about their extensional rheology, which plays a role in several methods of coatings applications. To gain some understanding of extensional flow behaviour, we subjected one of the associative polymers of Jenkins' study to extensional motion in a converging channel. To determine bulk stresses in the solutions, the pressure drop in the channel was measured; to determine changes in molecular structure, fluorescent probes were added to the fluids and simultaneous spectroscopic measurements were made.

2. EXPERIMENTAL METHOD AND RESULTS

The test section of the flow apparatus was a converging channel which had an inlet diameter of 8.7 mm and which smoothly converged over a length of 33 mm to an exit diameter of 0.42 mm. The change in diameter with axial distance was measured so that local velocities and strain rates could be determined from the measured discharge rate. Test fluids entered the channel from a pressurized reservoir and issued to atmosphere. Measurements were made of the overall pressure drop and of the flow rate. For the spectroscopic measurements, the quartz channel was installed in the sample chamber of an SLM fluorimeter and its excitation beam was focussed on the last 4 mm of the channel. Complete details of the experimental technique are given in reference 2.

A Newtonian baseline was established with a series of water-glycerine mixtures, yielding a single curve of pressure coefficient versus Reynolds number for Re from 10^{-2} to 10^4 (based on exit conditions). To put the associative polymer data in context, solutions of a neutral linear polymer – PEO with a molecular mass of about $3{\times}10^6$ – were tested first. Data for one solution are shown in Figure 1, along with data for water and for a Newtonian fluid having the same viscosity as the polymer solution. The extensional stresses expected for this PEO solution are evident from the larger pressure drops.

The associative polymer is composed of PEO oligomers ($M_n = 8000$, $M_w/M_n < 1.1$) connected by diisocyanate groups and terminated by hexadecyl substituents, to form a chain with a number-averaged molecular weight of 54000. ($M_w/M_n \sim 2$). Solutions were prepared with concentrations from 0.1 to 1.0% by weight. Two molecular fluorescence probes, dimethyl pyrenyl ether (dipyme) and pyrene, were separately added to 1% solutions to obtain information regarding the microenvironment under flow conditions.

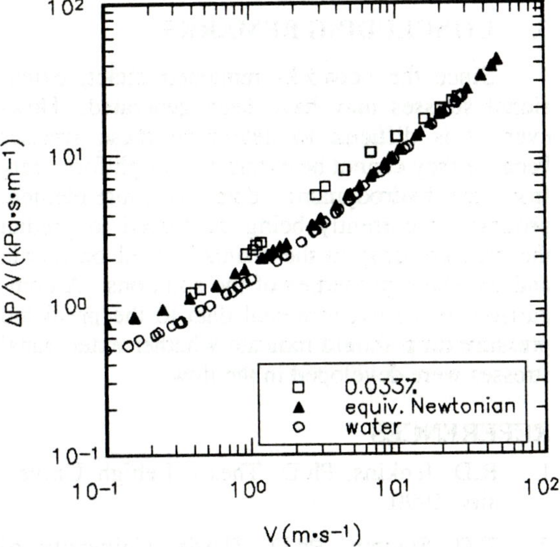

Figure 1. Hydrodynamic data for a linear, high MM PEO polymer (0.033% concentration) compared to water and to a Newtonian fluid of equivalent low-shear viscosity.

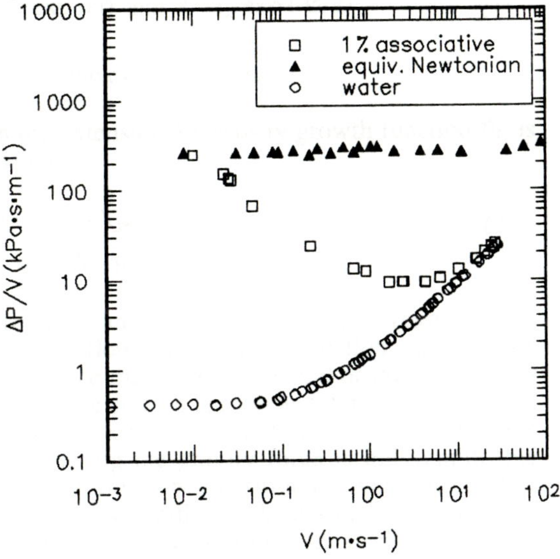

Figure 2. Hydrodynamic data for a hydrophobically-modified associative polymer (MM 54000, 1.0% w/w) compared to water and to a Newtonian fluid of equivalent low-shear viscosity.

The hydrodynamic data for the 1% solution are shown in Figure 2, along with data for water and for a Newtonian fluid having the same low-shear viscosity as the solution. The ordinate in the figure is the ratio of the pressure drop to the velocity. In terms of the pressure drop itself, the figure indicates that the drop departs from the Newtonian curve and increases only gradually with flow rate. The point of departure corresponds to a wall shear rate of 200 s^{-1}, which, as our own viscometric tests showed, is the critical shear rate when the viscosity drops off suddenly and the clusters (presumably) start to break up. With break-up occurring first near the wall in the channel, the fluid there had a lower viscosity and thus effectively lubricated the central flow and enhanced the flow rate. A further consequence of the thin boundary layer was that the bulk of the flow was subjected primarily to extensional motion.

The effect of this motion on the associative networks was detected from the fluorescent probes. These probes, hydrophobic like the hexadecyl end groups, solubilize primarily in the clusters formed by the end groups. Measurements showed that the hydrodynamic data were not altered by the addition of either probe. Pyrene and dipyme (which contains pyrene groups) are attractive probes because their vibronic fine structure (peak intensity ratio I_1/I_3) is very sensitive to the polarity of the probe environment. In addition, at high local concentrations, a new excimer emission I_e can be observed in the fluorescence spectrum along with a concomitant decrease in monomer emission I_m. The ratio I_e/I_m is a measure of the ease of excimer formation and is sensitive to the microfluidity of the local environment. Fluorescence results for dipyme are displayed in Figures 3 and 4. The values of I_1/I_3, measured at λ378nm/λ388nm, in Figure 3 confirm that the probe is in a micelle-like environment. This figure shows that I_1/I_3 did not change for extensional rates up to 1700 s^{-1}. The extensional rate is defined here as the average over the length of the channel, the maximum being about three times the average. It should be noted that the corresponding residence time of the fluid in the 4-mm optical detection length was about 0.2 ms. Similarly, Figure 4 illustrates that I_e/I_m, measured at λ495nm/λ398nm, also did not change with extensional rate. These probe data suggest that the

where η_s is the solvent viscosity and **D** is the rate-of-deformation tensor.

The polymer contribution to the total stress is taken as the sum of N components of stress. Then, the kth component of the polymer stress contribution for the Oldroyd-B model [5] is:

$$\tau_k + \lambda_k \tau_{k(1)} = 2\eta_k \mathbf{D} \tag{4}$$

where λ_k and η_k are the relaxation time and viscosity corresponding to the kth component of the stress. The uniaxial extensional stress growth function is given by:

$$\frac{\tau_k}{\dot{\varepsilon}} = \sum_{k=1}^{N} \left(\frac{2\eta_k}{1-2\lambda_k\dot{\varepsilon}}(1- \exp(-(1-2\lambda_k\dot{\varepsilon})t/\lambda_k)) + \frac{\eta_k}{1+\lambda_k\dot{\varepsilon}}(1- \exp(-(1+\lambda_k\dot{\varepsilon})\, t/\lambda_k))\right) \tag{5}$$

For the Giesekus model [6], the kth component of the polymer stress contribution is:

$$\tau_k + \lambda_k \tau_{k(1)} + \alpha_k \frac{\lambda_k}{\eta_k} \{\tau_k \cdot \tau_k\} = 2\eta_k \mathbf{D} \tag{6}$$

The uniaxial extensional stress growth function is given by:

$$\frac{\tau_p}{\dot{\varepsilon}} = \sum_{k=1}^{n} \left(\frac{4\eta_k(1-\exp(-tR_1/\lambda_k))}{(1-2\dot{\varepsilon}\lambda_k + R_1) - (1-2\dot{\varepsilon}\lambda_k - R_1)\exp(-tR_1/\lambda_k)} + \frac{2\eta_k(1-\exp(-tR_2/\lambda_k))}{(1+\dot{\varepsilon}\lambda_k + R_2) - (1+\dot{\varepsilon}\lambda_k - R_2)\exp(-tR_2/\lambda_k)} \right) \tag{7}$$

where $R_1 = \sqrt{1 - 4(1 - 2\alpha_k)\, \dot{\varepsilon}\lambda_k + 4(\dot{\varepsilon}\lambda_k)^2}$

and $R_2 = \sqrt{1 + 2(1 - 2\alpha_k)\, \dot{\varepsilon}\lambda_k + (\dot{\varepsilon}\lambda_k)^2}$

In the limit of small deformations the polymer stress contributions for each model reduce to the linear Maxwell model [4], and the η_k and λ_k values can be calculated from the small strain, oscillatory shear data. The η_k and λ_k values for the 4-mode Oldroyd-B and Giesekus models, and α_k values for the Giesekus model, given by Quinzani et al [3] for the PIB/PB/C14 fluid at 25°C, were shifted to 20°C. These parameters are shown in Table 1.

5. RESULTS AND DISCUSSION

The extensional stress growth functions of the PIB/PB/C14 fluid, plotted as the transient Trouton ratio against time, are shown in Figure 2, over a steady extensional rate range of 0.82 to 2.7 s^{-1}. The Trouton ratio rises rapidly to a value of 3, after which it increases further to very high values, of the order of 10^3. This strain-hardening effect is more pronounced the higher the extensional rate. However even at the highest strain ε (= $\dot{\varepsilon}t$) of 6 or more reached during the lowest extensional rate runs, no steady state was detected.

The reproducibility of the experiments is shown, for runs with $\dot{\varepsilon}$ = 2.0, in Figure 3 for both the horizontal and vertical configurations. The stress growth functions are almost identical during the entire duration of stretching for all the 4 runs presented.

The prediction of extensional stress growth functions by the 4-mode Oldroyd-B and Giesekus models are shown in Figure 4. The effect of the very viscous solvent is evident in these predictions where Trouton ratio rises rapidly to a value corresponding to three times the solvent viscosity. While both models predict identical growth function up to Trouton ratio of 3, the Oldroyd-B model predicts faster stress growth with time than does the Giesekus model and thus shows better agreement with the measured values. The Trouton ratio predicted by the Giesekus model near the end of all the experimental runs was an order of magnitude lower than the measured values. For the parameters shown in Table 1 for PIB/PB/C14 fluid, the Oldroyd-B model predicts infinite extensional viscosity for $\dot{\varepsilon} \geq 0.12$ s^{-1}, while the Giesekus model predicts steady-state Trouton ratio in the order of 1-2x10^3 for the extensional rates range obtained in our experiment. As no steady state has been measured for the fluid for any of the extensional rates tested, it was not possible to say whether or not the steady-state values predicted by the Giesekus model would represent the fluid well. It is also worth noting that in the range of extensional rates presented here, the second mode of polymer stress contribution dominates and the steady-state Trouton ratios are determined mainly by the very small α_2 value.

6. CONCLUSION

The extensional stress growth function of a constant viscosity, elastic fluid, used by Quinzani et al [3] has been measured at constant extensional rate from 0.82 to 2.7 s^{-1}. The fluid showed strain-hardening effect as expected of most polymer solution systems [4]. Predictions of the Oldroyd-B and Giesekus models, with 4 relaxation modes is reasonable compared to the measured stress growth results. While the Oldroyd-B model predicts infinite extensional viscosity for the extensional rate range measured, the Giesekus model predicts steady-state Trouton ratio of the order of 1-2x10^3, which is of the same magnitude as the value obtained at the end of the experiment.

REFERENCES

1. T. Sridhar, J. Non-Newtonian Fluid Mech., 35 (1990), 85-92.
2. T. Sridhar, V. Tirtaatmadja, D.A. Nguyen and R.K. Gupta, J. Non-Newtonian Fluid Mech., 40 (1991) 271-280.
3. L.M. Quinzani, G.H. McKinley, R.A. Brown and R.C. Armstrong, J. Rheology, 34(5), (1990) 705-748.
4. R.B. Bird, R.C. Armstrong and O. Hasseger, Dynamics of Polymeric Liquids, vol.1, Fluid Mechanics, Second Edition, Wiley-Interscience, New York, (1987).
5. J.G. Oldroyd, Proc. Royal Society, A200, (1950) 523-541.
6. H. Giesekus, J. Non-Newtonian Fluid Mech., 11, (1982) 69-109.

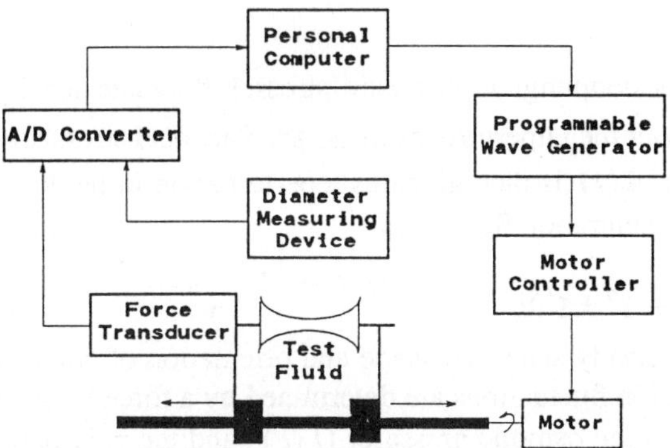

Figure 1 Schematic diagram of the filament stretching device used.

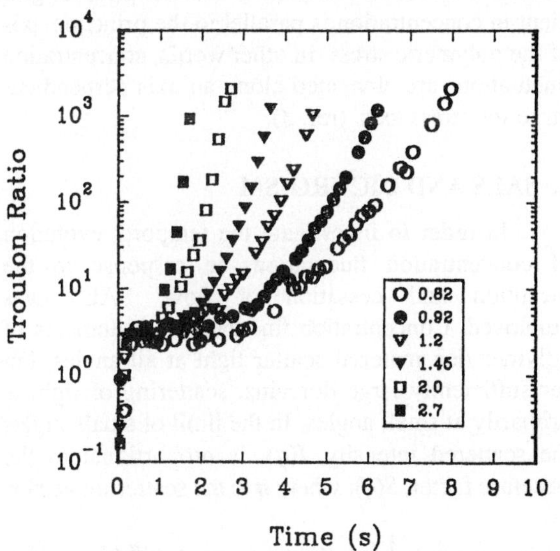

Figure 2 Transient Trouton ratio, at 20°C, at various extensional rates.

Table 1 Model Parameters for the solution of 0.31 wt% PIB in PB/C14, at 20°C.

Mode No	η_k (Pa s)	λ_k (s)	α_k
1	1.740	4.325	0.5
2	2.633	1.156	0.0001
3	2.603	0.172	0.001
4	1.901	0.0154	0.5
Solvent	12.75	–	–

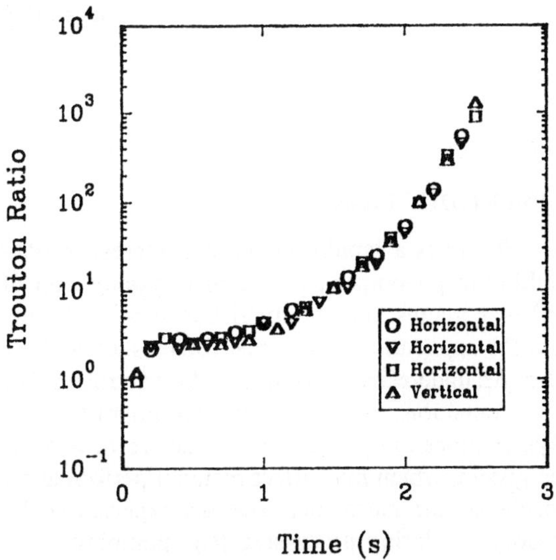

Figure 3 Transient Trouton ration for four experimental runs, with extensional rate of 2.0 s^{-1}.

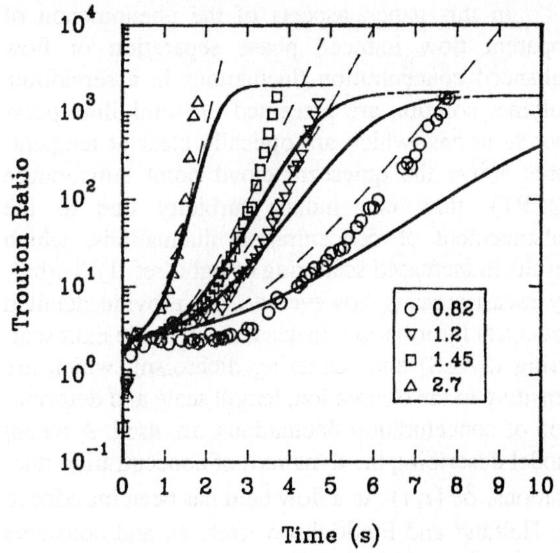

Figure 4 Transient Trouton ratio, compared with predictions by Giesekus model ——— and Oldroyd-B model - - - - -.

Theoretical and Applied Rheology, edited by P. Moldenaers and R. Keunings
Proc. XIth Int. Congr. on Rheology, Brussels, Belgium, August 17-21, 1992

Flow-Induced Structure and Dynamics of Concentration Fluctuations of Polymer Solutions

Jan W. van Egmond and Gerald G. Fuller

Department of Chemical Engineering
Stanford University
Stanford, CA 94305-5025

1. INTRODUCTION

There is a rapidly increasing interest in non-equilibrium phenomena in polymer systems and of particular interest are structural transitions resulting from the application of external forces such as flow fields. An understanding of flow induced structural or phase transitions is of fundamental importance in polymer processing. The phase structure of a polymer system, which may affect material performance, depends on the mechanical stresses experienced by the polymer during processing (e.g. pumping, mixing, extruding and molding) especially if phase transitions such as flow induced gelation, fibre formation, crystallization and phase separation (ref. 1) occur.

In this paper, aspects of the phenomenon of apparent flow induced phase separation or flow enhanced concentration fluctuations in a semidilute polymer solution are presented. In semidilute polymer solutions, which are optically clear at temperatures above the quiescent cloud point temperature (QCPT), flow can induce turbidity due to the enhancement of concentration fluctuations, which results in increased scattering of light (ref. 1). Turbidity measurements however do not provide detailed structural information. Instead, small angle light scattering (SALS) and scattering dichroism, which are sensitive to the orientation, length scale and deformation of concentration fluctuations, are used. A recent model describing the dynamics of concentration fluctuations, $\delta c(\boldsymbol{r}, t)$, in a flow field has been introduced by Helfand and Fredrickson (ref. 2), and considers the equation of motion for monomer concentration fluctuations. Central to the Helfand Fredrickson (HF) model is the coupling of mass transport to the concentration dependence of viscoelastic stresses. Due to this coupling, viscoelastic stresses fluctuate locally about the unperturbed stress, $\underset{\sim}{\tau}^o$. The local stress tensor, $\underset{\sim}{\tau}(\boldsymbol{r})$, is then given as a perturbation in the local concentration fluctuation:

$$\underset{\sim}{\tau} = \underset{\sim}{\tau}^o + \underset{\sim}{\tau}'\delta c \qquad (1)$$

At steady state, the shape and orientation of concentration fluctuations are determined by a force balance between osmotic pressure, $\Pi(c)$, and the polymeric stress:

$$\nabla \bullet \underset{\sim}{\tau}' = \nabla\Pi \qquad (2)$$

Equation (2) is a statement of the coupling between stress and structure and implies that the steepest gradient in concentration is parallel to the principle axis of the polymeric stress. In other words, concentration fluctuations are elongated along an axis perpendicular to the stress axis (ref. 3).

2. SALS AND DICHROISM

In order to investigate the temporal evolution of concentration fluctuations in response to the inception and cessation of flow, SALS was employed. Concentration fluctuations or domains of polymer rich material scatter light at all angles, but for sufficiently large domains, scattering of light is primarily at small angles. In the limit of small angles the scattered intensity, $I(\boldsymbol{q})$, is proportional to the structure factor, $S(\boldsymbol{q})$, where $\boldsymbol{q}$ is the scattering vector:

$$I(\boldsymbol{q}) \sim S(\boldsymbol{q}) \equiv \frac{1}{c}\int d\boldsymbol{r}\langle \delta c(\boldsymbol{r})\,\delta c(0)\rangle e^{-i\boldsymbol{q}\bullet\boldsymbol{r}} \qquad (3)$$

Since the structure factor is by definition the Fourier transform of the spatial correlation function of the concentration fluctuations, $\langle \delta c(\mathbf{r}) \delta c(0) \rangle$, we may infer the structure and dynamics of domains from the behavior of the scattering pattern.

Scattering dichroism is due to anisotropic scattering which results in anisotropy in the imaginary part of the refractive index tensor, $n''_{\alpha\beta}$ (ref. 4). Scattering dichroism thus gives a measure of the anisotropy and orientation of concentration fluctuations. From a generalized solution of Maxwell's equations, Onuki and Doi have derived a relation between the imaginary part of the refractive index tensor and the second moment in Fourier space of the structure factor (ref. 5):

$$n''_{\alpha\beta} \sim \int d\mathbf{q} \, (\delta_{\alpha\beta} - \hat{q}_\alpha \hat{q}_\beta) \, S(\mathbf{q}) \, . \tag{4}$$

For steady state, HF predicts a structure factor which depends on the fluctuating stress, $\underset{\sim}{\tau}'$ (ref. 2):

$$S(\mathbf{q}) \sim (A - \hat{q}\hat{q} : \underset{\sim}{\tau}'(\mathbf{q}))^{-1} , \tag{5}$$

where A depends on the osmotic pressure. Using Equation (5) scattering patterns can be interpreted as a map for the angular dependence of the fluctuating part of the stress. In the limit of small fluctuating stresses, substitution of Equation (5) into Equation (4) results in a linear relation between stress and dichroism:

$$n''_{\alpha\beta} \sim -\int d\mathbf{q} \, [\hat{q}_\alpha \hat{q}_\beta (\hat{q}\hat{q} : \underset{\sim}{\tau}'(\mathbf{q}))] \tag{6}$$

On the other hand, the stress optical rule (ref. 4) linearly relates intrinsic birefringence to the stress due to the segmental orientation of polymer molecules. Hence, it is possible in principle to obtain a measure of the local stress, $\underset{\sim}{\tau}(\mathbf{r})$, in Equation (1) by using a combination of birefringence and dichroism measurements.

3. RESULTS

In the present work, time dependent SALS and scattering dichroism have been utilized to investigate the dynamics of growth and orientation of concentration fluctuation enhancement for a 6% wt. solution of polystyrene (M_w = 1.86 x 10^6) in dioctyl phthalate (PS/DOP). All experiments were conducted over a range of temperatures between the QCPT (10^oC) and the theta temperature (22^oC), thus ensuring proximity of the system to the coexistence curve.

SALS patterns and dichroism measurements were conducted using an instrument described elsewhere in detail (ref. 6). The sample, illuminated by a 8mW HeNe laser, scatters the light onto a screen, from which scattering patterns were recorded at a rate of 2 frames/second by a 2D CCD array. Dichroism is determined from the anisotropic attenuation of the transmitted light, which is measured by a photodiode detector and a suitable polarization modulation scheme consisting of a polarizer and a rotating half wave plate. Various flow geometries were utilized. For simple shear flow, the flow-shear and the flow-vorticity plane were investigated using a concentric cylinder and parallel plate flow cell, respectively. In addition, a four roll mill allowed for investigation of uniaxial extensional flow.

Flow enhancement of concentration fluctuations is easily visualized by contour plots of the deviation of the structure factor, $\Delta S(\mathbf{q}, \nabla U) \equiv S(\mathbf{q}, \nabla U) - S(\mathbf{q}, 0)$, from the quiescent isotropic structure factor, $S(\mathbf{q}, 0)$. Here ∇U is the shear rate, $\dot{\gamma}$, or the extensional rate, $\dot{\varepsilon}$. In general, $\Delta S(\mathbf{q}, \nabla U)$ appears as a butterfly pattern consisting of two high intensity lobes, the orientation of maximum scattering being parallel to the direction of highest concentration gradient and thus perpendicular to the axis of elongation of concentration fluctuations (see Equation (2)). Figure 1 presents the steady state deviation in structure factor for a shear experiment at 16^oC in the flow-shear plane for a shear rate of $5s^{-1}$. The flow direction is from left to right while the direction of increasing velocity is from bottom to top. Similar steady state wide angle scattering patterns have been reported by Wu, *et al* for a 4% PS/DOP solution (ref. 7). Figure 2 presents the steady state structure factor for a simple shear experiment in the shear-vorticity plane. As expected, the angle of orientation of minimum scattering is perpendicular to the flow axis since in this geometry, the flow direction is also the principle axis of the stress tensor. Figure 3 represents the result for a uniaxial extensional flow experiment at 12^oC and an extensional rate, $\dot{\varepsilon}=5s^{-1}$, and here the axis of maximum scattering is parallel to the axis of extension, thus indicating that concentration fluctuations are elongated once again perpendicular to the principle stress axis.

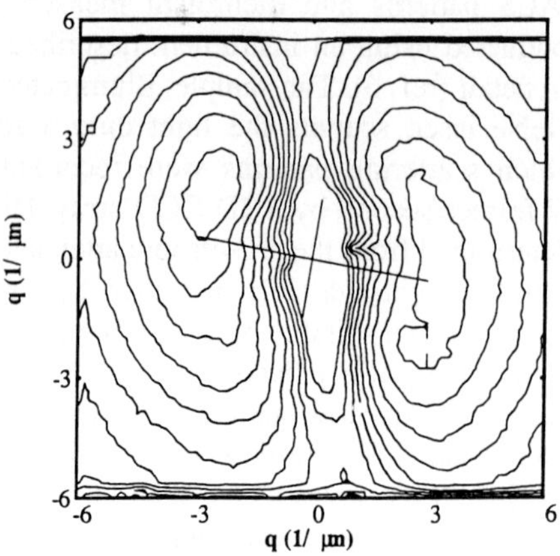

Figure 1. Contour plot of $\Delta S\,(\boldsymbol{q}, \dot{\gamma})$ for simple shear flow in couette geometry; T = 16°C, $\dot{\gamma} = 5\mathrm{s}^{-1}$.

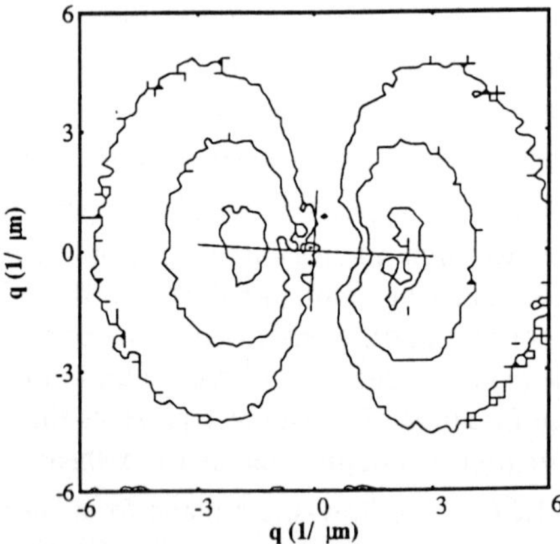

Figure 2. Contour plot of $\Delta S\,(\boldsymbol{q}, \dot{\gamma})$ for simple shear flow in parallel plate geometry; T = 16°C, $\dot{\gamma} = 5\mathrm{s}^{-1}$.

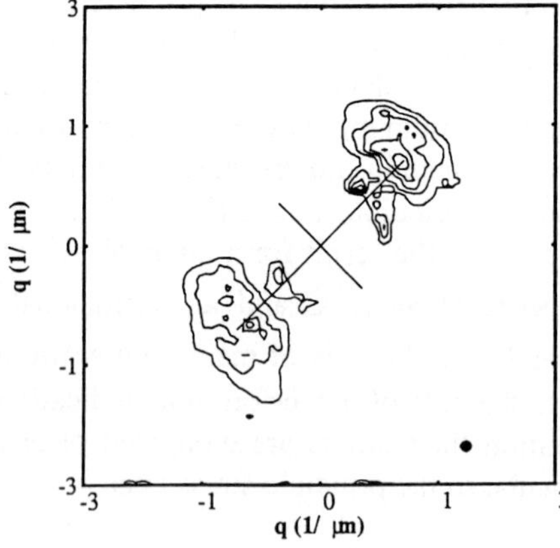

Figure 3. Contour plot of $\Delta S\,(\boldsymbol{q}, \dot{\gamma})$ for extensional flow in flow in four roll mill; T = 12°C, $\dot{\varepsilon} = 5\mathrm{s}^{-1}$.

4. DISCUSSION AND CONCLUSION

Using the Onuki-Doi relation between dichroism and structure factor, the dichroism, $\Delta n''_{calc}$, was calculated from the structure factor, and found to be consistent with dichroism measured from the transmitted light, $\Delta n''_{expt}$. This correspondence is a manifestation of the validity of the Onuki-Doi relation. Over the range of temperatures and shear rates considered, it was found that data for both $\Delta n''_{calc}$ and $\Delta n''_{expt}$ collapse to a monotonically increasing function of the Weissenberg number, $Wi \equiv \dot{\gamma}\tau_r$. Wi is the ratio of a characteristic polymer relaxation time, τ_r, to the time scale of the flow, which is taken here as the reciprocal of the shear rate.

Several interesting time dependent phenomena arising from the coupling between stress and structure are also observed. These include an overshoot in scattering intensity and dichroism on both inception and cessation of simple shear flow but not for uniaxial extensional flow. Moreover, for the flow-shear plane it is found that the orientation of deformed concentration fluctuations while always assuming a value of 135° to the flow direction on flow inception, decreases towards a limiting value of 45° with increasing *Wi*. In the flow-vorticity plane, the average orientation of concentration fluctuations is found to rotate from the vorticity axis to the flow axis. This behavior occurs when $Wi \gg 1$ and is thought to be related to shear thickening and the disappearance of the second normal stress coefficient.

References

1. Rangel-Nafaile, C., Metzner, A. B., Wissbrun, K. F., Macromolecules, 17, 1187 (1984).
2. Helfand, E. and Fredrickson, G.H., Phys. Rev. Lett., 62, 2468 (1989).
3. Larson, R. G., Rheol. Acta, submitted 1992.
4. Fuller, G.G., Annu. Rev. Fluid Mech., 22, 384 (1990).
5. Onuki, A. and Doi, M., J. Chem. Phys., 85, 1190 (1986).
6. van Egmond, J. W. and Fuller, G. G., J. Chem. Phys., submitted November 1991.
7. Wu. X.-L., Pine, D. J. and Dixon, Phys. Rev. Lett., 66, 2408 (1991).

Theoretical and Applied Rheology, edited by P. Moldenaers and R. Keunings
Proc. XIth Int. Congr. on Rheology, Brussels, Belgium, August 17-21, 1992

THERMALLY INDUCED STRUCTURE FORMATION IN POLYETHYLENE SOLUTIONS

L. AERTS [1], H. BERGHMANS [1], P. MOLDENAERS [2], J. MEWIS [2] AND M. KUNZ [3]

[1] Laboratory for Polymer Research, Katholieke Universiteit Leuven, Celestijnenlaan 200F, B-3001 Heverlee, Belgium

[2] Chemical Engineering Department, Katholieke Universiteit Leuven, De Croylaan 46, B-3001 Heverlee, Belgium

[3] Institut für Makromolekulare Chemie, Universität Freiburg, Hermann-Staudinger-Haus, Stefan-Meier-Str. 31, D—7800 Freiburg, Germany

1. INTRODUCTION

Polyethylene (PE) is one of the most studied crystallizable polymers. Morphological study on PE-crystals obtained from solution formed a base for the lamellar crystal model[1]. About a decade ago, the solution behaviour of PE regained interest after the gel-spinning method had been introduced as a route to ultra-high modulus PE-fibres[2]. In this method, a crystalline PE-decaline gel has been used as an intermediate. In 1946 already, Richards reported the phase behaviour of PE in different solvents[3]. In good solvents, only a melting temperature depression of PE is noticed. In poor solvents, a liquid-liquid (L-L) demixing region at the lower polymer concentration side is added to this scheme. In this work the morphology and rheological properties of several PE-solvent systems of moderately low concentration have been investigated and have been related to the phase behaviour of the system. The study tries to give more insight into the structure forming mechanisms in such gels, depending on the molecular weight of the PE and the solvent quality. The PE used ranged from High Density PE (HDPE) up to Ultra High Molecular Weight PE (UHMWPE). Decalin and diphenylether, respectively known as good and poor solvent, have been used as solvents. Morphology has been investigated by Transmission Electron Microscopy (TEM), while the rheological measurements have been carried out on a Rheometrics RMS 800 in parallel plate geometry.

2. PE IN A GOOD SOLVENT

When a 5 % (by weight) solution of PE in decalin is cooled to room temperature, only crystallization takes place. Oscillatory rheometry on these systems at a low strain level show an elastic, supensionlike behaviour. The loss tangent amounts to 0.1 . The storage moduli for a UHMWPE gel (10^4 Pa) and a comparable HDPE gel (2.10^2 Pa) differ almost two decades in absolute value. This was related to the mesh size of the crystalline network, evidenced by TEM analysis. On increasing the strain level, the dynamic moduli start to decrease. UHMWPE gels however show strain hardening from some tens of percent strain on.

3. PE IN A POOR SOLVENT

On cooling a moderately concentrated solution of PE in diphenylether, a L-L demixing preceeds the crystallization of PE. Under appropriate conditions of concentration and cooling rate, a continuous polymer rich phase with embedded droplets of solvent phase can be formed by L-L demixing. Further cooling results in the crystallization of the polymer from the concentrated phase and a rigid lamellar crystal network is formed throughout the system[4]. This is reflected in the rheological behaviour of a 5 % HDPE-diphenylether system. The storage modulus amounts 10^5 Pa, whereas the loss tangent is again 0.1 . The frequency and strain dependence of the dynamic moduli is very resemblant to the features of PE-decalin gels. Only the absolute values differ about three orders of magnitude.

4. DISCUSSION

From these observations a model for the network formation in crystalline PE-solvent systems is proposed. In this scheme a network is formed by the mutual adhesion of lamellae. Due to the intrinsic nature of the amorphous contacting regions, the loss tangent will always be the same, irrespective of the total number of interlamellar contacts. On stress or strain imposal the structure starts to break down because of the loosening of "sticking" lamellae. The much larger amount of tie molecules between the lamellae in case of the UHMWPE samples causes these systems to show strain hardening at higher strain level, in contrast to the HDPE samples, in which only breakdown of the structure can be observed.

REFERENCES

1. B. Wunderlich in "Macromolecular Physics",**1** and **2**, Academic Press,NY(1973-1976)
2. P. Smith and P.J. Lemstra, J. Mater. Sci.,**15**,505(1980)
3. R.B. Richards, Trans. Faraday Soc.,**42**,10(1946)
4. L. Aerts and H. Berghmans, submitted to Macromolecules

Theoretical and Applied Rheology, edited by P. Moldenaers and R. Keunings
Proc. XIth Int. Congr. on Rheology, Brussels, Belgium, August 17-21, 1992

RHEOLOGICAL BEHAVIOUR OF PECTIN LM/CALCIUM SOLUTIONS AND GELS.

M.A.V. AXELOS [1], C. GARNIER [1] and M. KOLB [2]

[1] Laboratoire de Physico-Chimie des Macromolécules, Institut de la Recherche Agronomique, BP 527, 44026 Nantes Cedex 03 (France)
[2] Laboratoire de Chimie Théorique, Ecole Normale Supérieure, 46 Allée d'Italie, Lyon Cedex (France)

1. INTRODUCTION

Pectin is a major anionic polysaccharide in plants. It consists of a linear backbone of randomly connected α (1-4) D galacturonic acid units and their methyl esters (ref. 1). It is a widely used gelling agent in the food industry. The most familiar example of a gelation process involving high-methoxyl (HM) pectin is the making of fruit jellies. Low-methoxyl (LM) pectin with degrees of esterification (DE) lower than 50%, gives physical gels in presence of divalent ions such as calcium. In this biopolymers network, the concept of single, covalent bonds is replaced by side-by-side associations of homogalacturonic acid regions of the polymer, through calcium mediation. The amount of methyl esters present limits the formation of such junction zones (ref. 2).

Systems undergo a sol-gel transition on decreasing the temperature or by changing the polymer-solvent interactions through the pH, the ionic strength or by adding different amount of calcium. Phase diagrams are established in order to describe the interdependence of all these factors and dynamic properties in the sol and gel phase are investigated using mechanical spectroscopy (ref. 3).

2. METHODS

Before the transition, constant shear viscosity measurements were performed with a Low-shear 40 Contraves Rheometer. After the transition, oscillatory stress measurements were performed using a Carri-Med CS-50 controlled stress rheometer with a cone-plate device. A deformation of 0.04 was maintained whatever the frequency range explored between 10^{-3} to 10 Hz. The kinetics of gel formation was followed, by measuring the storage and loss moduli, G' and G", at a fixed frequency until equilibrium was reached. Then the mechanical spectrum was recorded.

3. RESULTS

3.1. Sol - gel transition

From the rheological measurements performed close to the sol-gel transition, scaling exponents were determined independently for the viscosity, s=0.82(5), for the shear modulus, t=1.93(8), for the frequency-dependent modulus, Δ=0.71(2), and for the relaxation times below and above the transition, νz=2.67(12) and $\nu z'$=2.65(9). The exponents satisfy the scaling relations predicted by the theory and their numerical values agree with those from scalar elasticity percolation (ref. 4).

3.2. Fully set gels

On fully set gels, the mechanical spectra allow to point out that crosslinks in these gels may be considered as permanent. The magnitude of the equilibrium shear modulus were found to increase with pectin and calcium concentration and with the ionic strength, until the syneresis phase appears, and to decrease with the rise of the degree of esterification. The decrease of the pH leads to a different kinetics of formation and to a drop in the value of the shear modulus, G' may be reduced by a factor ten between pH 7 and 4.

REFERENCES

1. J.A. De Vries, F.M. Rombouts, A.G.J. Voragen and W. Pilnik, Carbohydr. Polym., 2 (1982) 25-33.
2. R. Kohn, Carbohydr. Res., 160 (1987) 343-353.
3. M.A.V. Axelos, Makromol. Chem., Macromol. Symp., 39 (1990) 323-328.
4. M.A.V. Axelos and M. Kolb, Phys. Rev. Lett., 64 (1990) 1457-1460.

Theoretical and Applied Rheology, edited by P. Moldenaers and R. Keunings
Proc. XIth Int. Congr. on Rheology, Brussels, Belgium, August 17-21, 1992

HYDRODYNAMIC INTERACTIONS OF POLYMER SOLUTIONS IN A GENERAL TWO-DIMENSIONAL FLOW

A.E. CHAVEZ, M. LOPEZ DE HARO AND O. MANERO

Instituto de Investigaciones en Materiales, UNAM, A. P. 70-360, Coyoacán 04510, México, D.F. (México)

1. INTRODUCTION

Hydrodynamic interactions are known to strongly influence the dynamic response of polymer solutions and this fact makes the representation of the hydrodynamic drag an important aspect in the calculations. In this investigation, effects of a general two-dimensional flow are incorporated explicitly in the frictional properties of the FENE model of a dilute polymer solution. The main objective is to develop a model to describe material properties and the variation of the mean square end-to-end distance R with flow-strength when effects of hydrodynamic interaction are present.

2. THE MODEL

We consider here a dilute suspension of flexible nonlinear dumbbells with finite spherical beads of radius 'a', immersed in a Newtonian solvent in steady homogeneous linear flow, which can be expressed in the form

$$\underline{V} = \underline{\underline{L}} \cdot \underline{R} \qquad (1)$$

$\underline{\underline{L}}$ defines the velocity gradient tensor for a general two-dimensional flow:

$$\underline{\underline{L}} = \frac{E}{2}\begin{bmatrix} 1+\lambda & 1-\lambda & 0 \\ -(1-\lambda) & -(1+\lambda) & 0 \\ 0 & 0 & 0 \end{bmatrix} \qquad (2)$$

where E is the magnitude of the local velocity gradient and λ a parameter which specifies the type of flow (ref. 1).

In a previous work (ref. 2), the assumptions and most important ideas about the model were presented. The hydrodynamic problem, including the calculation of the mobilities $\underline{\underline{A}}$ through the method of the induced forces, were made (ref. 3).

This approach gives analytical expressions for the mobilities and friction tensors beyond the Oseen approximation. It has been used in extensional flows where the velocity gradient tensor is diagonal (ref. 4).

In simple shear, diagonalization of this tensor renders expressions for the mobilities in term of the eigenvalues of $\underline{\underline{L}}$, i. e. $p = E\,\lambda^{1/2}$.

These are

$$\underline{\underline{A}}\,(6\pi\eta a) = \underline{\underline{I}} - \frac{3}{4}\frac{a}{R}\left(\underline{\underline{I}} + \underline{RR}\right) + -a(p/\nu)^{1/2}\left[9/10\begin{bmatrix}1&0&0\\0&1&0\\0&0&0\end{bmatrix}\right] + 3/4\left[7/15\,(Rx^2 + Ry^2)/R^2 - 6/5\right]\underline{\underline{I}} + +3/4\left[14/15 - 7/5\,(Rx^2 + Ry^2)/R^2\right]\underline{RR} \qquad (3)$$

3. RESULTS

The resulting system of algebraic equations for the components of $\langle\underline{RR}\rangle$, an averaged value of $\underline{RR}$ with respect to system configuration, shows that the shear flow limit can be numerically obtained. From this point, the computation of the influence of molecular size and state of flow in the mean square end-to-end distance and rheological properties is straightforward.

REFERENCES

1. N. Phan-Thien, O. Manero and L.G. Leal. Rheol Acta 23:151-162 (1984)
2. A.E. Chávez, M. López de Haro and O. Manero. J. Stat. Phys. 62:1255-1266 (1991)
3. D. Bedeaux and J.M. Rubí. Physical 144A (1987) 285-298
4. M. López de Haro, A. Pérez-Madrid and J. M. Rubí. J. Chem Phys. 88(12):7964-7969 (1988).

Theoretical and Applied Rheology, edited by P. Moldenaers and R. Keunings
Proc. XIth Int. Congr. on Rheology, Brussels, Belgium, August 17-21, 1992

ON DRAG-REDUCING SURFACTANT SOLUTIONS AND PUMPS

K. GASLJEVIC, B.A. JAROUX, AND E.F. MATTHYS

Dept. of Mechanical Engineering, University of California, Santa Barbara, CA 93106, USA

Drag-reducing surfactant solutions may be used to decrease friction losses in typical hydronic heating and cooling systems by about 50% [1]. We have conducted experiments to quantify the effect of these drag-reducing additives on the behavior of a typical pump, and we have also measured the effect of this pump on the fluid behavior in the pipe entrance region immediately after the pump. The fluid used in these experiments consists of 5000 ppm of Ethoquad T13-50 in deionized water with 4000 ppm of NaSal as counter-ion. This solution exhibits asymptotic drag reduction in pipes. Fig. 1 shows a comparison of the head and power characteristics (w.r.t. the pump design point) for water and for the solution in a Dayton 1P853 centrifugal pump. The figure shows that the head / flow characteristics are unchanged by the addition of the additives, but that the power required is reduced by about 7% (w.r.t. water). This energy saving is therefore available in addition to the savings due to drag reduction in the system pipes. Fig. 2 shows the pressure drop measured over 10-diameter long sections as a function of distance from the pump for several Reynolds numbers. (The flow rate was controlled by throttling downstream of the test section.) The pressure drop measured under fully-developed conditions (at 1000 diameters downstream) is also shown for reference (dashed lines) and is consistent with asymptotic drag reduction values. (The scatter in some of the pressure drop measurements is due to hole pressure errors.) The entrance region after the pump for Reynolds numbers increasing from 25000 to 150000 is seen to increase also from about 20 to 150 diameters. The latter length is significantly greater than for water but comparable to values found for polymer drag-reducers. In order to determine whether this entrance region is due primarily to degradation in the pump and subsequent reconstitution, or instead primarily to hydrodynamic effects, we have also conducted experiments in a blow-down (i.e. no pump) setup designed to minimize degradation before the test section. The results are shown for comparison. The pressure drops are seen to be only slightly higher with the pump (solid symbols) than without (open symbols), and the trends are very similar over the locations investigated. This in turn suggests that the entrance region is due primarily to hydrodynamic effects, and that the pump causes only minor degradation to the fluid even at these high Reynolds numbers. In conclusion, the use of pumps with surfactant solutions does not appear to lead to significant degradation or problems in the downstream pipes. Furthermore, the surfactant additives do also increase the efficiency of the pump, which results in additional energy savings beyond the usual drag reduction in pipes.

This work was supported by the California Institute for Energy Efficiency and the US National Science Foundation.

REFERENCES

1. K. Gasljevic and E.F. Matthys. Industrial Applications of Fluid Mechanics, ASME, N.Y., 1991, pp. 57-65.

FIG. 1

Power: Water
Power: Surfactant
Head: Water
Head: Surfactant
Re (pump) = 186000
RPM = 3450
Head/Head$_{des}$
Power/Power$_{wat\text{-}des}$
Flow/Flow$_{des}$

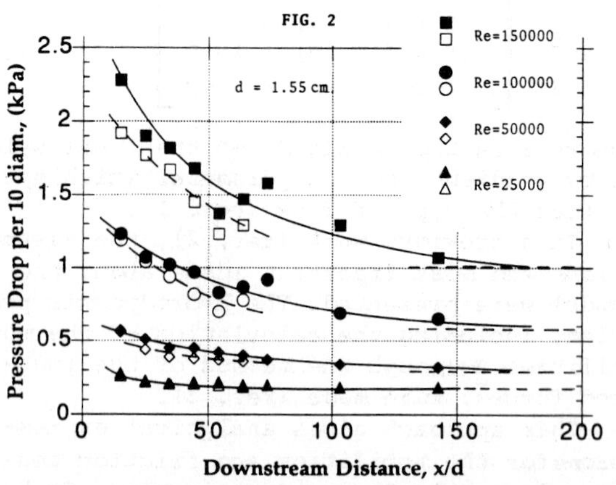

Theoretical and Applied Rheology, edited by P. Moldenaers and R. Keunings
Proc. XIth Int. Congr. on Rheology, Brussels, Belgium, August 17-21, 1992

STEADY AND DYNAMIC SHEAR CHARACTERISTICS OF SODIUM HYALURONATE SOLUTIONS

Y. GUO[1], T. JANKOWSKI[1], P. PRADIPASENA[1], C.K. RHA[1], J. BURNS[2] AND L. YU[2]

[1]Biomaterials Science and Engineering Laboratory, M.I.T., Cambridge, MA 02139
[2]Biopolymers Department, Genzyme Corp., Cambridge, MA 02139

1. INTRODUCTION

Sodium hyaluronate is widely distributed throughout various physiological systems. The rheological behaviour of sodium hyaluronate determines the functions it plays in physiological systems, dictates the formulation and processing of various products, and indicates the quality of the product and process.

Sodium hyaluronate used in this study was produced from <u>Streptococcus</u> <u>Zooepidemics</u> by fermentation.

2. METHODS

2.1 Intrinsic viscosity

Intrinsic viscosity was measured in a physiological buffer (pH=7) containing sodium phosphate monobasic (0.004 g/dl), sodium phosphate dibasic (0.056 g/dl) and sodium chloride (0.850 g/dl) at 25°C using a calibrated Cannon Fenske capillary viscometer (Cannon Instrument, State College, PA).

2.2 Molecular weight

Weight average molecular weight and molecular weight distribution was evaluated using a Multi-angle Laser Light Scattering instrument (Model Dawn-F Wyatt Technology Co., Santa Barbara, CA) in conjunction with HPLC and refractometer (Waters, Milford, MA).

2.3 Steady and dynamic viscosity

Steady and dynamic viscosity were measured in the physiological buffer at 25°C using Bohlin Rheometer system (Lund, Sweden).

2.4 Stress relaxation

A stress relaxation test was conducted under an applied constant small strain using the Bohlin Rheometer system (Lund, Sweden) and the buffer at 25°C.

3. SIGNIFICANCES

Sodium hyaluronate produced from <u>Streptococcus</u> <u>Zooepidemics</u> exhibits a large hydrodynamic volume as indicated by the value of intrinsic viscosity (16.4-33.3 dl/g). The molecular weight range was 900,000-2,500,000. Although there exist regions of local order along the backbone chain, sodium hyaluronate behaves as a free draining random coil in solution.

Sodium hyaluronate has three distinct interactive regimes: the intrinsic viscosity, the semi-dilute and the entanglement regimes as indicated by the reduced concentration (concentration x intrinsic viscosity) dependency of specific viscosity. The semi-dilute regime and the entanglement regime started from the reduced concentrations of 1.1 and 4.0, reflecting the onset of overlapping and entanglement, respectively. The slope or the interactive parameter in the entanglement regime is 3.9 for sodium hyaluronate. This value is similar to those for CMC and guar gum (4.0 and 3.9, refs. 1-2), and less than that for xanthan (4.2, ref. 3), but greater than those for dextran and carboxymethyl amylose (3.3, ref. 2). Thus the slope of the curve for interactive regimes is indicative of chain conformation and flexibility consequently the interactive potential of the polymer chain. Therefore, we represent the slope as the interactive parameter.

The molecular weight dependency of lower newtonian viscosity indicates that the lower newtonian viscosity is 3.8th power of the cMw

$\eta = 5.8 \times 10^{-23} (cMw)^{3.8}$.

The longest relaxation time was evaluated from the value of the critical reduced shear rate, and was also measured from the stress relaxation experiment. The results are the same for both cases, but they are one order of magnitude higher than that estimated from the Rouse model. This probably indicates the dominant effect of chain entanglement.

The cross-over frequency of sodium hyaluronate of Mw=2,500,000 is inversely proportional to the concentration square

$\omega_c = 0.12/c^2$.

REFERENCES

1. Y. Guo, C.K.Rha, M. Timonen, T. Vaara and A. Lehmussaari, XIth International Congress on Rheology, 17-21, August (1992).
2. E.R. Morris, A.N. Cutler, S.B. Ross-Murphy, D.A. Rees and J. Rice, Carbohydr. Polym., 1 (1981) 5-21.
3. G. Cuvelier and B. Launay, Advances in rheology 2. Fluids, Eds B.Mena, A. Garcia-Rejon and C. Rangel-Nafaile, P. 247, Univ. Nacional Autonoma de Mexico, Mexico.

Theoretical and Applied Rheology, edited by P. Moldenaers and R. Keunings
Proc. XIth Int. Congr. on Rheology, Brussels, Belgium, August 17-21, 1992

LONG-CHAIN HYDROPHOBIC DERIVATIVES OF HYDROXYPROPYL GUAR GUM (HPG): A RHEOLOGICAL STUDY IN SHEAR CONDITIONS

R. LAPASIN and S. PRICL

Istituto di Chimica Applicata e Industriale, Università di Trieste, Via Valerio 2, I-34127 Trieste, Italy.

1. INTRODUCTION

Hydrophobic polysaccharide derivatives constitute a new class of products for a variety of applications as fracturing fluids in EOR and rheological additives in paints. Low levels of hydrocarbon residues can be incorporated into guar gum ether, such as HPG, to yield high viscosity water-soluble products with peculiar non-Newtonian behavior (ref. 1).

2. EXPERIMENTAL

2.1 Materials

A sample of hydrophobic HPG (h-HPG, $M_w=7\cdot 10^5$, $MS=10^{-3}$), obtained by reaction of HPG (MS=1) with a mixture of long-chain aliphatic epoxides (C_{16}-C_{22}) in the presence of an alkaline catalyst, was supplied by Fratelli Lamberti S.p.A. (Albizzate, Italy). Aqueous systems were prepared at different polymers concentrations (C_p range 0.25 - 1.50 % w/w) and examined at different temperatures (T range 5 - 45°C).

2.2 Apparatus

Continuous and dynamic tests were performed with a rheometer Rotovisco Haake RV 100 (measuring system CV 100, coaxial cylinder device ZB 15). Normal stress analysis was performed with a rheometer Rotovisco Haake RV 20 (measuring system CV 20N, cone/plate device PK 30/4).

3. RESULTS AND DISCUSSION

3.1 Continuous shear behavior

For all the h-HPG systems examined, the flow curves $\tau(\dot\gamma)$ diverge at low shear rates with respect to those of the corresponding native polymer solutions. In particular, the appearance of a yield stress becomes evident with increasing C_p from both the $\tau(\dot\gamma)$ curves and the experimental relaxation data. Accordingly, the shear thinning behavior can be suitably described by the modified Cross model (ref. 2):

$$\tau = \tau_0 + \eta_\infty \dot\gamma + \frac{(\eta_0 - \eta_\infty)\dot\gamma}{1 + (\lambda\dot\gamma)^n}$$

The parameters η_0 and τ_0 are strongly dependent on C_p ($\eta_0 \propto C_p^{4.7}$, $\tau_0 \propto C_p^{4}$). An increase in temperature produces a τ_0 decrease and a crossing of the flow curves at high $\dot\gamma$.

The first normal stress difference N_1 is nearly independent of $\dot\gamma$ and higher than the shear stress in the shear rate range explored.

3.2 Oscillatory shear behavior

For all the h-HPG systems, the critical strain is close to 20% and lower than that of the corresponding HPG solutions. The profiles of the $G'(\omega)$ and $G''(\omega)$ curves do not modify sensibly with C_p and T. The frequency dependence of G' resemble that of a polysaccharide solution whereas G" shows a tendency to a plateau value for $\omega \to 0$.

3.3 Comparison between continuous and oscillatory material functions

The peculiar behavior of h-HPG aqueous systems with respect to polysaccharide solutions is confirmed by comparing the shear rate dependence of η and $N_1/2\dot\gamma^2$ with the frequency dependence of η' and G'/ω^2. These curves do not seem to converge in the low $\dot\gamma$ and ω regions.

REFERENCES

1. G. Molteni, C. Nicora, A. Cesàro and S. Pricl, Eur. Pat. 88121711.1, 1988.
2. R. Lapasin, S. Pricl and P. Esposito in: R.E. Carter (Ed.), Rheology of Food, Pharmaceutical and Biological Materials, Elsevier Applied Science, London, 1990, pp.122-132.

Theoretical and Applied Rheology, edited by P. Moldenaers and R. Keunings
Proc. XIth Int. Congr. on Rheology, Brussels, Belgium, August 17-21, 1992

AGGREGATE STRUCTURES IN WATER-SOLUBLE POLYMER SYSTEMS

Eleanor L. Meyer and Gerald G. Fuller

Department of Chemical Engineering, Stanford University, Stanford, CA 94305-5025

1.0 INTRODUCTION

The structures of polymers in solution and the effect of this structure on the solution flow properties is of interest in many systems. One such aqueous polymer system, xanthan gum solution, has been studied in our laboratory using the complementary techniques of birefringence and dichroism (ref. 1).

Xanthan gum is a very high molecular weight (M_w~$2x10^6$) biopolymer and is used extensively in industry as a thickening or stabilizing agent. Previous solution studies indicate that it has a rodlike conformation, but still has some degree of flexibility. In the native form, xanthan is thought to have either a single or double helix structure. Xanthan is also known to self-aggregate through entanglements and hydrogen bonding effects even at low concentrations. Many studies involving mechanical measurements or light scattering have found a critical concentration of aggregation of the xanthan chains. Unfortunately, mechanical techniques are limited in their ability to probe very low concentrations while light scattering measurements are restricted to examining static systems. The optical techniques of birefringence and dichroism have the advantage of the ability to probe very low concentrations in systems subjected to flow.

2.0 EXPERIMENTAL RESULTS

Using birefringence and dichroism optics we have been able to monitor xanthan dynamics under shear flow. The birefringence, due to intrinsic effects, allows us to examine segmental orientation, while the scattering dichroism allows us to interrogate the much larger aggregate orientation.

The results of experiments with these two techniques reveal two very different orientational and relaxation responses over the dilute/semidilute concentration range of 50 to 3000 ppm. The orientation angles of birefringence and dichroism with respect to the flow direction differ in both magnitude as well as their dependences on concentration. While birefringence data follow the usual result of decreasing orientation angle with increasing concentration, dichroism results indicate a very different and unexpected steady state behavior. Instead, dichroism orientation angles are low (in the flow direction) for low concentrations, but beyond a critical concentration rise rapidly. Clearly, two different and opposite orientational effects are occurring at the two length scales of the polymer segments and polymer aggregates.

The existence of a critical concentration also appears to be present in birefringence/dichroism relaxation. Interestingly, this critical concentration range coincides with that of the literature, but dichroism optical measurements confirm the existence of aggregates even below this concentration where many researchers believed aggregates to no longer be present. These results are further supported by previous work done by Southwick et al. (ref. 2) who studied time-dependent aggregation of xanthan in quiescent systems using light scattering. These researchers also found evidence for aggregates well below the critical concentration of aggregation cited in the literature.

These optical techniques will also be used to study the effect of the addition of a more flexible coil polymer, carboxymethyl cellulose, on the more rodlike xanthan dynamics.

1. Gerald G. Fuller, *Annual Review of Fluid Mechanics*, 22 (1990) 387-417.
2. Jeffrey G. Southwick, Hoosung Lee, Alexander M. Jamieson, and John Blackwell, *Carbohydrate Research*, 84 (1980) 287-295.

Theoretical and Applied Rheology, edited by P. Moldenaers and R. Keunings
Proc. XIth Int. Congr. on Rheology, Brussels, Belgium, August 17-21, 1992

RHEOLOGICAL CHARACTERIZATION OF POLYELECTROLYTE SOLUTIONS

P. SCHÜMMER and D. ASCHOFF

Institut für Verfahrenstechnik, RWTH Aachen, Turmstr. 46, D–W–5100 Aachen (Germany)

1. INTRODUCTION

The particular rheological properties of poly–electrolyte solutions are caused by the com–plicated electrostatic interactions between the macromolecules and the interactions between the polymer chains and the mobile ions. Even in highly dilute solutions where mechanical and hydrodynamical interactions may be neglected long–range Coulomb forces act on the particles. This has the macro–scopic effect of infinitely increasing the specific viscosity with decreasing concentra–tion (ref. 1). Furthermore changes in the shape of the polymer coils are expected to occur caused by the migration of the coun–terions into the solution. According to the early work of Katchalsky (ref. 2,3) the change in the shape of the polyion in the thermodynamical equilibrium compared to its uncharged reference state can be attributed to the change in the free energy of the electrostatic field of the molecule. The latter can be obtained by numerical integration of the titration curve. According to the theory of Manning (ref. 4,5) condension of the counterions will occur above a certain charge density on the polymer chain. This means that the free energy of the electrostatic field keeps constant above a critical degree of dissociation. In our experimental work we have tried to find out wether conclusions can be drawn from these thermodynamical theories on the rheological behaviour of the polyelectrolyte solutions.

2. METHODS AND RESULTS

Dilute and semi–dilute aqueous solutions of polyacrylic and polymethacrylic acid have been investigated in steady and oscillatory shear flow. The polyacrylic acid samples were quite polydisperse with viscosity average molecular weights between 30,000 and 800,000 g/mol. In contrast with these the polymethacrylic acid samples were characterized by a narrow molecular weight distribution having molecular weights between 87,000 and 1,600,000 g/mol. Up to the critical value of the charge density both, the shear viscosity and the real as well as the imaginary part of the complex viscosity showed a strong increase with the degree of dissociation. Above the critical value the rheological properties are nearly constant. This corresponds to Mannings theory of condensed counterions. The free electrostatical energy of the macromolecules has been calculated from the titration curves as a function of the polymer concentration and the degree of dissociation. This allows the determination of the specific viscosity as a function of the polymer concentration at constant dimensions of the polymer coil. Taking into account the electrostatical interactions between the macromole–cules (ref. 6,7) this dependence could be described quite well.

1. J. Cohen and Y. Rabin, J. Chem. Phys. 88 (1988) 7111–7116
2. A. Katchalsky, J. Polymer Sci., 7 (1951) 393–412
3. S. Lifson and A. Katchalsky, J. Polymer Sci., 13 (1954) 43
4. G.S. Manning, J. Chem. Phys., 51 (1969) 924–933
5. G.S. Manning and A. Holtzer, J. Chem. Phys. 77 (1973) 2206–2212
6. L. Onsager and R.M. Fuoss, J. Phys. Chem., 36 (1932) 2689–2778
7. W. Hess and R. Klein, Adv. Phys., 32 (1983) 173–283

Theoretical and Applied Rheology, edited by P. Moldenaers and R. Keunings
Proc. XIth Int. Congr. on Rheology, Brussels, Belgium, August 17-21, 1992

DESCRIPTION OF RHEOLOGICAL PROPERTIES OF CONCENTRATED POLYDIENEURETHANE SOLUTIONS USING THE FLUCTUATION ENTANGLEMENT THEORY

V.V. TERESHATOV and V.YU. SENICHEV

Institute of Technical Chemistry, Ural Branch of Russian Academy of Sciences, Bolshevistskaya 116, Perm 614600 (Russia)

As a result of a theoretical examination of the interchain hydrogen bond (HB) network in concentrated polydieneurethane solutions and using the theory of fluctuation entanglements, it was shown that the viscosity of polymer solutions with HB is proportional to the square of the concentration of elastically active chains N_h caused by HB.

The concentration of elastically active chains N_h is a function of volume fraction of a polymer in the solution φ_2^n. Thus, in the range of concentrated solutions the value of parameter m = 2n may be used as a measure of the physical activity of a liquid in the degradation of the secondary structure of polymers with the specific interaction.

The experimental verification of theoretical data was carried out on the example of oligodieneurethane (ODU) with molecular weight 4500. The transformer oil (TO) was used as plasticizer. The change in the viscosity of the system ODU-TO was studied over a wide range of plasticizer concentrations from 0 to 80% in the temperature interval from 23 to 85°C.

A good agreement was obtained between calculated and experimental viscosity dependences of the system studied on the TO concentration when the value of parameter n = 3.7 was established from experimental data on the variation in the initial modulus of crosslinked polydieneurethane during its swelling in TO.

Theoretical and Applied Rheology, edited by P. Moldenaers and R. Keunings
Proc. XIth Int. Congr. on Rheology, Brussels, Belgium, August 17-21, 1992

RHEO-OPTICAL DETECTION OF SHEAR INDUCED ORIENTATION AND DEFORMATION OF POLYMERS IN SOLUTION

M. ZISENIS, A. LINK, B. PRÖTZL and J. SPRINGER

Technische Universität Berlin, Institut für Technische Chemie, Fachgebiet Makromolekulare Chemie
D-1000 Berlin 12 (Germany)

1. INTRODUCTION

Light scattering experiments on macromole cules in solution subjected to shear flow enable the evaluation of a number of interesting molecular parameters and last not least the study of the rheology of polymer solutions. For this aim rheo-photometers have been constructed (refs.1-3) to detect the wide angle laser light scattering of polymer solutions in a Searle-type shear cell.

2. METHOD

The gap between the two concentric cylinders of the shear cell is passed by a polarised laser beam. Through a window in the stationary outer cylinder the scattered light is detected by a photomultiplier on the partial surface of a fictitious sphere. The shear dependent variation of the distribution of scattered light intensity gives direct information about the orientation and deformation behaviour of the macromolecules.

3. EXPERIMENTAL

Solutions subjected to rheooptical investigation have to meet several viscometric and optical conditions (refs.2,3). Up to now our investigations have intensively been performed on dilute solutions of high molecular weight polystyrene standards in good solvents (oligostyrene/toluene-compositions) in poor solvents (trans-decalin, dioctylphthalate) and in solvents of medium solvent power (low molecular weight polymethylvinylether/toluene-compositions).

4. RESULTS

4.1 Orientation in shear flow

Following the theory of Peterlin et al. (ref.4), from the scattered light distribution in the flow-plane, which is parallel to the vectors of flow-velocity and velocity-gradient, the orientation angle χ is estimated. The shear dependent orientation, extrapolated to zero-concentration, depends on molar mass, solvent power and viscosity. Compared to the theoretical models by Kuhn, Rouse and Zimm the experimental orientation tends to Zimm like behaviour with increasing shear rate (ref.5).
Our polymer-solvent-systems are in principle different from those used for investigation by streaming birefringence and estimation of the orientation angle is quite easy done with our rheo-photometer. In addition, the rheo-photometer allows measurements in Zimm-planes under shear. From these measurements information about the molecular structure of the macromolecules is obtained.

4.2 Deformation in shear flow

It was shown that conventional evaluation of LS-data from measurements in Zimm-planes under shear is severely distorted by the orientation process. Therefore, according to a proposal of Peterlin (ref.6) in 1957, measurements have been performed by variation of detection position and of the wavelength of the incident beam (ref.5). With the results of these measurements for the first time the 3-dimensional expansion of single macromolecules in solution and the shear induced deformation of its shape can be estimated. Though the observed deformation ratio obeys the predicted type of scaling law (refs.4,6), substantially lower deformation rates revealed.

REFERENCES

1. A. Wölfle, J. Springer, Coll. Polym. Sci., 262, 876 (1984).
2. D. Cleschinsky, H. Stock, J. Springer, Coll. Polym. Sci., 269, 1250 (1991).
3. A. Link, J. Springer, submitted for publication
4. A. Peterlin, W. Heller, M. Nakagaki, J. Chem. Phys., 28, 470 (1958).
5. A. Link, M. Zisenis, B. Prötzl, J. Springer, Analysis of Polymers, Hüthig & Wepf, Basel-Heidelberg-New York, in press.
6. A. Peterlin, J. Polym. Sci., 23, 189 (1957).

AUTHOR INDEX